Optical Shop Testing

Optical Shop Testing

Second Edition

Edited by
DANIEL MALACARA

A Wiley-Interscience Publication
John Wiley & Sons, Inc.
New York / Chichester / Brisbane / Toronto / Singapore

Wiley Series in Pure and Applied Optics

The Wiley Series in Pure and Applied Optics publishes outstanding books in the field of optics. The nature of these books may be basic ("pure" optics) or practical ("applied" optics). The books are directed towards one or more of the following audiences: researchers in university, government, or industrial laboratories; practitioners of optics in industry; or graduate-level courses in universities. The emphasis is on the quality of the book and its importance to the discipline of optics.

In recognition of the importance of preserving what has been written, it is a policy of John Wiley & Sons, Inc., to have books of enduring value published in the United States printed on acid-free paper, and we exert our best efforts to that end.

Copyright © 1992 by John Wiley & Sons, Inc.

All rights reserved. Published simultaneously in Canada.

Reproduction or translation of any part of this work beyond that permitted by Section 107 or 108 of the 1976 United States Copyright Act without the permission of the copyright owner is unlawful. Requests for permission or further information should be addressed to the Permissions Department, John Wiley & Sons, Inc.

Library of Congress Cataloging in Publication Data:
Optical shop testing/edited by Daniel Malacara.
    p. cm.—(Wiley series in pure and applied optics, ISSN 0277-2493)
   Includes bibliographical references and index.
   1. Optical measurements. 2. Interferometry. 3. Interferometers.
I. Malacara, Daniel, 1937– . II. Series.
QC367.059 1991
661.2—dc20   91-16856
ISBN 0-471-52232-5   CIP

Printed in the United States of America

10 9 8

To my wife and children
To my mother
To the memory of my father

Contributors

John H. Bruning *GCA Tropel Division, Fairport, New York*

Alejandro Cornejo-Rodríguez *Instituto Nacional de Astrofísica, Optica y Electrónica, Puebla, Pue Mexico*

Katherine Creath *Optical Sciences Center, University of Arizona, Tucson, Arizona*

Scott L. De Vore *6800 East Ventana Place, Tucson, Arizona*

Isaac Ghozeil *The Aerospace Corporation, Selseguido, California*

John E. Greivenkamp *Optical Sciences Center, University of Arizona, Tucson, Arizona*

Parameswaran Hariharan *CSIRO Division of Applied Physics National Measurement Laboratory, Linfield, Australia*

Daniel Malacara *Centro de Investigaciones en Optica, A.C. León, Mexico*

Zacarias Malacara *Centro de Investigaciones en Optica, A.C. León, Mexico*

Sham Mallick *Institud d'Optique Théorique et Appliquée, France*

Murty V. Mantravadi *2301 West 120th Street, Hawthorne, California*

Arquimedes Morales *Centro de Investigaciones en Optica, A.C. León, Mexico*

Abe Offner *100 Leeuwarden Road, Darien, Connecticut*

Jorge Ojeda-Castañeda *Instituto Nacional de Astrofísica, Optica y Electrónica, Puebla, Pue. Mexico*

Chandrasekhar Roychoudhuri *The Perkin Elmer Corporation, Applied Science and Technology Sector, Research, Danbury, Connecticut*

Walter T. Welford (deceased) *Blacket Laboratory, Imperial College, London, England*

James C. Wyant *Wyko Corporation, Tucson, Arizona*

Contents

Preface to the Second Edition   xv

Preface to the First Edition   xvii

Chapter 1.   Newton, Fizeau, and Haidinger Interferometers   1
M. V. Mantravadi

    1.1. Newton Interferometer, 1
    1.2. Fizeau Interferometer, 18
    1.3. Haidinger Interferometer, 36
    1.4. Absolute Testing of Flats, 43

Chapter 2.   Twyman–Green Interferometer   51
D. Malacara

    2.1. Introduction, 51
    2.2. Beam Splitter Plate, 53
    2.3. Coherence Requirements, 58
    2.4. Uses of a Twyman–Green Interferometer, 63
    2.5. Unequal-Path Interferometer, 69
    2.6. Variations from the Twyman–Green Configuration, 73
    2.7. Typical Interferograms and Their Analysis, 77
    2.8. Imaging of the Interference Pattern, 86

Chapter 3.   Common-Path Interferometers   95
S. Mallick

    3.1. Introduction, 95
    3.2. Burch's Interferometer Employing Two Matched Scatter Plates, 96
    3.3. Birefringent Beam Splitters, 99
    3.4. Lateral Shearing Interferometers, 103
    3.5. Double-Focus Interferometer, 108
    3.6. Saunders's Prism Interferometer, 110
    3.7. Point Diffraction Interferometer, 112

3.8. Zernike Tests with Common-Path Interferometers, 113
3.9. Measurement of the Optical Transfer Function, 113

Chapter 4. Lateral Shearing Interferometers 123
M. V. Mantravadi

4.1. Introduction, 123
4.2. Considerations Regarding Coherence Properties of the Light Source, 124
4.3. Brief Theory of Lateral Shearing Interferometry, 125
4.4. Evaluation of an Unknown Wavefront, 135
4.5. Lateral Shearing Interferometers in Collimated Light (White Light Compensated), 138
4.6. Lateral Shearing Interferometers in Convergent Light (White Light Compensated), 144
4.7. Lateral Shearing Interferometers Using Lasers, 151
4.8. Other Types of Lateral Shearing Interferometers, 159

Chapter 5. Radial, Rotational, and Reversal Shear Interferometers 173
D. Malacara

5.1. Introduction, 173
5.2. Radial Shear Interferometers, 173
5.3. Rotational Shear Interferometers, 191
5.4. Reversal Shear Interferometers, 198

Chapter 6. Multiple-Beam Interferometers 207
C. Roychoudhuri

6.1. Brief Historical Introduction, 207
6.2. Precision in Multiple-Beam Interferometry, 210
6.3. Multiple-Beam Fizeau Interferometer, 212
6.4. Fringes of Equal Chromatic Order, 221
6.5. Reduction of Fringe Interval in Multiple-Beam Interferometry, 225
6.6. Plane Parallel Fabry–Perot Interferometer, 225
6.7. Tolansky Fringes with Fabry–Perot Interferometer, 231
6.8. Multiple-Beam Interferometer for Curved Surfaces, 232
6.9. Coupled and Series Interferometers, 234
6.10. Holographic Multiple-Beam Interferometers, 237
6.11. Final Comments, 237

Chapter 7. Multiple-Pass Interferometers 247
P. Hariharan

7.1. Double-Pass Interferometers, 247
7.2. Multipass Interferometry, 255

Chapter 8. Foucault, Wire, and Phase Modulation Tests 265
J. Ojeda-Castañeda

8.1. Introduction, 265
8.2. Foucault or Knife-Edge Test, 265
8.3. Wire Test, 288
8.4. Platzeck–Gaviola Test, 295
8.5. Phase Modulation Tests, 299
8.6. Ritchey–Common Test, 309

Chapter 9. Ronchi Test 321
A. Cornejo-Rodríguez

9.1. Introduction, 321
9.2. Geometrical Theory, 322
9.3. Wavefront Shape Determination, 335
9.4. Physical Theory, 342
9.5. Practical Aspects of the Ronchi Test, 350
9.6. Some Related Tests, 353

Chapter 10. Hartmann and Other Screen Tests 367
I. Ghozeil

10.1. Introduction, 367
10.2. Theory, 370
10.3. Types of Screens, 374
10.4. Hartmann Test Implementation, 382
10.5. Data Reduction, 385
10.6. The Michelson and Gardner–Bennett Tests, 391
10.7. Hartmann Tests of the Future, 392
10.8. Summary, 392

Chapter 11. Star Tests 397
W. T. Welford

11.1. Principles of the Star Test for Small Aberrations, 398
11.2. Practical Aspects with Small Aberrations, 412
11.3. The Star Test with Large Aberrations, 420

Chapter 12. Null Tests Using Compensators 427
A. Offner and D. Malacara

- 12.1. Introduction and Historical Background, 427
- 12.2. The Dall Compensator, 430
- 12.3. The Shafer Compensator, 433
- 12.4. The Offner Compensator, 434
- 12.5. Other Null Tests for Concave Conicoids, 443
- 12.6. Compensators for Convex Conicoids, 446
- 12.7. Hindle-Type Tests, 450

Chapter 13. Interferogram Evaluation and Wavefront Fitting 455
D. Malacara and S. L. DeVore

- 13.1. Introduction, 455
- 13.2. Polynomial Wavefront Representation, 456
- 13.3. Wavefront Fitting and Data Analysis, 472
- 13.4. Fringe Digitization, 487
- 13.5. Fourier Analysis of Interferograms, 491
- 13.6. Direct Measuring Interferometry, 494

Chapter 14. Phase Shifting Interferometers 501
J. E. Greivenkamp and J. H. Bruning

- 14.1. Introduction, 501
- 14.2. Fundamentals Concepts, 502
- 14.3. Advantages of PSI, 504
- 14.4. Methods of Phase Shifting, 506
- 14.5. Detecting the Wavefront Phase, 510
- 14.6. Phase Unwrapping: Introduction, 514
- 14.7. Integrating Bucket Data Collection, 515
- 14.8. PSI Algorithms, 518
- 14.9. Phase Shift Callibration, 533
- 14.10. Error Sources, 536
- 14.11. Detectors and Spatial Sampling, 546
- 14.12. Phase Unwrapping, 551
- 14.13. Aspheres and Extended Range PSI Techniques, 553
- 14.14. Other Analysis Methods, 562
- 14.15. Computer Processing and Output, 568
- 14.16. Implementation and Applications, 571
- 14.17. Future Trends for PSI, 588

Chapter 15. Holographic and Speckle Tests 599
K. Creath and J. C. Wyant

15.1. Introduction, 599
15.2. Interferometers Using Real Holograms, 600
15.3. Interferometers Using Synthetic Holograms, 603
15.4. Two-Wavelength and Multiple-Wavelength Techniques, 612
15.5. Holographic Interferometry for Nondestructive Testing, 617
15.6. Speckle Interferometry and TV Holography, 627

Chapter 16. Moiré and Fringe Projection Techniques 653
K. Creath and J. C. Wyant

16.1. Introduction, 653
16.2. What is Moiré?, 654
16.3. Moiré and Interferograms, 658
16.4. Historical Review, 666
16.5. Fringe Projection, 668
16.6. Shadow Moiré, 671
16.7. Projected Moiré, 675
16.8. Two-Angle Holography, 675
16.9. Common Features, 676
16.10. Comparison to Conventional Interferometry, 677
16.11. Applications, 678
16.12. Summary, 681

Chapter 17. Contact and Noncontact Profilers 687
K. Creath and A. Morales

17.1. Introduction, 687
17.2. Stilus Profilers, 689
17.3. Scanning Probe Microscopes, 691
17.4. Optical Focus Sensors, 696
17.5. Interferometric Optical Profilers, 697

Chapter 18. Angle, Distance, Curvature, and Focal Length 715
Z. Malacara

18.1. Introduction, 715
18.2. Angle Measurements, 715
18.3. Distance Measurements, 725

    18.4. Radius of Curvature Measurements, 728
    18.5. Focal Length Measurements, 735

Appendix 1. An Optical Surface and Its Characteristics   **743**

Appendix 2. Some Useful Null Testing Configurations   **755**

Additional Bibliography   **763**

Index   **765**

Preface to Second Edition

The first edition of this book was published in 1978. Now, 13 years later, the state of the art in optical testing has changed and improved greatly. Far from decreasing, the interest in the field has been growing steadily. More than 600 research papers directly related to this field have been published in the optics journals in these 13 years. Old techniques have improved and new methods have been developed.

The purpose of this second edition is to update the first edition by revising some chapters and adding a few others. Fringe scanning techniques were just starting in 1978 and have now been developed to maturity, and even the name has changed to "heterodyning of phase shift interferometry." This chapter has been completely rewritten.

The methods for wavefront fitting and evaluation have also developed so much that a new chapter to cover this field is now included.

I have received many ideas from colleagues and friends for improving the book. One of these was to add an appendix where the appropriate tests for typical optical surfaces are suggested. This is now being included.

The discussion on holographic and moiré methods (covered in one chapter in the first edition) has now been split into two chapters to more fully describe the new techniques and procedures.

The chapter on parameter measurements has been completely rewritten to more fully describe these methods, which have recently received much attention in the literature.

As it is to be expected, some errors that were found in the first edition have been corrected here. I would like to acknowledge the support and encouragement of both the Institute of Optics of the University of Rochester, and the Centro de Investigaciones en Optica, in León, Mexico, where this work was carried out.

DANIEL MALACARA

León, Mexico.
January 1992

Preface to the First Edition

The purpose of this book is twofold. The first purpose is to bring together in a single book descriptions of all tests applicable in the optical shop to optical components and systems. In this way they can easily be compared and the most advantageous be chosen for use. Each chapter has been written by a specialist in the field who has used—and often improved—the test he discusses.

The second purpose has grown out of the changing nature of optical testing. For many years opticians in charge of the polishing and figuring of surfaces also carried the responsibility of testing. Generally, they used such semiquantitative but relatively simple tests as the Fizeau, the Foucault, and the Ronchi. These still have their places, but now precision and time are gained by employing electronic detectors, laser sources, and/or holograms as components of test instruments, and by using computers in the analysis of large quantities of data. Individuals participate in optical testing who are not necessarily opticians, but who have the training required to use the newer techniques. This book has been planned to help in that training and to assist the optician in his dialog with the new specialists who aid him. Together they can make ever better optical systems. The book can also serve as a textbook for a course in optical testing or interferometry, since the physical principles of each test are described.

Emphasis throughout has been placed on tests to be used during the construction phase of optical components and systems, rather than those for final evaluation. Thus, for instance, optical transfer measuring instruments are not described. The notation throughout the book has been made as uniform as possible. Mathematical formulas are used, but lengthy derivations available in the literature have not been repeated.

The book has fifteen chapters. The first seven chapters describe all the main interferometers useful in optical testing. Chapters 8 to 10 discuss the tests that can have a geometry as well as a physical interpretation, such as the Foucault, Ronchi, and Hartmann tests, including phase modulation tests (Zernike, Lyot, etc.) as a generalization of the physical theory of the Foucault test. Chapter 11 treats the well-known star test. Chapters 12 and 13 consider two very modern tools in optical testing, namely, the holographic and fringe scanning techniques. Chapter 14 describes compensators for the spherical aberration of systems like aspherical surfaces, used to obtain a null test. Finally, Chapter 15 considers the measurement of some parameters in optical systems, for example, radii of curvature, focal lengths, and angles.

Four appendices at the end of the book provide mathematical and optical background material very useful and even necessary in optical testing.

This book is the culmination of a very old dream that Alejandro Cornejo and I have long held. As a first step we compiled, together with Dr. Murty, an extensive bibliography on the subject, which was published in *Applied Optics* (May 1975). Now our idea has become a reality, thanks to all the contributors to this book.

I cannot conclude without mentioning some of the persons who offered me help in the editorial work. First I want to mention A. Cornejo for his encouragement and continuous help in revising manuscripts. I also thank Dr. Robert Noble and Mrs. Nel Noble for their great assistance in revising manuscripts and offering me many useful suggestions. The help of Mr. Eliezer Jara and Mrs. Zelma Jara, who prepared many drawings and pictures, is also greatly appreciated. The financial support and the enthusiasm of the Instituto Nacional de Astrofísica, Optica y Electrónica and of its director were of fundamental importance. Finally, I especially thank my wife Isabel for her encouragement and understanding.

DANIEL MALACARA

Tonantzitla, México
September 1977

1
Newton, Fizeau, and Haidinger Interferometers

M. V. Mantravadi

Newton, Fizeau, and Haidinger interferometers are among the simplest and most powerful tools available to a working optician. With very little effort these interferometers can be set up in an optical workshop for routine testing of optical components to an accuracy of a fraction of the wavelength of light. Even though these instruments are simple in application and interpretation, the physical principles underlying them involve a certain appreciation and application of physical optics. In this chapter we examine the various aspects of these interferometers and also consider the recent application of laser sources to them.

1.1. NEWTON INTERFEROMETER

We will take the liberty of calling any arrangement of two surfaces in contact illuminated by a monochromatic source of light a Newton interferometer. Thus the familiar setup to obtain Newton rings in the college physical optics experiment is also a Newton interferometer, the only difference being the large air gap as one moves away from the point of contact, as seen in Fig. 1.1. Because of this, it is sometimes necessary to view these Newton rings through a magnifier or even a low-power microscope. In the optical workshop we are generally concerned that an optical flat, one being fabricated, is matching the accurate surface of another reference flat or that a curved spherical surface is matching the correspondingly opposite curved spherical master surface. Under these conditions the air gap is seldom more than a few wavelengths of light in thickness. In the various forms of the Newton interferometer, we are mainly interested in determining the nonuniformity of this air gap thickness by observing and interpreting Newton fringes. A simple way to observe these Newton fringes is illustrated in Fig. 1.2. Any light source such as a sodium vapor lamp, low-pres-

---

Optical Shop Testing, Second Edition, Edited by Daniel Malacara.
ISBN 0-471-52232-5 © 1992, John Wiley & Sons, Inc.

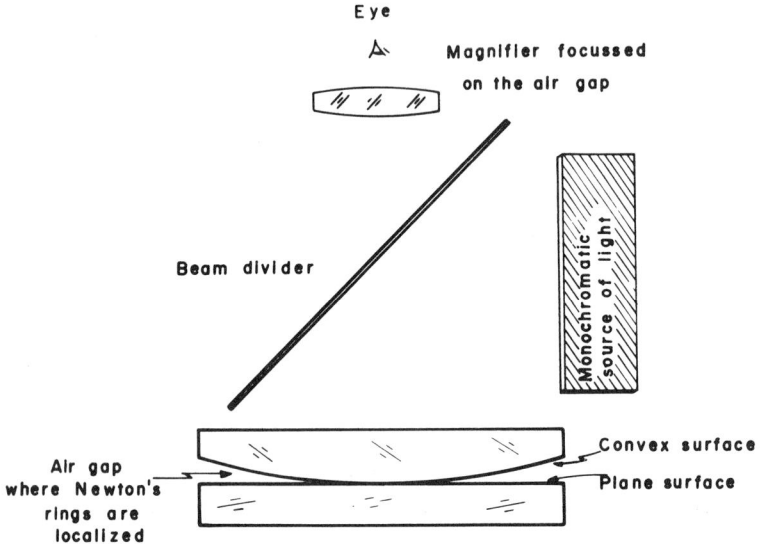

Figure 1.1. Illustration of the setup for Newton rings. A plano-convex lens about 1 or 2 meters in focal length is placed with its convex surface in contact with the plane surface of an optical flat and illuminated by monochromatic light.

sure mercury vapor lamp, or helium discharge lamp can be used in the setup. Under certain situations even an ordinary tungsten lamp can serve this purpose.

Let us first see what happens when two perfect optical flats are put in contact with each other and forming a small air wedge, as illustrated in Fig. 1.3. The air wedge is thin on the left and thick on the right. Generally, the thickness of the air gap, even on the thin side, will be not zero but finite; and unless one presses very hard on the edge, it is difficult to make this air gap on the thin side exactly zero. Hence we may imagine the two planes projected backward, as shown in Fig. 1.3, and meeting in a line of intersection. Let monochromatic light of wavelength λ be incident on the optical flat combination normally. Let α be the wedge angle between the two flats. Then the air gap at a distance x from the line of intersection of the two planes is αx, and the optical path difference is $2\alpha x$. To this we have to add an additional $\lambda/2$ because of the phase change of π taking place as a result of reflection on the bottom plane surface. Thus we have the optical path difference (OPD) at distance x given by $2\alpha x + \lambda/2$.

Hence the dark fringes may be represented by

$$2\alpha x = n\lambda \tag{1.1}$$

where n is an integer, and the bright fringes may be represented by

$$2\alpha x + \frac{\lambda}{2} = n\lambda. \tag{1.2}$$

1.1. NEWTON INTERFEROMETER

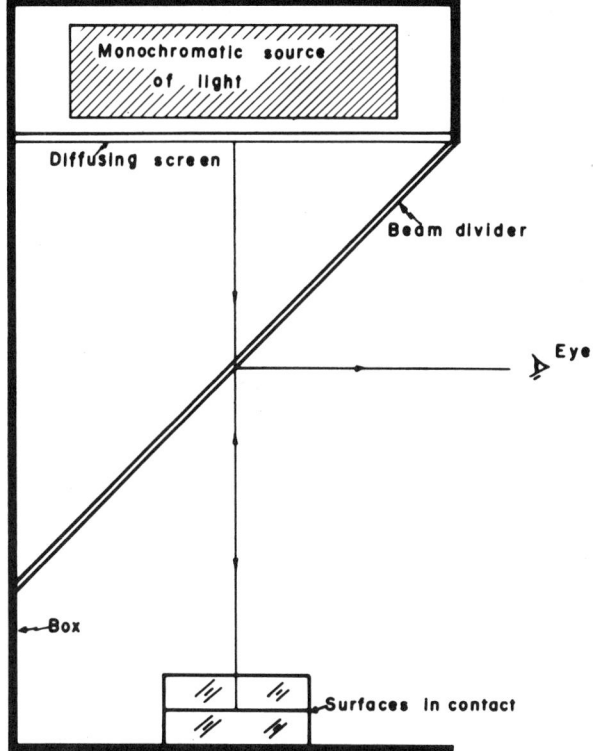

Figure 1.2. A simple arrangement to observe the Newton fringes in the optical workshop. With this arrangement plane and long radius spherical surfaces can be tested.

Each of these equations represents a system of equally spaced straight fringes, and the distance d between two consecutive bright or dark fringes is

$$d = \frac{\lambda}{2\alpha}. \tag{1.3}$$

Thus the appearance of the fringes is as shown in Fig. 1.3 when two good optical flats are put in contact with each other, forming a small air wedge, and are viewed in monochromatic light.

Now let us see what the appearance of Newton fringes is when one surface is optically flat while the other surface is not. Several situations are possible and in fact occur in actual practice. When one starts making a surface a plane, it does not turn out to be a plane on the first try; probably it becomes spherical with a long radius of curvature. It is necessary to test the surface from time to time with a reference flat to ascertain its deviation from flatness. Let us consider a spherical surface of large radius of curvature R in contact with the optical flat.

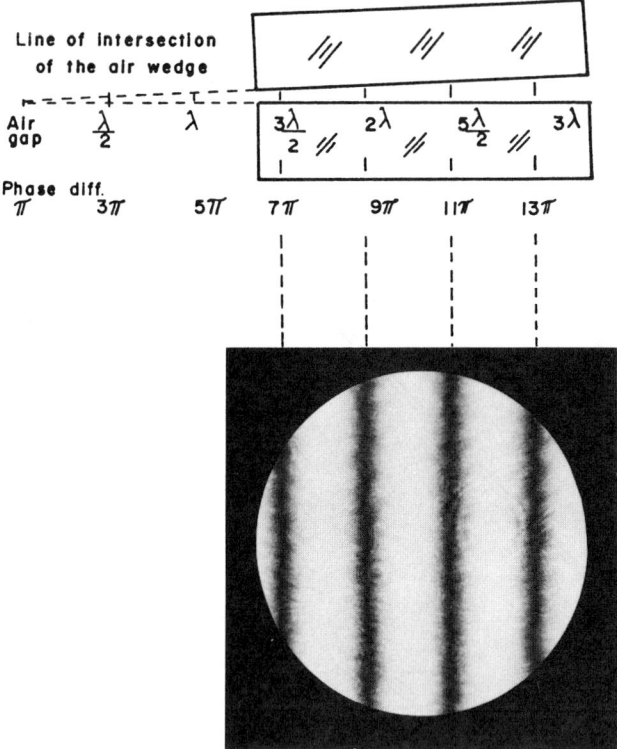

Figure 1.3. The principle of formation of straight, equally spaced fringes between two optically plane surfaces when the air gap is in the form of a wedge. The fringes are parallel to the line of intersection of the two plane surfaces.

Then the sag of the surface is given by $x^2/2R$, where x is the distance measured from the center of symmetry. Hence the OPD is given by $x^2/R + \lambda/2$, and the positions of dark fringes are expressed by

$$\frac{x^2}{R} = n\lambda. \tag{1.4}$$

Hence the distance of the nth dark fringe from the center is given by

$$x_n = \sqrt{nR\lambda}. \tag{1.5}$$

From this it is easy to show that the distance between the $(n + 1)$th and the nth fringe is given by

$$x_{n+1} - x_n = \sqrt{R\lambda}\left(\sqrt{n+1} - \sqrt{n}\right) \tag{1.6}$$

1.1. NEWTON INTERFEROMETER

and similarly the distance between the $(n + 2)$th and the $(n + 1)$th fringe is given by

$$x_{n+2} - x_{n+1} = \sqrt{R\lambda} \, (\sqrt{n + 2} - \sqrt{n + 1}). \tag{1.7}$$

From Eqs. (1.6) and (1.7) we can form the ratio

$$\frac{x_{n+1} - x_n}{x_{n+2} - x_{n+1}} \approx 1 + \frac{1}{2n}. \tag{1.8}$$

Thus it is seen that, when we look at fringes with large values of n, they appear to be almost equally spaced. Hence, when we are testing for the presence of curvature in the surface, it is desirable to manipulate the plates in such a way that we see the fringes with lower order n. In Fig. 1.4 the appearance of Newton fringes is shown when the maximum value of $x^2/2R$ is 2λ. Thus there will be four circular fringes in this situation. If the maximum value of $x^2/2R$ is $\lambda/2$, we have just one circular fringe. Thus, by observation of full circular fringes, we can detect a maximum error of $\lambda/2$ in the flatness of the surface. If the maximum error is less than $\lambda/2$, we have to adopt a different procedure. In this case the center of symmetry of the circular fringes is displaced sideways by suitable manipulation of the two components. Thus we obtain fringes in the aperture of the two surfaces in contact with a larger value of n; these fringes are arcs of circles, and their separations are almost but not exactly equal. Let us take as examples of maximum value $x^2/2R = \lambda/4$ and $\lambda/8$. Figures 1.5 and 1.6, respectively, illustrate the appearance of the fringes in these two cases. As can be inferred, the fringes become straighter and straighter as the value of R increases.

In the optical workshop it is necessary to know also whether the surface that

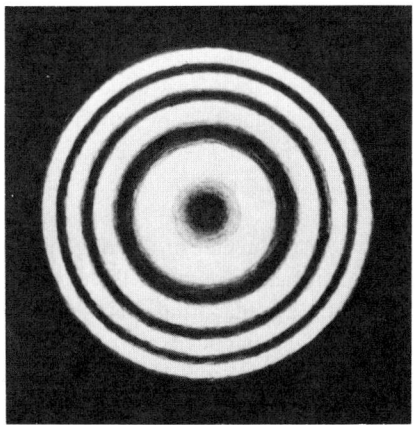

Figure 1.4. Appearance of the Newton fringes when a surface of long radius of curvature is kept on a good optical flat. This situation is for a surface deviating by 2λ from the plane at its maximum.

Figure 1.5. Appearance of the Newton fringes when a surface of long radius of curvature is kept on a good optical flat. This situation is for a surface deviating by $\lambda/4$ from the plane at its maximum. The center of symmetry of the fringes is outside the aperture of the surfaces, and hence only arcs of circles are seen.

Figure 1.6. Same as Fig. 1.5 except that the maximum error is $\lambda/8$.

is being tested is concave or convex with respect to the reference optical flat. This can be easily judged by several procedures. One simple method involves pressing gently near the edge of the top flat by means of a wooden stick or pencil. If the surface is convex, the center of the fringe system moves toward the point of application of pressure. If the surface is concave, the center of the fringe system moves away from the point of application of pressure, as shown in Fig. 1.7a.

A second very simple method is to press near the center of the ring system on the top flat, as shown in Fig. 1.7b. If the surface is convex, the center of the fringe is not displaced but the diameter of the circular fringes is increased.

Another method of deciding whether the surface is convex or concave involves the use of a source of white light. If slight pressure is applied at the center of the surfaces, the air gap at this point tends to become almost zero when the surface is convex. Hence the fringe at this point is dark, and the first

1.1. NEWTON INTERFEROMETER

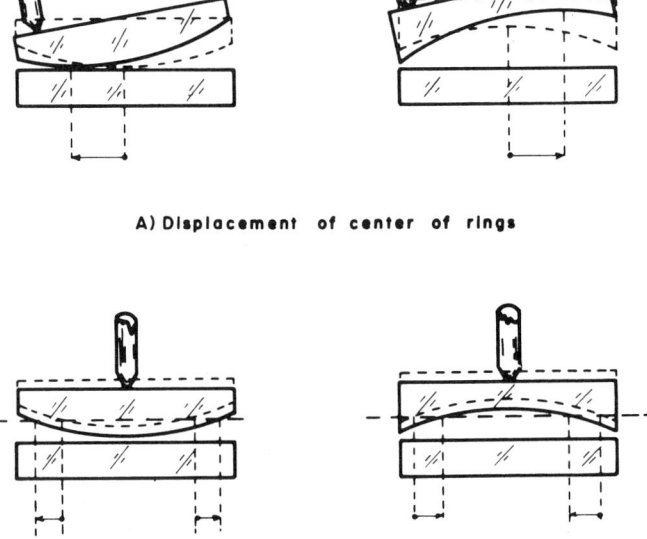

Figure 1.7. Two methods to determine whether the surface under test is convex or concave with respect to the reference surface: (a) by pressing near the edge, and (b) by pressing near the center of the top plate.

bright fringe will be almost colorless or white. The next bright fringe is tinged bluish on the inside and reddish on the outside. On the other hand, if the surface is concave, the contact is not a point contact but occurs along a circle, and the air gap thickness tends to zero along this circle. The dark fringe will be along this circle, and the sequence of colored fringes will be the same as before as one proceeds from the black fringe. This situation is illustrated in Figs. 1.8 and 1.9. This procedure is not very easy to perform unless the surfaces are clean and is not generally recommended.

A fourth and simpler procedure is based on the movement of the fringe pattern as one moves his eye from a normal to an oblique viewing position. Before explaining this procedure, it is necessary to find a simple expression for the optical path difference between two reflected rays at an air gap of thickness t and an angle of incidence θ. This is illustrated in Fig. 1.10, where it can be seen that

$$\text{OPD} = \frac{2t}{\cos \theta} - 2t \tan \theta \sin \theta = 2t \cos \theta. \qquad (1.9)$$

Thus the OPD at normal incidence, namely, $2t$, is always greater than the OPD at an angle θ for the same value of air gap thickness t. Using this fact, let us

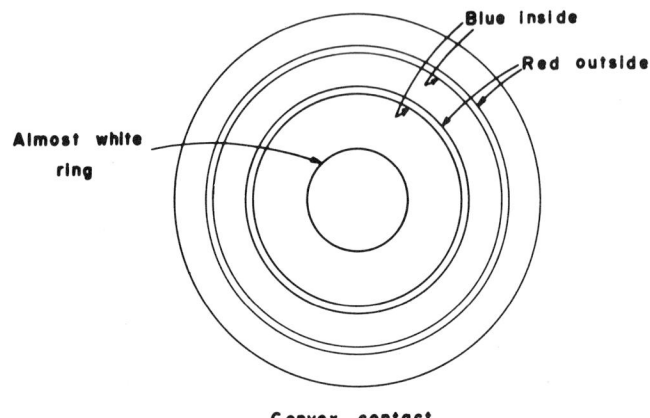

Figure 1.8. Convex contact and appearance of the colored fringes with white light illumination. Pressure is applied at center.

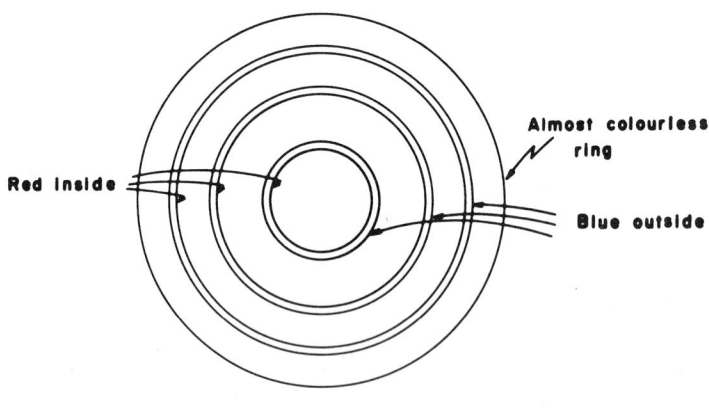

Figure 1.9. Concave contact and the appearance of the colored fringes with white light illumination. Pressure is applied at center.

see what happens when we have a convex contact between the two surfaces. The air gap increases as we go away from the point of contact. When we view the fringes obliquely, the OPD at a particular point is decreased, and consequently the fringes appear to move away from the center as we move our eye from normal to oblique position. The reverse of this situation occurs for a concave surface in contact with a plane surface.

We may consider many other situations where the surfaces are not plane or spherical. The nature and the appearance of such fringes when viewed are given in the usual manner in Table 1.1.

1.1. NEWTON INTERFEROMETER

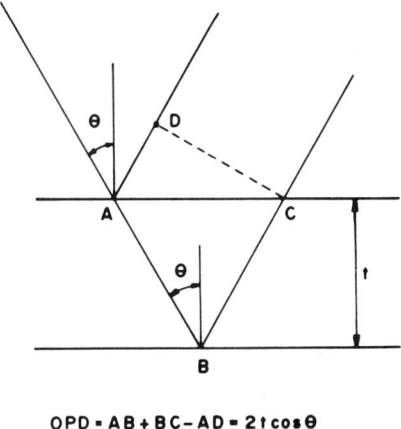

OPD = AB + BC − AD = 2t cos θ

Figure 1.10. Ray diagram for calculation of the optical path difference between two reflected rays from an air gap of thickness t and angle of incidence θ.

Table 1.1. Nature of Newton Fringes for Different Surfaces with Reference to a Standard Flat

S. No.	Surface type	Appearance of the Newton's fringes	
		Without tilt	With tilt
1	Plane		
2	Almost plane		
3	Spherical		
4	Conical		
5	Cylindrical		
6	Astigmatic (curvatures of same sign)		
7	Astigmatic (curvatures of opposite sign)		
8	Highly irregular		

We have mentioned that the reference surface is a flat surface against which a nearly plane surface that is being made is tested. By the same procedure, spherical or cylindrical surfaces having long radii of curvature can be tested. However, when such surfaces have very short radii of curvature, it is necessary to use special illumination, which will be discussed in Section 1.2 on the Fizeau interferometer.

1.1.1. Source Size Considerations

We now examine the effect of source size on the accuracy of interpretation of Newton fringes. The OPD given in Eq. (1.9) for an air gap of thickness t may be approximated to

$$\text{OPD} = 2t \cos \theta \approx 2t - t\theta^2 \qquad (1.10)$$

for small values of θ. Now, in the Newton interferometer we are interested in determining the variations of t. Hence, to reduce the influence of θ as much as possible, we should have

$$t\theta^2 \leq k\lambda \qquad (1.11)$$

where k is a fractional part and $k\lambda$ is the accuracy in which we are interested. Also, it is necessary that the contrast of the fringes be very good. For this the semiangle of illumination θ should be such that

$$t\theta^2 \leq \frac{\lambda}{4}. \qquad (1.12)$$

Now t itself is variable between the surfaces, kept one over the other. Generally speaking, however, the maximum value of t may not be more than a small multiple of λ. We may safely assume that the value of t never exceeds 6λ anywhere in the air gap, especially when the flats are clean and of better than $\lambda/4$ accuracy. Hence we have the simple rule that $\theta^2 \leq \frac{1}{24}$ or $\theta \leq 0.2$, which ensures good visibility of the fringes. This condition is always satisfied in the Newton setup. We may set $\lambda/20$ as the maximum accuracy to which the optical flat is to be assessed. Hence we may set

$$t\theta^2 \leq \frac{\lambda}{20} \quad \text{or} \quad 2\theta \leq 0.2. \qquad (1.13)$$

This condition automatically ensures good visibility of the fringes.

1.1. NEWTON INTERFEROMETER

From the foregoing analysis it can be seen that the illumination angle on the two flats in contact should not exceed 0.2 rad or 12° approximately. This is a sufficiently large angle of illumination, and hence most setups for the observation of Newton fringes do not use any collimating lens but employ an extended source of monochromatic light. In fact, to satisfy this condition, it is necessary to observe the fringes from a minimum distance that is roughly five times the diameter of the optical flats in contact. If the fringes are to be photographed, the camera lens is located at roughly the same minimum distance. To obtain higher accuracy, the distance from which the photograph is taken may be increased.

If the observing distance is not large enough, equal thickness fringes will not be observed. Instead, localized fringes will appear. These fringes are called localized because they seem to be located either above or below the air gap. The fringes are localized in the region where corresponding rays from the two virtual images of the light source intersect each other. It has been shown that this condition may be derived from the van Cittert–Zernike theorem. (Wyant 1978; Simon and Comatri 1987; Hariharan and Steel 1989).

1.1.2. Some Suitable Light Sources

For setting up a Newton interferometer, we require a suitable monochromatic source. Several sources are available and are convenient. One source is, of course, a sodium vapor lamp, which does not require any filter. Another source is a low-pressure mercury vapor lamp with a glass envelope to absorb the ultraviolet light. A third possible source is a helium discharge lamp in the form of a zigzag discharge tube and with a ground glass to diffuse the light. Table 1.2 gives the various wavelengths that can be used in these different spectral lamps.

1.1.3. Materials for the Optical Flats

The optical flats are generally made of glass, fused silica, or more recently developed zero expansion materials such as CerVit and ULE glass. Small optical flats less than 5 cm in diameter can be made of glass; they reach homogeneous temperature conditions reasonably quickly after some handling. It is preferable to make optical flats of larger sizes from fused silica or zero expansion materials. Table 1.3 gives relevant information regarding the materials commonly used for making optical flats.

When making a reference optical flat, it is necessary to consider carefully not only the material to be used but also the weight, size, testing methods, and many other important parameters (Primak 1984, 1989a, 1989b; Schulz and Schwider 1987).

Table 1.2. Characteristics, Such as Wavelength, of Various Lamps Suitable as Sources in Newton's Interferometer

Serial Number	Lamp Type	Wavelength(s) Normally Used (Å)	Remarks
1	Sodium vapor	5893	The wavelength is the average of the doublet 5890 and 5896. Warm-up time is about 10 min.
2	Low-pressure mercury vapor	5461	Because of other wavelengths present, the fringes must be viewed through the green filter, isolating the 5461-Å line. There is no warm-up time. Tube lights without fluorescent coating can be used.
3	Low-pressure helium discharge	5876	Because of other wavelengths present, a yellow filter must be used to view the fringes. There is no warm-up time.
4	Thallium vapor	5350	Characteristics are similar to those of the sodium vapor lamp. Warm-up time is about 10 min.
5	Cadmium vapor	6438	Red filter to view the fringes is required. Warm-up time is about 10 min.

1.1.4. Simple Procedure for Estimating Peak Error

Generally optical surfaces are made to an accuracy ranging from a peak error of 2λ on the lower accuracy side to $\lambda/100$ on the higher side. It is possible by means of the Newton interferometer to estimate peak errors up to about $\lambda/10$ by visual observation alone. Beyond that, it is advisable to obtain a photograph of the fringe system and to make measurements on this photograph. Figure 1.11 shows a typical interferogram as viewed in a Newton interferometer. Here we have a peak error much less than $\lambda/4$. Consequently, the top plate is tilted slightly to obtain the almost straight fringes. The central diametral fringe is observed against a straight reference line such as the reference grid kept in the Newton interferometer in Fig. 1.2. By means of this grid of straight lines, it is possible to estimate the deviation of the fringe from straightness and also the

1.1. NEWTON INTERFEROMETER

Table 1.3. Materials Used for Making Optical Flats and Their Properties

Serial Number	Material	Coefficient of Linear Expansion	Remarks
1	BK7, BSC	$75\text{-}80 \times 10^{-7}/°C$	These are borosilicate glasses that can be obtained with a high degree of homogeneity.
2	Pyrex	$25\text{-}30 \times 10^{-7}/°C$	This is also a borosilicate glass but has higher silica content. Several manufacturers make similar type glass under different brand names. This is a good material for making general quality optical flats and test plates.
3	Fused silica or quartz	$6 \times 10^{-7}/°C$	This is generally the best material for making optical flats. Different grades of the material are available, based mainly on the degree of homogeneity.
4	CerVit, Zerodur	$0\text{-}1 \times 10^{-7}/°C$	This material and similar ones made by different companies under different trade names have practically zero expansion at normal ambient temperatures.
5	ULE fused silica	$0\text{-}1 \times 10^{-7}/°C$	This is a mixture of silica with about 7% titania.

fringe spacing. If the fringe spacing is d and the peak deviation from straightness is k, the peak error of the flat is given by

$$\text{Peak error} = \left(\frac{k}{d}\right)\left(\frac{\lambda}{2}\right). \tag{1.14}$$

In Fig. 1.11 $k = 2.5$ mm and $d = 25$ mm; hence we can say that the peak error is $\lambda/20$. Even in this case it is desirable to know whether the surface is

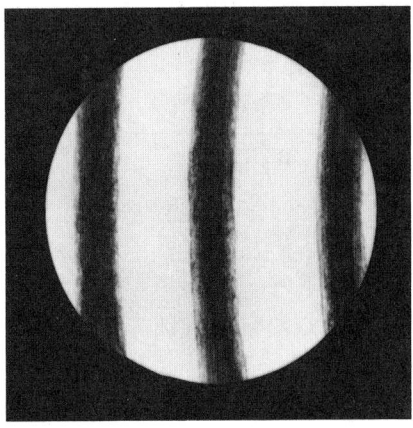

Figure 1.11. Newton fringes for an optical flat showing a peak error of $\lambda/20$.

convex or concave, and for this purpose we can use the procedure described earlier. The only difference is that we have to imagine the center of the fringe system to be outside the aperture of the two flats in contact.

1.1.5. Other Applications of the Newton Interferometer

Measurement of Spherical Surfaces. Probably one of the most common applications of the Newton interferometer is the testing of the faces of small lenses while they are being polished. A small test plate with the opposite radius of curvature is made to the necessary accuracy and then placed over the surface under test. A test plate is useful not only to detect surface irregularities but also to check the deviation of the radius of curvature from the desired value (Karow 1979).

The observation should be made in such a way that the light is reflected almost perpendicularly to the interferometer surfaces. Convex surfaces can be tested with the test plate shown in Fig. 1.12a, with a radius of curvature r in the upper surface given by

$$r = \frac{(N-1)(R+T)L}{NL + R + T}. \tag{1.15}$$

Concave surfaces can be tested as in Fig. 1.12b. The radius of curvature r of the upper surface is

$$r = \frac{(N-1)(R-T)L}{NL - R + T}. \tag{1.16}$$

It is important to remember that the fringes are localized very near the interferometer surfaces, and therefore the eye should be focused at that plane.

1.1. NEWTON INTERFEROMETER

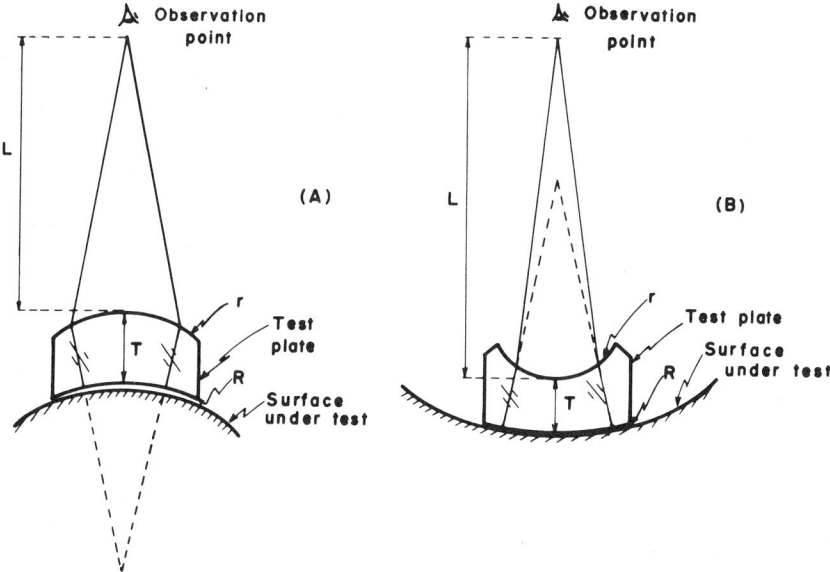

Figure 1.12. Test plates to test spherical surfaces with Newton fringes.

The radius of curvature is checked by counting the number of circular fringes. The relation between the deviation in the radius of curvature and the number of rings can be derived with the help of Fig. 1.13, where it can be shown that the distance ϵ between the two surfaces, measured perpendicularly to one of the surfaces, is given by

$$\epsilon = (r + \Delta r)\left\{1 - \left[1 - \frac{2(1 - \cos\theta)r\,\Delta r}{(r + \Delta r)^2}\right]^{1/2}\right\}. \quad (1.17)$$

If either Δr or the angle θ is small, this expression may be accurately represented by

$$\epsilon = (1 - \cos\theta)\Delta r. \quad (1.18)$$

Since the number n of fringes is given by $n = 2\epsilon/\lambda$, we can also write

$$\frac{n}{\Delta r} = \frac{2(1 - \cos\theta)}{\lambda}. \quad (1.19)$$

If D is the diameter of the surface, the angle θ is defined as $\sin\theta = D/2r$. Therefore a relation can be established between the increment per ring in the radius of curvature and the surface ratio r/D, as shown in Table 1.4.

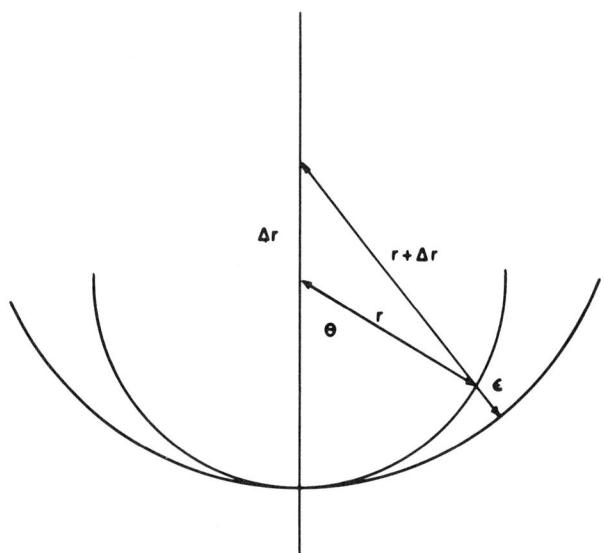

Figure 1.13. Geometry to find the separation between two spherical surfaces with different radii of curvature, measured along the radius of one of them.

Table 1.4. Radius of Curvature Increment per Fringe for Several Values of the Power Ratio r/D of the Spherical Surface Being Tested With Newton Fringes

r/D	$\Delta r/n$ (cm)
1.0	0.00020
2.0	0.00086
3.0	0.00195
4.0	0.00348
5.0	0.00545
6.0	0.00785
7.0	0.01069
8.0	0.01397
9.0	0.01768
10.0	0.02183
20.0	0.08736
30.0	0.19661
40.0	0.34970
50.0	0.54666
60.0	0.78712
70.0	1.07033
80.0	1.39665
90.0	1.77559
100.0	2.18144

1.1. NEWTON INTERFEROMETER

Measurement of Aspheric Profile. Malacara and Cornejo (1970) used the method of Newton fringes to determine the aspheric profile of a surface that deviates markedly from a spherical surface. This method is useful if the aspheric deviates from the nearest spherical by a few wavelengths of light (say, 10–20λ). The method consists in using a spherical test plate in contact with the aspherical surface and finding the position of the fringes by means of a measuring microscope. From these position values one can then obtain the actual air gap as a function of the distance, and a plot can be made and compared with the required aspheric plot. Figure 1.14 shows a typical schematic arrangement for this method.

It is important to consider that the surface under test probably does not have rotational symmetry. Therefore the measurements must be made along several diameters in order to obtain complete information about the whole surface. Instead of directly measuring the fringe positions with a microscope, a photograph can be taken and then the fringe positions measured with more conventional procedures.

If the reference surface is spherical, and the surface under test is aspherical (hyperboloid or paraboloid), the ideal fringe patterns will be those of a Twyman–Green interferometer for spherical aberration (see Chapter 2).

The reference surface may also be another aspherical surface that exactly matches the ideal configuration of the surface under test. This procedure is useful when a convex aspheric is to be made since a concave aspheric can be

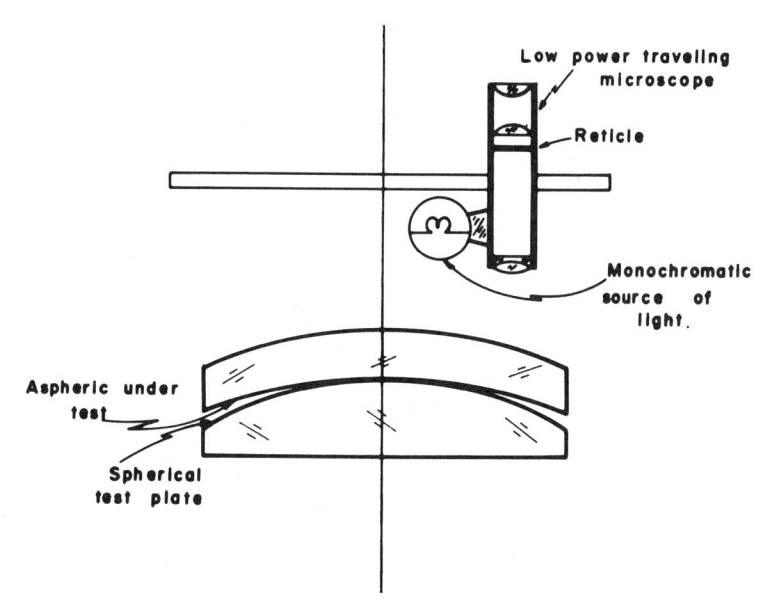

Figure 1.14. Schematic arrangement showing the method of measuring aspheric surfaces with a spherical test plate, using Newton fringes.

made and tested more easily than a convex surface. The advantage of this method is that a null test is obtained. It has the disadvantage that the relative centering of the surfaces is very critical because both surfaces have well-defined axes, and these must coincide while testing. This problem is not serious, however, because the centering can be achieved with some experience and with some device that permits careful adjustment.

When mathematically interpreting the interferograms, it should be remembered that the OPD is measured perpendicularly to the surfaces whereas the surface sagitta z (see Appendix 1) is given along the optical axis. Therefore the OPD is given by $2(z_1 - z_2) \cos \theta$, where $\sin \theta = Sc$.

Measurement of Flatness of Opaque Surfaces. Sometimes we encounter plane surfaces generated on such metal substrates as steel, brass, and copper. An optical flat made of glass should be put on top of such objects for viewing Newton fringes. It is not always the case that the metal object is in the form of a parallel plate. The plane surface may be generated on an otherwise irregular component, and hence some means of holding the component while testing becomes necessary. This can be avoided if we can put the object on top of the optical flat and observe the fringes through the bottom side of the flat. This sort of arrangement is shown in Fig. 1.15. Since most metal surfaces have reflectivities quite high compared to the value for a glass surface, the contrast of the fringes is not very good. To improve this situation, the optical flat is coated with a thin evaporated film of chromium or inconel having a reflectivity of about 30–40%. This brings about the formation of sharper, more visible fringes.

It is necessary to point out that, if the object is very heavy, it will bend the optical flat and the measurement will not be accurate. Therefore this kind of arrangement is suitable for testing only small, light opaque objects. In dealing with heavy objects, it is preferable to place the optical flat on top of the object.

1.2. FIZEAU INTERFEROMETER

In the Newton interferometer the air gap between the surfaces is very small, of the order of a few wavelengths of light. Sometimes it is convenient to obtain fringes similar to the ones obtained in the Newton interferometer, but with a much larger air gap. When the air gap is larger, the surfaces need not be cleaned as thoroughly as they must be before being tested in the Newton interferometer. However, the surfaces may become scratched if not cleaned properly.

We showed earlier that the angular size of the source to be used depends on the air gap. If, for instance, the air gap between the flats is 5 mm, the permissible value of 2θ is given by Eq. (1.12) and is

$$2\theta \leq 10^{-2} \text{ rad}, \tag{1.20}$$

1.2. FIZEAU INTERFEROMETER

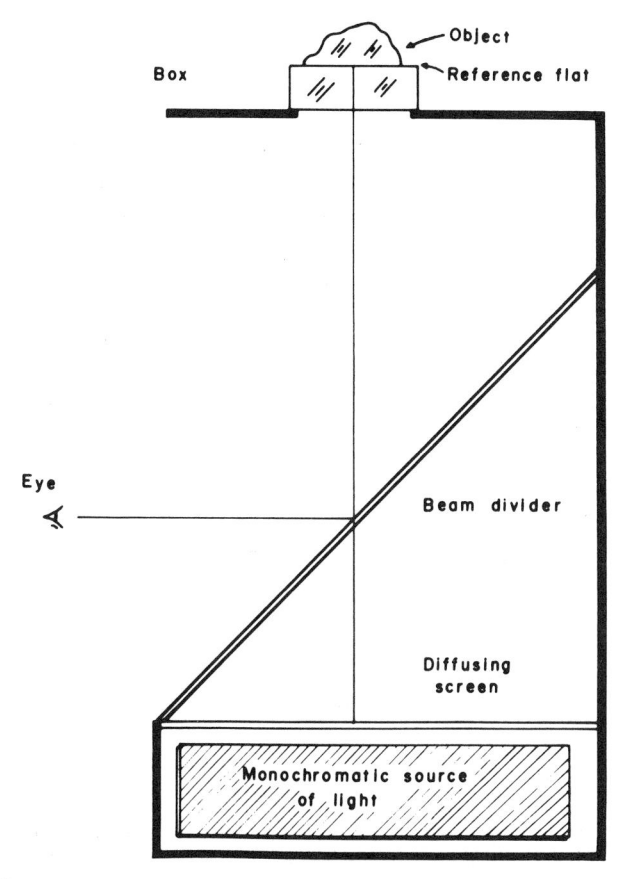

Figure 1.15. Schematic arrangement showing the method of testing opaque plane surfaces on irregular objects by placing them on top of the optical flats.

taking $\lambda = 5 \times 10^{-4}$ mm. Such a small angular source can be obtained generally by the use of a pinhole illuminated by a monochromatic source of light and located at the focus of a collimating lens or mirror. Thus, for example, a collimating lens of 250 mm focal length with a pinhole of 2.5 mm diameter will satisfy these requirements. It can be seen that, as we increase the air gap more and more, the pinhole becomes smaller and smaller.

1.2.1. The Basic Fizeau Interferometer

From the foregoing considerations it is seen that we should have a collimating system in a Fizeau interferometer. Figure 1.16 shows the schematic arrangement of a Fizeau interferometer using a lens for collimation. The optical flat

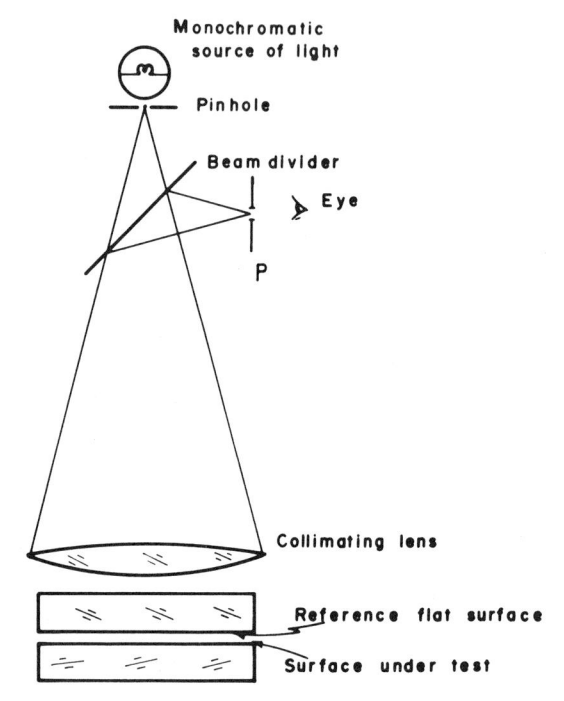

Figure 1.16. Schematic arrangement of a Fizeau interferometer using a lens for collimation of light.

that serves as the reference is generally mounted along with the lens and is preadjusted so that the image of the pinhole reflected by the reference surface falls on the pinhole itself. Either the back side of the flat is antireflection coated or (more conveniently) the reference optical flat is made in the form of a wedge (about 10–20 min of arc) so that the reflection from the back surface can be isolated. To view the fringes, a beam divider is located close to the pinhole. The surface under test is kept below the reference flat, and the air gap adjusted to the smallest value possible; then the air wedge is gradually reduced by manipulating the flat under test. When the air wedge is very large, two distinct images of the pinhole by the two surfaces can be seen in the plane P in Fig. 1.16. By making use of screws provided to tilt the flat under test, one can observe the movement of the image of the pinhole and can stop when it coincides with that of the reference flat. Then the observer places his eye at the plane P and sees, localized at the air gap, the fringes due to variation in the air gap thickness. Further adjustment, while looking at the fringes, can be made to alter the number and direction of the fringes. The interpretation of these fringes is exactly the same as that for Newton fringes.

Figure 1.17 is a schematic of a Fizeau interferometer using a concave mirror

1.2. FIZEAU INTERFEROMETER

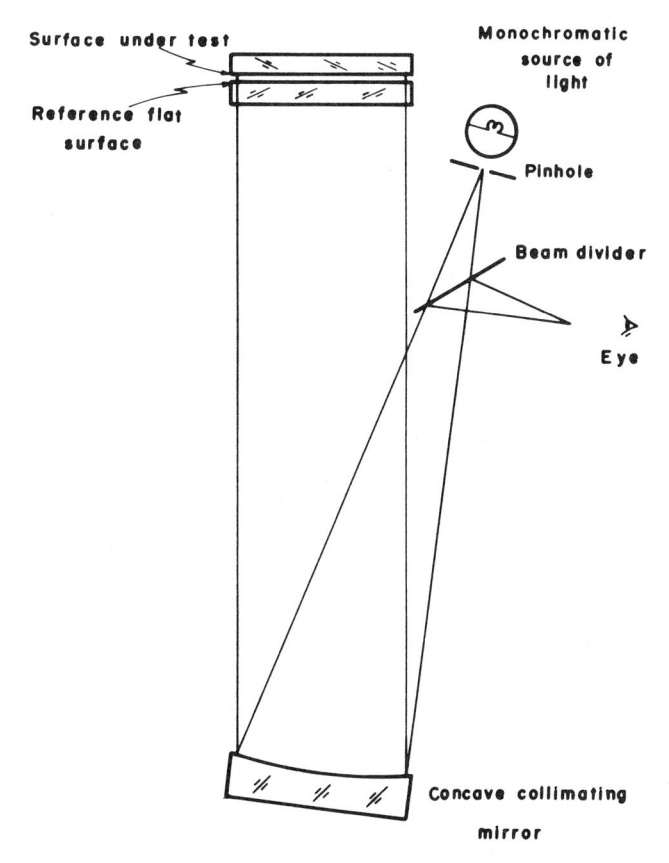

Figure 1.17. Schematic arrangement of a Fizeau interferometer using a concave mirror for collimation of light.

as the collimating element. If a long focal length is chosen for the concave mirror, a spherical mirror can be used. For shorter focal lengths an off-axis paraboloidal mirror may be required. Both the schemes of Figs. 1.16 and 1.17 may be arranged in either a vertical (upright and inverted) or a horizontal layout. In the vertical situation the optical flats are horizontal, whereas in the horizontal layout the optical flats stand on their edges.

1.2.2. Liquid Reference Flats

It is well known that a liquid surface can be used as a reference flat. Basically the liquid surface has a radius of curvature equal to that of the earth. If the radius of the earth is taken as 6400 km, the sag of the surface is (Grigor'ev et al. 1986; Ketelsen and Anderson 1988).

$$\frac{y^2}{2R} = \frac{y^2}{2 \times 6.4 \times 10^9} \text{ (mm)} \qquad (1.21)$$

where $2y$ is the diameter of the liquid surface considered. If we stipulate that this should not exceed $\lambda/100$ ($\lambda = 5 \times 10^{-4}$ mm), then

$$y^2 \leq 6.4 \times 10^4$$

or

$$2y \leq 512 \text{ mm}. \qquad (1.22)$$

Thus a liquid surface of about 0.5 m diameter has a peak error of only $\lambda/100$ as compared to an ideal flat. Therefore it has been a very attractive proposition to build liquid flats as standard references. In practice, however, there are many problems, mainly in isolating the disturbing influence of vibrations. It is also necessary to exclude the region near the wall of the vessel that holds the liquid and to make sure that no dust particles are settling down on the surface. Possible liquids that can be useful for the purpose are clear and viscous, such as glycerin, certain mineral oils, and bleached castor oil. Water is probably not suitable because of its low viscosity. Mercury may not be suitable because of its high reflectivity; the two interfering beams will have very unequal intensities, resulting in poor contrast of the fringes unless the surface under test is also suitably coated. However, mercury has been used as a true horizontal reference plane reflecting surface in certain surveying and astronomical instruments.

1.2.3. Testing Nearly Parallel Plates

In many applications glass plates having surfaces that are both plane and parallel are required. In such cases the small wedge angle of the plate can be determined by the Fizeau interferometer, and the reference flat of the interferometer need not be used since the fringes are formed between the surfaces of the plate being tested. If A is the angle of the wedge and N is the refractive index of the glass, the angle between the front- and back-reflected wavefronts is given by $2NA$, and hence the fringes can be expressed as

$$2NA = \frac{\lambda}{d} \qquad (1.23)$$

where d is the distance between two consecutive bright or dark fringes. Hence the angle A is given by

$$A = \frac{\lambda}{2Nd}. \qquad (1.24)$$

1.2. FIZEAU INTERFEROMETER

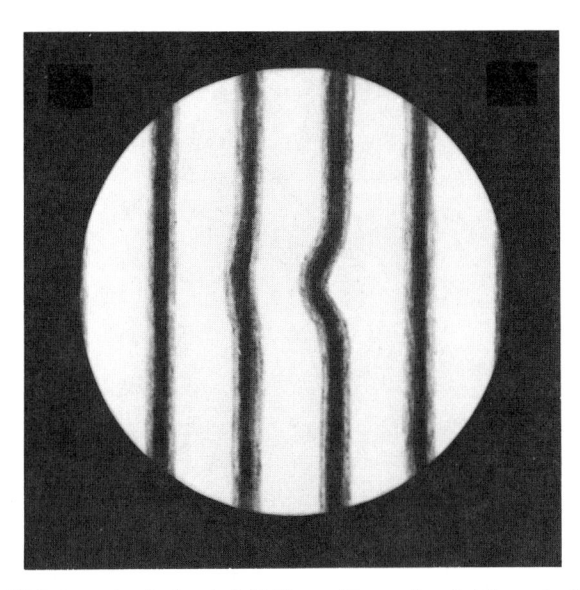

Figure 1.18. Kink formation in the straight Fizeau fringes of a slightly wedged plate, obtained by locally heating the plate. The kink is pointing toward the thin side of the wedge.

To determine the thinner side of the wedge, a simple method is to touch the plate with a hot rod or even with a finger. Because of the slight local expansion, the thickness of the plate increases slightly. Hence a straight fringe passing through the region will form a kink pointing toward the thin side, as shown in Fig. 1.18. For instance, if we take $N = 1.5$, $\lambda = 5 \times 10^{-4}$ mm, and $A = 5 \times 10^{-6}$ (1 s of arc), we get for d a value of about 33 mm. Hence a plate of 33 mm diameter, showing one fringe, has a wedge angle of 1 s of arc. If the plate also has some surface errors, we get curved fringes, indicating both surface and wedge errors. If the surfaces are independently tested and found to be flat, and even in this situation one is getting curved fringes, these should be attributed to variation of the refractive index inside the plate in an irregular manner. In fact, by combining the tests on the Newton interferometer and the Fizeau interferometer for a parallel plate, it is possible to evaluate the refractive index variation (inhomogeneity) (Murty 1963; Murty 1964a; Forman 1964).

1.2.4. Fizeau Interferometer for Curved Surfaces

Just as collimated light is employed for testing optical flats on the Fizeau interferometer, it is possible to use either divergent or convergent light for testing curved surfaces. Figure 1.19 shows an arrangement for testing a concave surface against a reference convex surface. The point source of light is located at the center of curvature of the convex reference surface. The concave surface

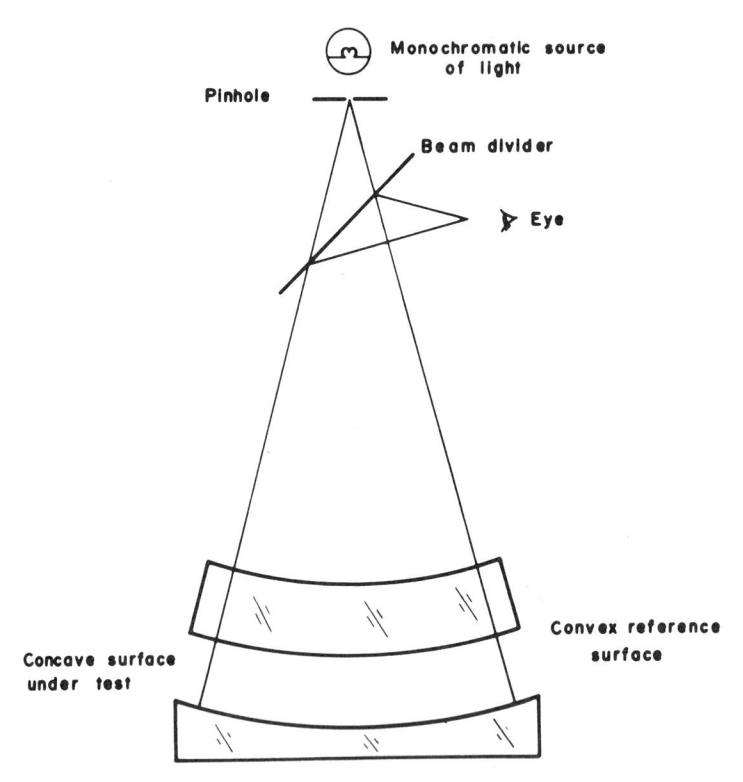

Figure 1.19. Fizeau interferometer setup for curved surfaces. Here the convex surface is the reference surface, and the concave surface is under test.

under test is adjusted until its center of curvature, too, almost coincides with the point source of light. The procedure is exactly the same as before except that to achieve the uniform air gap we have to provide some translational motion also (Moore and Slaymaker 1980).

The same setup can be used very easily for checking the uniformity of thickness (concentricity) of spherical shells. In this case the interfering beams are obtained from the front and back of the two spherical concentric surfaces. Figure 1.20 shows this setup for testing the concentricity of a spherical shell. If the radii of curvature are correct but the shell has a wedge (the centers of curvature are laterally displaced), we get straight fringes characteristic of the wedge. The hot rod or finger touch procedure described in Section 1.2.3 can be adopted to determine which side is thinner. If the two radii are not of proper value $(\overline{r_1 - r_2} \neq t$, where r_1 and r_2 are the two radii and t is the center thickness), the value of t is not constant over the entire shell. Hence we get circular fringes like Newton fringes. If in addition a wedge is present, the center of these circular fringes will be decentered with respect to the center of the shell. In this

1.2. FIZEAU INTERFEROMETER

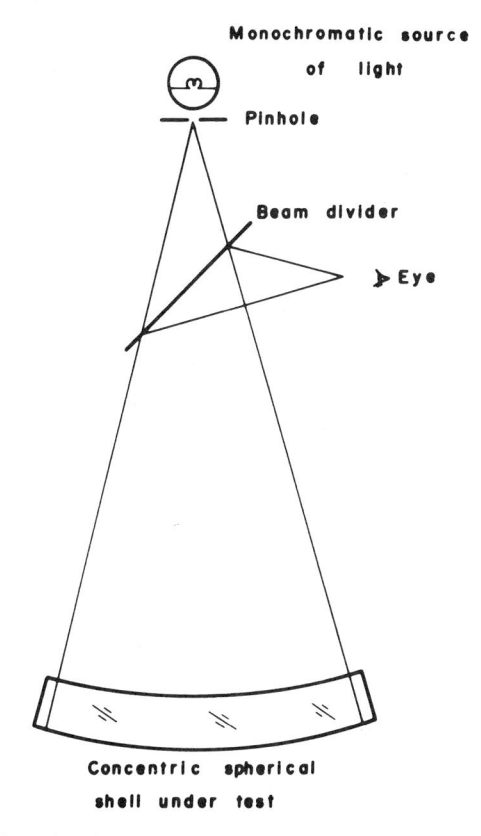

Figure 1.20. Fizeau interferometer setup for testing the concentricity of a spherical shell.

situation also we can adopt the hot rod or finger touch procedure to decide whether the shell is thin at the edge or at the center.

We can also have an arrangement for testing convex surfaces against a concave reference surface, as shown schematically in Fig. 1.21. Here we use a lens or a group of lenses at finite conjugate distances such that the point source of light is at one conjugate whereas the common center of curvature of the test surface and the reference surface is at the other conjugate. The concave reference surface is fixed to the instrument, while the convex surface under test is manipulated in the usual manner to obtain a uniform air gap.

1.2.5. Monochromaticity Requirement for the Source

In the Fizeau interferometer the air gap can be made quite small when plane surfaces are tested. Hence the total optical path difference involved may not exceed a few millimeters, and any small, low-pressure mercury vapor lamp can

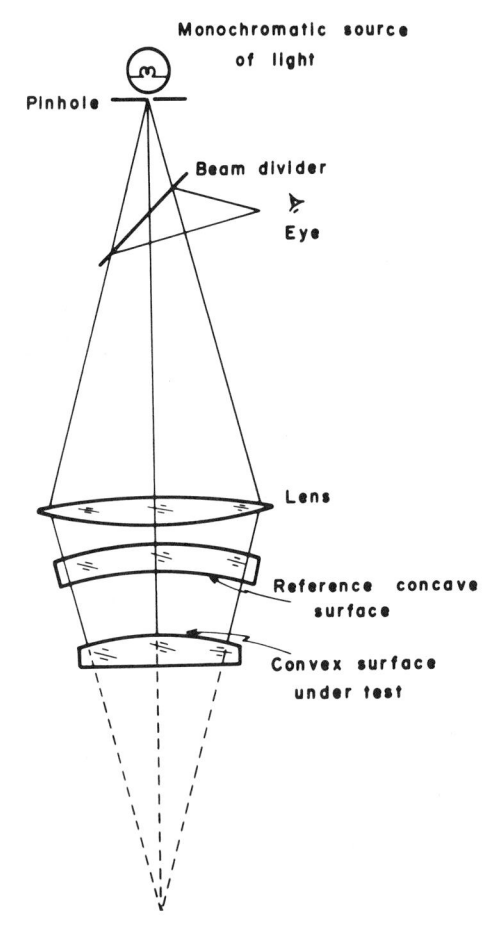

Figure 1.21. Fizeau interferometer setup for testing a convex surface against a concave reference surface.

be used with a green filter as the source of light. When testing for the wedge of thick plates of glass, there is a limitation on the thickness. For instance, a glass plate of 25 mm thickness gives rise to the equivalent of a 75-mm air gap between the front- and back-reflected wavefronts. For the lamp mentioned, this OPD is probably the maximum we can use. For plates of greater thickness, the contrast of the interference fringes is greatly reduced because the lamp does not give a very sharp spectral line. Similarly, the same situation (low contrast) occurs when thick glass shells are tested or when spherical test plates are tested with one test plate, and hence the air gap can be large for certain situations. This limitation can be eliminated, however, if we can use a source of very high monochromaticity. Fortunately such a source, the laser, has recently become

1.2. FIZEAU INTERFEROMETER

available. For our application the low-power (2-mW) helium–neon gas laser operating in a single mode TEM$_{00}$ and with a wavelength of emission at 6328 Å is ideal. With this as the source of light, we can tolerate an OPD of at least 2 m and obtain Fizeau fringes of high contrast. Even larger OPDs are possible provided that a properly stabilized laser is chosen and vibration isolation is provided for the instrument.

1.2.6. Fizeau Interferometer with Laser Source

We shall now describe a Fizeau interferometer using a source such as the helium–neon gas laser of about 2 mW power lasing at 6328 Å in the single mode. A schematic diagram is shown in Fig. 1.22. A very well corrected objective serves to collimate the light from the pinhole, illuminated by a combination of the laser and a microscope objective. Between the collimating objective and the pinhole (spatial filter), a beam divider is placed so that the fringes can be observed from the side. It is also desirable to provide a screen, upon which the Fizeau fringes are projected, to avoid looking into the instrument, as is normally done when conventional light sources are used. The laser has a high radiance compared to other sources, and a direct view may be dangerous to the eye under some circumstances. The reference plane surface is permanently adjusted so that the reflected image of the pinhole is autocollimated. The surface under test is adjusted until the image reflected from it also comes into coincidence with

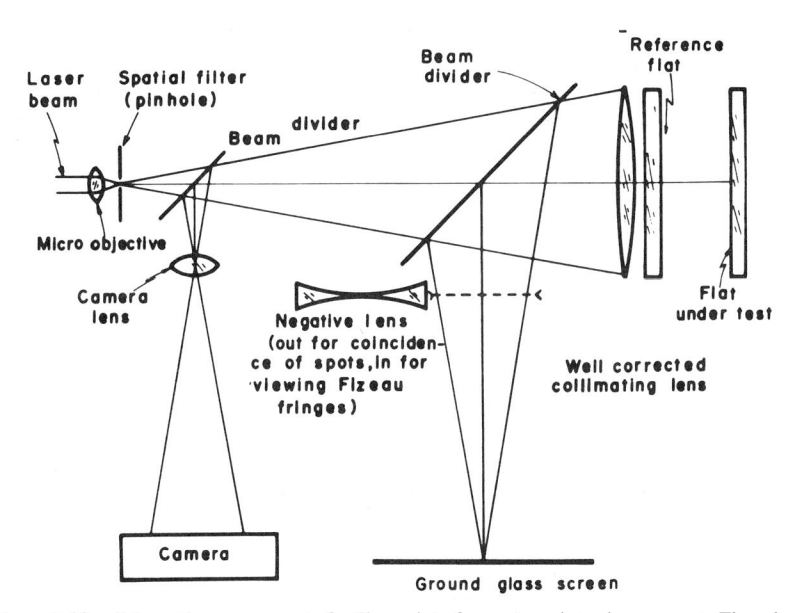

Figure 1.22. Schematic arrangement of a Fizeau interferometer using a laser source. The scheme shown here is for plane surfaces.

the pinhole. To facilitate preliminary adjustment, the screen is used to project the two pinhole images from the two reflecting plane surfaces. This is accomplished by removing the negative lens between the beam divider and the ground-glass screen. The pinhole image from the reference surface is at the center of the screen, whereas the one from the surface under test is somewhere on the screen; by manipulation of this surface, the two spots of light on the screen can be brought into coincidence. Then the negative lens is inserted in the path, and the Fizeau fringes are projected on the screen. These fringes can be further adjusted in direction and number as required. By the use of another beam divider, it is possible to divert part of the beam to a camera for taking a photograph of the fringe pattern. The whole instrument must be mounted on a suitable vibration-isolated platform.

This instrument can be used for various other applications that are normally not possible with conventional sources of light. We describe some such applications in the sections that follow. In addition, many possibilities exist for other applications, depending on the particular situations involved.

Several commercial Fizeau interferometers have been available for several years, but probably the two most widely known are the Zygo interferometer (Forman 1979), shown in Fig. 1.23, and the Wyko interferometer, shown in Fig. 1.24.

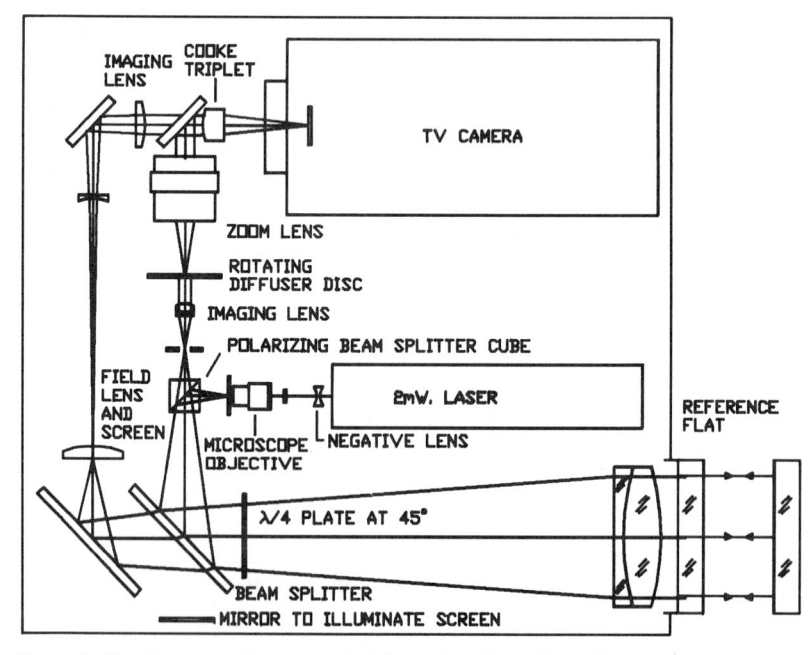

Figure 1.23. Fizeau interferometer manufactured by Zygo Corp. (Courtesy of Zygo, Corp.)

1.2. FIZEAU INTERFEROMETER

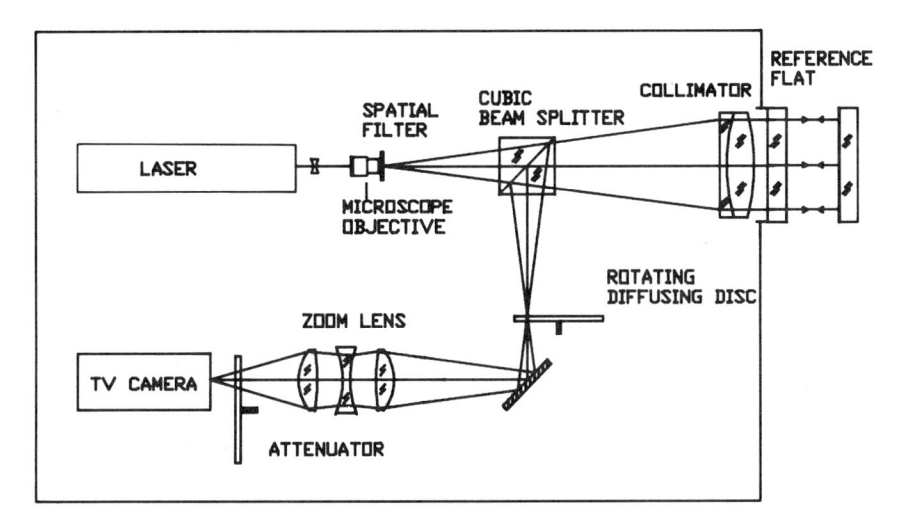

Figure 1.24. Fizeau interferometer manufactured by Wyko Corp. (Courtesy of Wyko, Corp.)

1.2.7. Multiple-Beam Fizeau Setup

If, instead of two-beam fringes, multiple-beam fringes of very good sharpness are required, the reference optical flat and the optical flat under test are coated with a reflecting material of about 80-90% reflectivity (see Chapter 6), such as aluminum or silver. If higher reflectivities are required, multilayer dielectric coatings can be applied. In fact the instrument may be provided with several reference flats having coatings of different reflectivities.

1.2.8. Testing the Inhomogeneity of Large Glass or Fused Quartz Samples

The sample is made in the form of a parallel plate. The surfaces should be made as flat as possible with a peak error of not more than λ. Then the plate is sandwiched between two well-made parallel plates of glass with a suitable oil matching the refractive index of the sample. This will make the small surface errors of the sample negligible, and only straight fringe deformation due to the inhomogeneity of the sample will be seen. If the sandwich is kept in the cavity formed by the two coated mirrors, very sharp dark fringes on a bright background are obtained. If, for instance, the maximum fringe deviation from straightness is k and the distance between two fringes is d, the optical path difference is $(k/d)\lambda$. Now the OPD due to the inhomogeneity ΔN and thickness t of the sample is given by $2\Delta N \cdot t$, and hence

$$\Delta N = \left(\frac{k}{d}\right)\left(\frac{\lambda}{2t}\right). \tag{1.25}$$

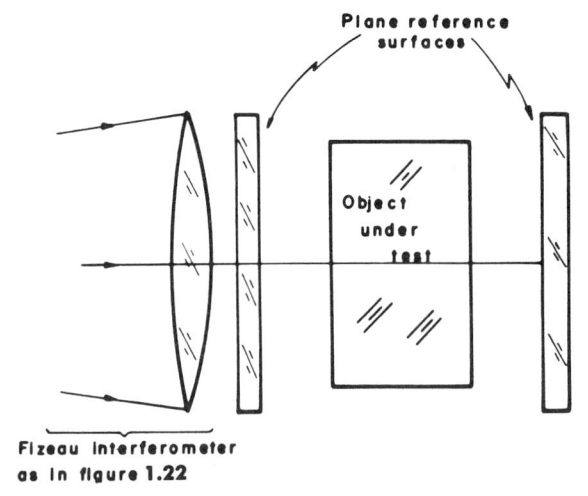

Figure 1.25. Schematic arrangement of a Fizeau interferometer for testing the homogeneity of solid samples of glass, fused quartz, and so on.

As an example, if $k/d = 0.25$, $\lambda = 6328$ Å, and $t = 50$ mm, we have $\Delta N = 1.6 \times 10^{-6}$. Thus a maximum variation of 1.6×10^{-6} may be expected in the sample for the direction in which it has been tested. Figure 1.25 shows the schematic arrangement of the Fizeau interferometer for the method just described.

1.2.9. Testing Cube Corner and Right-Angle Prisms

If the right angles of cube corner and right-angle prisms are exact without any error, they reflect an incident plane wavefront as a single emerging plane wavefront. Otherwise the reflected wavefront consists of several plane wavefronts. Thus it is possible to see the errors in the wavefronts reflected from such prisms. Because of the total internal reflection, the intensity of reflected light from these prisms is very high. The reference flat, if it is not coated, will give only a very low reflection, and hence fringes of poor contact result. On the other hand, if the reference flat is coated, a confusing system of fringes will appear because of multiple reflections, when there is any error in the right angle. Hence it is preferable to obtain effectively two-beam interference fringes. This can be done in two ways.

The reference flat, of course, is uncoated. To reduce the effective reflectivity of the right angle or cube prism, we can introduce a parallel plate of glass coated with a metallic film having a transmission between 20 and 30%. In this case the intensities of the two beams matched reasonably well, and we get good-contrast two-beam fringes. The coated plate between the prism and the uncoated reference flat should be tilted sufficiently to avoid the directly reflected beam. This method is shown schematically in Fig. 1.26.

1.2. FIZEAU INTERFEROMETER   31

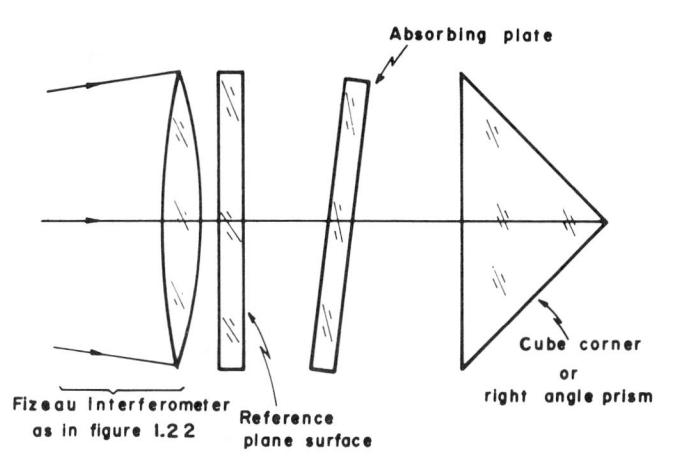

Figure 1.26. Schematic arrangement of a Fizeau interferometer for testing cube corner prisms and right-angle prisms. Here an absorbing plate is inserted between the prisms and the reference flat surface to equalize the intensities of the two interfering beams.

Another possible method is to reduce the reflectivity of one of the total reflecting surfaces. This can be done by constructing a special cell in which the prism is mounted, and behind one reflecting surface a thin layer of water or some other suitable liquid is in contact with the surface. Thus, in effect, the refractive index difference is reduced at one total internal reflecting surface, and hence the intensity of the wavefront reflected from the prism matches that of an uncoated flat. This method is shown schematically in Fig. 1.27. A very good cube corner prism will give rise to an interferogram like that shown in Fig.

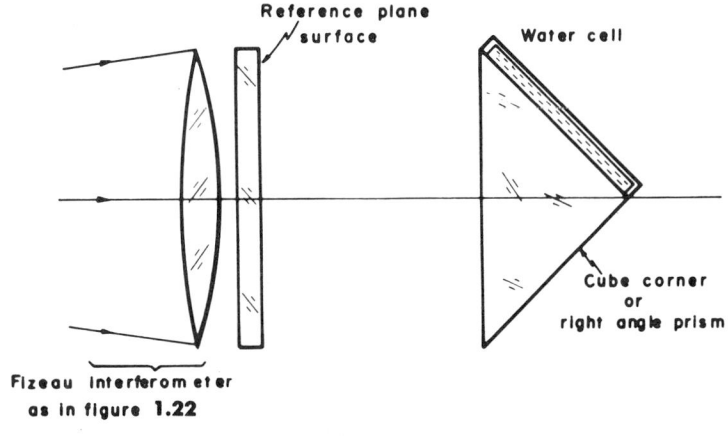

Figure 1.27. A scheme for reducing the intensity of reflected light from the cube corner prism and the right-angle prism. One of the total internally reflecting faces is brought into contact with water or some other liquid by the use of a cell behind it.

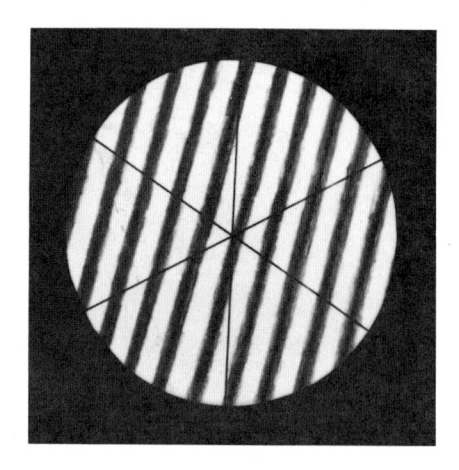

Figure 1.28. The interferogram of a very good cube corner prism. The reference flat surface is to be tilted slightly to obtain the straight fringes.

1.28. The fringes are straight throughout the aperture. A cube corner prism with angular errors produces an interferogram such as is shown in Fig. 1.29, in which the straight fringes abruptly change their direction. Figures 1.30 and 1.31 show similar situations for a right-angle prism of no error and of some angular error, respectively. If, in addition to angle errors, the surfaces are not flat or the glass is not homogeneous, an interferogram with curved fringes is obtained.

We describe here a brief method for obtaining the angular error in a right-angle prism. If the right angle has an error, the fringes look like those shown in Fig. 1.31 and can be manipulated to look like those in Fig. 1.32. If $2L$ is the width of the face of the prism, $\pi/2 \pm \epsilon$ is the angle of the prism, d is the distance between two successive fringes, k is the deviation of the fringe from the straight fringe after bending, N is the refractive index of the prism, and λ

Figure 1.29. The interferogram of a cube corner prism with angular errors.

1.2. FIZEAU INTERFEROMETER

Figure 1.30. The interferogram of a good right-angle prism.

Figure 1.31. The interferogram of a right-angle prism with a slight error in the 90° angle.

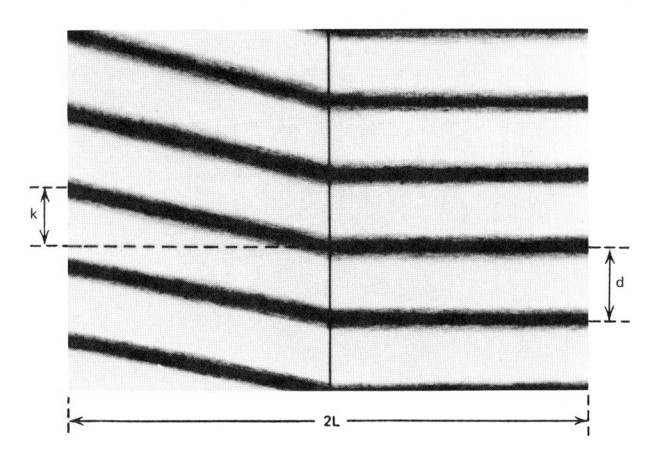

Figure 1.32. The interferogram of a right-angle prism with a slight error in the right angle. The fringes are adjusted so that they are perpendicular to the roof edge on one side.

is the wavelength used, the error is given by

$$\epsilon = \left(\frac{k}{d}\right)\left(\frac{\lambda}{4NL}\right). \tag{1.26}$$

For example, for a prism of 100-mm face width and $k/d = 0.25$, the error ϵ of the 90° angle is about 1 s of arc. In regard to the sign of the error, the hot rod or finger procedure described in Section 1.2.3 can be used.

1.2.10. Testing Concave and Convex Surfaces

The reference surface is again the uncoated flat surface that is part of the Fizeau interferometer. The collimated light from the instrument, after passing through the optical flat, is again focused by the use of another highly corrected lens. If the surface is concave, it is set up as shown in Fig. 1.33; if convex, as shown in Fig. 1.34. When the surface is spherical and the center of curvature coincides with the focus of the lens, a plane wavefront is reflected back. Hence we should obtain straight fringes due to interference of the two beams. If the optical reference flat and the spherical surfaces are coated with high reflecting material, we can get very sharp, multiple-beam Fizeau fringes. If the surfaces are not spherical but are aspheric, appropriate null lenses must be used in the interferometer. This setup can also be used to measure the radius of curvature if a length-measuring arrangement is provided.

Another interferometer, which may be considered as a Fizeau interferometer similar to the scheme shown in Fig. 1.19, was devised by Shack (Shack and Hopkins 1979; Smith 1979). The difference is that this scheme uses a He–Ne laser source to give very large coherence length. Hence, the separation between the convex reference surface and the concave surface under test can be very

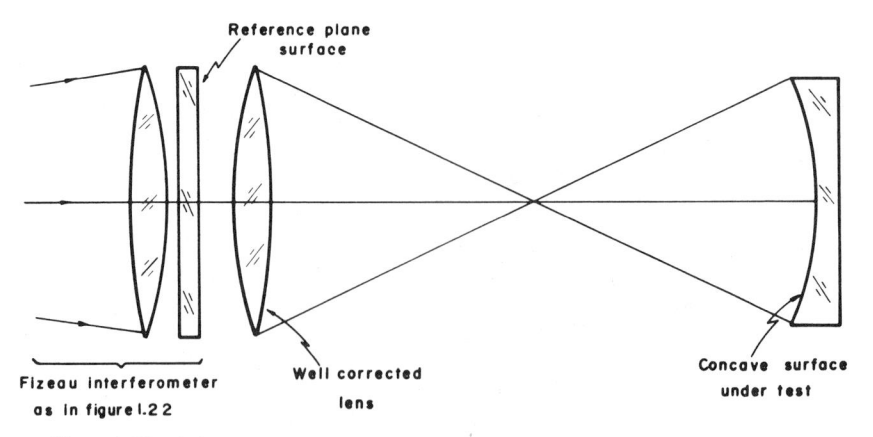

Figure 1.33. Schematic diagram of a Fizeau interferometer for testing a concave surface.

1.2. FIZEAU INTERFEROMETER

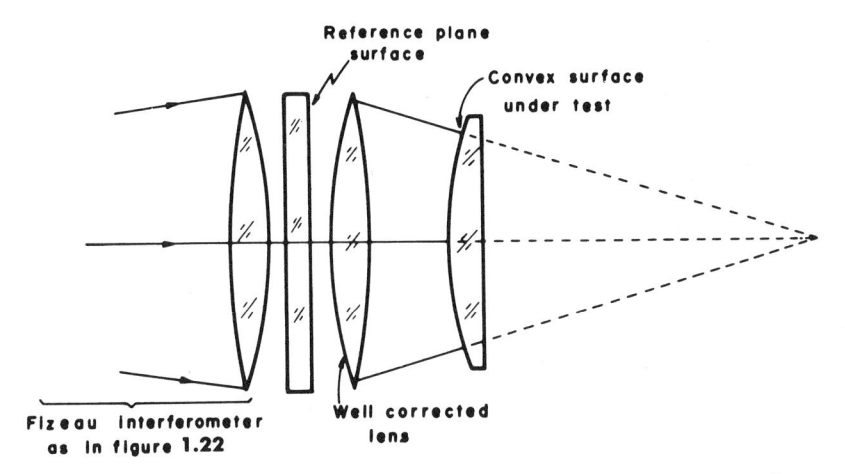

Figure 1.34. Schematic diagram of a Fizeau interferometer for testing a convex surface.

large (typically several meters). Also, the convex reference surface becomes part of the instrument and can be of very short radius of curvature. The scheme, in fact, incorporates the device in the form of a beam divider cube with one of the faces made into a convex spherical surface. The Shack interferometer is shown schematically in Fig. 1.35. It is possible to test a large aspherical surface with this interferometer if a suitable null corrector is inserted between the interferometer and the surface under test.

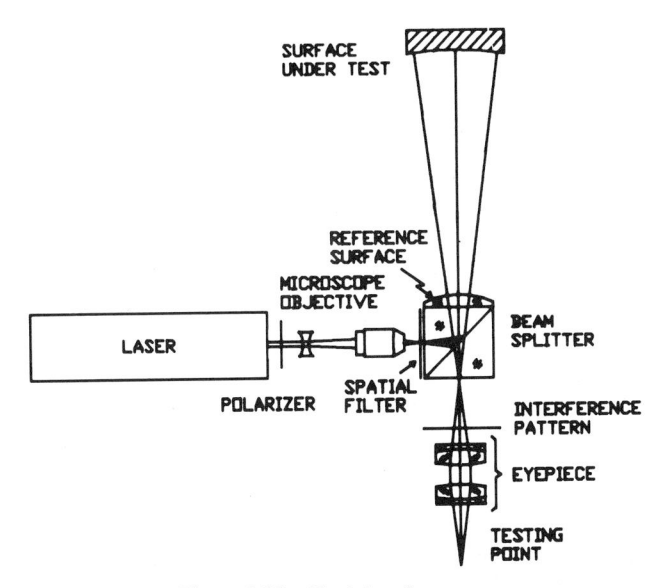

Figure 1.35. Shack interferometer.

1.2.11. Quality of Collimation Lens Required

We shall briefly examine the quality of collimating lens required for the Fizeau interferometer. Basically we are interested in determining the variation in air gap thickness. However, the OPD is a function of not only air gap thickness but also the angle of illumination, and at a particular point this is $2t \cos \theta$. The air gap t varies because of the surface defects of the flats under test, while the variation of θ is due to the finite size of the source and to the aberration of the collimating lens.

For Fizeau interferometers using conventional sources of light, the maximum air gap that is useful is 50 mm. Also, in this case we have to consider the size of the source and the aberration of the lens separately. The effect of the size of the source is mainly on the visibility of the Fizeau fringes. The excess optical path difference $t\theta^2$ should be less than $\lambda/4$ for good contrast of the Fizeau fringes, and the pinhole is chosen to satisfy this condition. The effect of the pinhole is uniform over the entire area of the Fizeau fringes. On the other hand, the effect of aberration in the collimating lens is not uniform. Thus we have to consider the angular aberration of the lens and its effect. If ϕ is the maximum angular aberration of the lens, then $t\phi^2$ should be less than $k\lambda$, where k is a small fraction that depends on the accuracy required in the instrument. Thus let us set $k = 0.001$ so that the contribution of $t\phi^2$ is 0.001λ. Taking a maximum value of $t = 50$ mm for the ordinary source situation, we have

$$\phi^2 \leq \frac{0.001\lambda}{t} \approx 10^{-8}$$

or

$$\phi \leq 10^{-4} \text{ radian.} \tag{1.27}$$

This angular aberration is quite large, being of the order of 20 sec of arc. Hence suitable lenses or mirror systems can be designed for the purpose (Taylor 1957, Yoder 1957, Murty and Shukla 1970).

1.3. HAIDINGER INTERFEROMETER

With the Newton and Fizeau interferometers we are basically interested in finding the variation in the air gap thickness. In these cases the fringes are referred to as fringes of equal thickness. If, however, the thickness of the air gap is uniform and it is illuminated by a source of large angular size, we get what are called fringes of equal inclination. These fringes are formed at infinity, and a

1.3. HAIDINGER INTERFEROMETER

suitable lens can be used to focus them on its focal plane. If the parallel gap is that of air, we have the simple relation $2t \cos \theta = n\lambda$, as given in Eq. (1.9), from which we can easily see that, for a constant value of t, we obtain fringes of equal inclination that are circles and are formed at infinity.

If the air gap is replaced by a solid plate such as a very good parallel plate of glass, Eq. (1.9) is modified slightly to include the effect of the refractive index N of the plate and becomes

$$2Nt \cos \theta' = n\lambda \tag{1.28}$$

where θ' is the angle of refraction inside the glass plate. For small values of θ', we may approximate this expression as

$$2Nt - \left(\frac{t}{N}\right)\theta^2 = n\lambda. \tag{1.29}$$

To see Haidinger fringes with simple equipment, the following method, illustrated in Fig. 1.36, may be adopted. A parallel plate of glass is kept on a black paper and is illuminated by the diffuse light reflected from a white card at 45°. At the center of the white card is a small hole through which we look at the plate. With relaxed accommodation our eyes are essentially focused at infinity, and we see a system of concentric circular fringes. For the light source we can use a sodium or even a fluorescent lamp.

In the situation where the laser is the source of light, there is a much higher

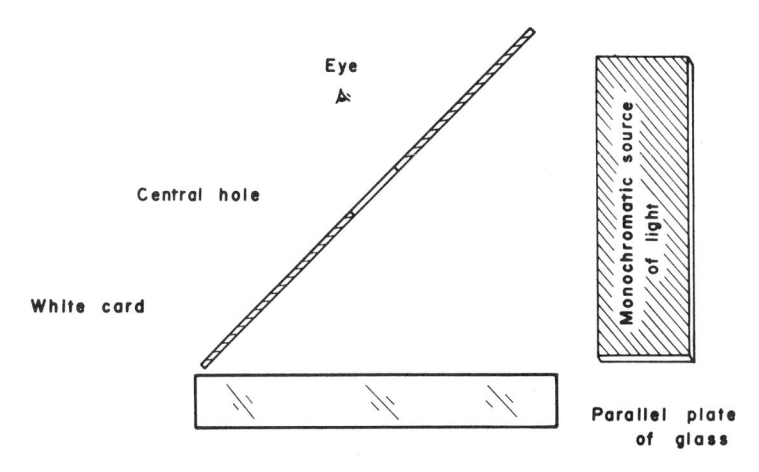

Figure 1.36. A simple arrangement to see Haidinger fringes for a nearly parallel glass plate.

limit for the value of t. Even though several meters can be used for t, we shall set $t = 1000$ mm. In this case, using $\lambda = 6.328 \times 10^{-4}$ mm, we get for ϕ an upper limit of 5 s of arc. Hence it is not difficult to design a collimating system to satisfy this condition.

Another aspect that is important, especially with large values of t, is the lateral shear one can get in the instrument. To avoid this, the autocollimated pinhole images must coincide with the pinhole itself. Similarly, if the collimating lens is not properly collimated, either a convergent or a divergent beam will emerge. The collimation may be accurately performed by using any of the various devices available, such as the plane parallel plate shearing interferometer (Murty 1964b).

A somewhat better method is to use a lens for focusing the system of Haidinger fringes on its focal plane. This requires a setup almost identical to that for the Fizeau interferometer. The only difference is that, instead of a pinhole, a wider aperture is used to have a large angular size for the source. The Haidinger fringes are then formed in the focal plane of the lens.

1.3.1. Applications of Haidinger Fringes

The Haidinger fringes may be used as a complementary test to that provided by the Fizeau interferometer. If we are testing a nearly parallel plate, we can find its wedge angle either by the Fizeau or by the Haidinger method. In the Haidinger method we look for the stability of the concentric fringes as we move our line of sight across the plate with a small aperture. If t is slowly varying, the center of the circular fringe system also appears to change. If t is decreasing, we are moving toward the thinner side of the wedge, and in this case the circular Haidinger fringes appear to expand from the center. On the contrary, the fringes appear to converge to the center if we are moving toward the thick side of the wedge. If we note how many times the center of the fringe system has gone through bright and dark cycles, we can also estimate the wedge angle in the same manner as for the Fizeau situation.

1.3.2. Use of Laser Source for Haidinger Interferometer

A helium–neon laser source of low power is very useful for this interferometer, as it is for the Fizeau instrument. It enables the fringes to be projected on a screen. In this case the laser can be made to give effectively a point source of light, and consequently the Haidinger fringes can be considered as the interference from two point sources that are coherent to each other. Hence it is possible to obtain the circular fringes even at a finite distance from the two coherent point sources, and no lens is needed to form the fringes in its focal plane. Figure 1.37 shows the two point images of a point source reflected from a glass plate having a wedge. For the purpose of analysis, it is sufficient for us to consider

1.3. HAIDINGER INTERFEROMETER

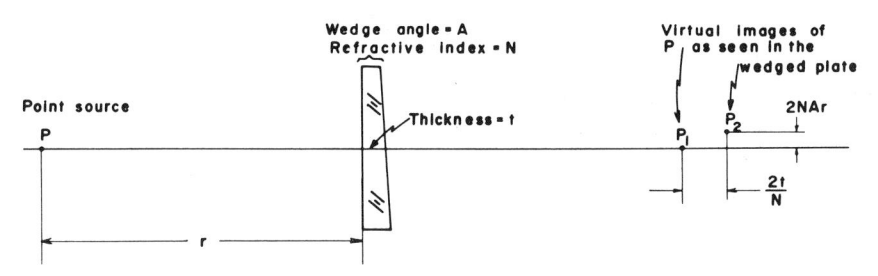

Figure 1.37. Various parameters related to the formation of two virtual coherent sources from a single point source by a wedged plate.

two point sources of light that are coherent to each other. Then, if we place a screen sufficiently far away and perpendicular to the line joining the two sources, we get a system of concentric circular fringes similar to Newton's rings and the center of the fringes is collinear with the two point sources. Also, for a glass plate of refractive index N, the distance between the virtual point sources is $2t/N$, where t is the thickness of the plate. Now, if the glass plate has a small wedge, the two virtual sources will also have a slight lateral displacement with respect to each other; this is given by $2NAr$, where A is the wedge angle and r is the distance of the point source from the wedged plate. These various parameters are illustrated in Fig. 1.37.

To apply this theory in practice, several methods are available. One method, proposed by Wasilik et al. (1971), is illustrated in Fig. 1.38. The laser beam is allowed to pass through a small hole in a white cardboard and is then incident on the glass plate under test. To provide some divergence for the laser beam, a negative or positive lens of about 50- to 100-mm focal length is introduced

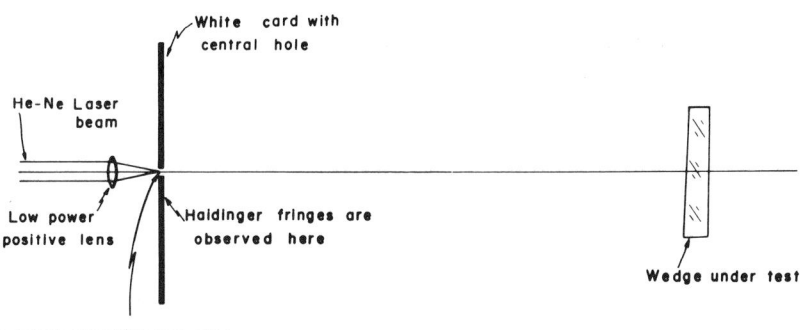

Figure 1.38. A schematic arrangement for observing the Haidinger fringes and measuring the displacement of the center. Here a laser beam is passed through a hole is a cardboard, and the Haidinger fringes are observed around the hole on the cardboard.

behind the cardboard centrally. The lens can be fixed in such a manner that it does not deviate the beam but only expands it slightly. This cardboard may be made specially, along with the lens, to fit on the laser. Several concentric circles with known spacing may be drawn on the cardboard for measuring purposes. The plate under test is kept on a platform that can be tilted. The plate is adjusted until the spot of laser light reflected from it goes back through the hole in the cardboard. In this situation concentric circular Haidinger fringes will be seen on the cardboard surrounding the hole. If the plate is free from the wedge, the center of the Haidinger fringe system coincides with the center of the hole. If the plate has a wedge, the center of the Haidinger fringe system is displaced with respect to the center of the hole. An approximate formula relating this displacement to the wedge angle of the wedged glass plate is as follows:

$$d = \frac{2N^2 r^2 A}{t} \qquad (1.30)$$

where d = displacement of the Haidinger fringe system,
A = wedge angle of the plate,
t = thickness of the glass plate,
N = refractive index of the glass plate,
r = distance of the point source from the glass plate.

For example, if $A = 1$ s of arc (5×10^{-6} rad), $N = 1.5$, $r = 1000$ mm, and $t = 10$ mm, we have $d = 2.25$ mm, which can be easily detected. Hence this method is quite sensitive and useful.

Another method is illustrated in Fig. 1.39. Here the laser beam passes through

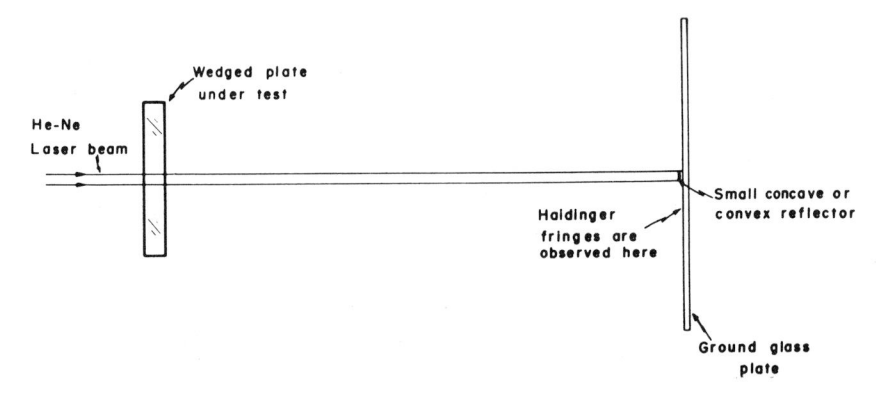

Figure 1.39. A schematic arrangement for observing the Haidinger fringes and measuring the displacement of the center. Here the laser beam is directed back into the wedged glass plate by a small concave or convex mirror on the ground glass. The Haidinger fringes are formed on the ground glass.

1.3. HAIDINGER INTERFEROMETER

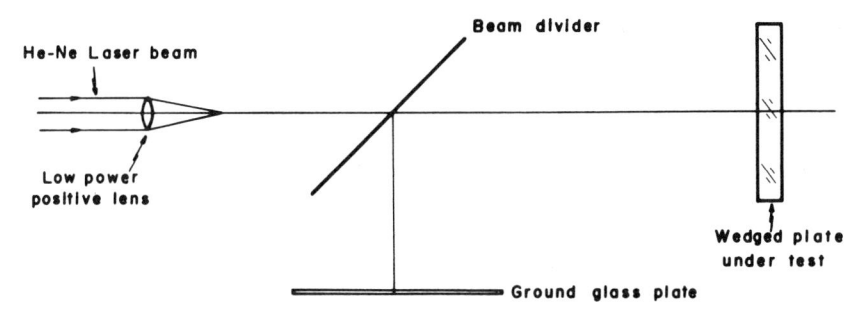

Figure 1.40. A schematic arrangement for observing the Haidinger fringes and measuring the displacement of the center. Here the fringes are observed on a ground glass, and by means of a beam divider, the central obscuration is avoided.

the wedged glass plate and falls on a specially prepared ground-glass plate in the center of which a small concave or convex reflector of about 50–100 mm radius of curvature is cemented. The size of the reflector should be slightly greater than the spot size of the laser. Thus the laser beam is reflected back onto the glass plate. The wedged plate is adjusted until the reflected spot from it coincides with the small reflecting mirror on the ground glass. Now Haidinger fringes can be seen on the ground glass, and the center of the fringe system is displaced with respect to the mirror on the ground glass. The same formula, Eq. (1.30), is valid for this case also.

A third method utilizes a beam divider, as shown in Fig. 1.40. The laser beam passes through the beam divider, after reflection from the wedged plate is again reflected at the beam divider, and finally falls on a ground-glass screen. The plate under test is adjusted until the laser spot reflected from it goes back on itself. After the position of the spot on the ground glass has been noted, a negative or positive lens is introduced into the laser beam close to the laser side. This widens the beam sufficiently so that circular Haidinger fringes can be seen on the ground-glass screen. The displacement of the center of the Haidinger fringe system is measured, and the same formula, Eq. (1.30), can be used for calculating the wedge angle A.

1.3.3. Other Applications of Haidinger Fringes

We have discussed earlier the application of Haidinger fringes for the determination of the very small wedge angle of a nearly parallel plate of glass. There are many types of prisms that can be reduced to equivalent parallel plates and hence can be tested for deviation from their nominal angles. A typical example is a right-angle prism with nominal angles of 90°, 45°, and 45°. In such a prism it is usually required that the 90° angle be very close to its nominal value and that the two 45° angles be equal to each other. In addition, all the faces of

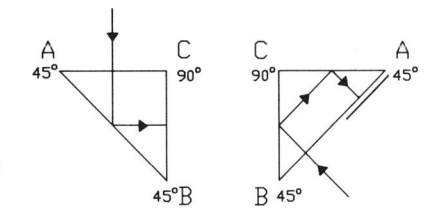

Figure 1.41. Schematic of the 45°–90°–45° prism to be equivalent parallel plate of glass.

the prism should be perpendicular to a base plane. If not, we say that the prism has pyramidal error that is objectionable in many applications. Figure 1.41 shows how the right-angle prism can be treated as an equivalent parallel plate with very small wedge angle. If the beam is incident first on the face *AC*, the beam returning after reflection from the face *BC* is nearly parallel to the one reflected from the face *AC*, and hence Haidinger fringes are seen as a result of the interference between these two beams. This arrangement checks the equality of the angles *A* and *B*. If there were no pyramidal error and the two angles are equal, the center of the Haidinger fringes will be exactly at the center of the beam spot. If the angles are equal but there is a pyramidal error, the center of the Haidinger fringes will be displaced vertically. If both errors are present, the center will be displaced both vertically and horizontally. The effect of the pyramidal error is to rotate the line of intersection of the two planes of the equivalent wedge so that it is neither vertical nor horizontal. If the beam is incident first on *AB*, the return beam reflected from the internal face of *AB* will be nearly parallel to the one reflected from *AB* externally, and hence we get again Haidinger fringes due to the interference of these two beams. This arrangement checks the exactness of the 90° angle of the angle *C*. If the center of the Haidinger fringes is not displaced horizontally, the 90° angle is exact; and, if in addition there is no vertical displacement, there is no pyramidal error. More details of this method may be found in the paper by Saxena and Yeswanth (1990).

Other examples of prisms that may be treated as equivalent parallel plates

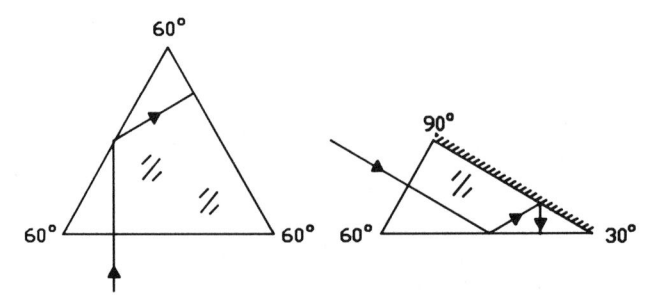

Figure 1.42. Schematics of two other prisms to be equivalent parallel plates of glass.

1.4. ABSOLUTE TESTING OF FLATS

are shown in Fig. 1.42. Readers may come across other examples depending on their particular situations.

The formula given in (1.29) is applicable in the situations described noting that now the displacement has two components—one in the vertical direction and one in the horizontal direction.

1.4. ABSOLUTE TESTING OF FLATS

Until now we have considered the testing of flats against a "perfect" flat taken as a reference. It is, however, often necessary to make a flat when a good reference flat is not available. In this case an alternative is to use a liquid flat, as mentioned in Section 1.2.2. Another possibility is to make three flats at the same time and test them with many combinations in order to obtain the absolute departure of the three surfaces with respect to an ideal flat.

Let us assume that we have the three surfaces in Fig. 1.43 and that they are tested in the three combinations shown in Figure 1.44. From these measurements we can determine the functions $g_{AB}(x, y)$, $g_{AC}(x, y)$ and $g_{BC}(x, y)$, defined as follows:

$$f_A(x, y) + f_B(-x, y) = g_{AB}(x, y),$$
$$f_A(x, y) + f_C(-x, y) = g_{AC}(x, y),$$
$$f_B(x, y) + f_C(-x, y) = g_{BC}(x, y), \tag{1.31}$$

where $f_A(x, y)$, $f_B(x, y)$, and $f_C(x, y)$ are the functions defining the deviations for the three surfaces. We then have a system of three equations with four unknowns: $f_A(x, y)$, $f_B(-x, y)$, $f_B(x, y)$, and $f_c(-x, y)$. The system has a solution over the whole plane only if we assume symmetry about the y axis, at

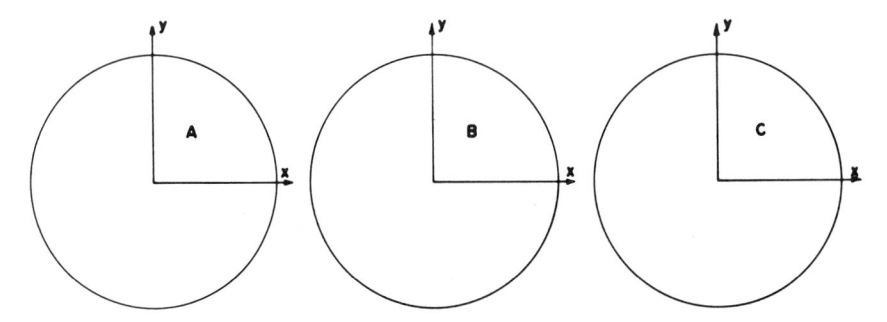

Figure 1.43. Three surfaces whose absolute deviations with respect to an ideal flat plane are to be determined.

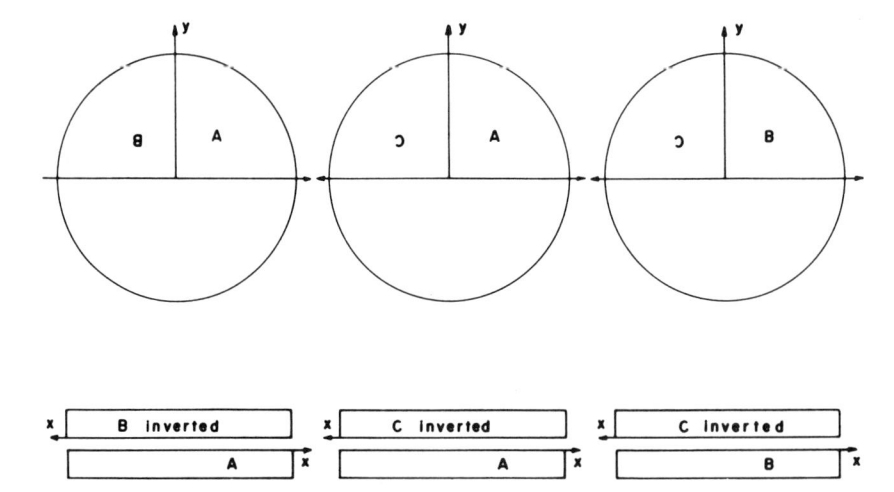

Figure 1.44. Three combinations used to test surfaces A, B, and C.

least for surface B, $f_B(x, y) = f_B(-x, y)$; otherwise a solution exists only at the y axis ($x = 0$), as follows:

$$f_A(0, y) = \frac{g_{AB}(0, y) + g_{AC}(0, y) - g_{BC}(0, y)}{2},$$

$$f_B(0, y) = \frac{g_{AB}(0, y) - g_{AC}(0, y) + g_{BC}(0, y)}{2},$$

$$f_C(0, y) = \frac{-g_{AB}(0, y) + g_{AC}(0, y) + g_{BC}(0, y)}{2}. \tag{1.32}$$

Without having to assume any kind of symmetry, information about the whole surface can be obtained with only three surfaces, but it is necessary to test them in more than three combinations that include rotations of one surface with respect to another. This problem is treated comprehensively by Schulz and Schwider (1976); their work also includes the testing of spherical surfaces.

The problem of the absolute testing and calibration of an optical flat has also been studied by Fritz (1983, 1984) and by Truax (1988).

REFERENCES

Forman, P. F., "A Note on Possible Errors Due to Thickness Variations in Testing Nominally Parallel Plates," *Appl. Opt.*, **3**, 646 (1964).

REFERENCES

Forman, Paul F., "The Zygo Interferometer System," *Proc. SPIE*, **192**, 41 (1979).

Fritz, Bernard S., "Absolute Calibration of an Optical Flat," *Proc. SPIE*, **433**, 123 (1983).

Fritz, Bernard S., "Absolute Calibration of an Optical Flat," *Opt. Eng.*, **23**, 379 (1984).

Grigor'ev, V. A., Ya. O. Zaborov, and P. P. Ivanov, "Use of a Liquid Mirror for the Calibration of an Interferometer," *Sov. J. Opt. Technol.*, **53**, 613 (1986).

Hariharan, P. and W. H. Steel, "Fringe Localization Depth: A Comment," *Appl. Opt.*, **28**, 29 (1989).

Karow, Hank H., "Interferometric Testing in a Precision Optics Shop: A Review of Testplate Testing," *Proc. SPIE*, **192**, 56 (1979).

Ketelsen, Dean A. and David S. Anderson, "Optical Testing with Large Liquid Flats," *Proc. SPIE*, **966**, 365 (1988).

Malacara, D. and A. Cornejo, "Testing of Aspherical Surfaces with Newton Fringes," *Appl. Opt*, **9**, 837 (1970).

Moore, Robert C. and F. H. Slaymaker, "Direct Measurement of Phase in a Spherical-Wave Fizeau Interferometer," *Appl. Opt.*, **19**, 2196 (1980).

Murty, M. V. R. K., "A Note on the Testing of Large Aperture Plane Parallel Plates of Glass," *Appl. Opt.*, **2**, 1337 (1963).

Murty, M. V. R. K., "Addendum to: A Note on the Testing of Large Aperture Plane Parallel Plates of Glass," *Appl. Opt.*, **3**, 784 (1964a).

Murty, M. V. R. K., "The Use of a Single Plane Parallel Plate as a Lateral Shearing Interferometer with a Visible Gas Laser Source," *Appl. Opt.*, **3**, 531 (1964b).

Murty, M. V. R. K. and R. P. Shukla, "Some Considerations of the Fizeau Interferometer," *Bull. Opt. Soc. India*, **4**, 13 (1970).

Primak, William, "Optical Flatness Standard," *Opt. Eng.*, **23**, 806 (1984).

Primak, William, "Optical Flatness Standard II: Reduction of Interferograms," *Proc. SPIE*, **954**, 375 (1989a).

Primak, William, "Optical Flatness Standard: Comment," *Opt. Eng.*, **28**, 934 (1989b).

Saxena, A. K. and L. Yeswanth, "Low Cost Method for Subarcsecond Testing of a Right Angle Prism," *Opt. Eng.*, **29**, 1516–1520 (1990).

Schulz, Gunter and J. Schwider, in *Progress in Optics*, vol. 13, E. Wolf, Ed., North-Holland, Amsterdam, 1976, Chap. IV.

Schulz, Gunter and J. Schwider, "Comments on the Paper 'Optical Flatness Standard,'" *Opt. Eng.*, **26**, 559 (1987).

Shack, Roland V. and George W. Hopkins, "The Shack Interferometer," *Opt. Eng.*, **18**, 226 (1979).

Simon, J. M. and S. A. Comatri, "Fringe Localization Depth," *Appl. Opt.*, **26**, 5125 (1987).

Smith, W. Scott, "Versatile Interferometer for Shop Use," *Proc. SPIE*, **192**, 13 (1979).

Taylor, W. G. A., "Spherical Aberration in the Fizeau Interferometer," *J. Sci. Instrum.*, **34**, 399 (1957).

Truax, Bruce E., "Absolute Interferometric Testing of Spherical Surfaces," *Proc. SPIE*, **966**, 130 (1988).

Wasilik, H., T. V. Biomquist, and C. S. Willet, "Measurement of Parallelism of the

Surfaces of a Transparent Sample Using Two-Beam Nonlocalized Fringes Produced by Laser," *Appl. Opt.*, **10,** 2107 (1971).

Wyant, James C., "Fringe Localization," *Appl. Opt.*, **17,** 1853 (1978).

Yoder, P. R., Jr., and W. W. Hollis, "Design of a Compact Wide Aperture Fizeau Interferometer," *J. Opt. Soc. Am.*, **47,** 858 (1957).

ADDITIONAL REFERENCES

Newton Interferometer

Blissinger, H. D., "How to Interpret an Interferogram," in *Optical Interferograms—Reduction and Interpretation.* ASTM Symposium, ASTM Tech. Publ. 666, A. H. Guenther and D. H. Liedberg, Eds., Am. Soc. for Test. and Mat., Florida, 1978.

Carman, P. D., "Control and Interferometric Measurement of Plate Flatness," *J. Opt. Soc. Am.*, **45,** 1009 (1955).

Einsporn, E., "Uber die Verbesserung der Gute Prüfung optischer Planschliffe" (About the Finesse, Correction, and Testing of Optical Flats), *Optik*, **7,** 174 (1950).

Emerson, W. B., "Determination of Planeness and Bending of Optical Flats," *J. Res. Nat. Bur. Stand*, **49,** 241 (1952).

Harper, D. C., "Preparation of Drawings for Optical Elements and Methods of Testing," *Appl. Opt.*, **9,** 527 (1970).

Glassman, A. T., "A Manual Method for Interpretation of Nominally Plane Wavefronts," in *Optical Interferograms—Reduction and Interpretation.* ASTM Symposyum, ASTM Tech. Publ. 666, A. H. Guenther and D. H. Liedberg, Eds., Am. Soc. for Test. and Mat., Florida, 1978.

Karlin, O. G. and V. A. Syutkin, "The Use of Spherical and Aspherical Test Glasses for Inspecting Aspherical Surfaces," *Soc. J. Opt. Technol.*, **39,** 156 (1972).

Kruizinga, Bob, "Testing of Cylindrical Optical Surfaces," *Proc. SPIE*, **656,** 183 (1986).

Landewehr, R., "Zur Messung der Ebenheit von reflecktierenden Flachen Mittels Interferenzen gleichet Dicke"(The measurement of Planes Reflecting Surfaces Using Fringes of Equal Thickness), *Optik*, **5,** 354 (1949).

Saunders, J. B., "Testing of Large Optical Surfaces with Small Test Plates," *J. Res. Nat. Bur. Stand*, **53,** 29 (1954).

Schulz, G., "Ein Interferenzverfahren zur absoluten Ebenheitsprüfung langs beliebiger Zentralschnitte" (An Interference Method for the Absolute Evenness Test along the Longitudinal Axis in a Central Plate), *Opt. Acta,* **14,** 375 (1967).

Schulz, G., "Interfentielle Absoluteprüfung zweier Flachen" (Absolute Interferometric Test for Two Surfaces), *Opt. Acta,* **20,** 699 (1973).

Schulz, G., J. Schwider, C. Hiller, and B. Kicker, "Establishing an Optical Flatness Standard," *Appl. Opt*, **10,** 929 (1971).

Schwider, J., "Ein Interferenzverfahren zur Absoluteprüfung von Planflächenermalen. II" (An Interference Method for the Absolute Test of Flats, II), *Opt. Acta*, **14**, 389, (1967).

Schwider, J., "Absolute Flachenprüfung durch Kombination Eines Normals mit Einem Kompensationshologramm," *Opt. Communic.*, **6**, 58 (1972).

Schwider, J., G. Schulz, R. Riekher, and G. Minkwitz, "Ein Interferenzverfaheren zur Absoluteprüfung von Planflächenermalen. I" (An Interference Method for the Absolute Test of Flats I), *Opt. Acta*, **13**, 103 (1966).

Fizeau Interferometer

Ashton, A. and A. C. Marchant, "Note on the Testing of Large Glass Panels," *Opt. Acta*, **14**, 203 (1967).

Augustyn, Walter H., "Aspheric Testing by Conventional Interferometry," *Proc. SPIE*, **220**, 71 (1980).

Barrell, H. and R. Marriner, "Liquid Surface Interferometry," *Nature*, **162**, 529 (1948).

Barrell, H. and R. Marriner, "A Liquid Surface Interferometer," *Br. Sci. News*, **2**, 130 (1949).

Barrell, H. and J. S. Preston, "An Improved Beam Divider for Fizeau Interferometers," *Proc. Phys. Soc.*, **B64**, 97 (1951).

Biddles, B. J., "A Non-contacting Interferometer for Testing Steeply Curved Surfaces," *Opt. Acta*, **16**, 137 (1969).

Bruce, C. F. and W. A. F. Cuningham, "Measurement of Angle by Interferometry," *Aust. J. Appl. Sci.*, **1**, 243 (1950).

Bunnagel, R., "Investigation of the Use of a Liquid Surface Mirror for a Flat Plane of Reference," *Z. Angew. Phys.*, **8**, 342 (1956).

Bunnagel, R., H. A. Oehring, and K. Steiner, "Fizeau Interferometer for Measuring Flatness of Optical Surfaces," *Appl. Opt.*, **7**, 331 (1968).

Chapham, P. B. and G. D. Dew, "Surface-Coated Reference Flats for Testing Fully Aluminized Surfaces by Means of the Fizeau Interferometer," *J. Sci. Instrum*, **44**, 899 (1967).

Collyer, P. W., "A Method of Sharpening Fizeau Fringes," *J. Opt. Soc. Am*, **41**, 285 (1951).

Dew, G. D., "A Method for the Precise Evaluation of Interferograms," *J. Sci. Instrum*, **41**, 160 (1964).

Dew, G. D., "The Measurement of Optical Flatness," *J. Sci. Instrum*, **43**, 409 (1966).

Dew, G. D., "Systems of Minimum Deflection Supports for Optical Flats," *J. Sci. Instrum.*, **43**, 809 (1966).

Dew, G. D., "Optical Flatness Measurement—The Construction and Use of Fizeau Interferometer," *N. P. L. Optical Metrology Report* No. 1, 1967.

Domenicali, Peter L., "Infrared Laser Interferometer," *Proc. SPIE*, **192**, 6 (1979).

Dukhopel, I. I. and I. Ye Urnis, "Selection of the Best Interferometer for Quality Control of Spherical Surfaces," *Soc. J. Opt. Technol*, **36**, 545 (1969).

Eliseev, Yu. V., O. A. Klochcova, Yu. P. Kontievskil, and Ya A. Perezhogin, "IKP-2 Interferometer for Testing Flatness," *Sov. J. Opt. Technol.*, **48**, 362 (1981).

Forman, Paul F., "Interferometric Examination of Lenses, Mirrors and Optical Systems," *Proc. SPIE*, **163**, 103 (1979).

Gates, J. W., "A Slow Motion Adjustment for Horizontal Interferometer Mirrors," *J. Sci. Instrum.*, **30**, 484 (1953).

Gates, J. W., "An Interferometer for Testing Sphericity," in *Optics in Metrology*, Pol Mollet, Ed., Pergamon, Oxford, 1960, p. 201.

Geary, Joseph M., "Interferometry on Grazing Incidence Optics," *Opt. Eng.*, **26**, 1225 (1987).

Geary, Joseph M. and Lawrence J. Parker, "New Test for Cylindrical Optics," *Opt. Eng.*, **26**, 813 (1987).

Gorshkov, V. A., V. S. Kryakhtunov, S. P. Nevrov, O. N. Fomin, and Yu. A. Sharov, "Interferometer for Testing the Shape of Plane Surfaces of Large Optical Elements," *Sov. J. Opt. Technol.*, **53**, 138 (1986).

Gubel, N. N., I. I. Dukhopel, Yu. A. Myasnikov, and I. Ye Urnis, "Interferometers for Inspecting Spherical Surfaces Subtended by Large Angles," *Sov. J. Opt. Technol.*, **40**, 27 (1973).

Harris, S. J., The Universal Fizeau Interferometer, Ph. D. Thesis, University of Reading, England, 1971.

Harris, J. S., R. L. Fusek, and J. S. Marcheski, "Stroboscopic Interferometer," *Appl. Opt.*, **18**, 2368 (1979).

Hayes, John, "Interferometry at 10.6 Micrometers," *Lasers and Applications*, Dec., 71 (1985).

Hodgkinson, I. J., "A Method for Mapping and Determining the Surface Defects Function of Pairs of Coated Optical Flats," *Appl. Opt.*, **8**, 1373 (1969).

Hunt, P. G., "Optical Cements: A Laboratory Assessment," *Opt. Acta*, **14**, 401 (1967.)

Kafri Oded and Kathi Kreske, "Comparison and Combined Operation of a Moire Deflectometer, Fizeau Interferometer, and Schlieren Device," *Appl. Opt.*, **27**, 4941 (1988).

Kafri Oded and Kathi Kreske, "Flatness Analysis of Hard Disks," *Opt. Eng.*, **27**, 878 (1988).

Komrakov, B. M. and V. A. Chudakova, "Testing the Shape of Aspherical Surfaces by Means of a Laser Interferometer," *Sov. J. Opt. Technol.*, **48**, 608 (1981).

Kontiyevskiy, Yu, P., O. A. Klochkova, and A. Ya Perezhogin, "An Improved Two-Beam Interferometer," *Soc. J. Opt. Technol.*, **35**, 559 (1968).

Laurent, M. L., "Sur Plusieurs Appareils d'Optique, Destines a Controler les Surfaces Planes: Paralleles, Perpendiculars et Obliques" (About Several Kinds of Optical Apparatus, Designed to Control Flat Surfaces: Parallel, Perpendicular, and Oblique). *C. R. Acad. Sci. (Paris)*, **94**, 134 (1883).

Magner, T. J., J. Zaniewski, S. Rice, and C. Fleetwood, "Fabrication and Testing of Off-Axis Parabolic Mirrors," *Opt. Laser Technol.*, **19**, 91 (1987).

Moreau, B. G. and R. E. Hopkins, "Application of Wax to Fine Ground Surfaces to Simulate Polish," *Appl. Opt.*, **8**, 2150 (1969).

Polster, H. D., "The Determination of the Absolute Contours of Optical Flats. II," *Appl. Opt.*, **7,** 977 (1968).

Primak, W., "The Determination of the Absolute Contours of Optical Flats," *Appl. Opt.*, **6,** 1917 (1967).

Puryaev, D. T. and N. L. Lazareva, "Interferometer for Testing the Shape of Concave Spherical Surfaces," *Sov. J. Opt. Technol.*, **49,** 156 (1982).

Schulz, G. and J. Schwider, "Precise Measurement of Planeness," *Appl. Opt.*, **6,** 1077 (1967).

Sen, D. and P. N. Puntambekar, "An Inverting Fizeau Interferometer," *Opt. Acta*, **12,** 137 (1965).

Sharma, G. K., "Making a Simple Fizeau-Interferometer for Workshop Use," *J. Optics (India)*, **13,** 82 (1984).

Shnurr, Alvin D. and Mann, Allen, "Optical Figure Characterization for Cylindrical Mirrors and Lenses," *Opt. Eng.*, **20,** 412 (1981).

Slomba, Albert F. and L. Montagnino, "Subaperture Testing for Mid-Frequency Figure Control on Large Aspheric Mirrors," *Proc. SPIE*, **429,** 114 (1983).

Truax, Bruce E., "Programmable Interferometry," *Proc. SPIE*, **680,** 10 (1986).

Truax, Bruce E., "Interferometers Eliminate Test Plates During Fabrication," *Appl. Opt.*, **27,** 2090 (1988).

Wright, J. J., "Technical Note—Simple Method for Improving Fringe Contrast When Testing Mirror Surfaces in the Fizeau Interferoscope," *Opt. and Laser Technol*, 327 (1981).

Yatagai, Toyohiko, Shigeru Inaba, Hideki Nakano, and Masane Suzuki, "Automatic Flatness Tester for VLSI," *Proc. SPIE*, **429,** 130 (1983).

Haidinger Interferometer

Bergman, T. G. and J. L. Thompson, "An Interference Method for Determining the Degree of Parallelism of (Laser) Surfaces," *Appl. Opt.*, **7,** 923 (1968).

Ford, D. L. and J. H. Shaw, "Rapid Method of Aligning Fabry-Perot Etalons," *Appl. Opt.*, **8,** 2555 (1969).

Hillenkamp, F., "Note on the Interference Method for Determining the Degree of Parallelism of Laser Surfaces," *Appl. Opt.*, **10,** 1982 (1971).

Raman, C. V. and V. S. Rajagopalan, "Haidinger's Fringes in Non-uniform Plates," *Philos Mag.*, Ser. 7, **29,** 508 (1939).

Raman, C. V. and V. S. Rajagopalan, "Haidinger Interference in Curved Plates," *J. Opt. Soc. Am.*, **29,** 413 (1939).

Roesler, F. L., "Mapping of High Quality Optical Flats without Reflection Coating," *J. Opt. Soc. Am.,* **52,** 471 (1962).

Roesler, F. L. and W. Traub, "Precision Mapping of the Uncoated Optical Flats," *Appl. Opt.*, **5,** 463 (1966).

Schonrock, O., "Testing Planeness of Surfaces by Haidinger's Rings," *Z. Instrumentenkd*, **59,** 31 (1939).

2

Twyman–Green Interferometer

D. Malacara

2.1. INTRODUCTION

The Twyman–Green interferometer is a modification of the Michelson interferometer used to test optical components. It was invented and patented by F. Twyman and A. Green (1916) for the testing of prisms and microscope objectives and was later adapted and applied to the testing of camera lenses (Twyman 1919). The first publications on this instrument were those of Twyman (1918b, 1920, 1921, 1922–1923). The instrument has been very useful and so popular that many review papers (Briers 1972) and books (Twyman 1957, Candler 1951, Horne 1972, Cook 1971, U.S. Department of Defense 1963) describe it in detail.

One of the basic Twyman–Green configurations is illustrated in Fig. 2.1. After the system has been illuminated with a quasi-monochromatic point light source, the light is collimated by means of lens L_1 in order to form flat wavefronts. The wavefront is then divided in amplitude by means of a beam splitter plate. After reflection, light from both mirrors M_1 and M_2 impinges again on the beam splitter. Two interference patterns are then formed, one going to lens L_2 and the other going back to the light source. Lens L_2 permits all of the light from the aperture to enter the eye so that the entire field can be seen. The observed fringes are of the equal thickness type.

It is easy to see that if the beam splitter is all dielectric, the main interference pattern is complementary to the one returning to the source; in other words a bright fringe in one pattern corresponds to a dark fringe in the other. This has to be so because of the conservation of energy principle, even though the optical path difference is the same for both patterns. Phase shifts upon reflection account for this complementarity. The case of an absorbing beam splitter has been treated by Parmigiani (1981).

It is interesting that Michelson (1918) did not consider the instrument appli-

---

Optical Shop Testing, Second Edition, Edited by Daniel Malacara.
ISBN 0-471-52232-5 © 1992, John Wiley & Sons, Inc.

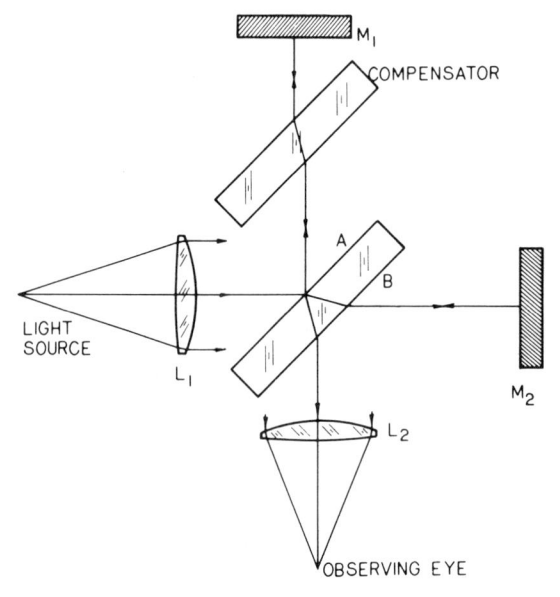

Figure 2.1. Basic Twyman–Green interferometer configuration.

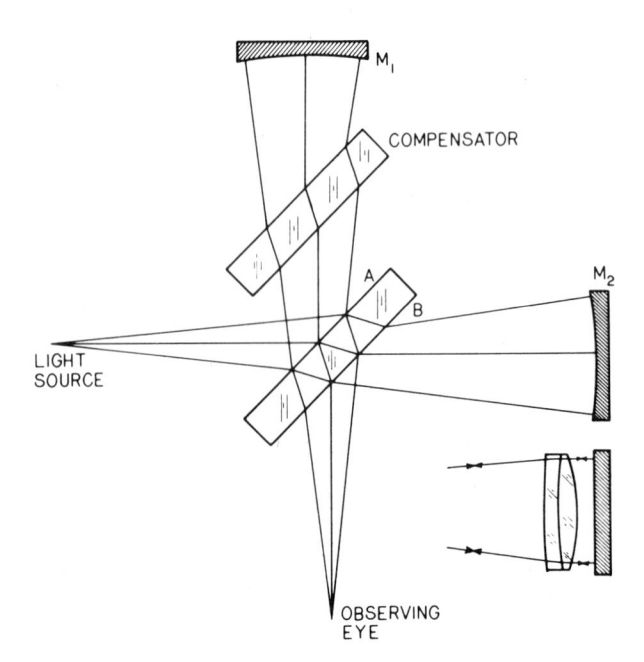

Figure 2.2. Twyman–Green interferometer (Williams type).

2.2. BEAM-SPLITTER PLATE

Figure 2.3. A general-purpose Twyman–Green interferometer.

cable to the testing of large optics, pointing out at the same time that the arrangement we now know as an unequal-path interferometer was impractical because of the lack of sufficiently coherent light sources. To answer Michelson's comments, Twyman (1918a) pointed out that the arrangement shown in Fig. 2.2 had been suggested in his patent (Twyman and Green 1916) for the testing of large mirrors or lenses. This procedure eliminates the need for a large collimator and beam splitter but unfortunately requires (for sources of limited coherence) a concave spherical mirror as large as the optical element under test. This kind of arrangement is often referred to as a Williams interferometer (Grigull and Rottenkolber 1967, De Vany 1965) because Burch (1940) attributed it to Williams. A Twyman–Green interferometer for general laboratory usage is shown in Fig. 2.3.

2.2. BEAM-SPLITTER PLATE

The beam-splitter plate is made in such a way that face A reflects the light in the appropriate amount by means of a partially reflective coating, but face B should not reflect any light. To avoid reflections on the second face a multilayer antireflection coating can be used. However, an easier solution is to place the plate at Brewster's angle and then to use a source of light with p polarization, which goes through the interface without any reflection. Still another solution is to introduce a wedge angle on the plate so that the unwanted reflected light escapes from the system. The contrast of the fringes does not depend on the reflectance R of face A; only the irradiance I on the fringe maxima is affected, since

$$I = 4I_0 RT, \tag{2.1}$$

where I_0 is the irradiance of the incident wavefront and T is the transmittance. If there is no absorption (nonmetallic coating), $T \simeq 1 - R$ and there is a maximum value of I for $R = T = \frac{1}{2}$.

The light going to the observer from mirror M_1 has traversed the beam splitter only once, whereas the light from mirror M_2 has gone through it three times. An interferometer that has more glass in one arm than in the other, as in this case, is said to be uncompensated. The interferometer can be compensated by inserting another piece of glass in front of mirror M_1, as shown in Figs. 2.1 and 2.2.

The importance of compensating an interferometer is clearly seen in the next section. Adjustable compensators for Williams configurations (Steel 1963) and Twyman–Green configurations (Steel 1962; Mertz 1959; Connes 1956) have been described in the literature.

We can show with a few algebraic steps and the law of refraction that a beam splitter or compensating plate shifts the optical axis laterally and parallel to itself by the following amount:

$$d = t \sin \theta \left(1 - \frac{\cos \theta}{(N^2 - \sin^2 \theta)^{1/2}} \right), \qquad (2.2)$$

where θ is the incidence angle, t is the plate thickness, and N is the refractive index.

2.2.1. Optical Path Difference Introduced by the Plate

As pointed out before, an interferometer is said to be uncompensated when it has more glass in one of its arms than in the other (a) because an optical component (lens or prism) is present in one arm, in order to test it or (b) because the light travels once through the beam splitter in one path and three times in the other path, and the compensating plate is absent. Both of these situations can be included in a general case in which an inclined plane glass plate is placed in one of the arms. The unfolded optical paths for both arms of the interferometer are shown in Fig. 2.4. Here we may see that the complete effect is equivalent to going through a system of two plates. The optical path difference (OPD) introduced by one passage through a glass plate is a function of the angle of incidence of the light, as shown in Fig. 2.5, yielding

$$\text{OPD} = N(\overline{AB}) + (\overline{BC}) - (\overline{AD}) \qquad (2.3)$$

and then

$$\text{OPD}(\varphi) = t_N(N \cos \varphi' - \cos \varphi). \qquad (2.4)$$

2.2. BEAM-SPLITTER PLATE

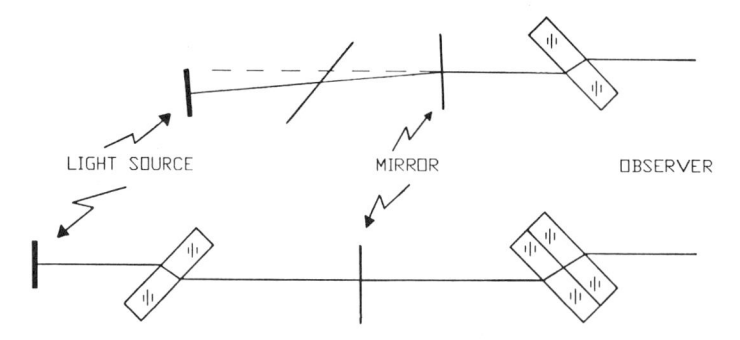

Figure 2.4. Light paths for both interfering beams in an uncompensated interferometer.

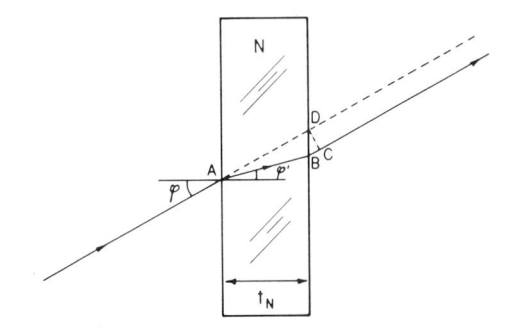

Figure 2.5. Optical path difference introduced by a plane parallel plate.

If the plates are inclined at an angle φ_0 with respect to the optical axis and the ray direction is defined by the angle θ and ψ as in Fig. 2.6, the OPD introduced by both passages may be computed by

$$\text{OPD}(\varphi) = t_N(N \cos \varphi'_1 + N \cos \varphi'_2 - \cos \varphi_1 - \cos \varphi_2) + 2t_0 \cos \theta, \quad (2.5)$$

where subscript 1 and 2 designate the first and second passages, respectively, through the plate. The last term corresponds to an additional OPD introduced by a shift t_0 of one of the mirrors along the optical axis. Angles φ_1 and φ_2 are obtained from angles θ, ψ, and φ_0 by means of the relations

$$\cos \varphi_1 = \cos \varphi_0 \cos \theta + \sin \varphi_0 \sin \theta \cos \psi,$$
$$\cos \varphi_2 = \cos \varphi_0 \cos \theta - \sin \varphi_0 \sin \theta \cos \psi. \quad (2.6)$$

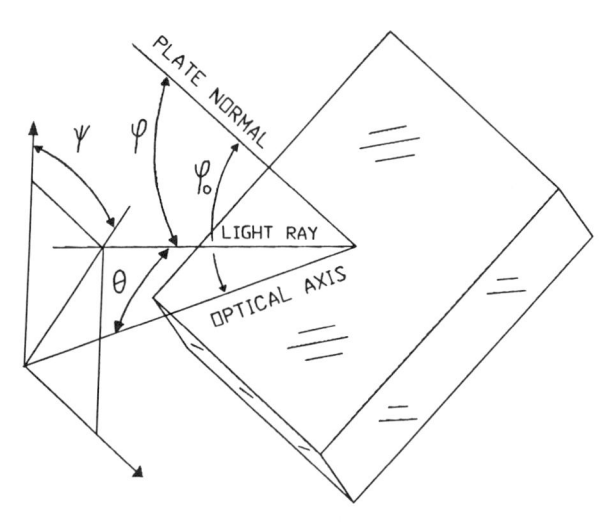

Figure 2.6. Light passing through an inclined plane parallel plate.

If the glass plate is normal to the optical axis, $\varphi_0 = 0$ and thus $\varphi_1 = \varphi_2 = \theta$ and $\varphi_1' = \varphi_2' = \theta'$. In this case Eq. (2.5) reduces to

$$\text{OPD}(\theta) = 2t_N(N \cos \theta' - \cos \theta) + 2t_0 \cos \theta. \tag{2.7}$$

2.2.2. Required Accuracy in the Beam Splitter

When constructing a Twyman–Green interferogram, it is very important to determine the required surface quality in each one of the two faces of the beam splitter for a desired interferometer accuracy. These may be found considering the following model of a beam splitter with three small localized defects with thicknesses δt_1, δt_2, and δt_3, as shown in Fig. 2.7.

In this model we use the fact that in the external reflection in defect 1, if the reflection is displaced a small distance t from the plate, then the optical path is reduced by an amount

$$\text{OPD} = 2t \cos \theta. \tag{2.8}$$

For beams reflected in the two faces of glass plate with plane and parallel faces, there is an optical path difference given by

$$\text{OPD} = 2Nt \cos \theta'. \tag{2.9}$$

Thus in the internal reflection at the localized defect 1, the optical path is increased. Upon transmission through a glass plate with thickness t, the additional

2.2. BEAM-SPLITTER PLATE

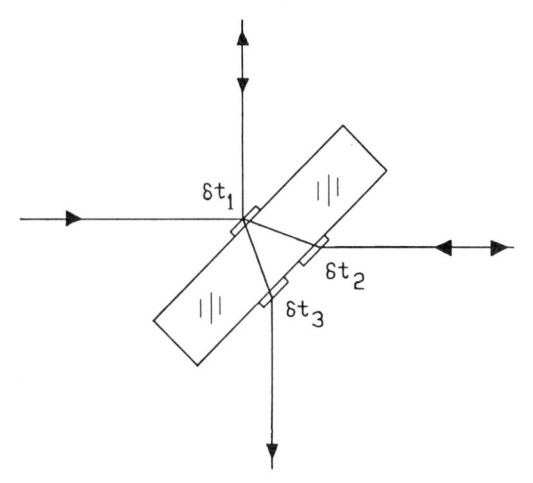

Figure 2.7. Effect of errors in the faces of the beam splitter of a Twyman–Green interferometer.

optical path introduced by the plate is

$$\text{OPD} = t[N \cos \theta' - \cos \theta]. \tag{2.10}$$

Thus the optical path (OP) introduced by the plate defects for the beam reflected from mirror A is

$$\delta \text{OP}_A = -2\delta t_1 \cos \theta + (\delta t_1 + \delta t_3)[N \cos \theta' - \cos \theta], \tag{2.11}$$

and the optical path introduced by the plate defects for the beam reflected from mirror B is

$$\delta \text{OP}_B = 2N\delta t_1 \cos \theta' + (\delta t_1 + \delta t_3 + 2\delta t_2)[N \cos \theta' - \cos \theta]. \tag{2.12}$$

Thus the difference between these two quantities is the OPD introduced by the plate errors:

$$\delta \text{OPD} = 2\delta t_1[\cos \theta + N \cos \theta'] + 2\, \delta t_2[N \cos \theta' - \cos \theta]. \tag{2.13}$$

As expected, this OPD does not depend on the error δt_3 since it is common for both beams. Of course, any error δt_2 is also an error δt_3 for some other ray.

If OPD is the desired interferometer accuracy, the required accuracy δt_1 on the plate face 1 is given by

$$\delta t_1 = \frac{\delta \text{OPD}}{2[\cos \theta + N \cos \theta']}, \tag{2.14}$$

and the required accuracy δt_2 on the plate face 2 is

$$\delta t_2 = \frac{\delta \text{OPD}}{2[N \cos \theta' - \cos \theta]}. \quad (2.15)$$

These results mean that roughly, the reflecting face must be polished with about twice the interferometer required accuracy, while the other face must have a quality of only about half the interferometer accuracy.

2.3. COHERENCE REQUIREMENTS

The size (spatial coherence) and monochromaticity (temporal coherence) of the light source must satisfy certain minimum requirements that depend on the geometry of the system, as described by Hansen (1955, 1984) and by Birch (1979). It is interesting to know that if the optical element under test has very steep reflections, the state of polarization of the light may change in the reflection, introducing changes in the contrast (Ferguson 1982). However, in most cases the important factor in the contrast is the coherence of the light source.

2.3.1. Spatial Coherence

There are two cases for which the collimated wavefront has ray lights spread over a solid angle with diameter 2θ, and hence the final accuracy of the interferometry or the contrast will be reduced:

1. The collimator has spherical aberration, in which case

$$\theta = \frac{\text{TA}}{f} \quad (2.16)$$

where TA is the maximum value of the transverse spherical aberration of the collimator at its best focus position. This aberration might limit the accuracy of the interferometer, unless the OPD remains constant with changes in the angle θ. Otherwise, given the maximum value of θ, the maximum change in the OPD should be smaller than the desired accuracy.

2. The light source is not a mathematical point but has a small diameter $2a$; then

$$\theta = \frac{a}{f} \quad (2.17)$$

where f is the focal length of the collimator.

2.3. COHERENCE REQUIREMENTS

Fringes with high contrast are obtained, using an extended thermal source, only if the OPDs for the two paths from any point of the source (θ) differ by an amount smaller than $\lambda/4$, according to the Rayleigh criterion. On the other hand radiometric considerations usually require as large a source as possible that will not degrade the contrast of the fringes.

When the beam splitter is not compensated by another identical glass plate, we may show that the maximum light source size has an elliptical shape. This is why the fringes are elliptical in an uncompensated Michelson interferometer. The shape and the size of the ellipse not only are functions of φ_0, θ, and ψ but also depend very critically on t_0.

The simpler case of a glass plate with its normal along the optical axis can be analyzed with more detail, as will be shown. The OPD is given by Eq. (2.7). As shown in Fig. 2.8, the value of the OPD changes with the value of θ, depending on the value of t_0. The maximum allowed value of the semidiameter θ of the light source, as seen from the collimator, is that which gives a variation of the OPD equal to $\lambda/4$. On the other hand the maximum allowed value of the angle θ due to spherical aberration of the collimator is that which gives a variation of the OPD equal to the accuracy desired from the interferometer.

When testing small optics using a nonmonochromatic light source, it is convenient to choose

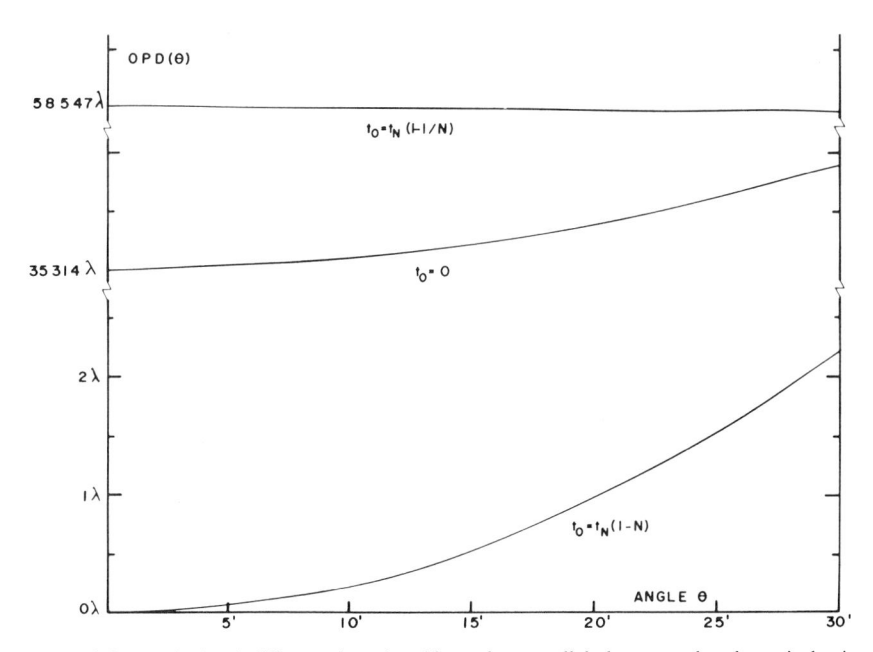

$$t_0 = t_N(1 - N) \qquad (2.18)$$

Figure 2.8. Optical path difference introduced by a plane parallel plate normal to the optical axis ($t_N = 2$ cm, $N = 1.52$, $\lambda = 589$ nm).

so that OPD(0°) = 0, but this situation will require an even smaller light source. It should be pointed out that, when testing large optics, the value of t_0 cannot be changed at will since in general it will be very large.

If an extended quasimonochromatic light is used, a good condition is

$$\frac{d\text{OPD}(\theta)}{d\theta} = 0 \tag{2.19}$$

yielding

$$t_0 = t_N\left(1 - \frac{1}{N}\right). \tag{2.20}$$

It is interesting to see from this equation that the interferometer is compensated to first order for the size of the source when the apparent distance of the image of the collimator (or the light source in a Michelson interferometer) is the same for both arms as seen by an observer, as pointed out by Steel (1962) and Slevogt (1954).

When the light source is extended and the interferometer is compensated in this manner, the fringes are localized near the mirror surfaces because of the way the image of the light source moves when one of the mirrors is tilted in order to obtain the fringes, as shown in Fig. 2.4. Thus the viewing system must be focused near the mirrors to see the fringes. When testing a lens, however, the fringes are to be observed at the pupil; therefore the mirror should be as close as possible to the lens. This is why a convex mirror with the longest possible radius of curvature is desirable (Steel 1966) when testing telescope objectives. On the other hand, the entrance pupil of a microscope objective is at infinity; hence the exit pupil is at the back focus. Dyson (1959) described an optical system like the one shown in Fig. 2.9, which images the mirror surface on the back focus of the microscope objective, where the fringes are desired. This system is very interesting because it has all Seidel aberrations equal to zero.

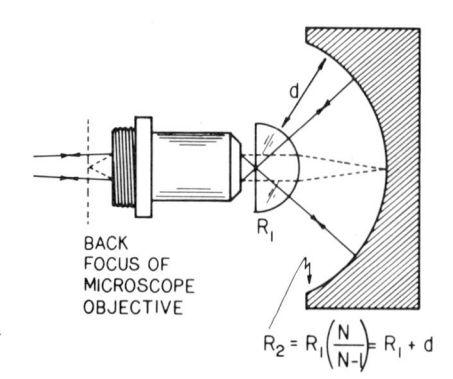

Figure 2.9. Dyson's system used to test microscope objectives.

2.3. COHERENCE REQUIREMENTS

The limitation on the size of a pinhole source was examined in a slightly different manner by Guild (1920–1921), as explained later. Imagine that the small source is greatly enlarged to form an extended source. Then fit an eyepiece in front of lens L_2 (see Fig. 2.1) to form a telescope. Under these conditions equal inclination fringes in the form of concentric rings, like the ones normally observed in the Michelson interferometer, are observed. If the mirrors are exactly perpendicular to their optical axes, the rings will be exactly centered. The ideal size of the source is that which allows only the central spot on the fringe system to be observed. The size of the central spot increases when the OPD (θ) reduces its dependence on θ by one of the adjustments described earlier, making possible the use of a larger source, although the effective size of the spot is then limited by the pupil of the observing eye or the camera.

In all the foregoing considerations, the two interfering wavefronts are assumed to have the same orientation, that is, without any rotations or reversals with respect to each other. In other words, if one of the beams is rotated or reversed, the other should also be rotated or reversed. A wavefront can be rotated 180° by means of a cube corner prism or a cat's-eye retroreflector formed by a convergent lens and a flat mirror at its focus. The wavefront can be reversed upon reflection on a system of two mutually perpendicular flat surfaces, for example, in a Porro prism. Murty (1964) showed that, if one of the wavefronts is rotated or reversed with respect to the other, then, to have fringes with good contrast and without phase shifts, the pinhole diameter $2a$ should satisfy the condition

$$2a \leq \frac{1.22\lambda f}{D} \tag{2.21}$$

so that diametrically opposite points over the wavefront are coherent to each other. Here, f and D are the collimator's focal length and diameter. Then $2a$ is extremely small and therefore an impractical size for some sources. However, there is no problem if a gas laser is used because its radiance and spatial coherence are extremely high.

2.3.2 Temporal Coherence

The OPD (θ) is given by Eq. (2.5) also imposes some minimum requirements on the monochromaticity of the light source. Considering first the case of an interferometer that is uncompensated because of the lack of a compensating plate or the presence of an optical system with lenses or prisms in one of the arms, from Eq. (2.5) we can write

$$[\Delta \text{OPD}(\theta)]_{\theta=0} = 2t_N \left(\frac{dN}{d\lambda}\right) \Delta\lambda \tag{2.22}$$

and, using again the Rayleigh criterion (ΔOPD $\leq \lambda/4$),

$$\Delta \lambda \leq \frac{\lambda}{8 t_N (dN/d\lambda)}. \tag{2.23}$$

Therefore, since the glass dispersion $(dN/d\lambda)$ is never zero, and t_N is also nonzero, the bandwidth $\Delta \lambda$ must not have a very large value if the interferometer is not compensated. If the interferometer is exactly compensated, white-light fringes can be observed when OPD is nearly zero; otherwise in a highly monochromatic light source such as a low vapor pressure lamp or (even better) a gas laser must be used.

If many different kinds of glasses are present in both arms of the interferometer, we may take a more general approach by considering that the interferometer is compensated for the bandwidth $\Delta \lambda$ of the light if the phase difference for the light following the two paths in the interferometer is independent of the wavelength. According to Steel (1962), we can say that, if each arm of the interferometer contains a series of optical components of thickness t and refractive index N, the phase difference for the two arms is

$$\phi = \frac{2\pi}{\lambda} \left(\sum_1 Nt - \sum_2 Nt \right). \tag{2.24}$$

This relative phase is independent of the wavelength when $d\phi/d\lambda = 0$, thus giving

$$\sum_1 \tilde{N} t = \sum_2 \tilde{N} t \tag{2.25}$$

where $\tilde{N}$ is the "group refractive index," defined by

$$\tilde{N} = N - \lambda \frac{dN}{d\lambda}. \tag{2.26}$$

Therefore the interferometer is compensated for the bandwidth $\Delta \lambda$ when the "group optical path" for both arms is the same. Steel (1962) pointed out that the compensation for the bandwidth of the light source can be examined by looking at the fringes, formed by a white-light source, through a spectroscope with its slit perpendicular to the fringes. The spectrum is crossed by the fringes, and their inclination shows the change of fringe position with wavelength. The fringes will be straight along the direction of dispersion if the bandwidth compensation is perfect. Otherwise the glass optical paths can be adjusted until the fringes show a maximum (zero slope) at the wavelength to be used. If the bandwidth of the light source is very large, a detailed balancing of the glass paths

2.4. USES OF A TWYMAN-GREEN INTERFEROMETER

has to be made, by using the same types and thicknesses of the glass on both arms.

If the OPD(0) is very large, as in the unequal-path interferometer (described in Section 2.5), the last term in Eq. (2.5) dominates, and we can write

$$\text{OPD}(0) = 2t_0 = m\lambda, \tag{2.27}$$

but from the Rayleigh criterion the order number m should not change from one end of the wavelength bandwidth to the other by more than $\frac{1}{4}$; thus

$$m\lambda = (m + \tfrac{1}{4})(\lambda - \Delta\lambda), \tag{2.28}$$

where $\Delta\lambda$ is the maximum allowed bandwidth. Thus we can write the approximation

$$\Delta\lambda \leq \frac{\lambda}{8t_0}. \tag{2.29}$$

Since the length of a train of waves with bandwidth $\Delta\lambda$ is equal to $\lambda^2/\Delta\lambda$, this condition is equivalent to saying that the OPD(0) should be smaller than one-fourth of the length of the wavetrain (or wavelength of the modulation).

A very interesting and practical case occurs when the light source is a gas laser, but this discussion is left to Section 2.5 on the unequal-path interferometer.

2.4. USES OF A TWYMAN-GREEN INTERFEROMETER

Many different kinds of optical components can be tested with this instrument. The simplest one to test is a plane parallel plate of glass. The OPD introduced by the presence of the glass plate is given by

$$\text{OPD} = 2(N - 1)t, \tag{2.30}$$

where t is the plate thickness and N is the refractive index. If the interferometer is adjusted so that no fringes are observed before introducing the plate into the light beam, all the fringes that appear are due to the plate. If the field remains free of fringes, we can say that the quantity $(N - 1)t$ is constant over all the plate. If straight fringes are observed, we can assume that the glass is perfectly homogeneous (N constant) and that the fringes are due to an angle ϵ between the two flat faces, given by

$$\epsilon = \frac{\alpha}{2(N - 1)}, \tag{2.31}$$

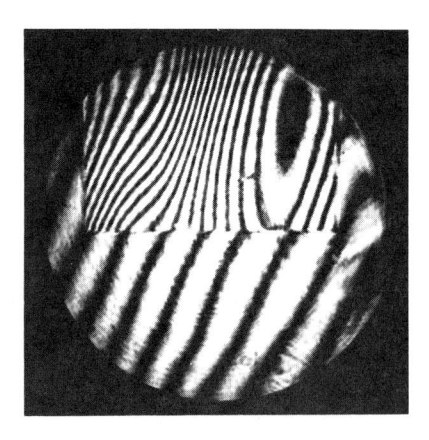

Figure 2.10. Testing a glass plate.

where α is a small angle between the two interfering wavefronts, which can be determined from

$$\alpha = n\lambda. \tag{2.32}$$

Here n is the number of interference fringes per unit length being observed.

The fringes, however, may not be straight but quite distorted, as in Fig. 2.10, because of bad surfaces or inhomogeneities in the index, since the only quantity we can determine is $(N - 1)t$. To measure the independent variations of N and t, we must complement this test with another made in a Fizeau interferometer, which measures the values of Nt (Kowalik 1978).

Many different kinds of material can be tested with this basic arrangement (Adachi et al. 1961, 1962; Masuda et al. 1962; Twyman and Dalladay 1921–1922).

2.4.1. Testing of Prisms and Diffraction Rulings

The Twyman–Green interferometer is a very useful instrument for testing prisms. Especially interesting is its application for testing the accuracy of the 90° angle between two of the faces of a right-angle (Porro) prism, a roof (Amici) prism, or a cube corner prism. As explained in Section 2.2.1, relative rotation or reversal of the wavefronts should be corrected, as shown in Fig. 2.11, if a gas laser is not used. The arrangements in Fig. 2.12 can be used when a gas laser source is employed.

Luneburg (1964) showed that the angular error ϵ in a roof face of a prism is

$$\epsilon = \frac{\alpha}{4KN \sin \theta}, \tag{2.33}$$

2.4. USES OF A TWYMAN–GREEN INTERFEROMETER

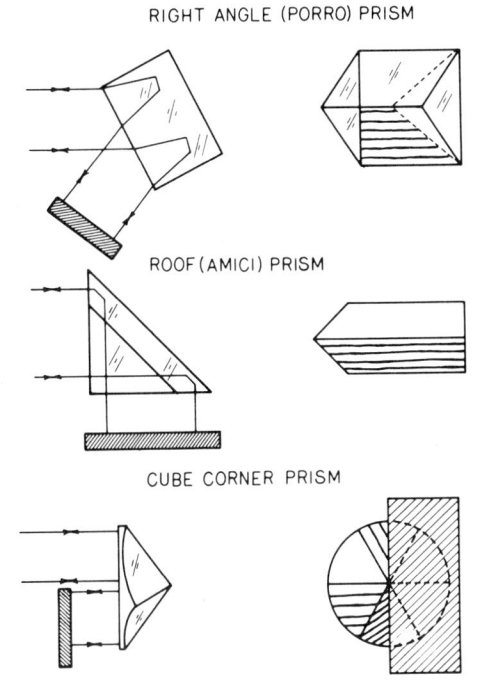

Figure 2.11. Testing some prisms.

where N is the refractive index of the material, α is the angle between the two exiting wavefronts in a single pass through the prism, θ is the angle between the roof edge and the incident beam, and K is the number of times the light is reflected on the roof face. For the arrangements shown in Figs. 2.11 and 2.12 we have the values in Table 2.1. The angle α is determined from Eq. (2.32),

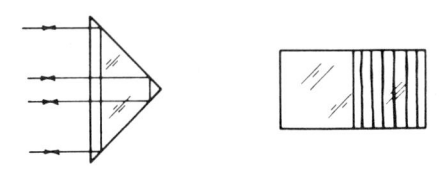

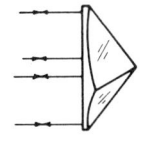

Figure 2.12. Testing some prisms with laser illumination.

Table 2.1 Values of Angle θ and of K for Prisms in Figs. 2.10 and 2.11

Prism	Figure	θ	sin θ	K
Porro	2.10	60°	0.8662	2
Amici	2.10	45°	$1/\sqrt{2}$	2
Cube corner	2.10	54.7°	$\sqrt{2/3}$	2
Porro	2.11	90°	1	1
Cube corner	2.11	54.7°	$\sqrt{2/3}$	1

but wth the interferometer adjusted in such a way that all the fringes in one of the faces are eliminated. Thomas and Wyant (1977) made a complete study of the testing of cube corner prisms.

A dispersive prism can also be tested, as shown in Fig. 2.13a. This arrangement of smoothly changing inhomogeneities in the glass may be compensated for by appropriately figuring the faces. An axicon may be tested in a Twyman–Green interferometer, using the method described by Fantone (1981), as well as reflaxicons (Hayes et al. 1981).

In 1935 Bisacre and Simeon suggested a method whereby a diffraction grating could be tested by means of a Twyman–Green interferometer. Unfortunately, they never published their work (Candler 1951). They used the arrangement shown in Fig. 2.13b. The interferometer is initially adjusted to obtain horizontal fringes in the first order. Then the grating is rotated to pass to the third order, in which the ghosts, if any, are stronger. If there are ghosts, the fringes have a sawtooth appearance. When the spacing between the horizontal

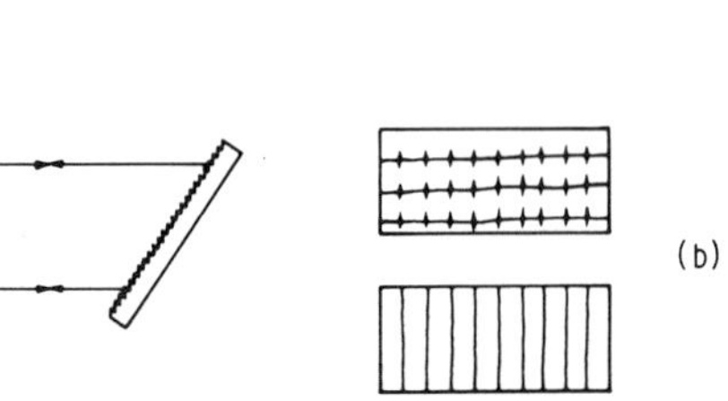

Figure 2.13. Testing a dispersive prism and a diffraction grating.

2.4. USES OF A TWYMAN-GREEN INTERFEROMETER

fringes is increased, the teeth become larger and larger until they form a system of vertical fringes. Jaroszewics (1986), using this interferometer, has also tested the spacing error of a plane diffraction grating.

2.4.2. Testing of Lenses

One of the early applications of the Twyman-Green interferometer was the testing of lenses and camera objectives (Twyman 1920; Smith 1926-1927), including the measurement of the chromatic aberration (Martin and Kingslake 1923-1924). Any of the arrangements in Fig. 2.14 can be used to test a convergent lens. A convex spherical mirror with its center of curvature at the focus of the lens is used for lenses with long focal lengths, and a concave spherical mirror for lenses with short focal lengths. A small, flat mirror at the focus of the lens can also be employed to great advantage, since the portion of the flat mirror being used is so small that its surface does not need to be very accurate. However, because of the spatial coherence requirements described in Section 2.3.1, the same arrangement or a cube corner prism must be employed on the other interferometer arm, if a laser is not used. Another characteristic of this method is that asymmetric aberrations like coma are canceled out, leaving only symmetric aberrations like spherical aberration and astigmatism.

When a lens is to be tested off-axis, it is convenient to mount it in a nodal lens bench, as shown schematically in Fig. 2.15. The lens L under test is mounted in a rotating mount so that the lens can be rotated about the nodal point N. Since the focal surface is usually designed to be a plane and not a sphere, mirror M is moved backward a small distance $\overline{FP}$ by pushing the mirror support

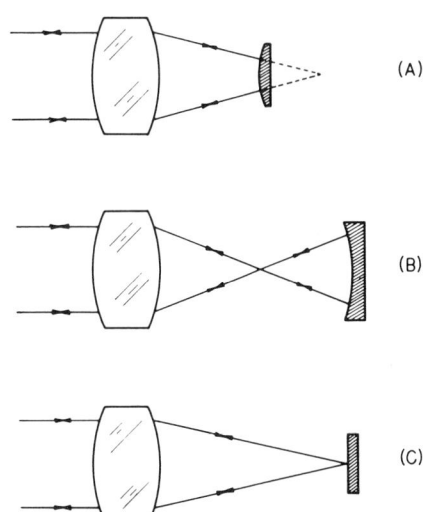

Figure 2.14. Three possible arrangements to test a lens.

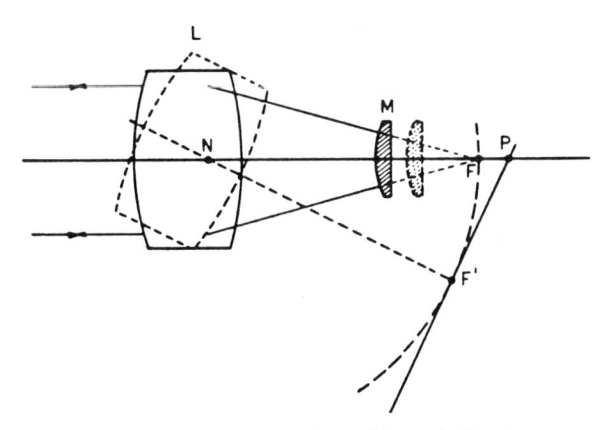

Figure 2.15. Testing a lens with a nodal bench.

against a metallic bar $\overline{F'P}$, fixed with respect to lens L. Interferograms obtained with lenses having third-order aberrations will be shown in Section 2.7.

Testing a large lens on the Twyman–Green interferometer requires the use of a beam splitter plate even larger than the lens. To avoid this difficulty, according to Burch (1940), Williams suggested using a concave mirror on one of the arms and the large lens with a flat mirror in front of it on the other arm, as shown in Fig. 2.2. This type of configuration was also suggested later by Hopkins (1962) for use with a gas laser in an unequal path configuration.

Complete small telescopes can also be tested with good results, as shown by Ostrovskaya and Filimonova (1969).

2.4.3. Testing of Microscope Objectives

Twyman (1920, 1922–1923) also used his instrument for the testing of microscope objectives with good success. The arrangement is essentially the one used for a lens, but a convex mirror cannot, in general, be employed because of the short focus of the objective. Since the microscope sometimes works at a finite tube length, a negative lens is added to change the collimated light and simulate a light source 16 cm away. This lens must be corrected for spherical aberration, but it is not necessary to correct it quasimonochromatic. As shown in Fig. 2.16, several arrangements can be used to test an objective, the most common being a spherical concave mirror with its center of curvature at the focus of the objective. A solid spherical reflector slightly thicker than a hemisphere can serve to simulate the presence of a cover glass.

A plane mirror at the focus of the objective can also be used, but in this case the wavefront is rotated 180°. Therefore we should either use a laser light source or rotate the wavefront on the other arm. This can be done by means of a cube corner prism or with another microscope objective with the same flat mirror

2.5. UNEQUAL-PATH INTERFEROMETER

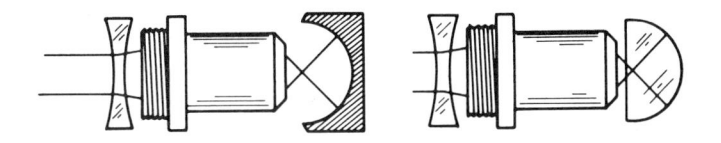

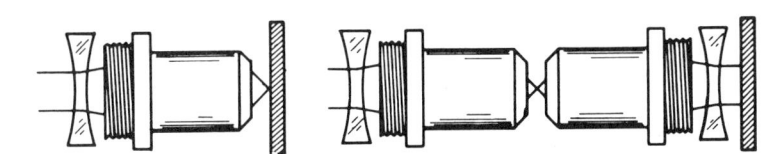

Figure 2.16. Testing microscope objectives.

arrangement. It should be pointed out that the interferogram in this case represents the difference between the aberrations of the two objectives.

A fourth arrangement is formed by two oppositely placed microscope objectives. In this case the interferogram represents the sum of the aberrations of the two objectives. However, when the aberrations to be measured are not small or the pinhole is not small, the best arrangement is the one shown in Fig. 2.9, which uses Dyson's system.

2.5. UNEQUAL-PATH INTERFEROMETER

In Section 2.3 we discussed the coherence requirements of a Twyman–Green interferometer and pointed out that, when a laser light source is used, extremely large OPDs can be introduced (Morokuma et al. 1963). In explaining this, let us first consider the spectrum of the light emitted by a laser. As shown in Fig. 2.17a, the light emitted by a gas laser usually consists of several spectral lines (longitudinal modes) (Sinclair and Bell 1969), spaced equally at a frequency interval $\Delta \nu$, given by

$$\Delta \nu = \frac{c}{2L}. \tag{2.34}$$

If the cavity length L of a laser changes for some reason (thermal expansion or contractions, mechanical vibrations, etc.), the lines move in concert along the frequency scale, preserving their relative distances $\Delta \nu$, but always with intensities inside the dotted envelope (power gain curve) in Fig. 2.17a.

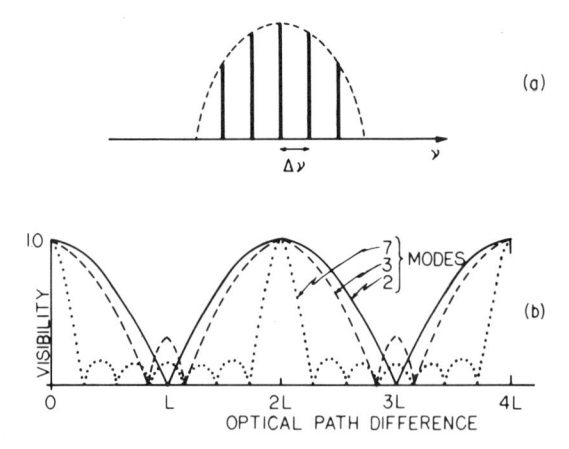

Figure 2.17. Visibility in the interferometer using a gas laser.

Lasers that have only one spectral line are called single-mode or single-frequency lasers. They produce a perfect unmodulated wavetrain, but because of instabilities in a cavity length L the frequency is also unstable. By the use of servomechanisms, however, single-frequency lasers with extremely stable frequencies are commercially produced. They are the ideal source for interferometry since an OPD as long as desired can be introduced without any loss in contrast. Unfortunately these lasers are very expensive and have very low power outputs (less than 1 mW). Even so, a 1-mW laser has a higher radiance than any other type of interferometric source.

It can be shown (Collier et al. 1971) that the theoretical visibility in an interferometer, when a laser source with several longitudinal modes is used, is as illustrated in Fig. 2.17b. Therefore, to have good contrast, the OPD(0) has to be near an integral multiple of $2L$; thus

$$\text{OPD}(0) = 2t_0 \simeq 2ML. \tag{2.35}$$

Hence lasers are very convenient for Twyman–Green interferometry provided that the mirrors in the interferometer can be adjusted to satisfy this condition. Because of mechanical instability, the laser cavity normally vibrates, producing a continuous instability in the frequency of the lines. This does not represent any serious problem for small OPD's, of the order of 1 m. If, however, the OPD is very large, of the order of 10 m or more, an almost periodic variation of the contrast is introduced, as shown by Batishko and Shannon (1972). To overcome the inconvenience of this effect they recommend taking photographs with exposures of the order of $\frac{1}{250}$ s. This exposure is fast enough to stop the vibration of the fringes but slow enough so that the contrast variation is integrated out.

2.5. UNEQUAL-PATH INTERFEROMETER

A laser with two longitudinal modes can be stabilized to avoid contrast changes by a method recommended by Bennet et al. (1973), Gordon and Jacobs (1974), and Balhorn et al. (1972).

Some suggestions for aligning and adjusting the unequal path interferometer had been given by Zielinski (1978, 1979).

2.5.1 Some Special Designs

With the advent of the laser, it became practical to use Twyman–Green interferometers with large optical path differences. Probably the first one to suggest this was Hopkins (1962). An instrument of this type following a Williams arrangement was made by Grigull and Rottenkolder (1967) for wind-tunnel observations and the testing of spherical mirrors.

A very versatile unequal-path interferometer for optical shop testing was designed by Houston et al. (1967). A schematic diagram of this interferometer is shown in Fig. 2.18. The beam splitter plate, which is at the Brewster angle, has a wedge angle of 2–3 min of arc between the surfaces. The reflecting surface of this plate is located to receive the rays returning from the test specimen in order to preclude astigmatism and other undesirable effects. A two-lens beam diverger can be placed in one arm of the interferometer. It is made of high-index glass, all the surface being spherical, and has the capability for testing a surface as fast as $f/1.7$. A null lens can be used to test an aspheric element, with the combination beam diverger and null lens spaced and aligned as depicted in Fig. 2.19 (see Chapter 15).

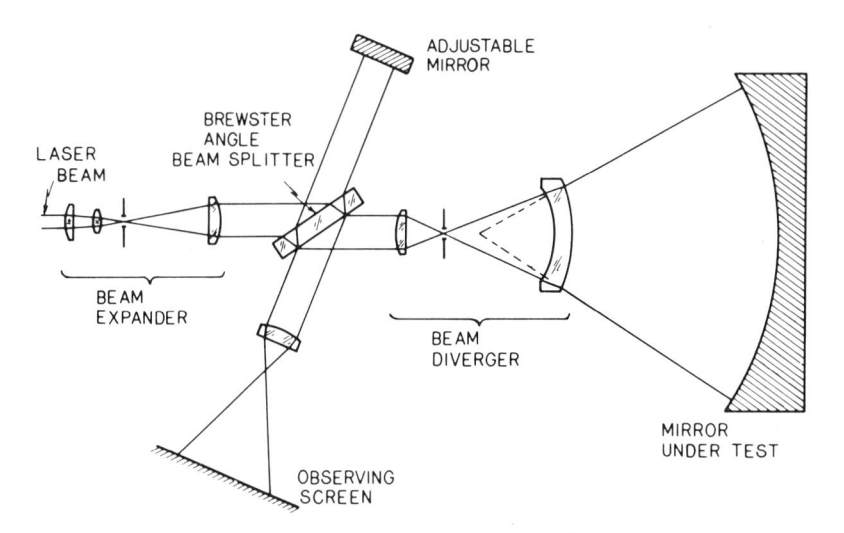

Figure 2.18. Houston's unequal-path interferometer.

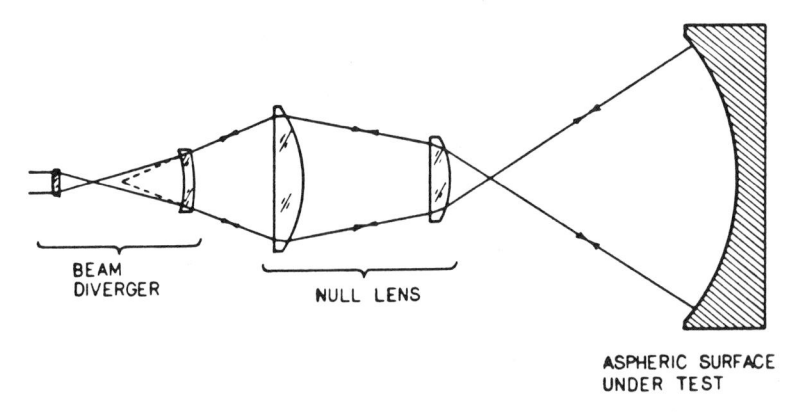

Figure 2.19. Null lens and lens diverger for unequal-path interferometer.

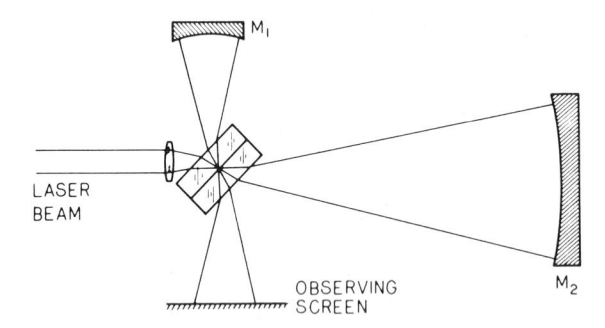

Figure 2.20. Kocher's unequal-path interferometer.

Another unequal-path interferometer was designed by Kocker (1972). This instrument, shown in Fig. 2.20, is quite similar to the Twyman–Green interferometer in Fig. 2.2. A significant feature is the use of an optically thick beam splitter substrate in the diverging beam. Such a plate introduces aberrations, but they are intentionally made equal on both arms. To a first approximation there is no effect on the fringe pattern, especially if the total thickness is kept small.

Bruning and Herriot (1970) designed an interferometer that reduces spurious fringes and noise to a very low level by the use of very interesting polarized light devices. Buin et al. (1969) reported a successful industrial use of unequal-path interferometers.

2.5.2. Improving the Fringe Stability

The unequal-path interferometer frequently has the problem that the fringes are very unstable due to vibrations on the surface under test. Two different solutions have been proposed to solve this problem. One solution is to use the arrange-

2.6. VARIATIONS FROM THE TWYMAN–GREEN CONFIGURATION

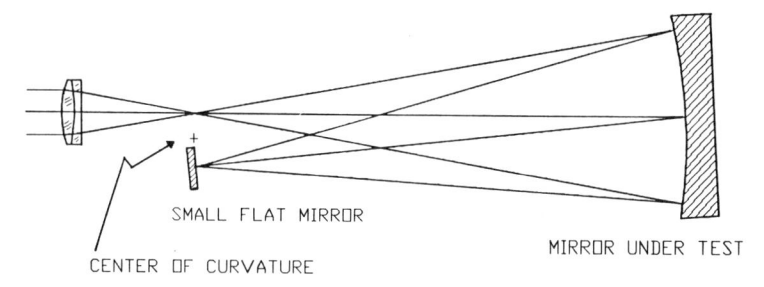

Figure 2.21. Stabilization of interference fringes by eliminating antisymmetric error in the wavefront, by means of a reflection in a small flat mirror.

ment in Fig. 2.21. The diverging beam of light going out from the interferometer illuminates the concave surface under test and then the reflected convergent beam gets reflected in a small flat mirror near the center of curvature. Then, the light returns to the mirror, but the returning wavefront has been rotated 180° with respect to the incident wavefront. This rotation of the wavefront eliminates all antisymmetric components of the wavefront error, leaving only the symmetric components. Thus, coma and tilts of the wavefront cannot be detected. Any mirror vibrations may be interpretated as tilt movements, but they cannot be detected with this arrangement.

The mirror vibrations may also be eliminated from the interferogram by introducing exactly the same vibrations in the reference wavefront. This is done with an arrangement described by Hardesty (1979).

2.6. VARIATIONS FROM THE TWYMAN–GREEN CONFIGURATION

Many variations of the Twyman–Green interferometer have been invented; interesting among them are some small, compact interferometers designed by Van Heel and Simons (1967) and by Basile (1979).

A carbon dioxide laser has been used as a light source of long wavelengths (10.6 μm) in a Twyman–Green interferometer (Munnerlyn et al. 1969; Lewandowski et al. 1986; Kwon et al. 1979) in order to measure unpolished or rough surfaces.

Another interesting development is the invention of a cheap interferometer whose defects are corrected by means of a hologram (Rogers 1970). First, a photograph is taken in a very imperfect and inexpensive interferometer, introducing a large tilt. This gives rise to an exceedingly fine set of fringes invisible to the naked eye but capable of being photographed. Then a second exposure is taken on the same photographic plate, after introducing into the interferometer the plate to be tested. A Moiré pattern appears on the developed plate, giving the contours of the surface quality of the plate under test.

Another interesting holographic Twyman–Green interferometer has been described by Chen and Breckinridge (1982). In this design a single holographic optical element combines the functions of a beam splitter, beam diverger, and null compensating lens. MacDonell (1979) designed a large aperture interferometer, using a holographic compensator.

2.6.1. Mach–Zehnder Interferometer

Frequently the Mach–Zehnder configuration shown in Fig. 2.22 has some advantages with respect to the Twyman–Green configuration. For example, if the sample under test has a large aberration, it is better to pass the beam of light only once through it. Another advantage is that this interferometer is automatically compensated because it has two beam splitters.

A commercial version of this interferometer configuration, manufactured by Wyko Corp., uses a pinhole in one of the branches to generate a perfect wavefront, in order to test the wavefront quality of the light source, as shown in Fig. 2.23. (Creath, 1987; Leung and Lange 1983). A Mach–Zehnder has also been

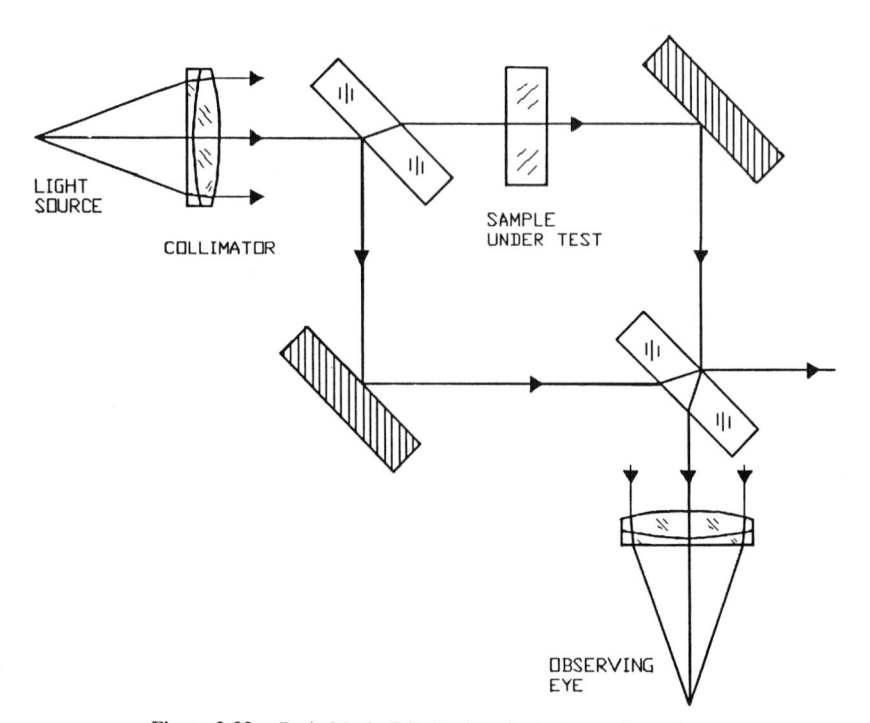

Figure 2.22. Basic Mach–Zehnder interferometer configuration.

2.6. VARIATIONS FROM THE TWYMAN–GREEN CONFIGURATION

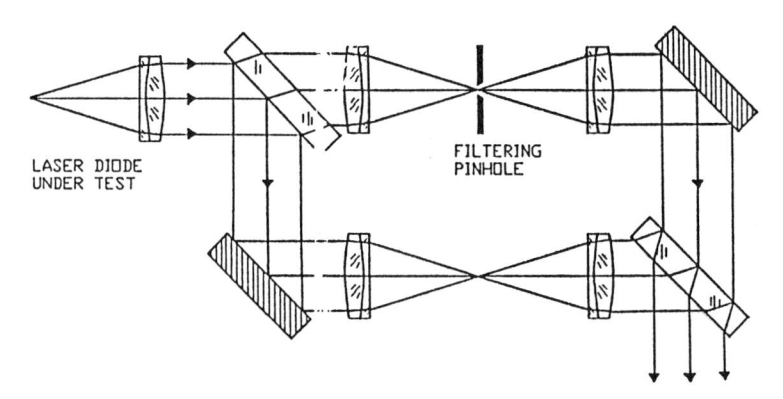

Figure 2.23. Mach–Zehnder interferometer used to test the wavefront quality of a laser diode.

used to test off-axis paraboloids (Gerth et al. 1978). Cuadrado et al. (1987) have described a method to align a Mach–Zehnder interferometer using equilateral hyperbolic zone plates, and Flack (1978) has analyzed the errors that result from a test section misalignment.

2.6.2. Interferometers with Diffractive Beam Splitters

Some two-beam interferometers may be thought of as modifications from the basic Twyman–Green interferometer. Some of them use diffraction gratings as beam splitters, as the one shown in Fig. 2.24, described by Molesini et al. (1984). The advantage of this particular configuration is that relatively large errors in the grating flatness may be tolerated. In the next section we will see some other interferometers using diffraction gratings as beam splitters.

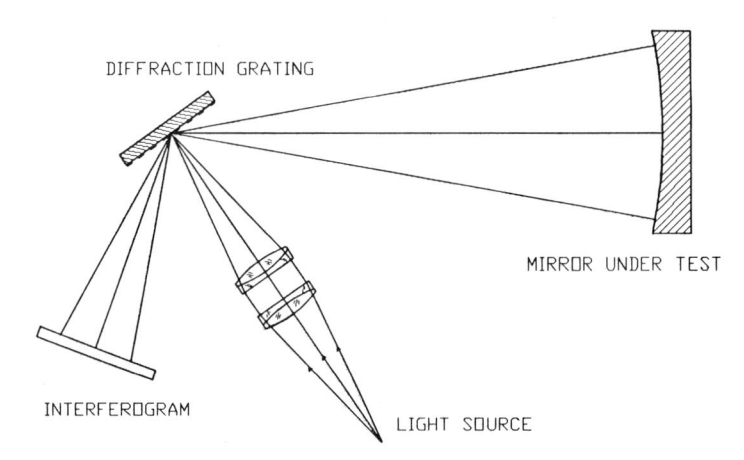

Figure 2.24. Interferometer using a diffraction grating beam splitter.

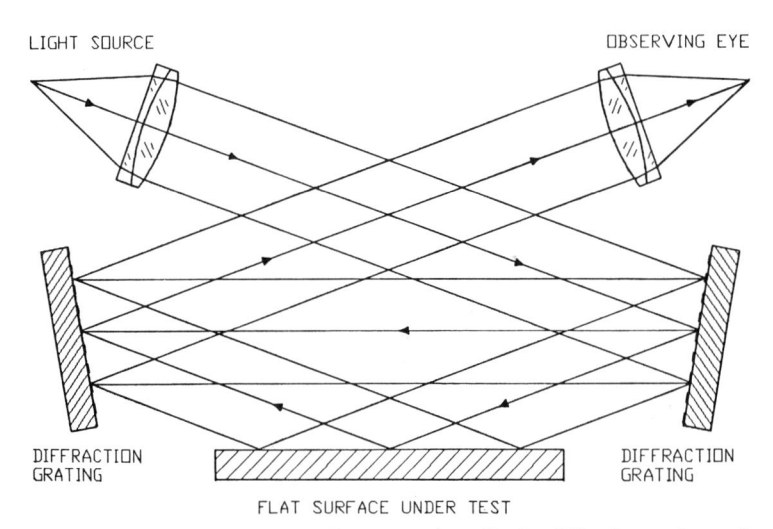

Figure 2.25. Oblique incidence interferometer using reflecting diffraction gratings as beam splitters.

2.6.3. Oblique Incidence Interferometers

Another kind of two-beam interferometers have triangular paths, so that one of the beams is obliquely reflected on the flat surfaces under test. It may be easily proven that under these conditions a small error with height h on the surface under test introduces an error equal to $2h \cos \theta$, where θ is the incidence angle. Thus, the interferometer is desensitized by a factor $\cos \theta$. Another consequence of the oblique incidence is that the reflectivity of the surface under test is greatly increased. Thus, an interferometer with oblique incidence is ideally suited for testing ground or mate flat surfaces, whose flatness and reflectivity are not good enough to be tested by conventional interferometry.

Oblique incidence interferometers have been described by Linnik (1942), Saunders and Gross (1959), Birch (1973, 1979), Hariharan (1975), and MacBean (1984). Some of these interferometers use diffraction gratings as beam splitters, as the one designed by Hariharan and shown in Fig. 2.25. Small ground and almost flat spherical surfaces may be tested with oblique incidence interferometers, as shown by Jones (1979).

2.6.4. Phase-Conjugating Interferometer

Phase-conjugating mirrors are very useful tools in interferometry. They eliminate the need for a perfect reference wavefront. A Twyman–Green interferometer as shown in Fig. 2.26, using a phase-conjugating mirror, has been described by Feinberg (1983) and Howes (1986a, 1986b). The phase-conjugating

2.7. TYPICAL INTERFEROGRAMS AND THEIR ANALYSIS

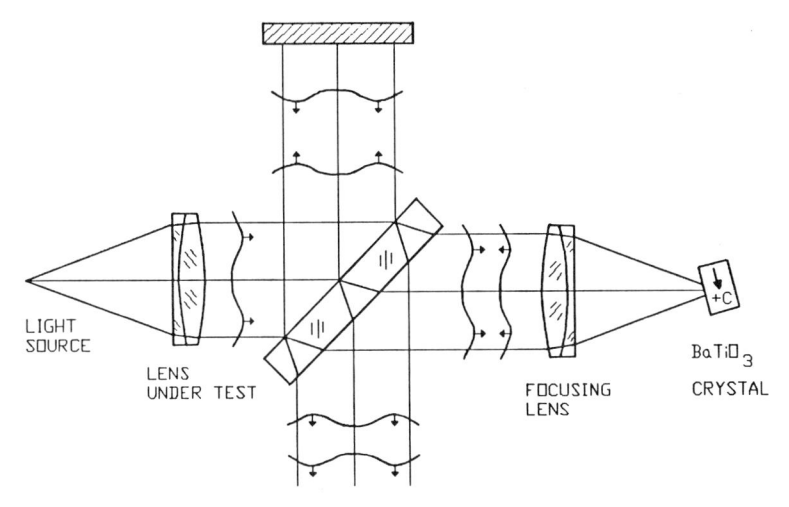

Figure 2.26. Twyman–Green interferometer with a phase-conjugating mirror.

mirror is formed by a BaTiO$_3$ crystal, with the C axis parallel to one of its edges and inclined 20° with respect to a plane perpendicular to the optical axis. The phase conjugation is obtained by four-wave mixing. These pumping beams are automatically self-generated from a 30-mW argon laser ($\lambda = 514.5$ nm) incident beam, by internal reflection at the crystal faces. Thus, it is a self-pumped phase-conjugating mirror.

The property of this self-conjugating mirror is that the wavefront incident of the mirror is reflected back along the same ray directions that the incident wavefront has. Thus the wavefront deformations change sign as shown in Fig. 2.26. Since the returning rays have the same directions as the incident rays, the quality of the focusing lens is not important. However, the quality of the light source collimator is important. Any wavefront distortions produced by this collimator will appear in the final interferogram, but duplicated. In other words the wavefront is not tested against a flat reference but against another wavefront, with deformations opposite in sign.

Then the lens under test is the collimator, and the sensitivity is the same as that in the common Twyman–Green interferometer, but with only a single pass through the lens. The advantage is that no perfect lenses are necessary. The disadvantage is that an argon laser is required.

2.7. TYPICAL INTERFEROGRAMS AND THEIR ANALYSIS

The description of interferograms due to primary aberrations will follow Kingslake's (1925–1926) treatment. The derivation of an aberrated wavefront with respect to a sphere with its center at the Gaussian image point is given by

$$\text{OPD} = A(x^2 + y^2)^2 + By(x^2 + y^2) + C(x^2 + 3y^2) \quad (2.36)$$
$$+ D(x^2 + y^2) + Ey + Fx,$$

where

$$A = \frac{\text{longitudinal third-order spherical aberration}}{4S^2F^2}, \quad (2.37)$$

$$B = \frac{\text{sagittal coma}}{S^2 lh}, \quad (2.38)$$

$$C = \frac{\text{sagittal astigmatism}}{2t^2}, \quad (2.39)$$

$$D = \frac{\delta l}{2l^2} \quad \text{(defocusing)}, \quad (2.40)$$

$$E = \frac{\delta h}{l} \quad \text{(tilt about } x \text{ axis)}, \quad (2.41)$$

$$F = \frac{\delta t}{l} \quad \text{(tilt about } y \text{ axis)}. \quad (2.42)$$

Here S is the semiaperture of the lens, f is the focal length, l is the distance from the back surface of the lens to the image, and h is the image height. The quantities δl is a defocusing term measured as a displacement of the image along the optical axis. The quantities δh and δt are wavefront tilts, produced by lateral displacement of the image along the y and x axis, respectively.

In polar coordinates (ρ, θ) Eq. (2.36) can also be written ($x = \rho \sin \theta$; $y = \rho \cos \theta$) as

$$\text{OPD} = A\rho^4 + B\rho^3 \cos \theta + C\rho^2 (1 + 2 \cos^2 \theta) \quad (2.43)$$
$$+ D\rho^2 + E\rho \cos \theta + F\rho \sin \theta.$$

For simplicity, in computing typical interferograms of primary aberrations, a normalized entrance pupil with unit semidiameter S was assumed.

Then the coefficients A, B, C, D, E, and F in Eq. (2.36) all represent the number of wavelengths of aberration. The interferograms were computed, and then drawings were made.

1. *Perfect Lens.* The patterns for a perfect lens without ($E = F = 0$) and with ($E = 5$) tilt are shown in Figs. 2.27a and 2.27b. A perfect lens with

2.7. TYPICAL INTERFEROGRAMS AND THEIR ANALYSIS

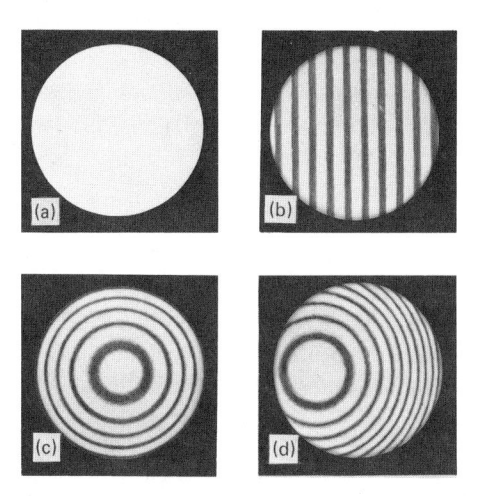

Figure 2.27. Interferograms for a perfect lens. (*a*) With no tilt or defocusing. (*b*) With tilt. (*c*) With defocusing. (*d*) With tilt and defocusing.

defocusing ($D = 5$) and with defocusing and tilt ($D = 5$, $E = 5$) is illustrated in Figs. 2.27*c* and 2.27*d*.

2. *Spherical Aberrations.* The patterns for pure spherical aberration were computed assuming that $A = 6$. They are shown at the paraxial focus ($D = 0$); without ($E = F = 0$) and with ($E = 5$) tilt are illustrated in Figs. 2.28*b* and 2.28*e*. The patterns at the marginal focus are obtained by setting in Eq. (2.43)

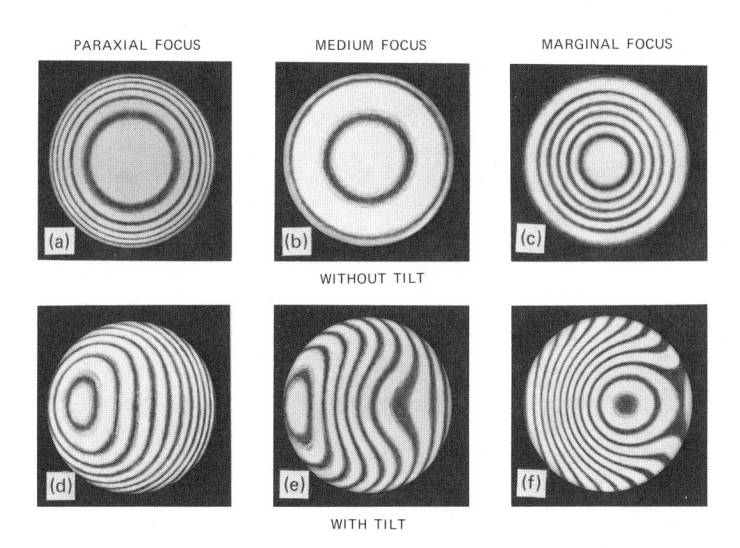

Figure 2.28. Interferograms for a lens with spherical aberration at the paraxial, medium, and marginal foci.

only A and D different from zero,

$$\frac{d\text{OPD}}{d\rho} = 4A\rho^3 + 2D\rho = 0. \tag{2.44}$$

Therefore we set $D = 2$, $A = 10$. These interferograms without ($E = F = 0$) and with ($E = 5$) tilt are shown in Figs. 2.28c and 2.28f.

3. *Coma.* All the patterns for coma were obtained using $B = 5$. Figure 2.29 shows them for the paraxial focus ($D = 0$), and Fig. 2.30 for a small defocusing ($D = 2$). In both figures the central pattern has no tilt ($E = F = 0$), and the surrounding pictures are for different tilt combinations ($E = \pm 3$, $F = \pm 3$).

Figure 2.29. Interferograms for a lens with coma, at the paraxial focus.

2.7. TYPICAL INTERFEROGRAMS AND THEIR ANALYSIS 81

Figure 2.30. Interferograms for a lens with coma and a small defocusing.

4. *Astigmatism.* All the patterns for astigmatism were computed for $C = 2$. If $D = 0$, we obtain the Petzval focus. The OPD for astigmatism can be written from Eq. (2.36) as

$$\text{OPD} = (C + D)x^2 + (3C + D)y^2. \qquad (2.45)$$

Therefore the sagittal focus is obtained for $C + D = 0$ and the tangential focus for $3C + d = 0$. The medium focus is obtained for $C + D = -(3C + D)$; hence $D = -2C$.

Figure 2.31 shows the patterns at the Petzval focus with tilts in all directions ($E = \pm 3$, $F = \pm 3$). Figures 2.32 to 2.34 show the patterns at the sagittal, medium, and tangential foci, respectively, also with tilts in all directions.

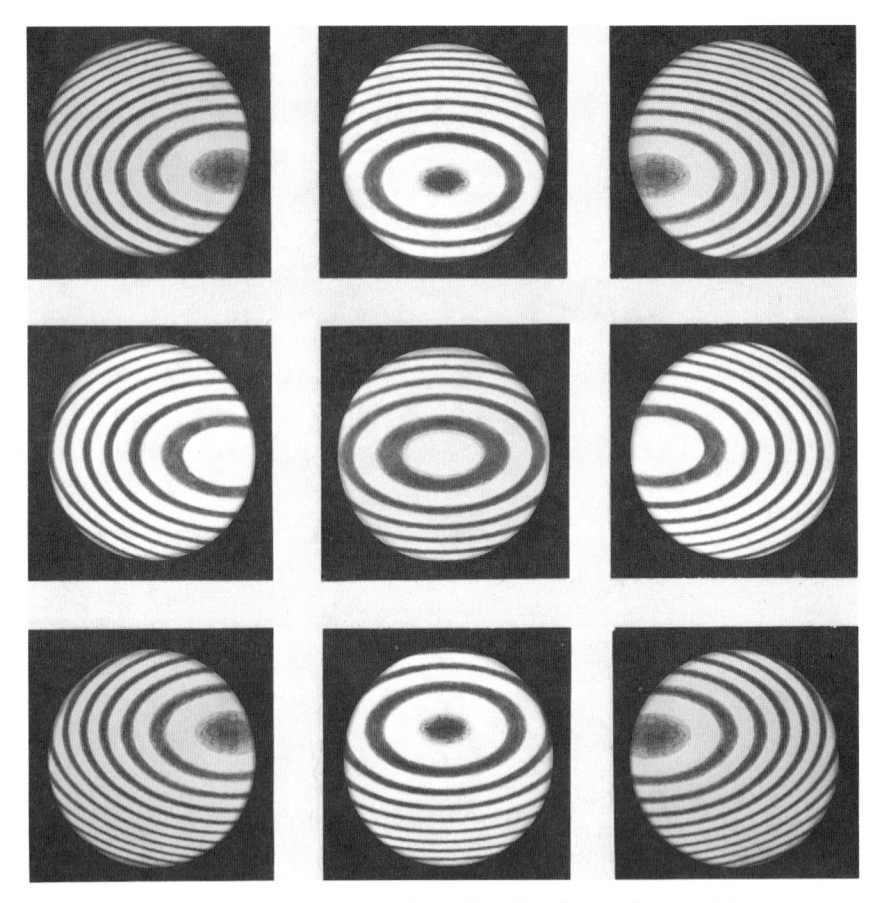

Figure 2.31. Interferograms for a lens with astigmatism, at the Petzval focus.

5. *Combined Aberrations.* Figure 2.35 shows the patterns for combined aberrations: spherical aberration plus coma in Fig. 2.35a, spherical aberration plus astigmatism in Fig. 2.35b, coma plus astigmatism ($B = 5$, $D = 2$) in Fig. 2.35c, and, finally, spherical aberration plus coma astigmatism in Fig. 2.35d.

Pictures of typical interferograms are shown in a paper by Maréchal and Dejonc (1950). These interferograms can be simulated by beams of fringes of equal inclination on a Michelson interferometer (Murty 1960) using the OPDs introduced by a plane parallel plate and cube corner prisms instead of mirrors, or by electronic circuits on a CRT (Geary et al. 1978; Geary 1979).

This type of interferogram was first analyzed by Kingslake (1926–1927). He measured the OPD at several points on the x and y axes, just by fringe counting.

2.7. TYPICAL INTERFEROGRAMS AND THEIR ANALYSIS

Figure 2.32. Interferograms for a lens with astigmatism, at the sagittal focus.

Then, solving a system of linear equations, he computed the OPD coefficients A, B, C, D, E, and F. Another method for analyzing a Twyman–Green interferogram was proposed by Saunders (1965). He found that the measurement of four appropriately chosen points is sufficient to determine any of the three primary aberrations. The points were selected as in Fig. 2.36, and then the aberration coefficients were computed as

$$A = \frac{128}{81r^2} [P_1 - P_9 + 2(P_8 - P_7)], \tag{2.46}$$

$$B = \frac{2}{3r^3} (P_2 - P_4 + 2P_6 - 2P_5), \tag{2.47}$$

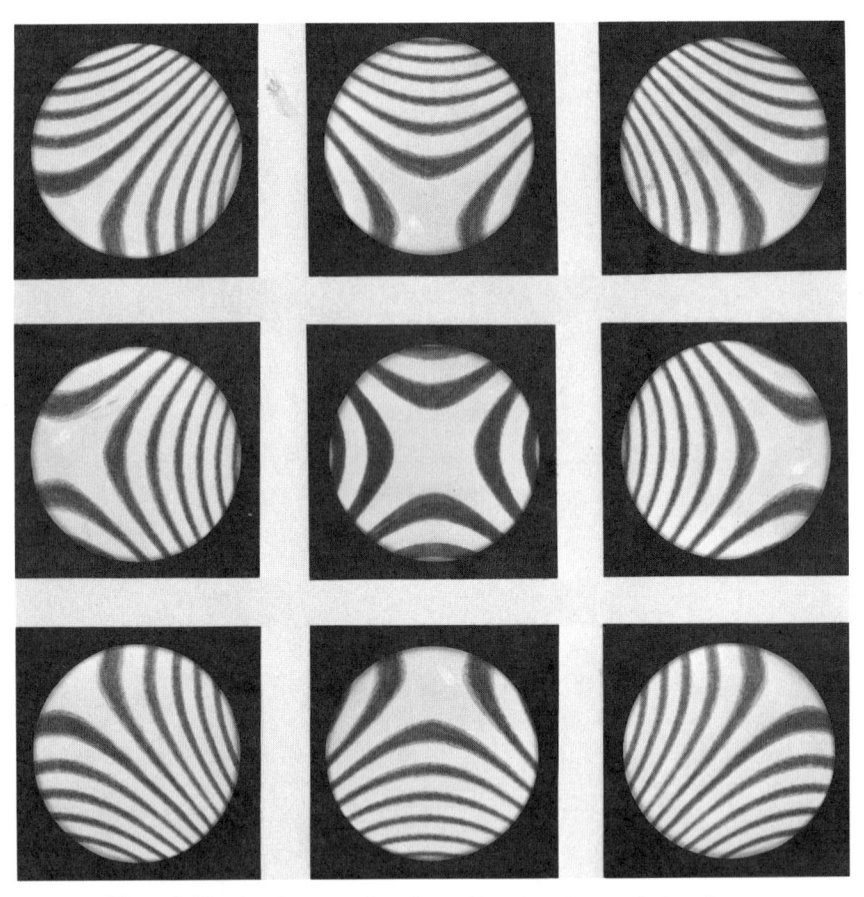

Figure 2.33. Interferograms for a lens with astigmatism, at the best focus.

$$C = \frac{1}{4r^2} (P_2 + P_4 - P_1 - P_3), \qquad (2.48)$$

where P_i is the interference order at point i.

If a picture of the interferogram is not taken, the aberration coefficients can be determined by direct reading on the interferogram setting, looking for interference patterns with different foci and tilts (Perry 1923–1924). To make these readings easier, some optical arrangements may be used to separate symmetrical and asymmetrical wavefront aberrations, as shown by Hariharan and Sen (1961).

2.7. TYPICAL INTERFEROGRAMS AND THEIR ANALYSIS

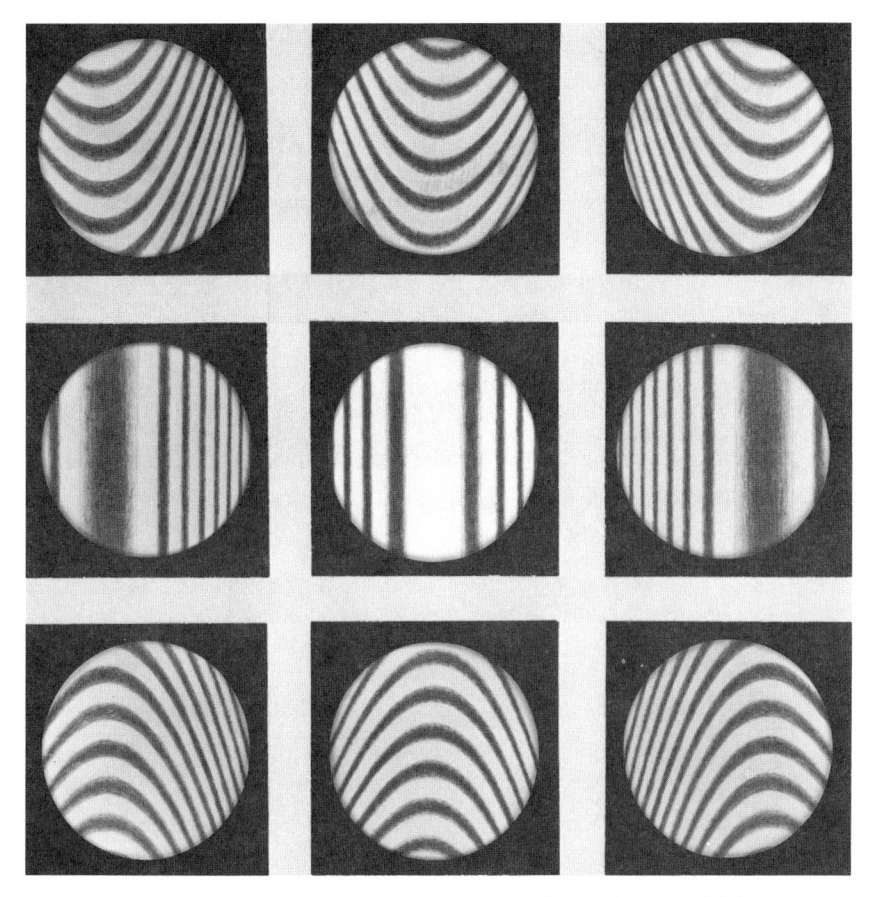

Figure 2.34. Interferograms for a lens with astigmatism, at the tangential focus.

2.7.1. Analysis of Interferograms of Arbitrary Wavefronts

The problem of determining the shape of a wavefront with arbitary shape from a Twyman–Green interferogram has been considered very generally and briefly by Berggren (1970) and more completely by Rimme et al. (1972). The procedure consists in measuring the positions of the fringes at many points over the interferograms and taking readings of the position (x, y) and the order of interference m. Since the measurements are taken at a limited number of points, an interpolation procedure must be adopted. The interpolation may be performed by the use of splines of a least-squares procedure, as described in detail in chapter 16.

If the wavefront is smooth enough, it is very convenient to express the final

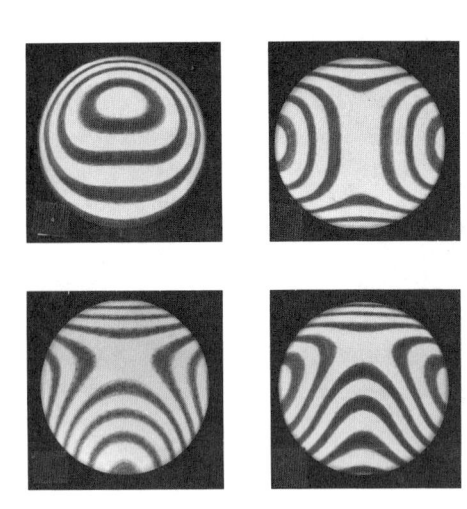

Figure 2.35. Interferograms for a lens with combined aberrations. (*a*) Spherical aberration with coma. (*b*) Spherical aberration with astigmatism. (*c*) Coma with astigmatism. (*d*) Spherical aberration with coma and astigmatism.

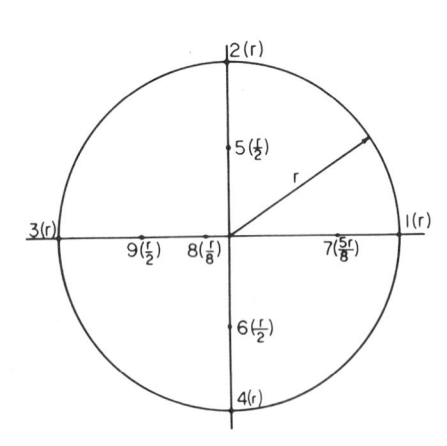

Figure 2.36. Distribution of reference points for evaluation of primary aberrations.

wavefront $W(x, y)$ in terms of a linear combination of Zernike polynomials. Then the process of removing or adding defocusing or tilts becomes much simpler. With the final results it is an extremely simple matter to plot level maps of the wavefront.

Another fundamentally different approach for finding the wavefront is the phase shifting procedure described in Chapter 13.

2.8. IMAGING OF THE INTERFERENCE PATTERN

An aberrated wavefront continuously changes its shape as it travels; therefore, if the optical system is not perfect, the interference pattern will also continuously change as the beam advances, as shown in Fig. 2.37. The errors of an

2.8. IMAGING OF THE INTERFERENCE PATTERNS

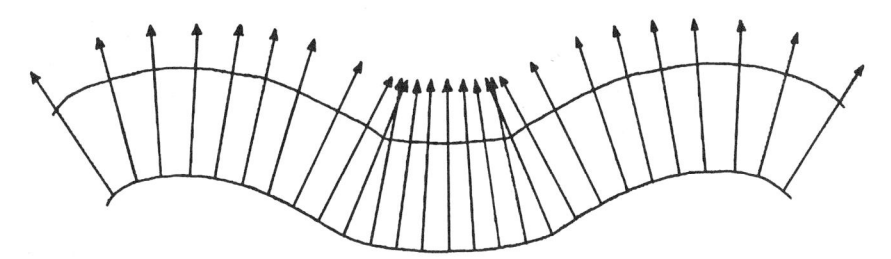

Figure 2.37. Wavefront shape changes as the wavefront travels.

instrument are represented by wavefront distortions on the pupil, and hence, the interferogram should be taken at that place.

When testing a lens with any of the configurations in Fig. 2.14, the wavefront travels twice through the lens, the second time after being reflected at the small mirror in front of the lens. If the aberration is small, the total wavefront deformation is twice the deformation introduced in a single pass through the lens. However, if the aberration is large, this is not so because the wavefront changes in its travel from the lens to the mirror and back to the lens. Then, the spot in the surface on which the defect is located is not imaged back onto itself by the concave or convex mirror and the ray will not pass through this defect the second time. Great confusion then results in regard to the interpretation of the interferogram, since the defect is not precisely duplicated by the double pass through the lens (Dyson 1959). It may be shown that the image of the lens is formed at a distance S from the lens, which is given by

$$S = \frac{2(F - r)^2}{2F - r} \qquad (2.49)$$

where F is the focal length and r is the mirror radius of curvature ($r > 0$ for a convex mirror and $r < 0$ for a concave mirror). We may see that the ideal mirror is convex and very close to the lens ($r \sim F$).

An adequate optical arrangement has to be used if the lens under test has a large aberration in order to image its pupil back on itself. Any auxiliary lenses or mirrors must preserve the wavefront shape. Some examples of these arrangements are in Fig. 2.38 (Malacara and Menchaca 1985).

However, for microscope objectives this solution is not satisfactory because the ideal place to observe the fringes is at the back focus. However, in this case, the Dyson system illustrated in Fig. 2.9 is an ideal solution. It is interesting to point out that Dyson's system may be used to place the self-conjugate plane at concave or convex surface while maintaining the concentricity of the surfaces.

The second problem is to image the interference pattern on the observing

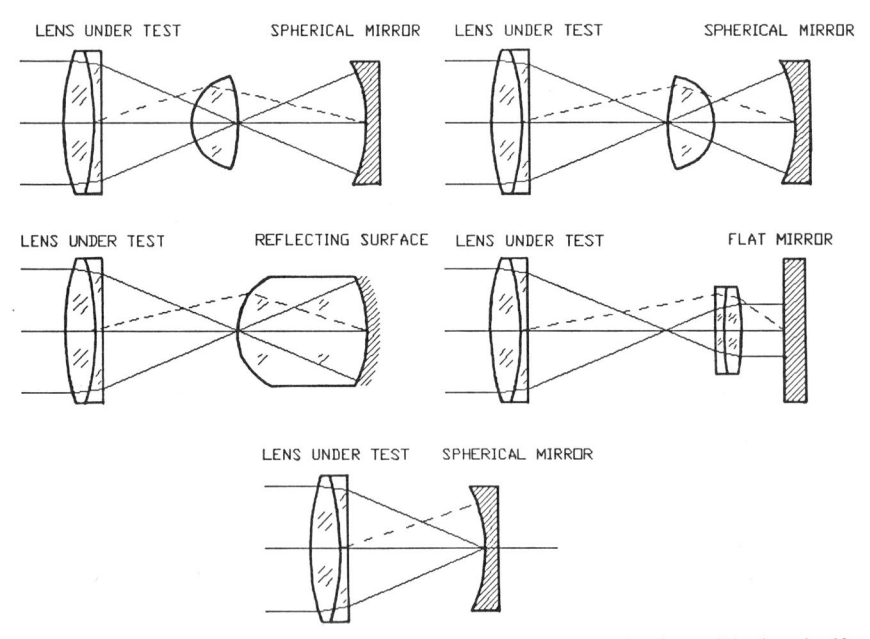

Figure 2.38. Five different optical arrangements to test a lens by imaging its pupil back on itself.

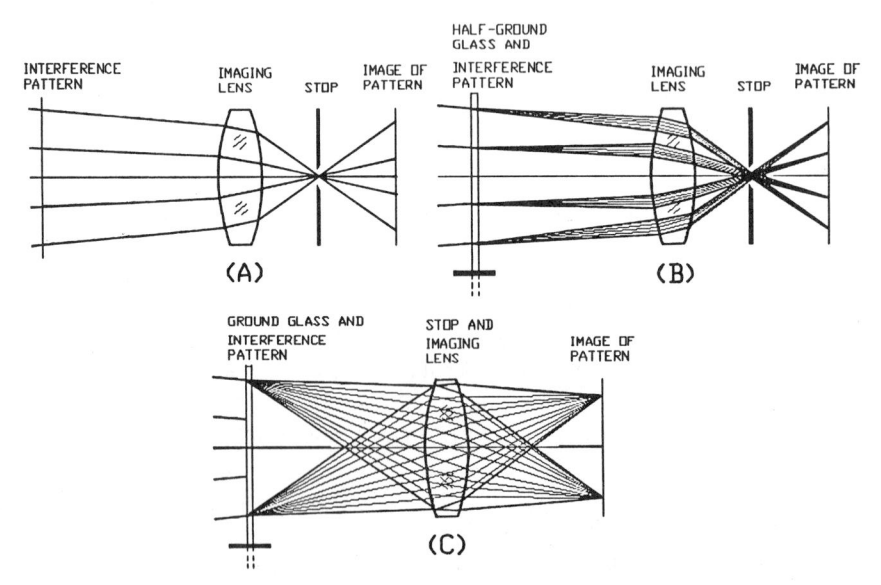

Figure 2.39. Imaging of the interferogram on the observation plane by means of a lens. (*a*) Without any rotating ground glass. (*b*) With a rotating half ground glass. (*c*) With a rotating fully ground glass.

detector, screen, or photographic plate. The imaging lens does not need to preserve the wavefront shape since it is generally placed after the beam splitter, and thus both interfering wavefronts pass through this lens. However, this lens has to be designed in such a way that the interference pattern is imaged without any distortion, assuming that the pupil of the system is at the closest image of the light source, as shown in Fig. 2.39a. A rotating ground glass in the plane of the interferogram might be useful sometimes in order to reduce the noise due to speckle and dust in the optical components. Ideally, this rotating glass should not be completely ground in order to reduce the loss of brightness and to keep the stop of the imaging lens at the original position, as shown in Fig. 2.39b. If the rotating glass is completely ground, the stop of the imaging lens should be shifted to the lens in order to use all available light, but then the lens must be designed taking this new stop position into consideration, as in Fig. 2.39c.

REFERENCES

Adachi, I., T. Masuda, and S. Nishiyama, "A Testing of Optical Materials by the Twyman Type Interferometer," *Atti Fond. Giorgio Ronchi Contrib. Ist. Naz. Ottica*, **16,** 666 (1961).

Adachi, I., T. Masuda, T. Nakata, and S. Nishiyama, "The Testing of Optical Materials by the Twyman Type Interferometer. III," *Atti Fond. Giorgio Ronchi Contrib. Ist. Naz. Ottica*, **17,** 319 (1962).

Balhorn, R., H. Kunzmann, and F. Lebowsky, "Frequency Stabilization of Internal-Mirror Helium-Neon Lasers," *Appl. Opt.*, **11,** 742 (1972).

Basile, G. "Two-Wave Interferometer for Use with a Monochromatic Source," *Appl. Opt.*, **18,** 422 (1979).

Batishko, C. R. and R. R. Shannon, "Problem in Large-Path Difference Laser Interferometry," *Appl. Opt.*, **11,** 195 (1972).

Bennett, S. J., R. E. Ward, and D. C. Wilson, "Comments on Frequency Stabilization of Internal Mirror He-Ne Lasers," *Appl. Opt.*, **12,** 1406 (1973).

Berggren, R., "Analysis of Interferograms," *Opt. Spectra.*, **4** (11), 22 (1970).

Birch, K. G., "Oblique Incidence Interferometry Applied to Non-Optical Surfaces," *J. Phys. E*, **6,** 1045 (1973).

Birch, K. G., "Interferometric Examination of Lenses and Other Components," *Proc. SPIE*, **163,** 112 (1979).

Briers, J. D., "Interferometric Testing of Optical Systems and Components: A Review," *Opt. Laser Technol.*, **4,** 28 (1972).

Bruning, J. H. and D. R. Herriot, "A Versatile Laser Interferometer," *Appl. Opt.*, **9,** 2180 (1970).

Buin, A. P., M. P. Semenova, and L. A. Kiryukhina, "Inspection of the Surface Quality of Large Scale Optical Components of an Unequal Arm Interferometer," *Sov. J. Opt. Technol.*, **36,** 720 (1969).

Burch, C. R., "The Williams Interferometer," *Monthly Not R. Soc.*, **100,** 488 (1940).

Candler, C., *Modern Interferometers*, Hilger and Watts, London, 1951, Chaps. 6 and 7.

Chen, Chungte W. and James B. Breckinridge, "Holographic Twyman-Green Interferometer," *Appl. Opt.*, **21**, 2563, (1982).

Collier, R. T., C. B. Burkhardt, and L. H. Lin, *Optical Holography*, Academic Press, New York, 1971, p. 146.

Connes, P., "Aumentation du Produit Luminosité X Resolution des Interférometres par l'Emploie d'une Difference de Marche Independente de l'Incidence," *Rev. Opt.*, **35**, 37 (1956).

Cook, A. H., *Interference of Electromagnetic Waves*, Clarendon Press, Oxford, 1971, Chaps. 2 and 4.

Creath, Katherine, "Wyko Systems for Optical Metrology," *Proc. SPIE*, **816**, 111 (1987).

Cuadrado, Javier M., Maria Va. Perez, and Carlos Gomez-Reino, "Equilateral Hiperbolic Zone Plates: Their Use in the Alignment of a Mach-Zehnder Interferometer," *Appl. Opt.*, **26**, 1527 (1987).

De Vany, A. S., "On Using a Williams Interferometer for Making a Divider Plate," *Appl. Opt.*, **4**, 365 (1965).

Dyson, J., "Unit Magnification Optical System without Seidel Aberrations," *J. Opt. Soc. Am.*, **49**, 713 (1959).

Fantone, Stephen D., "Simple Method for Testing an Axicon," *Appl. Opt.*, **20**, 3685 (1981).

Feinberg, J. "Interferometer with a Self-Pumped Phase Conjugating Mirror," *Opt. Lett.*, **8**, 569 (1983).

Ferguson, Thomas R., "Polarization Effects in Inteferograms of Conical Optical Elements," *Appl. Opt.*, **21**, 514 (1982).

Flack, Jr., Ronald D., "Mach-Zehnder Interferometer Errors Resulting from Test Section Misalignment," *Appl. Opt.*, **17**, 985 (1978).

Geary, Joseph M., "Real Time Interferogram Simulation," *Opt. Eng.*, **18**, 39 (1979).

Gerth, H. L., R. E. Sladky, M. J. Besik, and C. A. Washington, "Fabrication of Off Axis Parabolic Mirrors," *Opt. Eng.*, **17**, 588 (1978).

Gordon, S. K. and S. F. Jacobs, "Modification of Inexpensive Multimode Lasers to Produce a Stabilized Single Frequency Beam," *Appl. Opt.*, **13**, 231 (1974).

Grigull, V. and H. Rottenkolber, "Two Beam Interferometer Using a Laser," *J. Opt. Soc Am.*, **57**, 149 (1967).

Guild J., "Fringe Systems in Uncompensated Interferometers," *Proc. Phys. Soc.*, **33**, 40 (1920-1921).

Hansen, G., "Die Sichtbarkeit der Interferenzen beim Twyman Interferometer," *Optik*, **12**, 5 (1955).

Hansen, G., "On Twyman Interferometers," *Optik*, **67**, 79 (1984).

Hardesty, Chuck, "Vibration Insensitive Laser Unequal Path Interferometer (LUPI) Test," *Proc. SPIE*, **192**, 93 (1979).

Hariharan, P., "Improved Oblique Incidence Interferometer," *Opt. Eng.*, **14**, 257 (1975).

REFERENCES

Hariharan, P. and D. Sen, "The Separation of Symmetrical and Asymmetrical Wavefront Aberrations in the Twyman Interferometer," *Proc. Phys. Soc.*, **77,** 328 (1961).

Hayes, John, K. L. Underwood, John S. Loomis, Robert E. Parks et al., "Testing of Non-Linear Diamont-Turned Reflaxicons," *Appl. Opt.*, **20,** 235 (1981).

Hopkins, R. E., "Re-evaluation of the Problem of Optical Design," *J. Opt. Soc. Am.*, **52,** 1218 (1962).

Horne, D. F., *Optical Production Technology*, Adam Hilger, London, 1972, and Crane Russak, New York, 1972, Chap. 11.

Houston, J. B., Jr., C. J. Buccini, and P. K. O'Neill, "A Laser Unequal Path Interferometer for the Optical Shop," *Appl. Opt.*, **6,** 1237 (1967).

Howes, Walton L., "Lens Collimation and Testing Using a Twyman–Green Interferometer with a Self-Pumped Phase Conjugation Mirror," *Appl. Opt.*, **25,** 473 (1986a).

Howes, Walton L., "Large Aperture Interferometer with Phase-Conjugate Self-Reference Beam," *Appl. Opt.*, **25,** 3167 (1986b).

Jaroszewics, Zibgniew, "Interferometric Testing of the Spacing Error of a Plane Diffraction Grating," *Opt. Commun.*, **60,** 345 (1986).

Jones, Robert A., "Fabrication of Small Nonsymmetrical Aspheric Surfaces," *Appl. Opt.*, **18,** 1244 (1979).

Kingslake, R., "The Interferometer Patterns Due to the Primary Aberrations," *Trans. Opt. Soc.*, **27,** 94 (1925–1926).

Kingslake, R., "The Analysis of an Interferogram," *Trans. Opt. Soc.*, **28,** 1 (1926–1927).

Kocher, D. G., "Twyman–Green Interferometer to Test Large Aperture Optical Systems," *Appl. Opt.*, **11,** 1872 (1972).

Kowalik, W. "Interference Measurement of Continuous Heterogeneities in Optical Materials," *Appl. Opt.*, **17,** 2956 (1978).

Kwon, Osuk Y., James C. Wyant, and C. R. Hayslett, "Long-Wavelength Interferometer in the Optical Shop," *Proc. SPIE*, **192,** 88 (1979).

Linnik, V. P., "An Interferometer for Controlling Large Mechanical Details," *Comptes Rendus (Doklady) de l'Academie des Sciences de l''URSS*, **35,** 16 (1942).

Leung, Kang M. and Steve Lange, "Wavefront Evaluation on Laser Diodes Using a Phase Measurement Interferometer," *Proc. SPIE*, **429,** 27 (1983).

Lewandoswki, Jacques, Bernard Mongeau, Maurice Cormier, and Jean La-Pierre, "Infrared Interferometers at 10 μm," *J. Appl. Phys.*, **661,** 132 (1986).

Luneburg, R. K., *Mathematical Theory of Optics*, University of California Press, Berkeley, 1964, Appendix 2, p. 372.

MacBean, Myles D. A., "Oblique Incidence Interferometry of Rough Surfaces Using a Novel Dove-Prism Spectrometer," *Appl. Opt.*, **23,** 4024 (1984).

Malacara, Daniel and Menchaca, Carmen, "Imaging of the Wavefront Under Test in Interferometry," *Proc. SPIE*, **540,** 34 (1985).

Maréchal, A. and P. Dejonc, "Quelques Aspects de Franges de Twyman" (Some Aspects of Twyman Fringes), *Rev. Opt. Theor. Instrum.*, **29,** 430 (1950).

Martin, L. C., R. Kingslake, "The Measurement of Chromatic Aberration on the Hilger Lens Testing Interferometer," *Trans. Opt. Soc.*, **25,** 213 (1923–1924).

Masuda, T., S. Nishiyama, T. Nakata, and I. Adachi, "The Testing of Optical Materials by the Twyman Type Interferometer. II," *Atti Fond. Giorgio Ronchi Contrib. Ist. Naz. Ottica*, **17**, 197 (1962).

McDonell, M. M. and T. F. DeYoung, "Inexpensive Large Aperture Interferometer," *Proc. SPIE*, **192**, 145 (1979).

Mertz, L., *International Commission for Optics Conference*, Stockholm, 1959.

Michelson, A. A., "On the Correction of Optical Surfaces," *Astrophys. J.*, **47**, 283 (1918).

Molesini, G., G. Pedrini, and F. Quercioli, "Laser Unequal Path Interferometer Configurations by Grating Splitting at the Fourier Plane," *Opt. Eng.*, **23**, 646 (1984).

Morokuma, T., K. F. Neflen, T. R. Lawrence, and T. M. Klucher, "Interference Fringes with Long Path Difference Using He-Ne Laser," *J. Opt. Soc. Am.*, **53**, 394 (1963).

Munnerlyn, C. K., M. P. Givens, and R. E. Hopkins, "Interferometric Measurement of Optically Rough Surfaces," *IEEE J. Quantum Electron.*, **QE-5**, 359 (1969).

Murty, M. V. R. K., "Simulation of Primary Aberrations of Lens Using a Generalized Michelson Interferometer," *J. Opt. Soc. Am.*, **50**, 1089, (1960).

Murty, M. V. R. K., "Interference between Wavefronts Rotated or Reversed with Respect to Each Other and Its Relation to Spatial Coherence," *J. Opt. Soc. Am.*, **54**, 1187 (1964).

Ostrovskaya, M. A. and N. F. Filimonova, "Use of the Gas Laser for Interferometric Quality Control in Telescope Manufacture," *Sov. J. Opt. Technol.*, **36**, 563 (1969).

Parmigiani, F., "Phase Dependence of Michelson Interferometer Outputs on the Absorbing Beam Splitter Thickness," *Opt. Commun.*, **38**, 319 (1981).

Perry, J. W., "The Determination of Aberrations, as Expressed in Geometrical Optics, from the Indications of the Hilger Interferometer," *Trans. Opt. Soc.*, **25**, 97 (1923-1924).

Rimmer, M. P., D. M. King, and D. G. Fox, "Computer Program for the Analysis of Interferometric Test Data," *Appl. Opt.*, **11**, 2790 (1972).

Rogers, G. L., "The Equivalent Interferometer in Holography," *Opt. Acta*, **17**, 257 (1970).

Saunders, J. B., "Precision Method for Evaluating Primary Aberrations of Lenses with a Twyman Interferometer," *J. Res. Nat. Bur. Stand.*, **69C**, 251 (1965).

Saunders, James B. and Franz L. Gross, "Interferometer for Large Surfaces," *J. Res. Nat. Bur. Stand.*, **62**, 137 (1959).

Sinclair, D. C. and W. E. Bell, *Gas Laser Technology*, Holt, Rinehart & Winston, New York, 1969, Chap. 5.

Slevogt, H., "Zur geometrischen Optik der Zweistrahl-Interferometer," (About the Geometrical Optics of Two-Beam Interferometers), *Optik*, **11**, 366 (1954).

Smith, T., "The theory of the Lens-Testing Interferometer," *Trans. Opt. Soc.*, **28**, 104 (1926-1927).

Steel, W. H., "Adjustable Compensators for Two-Beam Interferometers," *Opt. Acta*, **9**, 111 (1962).

Steel, W. H., "The Compensation of a Williams Interferometer," *Opt. Acta*, **10**, 206 (1963).

Steel, W. H., "Two-Beam Interferometry," in *Progress in Optics*, Vol. 5, E. Wolf, Ed., North-Holland, Amsterdam, 1966, Chap. 3.

Thomas, D. A., and J. C. Wyant, "Determination of the dihedral angle errors of a corner cube from its Twyman–Green interferogram," *J. Opt. Soc. Am.*, **67**, 467 (1977).

Twyman, F., "Correction of Optical Surfaces," *Astrophys. J.*, **48**, 256 (1918a).

Twyman, F., "Interferometers for the Experimental Study of Optical Systems from the Point of View of the Wave Theory," *Philos. Mag.*, Ser. 6, **35**, 49 (1918b).

Twyman, F., British Patent (camera lens) 130224 (1919).

Twyman, F., "The Testing of Microscope Objectives and Microscopes by Interferometry," *Trans. Faraday Soc.*, **16**, 208 (1920).

Twyman, F., "An Interferometer for Testing Camera Lenses," *Trans. Opt. Soc.*, **22**, 174 (1920–1921); also appeared in *Philos. Mag.*, **42**, 777 (1921).

Twyman, F., "The Hilger Microscope Interferometer," *Trans. Opt. Soc.*, **24**, 189 (1922–1923).

Twyman, F., *Prism and Lens Making*, Hilger and Watts, London, Chaps. 11 and 12, 1957.

Twyman, F., and A. J. Dalladay, "Variation in Refractive Index near the Surfaces of Glass Melts," *Trans. Opt. Soc.*, **23**, 131 (1921–1922).

Twyman, F., and A. Green, British Patent (prisms and microscopes) 103832 (1916).

U.S. Department of Defense, *Military Handbook* 141 (Mil-HDBK-141), 1963, Sections 16 and 25.

Van Heel, A. C. S. and C. A. J. Simons, "Lens and Surface Testing with Compact Interferometers," *Appl. Opt.*, **6**, 803 (1967).

Zielinski, Robert J., "Unequal Path Interferometer Alignment and Use," *Opt. Eng.*, **18**, 479, (1979).

Zielinski, Robert J., "Unequal Path Interferometer Alignment and Use," *Proc. SPIE*, **153**, 51 (1978).

ADDITIONAL REFERENCES

Dorband, B. and H. J. Tiziani, "Auslegun von Kompensationsystem zur Interferometrischen Prüfung Asphärischer Flächen," *Optik*, **67**, 1 (1984).

Dutton, D., A. Cornejo, and M. Latta, "A Semiautomatic Method for Interpreting Shearing Interferograms," *Appl. Opt.*, **7**, 125, (1968).

Forsythe, G. E., "Generation and Use of Orthogonal Polynomials for Data-Fitting on a Digital Computer," *J. Soc. Indust. Appl. Math.*, **5**, 74 (1957).

Geary, Joseph M., David Holmes Holmes, and Zeringue Zeringue, "Real Time Interferogram Simulation," in *Optical Interferograms—Reduction and Interpretation*, ASTM Symposium, Am. Soc. for Test. and Mat. Florida, 1978.

Osuk, Kwon, J. C. Wyant, and C. R. Haystett, "Infrared Twyman–Green Interferometry," *J. Opt. Soc. Am.*, **67**, 1365 (1977).

Seligson, Joel L., C. A. Callari, J. E. Greivenkamp, and J. W. Ward, "Stability of Lateral-Shearing Heterodyne Twyman-Green Interferometer," *Opt. Eng.*, **23,** 353, (1984).

Simon, Juan M. and Maria C. Simon, "Wollaston Prism as a Beam Splitter in Convergent Light," *Appl. Opt.*, **17,** 3352 (1978).

Watrasiewics, B. M., "Turned Edge Fringes in the Twyman-Green Interferometer Due to Focusing Errors," *J. Sci. Instr.*, **42,** 897 (1965).

3

Common-Path Interferometers

S. Mallick

3.1. INTRODUCTION

In the general type of interferometer, such as the Twyman–Green or Mach–Zehnder, the reference and test beams follow widely separated paths and are, therefore, differently affected by mechanical shocks and temperature fluctuations. Thus, if suitable precautions are not taken, the fringe pattern in the observation plane is unstable and measurements are not possible. The problems are particularly acute when optical systems of large aperture are being tested. Most of the difficulty can be avoided by using so-called common-path interferometers in which the reference and test beams traverse the same general path. These interferometers have the additional advantage that they do not require perfect optical components (the master) of dimensions equal to those of the system under test for producing the reference beam. Furthermore, the path difference between the two beams in the center of the field of view is, in general, zero, making the use of white light possible.

In certain common-path interferometers the reference beam is made to traverse a small area of the optical system under test and is, therefore, unaffected by system aberrations. When this beam interferes with the test beam, which has traversed the full aperture of the optical system, explicit information about the system defects is obtained. However, in most common-path interferometers both the reference and test beams are affected by the aberrations, and interference is produced by shearing one beam with respect to the other. The information obtained in this case is implicit, and some computations are needed to determine the shape of the aberrated wavefront.

The beam splitting is brought about by amplitude division with the help of a partially scattering surface, a doubly refracting crystal, or a semireflecting surface. We consider a few examples of these instruments in this chapter.

Optical Shop Testing, Second Edition, Edited by Daniel Malacara.
ISBN 0-471-52232-5 © 1992, John Wiley & Sons, Inc.

3.2. BURCH'S INTERFEROMETER EMPLOYING TWO MATCHED SCATTER PLATES

Figure 3.1 is a schematic diagram of Burch's interferometer as applied to the testing of a concave mirror M (Burch 1953, 1962, 1969). A lens L forms an image of a small source S at S' on the mirror. The beam splitting is brought about by a weakly scattering plate R_1, which is located at the center of curvature C of the mirror. Two identical scatter plates R_1 and R_2 can be made, for example, by photographing a speckle pattern or by taking two replicas from a lightly ground surface (Houston 1970). The mirror M forms an image of R_1 at R_2, which is placed the other way round so that there is a point-for-point coincidence between R_2 and the image of R_1. A semireflecting plate B directs the light returning from the mirror M onto R_2.

A part of the light incident on the scatter plate R_1 passes through it without scattering and arrives at S'. Since this beam touches the mirror M only at a small region around S', it is not affected by the errors of the mirror surface. This beam acts as the reference beam. Some of the incident light is, however, scattered by R_1 and fills all of the aperture of M. This beam picks up the errors of the mirror and is the test beam.

Let us consider a ray incident at a point A on the scatter plate R_1. The directly transmitted ray (solid line in Fig. 3.1) follows the path $AS'A'$ and encounters at A' a scattering center that is identical to the one at A. This ray is scattered at A' and gives rise to a cone of rays. The rays scattered at A (dotted lines) fill the mirror M, arrive at the image A', and pass through R_2 without scattering. We have thus two mutually coherent beams emerging from R_2; one beam is directly transmitted by R_1 and scattered by R_2, and the second is scattered by R_1 and transmitted by R_2. An observer looking at the mirror surface through R_2 will see an interferogram between these two beams. If the mirror is free of any

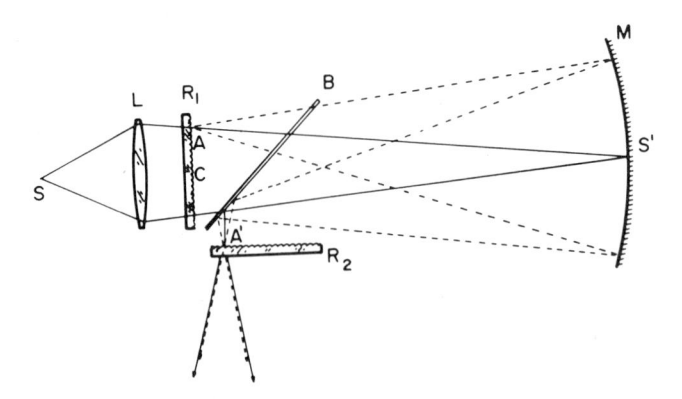

Figure 3.1. Burch's scatter plate interferometer, used for testing a concave mirror.

3.2. BURCH'S INTERFEROMETER WITH MATCHED SCATTER PLATES

error in the region of S', the interferogram will provide explicit information about the mirror aberrations, as in any separate-beam interferometer.

To obtain a permanent record of the interference pattern, a camera lens, placed after R_2, is used to form an image of the mirror surface on a photographic film. Each point of the film receives light from S' and from a conjugate point on the mirror surface; the interference effect (the light intensity) at the film will thus give information about the mirror aberration at the conjugate point.

The light that is directly transmitted by both R_1 and R_2 gives rise to a bright spot located at S' and is quite troublesome for visual observations. The light that is scattered by R_1 and again by R_2 gives a weak background and diminishes slightly the contrast of fringes. The dimensions of the source S should be such that its image S' remains localized within a fringe. If the fringes are quite broad in a certain region of the mirror (this is equivalent to saying that the mirror is almost free of aberrations in this region), the source image S' should be made to lie there.

A slight displacement of one of the scatter plates in its own plane, with respect to the image of the other, gives rise to a set of parallel straight-line fringes, and the mirror defects appear as distortions in the straightness of the fringes.

Scott (1969) used this interferometer to test a 91.5-cm, $f/4$ paraboloid and a 35-cm Gregorian secondary. The details of testing and of modifications made in Burch's original design are discussed in the article cited.

Figure 3.2 shows the principle for testing a converging lens.

Burch's interferometer is quite sensitive to vibrations taking the form of tilts about an axis normal to the line of sight or translations across the line of sight. To make the system insensitive to these vibrations, Shoemaker and Murty (1966) modified the setup by replacing the second scatter plate with a plane mirror and thus reimaging the first scatter plate point by point back on itself. This setup gives double sensitivity for even-order aberrations but cannot detect odd aberrations. An obvious great advantage is that only one scatter plate has to be made.

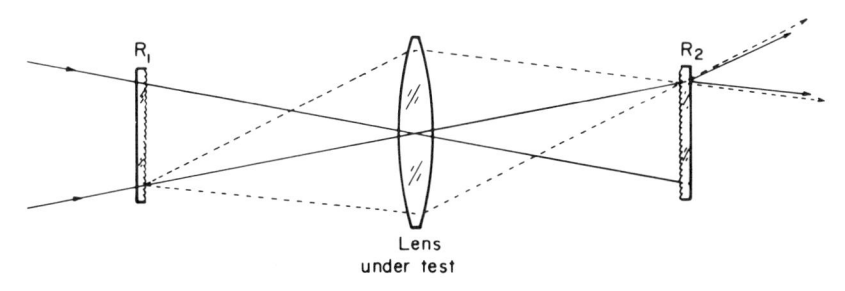

Figure 3.2. The principle of Burch's interferometer for testing a lens.

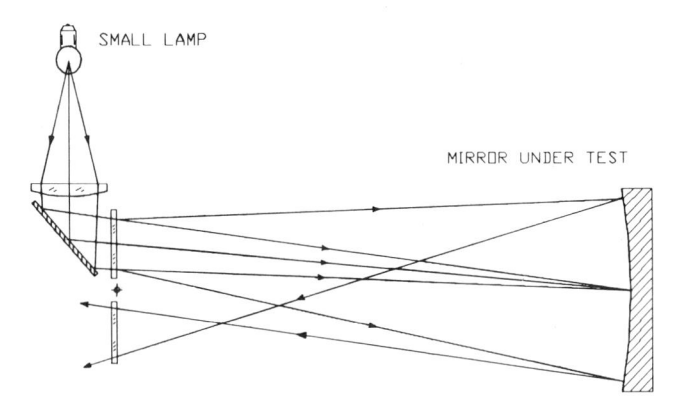

Figure 3.3. Burch interferometer with identical scattering plates.

Huang et al. (1989) have used some polarization elements in order to introduce a variable phase shift between the test and the reference beams and employ the fringe scanning method for determining the phase distribution over the test wavefront. Some further developments and modifications of the scatter plate interferometer are due to Su et al. (1984, 1986, 1987) in order to produce a null test for a concave conic surface and to Rubin (1980).

Two practical Burch interferometers for testing concave surfaces with a ratio of the radius of curvature to the diameter larger than about six, in order to be able to test them off-axis, are shown in Figs. 3.3 and 3.4. The light source is

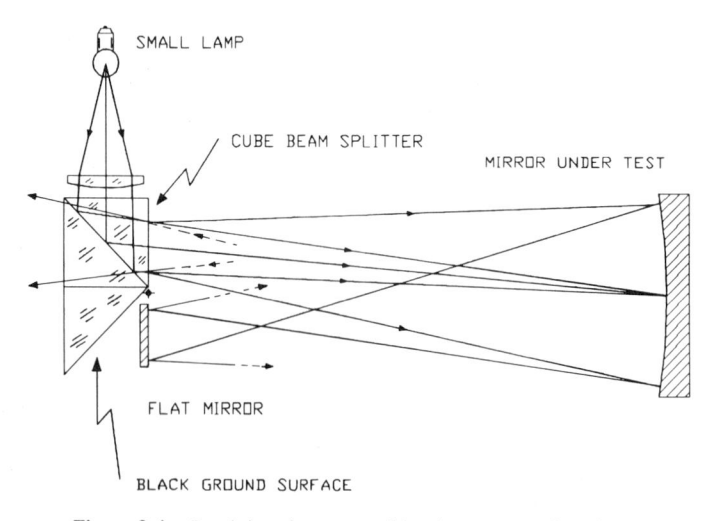

Figure 3.4. Burch interferometer with only one scattering plate.

3.3. BIREFRINGENT BEAM SPLITTERS

a small tungsten lamp. In the interferometer in Fig. 3.3 the two scattering plates have to be identical, but one rotated 180° with respect to the other. The scattering plate for the interferometer in Fig. 3.4 is made by fine grinding and then half polishing the front face of a cube beam splitter. Symmetrically placed with respect to the center of the curvature is a small flat mirror. In order to prevent an unwanted reflection from going to the eye of the observer, a small triangular prism with a ground and black painted surface is cemented to the cube.

3.3. BIREFRINGENT BEAM SPLITTERS

An important class of interferometers uses birefringent crystal elements as beam splitters. These interferometers are known as polarization interferometers (Françon and Mallick 1971). We discuss in this section three principal types of these beam splitters.

3.3.1. Savart Polariscope

A Savart polariscope consists of two identical uniaxial crystal plates with the optic axis cut at 45° to the plate normal (Fig. 3.5). The principal sections (plane containing the optic axis and the plate normal) of the two plates are crossed with each other. The optic axis of the first plate lies in the plane of the page and that of the second plate makes an angle of 45° with it; the dotted double arrow in the figure represents the projection of the optic axis on this plane. An incident ray is split by the first plate into two rays, the ordinary ray O and the

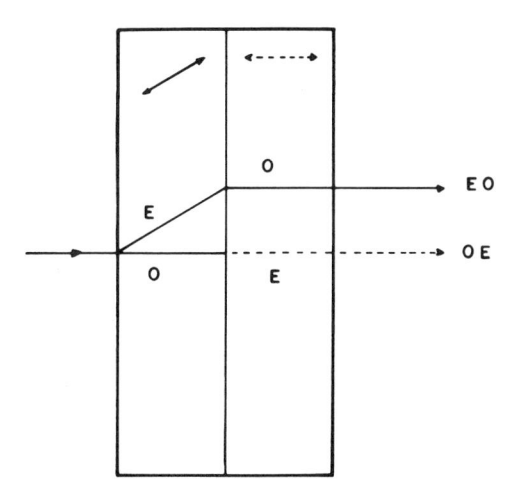

Figure 3.5. Beam splitting produced by a Savart polariscope. The figure is drawn for a polariscope made of a positive crystal (e.g., quartz).

extraordinary ray E. Since the second plate is turned through 90° with respect to the first one, the ordinary ray in the first plate becomes extraordinary in the second, and vice versa. The ray OE does not lie in the plane of the page, though it emerges parallel to its sister ray EO; the dotted line represents the projection of the ray path on this plane. The lateral displacements between the two rays, each produced by one of the two component plates, are equal and are in perpendicular directions. The total displacement between the emerging EO and OE rays produced by a Savart polariscope of thickness $2t$ is given by

$$d = \sqrt{2} \, \frac{n_e^2 - n_o^2}{n_e^2 + n_o^2} \, t \tag{3.1}$$

where n_o and n_e are the ordinary and extraordinary indices of refraction, respectively. A 1-cm-thick polariscope will produce a lateral displacement of 80 μm if it is made of quartz and a 1.5-mm displacement if made of calcite. In Fig. 3.5 if the incident ray is inclined to the plate normal, the two emerging rays are still parallel to the original ray, and their relative displacement remains practically unaltered.

The parallel emerging rays interfere in the far field (or in the back focal plane of a positive lens), and the interference pattern is similar to that produced in Young's experiment with the two mutually coherent sources separated by a distance equal to d. For small angles of incidence, the fringes are equidistant straight lines normal to the direction of displacement. The angular spacing of these fringes is as follows:

$$\text{Angular spacing} = \frac{\lambda}{d}. \tag{3.2}$$

The zero-order fringe corresponds to normal incidence and lies in the center of the field of view. With a Savart polariscope of 1 cm thickness and a lens of 10 cm focal length, the fringe spacing in yellow light is 2 mm for quartz and 0.1 mm for calcite.

The OE and EO rays emerging from the Savart polariscope vibrate in mutually orthogonal directions. To make them interfere, the vibration directions are set parallel to each other by means of a linear polarizer, the transmission axis of which is oriented at 45° to the orthogonal vibrations. This polarizer is, however, not sufficient for interference to take place. We know that a natural (unpolarized) ray of light is equivalent to two mutually incoherent components of equal amplitude vibrating in perpendicular directions. Thus the ordinary and the extraordinary rays produced by a crystal have no permanent phase difference between them. To make these rays mutually coherent, a polarizer is placed across the incident beam so that only a single component of the natural light is transmitted onto the crystal. The transmission axis of this polarizer is at 45° to the principal axes of the crystal.

3.3. BIREFRINGENT BEAM SPLITTERS

3.3.2. Wollaston Prism

A Wollaston prism (Fig. 3.6) consists of two similar wedges cemented together in such a way that the combination forms a plane parallel plate. The optic axes in the two component wedges are parallel to the external faces and are mutually perpendicular. A Wollaston prism splits an incident ray into two rays traveling in different directions; the lateral displacement between the rays is thus different at different distances from the Wollaston. The angular splitting α is given by the relation

$$\alpha = 2(n_e - n_o)\tan \theta \qquad (3.3)$$

where θ is the wedge angle. For most practical purposes α can be considered to be independent of the angle of incidence. For an angle $\theta = 5°$ the angular splitting is 6 min of arc for a Wollaston prism made of quartz and $2°$ for one made of calcite.

The path difference between the *OE* and the *EO* rays (Fig. 3.7) emerging at a distance x from the axis y-y' of the Wollaston prism is given by

$$\Delta = 2(n_e - n_o)x \tan \theta = \alpha x. \qquad (3.4)$$

The path difference is zero along the axis, where the thicknesses of the two component wedges are equal, and increases linearly with x. When a Wollaston prism is placed between two suitably oriented polarizers, one observes a system of straight-line fringes parallel to the edges of the component wedges and localized in the interior of the prism. (The fringes are perpendicular to the plane

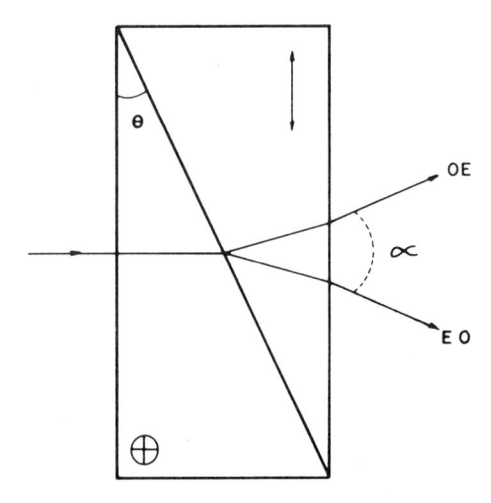

Figure 3.6. Beam splitting by a Wollaston prism made of a positive crystal.

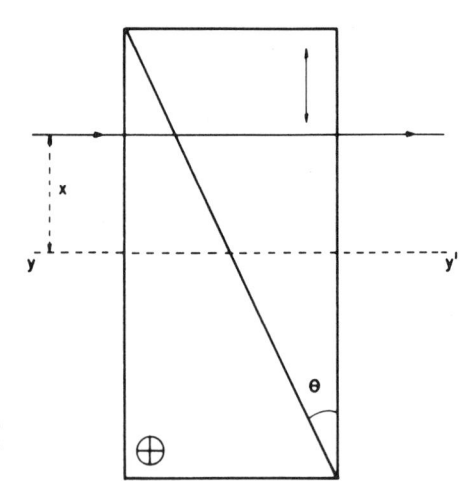

Figure 3.7. Path difference produced by a Wollaston prism between the two split-up rays is linearly related to x.

of the Fig. 3.7.) The path difference along the axis being zero, fringes are visible in white light. The fringe spacing is equal to

$$x_0 = \frac{\lambda}{2(n_e - n_o)\tan \theta}. \tag{3.5}$$

With $\theta = 5°$, $\lambda = 0.55$ μm, and $(n_e - n_o) = 9 \times 10^{-3}$ (quartz), there are approximately three fringes per millimeter. When the angle θ is very small (a few minutes of arc), the fringes are wide apart and the Wollaston prism can be used as a compensator. In this form the Wollaston prism is known as a Babinet compensator.

Relation (3.4) for the path difference between OE and EO rays is true for normal incidence (the angular splitting, being small, is neglected for the calculation of Δ). For nonnormal incidence a term proportional to the square of the angle of incidence is added to the right-hand side of Eq. (3.4). However, this term is negligible, for example, for a quartz prism of 10 mm thickness and for a case in which the angles of incidence remain less than $10°$. Some modified Wollaston prisms have been devised that can accept much larger angles of incidence.

3.3.3. Double-Focus Systems

A lens made of a birefringent crystal acts as a beam splitter. A parallel beam of light incident on such a lens will be split into an ordinary beam and an extraordinary beam, which come to focus at two different points (Fig. 3.8). The O and E images are displaced along the axis of the beam, in contrast to the case

3.4. LATERAL SHEARING INTERFEROMETERS 103

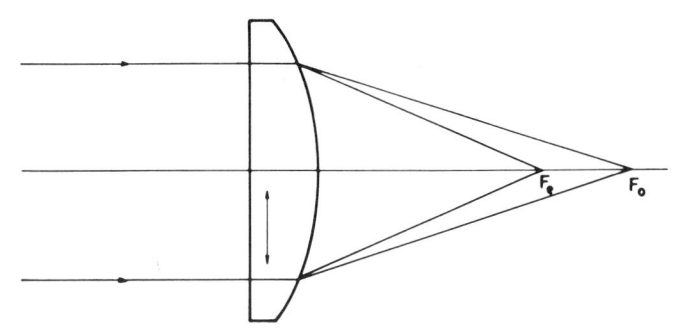

Figure 3.8. A birefringent lens splits up an incident beam into an ordinary and an extraordinary beam, which are brought to focus at two different points along the lens axis. The figure is drawn for a lens made of a positive crystal.

of a Savart polariscope or of a Wollaston prism, where the displacement is normal to the direction of the incident beam. Various types of compound lenses suitable for specific applications have been designed.

3.4. LATERAL SHEARING INTERFEROMETERS

3.4.1. Use of a Savart Polariscope

Lateral shearing interferometers using birefringent beam splitters have been widely used to study the aberrations of an optical system. We describe here an arrangement by Françon and Jordery (1953) in which a Savart polariscope is used to produce a lateral shear of the aberrated wavefront (Fig. 3.9). The lens L (or the mirror) under test forms an image S' of a small source S. The distance of L from the source is fixed by the conditions under which the lens is to be tested. The lens L_1 collimates the light coming from S' so that the Savart polariscope Q is traversed by a parallel beam of light. Two linear polarizers (not shown in the figure) are placed before and after the Savart Q. The combination of lenses L_1 and L_2 constitutes a low-power microscope that is focused on the test lens L. If the lens L is perfect, the wavefront Σ is plane, and the ordinary and the extraordinary wavefronts produced by the Savart will have a uniform path difference between them. The eye placed in the focal plane of L_2 will observe a uniform color (or a uniform intensity in the case of monochromatic light) in the entire field of view. In the presence of aberrations Σ will be deformed and the field of view will appear nonuniform. If the aberrations are large, a system of fringes will be observed. The nature and the magnitude of aberrations can be determined from the observed variations of color (or of intensity).

The far-field fringes of the Savart polariscope Q are located virtually in the

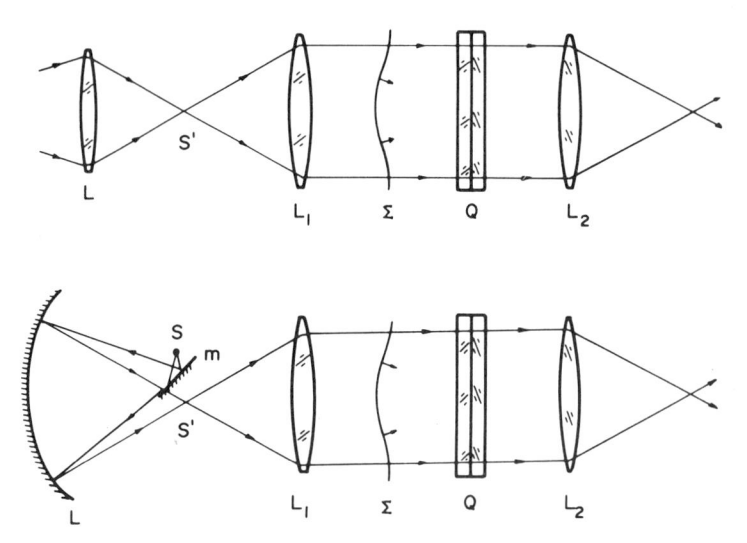

Figure 3.9. Interference arrangement employing a Savart polariscope Q for testing the optical system L.

plane of source image S'. The source size should be such that S' occupies a small fraction (say one fifth) of a fringe width.

The background color (or intensity) can be chosen by inclining the Savart Q about an axis parallel to the fringes. When the Savart is normal to the optical axis, that is, normal to the incident light, the zero-order fringe coincides with source image S' and the background will be dark (crossed polarizers). By inclining the Savart, S' can be made to coincide with a fringe of any desired color, which will then appear in the background. Instead of producing a uniform ground color in the field of view, we may produce a regular system of rectilinear fringes that are deformed in the region where the wavefront departs from the ideal form. Such fringes can be produced in a plane conjugate to the test lens by placing an additional Savart to the right of L_2.

To illustrate the principle of the method, we study the aspect of the field of view in the presence of primary spherical aberration. The ground is chosen to be of uniform intensity. The distance parallel to the optical axis between the aberrated wavefront Σ and the ideal wavefront (corresponding to the Gaussian image point) at a height h from the axis is given by

$$z = ah^4 \tag{3.6}$$

where a is a constant depending on the magnitude of the aberration. To determine the aspect of the field of view, we have to calculate the path difference between the two sheared wavefronts Σ_1 and Σ_2 produced by the Savart polari-

3.4. LATERAL SHEARING INTERFEROMETERS

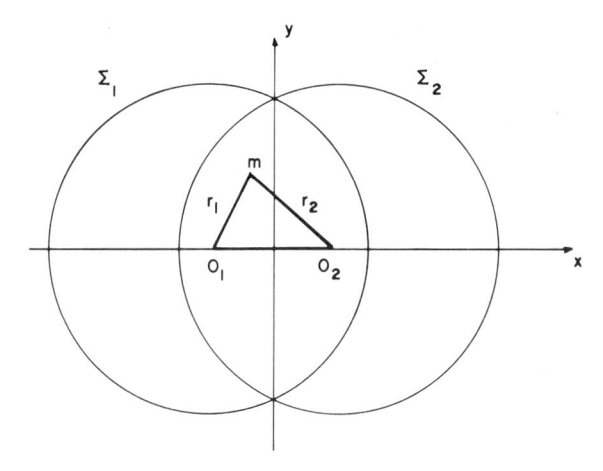

Figure 3.10. Representing the projection of the two sheared wavefronts on a plane normal to the optical axis of the system.

scope. Figure 3.10 represents Σ_1 and Σ_2 as projected on a plane perpendicular to the optical axis of the system (this is the plane of the ideal wavefront); O_1 and O_2 are the centers of Σ_1 and Σ_2, respectively. The coordinate system is so chosen that the x axis passes through O_1 and O_2 and the y axis is the right bisector of O_1-O_2. Now consider a point m (x, y) lying on the ideal plane wavefront; its distance from the aberrated wavefront Σ_1 is given by

$$z_1 = ar_1^4. \tag{3.7}$$

Similarly, the distance of point m from Σ_2 is given by

$$z_2 = ar_2^4. \tag{3.8}$$

The separation between Σ_1 and Σ_2 is, therefore,

$$z_1 - z_2 = a(r_1^4 - r_2^4) = 4a \times d\left(x^2 + y^2 + \frac{d^2}{4}\right) \tag{3.9}$$

where d is the shear between Σ_1 and Σ_2. The lines of equal path difference, $z_1 - z_2$, are represented in Fig. 3.11. The form of fringes for other aberrations can be determined similarly. Evidently, when the aberrations are small, no fringes will be seen; there will simply be small variations of intensity in the field of view.

A complete analysis of the lateral shearing interferogram can be carried out by a mathematical operation described by Saunders (1961, 1962) (see Chapter

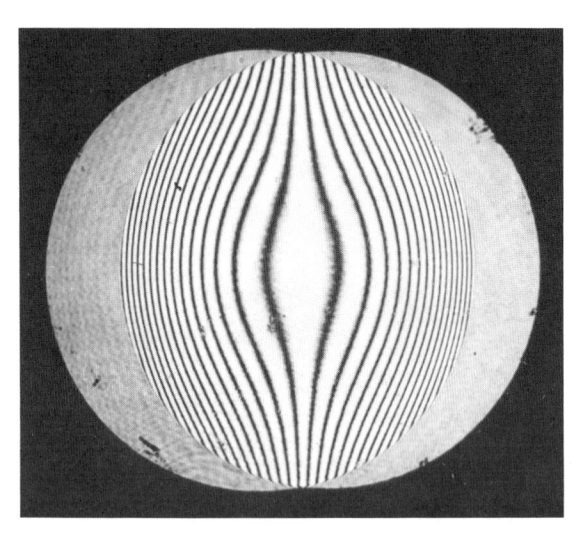

Figure 3.11. Lateral shearing interferogram of a wavefront distorted by spherical aberration of third order.

4). The method yields values of the deviations of the wavefront under test from a close fitting sphere. The reference sphere may be chosen statistically so that the results are the deviations from a best fitting surface.

3.4.2. Use of a Wollaston Prism

In the arrangement represented in Fig. 3.9, it is possible to use a Wollaston prism instead of the Savart polariscope. This prism is placed at the source image S'. The background intensity can be changed by translating the Wollaston prism laterally perpendicular to the optical axis. A system of rectilinear fringes can be produced in the background by shifting the Wollaston prism along the optical axis.

The size of the source in the setup of Fig. 3.9 is quite limited. It can be increased considerably, however, if the setup is modified so that the light passes twice through the Wollaston prism. Figure 3.12 illustrates such an arrangement. An image of the source S is formed on the Wollaston prism at the point S', which is near the center of curvature of the mirror M under test. A lens L forms an image of M on the observation screen M'. As usual, two polarizers are needed to complete the system; one may be placed between m and W and the second between W and L. If observations are to be made between parallel polarizers, a single polarizer, placed between W and L and covering all of the aperture of W, will suffice. If S' and S'' are symmetrically situated with respect to the central fringe of the Wollaston prism, the path difference between the

3.4. LATERAL SHEARING INTERFEROMETERS

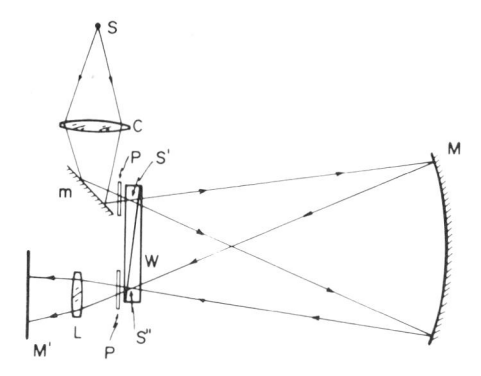

Figure 3.12. A double-pass compensated interferometer for testing the mirror M.

interfering beams is zero and the background appears uniformly dark/bright with crossed/parallel polarizers. The ground intensity can be varied by displacing W in a direction perpendicular to its fringes. A system of straight fringes will appear on the screen if W is displaced along the axis of the interferometer so that it is no longer located at the center of curvature of M.

Philbert (1958) and Philbert and Garyson (1961) employed this interferometer to control the homogeneity of optical glass (the glass plate is placed close to M) and to test spherical, paraboloidal, and plane mirrors during the process of figuring. To test a paraboloidal surface, the Wollaston prism is placed at the focus, and an auxiliary plane mirror is used to send back the parallel beam of light emerging from the paraboloid. A plane surface is tested by the arrangement represented in Fig. 3.13. During the final stages of figuring, the deviations from the perfect surface are quite small, and therefore the interferogram shows only slight variations in intensity. Under these conditions the as-

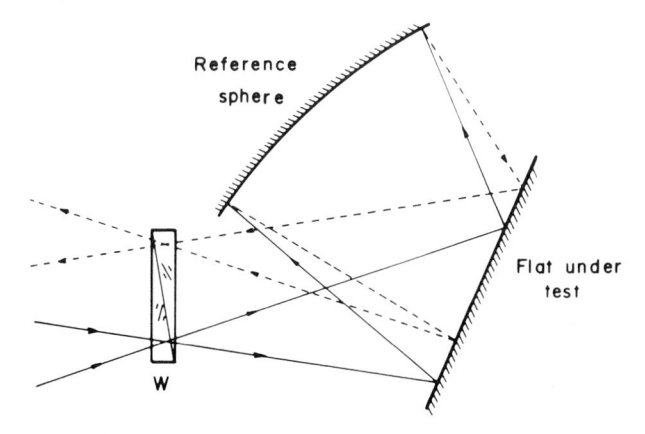

Figure 3.13. Setup for testing a flat surface.

pect of the field of view is similar to that observed in Foucault's knife-edge test.

To make the system insensitive to vibrations, Dyson (1963) used a small plane mirror near the Wollaston prism in order to form an image of it back on itself. Then, instead of a large prism, a small one is used since half of it is replaced by the small mirror.

3.5. DOUBLE-FOCUS INTERFEROMETER

Dyson (1957a, 1957b, 1970) devised an interferometer for the testing of optical components in which he employed a birefringent lens as a beam splitter (Fig. 3.14). The birefringent, double-focus lens L_1 is a symmetrical triplet, consisting of a central biconcave calcite lens and two biconvex glass lenses. The optic axis of calcite lies in the plane of the lens. The triplet is so designed as to have zero power for the ordinary ray and a focal length of a few centimeters for the extraordinary ray. As in the case of Burch's interferometer (Section 3.2), this arrangement gives explicit information about wavefront deformations since a part of the incident light is focused on a small region in the aperture of the system under test and acts as the reference beam.

The system under test in Fig. 3.14 is the concave mirror. The center of the triplet lens is located at the center of curvature of the mirror. A lens L_2 with its focus F_1 on the mirror surface is placed just to the right of the triplet. A quarter-wave plate with its principal axes at 45° to the optic axis of the calcite lens is also placed to the right of L_1. A collimated beam of linearly polarized light is incident from the left. The lens L_1 splits it up into an ordinary beam and an extraordinary beam. The O beam, undeviated by L_1, is brought to focus at F_1 by the lens L_2. An image of the source is thus formed at F_1. On its return journey the O beam is collimated by the lens L_2, and since its vibration direction has

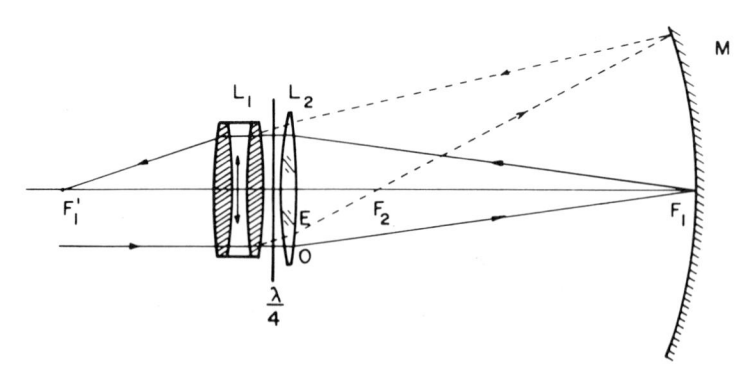

Figure 3.14. Dyson's double-focus interferometer.

3.5. DOUBLE-FOCUS INTERFEROMETER

been rotated through 90° because of the double passage through the quarter-wave plate, it is refracted to F'_1 by the lens L_1.

At its first passage the extraordinary beam is refracted by both the lenses L_1 and L_2 and converges to F_2, the focus of the combination L_1L_2. The beam then expands to fill the whole aperture of the mirror M. Because of the symmetry of the arrangement this beam, too, is brought to focus at F'_1. A semireflecting surface is placed to the left of the triplet so that the source (or the system of observation) can be placed outside the axis of the interferometer.

An observer receiving the light at F'_1 will see (a) a uniform disk of light, determined as to size by the angular aperture of the lens L_1 (reference field), and (b) the illuminated aperture of the mirror M (test field). These two fields will interfere (there is evidently an analyzer that sets the O and E vibrations parallel to each other), and in the absence of aberrations the resultant field will be of uniform intensity. If the triplet is slightly displaced laterally, so that its center no longer coincides with the center of curvature of the mirror, the field of view will be crossed with rectilinear fringes. When the triplet is displaced axially, circular fringes are observed. When the mirror has aberrations, these fringes are distorted. The aberrations can be deduced from these distortions in the same way as in any separate-path interferometer.

Dyson's interferometer is applicable to autostigmatic systems, that is, systems in which light diverging from a point in a particular plane is refocused to a point in the same plane to form an inverted image. Systems that are not autostigmatic can be converted to this form by the addition of one or more auxiliary components. To test a lens, for example, the scheme of Fig. 3.15 is employed. In Fig. 3.15a the lens is tested at infinite conjugates, and in Fig.

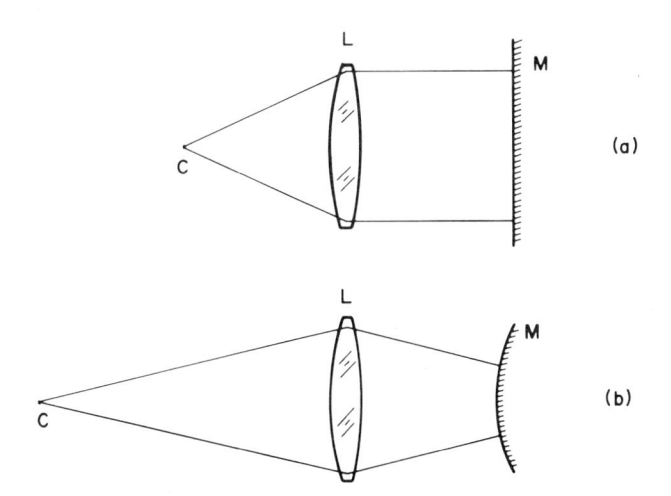

Figure 3.15. The lens L under test can be made autostigmatic by the addition of an auxiliary mirror M.

3.15b it is tested at finite conjugates. The focus C coincides with the center of the triplet. It may be noted that the system under test is not operating exactly under its correct conditions, as the test beam does not return along its original path. The arrangement gives the sum of the aberrations for two focal positions, one on each side of the desired position. The resultant error is often very small. Because of the aberrations of the triplet lens, optical systems of only moderate aperture—$f/5$, for example—can be tested with this interferometer.

The double-focus effect can also be produced with a Fresnel zone plate, which splits an incident beam into three beams converging to (or diverging from) different points on the optical axis. Common-path interferometers using single- or multiple-zone plates have been developed by many research workers for testing optical components [see, e.g., Murty (1963), Smart (1974), Lohmann (1985), Stevens (1988), and Huang et al. (1989)].

3.6. SAUNDERS'S PRISM INTERFEROMETER

Saunders (1967, 1970) described a lateral shearing interferometer in which the beam divider is made by cementing together the hypotenuse faces of two right-angle prisms, one of which is half silvered (Fig. 3.16). The faces B and B' are made highly reflecting. To obtain the zero-order fringe in the center of the field of view, the distances from the center of the beam-dividing surface to the two reflecting surfaces are made equal. If the two component prisms are identical, the two beams emerging from the face A' are mutually parallel. An angular shear between the beams can be introduced by rotating one prism relative to the other about an axis normal to the semireflecting surface. The direction of shear is approximately parallel to the vertex edges of the prisms. The shear can also be produced by making the angles α and α' of the two component prisms slightly different. This is usually the case when the prisms are not cut from a single

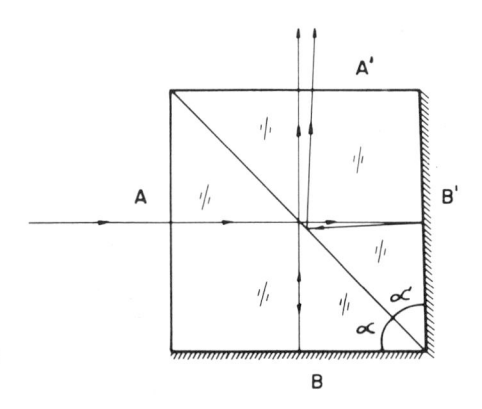

Figure 3.16. A beam splitter devised by Saunders.

3.6. SAUNDERS'S PRISM INTERFEROMETER

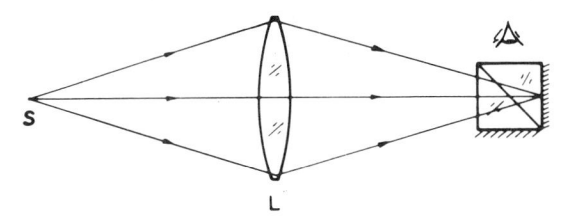

Figure 3.17. A lateral shearing interferometer using the beam splitter shown in Fig. 3.16.

large prism but are made separately. The angular shear is then equal to $2(\alpha - \alpha')$, and the direction of shear is perpendicular to the vertex edges.

Figure 3.17 shows an arrangement for testing a lens at finite conjugates. The prism is adjusted so that its back surface is approximately parallel to the image plane and near it, with the principal ray of light passing near the center of the prism. This adjustment should produce visible fringes. The fringe width is very large when the source image lies on the back surface of the prism. The fringe width can be decreased by moving the prism along the principal ray away from the source image. By translating the prism laterally parallel both to the image plane and to the direction of shear, any chosen fringe can be made to pass through any chosen point of the interferogram. The adjustments of Saunders's prism are similar to those of a Wollaston prism. To obtain high contrast fringes, the source size in the shear direction is kept small. The recommended size of the cube is 10–15 mm.

Saunders (1957) also studied a wavefront-reversing interferometer that employed a modified Kösters double-image prism. Figure 3.18 is a sketch of the arrangement for testing a lens with one conjugate at infinity. The base of the dividing prism is spherical, and its center of curvature, S_0, coincides with the image point at which the lens is to be tested. The observer's eye is located at S', the image of the source S. In this arrangement the part of the wavefront lying below the dividing plane of the prism appears to be folded onto the upper half after the second passage through the prism. When the dividing plane cuts

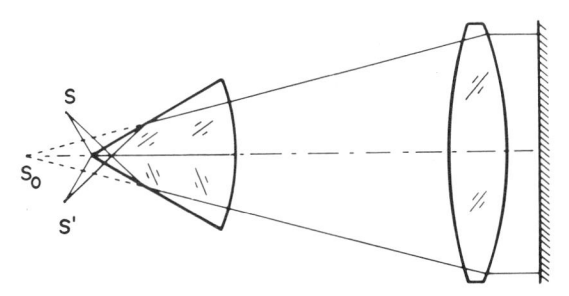

Figure 3.18. Saunders's wavefront-reversing interferometer.

through the center of the lens, the even-order aberrations are eliminated. However, when the dividing plane is adjusted to form an angle with the axis of the lens, the even-order terms are retained. Saunders gave different variations of this arrangement for determining different aberrations.

3.7. POINT DIFFRACTION INTERFEROMETER

Another interesting common-path interferometer is the so-called point diffraction interferometer, first described by Linnik in 1933, rediscovered by Smartt and Strong (1972), and more fully developed by Smart and Steel (1975). The principle of this interferometer is shown in Fig. 3.19. The wave to be examined is brought to a focus to produce an image, usually aberrated, of a point source. At the plane of that image an absorbing film is placed. This film contains a small pinhole or opaque disk in order to diffract the light and produce a spherical reference wavefront.

To produce an interferogram with good contrast, the wave passing through the film and the diffracted spherical wave should have the same amplitude. This is controlled by means of the filter transmittance and the pinhole or disk size. Also, the amplitude of the spherical wave depends on how much of the light in the image falls on the pinhole or disk, and this, in turn, depends on the aberrations of the wave and on the pinhole or disk position. Smartt and Steel (1975) advised using filter transmittances between 0.005 and 0.05, with a most common value of 0.01. The optimum size for the pinhole or disk is about the size of the Airy disk that the original wave would produce if it had no aberrations. To match the amplitudes of the two beams, Wu et al. (1984) use a clear pinhole in a polarizing sheet of vectograph film; rotation of a polarizer behind this sheet changes the amplitude of the beam transmitted by the film but not that of the diffracted beam.

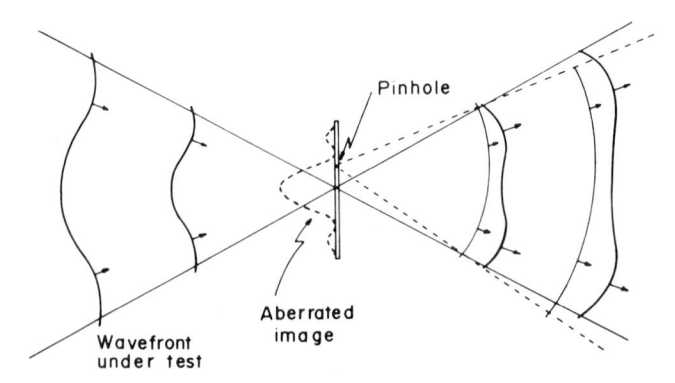

Figure 3.19. Principle used in the point diffraction interferometer.

The usual tilt and focus shift of the reference wavefront can be produced by displacing the diffracting point laterally and longitudinally, respectively. The point diffraction interferometer has been used with success to test astronomical telescopes and toric surfaces (Speer et al. 1979; Marioge et al. 1984). Smartt and Steel (1985) have developed a white-light interference microscope based on point diffraction interference principle.

3.8. ZERNIKE TESTS WITH COMMON-PATH INTERFEROMETERS

In any two-beam interferometer the irradiance in the interference pattern is a function of the phase difference between the two beams, as shown in Fig. 3.20. If the interferogram has many fringes, the irradiance goes through many maxima and minima of this function. However, if the wavefront is almost perfect and its deformations are smaller than half the wavelength of the light, the phase changes will not produce any variations in the irradiance.

These small wavefront errors may be easily detected if a bias in the phase difference is introduced by any means, so that it has a value equal to $\pi/2$ when the wavefront is perfect. Then, the interferometer sensitivity to small errors is very large. The Zernike test in the point diffraction interferometer from another point of view is studied in Chapter 8.

3.9. MEASUREMENT OF THE OPTICAL TRANSFER FUNCTION

3.9.1. Scanning Method

Common-path interferometers, particularly polarization interferometers, have been employed for measuring the transfer functions of optical systems. The transfer function of an optical system is given by the Fourier transform of its

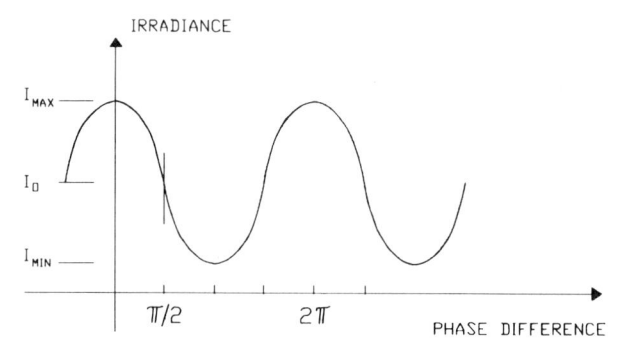

Figure 3.20. Irradiance versus phase difference in a two-beam interferometer.

spread function, the spread function being the intensity distribution in the image of a point source. There is no loss of generality if the point source is replaced by an incoherently illuminated line source, provided that the line source is considered along different azimuths. The optical transfer function (OTF), $H(N)$, is a complex function of the spatial frequency N and may be expressed as

$$H(N) = T(N) \exp[j\theta(N)] \tag{3.10}$$

where $T(N)$ represents the ratio of the contrast in the image to the contrast in the object, the object being a sinusoidal grating of frequency N, and $\theta(N)$ is the lateral shift of the image with respect to an ideal image (given by an aberration-free and diffraction-free system).

In practice, the object is not a line but a slit of finite width. The image is, then, the convolution of the slit function with the spread function of the optical system. Representing the Fourier transforms of the object and the image functions by $O(N)$ and $I(N)$, respectively, we have the relation

$$I(N) = O(N)H(N). \tag{3.11}$$

The Fourier transform of the object being known, the OTF can be determined if the Fourier transform of the image can be measured. To determine $I(N)$, the intensity distribution in the image may be measured by means of a narrow slit and its Fourier transform calculated numerically. A more direct method is to scan the image with a sinusoidal (or a nonsinusoidal) grating. If the transmission of the grating is represented by $\frac{1}{2}(1 + \cos 2\pi Nx)$, the light flux passing through the grating will be (neglecting the factor $\frac{1}{2}$)

$$\Phi(x_0) = \int_{-\infty}^{+\infty} i(x)[1 + \cos 2\pi N(x_0 - x)] \, dx \tag{3.12}$$

where $i(x)$ represents the intensity distribution in the image and x_0 is the distance of a line of maximum transmission from the origin of the image function. Equation (3.12) may also be written in the form

$$\begin{aligned}
\Phi(x_0) &= \int_{-\infty}^{+\infty} i(x) \, dx + \text{Re}\left\{\int_{-\infty}^{+\infty} i(x) \exp[j2\pi N(x_0 - x)] \, dx\right\} \\
&= \int_{-\infty}^{+\infty} i(x) \, dx + \text{Re}[\exp(j2\pi Nx_0)I(N)] \\
&= \int_{-\infty}^{+\infty} i(x) \, dx + \text{Re}\{\exp(j2\pi Nx_0)T'(N) \exp[\theta'(N)]\} \\
&= \int_{-\infty}^{+\infty} i(x) \, dx + T'(N) \cos[2\pi Nx_0 + \theta'(N)]
\end{aligned} \tag{3.13}$$

3.9. MEASUREMENT OF THE OPTICAL TRANSFER FUNCTION

where

$$T'(N) = T(N)|O(N)|$$

and

$$\theta'(N) = \theta(N) + \text{argument of } O(N). \tag{3.14}$$

From Eq. (3.13) we note that the output signal is modulated when the image is scanned by the grating, the amplitude of modulation being equal to $T'(N)$. The phase $\theta'(N)$ is determined by comparing the phase of the output signal with that of a reference signal.

Various types of scanning screens having sinusoidal and nonsinusoidal transmission functions have been produced and have been reviewed by Murata (1966). A system of two-beam fringes can be used as a scanning screen. Lohmann (1957) proposed the use of polarization interferometers for measuring the OTF at a particular frequency. Since fringes of variable frequency can easily be produced by polarization interferometers (Françon and Mallick 1971, Chapter 3), they can be used to determine the complete OTF curve (Mallick 1966; Prat 1966). The experimental setup is represented in Fig. 3.21, where Q is a birefringent system giving fringes of variable frequency in the far field; these are located virtually in the plane of the image $i(x)$. A quarter-wave plate with its principal axes at 45° to those of system Q and a rotating polarizer P_2 have the effect of giving a continuous lateral movement to the fringes. If the frequency of rotation of the polarizer is f, the output signal will have frequency $2f$. The reference signal is produced by a beam of linearly polarized light passing through the rotating polarizer P_2. The zero of the reference signal corresponds with the position of the rotating polarizer at which the center of a bright fringe falls on the origin of the image function.

In the preceding experiment the function of the slit and that of the sinusoidal fringes may be interchanged, in which case the optical system under test forms

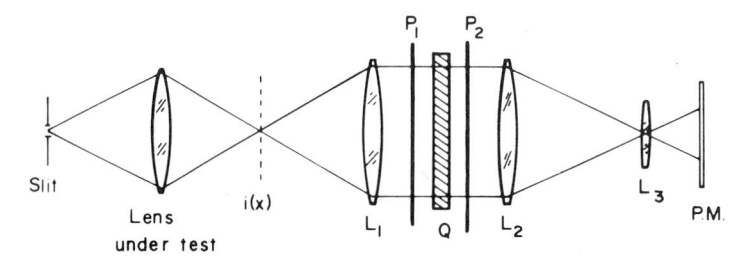

Figure 3.21. Experimental setup for measuring the OTF by a scanning method. The system under test forms an image of a slit source, and the sinusoidal fringes of variable frequency are used to scan the image.

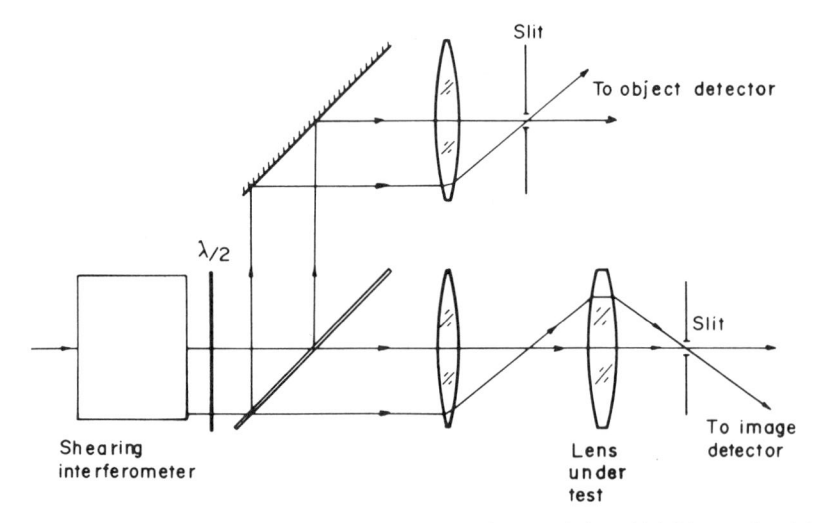

Figure 3.22. The system under test forms an image of a set of sinusoidal fringes of variable frequency, and a slit is used to scan this image. (After Steel 1967.)

an image of the fringe pattern and the slit is used to scan this image. Steel (1964) used such a disposition for measuring the OTF (Fig. 3.22). It has the advantage that a second slit and a photomultiplier can be used to monitor the object fringes and thus to measure the phase difference between the object and the image signals. The ratio of the modulations of the two signals and the phase difference between them give, respectively, the modulus and the phase of the optical transfer function.

3.9.2. Autocorrelation Method

The transfer function of an optical system is also given by the autocorrelation of the pupil function $f(x, y)$:

$$H(S) = \frac{1}{A} \int\!\!\!\int_{-\infty}^{+\infty} f\!\left(x + \frac{S}{2}, y\right) f^*\!\left(x - \frac{S}{2}, y\right) dx\, dy \qquad (3.15)$$

where

$$A = \int\!\!\!\int_{-\infty}^{+\infty} |f(x, y)|^2 \, dx\, dy \qquad (3.16)$$

3.9. MEASUREMENT OF THE OPTICAL TRANSFER FUNCTION

is a normalizing factor. Also, S represents the shear of the exit pupil expressed as a fraction of its radius; it is related to the spatial frequency N by the equation

$$S = \frac{\lambda}{n \sin \alpha} N \qquad (3.17)$$

where $n \sin \alpha$ is the numerical aperture of the optical system. The variables x and y are equal to the respective Cartesian coordinates divided by the radius of the exit pupil. The integral in Eq. (3.15) can be calculated if the aberrations of the optical system are known. This integration can be carried out experimentally by means of a shearing interferometer (Hopkins 1955). The optical system (the transfer function of which is to be measured) forms an image of a narrow, incoherently illuminated slit. This image is formed at infinity either by the test lens itself or by a well-corrected auxiliary lens that collimates the light leaving the test lens. The plane waveform is incident on a shearing interferometer that splits it into two wavefronts with a relative lateral shear S and with a phase difference ϕ between them. The two wavefronts interfere in a plane conjugate with the exit pupil, and the total flux in the interference pattern is given by

$$\Phi(\phi) = \int\int_{-\infty}^{+\infty} \left| f\left(x + \frac{S}{2}, y\right) + f\left(x - \frac{S}{2}, y\right) \exp(j\phi) \right|^2 dx\, dy$$

$$= \int\int_{-\infty}^{+\infty} \left| f\left(x + \frac{S}{2}, y\right) \right|^2 dx\, dy + \int\int_{-\infty}^{+\infty} \left| f\left(x - \frac{S}{2}, y\right) \right|^2 dx\, dy$$

$$+ 2 \operatorname{Re}\left[\exp(-j\phi) \int\int_{-\infty}^{+\infty} f\left(x + \frac{S}{2}, y\right) f^*\left(x - \frac{S}{2}, y\right) dx\, dy \right].$$

(3.18)

The first two terms on the right-hand side of Eq. (3.18) are each equal to A [Eq. (3.16)]. The integral in the third term is equal to $AH(S)$. Writing $H(S)$ equal to $T(S) \exp[j\theta(S)]$, we may express Eq. (3.18) in the form

$$\Phi(\phi) = 2A\{1 + T(S) \cos[\theta(S) - \phi]\}. \qquad (3.19)$$

If θ is varied linearly with time, the output flux will be modulated, the amplitude and the phase of modulation giving, respectively, the modulus and the argument of the transfer function for the frequency S.

Various types of shearing interferometers have been constructed to measure the OTF by this method. We discuss here an arrangement by Tsuruta (1963), who employed a Savart polariscope to shear the exit pupil (Fig. 3.23). A narrow split is placed in the front focal plane of the test lens L. A modified polariscope Q of variable thickness produces a variable shear of the exit pupil of L. An additional Savart Q', oriented at 180° with respect to Q, is used to produce a zero shear of the exit pupil and thus to make measurements of the OTF for zero frequency. When Q' is oriented parallel to Q, the shears of the two add together. A soleil compensator C is used to vary ϕ. The afocal system consisting of lenses L_1 and L_2 serves simply to decrease the width of the parallel beam leaving the test lens. The exit pupil of L is imaged on a matte plate M; it is on this plate that the interferogram is formed. A collecting lens concentrates the light on a photomultiplier. When measurements are made for high spatial frequencies (i.e., when the shear of the exit pupil is large), the slit width has to be correspondingly small. The slit width should be such that two points of the exit pupil separated by a distance equal to the shear have a high degree of coherence. In the experiment of Tsuruta the slit width was 2 μm. Since the path difference produced by a Savart polariscope is zero for normal incidence, there is no need to use light of high temporal coherence. Tsuruta used a tungsten lamp in conjunction with an interference filter having a bandwidth of 100 Å.

The setup of Fig. 3.23 is similar to that of Fig. 3.9 and can be used to measure the aberrations of the test lens L. The intensity distribution on the matte plate M is nothing but a lateral shearing interferogram of the wavefront emerging from the lens L.

The autocorrelation method can be shown to be equivalent to the scanning method by the following consideration. In the experimental setup of Fig. 3.23 the image of the slit source as formed by the lens under test, and the fringes of the Savart polariscope lie in the same plane at infinity. Varying ϕ by means of the compensator has the effect of displacing the fringe system with respect to the slit image. Consequently, if the flux is measured in the focal plane of a well-corrected lens placed after the compensator in Fig. 3.23, we obtain the OTF of the lens L by the scanning method. Thus the two methods are equivalent; the only difference is that in the autocorrelation method flux is measured

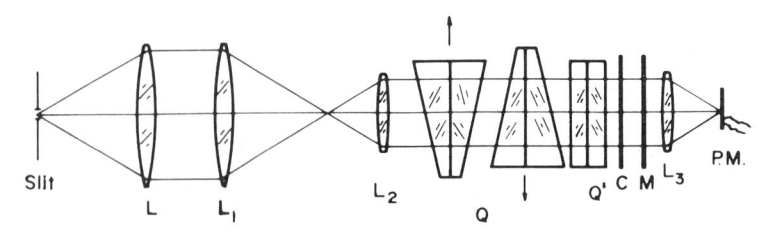

Figure 3.23. Experimental setup for measuring the OTF by the autocorrelation method.

in a plane conjugate to the exit pupil of the test lens, whereas in the scanning method flux is measured in a plane conjugate to the slit source.

REFERENCES

Burch, J. M., "Scatter Fringes of Equal Thickness," *Nature*, **171,** 889 (1953).

Burch, J. M., "Scatter-Fringe Interferometry," *J. Opt. Soc. Am.*, **52,** 600 (1962).

Burch, J. M., "Interferometry with Scattered Light," in *Optical Instruments and Techniques*, J. Home Dickson, Ed., Oriel Press, England, 1969, p. 213.

Dyson, J., "Common-Path Interferometer for Testing Purposes," *J. Opt. Soc. Am.*, **47,** 386 (1957a).

Dyson, J., "Interferometers," in *Concepts of Classical Optics*, John Strong, Ed., W. H. Freeman, San Francisco, 1957b, Appendix B p. 377.

Dyson, J., "Very Stable Common-Path Interferometers and Applications," *J. Opt. Soc. Am.*, **53,** 690 (1963).

Dyson, J., *Interferometry as a Measuring Tool*, Machinery Publishing Co. Brighton, 1970.

Françon, M. and M. Jordery, "Application des Interferences par Double Réfraction a l'Etude des Aberrations," *Rev. Opt.*, **32,** 601 (1953).

Françon, M. and S. Mallick, *Polarization Interferometers*, Wiley, New York, 1971.

Hopkins, H. H., "Interferometric Methods for the Study of Diffraction Images," *Opt. Acta*, **2,** 23 (1955).

Houston, J. B., Jr., "How to Make and Use a Scatterplate Interferometer," *Opt. Spectra*, **4**(6), 32 (1970).

Huang Juneji, Nagaaki Ohyama, and Toshio Honda, "A Null Test of Conic Surfaces in Zone Plate Interferometer," *Opt. Commun.*, **72,** 17 (1989).

Juneji Huang, Toshio Honda, Nagaaki Ohyama, and Jumpei Tsujiuchi, "Fringe Scanning Scatter Plate Interferometer," *Opt. Commun.*, **68,** 235 (1988).

Linnik, W., "Simple Interferometer to Test Optical Systems," Comptes Radnus de l'Académie des Sciences d l'U.R.SS. 1, 208 (1933). Abstract in *Z. Instrumentenkd.*, **54,** 463 (1934).

Lohmann, A., "Zur Messung des Opticachen Übertragungsfaktors," *Optik*, **14,** 510 (1957).

Lohmann, A. W., "An Interferometer with a Zone Plate Beam-Splitter," *Opt. Acta*, **32,** 1465 (1985).

Mallick, S., "Measurement of Optical Transfer Function with Polarization Interferometer," *Opt. Acta*, **13,** 247 (1966).

Marioge Jean-Paul, B. Bonino, F. Bridou, P. Fournet et al., "La Fabrication et le Controle de Surfaces Toriques," *J. Opt. (Paris)*, **15,** 286 (1984).

Murata, K., "Instruments for the Measuring of Optical Transfer Functions," in *Progress in Optics*, Vol. 5, E. Wolf, Ed., North-Holland, Amsterdam, 1966, p. 201.

Murty, M. V. R. K., "Common Path Interferometer Using Fresnel Zone Plates," *J. Opt. Soc. Am.*, **53,** 568 (1963).

Philbert, M., "Applications Métrologiques de la Strioscopie Interférentielles," *Rec. Opt.*, **37,** 598 (1958).

Philbert, M. and M. Garyson, "Réalisation et Controle par Strioscopie Interférentielle de Miroirs Plans, Sphériques et Paraboliques," in *Optical Instruments and Techniques*, K. J. Habell, Ed., Chapman and Hall, London, 1961, p. 352.

Prat, R., "Spectrométrie des Fréquences Spatiales et Cohérence," *Opt. Acta*, **13,** 73 (1966).

Rubin, Lawrence F., "Scatter Plate Interferometry," *Opt. Eng.*, **19,** 815 (1980).

Saunders, J. B., "The Kösters Double-Image Prism," in *Concepts of Classical Optics*, John Strong, Ed., W. H. Freeman, San Francisco, 1957, Appendix C, p. 393.

Saunders, J. B., "Measurement of Wavefronts without a Reference Standard I; The Wavefront-Shearing Interferometer," *J. Res. Nat. Bur. Stand.*, **65B,** 239 (1961).

Saunders, J. B., "Measurement of Wavefronts without a Reference Standard II: The Wavefront-Reversing Interferometer," *J. Res. Nat. Bur. Stand.*, **66B,** 29 (1962).

Saunders, J. B., "A Simple, Inexpensive Wavefront Shearing Interferometer," *Appl. Opt.*, **6,** 1581 (1967).

Saunders, J. B., "A Simple Interferometric Method for Workshop Testing of Optics," *Appl. Opt.*, **9,** 1623 (1970).

Scott, R. M., "Scatter Plate Interferometry," *Appl. Opt.*, **8,** 531 (1969).

Shoemaker, A. H. and M. V. R. K. Murty, "Some Further Aspects of Scatter-Fringe Interferometry," *Appl. Opt.*, **5,** 603 (1966).

Smartt, R. N. and J. Strong, "Point-Diffraction Interferometer" (abstract only), *J. Opt. Soc. Am.*, **62,** 737 (1972).

Smartt, Raymond N., "Zone Plate Interferometer," *Appl. Opt.*, **13,** 1093 (1974).

Smartt, R. N. and W. H. Steel, "Theory and Application of Point-Difference Interferometers," Proceedings of the ICO Conference on Optical Methods in Scientific and Industrial Measurements, Tokyo, 1974; *Jap. J. Appl. Phys.*, **14,** Suppl. 1, 351 (1975).

Smartt, Raymond N. and William H. Steel, "Point Diffraction Interferometer," *Appl. Opt.*, **24,** 1402 (1985).

Speer, Robert J., M. Crisp, D. Turner, S. Mrowka, and K. Tregidjo, "Grazing Incidence Interferometry: The Use of the Linnik Interferometer for Testing Image-Forming Reflection Systems," *Appl. Opt.*, **18,** 2003 (1979).

Steel, W. H., "A Polarization Interferometer for the Measurement of Transfer Functions," *Opt. Acta*, **11,** 9 (1964).

Steel, W. H., *Interferometry*, Cambridge University Press, London, 1967, p. 250.

Stevens, R. F., "Zone Plate Interferometers," *J. Mod. Optics (Form. Opt. Acta)*, **35,** 75 (1988).

Su, Der-chin, Toshio Honda, and Jumpei Tsujiuchi, "A Simple Method of Producing Accurately Symmetrical Scatter Plates," *Opt. Commun.*, **49,** 161 (1984).

Su, Der-chin, Nagaaki Ohyama, Toshio Honda, and Jumpei Tsujiuchi, "A Null Test of Aspherical Surfaces in Scatter Plate Interferometer," *Opt. Commun.*, **58,** 139 (1986).

Su, Der-chin, Toshio Honda, and Jumpei Tsujiuchi, "Some Advantages of Using Scatter Plate Interferometer in Testing Aspheric Surfaces," *Proc. SPIE.*, **813,** 217 (1987).

Tsuruta, T., "Measurement of Transfer Functions of Photographic Objectives by Means of a Polarizing Shearing Interferometer," *J. Opt. Soc. Am.*, **53,** 1156 (1963).

Wu, S. T., Ch. L. Xu, and Zh. J. Wang, "New Application of Point Diffraction Interferometer—Polarization Fringe Scanning PDI," in *ICO 13 Conference Digest, Optics in Modern Science and Technology*, H. Ohzu, Ed., Reidel, Dordrecht, 1984, p. 458.

ADDITIONAL REFERENCES

Aggarwal, A. K. and Susnil K. Kaura, "Further Applications of Point Diffraction Interferometer," *J. Optics (Paris)*, **17,** 135 (1986).

Harris, W., S. Mrowka, and R. J. Speer, "Linnik Interferometer: Its Use at Short Wavelengths," *Appl. Opt.*, **21,** 1155 (1982).

Harris, W., R. J. Speer, and V. Stanley, "Linnik Point Difraction Interferometer of Increased Sensitivity for the Measurement of Wavefront Error," *Proc. SPIE*, **235,** 122 (1980).

James, W. E. and P. Hariharan, "Line Difraction Test," *Appl. Opt.*, **25,** 3806 (1986).

Je Chang H., Won H. Lee, Jin H. Kwon, and Ok. S. Choe, "Speckle-Averaging Scatter Plate Interferometry," *Appl. Opt.*, **24,** 2042 (1985).

Koliopoulos, Chris L., D. Kuon, R. Shagam, J. C. Wyant, and C. R. Hayslett, "Infrared Point-Diffraction Interferometer," *Opt. Lett.*, **3,** 118 (1978).

Ohyama, Nagaaki, Ikuo Yamaguchi, Isao Ichimura, Toshio Honda, et al., "A Dynamic Zone-Plate Interferometer for Measuring Aspherical Surfaces," *Opt. Commun.*, **54,** 257 (1985).

Ohyama, Nagaaki, Isao Ichimura, Ikuo Yamaguchi, Toshio Honda, et al., "The Dynamic Zone Plate Interferometer for Measuring Aspherical Surfaces (II)," *Opt. Commun.*, **56,** 369 (1986).

Quercioli, F. and G. Molesini, "Contrast Reversal with a Point-Difraction Interferometer with a Carrier Frequency," *Opt. Commun.*, **35,** 303 (1980).

Rabinovich, V. B., "Method of Centering the Objective Lenses of a Linnik Microinterferometer," *Sov. J. Opt. Technol.*, **52,** 615 (1985).

Rabinovich, V. B., "Effect of Reference Mirror Adjustment Errors on Fringe Contrast in a Linnik Microinterferometer," *Sov. J. Opt. Technol.*, **53,** 475 (1986).

Rubin, Lawrence F., "Scatterplate Interferometry," *Proc. SPIE*, **192,** 27 (1979).

Rubin, Lawrence F. and C. Wyant James, "Energy Distribution in a Scatterplate Interferometer," *J. Opt. Soc. Am.*, **69,** 1305 (1979).

Rubin, Lawrence F. and U. Kuon, "Infrared Scatter Plate Interferometry," *Appl. Opt.*, **19,** 3219 (1980).

Rubin, Lawrence F., "Null Testing in a Modified Scatterplate Interferometer," *Appl. Opt.*, **19,** 1634 (1980).

Shimano, T., Nagaaki Ohyama, Jumpei Tsujiuchi, and Toshio Honda, "Analysis of Systematic Errors in the Dynamic Zone Plate Interferometer," *Opt. Commun.*, **64,** 1 (1987).

Smartt, Raymond N., "Special Applications of the Point-Difraction Interferometer," *Proc. SPIE*, **192,** 35 (1979).

Speer, Robert J. and Michael Chrisp, "All Grazing Incidence Interferometer for Testing and in situ Alignment of Imaging X-Ray Optics," *Proc. SPIE*, **184,** 172 (1979).

Su, Der-chin, Toshio Honda, and Jumpei Tsujiuchi, "Aperture of Scatter Plate and its Effects on Fringe Contrast in a Scatter Plate Interferometer," *Opt. Commun.*, **50,** 137 (1984).

Su, Der-chin, Toshio Honda, Jumpei Tsujiuchi, and Nagaaki Ohyama, "Symmetry Error of Scatter Plate and Its Effect on Fringe Contrast in a Scatter Plate Interferometer," *Opt. Commun.*, **52,** 157 (1984).

Underwood, Katle, James C. Wyant, and Chris L. Koliopoulos, "Self-Referencing Wavefront Sensor," *Proc. SPIE*, **351,** 108 (1982).

Wanzhi, Zhou and Lu Zhenwu, "Optical Testing Using a Point Difraction Holographic Interferometer," *Proc. SPIE*, **673,** 289 (1986).

4

Lateral Shearing Interferometers

M. V. Mantravadi

4.1. INTRODUCTION

Lateral shearing interferometry is an important field of interferometry and has been used extensively in diverse applications such as the testing of optical components and systems and the study of flow and diffusion phenomena in gases and liquids. Basically the method of lateral shearing interferometry consists of displacing the defective wavefront laterally by a small amount and obtaining the interference pattern between the original and the displaced wavefronts. Figure 4.1 schematically illustrates the principle. If the wavefront is nearly a plane, the lateral shear is obtained by displacing the wavefront in its own plane. If the wavefront is nearly spherical, the lateral shear is obtained by sliding the wavefront along itself by rotation about an axis passing through the center of curvature of the spherical wavefront.

There are many physical arrangements to obtain lateral shear. In general, in this chapter, we discuss arrangements that can be obtained by the use of beam dividers, which divide the amplitude of the incident wavefront but do not change the shape of the wavefront. This means that plane surfaces coated with semireflecting material are used as beam dividers. Several arrangements to obtain lateral shear will be described in this chapter mainly to show that with available components one can easily fashion a workable lateral shearing interferometer in one's laboratory or optical workshop. Another important consideration in the design of lateral shearing interferometers is the nature of the light source. From the point of view of lateral shearing interferometry, the sources can be classified into two categories: (*a*) laser sources, such as the helium–neon gas laser giving a 6328-Å light beam of very high spatial and temporal coherence, and (*b*) all other sources, such as gas discharge lamps, which are temporally coherent to some extent but not spatially coherent.

Optical Shop Testing, Second Edition, Edited by Daniel Malacara.
ISBN 0-471-52232-5 © 1992, John Wiley & Sons, Inc.

124   LATERAL SHEARING INTERFEROMETERS

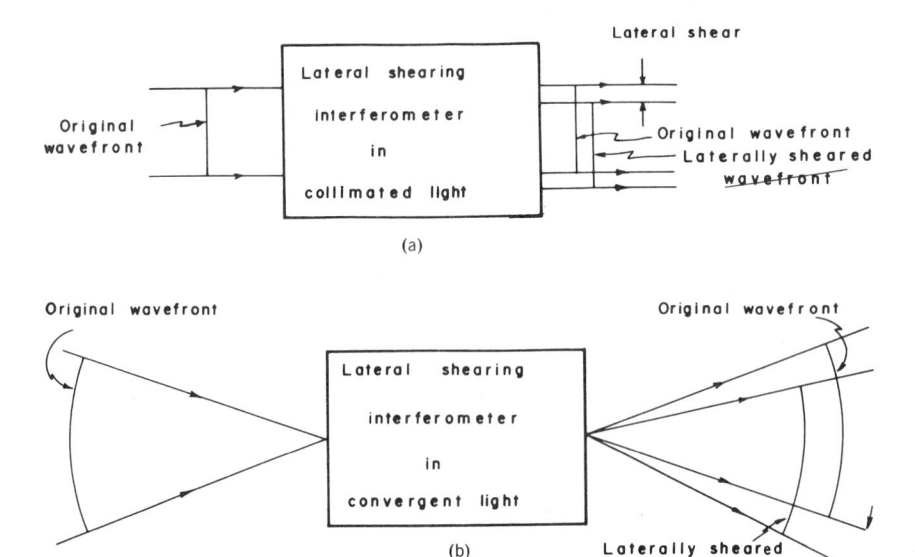

Figure 4.1. Schematic diagram illustrating lateral shearing interferometry in (*a*) collimated light and (*b*) convergent light.

4.2. CONSIDERATIONS REGARDING COHERENCE PROPERTIES OF THE LIGHT SOURCE

Figure 4.2 schematically illustrates the arrangement of a lateral shearing interferometer in which a shear takes place for a nearly plane wavefront obtained from the collimating lens. Let the full width of the wavefront be d and the amount of lateral shear S. If the focal length of the collimating lens used is f,

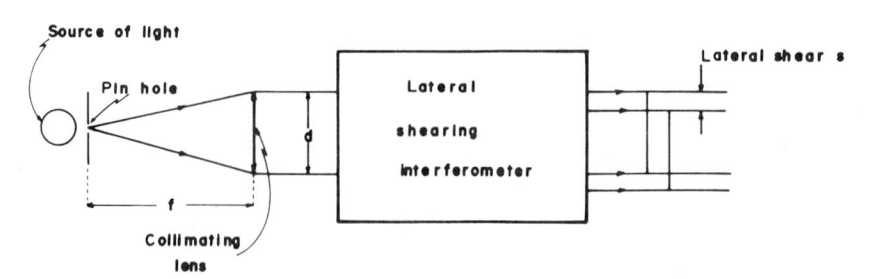

Figure 4.2. Schematic diagram indicating the various parameters for the consideration of the size of pinhole to be used in a lateral shearing interferometer.

4.3. BRIEF THEORY OF LATERAL SHEARING INTERFEROMETRY

there is full spatial coherence across the width of the wavefront when the size of the source is equal to the width of the central diffraction maximum (Airy disk) corresponding to the f number of the particular collimating lens. Thus the order of magnitude of the size of the pinhole to be used over the source to achieve spatial coherence is given by $(\lambda/d)f$, where λ is the wavelength of the particular spectral line of the source that is to be used. However, since the lateral shear is S, the spatial coherence should be sufficient so that interference can be observed between parts of the wavefront separated by the distance S, which is less than d. Hence the source (pinhole) size can be $(\lambda/d)f(d/S) = (\lambda/S)f$. Thus the pinhole size chosen is some multiple of the diffraction-limited pinhole size.

As an example let us assume that we are using a mercury discharge lamp as the source of light and that the green line (5461 Å) is isolated by means of a filter. If we are using a collimating lens of f number = 5, then, assuming a shear ratio S/d of 0.1, the pinhole must be of the order of 25 μm. This is an extremely small pinhole, and generally very little intensity can be obtained in the fringe pattern unless the source itself is very intense. Hence a source such as a high-pressure mercury arc must be used. Since such sources have poor temporal coherence, even after a spectral line suitable for the purpose has been isolated by means of a filter, there is always a need for compensating the two optical paths in an interferometer used as a lateral shearing instrument. This is sometimes referred to as white-light compensation, and when white light is used a lateral shearing interferogram in which the central fringe is achromatic and the other fringes are colored is obtained.

Until the gas laser came into general use, all lateral shearing interferometers were designed for white-light compensation. Now it is possible to devise lateral shearing interferometers in which the light paths of the two interfering beams are of unequal length (uncompensated). A laser source having a high degree of spatial and temporal coherence is, however, necessary for this purpose. A helium–neon laser giving out the 6328 Å line is the most useful source of light for many of these applications. Whereas a lateral shearing interferometer designed for white-light compensation can always be used with a laser, a lateral shearing interferometer designed for laser use and hence probably having unequal optical paths cannot give a visible or recordable interference fringe pattern with less coherent sources of light.

4.3. BRIEF THEORY OF LATERAL SHEARING INTERFEROMETRY

Figure 4.3 shows the original wavefront and also the laterally sheared wavefront. The wavefront is considered nearly plane so that wavefront errors may be small deviations from this plane. The wavefront error may be expressed as

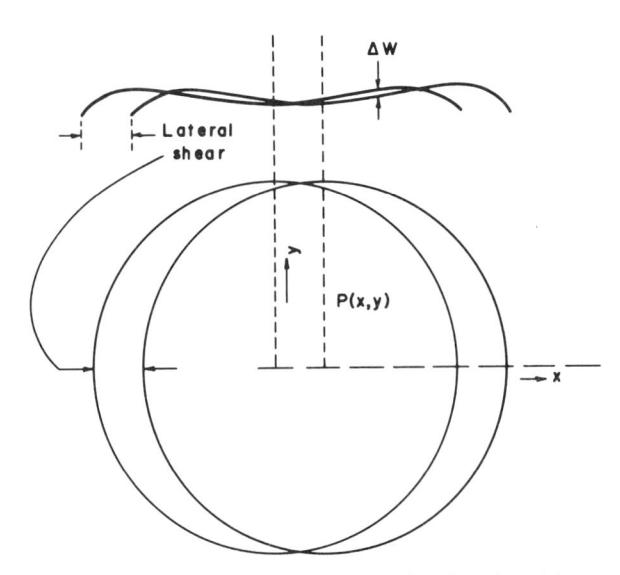

Figure 4.3. Schematic diagram illustrating (in plan and elevation) the original and the sheared wavefronts. A circular aperture is assumed. The lateral shear fringes appear in the common area of the two wavefronts.

$W(x, y)$, where (x, y) are the coordinates of the point P. When this wavefront is sheared in the x direction by an amount S, the error at the same point for the sheared wavefront is $W(x - S, y)$. The resulting path difference ΔW at P between the original and the sheared wavefronts is $W(x, y) - W(x - S, y)$. Thus, in lateral shearing interferometry, it is the quantity, ΔW, that is determined, and when S is zero, there is no path difference anywhere in the wavefront area and consequently no error can be seen, however large it may be. Now, the path difference ΔW may be obtained at various points on the wavefront from the usual relationship:

$$\Delta W = n\lambda \tag{4.1}$$

where n is the order of the interference fringe and λ is the wavelength used. It is of interest that, when S is small, Eq. (4.1) may be written as

$$\left(\frac{\partial W}{\partial x}\right) S = n\lambda. \tag{4.2}$$

Thus the information obtained in the lateral shearing interferometer is ray aberration $(\partial W/\partial x)$ in angular measure. The equation becomes more exact as $S \to 0$, but we also have seen that the sensitivity decreases as $S \to 0$. Thus we

4.3. BRIEF THEORY OF LATERAL SHEARING INTERFEROMETRY

must arrive at some compromise for the proper value of S if Eq. (4.2) is to be used exactly.

Let us now consider some specific situations.

Defocusing. The wavefront error for defocusing may be represented as

$$W(x, y) = D(x^2 + y^2). \tag{4.3}$$

This situation corresponds to a slight defocusing of the optical system so that, instead of a perfectly plane wavefront emerging from it, a slightly concave or convex spherical wavefront of very long radius of curvature emerges. Hence in this situation

$$\Delta W = 2DxS = n\lambda. \tag{4.4}$$

Equation (4.4) represents a system of straight fringes that are equally spaced and perpendicular to the x direction (direction of shear). This situation is shown in Fig. 4.4a. The straight fringes appear in the common area of the two intersecting wavefronts. If there is no defocusing ($D = 0$), the whole common area is filled with no fringes and appears to be of uniform intensity.

Tilt. When the wavefront is laterally sheared, normally we assume that the new wavefront is not tilted with respect to the original wavefront. In certain arrangements, however, it is possible to obtain a known amount of tilt between the two wavefronts. In such cases it is usual practice to obtain the tilt in the direction orthogonal to that of the lateral shear. The optical path difference associated with this tilt may be represented as a linear function of the y coordinate. Thus, in the case of tilt alone,

$$\Delta W = Ey = n\lambda \tag{4.5}$$

where E is the angle of tilt between the original and the sheared wavefronts, and the line of intersection of these wavefronts is parallel to the x axis. If defocusing and tilt are simultaneously present, the optical path difference is given by

$$\Delta W = 2DxS + Ey = n\lambda, \tag{4.6}$$

and this equation represents a system of straight fringes that are parallel neither to the x axis nor to the y axis. Only when either D or E is zero are they parallel to either the x axis or the y axis, respectively.

It is important to notice the difference between the two situations given by Eqs. (4.4) and (4.6). When there is no defocusing ($D = 0$), Eq. (4.4) gives a

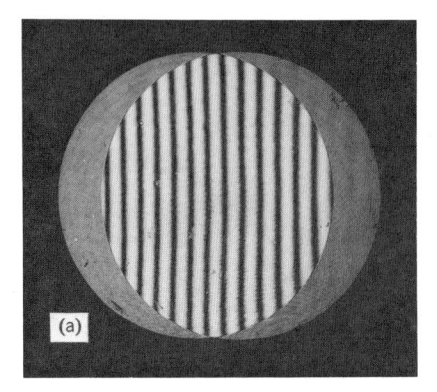

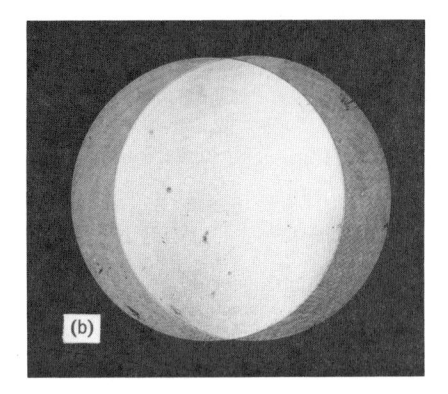

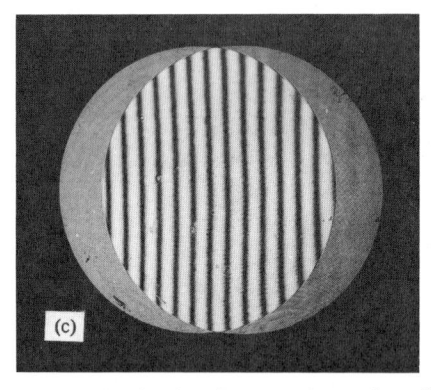

Figure 4.4. Sequence of lateral shearing interferograms for an aberrationless wavefront as one passes through the focus. The central fringeless pattern is obtained when there is no defocusing. The patterns on either side are due to slight defocusing in either direction by the same amount. (*a*) Inside the focus. (*b*) At the focus. (*c*) Outside the focus.

uniform or fringe-free field, while Eq. (4.6) gives a system of straight fringes parallel to the x axis. Thus, when an optical system is being collimated with respect to the point source of light, we go through the region of the focus. If we use a lateral shearing interferometer without tilt, the sequence of lateral shearing interferograms as we pass through the focal region will be as shown in Fig. 4.4. On the other hand, if we use a lateral shearing interferometer that can also introduce tilt for the same purpose, the sequence of interferograms will be as shown in Fig. 4.5. Thus in the latter cases it is possible to detect slight defocusing, because we are looking for a change in the direction of fringes rather than an absence of fringes. Therefore the ability to introduce tilt in addition to lateral shear is a distinct advantage in certain situations, and we shall see later how this feature can be used in different arrangements.

4.3. BRIEF THEORY OF LATERAL SHEARING INTERFEROMETRY   129

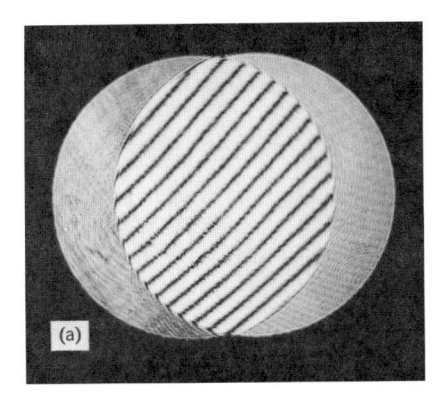

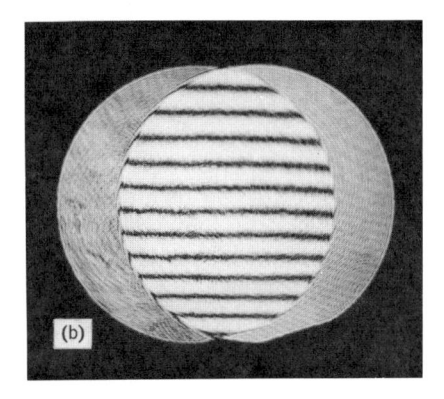

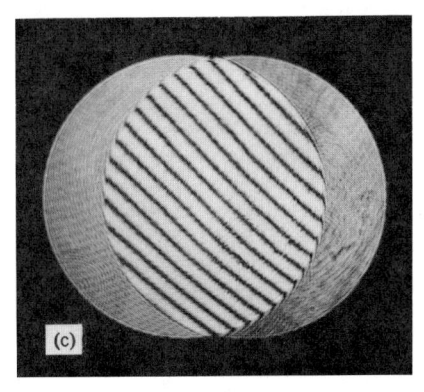

Figure 4.5. Sequence of lateral shearing interferograms of an aberrationless wavefront as one passes through the focus. In this case, however, a certain amount of tilt orthogonal to the direction of shear is introduced. At the focus the fringes are parallel to the direction of shear, whereas inside and outside the focus they are inclined. (*a*) Inside the focus. (*b*) At the focus. (*c*) Outside the focus.

4.3.1. Considerations of Lateral Shear in Relation to Primary Aberrations

Primary Spherical Aberration. The wavefront error for primary spherical aberration may be expressed as

$$W(x, y) = A(x^2 + y^2)^2. \tag{4.7}$$

Thus the shearing interferogram may be obtained from the equation

$$\Delta W = 4A(x^2 + y^2)xS = n\lambda \tag{4.8}$$

when there is no defocusing term. The system of fringes can be obtained from the following equation when defocusing is also present:

$$\Delta W = [4A(x^2 + y^2)x + 2Dx]S = n\lambda. \qquad (4.9)$$

Both Eqs. (4.8) and (4.9) are cubic, and consequently the fringes are cubic curves. Figure 4.6 shows the shape of fringes as we go through the focus in the presence of primary spherical aberration in the original wavefront. If, in addition to the preceding, we also have tilt, the equation for the fringes is given by

$$\Delta W = [4A(x^2 + y^2)x + 2Dx]S + Ey = n\lambda. \qquad (4.10)$$

When the primary spherical aberration is very small, and also when defocusing is absent, Eq. (4.10) may be approximated by

$$\Delta W = 4Ax^3 S + Ey = 0 \qquad (4.11)$$

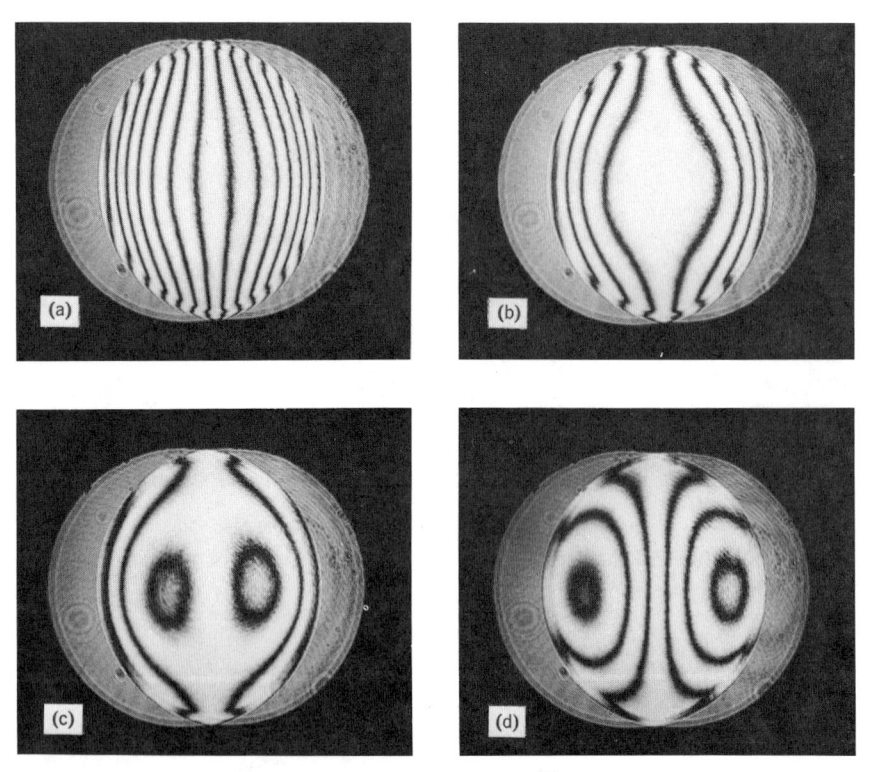

Figure 4.6. Typical lateral shearing interferograms in the presence of primary spherical aberration due to various amounts of defocusing. The second fringe pattern occurs when there is no defocusing. (*a*) Inside the focus. (*b*) At the focus. (*c*), (*d*) Outside the focus.

4.3. BRIEF THEORY OF LATERAL SHEARING INTERFEROMETRY

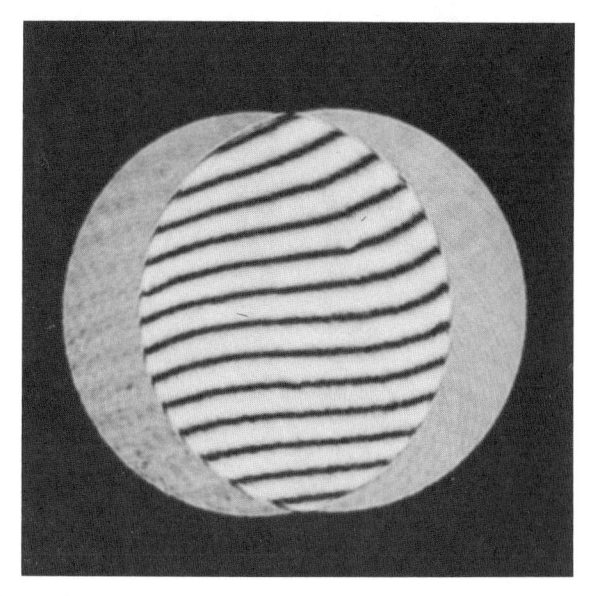

Figure 4.7. Typical lateral shearing interferogram in the presence of primary spherical aberration when there is a small tilt in the orthogonal direction. Note the characteristic shape of the fringe.

for the central fringe close to the x axis. This equation for the central fringe gives the characteristic ⌐⌐ -shaped curve by which very small amounts of spherical aberration can be visually detected. Typical fringes are shown in Fig. 4.7.

Primary Coma. The wavefront error for primary coma may be expressed as

$$W(x, y) = By(x^2 + y^2). \tag{4.12}$$

In view of the unsymmetrical nature of the aberration, the shape of the lateral shear fringes is different, depending on whether the shear is in the x direction, the y direction, or some other direction.

Let us first consider the case in which the shear is in the x direction. Then the shape of the fringes can be found from the following equation:

$$\Delta W = 2BxyS = n\lambda. \tag{4.13}$$

The fringes represented by Eq. (4.13) are rectangular hyperbolas with the asymptotes in the x and y directions. The effect of defocusing is to add another term $(2DxS)$ to Eq. (4.13), and this results in displacement of the center of the system of rectangular hyperbolas in the y direction. Figure 4.8 shows the system

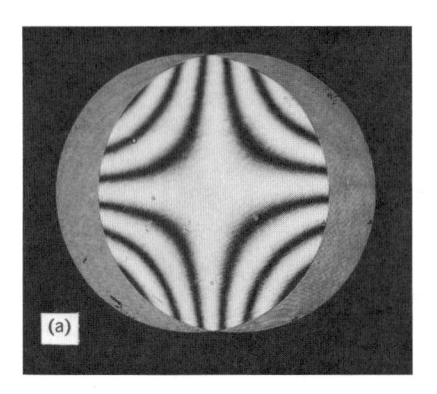

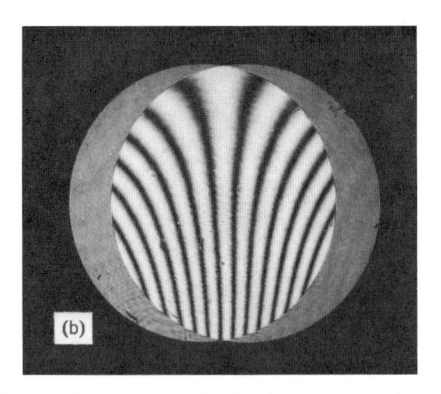

Figure 4.8. Lateral shearing interferograms of a wavefront suffering from primary coma. The direct of shear is in the sagittal direction. (*a*) Centered rectangular hyperbolas are obtained in this case, where there is no defocusing. (*b*) This interferogram is due to small defocusing, which causes displacement of the center of the rectangular hyperbolas in a direction perpendicular to the direction of shear.

of fringes under the conditions just discussed. If, in addition, there is tilt also in the orthogonal direction, the center of the hyperbolic fringes will be shifted in some other direction inclined to both the x and y axes.

Next, let us consider the situation in which the shear is in the y direction. Then, if the shear is T, the shape of the fringes can be found from the following equation:

$$\Delta W = B(x^2 + 3y^2)T = n\lambda. \tag{4.14}$$

In this case the fringes form a system of ellipses with a ratio of major to minor axes of $\sqrt{3}$. Also, the major axis is parallel to the x axis. The effects of defocusing and tilt are similar to those prevailing in the situation discussed earlier. Lateral shear fringes in the presence of coma with the shear parallel to the y axis are shown in Figs. 4.9*a* and 4.9*b* in the absence and the presence, respectively, of defocusing.

Primary Astigmatism. The wavefront error for primary astigmatism may be expressed as

$$W(x, y) = C(x^2 - y^2). \tag{4.15}$$

The situation in the case of astigmatism is peculiar because, if the lateral shear is in either the x or the y direction, we get straight fringes orthogonal to the direction of shear. Hence we can easily mistake an astigmatic wavefront for a

4.3. BRIEF THEORY OF LATERAL SHEARING INTERFEROMETRY

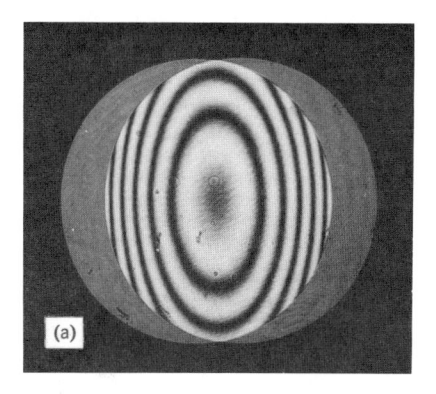

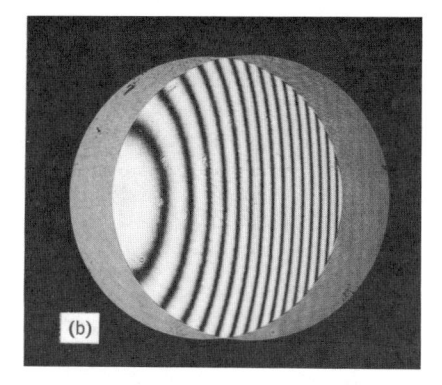

Figure 4.9. Lateral shearing interferograms of a wavefront suffering primary coma when the direction of shear is in the tangential direction. (*a*) This figure corresponds to a situation when there is no defocusing and hence the center of the elliptical fringes is at the center of the interferogram. (*b*) This figure corresponds to a situation where there is a certain amount of defocusing, resulting in the displacement of the center of the elliptical fringes in a direction parallel to the direction of shear.

true spherical wavefront. However, the introduction of defocusing changes the situation and reveals the presence of astigmatism. In the presence of defocusing, the fringe system for lateral shear in the x direction is given by

$$\Delta W = (2Dx + 2Cx)S = n\lambda \quad (4.16)$$

and for lateral shear in the y direction by

$$\Delta W = (2Dy - 2Cy)T = n\lambda. \quad (4.17)$$

If S and T are assumed to be of the same magnitude, there are two values of D, namely, $\pm C$, at which the lateral shearing interferogram shows no fringes. These two positions correspond to the usual tangential and sagittal foci of the astigmatic wavefront. Another way of detecting astigmatism is to see whether there is the same number of fringes at a given focal setting for both directions of lateral shear. Figure 4.10 shows this situation for the presence of astigmatism.

Another method whereby astigmatism can be detected involves the use of lateral shear in a general direction. In this case the system of fringes may be obtained from the following equation:

$$\Delta W = 2(D + C)xS + 2(D - C)yT = n\lambda. \quad (4.18)$$

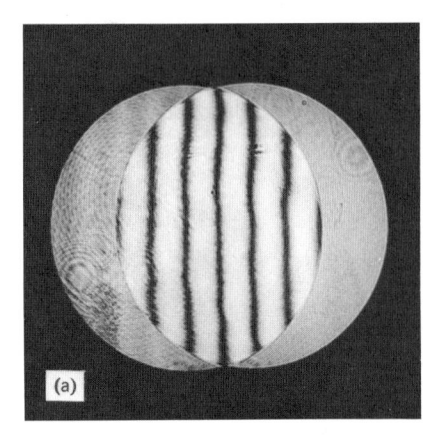

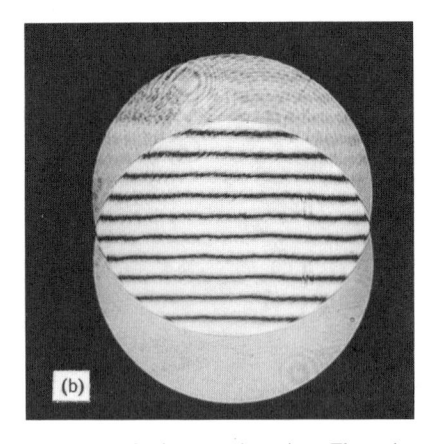

Figure 4.10. Lateral shearing interferograms in the presence of primary astigmatism. The astigmatism is perceived by noting the different number of straight fringes present when the shear direction at a given focal setting is (*a*) in a sagittal direction, and (*b*) in a tangential direction.

Equation (4.18) represents a system of equally spaced straight fringes, the slope of which is given by $(C + D)S/(C - D)T$. Thus, by changing the azimuth of the direction of shear and noting the corresponding direction of the fringes, it is possible to find the particular direction the slope of which deviates most from the orthogonal direction. Figure 4.11 illustrates this aspect of astigmatism in relation to lateral shearing interferometry.

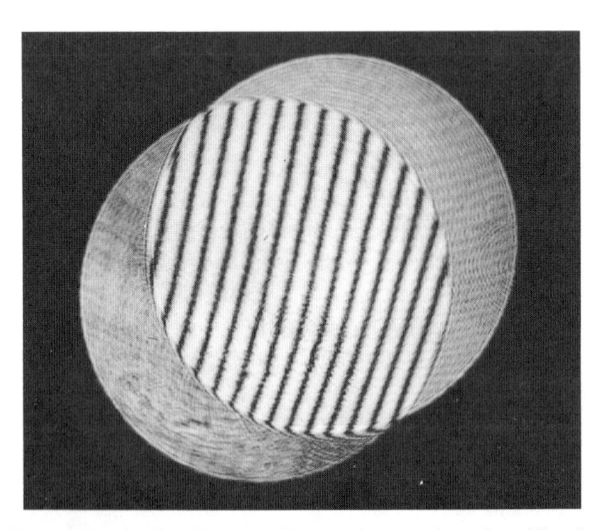

Figure 4.11. Lateral shearing interferogram due to primary astigmatism. The direction of shear is halfway between the sagittal and tangential directions. The presence of astigmatism is indicated by the inclination of the direction of straight fringes with respect to the direction of shear.

4.4. EVALUATION OF AN UNKNOWN WAVEFRONT

Curvature of Field and Distortion. Curvature of field is a displacement of focus longitudinally, and hence it can be treated merely as a defocusing situation. Distortion, being a linear function of the pupil variable y, is not generally detected.

Chromatic Aberration. Longitudinal chromatic aberration is nothing but a change of focus for different wavelengths. Hence, by changing the source of light or using different wavelengths from the same source, one can detect a change in the number of fringes due to defocusing and can thereby detect longitudinal chromatic aberration. Lateral chromatic aberration is again a linear function of the pupil variable y and is not generally detected.

The foregoing brief account of what happens when pure aberrations are present in the wavefront helps one to judge an optical system by simple and quick inspection utilizing lateral shearing interferometry.

4.4. EVALUATION OF AN UNKNOWN WAVEFRONT

We shall now see how it is possible to determine the shape of a wavefront from a lateral shearing interferogram of the wavefront in question. One method, proposed by Saunders (1961, 1970), estimates the order of interference at equally spaced points along a diameter and then evaluates the wavefront as in Fig. 4.12 by setting $W_1 = 0$ and $W_2 = \Delta W_1$, $W_3 = \Delta W_1 + \Delta W_2$, and so on, thus obtaining the wavefront by a summation of the lateral shear measurements ΔW_i. This procedure was extended to two dimensions by Saunders and Bruning (1968) and later by Rimmer (1972) and Nyssonen and Jerke (1973).

Another procedure assumes that the unknown wavefront $W(x, y)$ is a smooth function that can be represented by a polynomial; the interferometer function $\Delta W(x, y)$ is expressed in terms of the coefficients of the polynomial. Many values of ΔW are found from measurements of the fringe positions, and from those the coefficients of the wavefront are computed. Malacara (1965a), Murty and Malacara (1965), and Dutton et al. (1968) developed this method in one dimension to find the wavefront shape along a diameter parallel to the shear. Malacara and Mendez (1968) particularized the method for surfaces with rotational symmetry. In general, the polynomial method is very good, especially after being extended to two dimensions by Rimmer and Wyant (1975) in the following manner.

The wavefront is represented by the function $W(x, y)$, which we assume can be represented by a two-dimensional polynomial of degree k, of the form

$$W(x, y) = \sum_{n=0}^{k} \sum_{m=0}^{n} B_{nm} x^m y^{n-m} \tag{4.19}$$

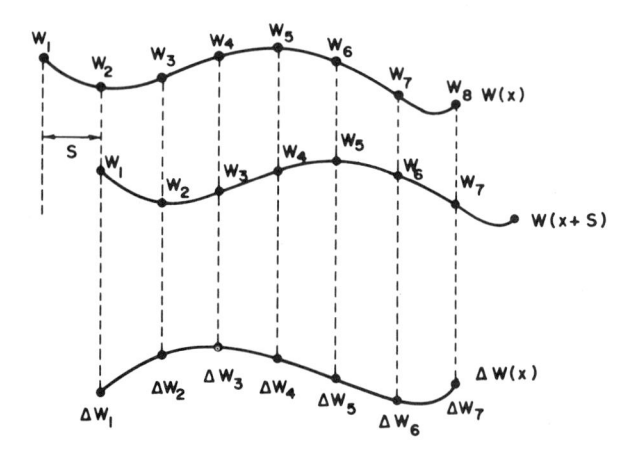

Figure 4.12. Procedure to find a wavefront from its lateral shearing interferogram.

which contains $N = (k + 1)(k + 2)/2$ terms. If we want to reconstruct the whole wavefront, we take two mutually perpendicularly sheared interferograms, represented by S and T. The sheared wavefronts will be given by

$$W(x + S, y) = \sum_{n=0}^{k} \sum_{m=0}^{n} B_{nm}(x + S)^m y^{n-m} \qquad (4.20)$$

and

$$W(x, y + T) = \sum_{n=0}^{k} \sum_{m=0}^{n} B_{nm} x^m (y + T)^{n-m} \qquad (4.21)$$

respectively. But, using the binomial theorem, we obtain

$$(x + S)^m = \sum_{j=0}^{m} \binom{m}{j} x^{m-j} S^j \qquad (4.22)$$

where the binomial factor is

$$\binom{m}{j} = \frac{m!}{(m-j)!j!}. \qquad (4.23)$$

Equations (4.20) and (4.21) can be transformed into

$$W(x + S, y) = \sum_{n=0}^{k} \sum_{m=0}^{n} \sum_{j=0}^{m} B_{nm} \binom{m}{j} x^{m-j} y^{n-m} S^j \qquad (4.24)$$

4.4. EVALUATION OF AN UNKNOWN WAVEFRONT

and

$$W(x, y + T) = \sum_{n=0}^{k} \sum_{m=0}^{n} \sum_{j=0}^{n-m} B_{nm} \binom{n-m}{j} x^m y^{n-m-j} T^j. \quad (4.25)$$

Observing that these two functions become equal to $W(x, y)$ when $j = 0$, and rearranging the sums, we can obtain the following two shearing interferogram equations:

$$\Delta W_S = W(x + S, y) - W(x, y) = \sum_{n=0}^{k-1} \sum_{m=0}^{n} C_{nm} x^m y^{n-m} \quad (4.26)$$

and

$$\Delta W_T = W(x, y + T) - W(x, y) = \sum_{n=0}^{k-1} \sum_{m=0}^{n} D_{nm} x^m y^{n-m} \quad (4.27)$$

where

$$C_{nm} = \sum_{j=1}^{k-n} \binom{j+m}{j} S^j B_{j+n, j+m} \quad (4.28)$$

and

$$D_{nm} = \sum_{j=1}^{k-n} \binom{j+n-m}{j} T^j B_{j+n, m}. \quad (4.29)$$

This result was given by Rimmer and Wyant (1975). The values of C_{nm} and D_{nm} may be obtained from the interferograms by means of a least-squares fitting of the measured values of ΔW_S and ΔW_T to functions (4.26) and (4.27), respectively. There are $M = k(k + 1)/2$ coefficients C_{nm} and D_{nm}, and from them we have to determine the N wavefront coefficients B_{nm}. Expression (4.28) represents a system of M equations with M unknowns, where the unknowns are all B_{nm} coefficients with the exception of B_{n0}. Similarly, expression (4.29) represents a system of M equations with M unknowns, where these unknowns are all B_{nm} coefficients with the exception of B_{nn}. If the wavefront has rotational symmetry, $B_{n0} = B_{nn} = 0$ for all values of n, it becomes sufficient to use either expression (4.28) or (4.29) and hence only one interferogram is necessary. If m is different from n and also different from zero, the value of B_{nm} is derived from both expressions and therefore an average of the two values can be taken, since they may be different because of numerical rounding errors.

Many alternative approaches to computing the wavefront of a lateral shearing interferometer have been devised, for example, by Gorshkov and Lysenko

(1980) and many others, but one of the most interesting is based on the Zernike polynomials. The first description of this method is due to Rimmer and Wyant (1975) and was later revised by Korwan (1983). The interferogram function ΔW as well as the wavefront W is expressed as a linear combination of Zernike polynomials.

4.5. LATERAL SHEARING INTERFEROMETERS IN COLLIMATED LIGHT (WHITE LIGHT COMPENSATED)

We first consider arrangements in which ordinary light sources are used, and hence white-light compensation is necessary. Here again we have two basic situations, namely, lateral shear in a collimated beam and lateral shear in a convergent beam. However, it is always possible to convert one type into the other by the use of a well-corrected lens.

4.5.1. Arrangements Based on the Jamin Interferometer

Figures 4.13, 4.14, and 4.15 show schematic arrangements of the Jamin interferometer modified by Murty (1964b) to serve as lateral shearing interferometers. The usual extended source of light is replaced by a small pinhole and the lens that is to be tested. The lens acts as the collimating lens, and this nearly collimated beam of light enters the interferometer. If the two plates of glass on either side are exactly parallel to each other, the two beams of light emerge without any lateral shear. To introduce lateral shear into this system, two methods are available. One method is to rotate one of the parallel plates of glass about an axis as shown in Fig. 4.13. In this case the shear is in the direction

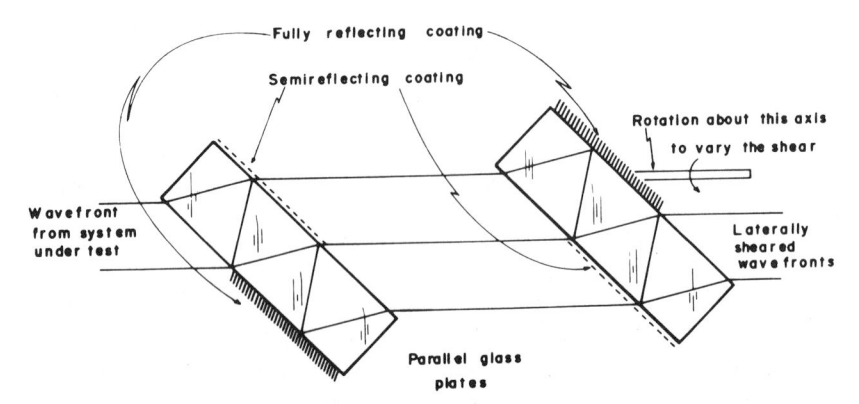

Figure 4.13. Schematic arrangement showing the modification of the Jamin interferometer into a lateral shearing interferometer. The lateral shear can be varied by rotating the end glass plate as shown in the figure.

4.5. LATERAL SHEARING INTERFEROMETERS IN COLLIMATED LIGHT

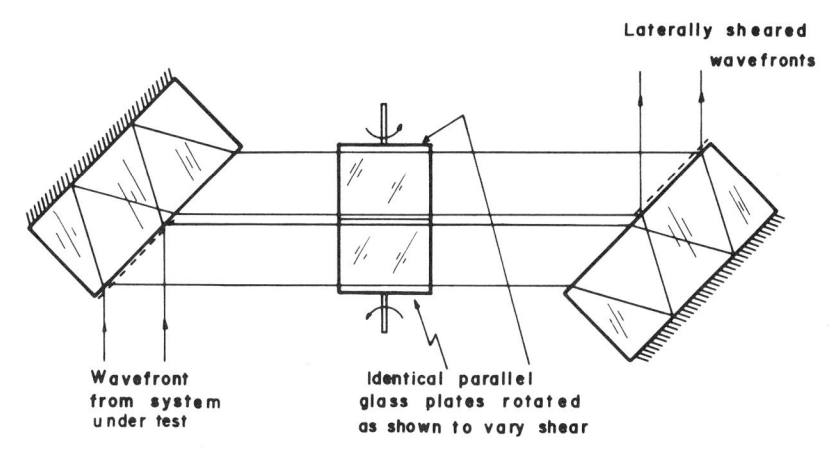

Figure 4.14. Another modification of the Jamin interferometer into a lateral shearing interferometer. In this modification the end glass plates are adjusted permanently to be parallel to each other. Lateral shear is achieved by rotating the two identical glass blocks in the two beams as indicated, in opposite directions.

perpendicular to the plane of the paper. The second method is shown in Fig. 4.14. In this arrangement the two blocks of glass of the Jamin interferometer are adjusted exactly parallel to each other; this can be done by observing the zero-order fringe with a broad source of light. Then in the two beams two identical parallel plates of glass are arranged in such a way that they can be rotated about the axes, as shown, in opposite directions by the same amount. This can be accomplished by the use of a simple gear arrangement. The Jamin plates should be big enough to accommodate the beams displaced by lateral shear. In the second arrangement it is possible to introduce a fixed tilt in an orthogonal direction by making the end Jamin plates slightly wedge shaped (by a few seconds of arc).

Another arrangement, which is a cyclical form of the Jamin interferometer, is shown in Fig. 4.15. In this only one parallel plate is sufficient, and the light path is folded by the right-angle prism or a set of two plane mirrors is adjusted to be at right angles to each other. To obtain lateral shear in this arrangement, it is necessary to use only one plane parallel glass plate. Since the two beams pass through it from either side, one beam is lifted up and the other pushed down so that the usual lateral shear is obtained. This seems to be a very convenient lateral shearing interferometer for testing lenses of small aperture. It is also possible to introduce tilt in the orthogonal direction by making the right angle of the right-angle prism or mirror slightly in error. In fact, this interferometer may be used for testing the accuracy of the right angle of a right-angle prism or mirror combination and also for detecting errors of the right angles in a cube corner prism.

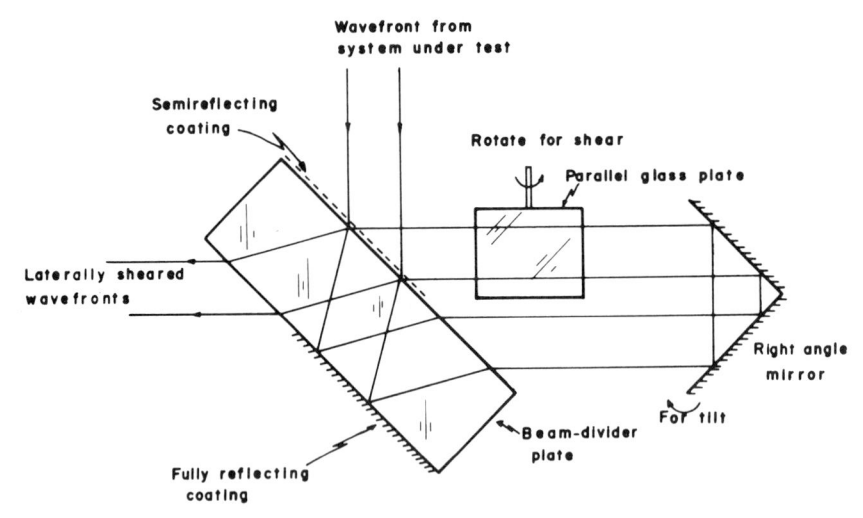

Figure 4.15. Another modification of the Jamin interferometer into a lateral shearing interferometer. The arrangement may be considered to be a folded version of the conventional Jamin interferometer. To achieve lateral shear, only one glass block, which is rotated as indicated, is needed.

4.5.2. Arrangements Based on the Michelson Interferometer

The Michelson interferometer, when adjusted for the zero-order interference position, is already compensated for white light. The conventional Michelson interferometer with plane mirrors as end reflectors cannot give lateral shear in collimated light. However, if we replace these plane mirrors by right-angle prisms or cube corner prisms, it is possible to obtain lateral shear (Kelsall 1959). The arrangement is shown in Fig. 4.16. For our discussion we may assume that

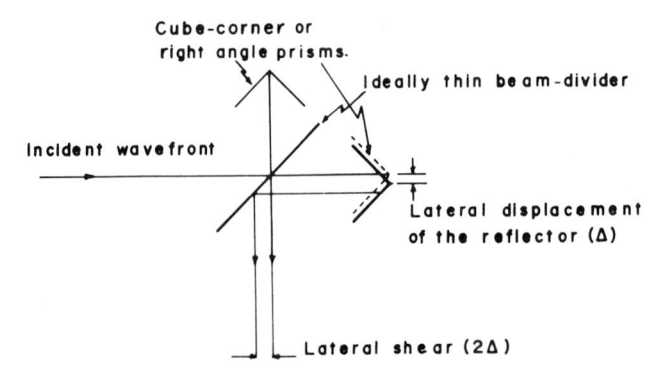

Figure 4.16. Schematic diagram of the modification of the Michelson interferometer into a lateral shearing interferometer. The end reflectors are either right-angle prisms or cube corner prisms.

4.5. LATERAL SHEARING INTERFEROMETERS IN COLLIMATED LIGHT

the two right-angle prisms are of identical size and material. When their edges are lined up parallel to each other, and when they are at exactly the same distance from the beam divider, we get no shear, provided that the virtual image of one prism coincides with the other prism as viewed in the beam divider. If one of the prisms is displaced laterally by some amount, the wavefront is sheared by twice this amount. If a very large amount of shear is wanted, the beam divider plate must be large enough to accommodate this. Similarly the right-angle prisms (at least one of them) also must be of sufficient size. Tilt is possible if the right-angle prism is tilted perpendicularly to the direction of shear. If cube corner prisms are used, no tilt is possible and only pure lateral shear is obtained.

4.5.3. Arrangements Based on a Cyclic Interferometer

A cyclic interferometer is one in which the two coherent light beams travel in both directions and finally emerge to interfere. In this sense the modification of the Jamin interferometer described previously and shown in Fig. 4.15 may also be considered as a cyclic interferometer. We shall see how a typical cyclic interferometer such as a triangular-path interferometer (Hariharan and Sen 1960) can be used as a lateral shearing interferometer. Figure 4.17 shows a typical arrangement for obtaining lateral shear in collimated light. Here there are two ways of obtaining lateral shear. One is to translate one of the plane mirrors, as shown in Fig. 4.18. The other method is similar to what was done in Fig. 4.15.

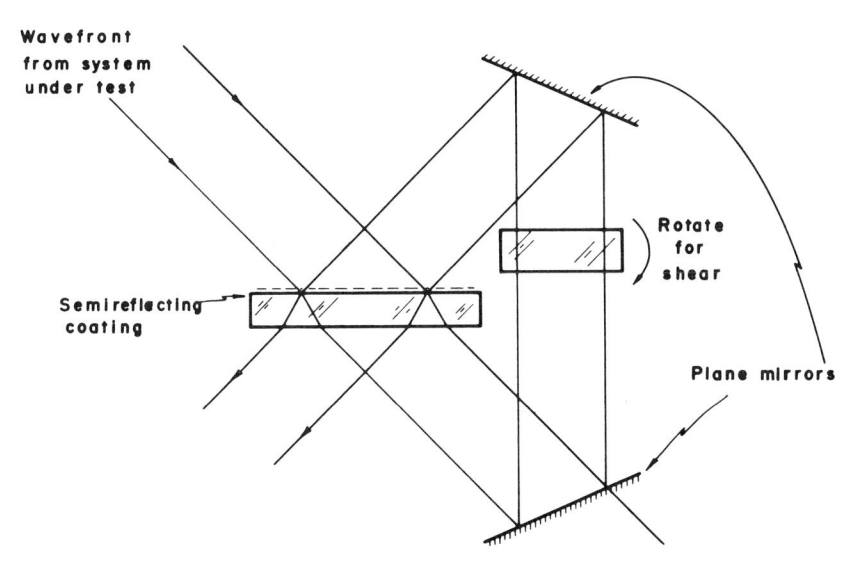

Figure 4.17. Schematic arrangement of a triangular-path cyclic interferometer, which can be used as a lateral shearing interferometer. One method of achieving lateral shear is to rotate the parallel glass plate as indicated.

142　　　　　　　　　　　　　　　　　　　　　　LATERAL SHEARING INTERFEROMETERS

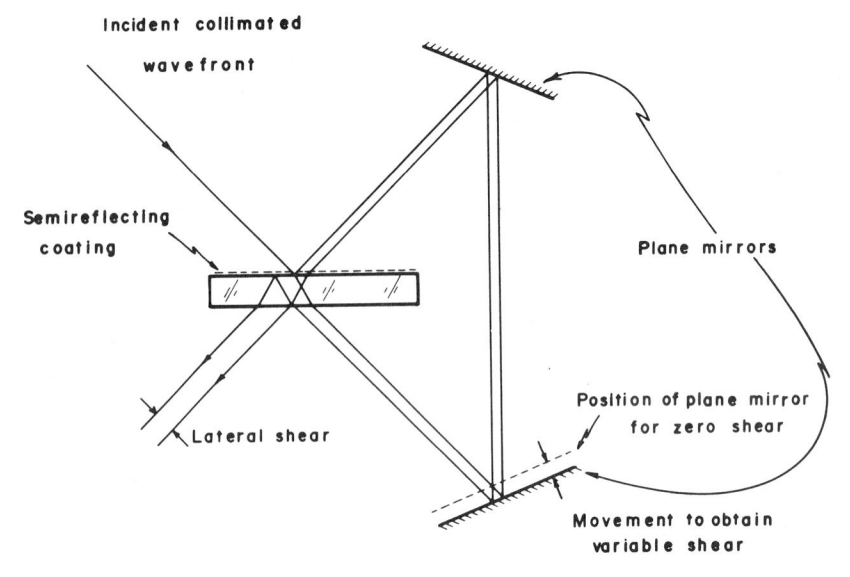

Figure 4.18. Schematic diagram of a triangular-path interferometer in which lateral shear is obtained by parallel displacement of one of the mirrors.

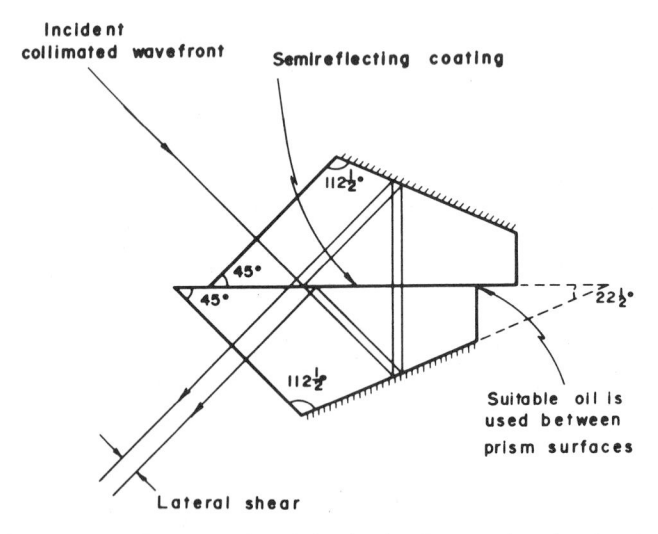

Figure 4.19. Schematic diagram of a lateral shearing interferometer based on the triangular-path interferometer. Here, two half-pentaprisms are utilized; by the displacement of one prism over the other, lateral shear can be achieved. This is suitable for testing small aperture optical systems.

4.5. LATERAL SHEARING INTERFEROMETERS IN COLLIMATED LIGHT

In the symmetrical instrument we may introduce a plane parallel plate of glass, and rotation of the glass plate then produces lateral shear. This is also a very easy instrument to construct and is very insensitive to vibration and other effects in the laboratory. An arrangement using two half-pentaprisms is shown in Fig. 4.19. Translation of one prism parallel to the beam-dividing surface results in variable shear.

Another cyclic shearing interferometer to be used with convergent light has been described by Kanjilal et al. (1984) and by Kanjilal and Puntambekar (1984).

4.5.4. Arrangements Based on the Mach–Zehnder Interferometer

A schematic arrangement of the Mach–Zehnder interferometer is shown in Fig. 4.20. There are two beam dividers and two plane reflectors. To obtain lateral shear in collimated light, one has to use one of two glass plane parallel plates of the same thickness and material in each beam of the instrument. The two plates must be arranged in such a way that they rotate as shown in Fig. 4.21. Thus, in principle, the Mach–Zehnder interferometer may also be modified for use as a lateral shearing interferometer in collimated light. However, in prac-

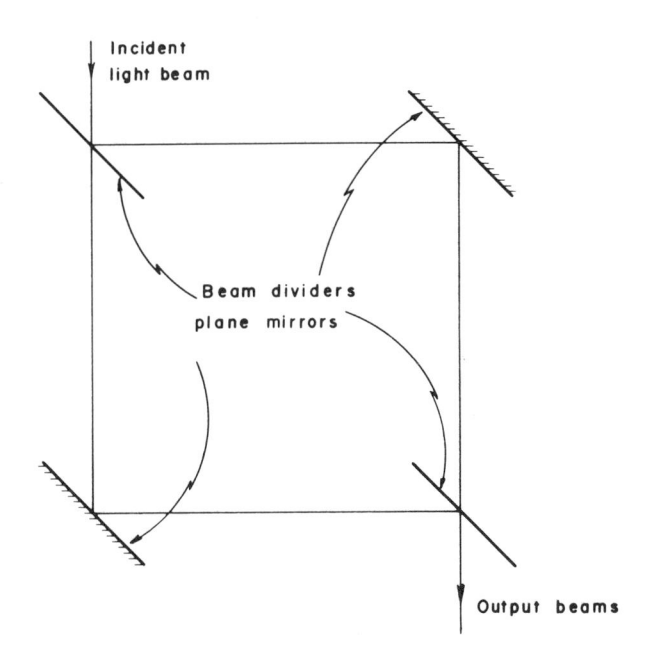

Figure 4.20. Schematic diagram of an idealized Mach–Zehnder interferometer using thin beam splitters.

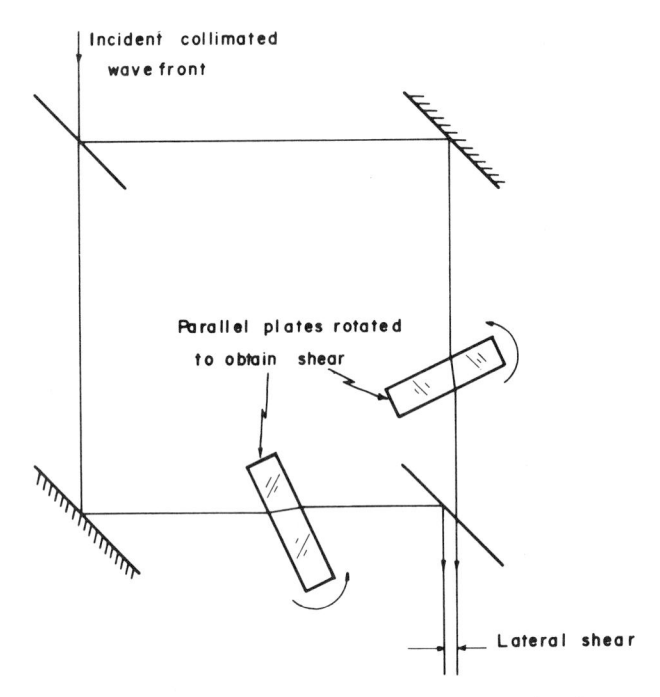

Figure 4.21. Schematic diagram illustrating how the Mach–Zehnder interferometer is modified into a lateral shearing interferometer. Two identical glass blocks are present in each path of the interferometer, and to obtain lateral shear they are rotated in the directions shown by equal amounts.

tice, it is a cumbersome device to adjust, especially if all the elements are separately mounted and each has its own tilting screws. We shall see later that the Mach–Zehnder interferometer is more useful for obtaining lateral shear in a convergent beam of light.

4.6. LATERAL SHEARING INTERFEROMETERS IN CONVERGENT LIGHT (WHITE LIGHT COMPENSATED)

4.6.1. Arrangements Based on the Michelson Interferometer

The basic scheme for all lateral shearing arrangements using convergent light is shown in Fig. 4.22. The converging, nearly spherical wavefront is directed to its center of curvature; in this region is located the lateral shearing interferometer. For this reason the size of the lateral shearing interferometer used in convergent light is always very small. Figure 4.23 shows the idealized thin beam divider type Michelson interferometer, into which a converging beam of light from the optical system under test is passing. If the two plane mirrors are

4.6. LATERAL SHEARING INTERFEROMETERS IN CONVERGENT LIGHT 145

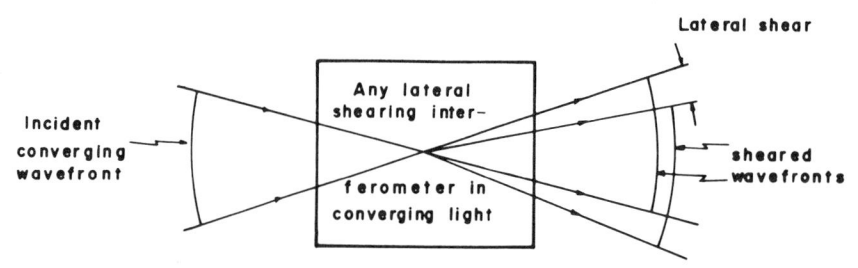

Figure 4.22. Schematic diagram of any lateral shearing interferometer introducing lateral shear in the convergent beam.

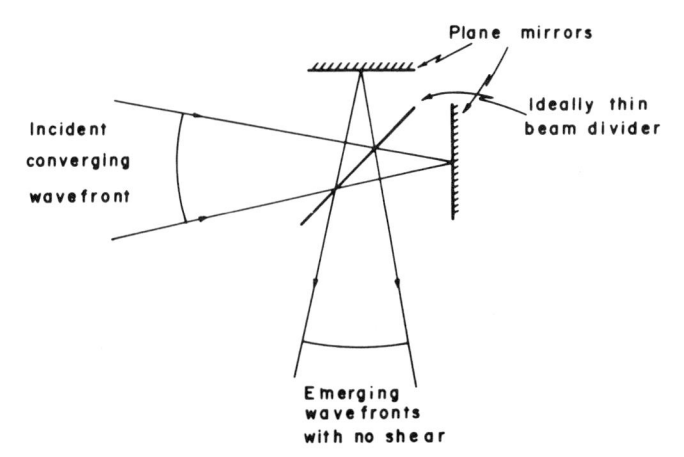

Figure 4.23. Schematic diagram of an idealized Michelson interferometer used as a lateral shearing interferometer. The diagram shows the two plane mirrors exactly symmetrical with respect to the thin beam divider, and the convergent beam is focusing on the two mirrors. In this situation the amount of lateral shear is zero.

symmetrical with respect to the beam divider, and if they are at equal distances from it, there is no lateral shear in the two diverging spherical wavefronts that emerge, and hence one sees a broad zero-order fringe in the whole field. Now suppose that the center of curvature of the converging beam is coincident with the centers of the two end reflectors and that one of the end reflectors is rotated through a small angle about a transverse axis passing through the center of curvature. Then lateral shear will result, as indicated in Fig. 4.24. If the wavefront has some aberration, it is made visible in the shearing pattern. If the center of curvature falls slightly outside the plane reflectors, defocusing results and we get the usual straight fringes in the absence of any aberration. It is interesting that it is not possible to obtain tilt in the orthogonal direction.

In actual practice this interferometer is made by taking two identical right-

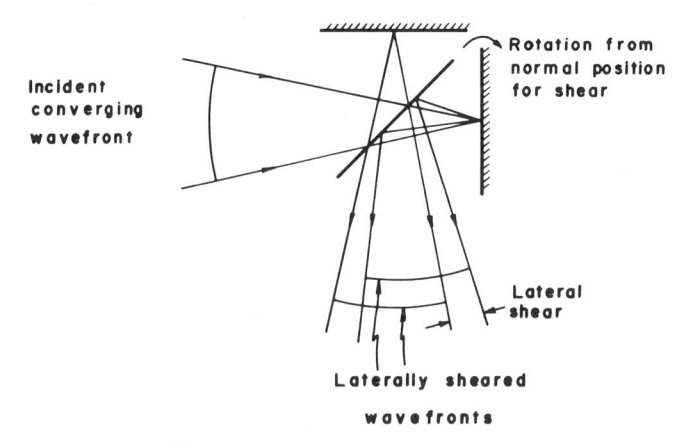

Figure 4.24. Schematic diagram of an idealized Michelson interferometer used as a lateral shearing interferometer. Tilting one of the plane mirrors about an axis passing through the point of convergence of the wavefront introduces lateral shear as shown in the figure.

angle prisms and cementing them together to give a proper amount of shear. This is shown in Fig. 4.25, which represents the practical solid version of the device. Lenouvel and Lenouvel (1938) were probably the first to devise this interferometer. More recently, Murty (1969), Saunders (1970), and several others have found this interferometer to be very useful and simple and adapted to the testing of optical systems. It is also possible to make a similar interferometer

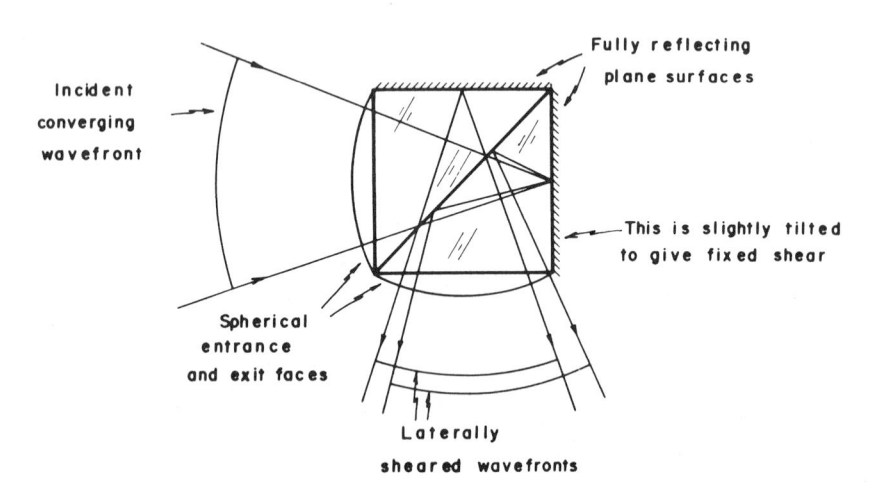

Figure 4.25. Schematic diagram of a practical lateral shearing interferometer in the form of a cemented cube. The plane surfaces are arranged in such a way that there is a fixed amount of shear. The entrance and the exit faces may be given convex spherical shapes so that the wavefronts enter and exit normally.

4.6. LATERAL SHEARING INTERFEROMETERS IN CONVERGENT LIGHT

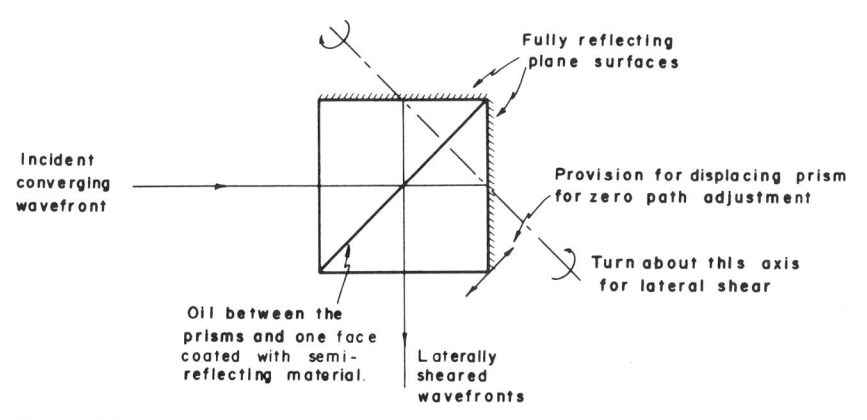

Figure 4.26. Schematic diagram of a lateral shearing interferometer based on the Michelson interferometer in which the lateral shear can be varied by rotating one or both prisms about an axis as shown in the figure.

in which one can vary the shear (Murty 1970) as shown in Fig. 4.26. In this case a suitable oil is used between the two hypotenuse faces and a suitable mechanism to rotate one of the prisms about an axis (see Fig. 4.26). The entrance and exit faces may be made spherical so that the light enters and exits almost normal to the surfaces. This interferometer has been used to measure cryogenic laser fusion targets (Tarvin et al. 1979).

4.6.2. Arrangements Based on the Mach–Zehnder Interferometer

Let us consider the idealized Mach–Zehnder interferometer with thin beam dividers as shown in Fig. 4.27. The converging wavefront is shown focused on the two plane mirrors. If one of these mirrors is rotated about a transverse axis passing through this focus, a lateral shear results between the two beams emerging from the interferometer. Again, if the second beam divider is tilted about an axis as shown in Fig. 4.27, a tilt is introduced that is orthogonal to the direction of shear. The introduction of this tilt is an important advantage of this interferometer. It is also possible to bring the focus to the second beam-dividing surface, in which case the roles of the plane mirror and the beam divider are interchanged.

An instrument that is made using two beam dividers and two plane mirrors, each separately mounted on a suitable base and capable of all possible adjustments, is extremely difficult to adjust quickly. Therefore we give here some examples of prealigned and preadjusted devices based on the Mach–Zehnder interferometer. Most of them use solid prisms of glass. Figure 4.28 shows almost a literal solidification of Fig. 4.27, devised by Saunders (1965), which fills the space between the beam dividers and the plane mirrors with glass. By

148 LATERAL SHEARING INTERFEROMETERS

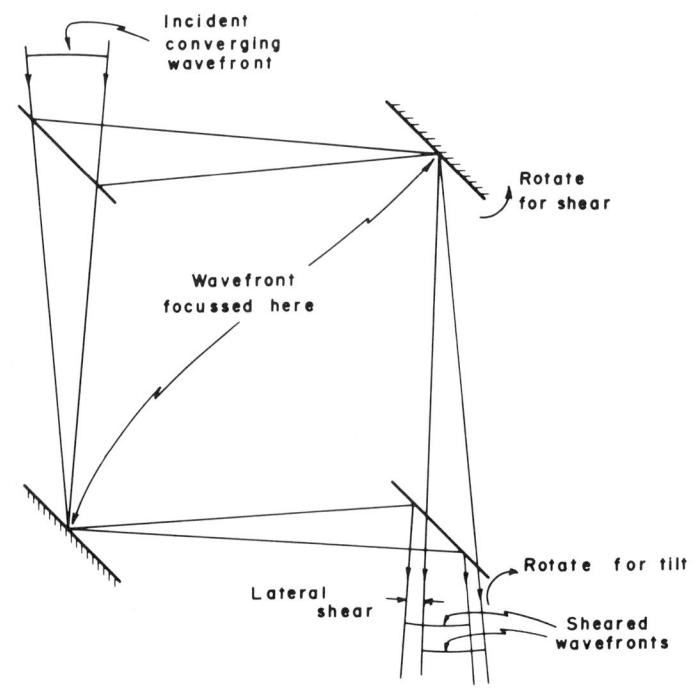

Figure 4.27. Schematic diagram of a lateral shearing interferometer based on the Mach–Zehnder interferometer. In this case the converging wavefront focuses on the two plane mirrors as indicated.

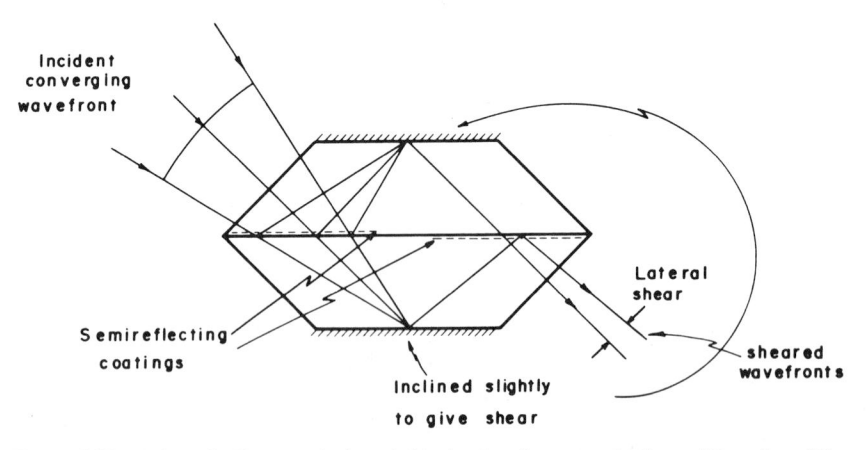

Figure 4.28. Schematic diagram of a lateral shearing interferometer that is a solid version of the one shown in Fig. 4.27.

4.6. LATERAL SHEARING INTERFEROMETERS IN CONVERGENT LIGHT

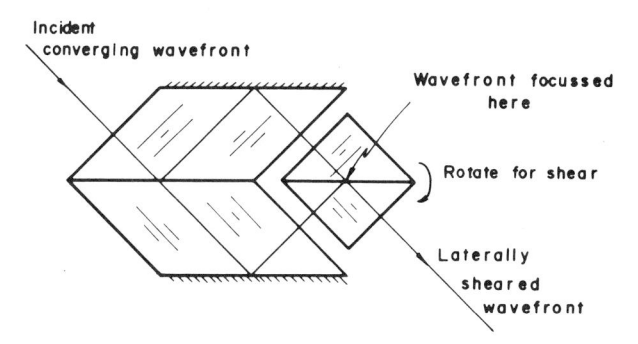

Figure 4.29. Schematic arrangement of a lateral shearing interferometer based on the Mach–Zehnder interferometer in which the lateral shear can be varied. The convergent beam is focused at the center of the cube beam divider, and rotation of the cube about its center gives variable lateral shear.

making the surfaces on which the convergent light focuses parallel to each other but not parallel to the beam divider, it is possible to obtain a fixed amount of lateral shear. To obtain two identical prisms of the shape shown in Fig. 4.28, one normally starts with one prism and afterwards cuts it into two. Tilt about an orthogonal direction is generally possible in this arrangement if the cement used is slightly wedged. An arrangement by Murty (unpublished) in which lateral shear is variable is shown in Fig. 4.29. Here two rhomboidal prisms are cemented with a beam-dividing surface, and the second beam divider is in the form of a cube. Now the center of curvature of the incident wavefront is focused at the center of the cube beam divider. By rotating the cube divider about an axis passing through its center, it is possible to vary the shear to some extent. A fixed tilt can be obtained by a suitable tilt of the reflecting surfaces.

Another arrangement of a lateral shearing interferometer by Murty (unpublished) based on the Mach–Zehnder interferometer is shown in Fig. 4.30. We may imagine two plano-convex spherical lenses such that their center thicknesses are roughly half their radii of curvature. The apex portions of both lenses are made flat, having a small area and coated with a suitable reflecting material. It is possible to make these flat surfaces in such a way that they are slightly inclined to the plane of the beam-dividing surface. This arrangement may be considered almost the same as the one shown in Fig. 4.28. Because of the spherical entrance and exit faces, a beam of large numerical aperture can be easily passed.

Another arrangement, devised by Saunders (1964a), that is a modification of the Mach–Zehnder interferometer is shown in Fig. 4.31. The prisms can be cut from a single prism of angles $\alpha = 20°$, $\beta = 60°$, and $\gamma = 100°$. The converging wavefront is focused on the inclined faces inside the prisms. If we

150   LATERAL SHEARING INTERFEROMETERS

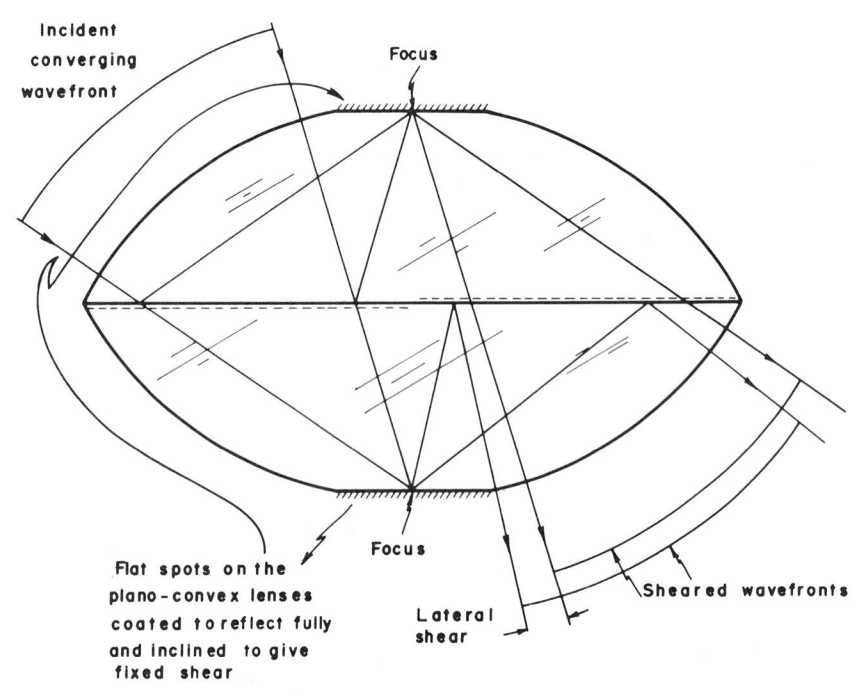

Figure 4.30. Schematic diagram of a lateral shearing interferometer fabricated from two thick plano-convex lenses.

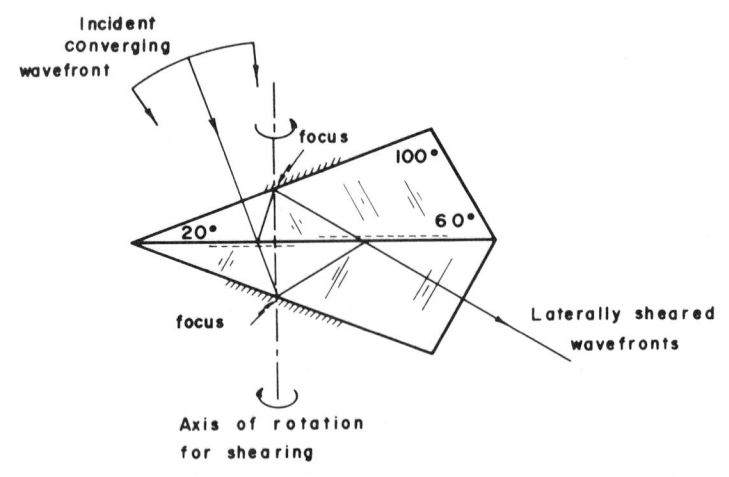

Figure 4.31. Schematic diagram of a lateral shearing interferometer consisting of two prisms. The convergent beam is focused on the inclined faces, and lateral shear is obtained by rotating the two prisms above the axis as indicated in the diagram. The same instrument may be made as a fixed shear device by cementing.

4.7. LATERAL SHEARING INTERFEROMETERS USING LASERS

imagine a line joining the two foci, lateral shear can be obtained by rotation of the two prisms about this line in opposite directions. This arrangement can very easily be made into a fixed shear instrument. By cementing, on the entrance and exit faces, proper plano-convex lenses, the instrument is made completely free of aberrations of its own. It is also possible to provide tilt by suitable wedging of the cement and by coating the beam divider on both prism faces, covering only half the surface of each.

4.7. LATERAL SHEARING INTERFEROMETERS USING LASERS

As pointed out earlier, any white-light-compensated lateral shearing interferometer works with a laser source, but it is possible to devise interferometers that are simple in construction and use when the compensation requirement is eliminated. The most convenient laser is the helium–neon laser giving a few milliwatts of power in the 6328-Å line. One of the simplest lateral shearing interferometers is a plane parallel plate devised by Murty (1964a). Figure 4.32 shows the schematic arrangement of the interferometer. The light from the laser is focused by a suitable microscope objective on a pinhole located at the focus of a collimating lens. The collimated beam of light is incident on the plane parallel plate, which is normally used without any coating on either surface. The light is reflected from the front and the back of the plate, and because of the thickness of the plate there is a lateral shear. The lateral shear S for a plate of thickness t and refractive index N, and for an angle of incidence i, is given

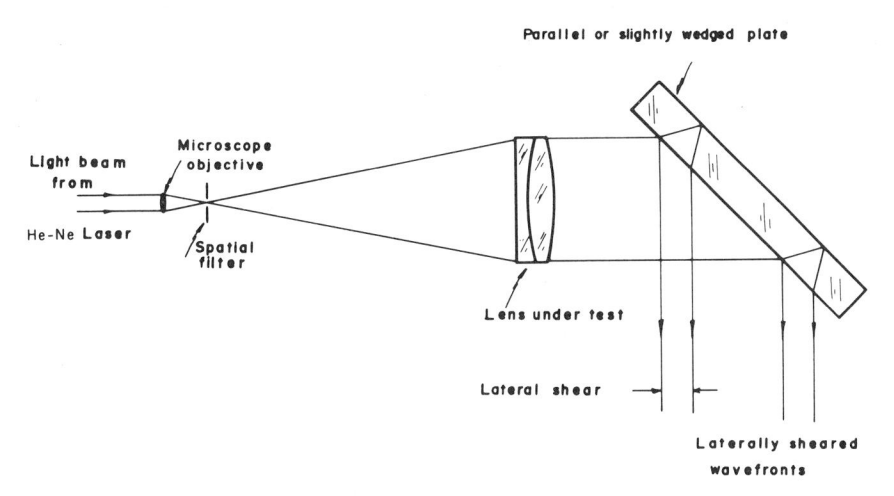

Figure 4.32. Schematic diagram of a lateral shearing interferometer using a laser source and a plane parallel plate of glass.

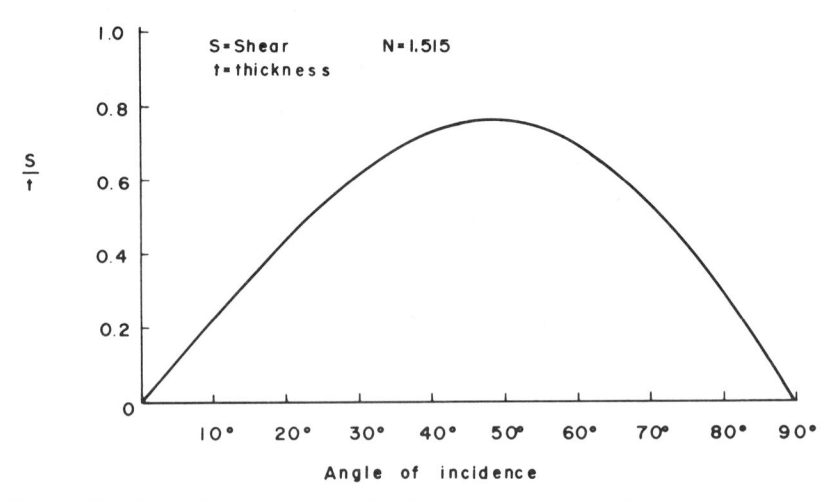

Figure 4.33. Plot of S/t versus angle of incidence for a typical borosilicate crown glass plate. From the plot it is seen that any angle of incidence up to a maximum of 50° is convenient.

by (Malacara 1965a)

$$S = t \sin 2i(N^2 - \sin^2 i)^{-1/2}. \qquad (4.30)$$

For glass of $N = 1.515$ at 6328 Å, Fig. 4.33 shows the plot of S/t versus i. From this it is seen that S/t has a maximum value of about 0.8, corresponding to an angle of incidence of 50°. Therefore a 45° angle of incidence is quite convenient to use in a practical setup.

It is possible to increase the intensity of the fringe pattern by coating the front and back surfaces. If this is done, the secondary reflection has enough intensity to show faint secondary shearing patterns. Therefore it may be best to use an uncoated plate. Figure 4.34 shows the modification of the arrangement

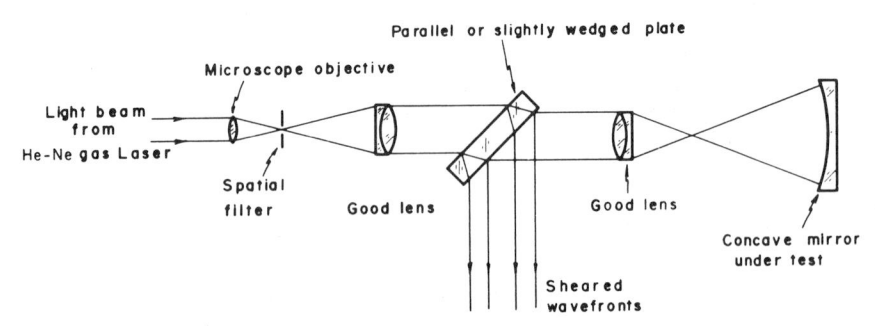

Figure 4.34. Schematic diagram of a parallel plate interferometer for use in testing large concave mirrors.

4.7. LATERAL SHEARING INTERFEROMETERS USING LASERS

used for testing a large concave mirror. If the mirror is not spherical, a suitable null correcting system may be inserted into the system. It is also possible to introduce a fixed amount of tilt orthogonal to the shear direction if the shearing plate is made slightly wedge shaped. The line in which the two planes of the wedge intersect is parallel to the plane of the paper in Fig. 4.34.

Another modification (Hariharan 1975) of the basic parallel plate interferometer is shown in Fig. 4.35. Here, to obtain the two plane reflecting surfaces, two separate plates are used. The inside surfaces of these plates are uncoated so that about 4% reflection occurs at these two surfaces. The outer surfaces of the two plates are treated with very high quality antireflection coatings so that they do not reflect any light. One of the plates is mounted on a platform that can be moved so that the air gap is variable and, hence variable shear is obtained. Provision may also be made on the plate mounting so that it can be rotated orthogonally to introduce tilt. Since the reflecting surfaces are separate and not solidly connected as in a single parallel plate, the stability of the fringe system depends on the stability of the mechanical mounts and special design.

The lateral shearing interferograms in Figs. 4.4 to 4.11 were obtained with the laser interferometer shown in Fig. 4.32. Even though only an uncoated glass plate is used, it is possible to see a sufficiently intense fringe pattern projected on a ground glass in a dimly illuminated room, and an exposure of only a fraction of a second is required to photograph the fringes.

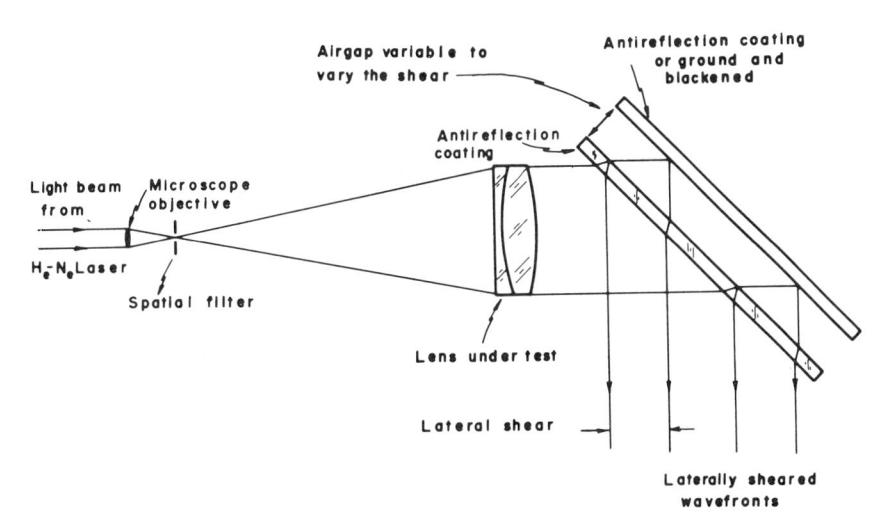

Figure 4.35. Modification of the parallel plate interferometer by the use of two separate glass plates. The reflections from the inner surface are utilized for the lateral shear, while the outer surfaces are treated with antireflection coatings. The lateral shear is varied by moving the back plate to change the air gap.

4.7.1 Other Applications of the Parallel Plate Interferometer

One of the most useful applications of the parallel plate interferometer is for checking the collimation of a lens. In Fig. 4.32, if the pinhole is not located at the focus of the lens, the resulting beam is slightly divergent or convergent. Hence the shearing interference pattern is as shown in Figs. 4.4a and 4.4c. Only when the pinhole coincides with the focus of the collimating lens will the common area of the two sheared apertures be free of any fringes, as seen in Fig. 4.4b. However, a better judgment can be made of the exact collimation if a wedged plate is used for the purpose of indicating it. In this case the plate is first used normal to the beam emerging from the collimating lens. The reflected beam shows the Fizeau fringes indicating the direction of the wedge. Since in this position the shear is zero, slight decollimation does not matter. The parallel plate is rotated in its own plane until the Fizeau fringes are horizontal. Then the plate is turned about a vertical axis so that the angle of incidence is about 45°. If the pinhole is slightly outside the focus, the shear fringes will not be horizontal but will be inclined. By moving the lens longitudinally one can know when one is going through the focus by observing the horizontal fringes. The sequence of the fringe patterns in passing through the focus is like that shown in Fig. 4.5.

Another application of the parallel plate interferometer is for the determination of inhomogeneity of solid glass samples. The sample may be prepared in the form of a parallel piece of glass and sandwiched between two very good plane plates, using a suitable oil to contact. This sandwich is placed between the collimating lens and the shearing plate, as shown in Fig. 4.36. Since a good

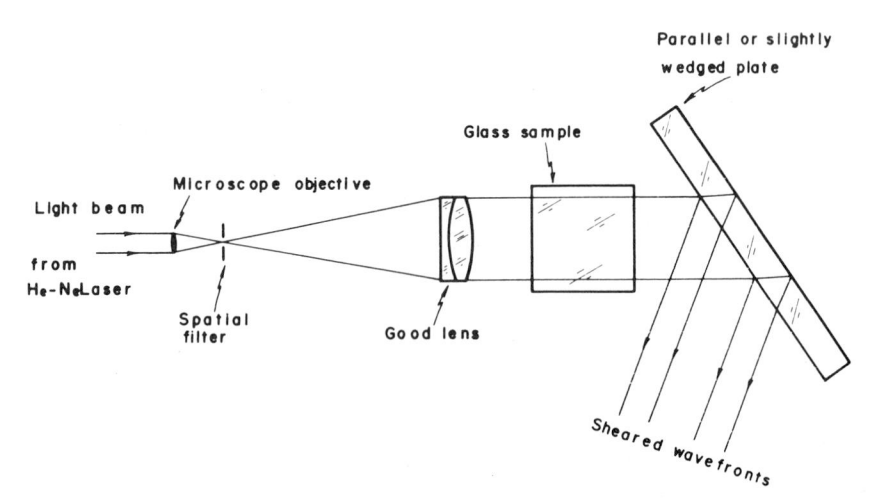

Figure 4.36. Schematic arrangement of a parallel plate interferometer for testing the homogeneity of a glass sample.

4.7. LATERAL SHEARING INTERFEROMETERS USING LASERS

plane wavefront is passing through the sample, any inhomogeneity distorts this wavefront, which when laterally sheared shows the inhomogeneity. This method can very easily reveal variation of the refractive index inside the material. The same region between the collimating lens and the shear plate can be used for flow studies, diffusion studies, and other research.

This application of the parallel plate interferometer for testing the collimation or alignment of lenses has been studied by Dickey and Harder (1978), Grindel (1986), and Sirohi and Kothiyal (1987a, 1987b). Lens parameters, like the focal length or the refractive index, may also be measured with this interferometer (Kasana and Rosenbruch 1983a, 1983b; Murty and Shukla 1983).

The evaluation of infrared materials has been done with this kind of interferometer, using a CO_2 laser, a parallel plate made out of ZnS, and a phosphor screen to observe the fringes (Venkata and Judal 1987).

Another application of the parallel plate interferometer involves the measurement of the surface imperfections of a large concave spherical or aspherical surface, with the instrument shown in Fig. 4.37, designed by Malacara (1965a). The system contains two well-corrected lenses, the first of which collimates the beam well. The shearing plate is located between the two lenses. An additional (unpublished) application of this instrument by Murty and Shukla for the measurement of the radius of curvature of a spherical concave surface is shown in Fig. 4.37. For concave surfaces it is necessary to move the second lens toward the right so that its focus coincides with the vertex of the concave mirror. Then the lens is moved left so that its focus coincides with the center of curvature of the concave mirror. In both settings the wavefront reflected back is a plane, and consequently one sees either a blank field or a field containing horizontal fringes, depending on whether a perfect parallel plate or a wedged plate is used. The displacement of the lens between these two positions must be accurately measured by the use of any of several techniques. Figure 4.38 shows the corre-

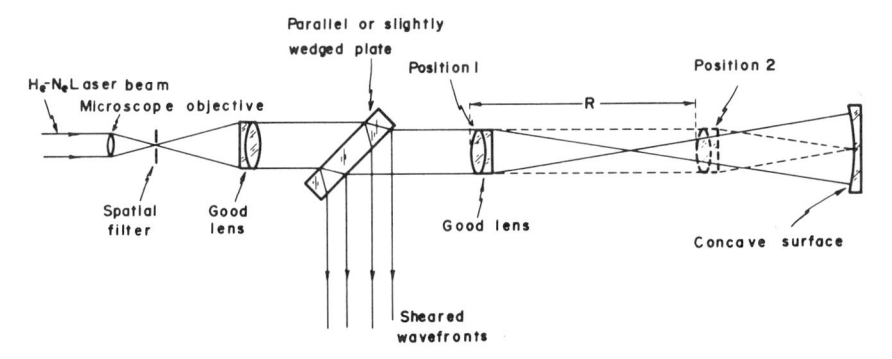

Figure 4.37. Schematic arrangement of a parallel plate interferometer for measuring the radius of curvature of a concave spherical surface.

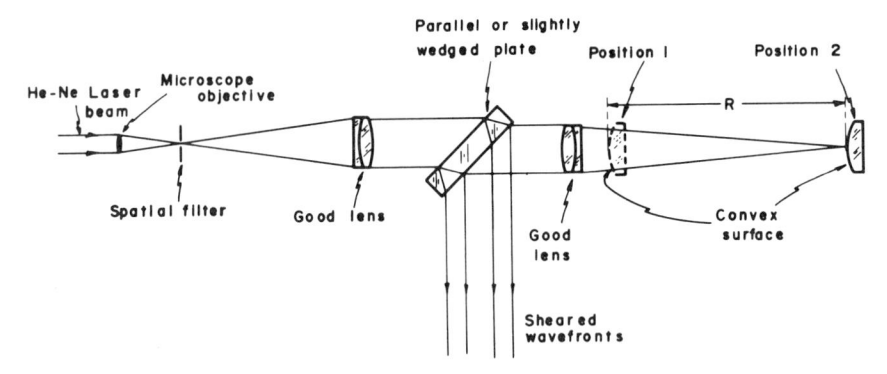

Figure 4.38. Schematic diagram of a parallel plate interferometer for measuring the radius of curvature of a convex spherical surface.

sponding arrangement for convex surfaces. In this case the longest radius of curvature that can be measured is slightly less than the focal length of the second lens. The collimating lenses in these applications have to be very good, and several designs have been suggested by Malacara (1965b), among others.

Another application of the interferometer is for the determination of the refractive index of nearly parallel plates of glass or liquids contained in a parallel-sided glass cell. The basic arrangement is similar to the one shown in Fig. 4.34 in which we use two well-corrected lenses to obtain the sharp and well-corrected focal point. The second lens is adjusted for the two positions corresponding to the two retroreflecting situations from the two faces of the parallel plate as shown in Fig. 4.39. The difference between the two positions is t/N where t is the thickness of the plate and N is the refractive index. If the thickness of the plate is independently measured, the refractive index is determined easily. It is possible to determine the refractive index to the fourth decimal place without much difficulty. Higher accuracies may involve taking into account the spherical aberration of the parallel plate and also controlling the temperature within narrow limits.

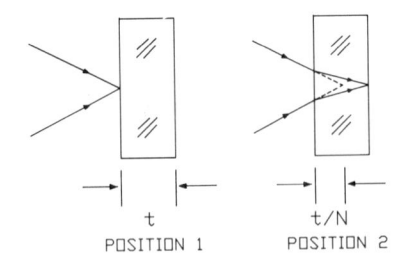

Figure 4.39. Schematic for the determination of the refractive index of a parallel plate.

4.7. LATERAL SHEARING INTERFEROMETERS USING LASERS

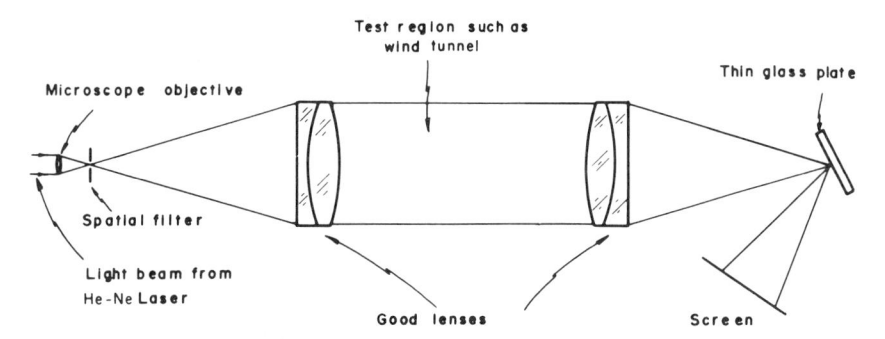

Figure 4.40. Lateral shearing interferometer utilizing a thin parallel plate at the point of convergence.

A very small, thin parallel plate (Tanner 1965) can be used at the focal spot of the system, as shown in Fig. 4.40. Laser light is collimated by the first lens, and after leaving sufficient space in the collimated region, the light is again focused by the second lens. At a position very close to the focus, a thin glass plate is set at about a 45° angle of incidence. As can be easily seen from Fig. 4.41, it is not possible to obtain pure lateral shear because of the longitudinal separation between the two images, seen as reflections from the two surfaces of the plate. The result is that, even for a well-corrected optical system, the lateral shear fringes are slightly curved, as shown in Fig. 4.42. Ideally this system requires an extremely thin plate of glass with a wedge between the surfaces. Alternatively, the system may be made in the form of an air wedge. Even

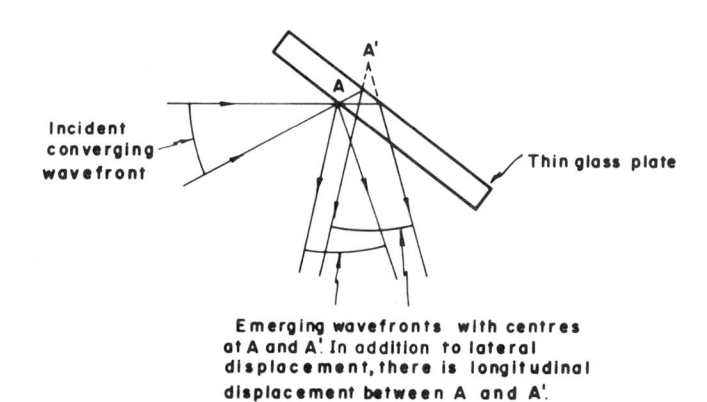

Figure 4.41. Ray diagram indicating that pure lateral cannot be obtained when the thin glass plate is located at the focus.

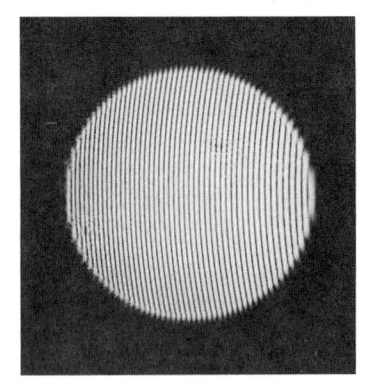

Figure 4.42. Typical appearance of the lateral shear fringes in interferometer shown in Fig. 4.39.

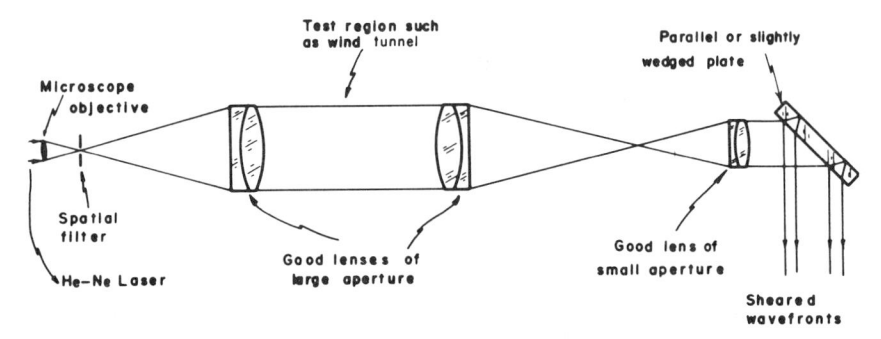

Figure 4.43. Schematic arrangement of the parallel plate interferometer for large aperture wind-tunnel applications.

then only a very limited amount of lateral shear is obtained, and it is better to use a system designed for large aperture wind tunnel applications, homogeneity measurements, and so forth, as shown in Fig. 4.43.

The parallel plate interferometer may be modified to obtain lateral shear directly in the divergent or convergent beam (Malacara et al. 1975), as shown in Fig. 4.44. However, there is always a certain amount of radial shear, along with lateral shear, in this modification, which hence has probably only limited applications.

The parallel plate interferometer has been modified by Schwider (1980) in order to make possible the fringe observation using white light. His method, called superposition fringe, is based on the chromatic compensation by means of a Fabry–Perot interferometer, placed in the collimated beam entering the shearing plate. This shearing plate is really formed by two plates, like in Fig. 4.35.

4.8. OTHER TYPES OF LATERAL SHEARING INTERFEROMETERS   159

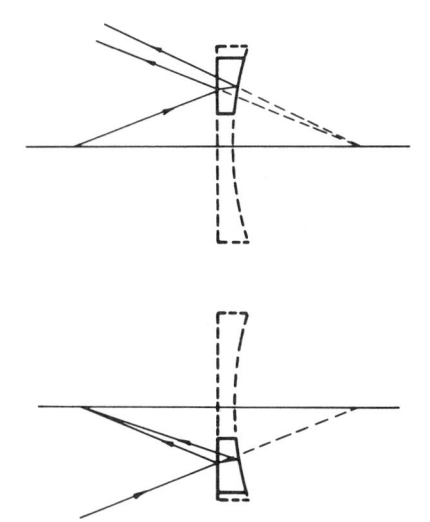

Figure 4.44. Modification of the parallel plate interferometer for convergent or divergent wavefronts.

4.8. OTHER TYPES OF LATERAL SHEARING INTERFEROMETERS

We have considered so far only interferometers in which a semireflecting surface acts as a beam divider to give amplitude division. There are several other types of interferometers to obtain lateral shear, in which other optical principles are used. We treat these briefly here because some of them are discussed in more detail in other chapters.

4.8.1. Lateral Shearing Interferometers Based on Diffraction

Let us consider a transmission type diffraction grating with a periodicity of d. Let us assume that a convergent beam of light is incident on the grating in such a way that the central ray of the beam is perpendicular to the grating and the converging focus of the beam coincides with the grating plane. Let the cone angle of the converging beam of light be 2α. Using the usual formula for the diffraction grating, we can express θ, the angle of diffraction of the central ray in first order, as

$$\sin \theta = \frac{\lambda}{d}. \tag{4.31}$$

If the value of d is chosen properly, the zero-order and first-order beams appear as intersecting circles, as shown in Fig. 4.45. Of course, higher-order beams

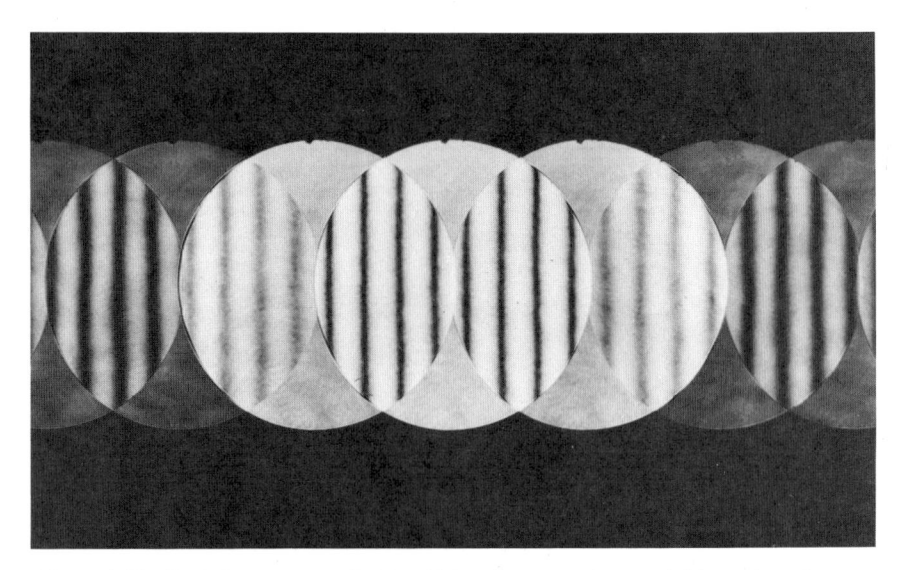

Figure 4.45. Typical appearance of a Ronchi interferogram when a suitable grating of proper spacing is chosen for the Ronchi grating interferometer. The two first-order beams just touch each other and pass through the center of the zero-order beam.

will also be present, but the higher the order the weaker is the beam irradiance. It is, however, possible to obtain gratings in which only zero-order and first-order beams occur. It is presumed that the aberration of the original wavefront is preserved in the zero-order and first-order beams; this is true when moderate angles of diffraction are used. Also, to avoid confusion, it is necessary that the two first-order beams not overlap. Hence, to satisfy this condition, we should have

$$\theta \geq \alpha \qquad (4.32)$$

which means that

$$d \leq 2\lambda(f \text{ number}). \qquad (4.33)$$

In the limit one would like to have the beams just touching each other as shown in Fig. 4.45, and in this case the equality sign is to be considered in expression (4.33). Therefore the choice of grating depends on the f number of the system. For instance, taking a system of f number = 5 and $\lambda = 0.5$ μm, we get 5 μm for the value of d. Hence a grating having 200 rulings per millimeter must be used in this situation. Another point to be mentioned in connection with the grating interferometer is that it is white light compensated, and hence in prin-

4.8. OTHER TYPES OF LATERAL SHEARING INTERFEROMETERS

ciple ordinary sources of light can be employed provided that a pinhole of proper size has been used to obtain a small effective source. Of course it is also possible to use this grating interferometer with helium–neon laser source, which is quite convenient. The gratings can be made to have various periodicities by recording interference fringes between plane wavefronts obtained from a laser source. The angle between the plane wavefronts is changed to obtain different periodicities. Then these gratings can be used for the purpose described earlier. This type of interferometer is due to Ronchi (1923), who also gave an extensive review of the subject (Ronchi 1964).

As can be seen from the foregoing description, the convenient amount of lateral shear is roughly equal to half the diameter of the beam. Hence, with the single grating, it is not possible to obtain a lesser amount of lateral shear without confusion. However, by the use of a double-frequency grating devised by Wyant (1973), any small amount of shear can be obtained. The grating has two distinct frequencies recorded on it. The lower frequency is chosen so that the zero and first orders are distinctly separated, and the higher frequency part of the grating gives two first-order beams that are sheared with respect to the other first-order beams. The appearance of the beams is shown in Fig. 4.46. By inserting another grating identical with the original one and adjusting it to be orthogonal to the first grating, it is possible to obtain a simultaneous shear pattern in both tangential and sagittal directions, as shown in Fig. 4.47. Figure 4.48 shows a typical lateral shearing pattern of a converging wavefront obtained when two identical double-frequency gratings are used orthogonal to each other and in contact. It is also possible to record the two orthogonal double-frequency gratings on the same photographic plate.

A technique by Rimmer and Wyant (1975) to obtain variable shear utilizes two crossed gratings of the same frequency. In this case a small rotation of one grating gives rise to variable shear. Figure 4.49 shows the appearance of the shearing patterns, which may be compared to those of Fig. 4.47. This system

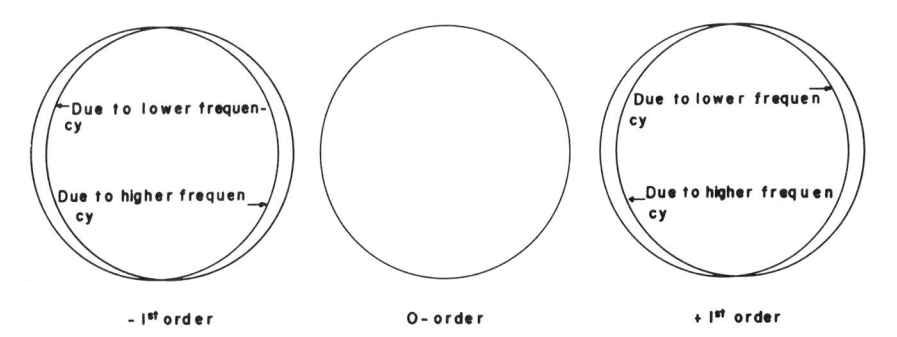

Figure 4.46. Typical appearance of the interferogram and the diffracted apertures when a double-frequency grating is used.

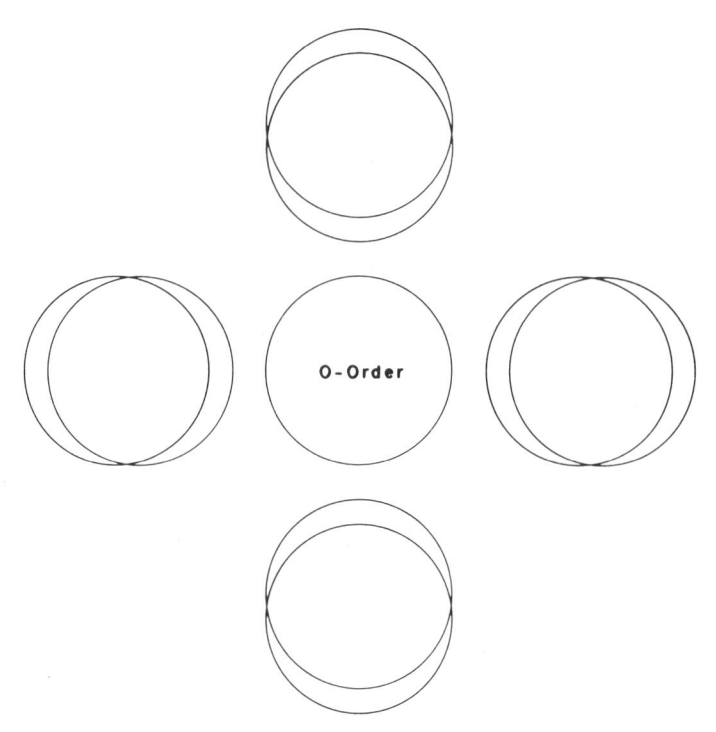

Figure 4.47. Appearance of the lateral shearing interferogram when two double-frequency gratings, orthogonal to each other, are used, so that lateral shear is obtained simultaneously in the sagittal and tangential directions.

is very useful when it is desirable to vary the shear. This principle was later applied to obtain a null test of an aspheric surface by Malacara and Mallick (1976).

A variation of the double-frequency grating interferometer, using two slightly displaced off-axis zone plates, has been designed by Joenathan et al. (1984).

A lateral shearing interferometer with a continuously varying amount of lateral shear has been described by Schwider (1984). The shear is obtained with two plane diffraction gratings in a collimated beam, one after the other, forming an angle between them. The interference takes place between the beam with zero order in the first grating and first order in the second grating with the beam with first order in the first grating and zero order in the second grating.

4.8.2. Lateral Shearing Interferometers Based on Polarization

Any birefringent material gives rise to two orthogonally polarized beams when a beam of unpolarized light is incident on it. However, these two orthogonally

4.8. OTHER TYPES OF LATERAL SHEARING INTERFEROMETERS

Figure 4.48. A typical double-frequency grating interferogram. (From Wyant 1973.)

polarized beams do not show any observable interference pattern, since they are mutually incoherent. However, if a beam of polarized light falls on such a material, the resulting orthogonally polarized beams are mutually coherent and can be made to show observable fringes upon interfering. There are many types of polarizing prisms, such as the Wollaston, in which it is possible to obtain two orthogonally polarized beams from a plane polarized incident beam. Many interferometers, some of them of the lateral shearing type, have been devised utilizing such double-image prisms. These are treated in more detail in Chapter 3.

Examples of this kind of interferometer, using a liquid crystal wedge as a polarizing element, have been described by Murty and Shukla (1980) and by Komissaruk and Mende (1981).

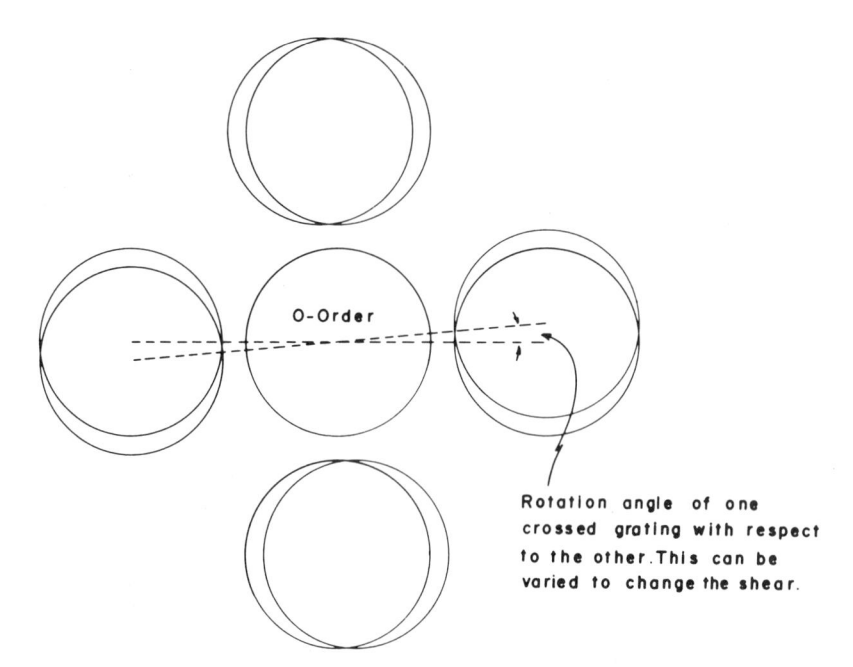

Figure 4.49. Another method of obtaining simultaneous lateral shear in both tangential and sagittal directions. Here two crossed gratings of the same frequency are in contact with each other, and the shear can be varied by rotating one grating with respect to the other. (From Wyant, 1973.)

Another interesting interferometer based on a Babinet compensator, has been described by Saxena (1979), as shown in Fig. 4.50. This interferometer is similar to the one described by Mallick in Chapter 3 on common-path interferometers, using a Wollaston compensator in double pass. As we know, a Babinet compensator is the thinner version of the Wollaston compensator. The lateral shear S introduced by this compensator is given by

$$S = 2R(N_e - N_o)\tan \alpha \tag{4.34}$$

where R is the radius of curvature of the concave surface under test, N_e and N_o are the extraordinary and ordinary refractive indices, respectively, and α is the wedge angle in the compensator. A typical Babinet compensator is made out of quartz ($N_e = 1.553305$ and $N_o = 1.544195$) and with an angle $\alpha = 7.5°$.

If the Babinet compensator between the two polaroids is observed with a microscope, a pattern of straight, parallel, and equidistant fringes is observed, resembling a Ronchi ruling. The two important differences are that, unlike in the Ronchi ruling, in this compensator (a) the pattern profile is sinusoidal and (b) any two contiguous bright lines have their relative phase different by 180°. The Ronchi ruling producing the same interference pattern would have a slit

4.8. OTHER TYPES OF LATERAL SHEARING INTERFEROMETERS

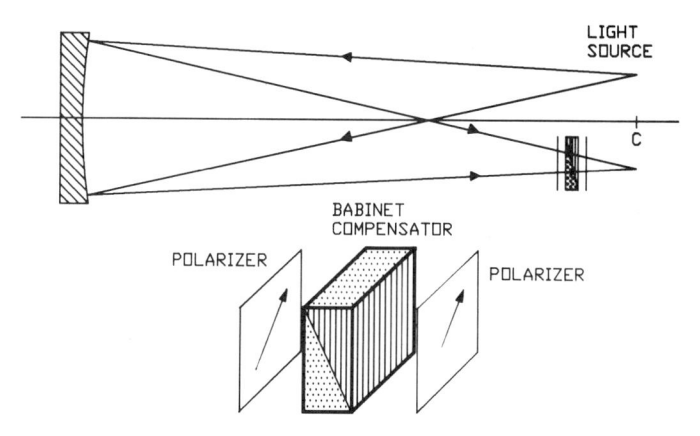

Figure 4.50. Interferometer using a Babinet compensator.

separation equal to the separation between two lines with the same phase. For the compensator just described, this slit separation is equal to 110 lines per inch.

Later, this interferometer was modified by Saxena and Jayarajan (1981) and by Saxena and Lancelot (1982) by using two crossed Babinet compensators. With this modification the sensitivity of the interferometer is doubled. Also, the sensitivity of the interferometer to azimuth variations in the orientation of the compensator is reduced.

The crossed Babinet compensator arrangement is illustrated in Fig. 4.51. The interferometer has two orthogonal lateral shears, produced by each compensator. Thus, the fringe pattern may be described by the following relation:

$$\frac{\partial W}{\partial x} S + \frac{\partial W}{\partial y} T = n\lambda \tag{4.35}$$

where W is the wavefront deformation, including defocusing and tilts, and S and T are the two orthogonal shears.

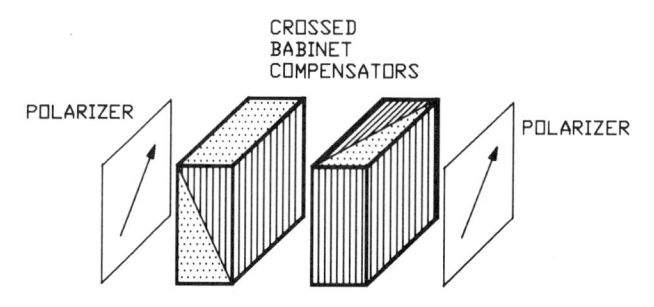

Figure 4.51. Interferometer using two crossed Babinet compensators.

The two compensators may be placed in close contact to each other, but they may also be separated by a small distance d. In this case the defocusing term will be different for each one of the two orthogonal lateral shears. Then, if the defocusing is referred to the point in the middle between the two compensators, the fringe pattern becomes

$$\left(\frac{\partial W}{\partial x} + \frac{d}{2R^2}x\right)S + \left(\frac{\partial W}{\partial y} - \frac{d}{2R^2}y\right)T = n\lambda. \tag{4.36}$$

In this case the fringes for spherical aberration become ⌐⌐ shaped as in Fig. 4.7.

REFERENCES

Dickey, F. M. and T. M. Harder, "Shearing Plate Optical Alignment," *Opt. Eng.*, **17**, 295 (1978).

Dutton, D., A. Cornejo, and M. Latta, "A Semiautomatic Method for Interpreting Shearing Interferograms," *Appl. Opt.*, **7**, 125 (1968).

Gorshkov, V. A. and V. G. Lysenko, "Study of Aspherical Wavefronts on a Lateral Shearing Interferometer," *Sov. J. Opt. Technol.*, **47**, 689 (1980).

Grindel, M. W., "Testing Collimation Using Shearing Interferometry," *Proc. SPIE*, **680**, 44 (1986).

Hariharan, P., "Simple Laser Interferometer with Variable Shear and Tilt," *Appl. Opt.*, **14**, 1056 (1975).

Hariharan, P., "Lateral and Radial Shearing Interferometers: A Comparison," *Appl. Opt.*, **27**, 3594 (1988).

Hariharan, P. and D. Sen, "Cyclic Shearing Interferometer," *J. Sci. Instrum.*, **37**, 374 (1960).

Joenathan, C., R. K. Mohanty, and R. S. Sirohi, "Lateral Shear Interferometry with Holo Shear Lens," *Opt. Commun.*, **52**, 153 (1984).

Kanjilal, A. K., P. N. Puntambekar, and D. Sen, "Compact Cyclic Shearing Interferometer: Part One," *Opt. Laser Technol.*, **16**, 261 (1984).

Kanjilal, A. K. and P. N. Puntambekar, "Compact Cyclic Shearing Interferometer: Part Two," *Opt. Laser Technol.*, **16**, 311 (1984).

Kasana, R. S. and K. J. Rosenbruch, "Determination of the Refractive Index of a Lens Using the Murty Shearing Interferometer," *Appl. Opt.*, **22**, 3526 (1983a).

Kasana, R. S. and K. J. Rosenbruch, "The Use of a Plane Parallel Glass Plate for Determining the Lens Parameters," *Opt. Commun.*, **46**, 69 (1983b).

Kelsall, D., Thesis, University of London, 1959; *Proc. Phys. Soc.*, **73**, 465 (1959).

Komissaruk, V. A. and N. P. Mende, "A Polarization Interferometer with Simplified Double-Refracting Prisms," *Opt. Laser Technol.*, **13**, 151 (1981).

Korwan, D., "Lateral Shearing Interferogram Analysis," *Proc. SPIE*, **429**, 194 (1983).

Lenouvel, L. and F. Lenouvel, "Etude des Faisceaux Convergents" (Convergent Beams Study), *Rev. Opt. Theor. Instrum.*, **17**, 350 (1938).

Malacara, D., Testing of Optical Surfaces, Ph.D. Thesis, Institute of Optics, University of Rochester, New York, 1965a.

Malacara, D., "Two Lenses to Collimate Red Laser Light," *Appl. Opt.*, **4**, 1652 (1965b).

Malacara, D. and S. Mallick, "Holographic Lateral Shear Interferometer," *Appl. Opt.*, **15**, 2695 (1976).

Malacara, D. and M. Mendez, "Lateral Shearing Interferometry of Wavefronts Having Rotational Symmetry," *Opt. Acta.*, **15**, 59 (1968).

Malacara, D., A. Cornejo, and M. V. R. K. Murty, "A Shearing Interferometer for Convergent or Divergent Beams," *Bol. Inst. Tonantzintla*, **1**, 233 (1975).

Murty, M. V. R. K., "The Use of a Single Plane Parallel Plate as a Lateral Shearing Interferometer with a Visible Gas Laser Source," *Appl. Opt.*, **3**, 531 (1964a).

Murty, M. V. R. K., "Some Modifications of the Jamin Interferometer Useful in Optical Testing," *Appl. Opt.*, **3**, 535 (1964b).

Murty, M. V. R. K., "Fabrication of Fixed Shear Cube Type Shearing Interferometer," *Bull. Opt. Soc. India*, **3**, 55 (1969).

Murty, M. V. R. K., "A Compact Lateral Shearing Interferometer Based on the Michelson Interferometer," *Appl. Opt.*, **9**, 1146 (1970).

Murty, M. V. R. K. and D. Malacara, "Some Applications of the Gas Laser as a Source of Light for the Testing of Optical Systems," Proceedings of the Conference on Photographic and Spectroscopic Optics, Tokyo and Kyoto, 1964, *Jap. J. Appl. Phys.*, **4**, Suppl. 1, 106 (1965).

Murty, M. V. R. K. and R. P. Shukla, "Liquid Crystal Wedge as a Polarizing Element and its Use in Shearing Interferometry," *Opt. Eng.*, **19**, 113 (1980).

Murty, M. V. R. K. and R. P. Shukla, "Parallel Plate Interferometer for the Precise Measurement of Radius of Curvature of a Test Plate & Focal Length of a Lens System," *Ind. J. Pure Appl. Phys.*, **21**, 587 (1983).

Nyssonen, D. and J. M. Jerke, "Lens Testing with a Simple Wavefront Shearing Interferometer," *Appl. Opt.*, **12**, 2061 (1973).

Rimmer, M. P., "A Method for Evaluating Lateral Shearing Interferograms," *Itek Corp. Internal Report* No. 72-5802-1, 1972.

Rimmer, M. P. and J. C. Wyant, "Evaluation of Large Aberrations Using a Lateral-Shear Interferometer Having Variable Shear," *Appl. Opt.*, **14**, 142 (1975).

Ronchi, V., *Ann. Sc. Norm. Super. Pisa*, **15** (1923).

Ronchi, V., "Forty Years of History of a Grating Interferometer," *Appl. Opt.*, **3**, 437 (1964).

Saunders, J. B., "Measurement of Wavefronts without a Reference Standard I: The Wavefront Shearing Interferometer," *J. Res. Nat. Bur. Stand.*, **65B**, 239 (1961).

Saunders, J. B., "Wavefront Shearing Prism Interferometer," *J. Res. Nat. Bur. Stand.*, **68C**, 155 (1964a).

Saunders, J. B., "Interferometer Test of the 26-Inch Refractor at Leander McCormick Observatory," *Astron. J.*, **69**, 449 (1964b).

Saunders, J. B., "Some Applications of the Wavefront Shearing Interferometer," Proceedings of the Conference on Photographic and Spectroscopic Optics, Tokyo and Kyoto, 1964, *Jap. J. Appl. Phys.*, **4**, Suppl. 1, 99 (1965).

Saunders, J. B., "A Simple Interferometric Method for Workshop Testing of Optics," *Appl. Opt.*, **9**, 1623 (1970).

Saunders, J. B. and R. J. Bruning, "A New Interferometric Test and Its Applications to the 84-Inch Reflecting Telescope at Kitt Peak National Observatory," *Astron. J.*, **73**, 415 (1968).

Saxena, A. K., "Quantitative Test for Concave Aspheric Surfaces Using a Babinet Compensator," *Appl. Opt.*, **18**, 2897 (1979).

Saxena, A. K. and A. P. Jayarajan, "Testing Concave Aspheric Surfaces: Use of two Crossed Babinet Compensators," *Appl. Opt.*, **20**, 724 (1981).

Saxena, A. K. and J. P. Lancelot, "Theoretical Fringe Profiles with Crossed Babinet Compensators in Testing Concave Aspheric Surfaces," *Appl. Opt.*, **21**, 4030 (1982).

Schwider, J., "Superposition Fringe Shear Interferometry," *Appl. Opt.*, **19**, 4233 (1980).

Schwider, J., "Continuous Lateral Shearing Interferometer," *Appl. Opt.*, **23**, 4403 (1984).

Sirohi, R. S. and M. P. Kothiyal, "Double Wedge Plate Shearing Interferometer for Collimation Test," *Appl. Opt.*, **26**, 4054 (1987a).

Sirohi, R. S. and M. P. Kothiyal, "On Collimation of a Laser Beam," *Proc. SPIE*, **813**, 205 (1987b).

Tanner, L. H., "Some Laser Interferometers for Use in Fluid Mechanics," *J. Sci. Instrum.*, **42**, 834 (1965).

Tarvin, J. A., R. D. Sigler, and G. E. Busch, "Wavefront Shearing Interferometer for Cryogenic Laser-Fusion Targets," *Appl. Opt.*, **18**, 2971 (1979).

Venkata, B. and D. P. Juyal, "A 10 μm CO2 Laser Interferometer Using Shearing Technique," *J. Optics (India)*, **16**, 31 (1987).

Wyant, J. C., "Double Frequency Grating Lateral Shear Interferometer," *Appl. Opt.*, **12**, 2057 (1973).

ADDITIONAL REFERENCES

Adachi, M. and K. Yasaka, "Roughness Measurement Using a Shearing Interference Microscope," *Appl. Opt.*, **25**, 764 (1986).

Ashton, A. and A. C. Marchant, "A Scanning Interferometer for Wavefront Aberration Measurements," *Appl. Opt.*, **8**, 1953 (1969).

Bates, W. J., "A Wavefront Shearing Interferometer," *Proc. Phys. Soc.*, **59**, 940 (1947).

Briers, J. D., "Prism Shearing Interferometer," *Opt. Technol. J.*, **1**, 196 (1969).

Briers, J. D., "Self-Compensation of Errors in a Lateral Shearing Interferometer," *Opt. Commun.*, **4**, 69 (1971).

Briers, J. D., "Interferometric Testing of Optical Systems and Components: A Review," *Opt. Laser Technol.*, **4**, 28 (1972).

ADDITIONAL REFERENCES

Brown, D. S., "A Shearing Interferometer with Fixed Shear and Its Application to Some Problems in the Testing of Astro-Optics," *Proc. Phys. Soc.*, **B67,** 232 (1954).

Brown, D. S., "The Application of Shearing Interferometry to Routine Optical Testing," *J. Sci. Instrum.*, **32,** 137 (1955).

Bryngdahl, O., "Applications of Shearing Interferometry," in *Progress in Optics*, Vol. 4, E. Wolf, Ed., North-Holland, Amsterdam, 1964, Chap. II, p. 39.

De Vany, A. S., "Some Aspects of Interferometric Testing and Optical Figuring," *Appl. Opt.*, **4,** 831 (1965).

De Vany, A. S., "Quasi-Ronchigrams as Mirror Transitive Images of Shearing Interferograms," *Appl. Opt.*, **9,** 1477 (1970).

De Vany, A. S., "Using a Murty Interferometer for Testing the Homogeneity of Test Samples of Optical Materials," *Appl. Opt.*, **10,** 1459 (1971).

De Vany, A. S., "Scanning Murty Interferometer for Optical Testing," *Appl. Opt.*, **11,** 1467 (1972).

Dickey, F. M., J. R. White, R. H. Allen, and R. L. Sledge, "Common Collimator Approach to Alignment and Performance Testing of Laser Ranger Designators and Thermal Imaging Systems," *Opt. Eng.*, **20,** 765 (1981).

Donath, E. and W. Carlough, "Radial Shearing Interferometer," *J. Opt. Soc. Am.*, **53,** 395 (1963).

Drew, R. L., "A Simplified Shearing Interferometer," *Proc. Phys. Soc.*, **B64,** 1005 (1951).

Flack, R. D., Jr., "Shearing Interferometer Inaccuracies due to a Misaligned Test Section," *Appl. Opt.*, **17,** 2873 (1978).

Freischlad, K., "Sensitivity of Heterodyne Shearing Interferometer," *Appl. Opt.*, **26,** 4053 (1987).

Ganesan, A. R., D. K. Sharma, and M. P. Kothiyal, "Universal Digital Speckle Shearing Interferometer," *Appl. Opt.*, **27,** 4731 (1988).

Gates, J. W., "Reverse Shearing Interferometry," *Nature*, **176,** 359 (1955).

Gorshkov, V. A., V. S. Kryakhtunov, and O. N. Fomin, "The Inters Lateral Shearing Interferometer for Inspecting the Surface Shape of Large Optical Components," *Sov. J. Opt. Technol.*, **47,** 77 (1980).

Haberland, E., "Uber Linsenfehler fur shiefe Buschel" (Upon the Lens Aberration with Oblique Bundles), *Z. Phys.*, **24,** 285 (1924).

Hardy, J. W. and A. J. MacGovern, "Shearing Interferometry: A Flexible Technique for Wavefront Measurement," *Proc. SPIE*, **816,** 180 (1987).

Hariharan, P., W. H. Steel, and J. C. Wyant, "Double Grating Interferometer with Variable Lateral Shear," *Opt. Commun.*, **11,** 317 (1974).

Hung, Y. Y. and C. Y. Liang, "Image-Shearing Camera for Direct Measurement of Surface Strains," *Appl. Opt.*, **18,** 1046 (1979).

Joenathan, C., R. K. Mohanty, and R. S. Sirohi, "Fringe Sharpening in Lateral Shear Interferometry," *Opt. Laser Technol.*, **17,** 310 (1985).

Joenathan, C., V. Parthiban, and R. S. Sirohi, "Shear Interferometry with Holographic Lenses," *Opt. Eng.*, **26,** 359 (1987).

Keenan, P. B., "Pseudo-Shear Interferometry," *Proc. SPIE*, **429,** 2 (1983).

Kelly, J. G. and R. A. Hargreaves, "A Rugged Inexpensive Shearing Interferometer," *Appl. Opt.*, **9**, 948 (1970).

Kolyshkina, L. L., S. I. Kromin, and V. N. Shekhtman, "Identification of the Interference Fringes on a Lateral-Shear Interferogram," *Sov. J. Opt. Technol.*, **53**, 617 (1986).

Komissaruk, V. A., "The Displacement Interferogram in the Case of a Wavefront Having Rotational Symmetry," *Sov. J. Opt. Technol.*, **36**, 456 (1969).

Komissaruk, V. A. and V. P. Martynov, "Use of Iceland Spar and Glass Prism in a Polarization Interferometer," *Sov. J. Opt. Technol.*, **47**, 311 (1980).

Kothiyal, M. P. and C. Delisle, "Shearing Interferometer for Phase Shifting Interferometry with Polarization Phase Shifter," *Appl. Opt.*, **24**, 4439 (1985).

Kwon, O. Y., "Infrared Lateral Shear Interferometers," *Appl. Opt.*, **19**, 1225 (1980).

Langenbeck, P., "Improved Collimation Test," *Appl. Opt.*, **9**, 2590 (1970).

Langenbeck, P., "Modifying a Shear Interferometer to Obtain a Neutral Reference Beam," *Appl. Opt.*, **9**, 2590 (1970).

Langenbeck, P., "Modifying a Shear Interferometer to Obtain a Neutral Reference Beam," *J. Opt. Soc. Am.*, **61**, 172 (1971).

Lee, P. H. and G. A. Woolsey, "Versatile Polarization Interferometer," *Appl. Opt.*, **20**, 3514 (1981).

Lewandowski, J., "Lateral Shear Interferometer for Infrared and Visible Light," *Appl. Opt.*, **28**, 2373 (1989).

Lohmann, A. and O. Bryngdahl, "A Lateral Wavefront Shearing Interferometer with Variable Shear," *Appl. Opt.*, **6**, 1934 (1967).

McLaughlin, J. L. and B. A. Horwitz, "Real-Time Snapshot Interferometer," *Proc. SPIE*, **680**, 35 (1986).

Matsuda, K., "Lateral Shear Interferometer Using Twin Three-Beam Holograms," *Appl. Opt.*, **19**, 2643 (1980).

Matsuda, K. and M. Namiki, "Holographic Lateral Shear Interferometer for Differential Interference Contrast Method," *J. Optics (Paris)*, **11**, 81 (1980).

Matsuda, K., "Holographic Lateral Shear Interferometers with Twin Three-Beam Holograms," *Opt. Laser Technol.*, **12**, 305 (1980).

Murthy, K. R., R. S. Sirohi, and M. P. Kothiyal, "Speckle Shearing Interferometry: A New Method," *Appl. Opt.*, **21**, 2865 (1982).

Murthy, Krishna R., R. S. Sirohi, and M. P. Kothiyal, "Speckle Shear Interferometry Using a Split Lens," *J. Optics (India)*, **11**, 9 (1982).

Murty, M. V. R. K., "Interferometry Applied to Testing of Optics," *Bull. Opt. Soc. India*, **1**, 29 (1967).

Murty, M. V. R. K., "A Simple Method of Introducing Tilt in the Ronchi and Cube Type of Shearing Interferometers," *Bull. Opt. Soc. India*, **5**, 1 (1971).

Murugov, V. M., G. P. Okutin, and I. Pankratov, "Shearing Interferometers Having a Wedge-Shaped Plate," *Sov. J. Opt. Technol.*, **49**, 672 (1982).

Ohlidal, I. and K. Navratil, "Analysis of the Basic Statistical Properties of Randomly Rough Curved Surfaces by Shearing Interferometry," *Appl. Opt.*, **24**, 2690 (1985).

Patorski, K., "Heuristic Explanation of Grating Shearing Interferometry Using Incoherent Illumination," *Opt. Acta*, **31**, 33 (1984).

Patorski, K., "Modified Double Grating Shearing Interferometer," *Opt. Appl.*, **14**, 149 (1984).

Patorski, K., "Talbot Interferometry with Increased Shear," *Appl. Opt.*, **24**, 4448 (1985).

Patorski, K., "Grating Shearing Interferometer with Variable Shear and Fringe Orientation," *Appl. Opt.*, **25**, 4192 (1986).

Patorski, K., "Talbot Interferometry with Increased Shear: Further Considerations," *Appl. Opt.*, **25**, 1111 (1986).

Patorski, K., "Generation of the Derivative of Out-of-Plane Displacements Using Conjugate Shear and Moire Interferometry," *Appl. Opt.*, **25**, 3146 (1986).

Patorski, K., "Conjugate Lateral Shear Interferometry and its Implementation," *J. Opt. Soc. Am. A*, **3**, 1862 (1986).

Patorski, K. and A. Ulinowicks, "Conjugate Lateral Shear Interferometry: Further Considerations," *Appl. Opt.*, **26**, 4506 (1987).

Patorski, K., "Talbot Interferometry with Increased Shear: Part 3," *Appl. Opt.*, **27**, 3875 (1988).

Patorski, K., "Shearing Interferometry and the Moire Method for Shear Strain Determination," *Appl. Opt.*, **27**, 3567 (1988).

Saunders, J. B., "Measurement of Wavefronts without a Reference Standard. II: The Wavefront Reversing Interferometer," *J. Res. Nat. Bur. Stand.*, **66B**, 29 (1962).

Saunders, J. B., "A Simple Inexpensive Wavefront Shearing Interferometer," *Appl. Opt.*, **6**, 1581 (1967).

Schwider, J. and R. Burow, "Wave Aberrations Caused by Misalignments of Aspherics, and their Elimination," *Opt. Appl.*, **9**, 33 (1979).

Sen, D. and P. N. Puntambekar, "Shearing Interferometer for Testing Corner Cubes and Right Angle Prisms," *Appl. Opt.*, **5**, 1009 (1966).

Sen, B. and D. Sen, "Interference with Beams Sheared in Orthogonal Axes," *Opt. Laser Technol.*, **17**, 315 (1985).

Seligson, J. L., C. A. Callari, J. E. Greivenkamp, and J. W. Ward, "Stability of a Lateral-Shearing Heterodyne Twyman-Green Interferometer," *Opt. Eng.*, **23**, 353 (1984).

Shaker, C., P. B. Godbole, and B. N. Gupta, "Shearing Interferometry Using Holo-Lenses," *Appl. Opt.*, **25**, 2477 (1986).

Shekhtman, V. N., "Construction of Light Wavefront from Lateral Shearing Interferograms," *Sov. J. Opt. Technol.*, **49**, 601 (1982).

Stumpf, K. D., "Real Time Interferometer," *Proc. SPIE*, **153**, 42 (1978).

Sweatt, W. C., "Rotatable Shear Plate Interferometer," *Opt. Eng.*, **29**, 1157 (1990).

Takezaki, J., S. Toyooka, H. Nishida, and H. Kobayashi, "Automatic Processing of Fringes Obtained by Shearography," *Proc. SPIE*, **813**, 39 (1987).

Thomas, D. A. and J. C. Wyant, "High Efficiency Grating Lateral Shear Interferometer," *Opt. Eng.*, **15**, 477 (1976).

Van Rooyen, E., "Design for a Variable Shear Prism Interferometer," *Appl. Opt.*, **7,** 2423 (1968).

Van Rooyen, E. and V. H. G. Houten, "Design of a Wavefront Shearing Interferometer Useful for Testing Large Aperture Optical Systems," *Appl. Opt.*, **8,** 91 (1969).

Weingardner, I. and H. Stenger, "A Simple Shear-Tilt Interferometer for the Measurement of Wavefront Aberrations," *Optik*, **70,** 124 (1985).

Wyant, J. C., "Interferometer for Measuring Power Distribution of Ophthalmic Lenses," *Itek Corp. Internal Report* No. OLTN 70-5, 1971.

Wyant, J. C., "Use of an AC Heterodyne Lateral Shear Interferometer with Real-Time Wavefront Correction Systems," *Appl. Opt.*, **14,** 2622 (1975).

Wyant, J. C. and F. D. Smith, "Interferometer for Measuring Power Distribution of Ophthalmic Lenses," *Appl. Opt.*, **14,** 1607 (1975).

Yatagai, T. and T. Kanou, "Aspherical Surface Testing with Shearing Interferometer Using Fringe Scanning Detection Method," *Proc. SPIE*, **429,** 136 (1983).

Yatagai, T. and T. Kanou, "Aspherical Surface Testing with Shearing Interferometer Using Fringe Scanning Detection Method," *Opt. Eng.*, **23,** 357 (1984).

5

Radial, Rotational, and Reversal Shear Interferometers

D. Malacara

5.1. INTRODUCTION

Although the most popular shearing interferometer is the lateral shearing instrument, other types are equally useful. In this chapter we examine radial, rotational, and reversal shear interferometers, whose basic wavefront operations are illustrated in Fig. 5.1. Many review papers (Briers 1972; Fouéré and Malacara 1975; Murty 1967) and books (Baird and Hanes 1967; Bryngdahl 1965; Steel 1966) have very good general descriptions of them. The radial shear interferometer produces two interfering wavefronts with identical deformations, but one of the wavefronts is contracted or expanded with respect to the other. The rotational shear interferometer produces two identical wavefronts, with one of them rotated with respect to the other. The reversal shear interferometer produces two wavefronts, in which deformations on one wavefront are symmetrical with respect to those on the other wavefront, with a diameter as axis of symmetry.

For the analysis of these shearing operations we can assume a completely general wavefront function given by (see Appendix 3)

$$W(\rho, \theta) = \sum_{n=0}^{k} \sum_{l=0}^{n} \rho^n (a_{nl} \cos l\theta + b_{nl} \sin l\theta) \tag{5.1}$$

where $(n - l)$ is given (n and l have the same parity) and generally $l \leq n$.

5.2. RADIAL SHEAR INTERFEROMETERS

Radial shear interferometers perform the basic operations illustrated in Fig. 5.2. One of the wavefronts is contracted or expanded with respect to the other. The

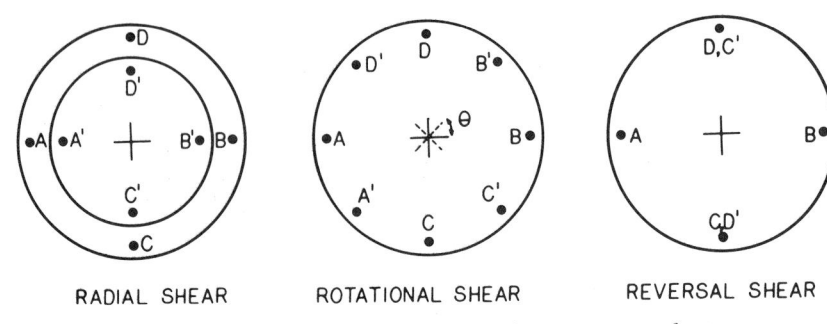

Figure 5.1. Three methods to obtain shear between two wavefronts.

interferometer may be thought of as an optical system producing two images of an object at A, with different magnifications, at the location A'. These two images must coincide at A' if a defocusing term is to be avoided.

As described in Chapter 2 on Twyman–Green interferometers, the interferogram to be analyzed has to be an image of the exit pupil of the system under test, especially when the wavefront deviations from the spherical shape are large. If the wavefront is almost spherical, this condition is not necessary. In the case of the radial shear interferometer, both wavefronts are deformed. Thus, both wavefronts in the interferogram must be images of the pupil. When the shear is very large, only the smaller wavefront needs to be an image of the exit pupil of the system. Fortunately, this condition may frequently be satisfied since, as described by Steel (1984), all radial shear interferometers have a second pair of conjugates, B and B', as shown in Fig. 5.2, with the same shear ratio as the images A and A'. Then, ideally, the exit pupil of the system under test must be located at B and the interferogram should be analyzed at B'.

Steel (1984) also showed that by reversing the direction of the light a radial

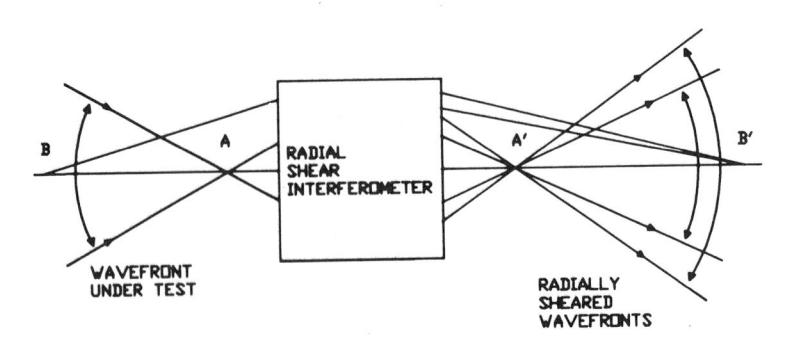

Figure 5.2. Schematic block for a radial shear interferometer.

5.2. RADIAL SHEAR INTERFEROMETERS

shear interferometer with the same shear is also obtained. Thus, any system may be used in four ways to produce the same shear.

These types of interferometers directly represent the wavefront deformations when the shear is large, but even if the shear is small, they are simpler to interpret than lateral shear interferometers, and unlike them, they provide information in all directions, not only in one. Then, only one interferogram is needed. Radial and lateral shear interferometers have been compared in detail by Hariharan (1988).

The lower sensitivity of radial shear interferometers makes them ideal for testing wavefronts with a high degree of asphericity because the number of fringes is smaller, instead of the typical approach of using a long-wavelength or two-wavelength interferometry.

When the surface under test has a central hole, the shear has to be small, otherwise the smaller of the two wavefronts would be over the hole of the larger wavefront. This small shear reduces the sensitivity, but this disadvantage may be compensated for by the use of phase shifting techniques (Hariharan et al. 1986).

We can assume that one of the wavefronts is contracted in the ratio ($S_c \leq 1$) defined by

$$S_c = \frac{\rho}{\rho'} \tag{5.2}$$

where ρ is the ratio of the radial distance of a point in the interference pattern to the maximum radius of the uncontracted wavefront. Similarly, ρ' is the ratio of the same radial distance to the maximum radius of the contracted wavefront. Then, for the contracted wavefront, we can write

$$W(\rho', \theta) = \sum_{n=0}^{k} \sum_{i=0}^{n} \rho'^n (a_{nl} \cos l\theta + b_{nl} \sin l\theta). \tag{5.3}$$

In a similar manner, if the other wavefront is expanded in the ratio ($S_e \geq 1$) given by

$$S_e = \frac{\rho}{\rho''}, \tag{5.4}$$

we can write

$$W(\rho'', \theta) = \sum_{n=0}^{k} \sum_{l=0}^{n} \rho''^n (a_{nl} \cos l\theta + b_{nl} \sin l\theta), \tag{5.5}$$

and hence the interference pattern will be given by the optical path difference (OPD):

$$\text{OPD} = W(\rho', \theta) - W(\rho'', \theta)$$

$$= \sum_{n=0}^{k} \sum_{l=0}^{n} (1 - R^n)\rho'^n(a_{nl} \cos l\theta + b_{nl} \sin l\theta) \quad (5.6)$$

where the effective radial shear R is defined as

$$R = \frac{S_c}{S_e}. \quad (5.7)$$

The sensitivity σ of a radial shear interferometer relative to that of a Twyman–Green interferometer is given by

$$\sigma = \frac{d\text{OPD}/d\rho'}{dW(\rho', \theta)/d\rho'}$$

$$= \frac{\sum_{n=0}^{k} \sum_{l=0}^{n} n\rho'^{n-1}(a_{nl} \cos l\theta + b_{nl} \sin l\theta)(1 - R^n)}{\sum_{n=0}^{k} \sum_{l=0}^{n} n\rho'^{n-1}(a_{nl} \cos l\theta + b_{nl} \sin l\theta)}. \quad (5.8)$$

but if only one aberration (n, l) is present, the relative sensitivity can be expressed as

$$\sigma_{n,l} = 1 - R^n \quad (5.9)$$

and is plotted in Fig. 5.3 for some aberrations. We can see that a moderate effective radial shear R equal to 0.5 gives a very high relative sensitivity. It should be noted that the analysis here is related to the ρ' scale, since this is the scale of the interference fringe pattern.

If the expansion of one of the wavefronts becomes infinite ($S_e \to \infty$, $R \to 0$), we have a kind of radial shear interferometer that is said to have exploded shear. This topic, however, is treated in Chapter 3 on common-path interferometers.

Radial shear interferograms are basically identical to the Twyman–Green interferograms studied in Chapter 2, especially if only pure aberrations are present. Of special interest are wavefronts with rotational symmetry because they are obtained during the testing of aspherical rotationally symmetric surfaces, for example, an astronomical mirror. The procedure for computing the wave-

5.2. RADIAL SHEAR INTERFEROMETERS

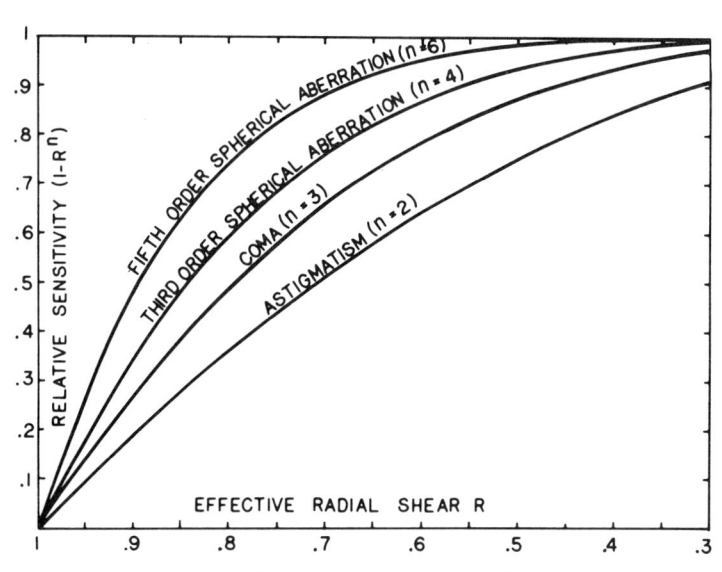

Figure 5.3. Relative sensitivity σ versus radial shear R.

front from a radial shear interferogram under these conditions has been given by Malacara (1974) and by Honda et al. (1987).

However, in general the wavefront is not rotationally symmetric, and even some lateral shear may be present. Then, the wavefront may be computed by an iterative procedure as described by Kohler and Gamiz (1986). In the first iteration the reference wavefront is assumed to be perfectly flat, and the interferogram is sampled and fitted to a polynomial. The result of this evaluation is then used in the second iteration, with a better estimation of the reference wavefront. This procedure produces a very accurate result, limited only by the sample spacing, reading errors, and the quality of the wavefront fitting.

5.2.1. Single-Pass Radial Shear Interferometers

In this kind of interferometer, the light goes through the instrument only once and two radially sheared interfering wavefronts are produced, as illustrated in Fig. 5.2. If the interferometer is illuminated by a small, circular extended source, the degree of coherence g_{12} between any two points on the wavefront is given (Hariharan and Sen 1961a; Murty 1964b) by

$$g_{12} = \frac{2J_1[(2\pi/\lambda)\alpha d]}{(2\pi/\lambda)\alpha d} \qquad (5.10)$$

where 2α is the angular diameter of the source as seen from the wavefront under consideration and d is the distance between the two points on the same wavefront.

For a radial shear interferometer the distance d between two points on the wavefront that interfere at a point on the interference pattern can be shown to be given by

$$d = \rho' - \rho'' = \rho'(1 - R). \tag{5.11}$$

Therefore we obtain for those two points (Hariharan and Sen 1961a)

$$g_{12} = \frac{2J_1[(2\pi/\lambda)(1 - R)\rho'\alpha]}{(2\pi/\lambda)(1 - R)\rho'\alpha}. \tag{5.12}$$

Since the visibility of the pattern is directly proportional to g_{12}, a fringe pattern like that in Fig. 5.4 will be obtained. The first minimum on the visibility is obtained when the argument x of $J_1(x)$ is equal to 1.22π. Hence, for good visibility over the whole pattern, the circular light source must have an angular semiconductor α, as seen from the wavefront (entrance pupil of the interferometer), smaller than a certain value given by

$$\alpha \leq \frac{1.22\lambda}{(1 - R)D} \tag{5.13}$$

where D is the entrance pupil diameter.

A radial shear interferometer is said to be compensated for the imperfect monochromaticity of the light (wavelength bandwidth) when the optical paths for the two interfering beams involve the same glass and air paths. Interferometers of this kind will now be discussed in this section.

Radial shear interferometers were first considered by Brown (1959), who

Figure 5.4. Fringe visibility changes in a radial shear interferometer with a large source. (From Hariharan and Sen 1961a.)

5.2. RADIAL SHEAR INTERFEROMETERS

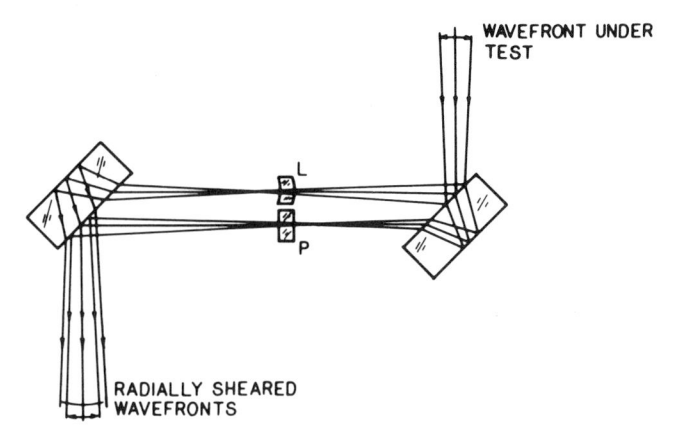

Figure 5.5. Brown's radial shear interferometer.

described the instrument shown in Fig. 5.5 (Brown 1962). Basically, this is a Jamin interferometer, but it uses convergent light and has a small meniscus lens in one of the beams. A compensating parallel plate is placed in the other beam.

Another of the first radial shear interferometers, designed by Hariharan and Sen (1961b), is shown in Fig. 5.6. It consists of a plane parallel beam splitter plate P and two plane mirrors M_1 and M_2. The radial shear is produced by two lenses L_1 and L_2, which are placed in such a manner that their foci are at the

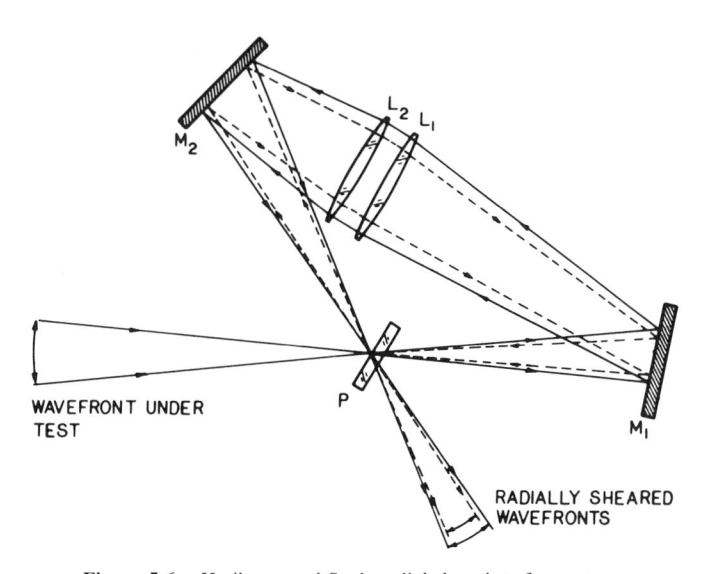

Figure 5.6. Hariharan and Sen's radial shear interferometer.

face of the beam splitter. The radial shear is produced when the two lenses have different focal lengths f_1 and f_2 ($f_2 > f_1$), and it is given by

$$R = \frac{f_1}{f_2}. \tag{5.14}$$

The two lenses could be replaced by a single lens, but two lenses serve to facilitate the elimination of aberrations produced by the lenses.

The visibility is equal to 1 only if the two beams have the same irradiance; but since the two beams are differently expanded, this is possible only if the reflectance $\mathcal{R}$ and the transmittance $\mathcal{T}$ of the beam splitter satisfy the equation

$$\frac{\mathcal{R}}{\mathcal{T}} = \left(\frac{f_2}{f_1}\right)^2. \tag{5.15}$$

Hariharan and Sen (1962) successfully used this interferometer to test microscope objectives.

Murty (1964a) suggested several arrangements to produce radial shear. One of them is based on the Mach–Zehnder interferometer, using telescopic systems S_1 and S_2 on each of the arms, as shown in Fig. 5.7. The effective radial shear F is given by $1/M^2$, where M is the magnification of each of the telescopes.

Another system is a cyclical interferometer (shown in Fig. 5.8) that resem-

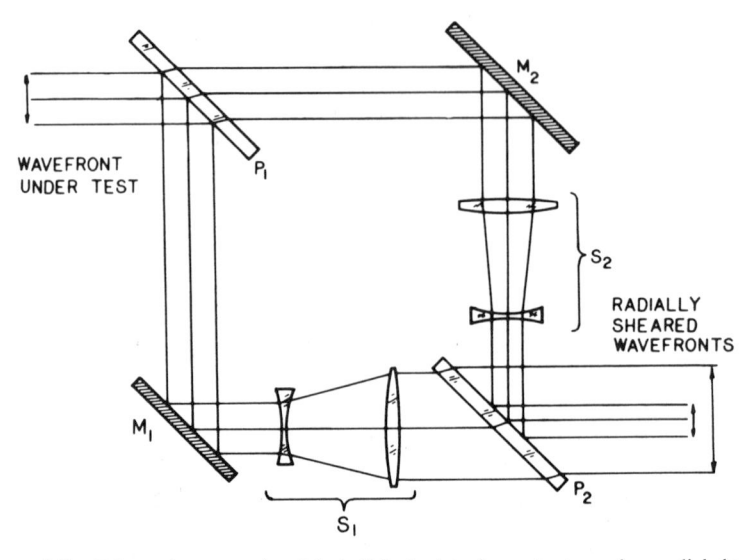

Figure 5.7. Telescopic systems in a Mach–Zehnder interferometer to produce radial shear.

5.2. RADIAL SHEAR INTERFEROMETERS

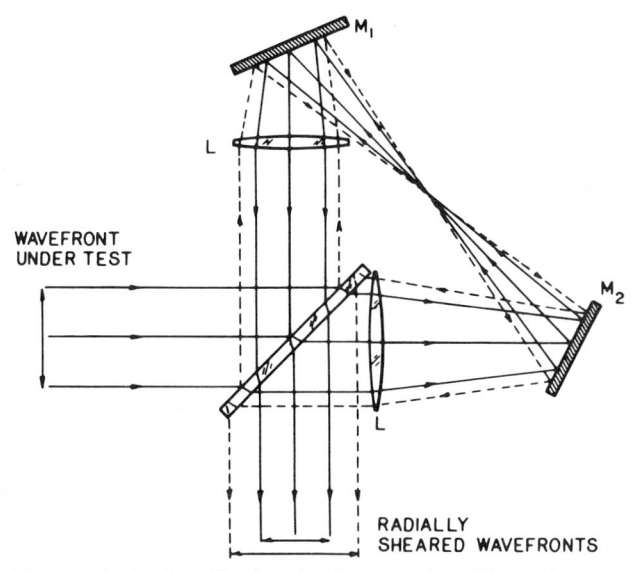

Figure 5.8. Cyclic radial shear interferometer for collimated beams.

bles the interferometer designed by Hariharan and Sen. Here, however, the light entering the interferometer should be collimated.

A very practical and interesting interferometer, also described by Murty (1964a), is based on the contraction and expansion of the numerical aperture by a hemispherical lens, as illustrated in Fig. 5.9. Using this principle and the basic cyclic configuration, Murty designed the interferometers shown in Figs.

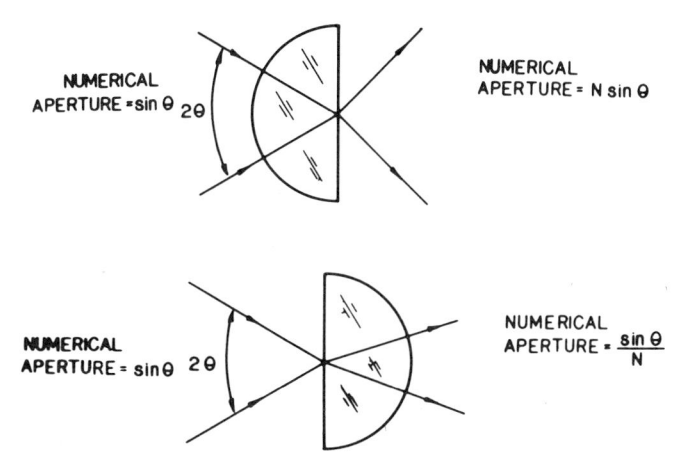

Figure 5.9. Contraction and expansion of the numerical aperture by a hemispherical lens.

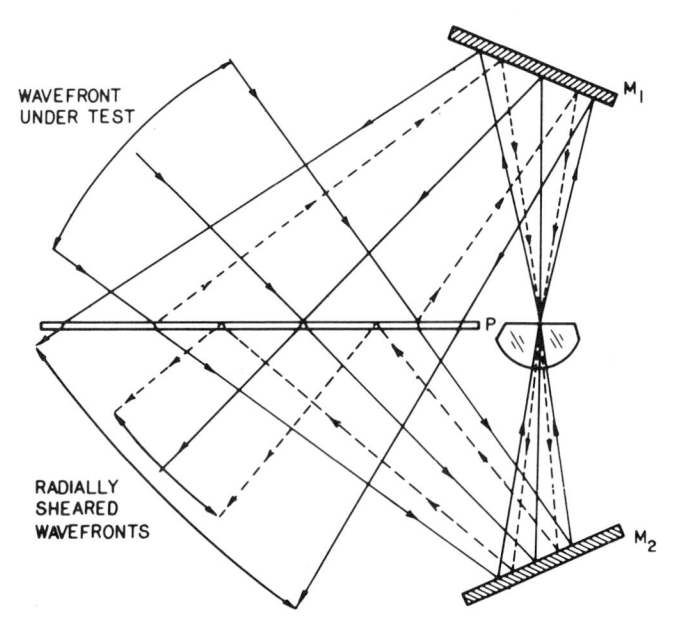

Figure 5.10. Murty's cyclic radial shear interferometer.

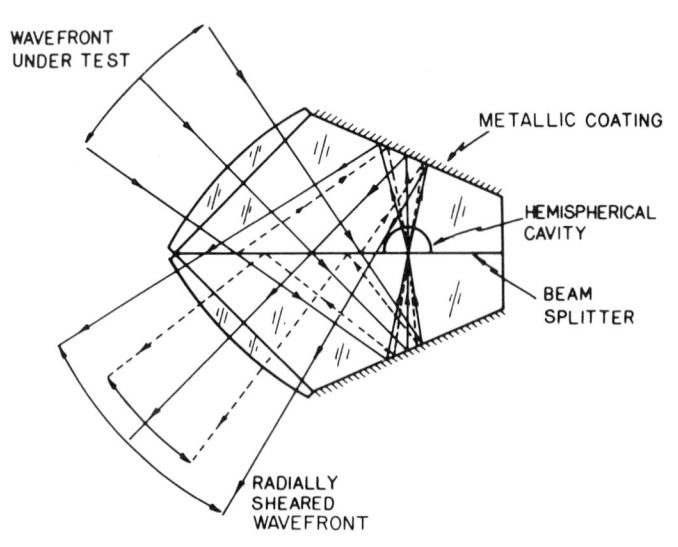

Figure 5.11. Murty's solid radial shear interferometer.

5.2. RADIAL SHEAR INTERFEROMETERS

5.10 and 5.11. The hemispherical cavity in the second interferometer may be emptied or filled with oil, in order to obtain the desired radial shear. Some very unconventional radial shear interferometers with discontinuous stepping wavefronts have been described by Bryngdahl (1970, 1971).

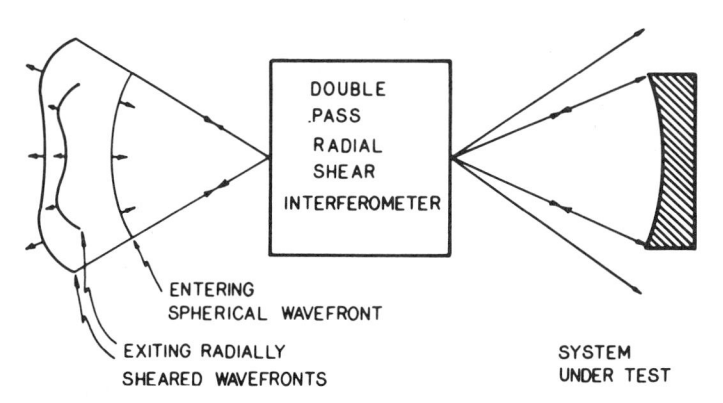

5.2.2. Double-Pass Radial Shear Interferometers

The schematic block for a double-pass radial shear interferometer is shown in Fig. 5.12. It must be recalled that the single-pass interferometer in Fig. 5.2 produces two interference patterns, one formed by the light passing through the interferometer and another by the light that is reflected back from the instrument in order to preserve the total amount of energy. If the reflected pattern were present in the double-pass interferometer, the observed pattern would be extremely complicated because four returning beams instead of two would be interfering. This problem is avoided if the two sheared interfering wavefronts illuminating the system under test after the first pass do not interfere with each other. This can be achieved if the two wavefronts are produced with orthogonal polarizations, but are nevertheless coherent to each other. As pointed out by Brown (1959) and later by Steel (1965), the spatial coherence requirements are greatly relaxed on double-pass interferometers because they are essentially compensated for the size of the source. Although the surface cannot be made extremely large—only about three times (Brown 1959) as large as that in the single-pass interferometer—this provides about ten times more light.

Two interferometers of this type have been designed by Steel to test microscope objectives. One of these is shown in Fig. 5.13 (Steel 1965). The radial shear is produced by two birefringent systems, each formed by two calcite components between two glass lenses. The glass is Schott La K11, chosen to match

Figure 5.12. Double-pass radial shear interferometer.

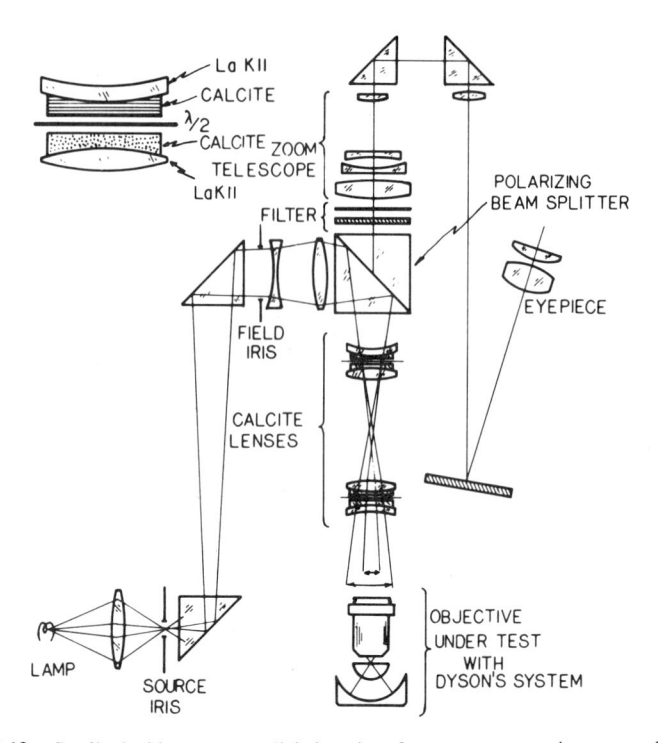

Figure 5.13. Steel's double-passage radial shear interferometer to test microscope objectives.

the ordinary index of calcite and to correct the chromatic aberration of the lens. The calcite lenses are designed to have small off-axis errors, are equiconcave, and are divided into two parts by plane surfaces. The optical axes of the two halves are at 90° with respect to each other. The plane of polarization between the halves is rotated 90° by means of a half-wave plate, so that the ordinary ray in the first half remains an ordinary ray in the second half. The two birefringent systems have a relative orientation such that the ordinary ray of the first system becomes the extraordinary ray of the second system, and vice versa.

The two birefringent systems are adjusted to satisfy the following conditions: (a) the apparent point of divergence of the two radially sheared wavefronts is at the proper distance (16 cm) from the microscope objective, and (b) the focal plane of the whole birefringent system coincides with the exit pupil of the microscope objective, which also coincides with the back focal plane of this objective. For the reasons explained in Chapter 2, a Dyson system is used in front of the microscope objective. Since the fringes must be observed at the exit pupil of the objective, a telescope is used to look at them.

The second double-passage radial shear interferometer designed by Steel (1966) is illustrated in Fig. 5.14. It is similar to the Hariharan and Sen inter-

5.2. RADIAL SHEAR INTERFEROMETERS

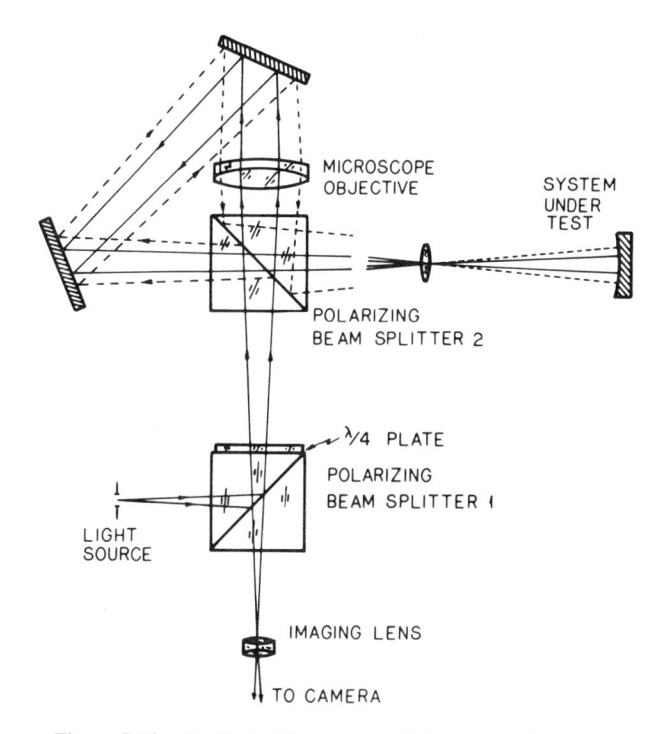

Figure 5.14. Steel's double-passage radial shear interferometer.

ferometer except that the two lenses are replaced by a single lens from a low-power microscope objective.

The combination of the polarizing beam splitter 1 and the $\lambda/4$ plate forms a source of circularly polarized light. When the light returns from the instrument after being reflected on the system under test, the direction of rotation of the circularly polarized light is reversed (assuming a perfect system). Therefore the returning light passes through the prism to go to the camera. If the system under test is imperfect, an optical path difference between the two radially sheared and orthogonally polarized beams is introduced. Then the returning light will not be circularly but elliptically polarized, giving rise to dark zones (fringes) on the camera.

5.2.3. Laser Radial Shear Interferometer

The radial shear interferometers so far described are of the equal-path type, with white-light compensation. This is necessary when conventional light sources are used. When a laser is employed, the two beams do not need to have the same optical paths.

With laser light the design of the interferometer is greatly simplified, but

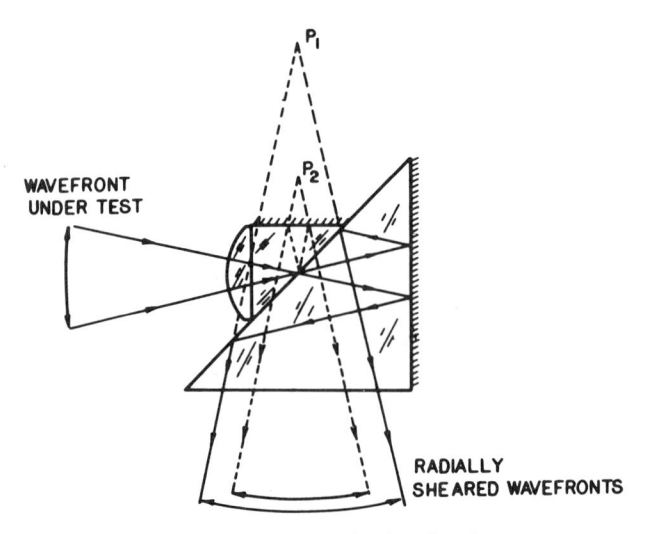

Figure 5.15. Som's laser radial shear interferometer.

some new problems are introduced. The main problem is the presence of many spurious fringes over the desired interference pattern due to reflections on lens surfaces and glass plates. All reflections produce interference fringes because of the long coherence length of laser light.

Probably the first laser radial shear interferometer was designed by Som (1970); it is illustrated in Fig. 5.15. There is, however, a great problem with this design since the virtual points of divergence P_1 and P_2 for the two wavefronts do not coincide, as pointed out by Murty and Shukla (1973). Hence a perfect system produces a system of concentric circular fringes, similar to Newton rings, and it is difficult to analyze the interferogram with this as a reference. Murty points out that for easier analysis a perfect optical system under test must give either a fringe-free field or a set of straight fringes.

To eliminate this problem, Murty and Shukla (1973) modified Sam's design and proposed the interferometer shown in Fig. 5.16, in which one of the reflecting surfaces is spherical. If a and b are the distances from the concave and plane mirrors, repectively, to the center of the beam-dividing surface, the radius of curvature r of the reflecting surface must be

$$r = \frac{(2b - a)a}{b - a}. \tag{5.16}$$

The effective radial shear R is then

$$R = \frac{a}{2b - a}. \tag{5.17}$$

5.2. RADIAL SHEAR INTERFEROMETERS

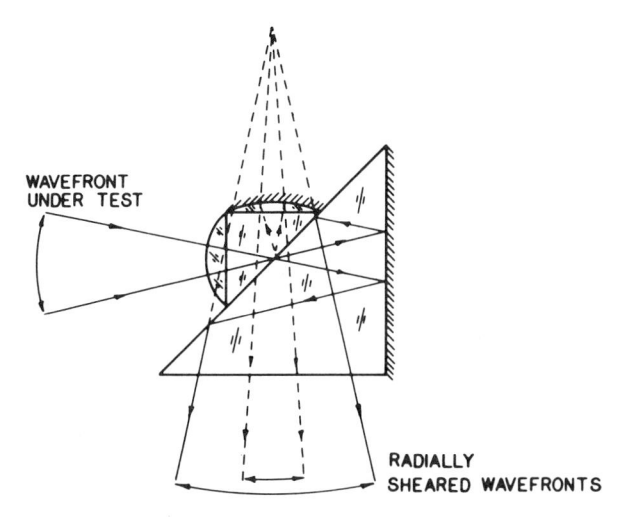

Figure 5.16. Murty's laser radial shear interferometer.

This interferometer can be fabricated very easily from a solid cube beam splitter. Ideally, the reflecting surface should be a hyperboloid of revolution, and thus a spherical surface introduces a small amount of spherical aberration. Murty et al. (1975) showed that the wavefront spherical aberration is given by

$$\text{OPD} = \frac{-Na^4 b^2 \alpha^4}{(b-a)^2 r^3} \tag{5.18}$$

where N is the refractive index of the glass and α is the numerical aperture. Thus it is necessary to reduce a as much as possible. If the numerical aperture is small, very high accuracies can be obtained.

Malacara et al. (1975) designed a lateral shearing interferometer for converging or diverging laser beams that uses a plano-concave prismatic glass plate. Like the interferometer just described, this instrument also produces some radial shear together with the lateral shear.

Hariharan et al. (1984) has used an interferometer based on the instrument designed by Murty and Shukla (1973) in order to perform phase shifting interferometry.

Another of the early laser radial shear interferometers was designed by Steel (1970); it is illustrated in Fig. 5.17. The two beams are split and recombined at the two surfaces of a prism, thus eliminating the possibility of unwanted fringes on the second face of a beam splitter. The diameter of one of the beams is reduced in size by means of a telescope with 5.5 magnification. The advantage of using glass spheres is that they do not need to be squared to the beam: a disadvantage is that they introduce spherical aberration. It must be pointed

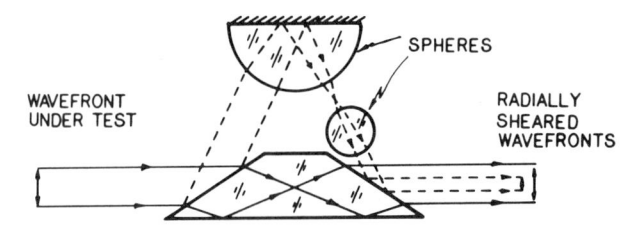

Figure 5.17. Steel's laser radial shear interferometer.

out that, simultaneously with the radial shear, this interferometer also produces reversal shear, since one of the wavefronts is reversed with respect to the other.

A very simple holographic radial shear interferometer, shown in Fig. 5.18, was devised by Fouéré and Malacara (1974) and Fouéré (1974). The first step in making this interferometer is to fabricate a Gabor zone plate by photographing the interference between a convergent and a flat wavefront. The numerical aperture of this plate is the numerical aperture that the interferometer will accept. The second step is to illuminate the Gabor zone plate with a convergent wavefront. If the point of convergence corresponds exactly to the focus of the zone plate, the diffracted +1 order beam emerges as a parallel beam. Let us now place a photographic plate behind the zone plate and make an exposure. By developing the photographic plate we obtain a hologram. The convergent beam (order 0) can be regarded as the reference beam, and the diffracted parallel beam (order +1) as the object beam.

Considering now Fig. 5.18b, if we place the hologram exactly in its original

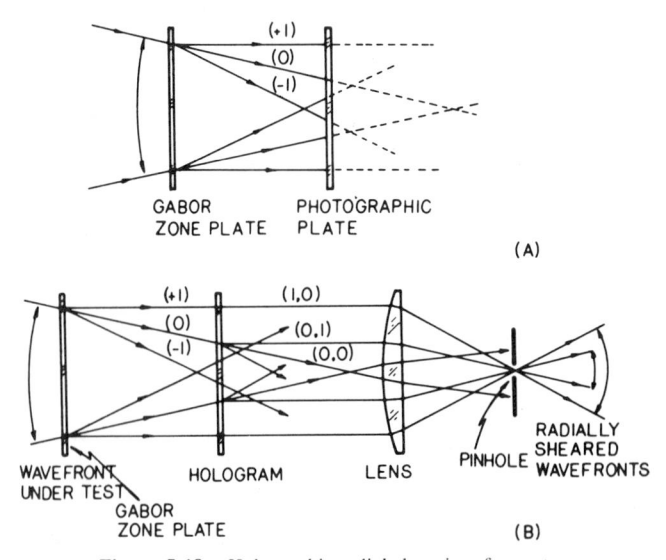

Figure 5.18. Holographic radial shear interferometer.

5.2. RADIAL SHEAR INTERFEROMETERS

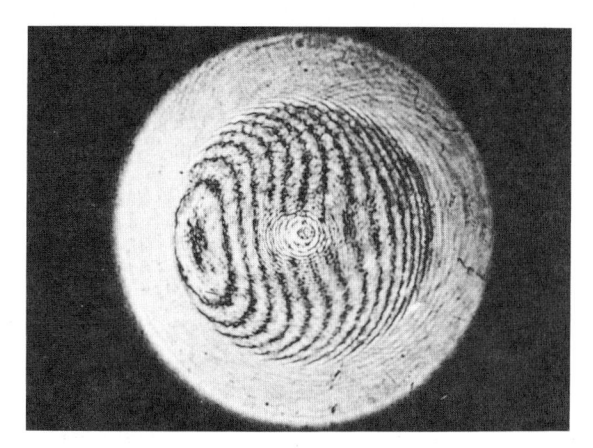

Figure 5.19. Interferogram obtained in holographic interferometer. (From Fouéré and Malacara 1974.)

position, by reconstruction we obtain an emergent parallel beam (0, 1). The incident parallel beam goes through the hologram as the parallel beam (1, 0). Thus, these two beams emerge parallel but with different magnifications, achieving a radial shear interferometer. All undesired diffracted beams are filtered out by means of a lens and a pinhole at its focus. It may be shown that all imperfections on the glass plates are automatically canceled out. Tilt and defocusing in this interferometer can very easily be obtained by small lateral and longitudinal movements, respectively, of the Gabor zone plate. Figure 5.19 shows an interferogram obtained with this interferometer.

Several variations of this basic configuration of two-zone plates have been proposed by Smartt and Hariharan (1985), in order to have also rotational shear or to produce an exiting spherical, instead of flat wavefront.

5.2.4. Thick-Lens Radial Shear Interferometers

Steel (1975) and Steel and Wanzhi (1984) have described an interesting class of laser interferometers, called thick-lens radial shear interferometers. Figure 5.20 shows some designs. One of the beams is obtained from the direct beam, going through the thick lens. The other beam is obtained with two internal reflections, first on the second lens face and then on the first lens face. If the direct beam has an irradiance I_1, the internally reflected beam has an irradiance I_2 given by

$$I_2 = \frac{R_1 R_2}{S^2} I_1 \tag{5.19}$$

where S is the radial shear, defined by the ratio of the diameter of the internally reflected beam to the diameter of the direct beam. The maximum constant of

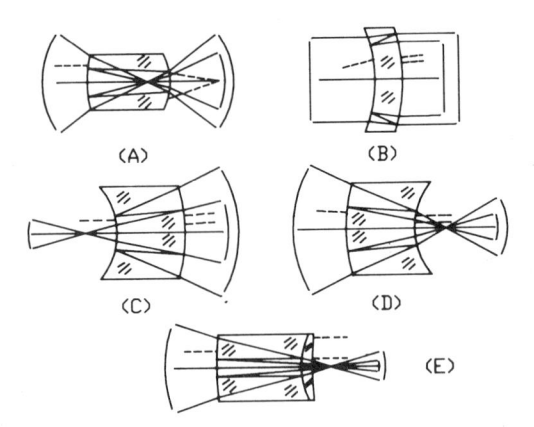

Figure 5.20. Thick-lens laser radial shear interferometers.

the fringes is obtained when $R_1 R_2 = S^2$. This is possible only if $S < 1$, in other words, when the direct beam has the smallest diameter. Steel and Wanzhi suggest keeping the contrast to a low value of the order of 0.8, to reduce the disturbing effects of higher-order ghosts due to secondary internal reflections. Unfortunately, these reflections arrive at the same focus, and thus they cannot be eliminated by a spatial filter.

Figure 5.20 shows some thick-lens interferometers. The first one, Fig. 5.20a, is a radial as well as a rotational shear interferometer; the rest are pure radial shearing. The interferometer in Fig. 5.20b may be used to test large collimated beams. In the interferometer in Fig. 5.20c the direct beam has a diameter smaller than the reflected beam. Thus, the contrast of the fringes cannot be controlled. The system in Fig. 5.20e may be used in convergent beams with the exit pupil located far away.

The reflecting radial shear interferometer described by Wanzhi (1984) is a special case of the thick lens interferometer, working in a reflection mode, as shown in Fig. 5.21a and 5.21b. The fringe contrast is much better than in the

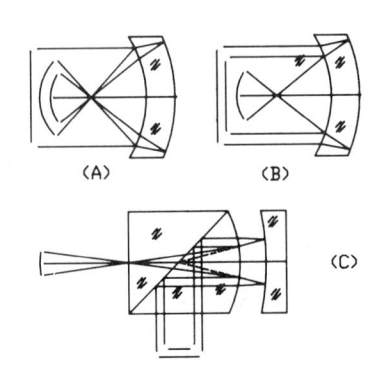

Figure 5.21. Reflection laser radial shear interferometers.

5.3 ROTATIONAL SHEAR INTERFEROMETERS

In rotational shear interferometers identical wavefronts are made to interfere, but one of them is rotated with respect to the other by a certain angle about their common optical axis. They are useful for detecting and evaluating non-symmetrical aberrations, like coma and astigmatism. Many procedures have been developed for the interpretation of rotational shearing interferograms. For example, Golikov et al. (1981) have described a method using cylindrical coordinates.

Let a wavefront be represented by $W(\rho, \theta)$. A rotational shear interferometer is one that performs the operation of rotating one wavefront with respect to the other, to give an interferogram defined by

$$\text{OPD}(\rho, \theta) = W\left(\rho, \theta - \frac{\phi}{2}\right) - W\left(\rho, \theta + \frac{\phi}{2}\right) \tag{5.20}$$

where ϕ is the rotation of one wavefront with respect to the other. If the wavefront is given by general expression (5.1), we obtain

$$\text{OPD}(\rho, \theta) = \sum_{n=0}^{k} \sum_{l=0}^{n} \rho^n \left\{ a_{nl} \left[\cos l\left(\theta - \frac{\phi}{2}\right) - \cos l\left(\theta + \frac{\phi}{2}\right) \right] \right.$$
$$\left. + b_{nl} \left[\sin l\left(\theta - \frac{\phi}{2}\right) - \sin l\left(\theta + \frac{\phi}{2}\right) \right] \right\} \tag{5.21}$$

where n and l are both even or both odd. It is interesting that all terms for $l = 0$ cancel out, rendering the rotational shear interferometers insensitive to rotationally symmetric wavefronts, as could be expected. Therefore the sums in this expression can be started from $n = l = 1$. If we now assume that the aberrations are produced by an axially symmetric optical system, we have the wavefront symmetric about the tangential (y–z) plane and therefore all coefficients b_{nl} become zero. Thus

$$\text{OPD}(\rho, \theta) = \sum_{n=1}^{k} \sum_{l=1}^{n} \rho^n a_{nl} \left[\cos l\left(\theta - \frac{\phi}{2}\right) - \cos l\left(\theta + \frac{\phi}{2}\right) \right]$$
$$= \sum_{n=1}^{k} \sum_{l=1}^{n} 2\rho^n a_{nl} \sin l\theta \sin \frac{l\phi}{2}. \tag{5.22}$$

The only two primary aberrations contained in this expression are astigmatism ($n = 2$, $l = 2$) and coma ($n = 3$, $l = 1$), in addition to a tilt of the wavefront about the x axis ($n = 1$, $l = 1$) that we may ignore, thus obtaining

$$\text{OPD}(\rho, \theta) = 2a_{22}\rho^2 \sin 2\theta \sin \phi + 2a_{31}\rho^3 \sin \theta \sin \frac{\phi}{2} \quad (5.23)$$

which can also be written as

$$\text{OPD}(\rho, \theta) = 2a_{22}\rho^2 \cos 2\left(\theta + \frac{\pi}{4}\right) \sin \phi$$
$$+ 2a_{31}\rho^3 \cos\left(\theta + \frac{\pi}{2}\right) \sin \frac{\phi}{2}. \quad (5.24)$$

Observing this expression, we can see that the sensitivity σ_{ast} for astigmatism with respect to that of a Twyman–Green interferometer is given (Murty and Hagerott 1966) by

$$\sigma_{\text{ast}} = 2 \sin \phi, \quad (5.25)$$

but it is also important to note that the interference pattern is rotated 45° with respect to the Twyman–Green pattern, as shown in Fig. 5.22, and that the rotational shear interferometer is insensitive to real defocusing. The apparent focus is such that the observed pattern is similar to the Twyman–Green pattern at the intermediate focus, between the tangential and sagittal foci.

The relative sensitivity σ_{coma} for coma is given (Murty and Hagerott 1966)

Figure 5.22. Astigmatism pattern in a rotational shear interferometer with maximum sensitivity ($\phi = 90°$) ($\alpha_{22} = 2\lambda$).

5.3. ROTATIONAL SHEAR INTERFEROMETERS

Figure 5.23. Coma pattern in a rotational shear interferometer with maximum sensitivity ($\phi = 180°$) ($\alpha_{31} = 5\lambda$).

by

$$\sigma_{\text{coma}} = 2 \sin \frac{\phi}{2}, \tag{5.26}$$

and the interference pattern is rotated 90° with respect to the corresponding Twyman–Green interferogram, as shown in Fig. 5.23. No defocusing appears, nor can it be introduced in this pattern.

The relative sensitivities σ_{ast} and σ_{coma} for astigmatism and coma, respectively, are plotted in Fig. 5.24 which shows that the ability of a rotational shear interferometer to detect astigmatism and coma depends on the amount of rotational shear ϕ. We can see that the coma can be isolated at $\phi = 180°$, but the astigmastism can never be isolated. However, the coma can be eliminated and

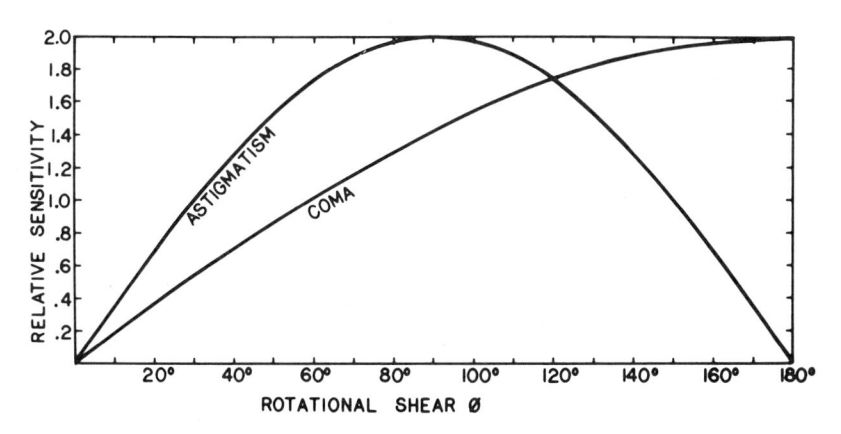

Figure 5.24. Relative sensitivity for astigmatism and coma in a rotational shear interferometer.

the astigmatism doubled by the use of a small, flat mirror to allow the wavefront to go twice through the optical system. The wavefront goes first to the system under test (concave mirror), proceeds next to the small, flat mirror near the center of curvature of the concave mirror, and returns through the same path.

It is important to point out the difference between the two similar processes in which either the symmetrical (power of cos θ is even) or the antisymmetrical (power of cos θ is odd) components of the wavefront are isolated. To isolate the asymmetrical components, the wavefront is made to interfere with an image of itself that is identical but is rotated 180°, in order to obtain the difference between them. To isolate the symmetrical components, the wavefront is rotated 180° and then passed again through the optical system under test, in order to duplicate the symmetrical errors and to eliminate the asymmetrical ones. Then the wavefront is made to interfere with an unaberrated wavefront. These two processes have been exploited by Hariharan and Sen (1961c) in a single arrangement, as described in Chapter 7 of this book.

As pointed out by Murty and Hagerott, it is interesting to consider the testing of a ribbed lightweight mirror. When such a mirror is polished, the region of the face directly on top of the rib is low, whereas the region between the ribs is high. The astigmatism of a mirror of this kind ($n = 2$, $l = m$) produces on a rotational shear interferometer on OPD given by

$$\text{OPD} = 2\rho^n a_{nl} \sin m\theta \sin \frac{m\phi}{2} \tag{5.27}$$

where m is the number of ribs supporting the faceplate. It should be noticed that $n = m$ are both even or both odd. The maximum relative sensitivity is 2 and occurs for $\phi = 180°$. Figure 5.25 shows the interferogram for a mirror with four ribs.

Figure 5.25. Pattern for a mirror with four ribs in a rotational shear interferometer with maximum sensitivity ($\phi = 45°$) ($\alpha_{24} = 4\lambda$).

5.3.1. Source Size Uncompensated Rotational Shear Interferometers

The degree of coherence g_{12} of a uniform circular source with angular diameter 2α is given in Eq. (5.10). Thus, to maintain good fringe contrast, the pinhole of a rotational shear interferometer must be of the proper size. If a point of the wavefront is sheared through an angle ϕ, the distance in the wavefront between two interfering points is given by

$$d = 2\rho \sin \frac{\phi}{2}. \tag{5.28}$$

Substituting this value in Eq. (5.10), we obtain

$$g_{12} = \frac{2J_1[(4\pi/\lambda)\alpha\rho \sin (\phi/2)]}{(4\pi/\lambda)\alpha\rho \sin (\phi/2)}. \tag{5.29}$$

For the reasons given in Section 5.2.1, to have good contrast on the interference pattern, the pinhole must be smaller than certain value given by Murty and Hagerott (1966) as

$$\alpha \leq \frac{1.22\lambda}{2D \sin (\phi/2)} \tag{5.30}$$

where D is the interferometer entrance pupil diameter, from which the angular diameter 2α of the source is measured.

Murty and Hagerott also designed the rotational shear interferometer shown in Fig. 5.26. It is a Jamin interferometer with two identical Dove prisms between the glass plates. A microscope objective L_1 collimates the light from a point source, and the pattern is observed by looking through the microscope objective L_2. One of the wavefronts is rotated an angle 2α by rotating one of the Dove prisms an angle α.

An ordinary Twyman–Green interferometer can be converted to a rotation shear interferometer, in order to test the illuminating wavefront, by means of several procedures. A 180° rotational shear is achieved, as suggested by Murty (1964c), by replacing the mirror in one of the arms with a cat's-eye retroreflector or a cube corner prism. Armitage and Lohmann (1965) suggested using two roof prisms as reversing prisms, instead of the flat mirrors. The magnitude of the rotational shear ϕ is changed by rotating one of the prisms about its optical axis by an amount $\phi/2$. The interferometer proposed by Armitage and Lohmann is illustrated in Fig. 5.27. The wavefront under test is collimated by the lens L_1, or a pinhole is used if L_1 is the lens to be tested. The pair of lenses L_2 and L_3 forms an image of the entrance pupil at L_1 on the observing screen. To obtain good coherence, the state of polarization of both beams must be identi-

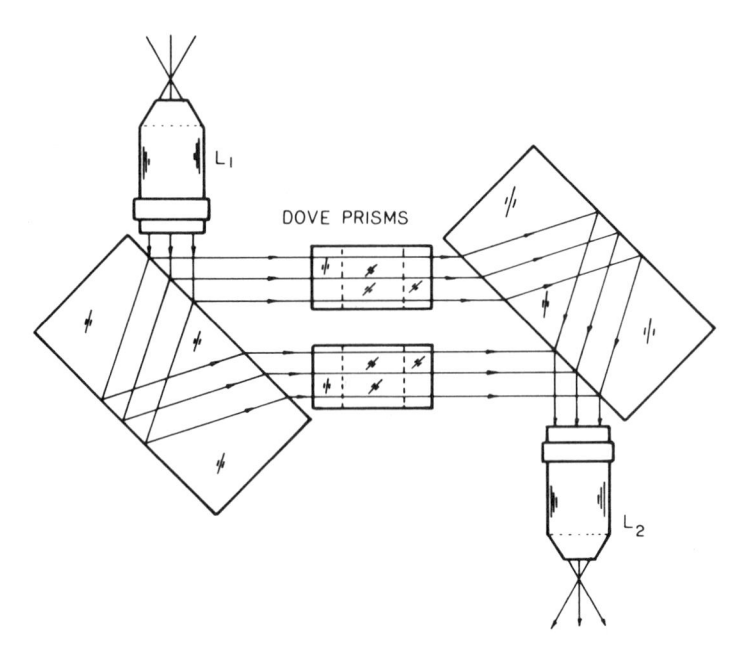

Figure 5.26. Murty and Hagerott's rotational shear interferometer.

cal, but this is not so when the rotating roof prism is rotated an angle $\phi/2$ because the roof prisms modify the state of polarization according to their positions. To solve this problem a polarization coupling is used as follows. Two polarizers are included, one before and one after the interferometer, both at 0°. Also, two quarter-wave plates are placed between the main interferometer body and the rotating roof prism. One plate is fixed at 45° to the main body of the

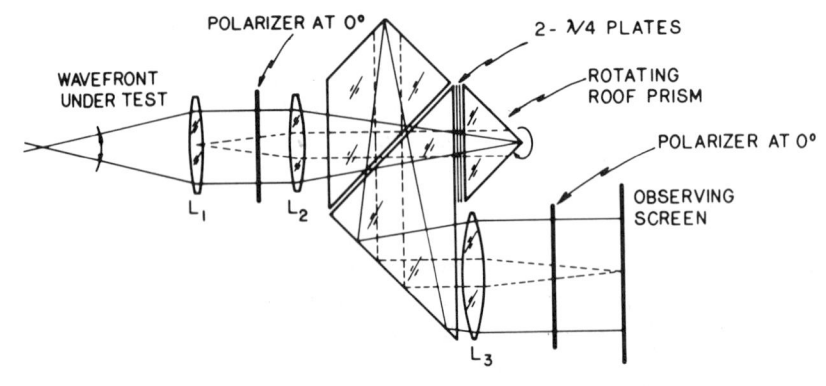

Figure 5.27. Armitage and Lohmann's rotational shear interferometer.

5.3. ROTATIONAL SHEAR INTERFEROMETERS

interferometer, and the other plate is fixed to the roof prism at 45° to the line of intersection of the roof planes. Between the two quarter-wave plates the light will be circularly polarized. Within the roof prism it will be linearly polarized and parallel to the roof ridge. Upon returning into the main body, the old direction of linear polarization is restored, independently of the rotational position of the roof prism.

The problem with the interferometer just described is that the use of the polarizers causes an appreciable loss of light. An alternative solution has been employed by Roddier et al. (1978, 1989) in the interferometer shown in Fig. 5.28. There are two roof prisms, one is fixed, and the other is rotatable about its optical axis, as in the interferometer by Armitage and Lohmann. Let us assume for the time being that the light entering the interferometer is linearly polarized, with the plane of polarization either parallel or perpendicular to the fixed roof ridge. If the rotatable roof prism is rotated an angle with respect to the first prism, we may decompose the electric vector in two orthogonal components, one parallel and one perpendicular to the roof ridge of this second prism. The phase shift upon reflection for these two components will be different in general, so, they will recombine after leaving the prism producing a different polarization state from that of the entering light beam. The only way to preserve the incident polarization state is to have a phase shift difference between these two components equal to 180°. The electrical component perpendicular to the roof edge comes out of the prism in opposite direction because of the two reflections. The net effect is that the linear polarization and its direction is thus preserved.

Since the total shift upon reflection for the two electrical components is about 74°, an additional phase shift of about 106° is required. This is obtained by cementing to the entrance face (hypotenuse) of the rotatable prism a phase plate with a phase retardation of 53°, with its principal axes parallel and perpendicular to the roof ridge.

We assumed at the beginning that the light entering the interferometer was linearly polarized, with the plane of polarization either parallel or perpendicular to the fixed roof ridge. If the light is unpolarized, the components parallel and perpendicular to the roof ridge of the fixed prism will produce different inter-

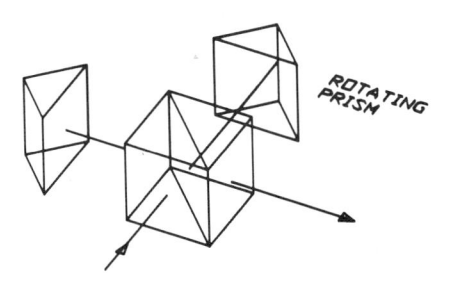

Figure 5.28. Rotational shear interferometer with phase compensation.

ference patterns. The same interference pattern is produced with these two components if they come out of the fixed prism in phase, as when they entered the prism. This goal is achieved by cementing another phase plate to the fixed prism. Then, unpolarized light may be used.

Another interferometer suggested by Armitage and Lohmann (1965) is based on the Sagnac or cyclic interferometer. Here the rotational shearing angle ϕ is produced by a Dove prism rotation of $\phi/4$ within the closed loop of the interferometer.

Although not really intended as a rotational shear interferometer, the inverting Fizeau interferometer designed by Sen and Puntambekar (1965, 1966) produces a rotational shear of 180°. This instrument and its adaptation to test spherical surfaces (Puntambekar and Sen 1971) are described in Chapter 7.

5.3.2. Source Size Compensated Rotational Shear Interferometers

A large light source cannot in general be used because of the coherence problems explained in the preceding section. The only way to have good contrast with a large source is to make the two images of the source coincide in position and orientation. At the same time the images of the object under test must be shared with respect to each other. One solution is to shear the wavefronts before they reach the object, with an equal magnitude but in an opposite sense to the main shear, which is to take place after the object. These two shearing operations cancel each other as far as coherence is concerned, but only the second shear takes place with the information about the object. These considerations, equivalent to the ones made in Section 5.2.2 for radial shear interferometers, were also advanced by Armitage and Lohmann (1965) in suggesting several compensated interferometers. The systems they proposed consist basically of two identical interferometers placed symmetrically one after the other, with the object under test between them.

As explained in Section 5.2.2, the same compensation can be obtained by using a rotational shear interferometer in a double-pass configuration.

5.4. REVERSAL SHEAR INTERFEROMETERS

The reversal of a wavefront about a reversing axis is illustrated in Fig. 5.29 where point P goes to point P', according to the transforming equations

$$\rho' \sin \theta' = \rho \sin \theta \tag{5.31}$$

and

$$\rho' \cos \theta' = S - \rho \cos \theta. \tag{5.32}$$

5.4 REVERSAL SHEAR INTERFEROMETERS

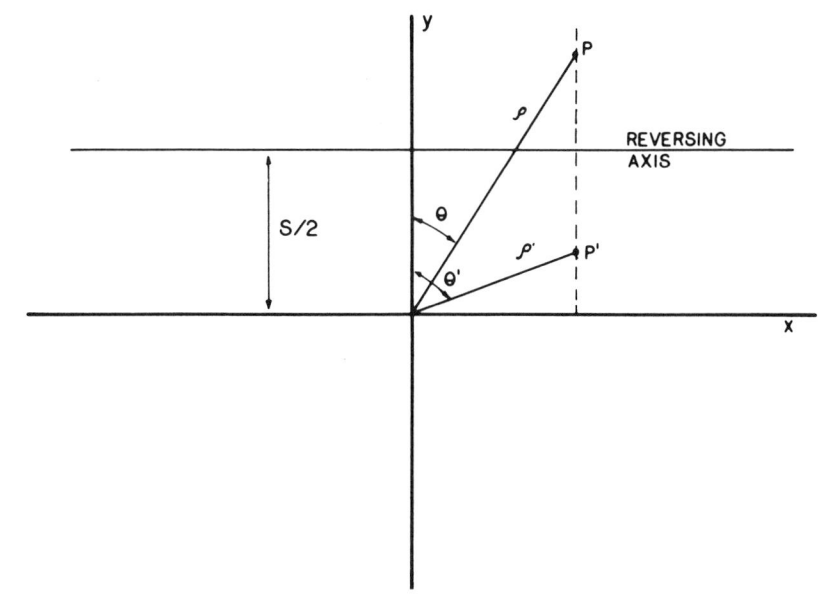

Figure 5.29. Reversal of a wavefront.

It is easy to see that this reversing about an arbitrary axis is equivalent to a reversion about the x axis followed by a lateral shear S in the y direction. If the wavefront is defined by Eq. (5.1), the interferogram in a reversal shear interferometer is given by

$$\text{OPD} = W(\rho, \theta) - W(\rho', \theta'). \tag{5.33}$$

Let us consider a wavefront that has only primary aberrations as follows:

$$W(\rho, \theta) = a_{20}\rho^2 + a_{40}\rho^4 + a_{31}\rho^3 \cos\theta + a_{22}\rho^2 \cos 2\theta + a_{11}\rho \cos\theta \tag{5.34}$$

where the aberrations being represented are defocusing, spherical aberration, coma, astigmatism, and a tilt about the x axis, respectively.

We can see that the reversal shear interferometer has no sensitivity to symmetric aberrations like defocusing (a_{20}), spherical aberration (a_{40}), and astigmatism (a_{22}) if the axis of reversion coincides with the x axis ($S = 0$). The interferometric pattern and the sensitivity of this interferometer to symmetrical aberrations are, however, identical to those of a lateral shearing interferometer with shear S when the axis of reversion is shifted a distance $S/2$, as in Fig. 5.29.

$$\text{OPD} = 2a_{31}\rho^3 \cos\theta + 2a_{11}\rho \cos\theta. \tag{5.35}$$

Thus the relative sensitivity of the reversal shear interferometer without any lateral shear ($S = 0$) to antisymmetric aberrations, like coma and tilt about the x axis, is equal to 2.

If the interferometer is uncompensated for the size of the light source, that is, if the reversion affects not only the object under test but also the light source, the contrast in the fringe pattern is given by Eq. (5.10) (Murty 1964c). The distance for any two interfering points is $(2\rho \cos\theta - S)$. Therefore its maximum value d is

$$d = D + S \tag{5.36}$$

where D is the waveform diameter. Then the maximum angular diameter α of the pinhole will be

$$\alpha \leq \frac{1.22\lambda}{D + S}. \tag{5.37}$$

An uncompensated reversal shear interferometer of this type is represented by a Twyman–Green interferometer with a right-angle prism used as a roof prism in one of the arms.

5.4.1. Some Reversal Shear Interferometers

In this section three different versions of a prism reversal shear interferometer are described. They are variations of a basic prism system invented by Kösters (1934), who pointed out that any combination of two exactly similar prisms can be used as an interferometer in the way he described, as long as one of the angles of the prism adjoining the common face is exactly half the angle that is opposite this common face.

One of the two systems suggested by Gates (1955b), based on Kösters's prism, is illustrated in Fig. 5.30; it uses two 30°–60°–90° prisms. The intersection of the plane defined by the beam-splitting surface with the optical system defines the axis of reversion. Lateral shear S is produced if the axis of reversion does not coincide with a diameter of the optical system under test. A tilting of the mirror under test provides a control of the separation of the interference fringes. A lens can be tested if an autocollimating flat mirror is used behind the lens.

The interferometer is compensated for use with white light because the optical paths are exactly equal. It is also compensated for the size of the light source, but in practice this size is limited to about 0.5 mm in diameter by small errors in construction, as pointed out by Gates.

5.4 REVERSAL SHEAR INTERFEROMETERS

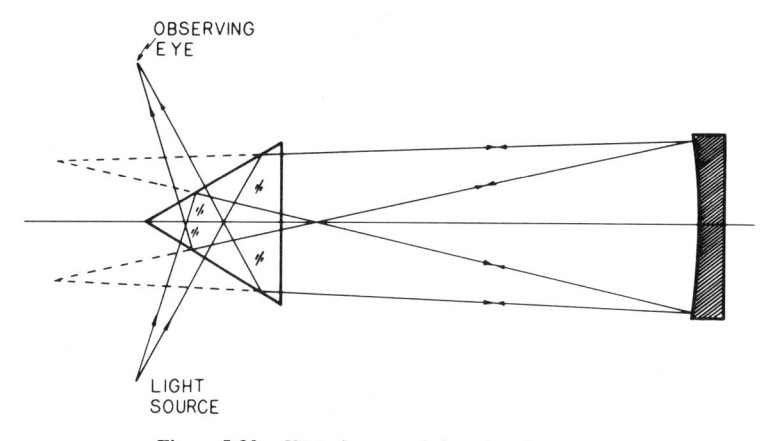

Figure 5.30. Köster's reversal shear interferometer.

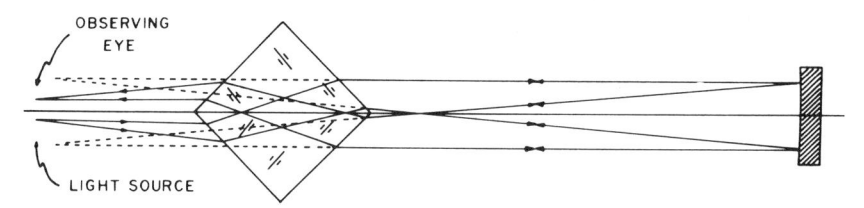

Figure 5.31. Gates' reversal shear interferometer.

Using the same principle, Gates (1955a) also suggested using the popular beam splitter cube, as shown in Fig. 5.31. This, however, has the disadvantage that only optical systems with small numerical apertures can be tested.

Two common disadvantages of both the instruments described by Gates are (a) that the virtual light source and image do not coincide but have some small separation, and (b) that some aberrations are introduced on the plane exit face of the prism. These problems are not present in a prism system invented by Saunders (1955), shown in Fig. 5.32. The use of this instrument for the testing of optical systems, as described by Saunders (1962), is illustrated in Fig. 5.33. Many other applications to metrology have also been found (Saunders 1960; Strong 1958). The construction method has been very well described by Saunders (1957).

Another interesting reversal shear interferometer to test lenses was described by Waetzman (1912) and later by Murty (1964b). It is illustrated in Fig. 5.34. This instrument, which has also been used to test right-angle and cube corner prisms by Murty (1964b) and Sen and Puntambekar (1966), produces reversal and lateral shears. It is interesting to observe that the asymmetric aberrations are canceled out because of the double passage through the lens and that the

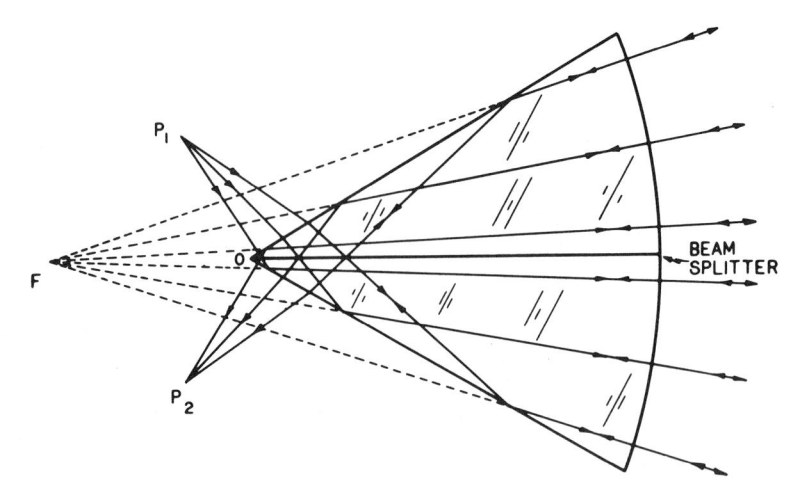

Figure 5.32. Saunders's prism system.

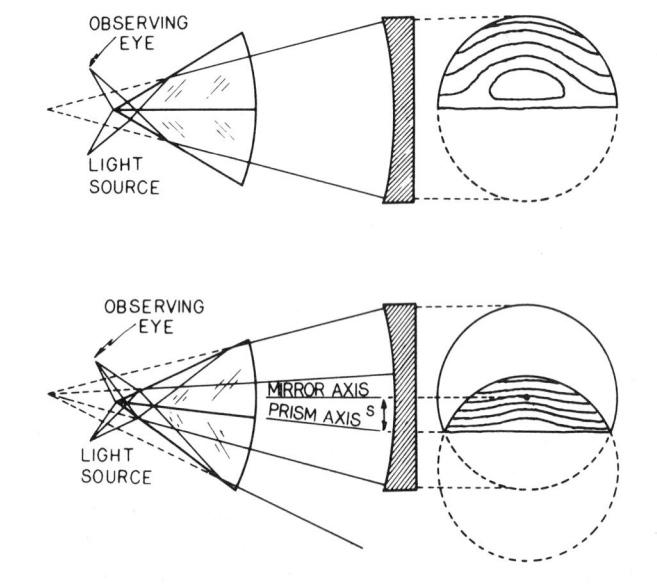

Figure 5.33. Saunders's reversal shear interferometer.

reversal interferometer is not sensitive to symmetric aberrations. Since the symmetric aberrations are detected by means of lateral shear only, this instrument can more properly be considered a lateral shearing interferometer.

Holographic techniques may also be used to obtain a reversal shear interferometer, as shown by Partiban et al. (1987, 1988).

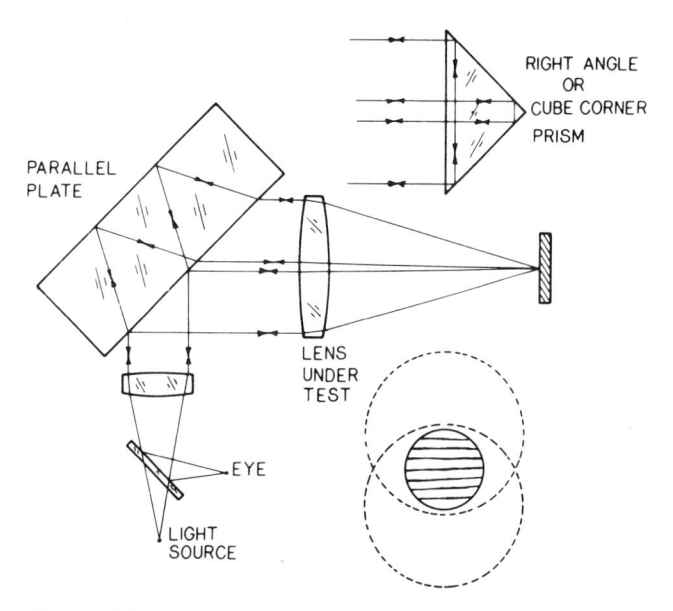

Figure 5.34. Modified Jamin interferometer to test lenses and prisms.

REFERENCES

Armitage, J. D. and A. Lohmann, "Rotary Shearing Interferometry," *Opt. Acta*, **12,** 185 (1965).

Baird, K. M. and G. R. Hanes, in *Applied Optics and Optical Engineering*, Vol. 4, R. Kingslake, Ed., Academic Press, New York, 1967, p. 336.

Briers, J. D., "Interferometric Testing of Optical Systems and Components: A Review," *Opt. Laser Technol.*, **4,** 28(1972).

Brown, D. S., "Radial Shear Interferograms," *Interferometry N.P.L. Symposium* No. 11, Her Majesty's Stationary Office, London, 1959, p. 253.

Brown, D. S., "Radial Shear Interferometry," *J. Sci. Instrum.*, **39,** 71 (1962).

Bryngdahl, O., in *Progress in Optics*, Vol. IV, E. Wolf, Ed., North-Holland, Amsterdam, 1965, Chap. II.

Bryngdahl, O., "Reversed-Radial Shearing Interferometry," *J. Opt. Soc. Am.*, **60,** 915 (1970).

Bryngdahl, O., "Shearing Interferometry with Constant Radial Displacement," *J. Opt. Soc. Am.*, **61,** 169 (1971).

Fouéré, J. C., "Holographic Interferometers for Optical Testing," *Opt. Laser Technol.*, **6,** 181 (1974).

Fouéré, J. C. and D. Malacara, "Holographic Radial Shear Interferometer," *Appl. Opt.*, **13,** 2035 (1974).

Fouéré, J. C. and D. Malacara, "Generalized Shearing Interferometry," *Bol. Inst. Tonantzintla*, **1**, 227 (1975).

Gates, J. W., "Reverse Shearing Interferometry," *Nature*, **176**, 359 (1955a).

Gates, J. W., "The Measurement of Comatic Aberrations by Interferometry," *Proc. Phys. Soc.*, **B68**, 1065 (1955b).

Golikov, A. P. M., L. Gurari, and S. I. Prytkov, "Interpretation of Rotational-Shearing Interferograms," *Sov. J. Opt. Technol.*, **48**, 676 (1981).

Hariharan, P. and D. Sen, "Effects of Partial Coherence in Two Beam Interference," *J. Opt. Soc. Am.*, **51**, 1307 (1961a).

Hariharan, P. and D. Sen, "Radial Shearing Interferometer," *J. Sci. Instrum.*, **38**, 428 (1961b).

Hariharan, P. and D. Sen, "The Separation of Symmetrical and Asymmetrical Wavefront Aberrations in the Twyman Interferometer," *Proc. Phys. Soc.*, **77**, 328 (1961c).

Hariharan, P. and D. Sen, "Interferometric Measurements of the Aberrations of Microscope Objectives," *Opt. Acta*, **9**, 159 (1962).

Hariharan P., B. F. Oreb, and Z. Wanzhi, "A Digital Radial Shear Interferometer for Testing Aspheric Surfaces," *Proc. 13th. Congr. ICO.*, Sapporo, Japan, 464, (1984).

Hariharan P., B. F. Oreb, and Z. Wanzhi, "Measurement of Aspheric Surfaces Using a Microcomputer-Controlled Digital Radial-Shear Interferometer," *Opt. Acta*, **31**, 989 (1984).

Hariharan P., B. F. Oreb, and Z. Wanzhi, "Digital Radial Shearing Interferometry: Testing Mirrors with a Central Hole," *Opt. Acta*, **33**, 251 (1986).

Hariharan P., "Lateral and Radial Shearing Interferometers: A Comparison," *Appl. Opt.*, **27**, 3594 (1988).

Honda T., J. Huang, J. Tsujiuchi, and J. C. Wyant, "Shape Measurement of Deep Aspheric Optical Surfaces by Radial Shear Interferometry," *Proc. SPIE*, **813**, 351 (1987).

Horton R. F., "Design of a White Light Radial Shear Interferometer for Segmented Mirror Control," *Opt. Eng.*, **27**, 1063 (1988).

Kohler D. R. and V. L. Gamiz, "Interferogram Reduction for Radial-Shear and Local-Reference-Holographic Interferograms," *Appl. Opt.*, **25**, 10, (1986).

Kösters, W., "Interferenzdoppelprisma für Messwecke," German Patent 595211 (1934).

Malacara, D., "Mathematical Interpretation of Radial Shearing Interferometers," *Appl. Opt.*, **13**, 1781 (1974).

Malacara, D., A. Cornejo, and M. V. R. K. Murty, "A Shearing Interferometer for Convergent or Divergent Beams," *Bol. Inst. Tonantzintla*, **1**, 233 (1975).

Murty, M. V. R. K., "A Compact Radial Shearing Interferometer Based on the Law of Refraction," *Appl. Opt.*, **3**, 853, (1964a).

Murty, M. V. R. K., "Some Modifications of the Jamin Interferometer Useful in Optical Testing," *Appl. Opt.*, **4**, 535 (1964b).

Murty, M. V. R. K., "Interference between Wavefronts Rotated or Reversed with Respect to Each Other and Its Relation to Spatial Coherence," *J. Opt. Soc. Am*, **54**, 1187 (1964c).

Murty, M. V. R. K., "Interferometry Applied to Testing to Optics," *Bull. Opt. Soc. India*, **1**, 29 (1967).

Murty, M. V. R. K. and E. C. Hagerott, "Rotational Shearing Interferometry," *Appl. Opt.*, **5**, 615 (1966).

Murty, M. V. R. K. and R. P. Shukla, "Radial Shearing Interferometers Using a Laser Source," *Appl. Opt.*, **12**, 2765 (1973).

Murty, M. V. R. K., R. P. Shukla, and A. Cornejo, "Aberration in a Radial Shearing Interferometer Using a Laser Source," *Indian J. Pure Appl. Phys.*, **13**, 384 (1975).

Parthiban V., C. Joenathan, and R. S. Sirohi, "Inversion and Folding Shear Interferometers Using Holographic Optical Elements," *Proc. SPIE*, **813**, 211 (1987).

Parthiban V., C. Joenathan, and R. S. Sirohi, "Simple Inverting Interferometer with Holoelements," *Appl. Opt.*, **27**, 1913 (1988).

Puntambekar, P. N. and D. Sen, "A Simple Inverting Interferometer," *Opt. Acta*, **18**, 719 (1971).

Roddier C., F. Roddier, and J. Demarcq, "Compact Rotational Shearing Interferometer for Astronomical Application," *Opt. Eng.*, **28**, 66 (1989).

Roddier F., C. Roddier, and J. Demarcq, "A Rotation Shearing Interferometer with Phase-Compensated Roof Prisms," *J. Optics (Paris)*, **9**, 145 (1978).

Saunders, J. B., "Inverting Interferometer," *J. Opt. Soc. Am.*, **45**, 133 (1955).

Saunders, J. B., "Construction of a Kösters Double-Image Prism," *J. Res. Nat. Bur. Stand.*, **58**, 21 (1957).

Saunders, J. B., in *Optics in Metrology*, P. Mollet, Ed., Pergamon, Oxford, 1960, p. 227.

Saunders, J. B., "Measurement of Wavefronts without a Reference Standard, 2: The Wavefront Reversing Interferometer," *J. Res. Nat. Bur. Stand.*, **66B**, 29 (1962).

Sen, D. and P. N. Puntambekar, "An Inverting Fizeau Interferometer," *Opt. Acta*, **12**, 137 (1965).

Sen, D. and P. N. Puntambekar, "Shearing Interferometers for Testing Corner Cubes and Right Angle Prisms," *Appl. Opt.*, **5**, 1009 (1966).

Smartt, R. N. and W. H. Steel, "Zone-Plate Radial-Shear Interferometers. A Study of Possible Configurations," *Opt. Acta*, **32**, 1475 (1985).

Som, S. C., "Theory of a Compact Radial Shearing Laser Interferometer," *Opt. Acta*, **17**, 107 (1970).

Steel, W. H., "A Radial Shear Interferometer for Testing Microscope Objectives," *J. Sci. Instrum.*, **42**, 102 (1965).

Steel, W. H., in *Progress in Optics*, Vol. V, E. Wolf, Ed., North-Holland, Amsterdam, 1966, Chap. III.

Steel, W. H., "A Radial-Shear Interferometer for Use with a Laser Source," *Opt. Acta*, **17**, 721 (1970).

Steel, W. H., "A Simple Radial Shear Interferometer," *Opt. Commun.*, **14**, 108 (1975).

Steel, W. H. and Z. Wanzhi, "A Survey of Thick-Lens Radial-Shear Interferometers," *Opt. Acta*, **31**, 379, (1984).

Strong, J., *Concepts of Classical Optics*. W. H. Freeman, San Francisco, 1958, Appendix C.

Waetzman, E., "Interferenzmethode sur Untersuchung der Abbildungsfehler optischer Systeme," (Interference Method for Determination of Aberrations of Optical Systems). *Ann. Phys.*, **39,** 1042 (1912).

Wanzhi, Z., "Reflecting Radial Shear Interferometer," *Opt. Commun.*, **49,** 83 (1984).

Wanzhi, Z., "Reflecting Radial-Shear Interferometers with an Air-Spaced System," *Opt. Commun.*, **53,** 74 (1985).

ADDITIONAL REFERENCES

Breckinridge, J. B., "A White-Light Amplitude Interferometer With 180-Degree Rotational Shear," *Opt. Eng.*, **17,** 156 (1978).

Gorshkov, V. A., O. N. Formin, and S. N. Gorlov, "Photoelectric Interferometer for Inspecting the Surface Shape of Large Optical Components," *Sov. J. Opt. Technol.*, **53,** 598 (1986).

Horner, J. L., "Additional Property of Interferometer Symmetry," *Appl. Opt.*, **17,** 505 (1978).

Joenathan, C., C. S. Narayanamurthy, and R. S. Sirohi, "Radial and Rotational Slope Contours in Speckle Shear Interferometry," *Opt. Commun.*, **56,** 309 (1986).

Joshi, D. K., P. N. Puntambekar, and D. Sen, "Interference with Approximately Constant Radial and Azimuthal Shears," *Opt. Acta*, **33,** 653 (1986).

Kanjilal, A. K. and D. Sen, "Polarizing Interferometer with Constant Radial and Azimuthal Shears," *Opt. Laser Technol.*, **18,** 151 (1986).

Murthy, K. R., R. K. Mohanty, R. S. Sirohi, and M. P. Kothiyal, "Radial Speckle Shearing Interferometer and its Engineering Applications," *Optik*, **67,** 85 (1984).

Ru, Q. S., N. Ohyama, T. Honda, and J. Tsujiuchi, "Constant Radial Shearing Interferometry with Circular Gratings," *Appl. Opt.*, **28,** 3350 (1989).

Truax, B. E., "A Phase Measuring Radial Shear Interferometer for Measuring the Wavefronts of Compact Disc Laser Pickups," *Proc. SPIE*, **661,** 74 (1986).

6
Multiple-Beam Interferometers

C. Roychoudhuri

6.1. BRIEF HISTORICAL INTRODUCTION

The historical origin of multiple-beam interference is found as early as 1836, when Airy derived an expression for the multiple-beam interference pattern that would be produced by a plane parallel plate. However, the idea remained unexploited since high-reflectance coatings were not available and uncoated glass plate has a reflectance of only 0.04. It can produce good visibility fringes but of a two-beam type in reflection; no recognizable fringes can be seen in transmission.[†] Then Fizeau (1862a, 1862b) devised his celebrated interferometer, which now bears his name (see Chapter 1). His invention led to the idea of studying surface topography by optical interferometer (Laurent 1883). The interferometer, being formed by two uncoated glass plates, gives interference fringes of a two-beam (cosine) type in reflection that contour the surface topography. Then Boulouch (1893) revived Airy's (1836) derivation of multiple-beam interference and demonstrated that with increasing reflectance of the Fizeau surfaces the multiple-beam fringes, both in reflection and in transmission, become increasingly sharper.

A few years later Fabry and Perot (1897) recognized the potential of the interferometer that consists of two plane parallel surfaces of high-reflection coating with variable separation, now known as the Fabry–Perot interferometer. Today it is still one of the most compact and highest resolving power spectrometers (Cook, 1971). To exploit this capability for high resolving power, the Fabry–Perot plates were usually used with a large separation that obscured the great potential of this instrument for mapping surface microtopography with

---

[†]The relative strengths of the first three reflected beams are 0.04, 0.037, and 0.000059, and those of the transmitted beams are 0.92, 0.0015, and 0.0000024.

Optical Shop Testing, Second Edition, Edited by Daniel Malacara.
ISBN 0-471-52232-5 © 1992, John Wiley & Sons, Inc.

ultrahigh local precision. Then in 1913 Benoit, Fabry, and Perot used a Fizeau interferometric arrangement with coated surfaces for their determination of the standard meter. They also missed the "optimum conditions" under which this multiple-beam Fizeau interferometer could be used for precision surface testing. These were provided and demonstrated by Tolansky (1944), although he himself noted (Tolansky 1948a, pp. 7, 184) from a private communication to him from Williams that Adams Hilger Ltd. of England had been using multiple-beam Fizeau fringes for optical flat testing for some years. A lucid and detailed analysis and application of a multiple-beam interferometer of the Fizeau type can be found in Tolansky's books (1948a, 1948b, 1948c, 1960, 1966, 1968). Section 6.3 summarizes the essentials of the multiple-beam Fizeau interferometer.

To obtain sharp fringes as well as localized precision with the multiple-beam Fizeau, the proper conditions of Tolansky (next section) require a very small separation and wedge angle between the mirrors. This mandates a high-quality test flat to give only a few narrow fringes of $\lambda/2$ interval[†] in the entire field of view, leaving most of the surface without any topographic information. The next year Tolansky (1945a, 1945b) developed the so-called fringes of equal chromatic order (FECO) interferometry, where the illumination is by white light, rather than quasimonochromatic light, and one observes small sections of the test surface successively through a prism spectrometer that displays fringes of equal chromatic order. The advantages of FECO over multiple-beam Fizeau are the greater precision of the former and its capability for distinguishing "hills" from "valleys" in a straightforward manner from the direction of bending of the colored fringes. Again, a good expression of the field is given in Tolansky's book (1948a). Contributions by Koehler (Koehler 1953, 1955a, 1955b; Koehler and Eberstein 1953; Koehler and White 1955) are also worth mentioning. We describe FECO interferometry briefly in Section 6.4, where we also mention a somewhat different but very useful technique of illuminating the Fizeau interferometer by two different wavelengths that can be chosen suitably from two different calibrated monochromator (Shaalan and Little 1975).

Later, several persons introduced various techniques to reduce the fringe interval of $\lambda/2$ so that microtopographic information can be obtained from otherwise fringe-free regions, using monochromatic or quasimonochromatic illumination. The first of these techniques, by Saunders (1951), introduces optical path differences into the interference film (formed by the two mirrors) by changing the air pressure by controlled amounts and taking a multiple photographic exposure of the fringe systems. Another technique to reduce the contour interval is to use a suitably chosen discrete set of wavelengths for illumination (Herroitt 1961; Schwider 1968; Pilston and Steinberg 1969). A somewhat simpler technique by Murty (1962) employs regular quasimonochromatic illumination but

---

[†]In some multiple-pass interferometers the interval between consecutive fringes is less than $\lambda/2$; see Chapter 7.

6.1. BRIEF HISTORICAL INTRODUCTION

a spatially separated set of pinholes. These techniques for reducing the fringe interval are described briefly in Section 6.5.

Some parallel development has also taken place in using the conventional Fabry–Perot interferometer with the plates parallel (rather than wedged) for deriving surface microtopography (Benedetti-Michelangeli 1968; Hodgkinson 1969) and for precision measurement of thin-film thickness (Schulz 1950a, 1950b). See Section 6.6 for methods of using a Fabry–Perot interferometer for testing.

A somewhat different set of multiple-beam fringes, produced by a Fabry–Perot illuminated by a point source but lacking the conventional fringe-focusing lens after the Fabry–Perot plates (see Figs. 6.2 and 6.17), was first observed and used by Tolansky (1943, 1946). The utility of these "Tolansky fringes" is described in Section 6.7.

The development of the spherical Fabry–Perot interferometer (Connes 1958), of general curved mirror laser cavities (Boyd and Gordon 1961; Fox and Li 1961), and of the scanning spherical Fabry–Perot (Herroitt 1963) paved the way for an elegant technique for testing curved surfaces with multiple-beam fringes preserving the localized precision (Herroitt 1966).

A similar spherical wave interferometer was also developed independently by Perkin-Elmer of the United States (Heintze et al. 1967) and SIRA (1967) of the United Kingdom (Biddles 1969). A strong desire to test many spherical surfaces by multiple-beam fringes against a single surface or a few master surfaces, coupled with the existing knowledge of Williams' (1950) interferometer, paved the way for this development. We describe such multiple-beam interferometers with curved surfaces in Section 6.8.

Other developments worthy of mention in the field of multiple-beam interferometry are the so-called dual interferometry and holographic multiple-beam interferometry. In the former, one combines the multiple-beam fringes with another wavefront in a dual interferometric setup to utilize a live Moiré technique or to control the background contrast to advantage, especially while testing opaque surfaces in reflection by multiple-beam Fizeau (Langenbeck 1968; Pastor and Lee 1968). This is described in Section 6.9. The latter development can be exploited when one has already chosen to use holographic interferometry by modifying the system from two-beam to multiple-beam interferometry, as suggested by Matsumoto (1969) and Bryngdahl (1969). This is mentioned briefly in Section 6.10. For a detailed exposition of Moiré and holographic interferometry see Chapter 12.

Readers with broader interest on interferometry should consult the following review articles: Vrabel and Brown (1975), Malacara et al. (1975), Briers (1972), Koppelmann (1969), Baird (1967), Baird and Hanes (1967), and Kuhn (1951). For a good review on precision interferometric testing, see Schulz and Schwider (1976). Readers interested in Fabry–Perot interferometer as a high-resolution spectrometer should consult the recent books by Hernandez (1986) and Vaughan (1989).

6.2. PRECISION IN MULTIPLE-BEAM INTERFEROMETRY

All the preceding chapters of this book dealing with various interferometric tests of optical components have one thing in common: The final fringe pattern, contouring the surface or the wavefront under test, is formed by interference between two wavefronts. Then the recorded intensity variation follows the $\cos^2 \phi$ or $(1 + \cos 2\phi)$ type curve shown in Fig. 6.1a. Such fringes are said to have the so-called fringe quality, the finesse number, equal to 2. This is understood from the definition of finesse, which is the ratio of the fringe interval to the width of the fringe at half its height (Fig. 6.1). In general, the fringe interval being $\lambda/2$, visual observation of two-beam cosine fringes can rarely achieve an error estimation of precision better than $\lambda/20$.

In contrast, multiple-beam fringes are extremely sharp and can have local finesse as high as 300 with surfaces of about 99% reflectance. The Airy curve of Fig. 6.1b is a typical example that shows the essential nature of most multiple-beam fringes. In this example the finesse of the fringes is 50, and simple measurements with such fringes can reveal surface microtopography with a precision close to $\lambda/500$. It is this high-resolution capability, achievable with simple laboratory equipment, that has made multiple-beam interferometry so popular.

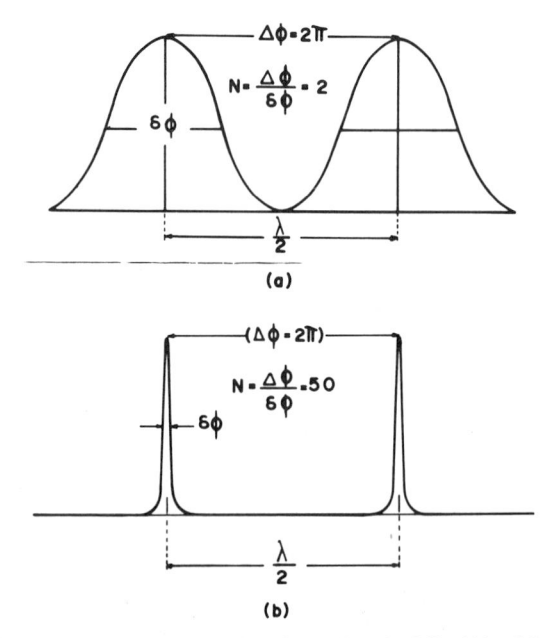

Figure 6.1. The finesse, the ratio of the fringe interval to the full width of the fringe at half its height, (a) for two-beam cosine fringes, and (b) for multiple-beam Fabry–Perot fringes. For example, the finesse of (b) is 50.

6.2. PRECISION IN MULTIPLE-BEAM INTERFEROMETRY

It must be realized, however, that to obtain multiple-beam fringes the interferometer plates must be coated with high-reflectance coating. This unquestionably interrupts the polishing process for a period of time, in contrast to two-beam interferometry, in which the surfaces under test can be simply cleaned and tested directly as frequently as the polisher wishes. For this reason multiple-beam interferometric tests are applicable chiefly to the field of thin-film technology (Bennett and Bennett 1967; Eastman 1975). The technique is also applied with relative ease (say, instead of electron microscopy) for special precision testing of the surface roughness of high-quality optical surfaces (Koehler 1955b; Koehler and White 1955; Hodgkinson 1970), but here one must be careful to choose a surface coating material (usually silver is selected; see Tolansky 1960) that contours the substrate as closely as possible and also to analyze the phase change on reflection from the coating substance (Tolansky 1948a; Koehler 1953). These problems usually do not arise in two-beam interferometry. Multiple-beam interferometry is not advisable for routine shop testing unless the surface to be tested is definitely better than $\lambda/20$.

Of course, multiple-beam interferometry is not the only method of obtaining high spatial precision. If fringe sharpening is the main criterion, one can simply image the two-beam Fizeau fringes on a high-resolution vidicon camera and observe the fringes on closed circuit television. The fringes can be artificially sharpened through electronic control of the contrast. The precision by simple visual observation can certainly exceed $\lambda/50$, and an on-line computer analysis may give ultrahigh precision. Fringe sharpening can also be achieved through nonlinear photography. Precision densitometric traces of a regular photographic record of two-beam fringes or a direct photoelectric scanning of the fringe field, when suitably analyzed, can also give very high precision (Roesler 1962; Roesler and Traub 1966; Dew 1964, 1966). A more recent technique, oscillating mirror interferometry (see Chapter 13) with a simple two-beam instrument but with an electronic analyzing system, can also achieve a precision of $\lambda/1000$ (Raymond 1970; Bruce and Sharpless 1975). In fact, Moss et al. (1971) and Logan (1973) stated that their two-beam heterodyne (Michelson) interferometry, used in a gravitational wave detection system, can detect a displacement of a mirror with a precision of 10^{-6} Å. A somewhat different technique of detecting an optical path variation down to 10^{-5} Å through the measurement of the beat signal from a three-longitudinal-mode three-mirror laser cavity has also been reported in the literature (Boersch et al. 1974).

Here one should remember that the overall thermal and mechanical stability of the interferometer assembly must be higher by an order of magnitude than the precision expected from the system, unless the very purpose of the testing is to measure the relative "instability" (Dyson 1968). The ultimate precision in interferometry is limited by the noise inherent in photoelectric detection (Hanes 1959, 1963; Hill and Bruce 1962; Raymond 1970) and also by the diffraction phenomena that lead to the "optical uncertainty principle," analogous to Heisenberg's uncertainty principle (Heisenberg 1949), discussed by Tolansky and Emara (1955), Thornton (1957), Koppelmann (1966), and Lang and Scott (1968). The very high resolution, on the order of an angstrom or a fraction thereof, that is obtained through multiple-beam interferometry is in the longitudinal, not in the lateral, direction. The resolution in the lateral direction is determined by the wavelength of the radiation due to diffraction.

The choice of multiple-beam interferometry should be guided by the following considerations: (a) the time available for the test, (b) the maximum precision of the test that is really necessary for the particular job the test surface is designed for, and (c) the equipment available in the laboratory.

6.3 MULTIPLE-BEAM FIZEAU INTERFEROMETER

6.3.1. Conditions for Fringe Formation

A multiple-beam Fizeau interferometer constitutes a very thin wedge-shaped film formed by two sufficiently plane and reflecting surfaces. Any incident beam falling on this film will, in general, produce a series of increasingly diverging beams (Fig. 6.3). These multiply reflected beams cannot be superposed in any plane with exactness. However, within certain approximations multiple-beam fringes are formed. Because the necessary conditions are more easily appreciated after following the process of formation of the ideal multiple-beam fringe by a Fabry–Perot interferometer, we shall briefly digress to describe fringe formation by this instrument.

The Fabry–Perot interferometer consists of two perfectly plane and parallel mirrors (Fig. 6.2). Naturally, a single incident wave will produce a series of waves by multiple reflection, and all of them can be superposed to form ideal

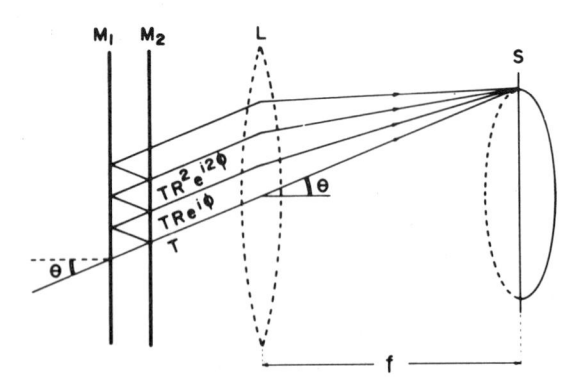

Figure 6.2. Fringe formation by a Fabry–Perot interferometer consisting of a pair of plane parallel mirrors M_1, M_2 followed by a fringe focusing lens L. S is the fringe plane in the focal plane of L.

6.3. MULTIPLE-BEAM FIZEAU INTERFEROMETER

Airy fringes (Tolansky 1948a) at the focal plane of a lens. The irradiance distribution is given by

$$I(\phi) = \left| \sum_{n=0}^{\infty} TR^n e^{in\phi} \right|^2$$

$$= \left(\frac{T}{1-R} \right)^2 \frac{1}{1 + (4N_R^2/\pi^2) \sin^2 \phi/2}, \quad (6.1)$$

with the ideal reflective finesse,

$$N_R = \frac{\pi \sqrt{R}}{1-R}, \quad (6.2)$$

where T and R are the intensity transmittance and reflectance, respectively, of both mirrors, and ϕ is the total effective phase or optical path delay between any two consecutive wavefronts.

Although the series in Eq. (6.1) is infinite, in practice the total effective number of interfering beams is finite because of the fact that the energy carried by the nth transmitted beam, T^2R^{2n}, becomes negligible for sufficiently large n. [See the solid curves of Fig. 6.4, which indicates the number of superposed beams required to form a multiple-beam fringe, for a particular reflectance, within 1% of the ideal (infinite sum) of the Airy curve.] This is a useful practical point to remember for any type of multiple-beam interferometry, as we shall soon see. If it were really necessary in practice to superpose the infinite sequence of multiply reflected beams to form the ideal Airy fringe, the interferometrist would have to wait throughout eternity to observe it (Roychoudhuri 1975)!

The next point of practical interest in multiple-beam Fabry–Perot fringe formation is that the phase delay ϕ between any two consecutive wavefronts is constant. Then, once ϕ has been determined, the entire series of multiply reflected beams combines to form the appropriate part of the ideal Airy fringe. Thus, when ϕ is either $2n\pi$ or $(2n+1)\pi$ (n any integer), the entire series of beams, when superposed, either combines constructively to form a bright fringe or adds destructively to form a dark fringe.

This is where the major point of departure appears between a plane parallel Fabry–Perot and a wedge-shaped Fizeau interferometer. The phase delay between the consecutive wavefronts produced by the wedged Fizeau mirrors is a progressively increasing quantity, rather than a constant one as in a parallel Fabry–Perot. The other difference is the spatial walk-off of the beams with multiple reflection (in Fizeau interferometry) that cannot be compensated for perfectly by any focusing or imaging device. Both these effects are displayed in Fig. 6.3.

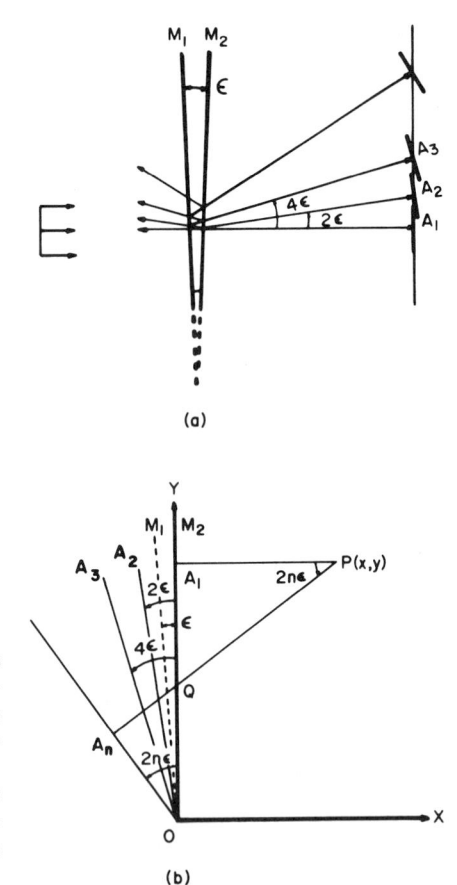

Figure 6.3. (a) Multiply reflected beams produced by a Fizeau interferometer consisting of a pair of "plane" mirrors M_1, M_2 with a wedge angle ϵ between them when illuminated by a collimated wavefront parallel to M_2. (b) Geometrical construction to aid in the computation of the relative path difference between multiply reflected beams produced by a Fizeau interferometer.

The progressive phase delay is easily derived by using Brossel's (1947) very general but elegant method, developed to compute the intensity distribution and localization of Fizeau fringes (see also Born and Wolf 1975, p. 286). We shall consider the particular case of perpendicular illumination of one of the mirrors by a collimated beam that is used in many optical shops (Figs. 6.3 and 6.6). Two partially transmitting mirrors M_1 and M_2 form a very small wedge angle ϵ, whose apex is at O. An incident plane wavefront, parallel to the mirror M_2, produces a series of beams due to multiple reflection. The arrow heads in Fig. 6.3a correspond to the "center" of the incident wavefront and hence demonstrate the beam walk-off defect of the Fizeau interferometer. Figure 6.3b shows the position of the multiply reflected wavefronts relative to the incident one at the plane of the mirror M_2 (Y plane). The choice of such a diagram to compute the phase delay (Brossel 1947; Tolansky 1948a) was based on the fact that none of the multiply reflected wavefronts suffers any phase delay relative to the in-

6.3. MULTIPLE-BEAM FIZEAU INTERFEROMETER

cident one (OA_1) at the apex of the wedge O, formed by the two Fizeau mirrors. The tilts of the multiply reflected wavefronts are designed by lines $OA_2(\angle 2\epsilon)$, $OA_3(\angle 4\epsilon)$, $\cdots$, $OA_n(\angle 2n\epsilon)$. If the effect of superposition is observed in transmission at a general point $P(x, y)$, the path difference between the first and the nth wavefronts is

$$d_n(x, y) = PA_n - PA_1. \tag{6.3}$$

But

$$PA_n = PQ + QA_n = PQ + OQ \sin 2n\epsilon$$
$$= PQ + (y - QA_1) \sin 2n\epsilon = x \cos 2n\epsilon + y \sin 2n\epsilon.$$

Then

$$d_n(x, y) = x(\cos 2n\epsilon - 1) + y \sin 2n\epsilon. \tag{6.4}$$

If the fringes are observed on mirror surface M_2 ($x = 0$), as is customarily done in optical shops through an imaging device, the expression for the phase difference simplifies to

$$d_n(0, y) = 2tn - \tfrac{4}{3} n^3 \epsilon^2 t \tag{6.5}$$

where we have used $t = y\epsilon$, the separation between the plates at y and the first two terms of the series expansion of $\sin 2n\epsilon$, considering $2n\epsilon$ to be still a small angle. For a parallel plate Fabry–Perot with orthogonal collimated illumination, the phase difference between the first and nth wavefronts is $2tn$. Then the nth beam of a Fizeau interferometer lags behind in phase, compared to $2tn$, by

$$\delta d_n = \tfrac{4}{3} n^3 \epsilon^2 t. \tag{6.6}$$

This lesser delay in the Fizeau than in the Fabry–Perot arithmetic series ($2tn$) is the single most important restriction against forming high-quality symmetric and sharp fringes, as was first realized by Tolansky. (This problem is avoided in white-light FECO interferometry, described in the next section, where the mirrors are set as nearly parallel as possible.) In particular, if

$$\frac{4}{3} n^3 \epsilon^2 t = \frac{\lambda}{2}, \tag{6.7}$$

the nth beam will add destructively rather than constructively to the first one. Hence, to obtain symmetric and sharp Fizeau fringes, one should restrict oneself to the Tolansky inequality,

$$\frac{4}{3} n^3 \epsilon^2 t < \frac{\lambda}{2}. \qquad (6.8)$$

Here the most rapidly varying quantity is n, the number of interfering beams. But we cannot set a low value for it because the very purpose of a multiple-beam Fizeau is to produce high-quality, high-finesse fringes that require superposition of a large number of regularly delayed beams; this number, in turn, depends on the reflectance of the surfaces. The broken curve in Fig. 6.4 shows how the finesse N_R (the ratio of the fringe interval to the full width of a fringe at half the height from its peak) of Fabry–Perot fringes increases with increasing reflectance. The solid curve in the same figure shows the effective number of beams M added by a computer to obtain a peak transmittance within 0.1% of

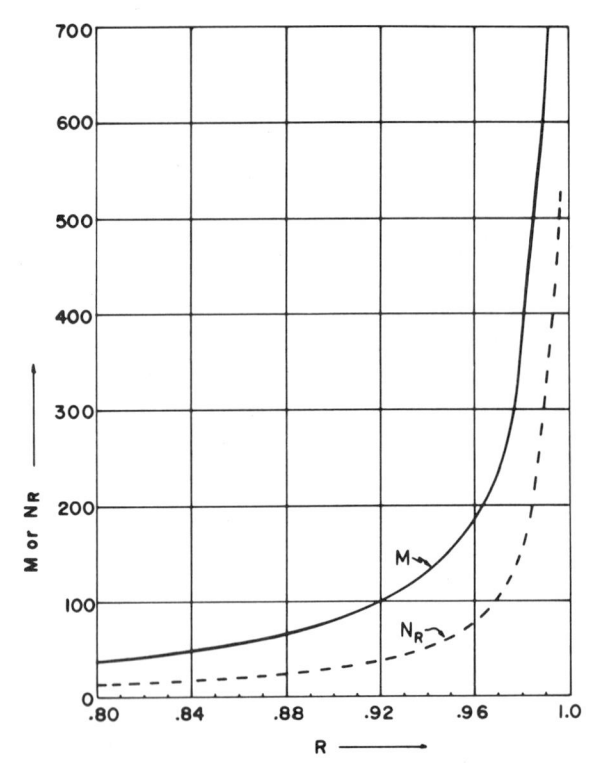

Figure 6.4. The number of interfering beams M that gives a value for the ideal Airy curve within 0.1% plotted against the reflectance R (continuous curve). The discontinuous curve shows the reflective finesse N_R plotted against the reflectance R.

6.3. MULTIPLE-BEAM FIZEAU INTERFEROMETER

the ideal Airy value [infinite sum of Eq. (6.1)] for different reflectances. It can be seen that, for a reflectance below 90%, the effective number of beams M that are superposed to obtain a finesse N_R is roughly equal to $2N_R$. For higher reflectivity the required value of M steadily increases.

These curves can be exploited for Fizeau interferometry in the following manner. If the surfaces to be tested have a given reflectance R (or equivalently, $R = \sqrt{R_1 R_2}$ when the reflectances are different for the surfaces), or one requires a desired finesse N_R such that the reflectance is R, then the curves of Fig. 6.4 indicate the value of M, the effective number of beams that are superposed. Then, to maintain the Tolansky inequality, one substitutes M for n in Eq. (6.8) and reduces the values of ϵ and t accordingly. Tolansky (1948a) gave some typical values of n, ϵ, and t for obtaining symmetric Fizeau fringes. The lowest limit of t, the plate separation, is usually determined by omnipresent dust particles whose average size is generally of the order of 1 μm. The lowest limit for the wedge angle ϵ is determined by the minimum number of fringes one wants within the field of view. The smaller the angle, the larger is the spatial separation between two adjacent fringes.

When the Tolansky inequality [Eq. (6.8)] is not satisfied (in other words, under the most general condition), the intensity profile of each Fizeau fringe at high reflectance is an asymmetric and composite one and constitutes a primary maximum followed by a series of secondary maxima on the side of the thicker wedge. The intensity of these secondary maxima rapidly decreases in the same direction as the increasing wedge separation. The primary maximum is broadened and is no longer symmetric like the Fabry–Perot fringes; also, its peak shifts toward the opening of the wedge with reduced height (Kinosita 1953). Figure 6.5 shows a qualitative representation of a general Fizeau fringe profile plotted against one due to Fabry–Perot. These general Fizeau fringes show a remarkable similarity to Tolansky fringes (Fig. 6.17) produced through a plane parallel Fabry–Perot but illuminated by various tilted rays from a point source. Such composite fringes will not destroy precision testing if the smooth edge of the fringes is used (Polster 1969). Characteristic Fizeau fringes are formed not only on the plane of the thin wedged film but also at various well-defined distances on planes named after Feussner (Barakat et al. 1965).

There is another important constraining parameter of a Fizeau interferometer: the lateral displacement Δ_n of the nth beam due to the wedge angle ϵ (Tolansky 1948a),

$$\Delta_n = 2n^2 t\epsilon. \qquad (6.9)$$

This implies that each multiple-beam fringe formed by a Fizeau interferometer always gives the contour map of the average surface topography over a region spanned by Δ_n, rather than a precise geometrical point-by-point mapping.

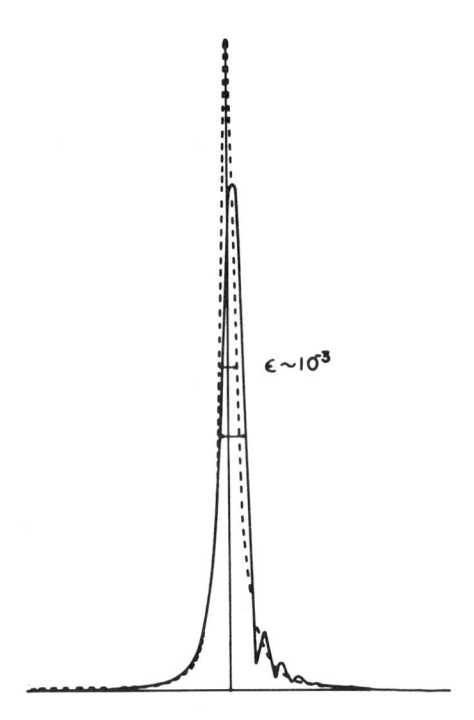

Figure 6.5. A composite Fizeau fringe (continuous curve) produced by a pair of flat wedged ($\sim 10^{-3}$ radian) mirrors compared with a symmetric Fabry–Perot fringe (discontinuous curve) produced by the same mirrors when they are perfectly parallel.

6.3.2. Fizeau Interferometry

The essential components of a Fizeau interferometer arrangement are shown in Fig. 6.6. A monochromatic or quasimonochromatic collimated beam illuminates the coated flats (M_1, M_2) forming a wedge. The fringes can be observed both in transmission (O_T) and in reflection (O_R). The fringes in transmission are sharp and bright, in very high contrast against an almost dark background, and those in reflection are exactly complementary (by simple energy conser-

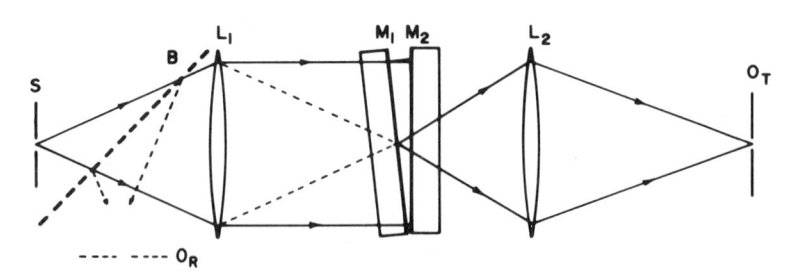

Figure 6.6. The essential components of Fizeau interferometric arrangement: S, point source, B, beam splitter to observe fringes in reflection (O_R); L_1, collimator; M_1, M_2, Fizeau mirrors, the wedge and separation very much exaggerated; L_2, observing lens, which images the localized fringes between the mirrors at the observation plane O_T.

6.3. MULTIPLE-BEAM FIZEAU INTERFEROMETER

vation), that is, they are dark fringes against a bright background (Fig. 6.7). The fringes in reflection, however, may have very poor contrast when the reflection coating (such as a silver layer) has a very high absorption coefficient (Tolansky 1948a). All Fizeau fringes have the same fringe interval of $\lambda/2$, just as do the multiple-beam Fabry–Perot or regular two-beam fringes. This is the reason why sharp Fizeau fringes leave most of the surface area under test with-

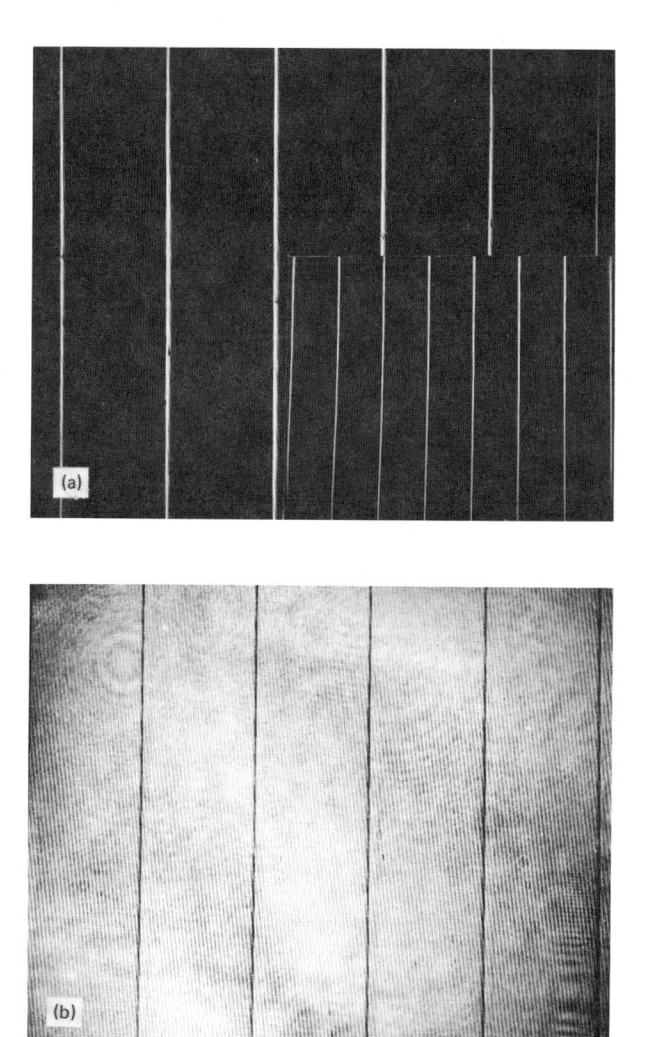

Figure 6.7. Fizeau fringes in transmission and in reflection. (*a*) Fringes in transmission constitute sharp, bright lines against a dark background. Insert: the same transmission fringes but without the collimating lens L_1 of Fig. 6.6. (*b*) Fringes in reflection constitute complementary fringes of dark lines against a bright background. (Fine fringes are due to back reflection from the beam splitter.)

out much available information. Alternatives to this arrangement are discussed in the following sections.

Since the surface contour fringes show the total optical path variation of both flats, one of them should be a very high quality reference flat so that the fringes can be interpreted as contours due only to the flat being tested (Clapham and Dew 1967). One way of obtaining an extremely high quality flat is to use a liquid surface, like liquid mercury. A detailed description of such a Fizeau interferometer has been given by Bünnagel et al. (1968). However, deviation of a surface from absolute planeness can also be obtained without the use of an absolute reference flat. The method (Schulz 1967; Schulz and Schwider 1967; Schulz et al. 1971) requires the use of three flats, none of which is a high-quality reference flat.

Because of the Tolansky inequality [Eq. (6.8)] good-quality Fizeau fringes are obtained more easily if the separation between the plates, t, is set to a very small value. However, this is not an absolute necessity, as has been demonstrated by Moos et al. (1963), who used a highly collimated laser beam to obtain the surface contours with a plate separation as large as 20 cm. Of course, the tilt must be very small; in their case it was less than 10^{-4} rad. This point is worth remembering for shopwork because a large separation ensures against developing scratches or spoiling the reflection coating by physical contact. A large separation requires a high-quality mounting to maintain a stable relative alignment.

In an interesting example of off-axis illumination for improving fringe sharpening, described by Langenbeck (1970), the incident angle is chosen in such a way that the incident beam is first reflected toward the apex of the Fizeau plates and then, after a controllable finite number of reflections, is reflected away from the apex. This can be appreciated from Fig. 6.8, where, if the illumination follows the direction of the ray SA_1, after a few reflections it becomes perpendicular to mirror M_2 at A_n and starts retracing its path. In this manner sharper Fizeau fringes can be formed in special situations where a relatively large sep-

Figure 6.8. Fringe sharpening in a Fizeau interferometer. For a given wedge between the Fizeau mirrors M_1, M_2, a suitable choice of the direction of illumination SA_1 can reduce the range of beam walk-off and increase the effective number of interfering beams to sharpen the fringes.

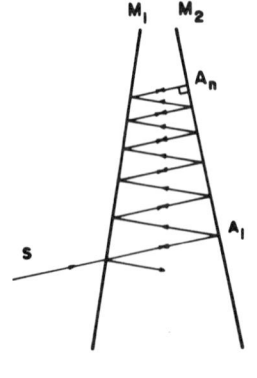

aration between Fizeau mirrors is required by a particular test object, such as a corner cube (Langenbeck 1970).

Eastman and Baumeister (1974) designed a regular Fizeau interferometer with one of the mirrors mounted on a piezoelectric scanning device. Instead of photographic recording, the fringes can be detected and analyzed electronically. The reported accuracy is about 20 Å, but techniques to improve the measurement precision are probably available. Fizeau interferometry with illumination of two wavelengths (Shaalan and Little 1975) is discussed briefly in the next section.

6.4. FRINGES OF EQUAL CHROMATIC ORDER

In 1945 Tolansky developed a new technique of surface microtopography using white-light illumination. The interferometer has evolved from the multiple-beam Fizeau interferometer, where the fringes are formed by a thin, wedge-shaped film bounded by two highly reflecting surfaces. This wedge reduces the fringe sharpness (finesse) because of beam walk-off (Fig. 6.3). Therefore Tolansky (1945a, 1945b) set the surfaces as parallel as possible, increasing the inherent finesse. With this arrangement a collimated and orthogonal illumination with white light will produce a channeled spectrum (Jenkins and White 1957) of all the wavelengths, λ_i, that satisfy the relation

$$2t = n_i \lambda_i. \tag{6.10}$$

However, to see the spectrum one needs to use a spectrograph for dispersion. When a narrow section of an ideal interference film (t constant) is imaged through a spectrograph, a series of laterally separated and parallel straight fringes (the channeled spectrum) that satisfy Eq. (6.10) is displayed. If there is a variation in t within the imaged section of the film, each point is passing a group of waves corresponding to the local value of t. The spectrographic image then consists of nonstraight "fringes," each one of which shifts along the wavelength scale, keeping t/λ constant. As stated earlier, the name fringes of equal chromatic order (FECO) is used since along each fringe the order number n is constant.

The basic interferometric setup is illustrated in Fig. 6.9. Figure 6.10a shows some FECO fringes (black and white reproduction) of a cleaved mica surface, the topographic variation of which is shown in Fig. 6.10b (Tolansky 1945a). Determination of the "hills" and "valleys" of a surface microtopography becomes very simple since fringes are convex to the violet on the "hills" and are concave to the violet on the "valleys." The steepness of the incline at a local region of a surface is given by the number of fringes intersected per unit length

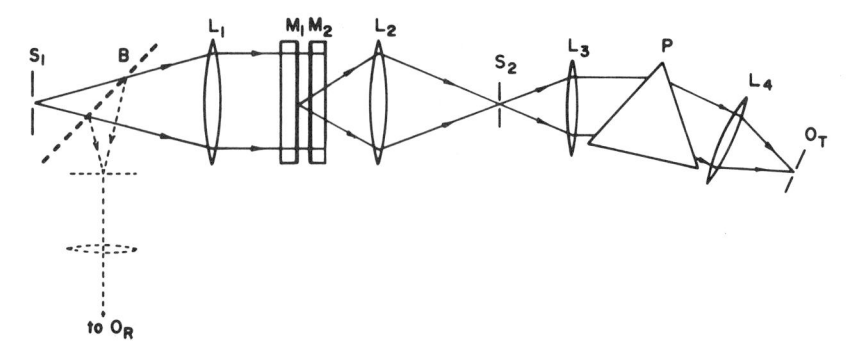

Figure 6.9. The essentials of the interferometric arrangements for observing fringes of equal chromatic order (FECO) either in transmission (O_T) or in reflection (O_R): S_1, white-light point source; B, beam splitter to aid observation in reflection; L_1, collimating lens; M_1, M_2, parallel mirrors forming the interferometer, one of which is being compared to the other as a reference; L_2, lens that images the channeled spectrum from a small section of M_1–M_2 on the spectrographic slit S_2 (slit S_2, lenses L_3 and L_4, and prism P form the spectrograph); O_T, observation plane for fringes in transmission.

of the vertical section of the fringes (Fig. 6.10). For a detailed exposition of FECO interferometry the reader is referred to Tolansky (1948a).

Fringes of equal chromatic order interferometry find a very useful application in the accurate determination of thin-film thickness and in surface roughness measurement (Bennett and Bennett 1967; Bennett 1976; Eastman 1975). The steadily increasing demand for very high quality optical surfaces in modern optical technology has motivated the development of FECO interferometry for studying surface roughness and has led to a better understanding of the mechanism of polishing (Koehler and White 1955; Koehler 1955b; Vinokurov et al. 1962; Hodgkinson 1970). One should be careful, however, to consider the problem of the phase change on reflection due to the material coated on the substrate (Tolansky 1948a; Koehler 1953; Schulz 1951a) and also the dependence of the phase change on the thickness of the thin-film material (Schulz and Scheibner 1950). It has been observed that the original surface microtopography is more faithfully contoured by a metallic silver film (Tolansky 1948a) than by a dielectric film.

Following the lead of Tolansky, Shaalan and Little (1975) reported a different technique of exploiting white-light fringes for the study of surface microtopography. To exploit the full advantage one should use an illumination for the Fizeau interferometer that has two wavelengths simultaneously present from two calibrated monochromators (like constant deviation spectrometers). Figure 6.11 shows the experimental arrangement. With this technique the direction of height of crystal cleavage steps or of thin films can be measured with great

6.4. FRINGES OF EQUAL CHROMATIC ORDER

(a)

(b)

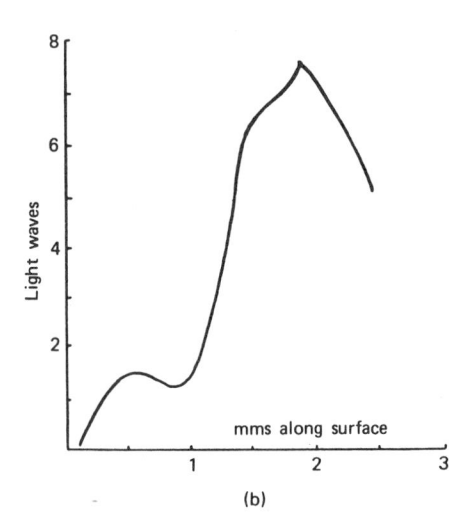

Figure 6.10. (a) Fringes of equal chromatic order (in black and white) from a freshly cleaved and silvered mica surface tested against a reference flat mirror. The dense fringes at the central region along the vertical direction indicate a sharp ridge. (b) A quantitative plot of the surface height variation along the vertical direction depicted by the fringes of (a). (From Tolansky 1970).

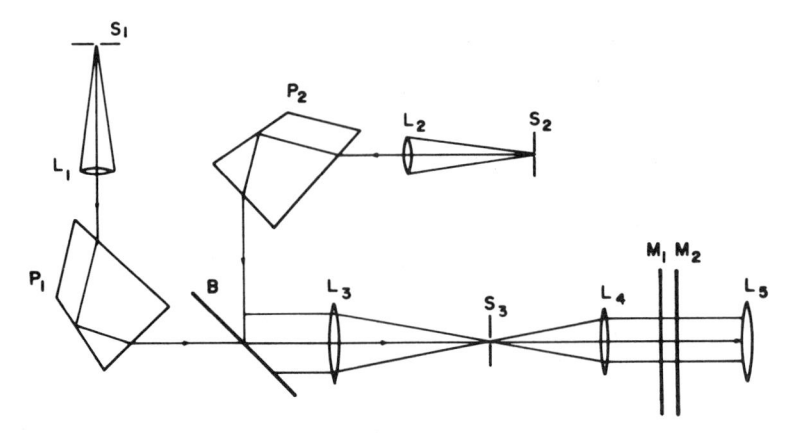

Figure 6.11. Fizeau interferometry with white light, using two calibrated monochromators: S_1, L_1, P_1 and S_2, L_2, P_2, the two monochromators; B, beam splitter; L_3, common spectrum-forming lens for both monochromators; S_3, entrance split for the Fizeau interferometer (M_1–M_2); L_4, collimator for the interferometer; L_5, observing lens.

facility. Any region of the surface under test can be scrutinized by first choosing one of the wavelengths to form a fringe in the region of interest, and then adjusting the second monochromator to produce a different colored but matched fringe on the other side of the step (Fig. 6.12). Since the two wavelengths are known, the direction and the height of the step are determined easily using Eq. (6.10); but one should take into account the phase change on reflection, which is not explicit in this equation (Shaalan and Little 1975).

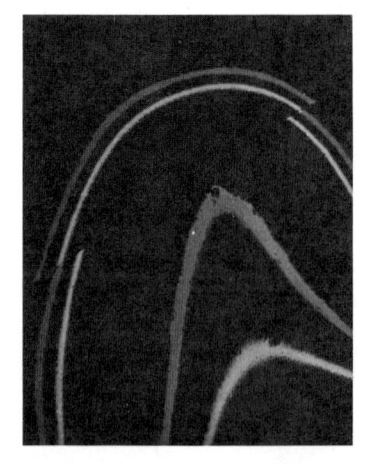

Figure 6.12. Multiple fringes of variable chromaticity meeting across a step. The lighter fringe corresponds to the red and the darker one to the green radiation as they appear in the original, which was in color. (From Shaalam and Little 1975).

6.5. REDUCTION OF FRINGE INTERVAL IN MULTIPLE-BEAM INTERFEROMETRY[†]

In most regular interferometry with monochromatic or quasimonochromatic illumination, the fringe interval is $\lambda/2$, the fringes being contours of equal optical thickness. This is the reason why in multiple-beam interferometry, the fringes being very sharp and of high contrast, most of the surface area under test produces almost no information about the surface topography. Even in the case of interferometry with white light, discussed in Section 6.4, only a small surface area is imaged on the spectrometer slit. One can scan the surface of the test object region by region. However, this procedure is somewhat tedious.

A simpler solution (Saunders 1951), using "pressure scanning" with normal quasimonochromatic illumination, gives sharp Fizeau fringes but obtains a fringe interval less than $\lambda/2$. The basic technique requires placing the entire interferometer in an airtight chamber and taking a series of exposures of the fringe system, the fringes being shifted between every exposure by the desired amount. The shift is achieved by a controlled amount of change in the optical path through a change in air pressure. The same objective can be achieved a bit more conveniently by mounting one of the mirrors on piezoelectric scanning devices and applying a suitable staircase voltage to the piezoelectric (Roychoudhuri 1974). This eliminates the necessity of putting the entire interferometer within an airtight system.

In a different solution demonstrated by Herriott (1961), one illuminates the interferometer simultaneously with a discrete set of wavelengths obtained through a monochromator whose entrance illumination consists of a set of spatially separated slits, instead of a single slit. Later Murty (1962) suggested a simpler solution, whereby one can use a regular quasimonochromatic (or monochromatic) source with a set of spatially separated pinholes to illuminate the interferometer. A multiple set of fringes with a reduced fringe interval is provided. A simple computation to achieve the desired fringe interval has been given by Murty (1962). Schwider (1968) also used Herriott's idea of multiple-wavelength illumination but employed the channeled spectra from a suitable Fabry–Perot. The flexibility of the test is greatly increased if one uses a tunable laser source, as described by Pilston and Steinberg (1969). With the continuously tunable dye lasers currently available, such tests can be carried out very conveniently and rapidly.

6.6 PLANE PARALLEL FABRY–PEROT INTERFEROMETER

It is somewhat surprising that, even though the capabilities of the plane parallel Fabry–Perot (1897) interferometer as a high-resolution spectroscopic instru-

---

[†]The fringe interval can also be decreased by multipass interferometry, discussed in Chapter 7.

ment were well understood, the interferometer was not used for surface measurements until fairly recently (Schulz 1950a, 1950b). The standard techniques that evolved for its utility as a spectroscopic instrument require (a) illumination by an extended source to obtain fringes of sufficient brightness and (b) a reasonable separation between the plates to provide sufficient resolving power. These two general conditions probably inhibited the use of this interferometer for surface topography because the fringes formed under these conditions cannot describe the local defects region by region; rather, every part of the fringes is characteristic of the overall defects of the entire surface (Chabbal 1953, 1958). In the following discussion we describe briefly, in chronological order, a series of uses of the Fabry–Perot interferometer for surface testing.

6.6.1. Measurement of Thin-Film Thickness

The thickness of thin evaporated films can be measured to within an accuracy of ± 15 Å (or better) by using conventional Fabry–Perot fringes (Schulz 1950a, 1950b). The method exploits the fact that a simple sodium vapor lamp emits a close doublet (5890 and 5896 Å) that forms, in general a pair of closely spaced fringes and, in particular, consonance and dissonance of the pair of fringes, depending on the separation between the plates. Then a knowledge of the wavelengths and plate separations for suitable positions of the pair of fringes on the two different layers of the plate gives one the thickness of the thin film. The precise method of quantitative analysis has been described by Schulz (1950a, 1950b). The advantage of the method over the multiple-beam Fizeau is that the interferometer-forming surfaces do not have to be pressed together so closely as to endanger the high-quality surfaces and/or reflection coatings. [But here it should be remembered that Fizeau fringes can also be formed with large separation of the plates but with a very small wedge angle (Moos et al. 1963).]

Schulz (1951a, 1951b) extended this technique to measure the phase change on reflection from a thin film, its dependence on the wavelength and the thickness of the material, and also the absorption coefficient of the film.

6.6.2. Surface Deviation from Planeness

Benedetti-Michelangeli (1968) developed a method to measure the local defects of plane surfaces by using a narrow collimated laser beam to illuminate the local region of the Fabry–Perot interferometer formed by the test and the reference plate. The computation of the defect is carried out by using the standard relation for the diameter of the Fabry–Perot fringes with the interferometer parameters as explained in this paper. The method of illumination consists of a narrow collimated beam incident on the parallel plates at an angle that can be varied in a smooth manner. The reported accuracy of the flatness variation is $\lambda/400$, but, as has been rightly claimed, the potential limit is much higher.

6.6. PLANE PARALLEL FABRY–PEROT INTERFEROMETER

Again, one advantage of this method is that the plates under test do not have to be pressed together so closely as to endanger the surfaces. [An important virtue of this short paper of Benedetti-Michelangeli (1968) is that it describes briefly but succinctly all the important factors that influence the fringe qualities of the Fabry–Perot.] A more accurate method is to exploit the steep slope of an Airy transmission curve and use a narrow collimated illumination that is orthogonal to the parallel plates. The regional defects are obtained by changing the position of the narrow collimated beam. An accuracy of $\lambda/1000$ can be obtained by this method (Koppelmann and Drebs 1961). A major defect of either of these methods is that the minimum area for which the average defect can be determined is limited by the spatial size of the collimated scanning beam, which can be scarcely smaller than a millimeter. Methods for determining point-by-point defects of surfaces are described in the following paragraphs.

It appears that Hodgkinson (1969) was the first one to exploit the full potentiality of the Fabry–Perot interferometer for the study of surface defects. Most multiple-beam interferometric tests for surface topography suffer from the general disadvantage that they do not record the complete topography of the entire surface at a time; rather, they sample it only along the narrow fringes, leaving most of the surface without any available information, as mentioned before. Techniques to reduce the fringe interval (Section 6.5) alleviate the problem only partially. Hodgkinson (1969) developed a method of recording an integrated interferogram with the transmitted wavefront by slowly moving one of the mirrors of the Fabry–Perot parallel to itself. The illuminating beam being an orthogonal collimated monochromatic radiation, the transmission of the interferometer at every point is proportional to the local plate separation. All the topographic information, as well as the defect distribution function (Chabbal 1958), can be obtained from the transmission characteristics of this integrated interferogram when it is properly developed (Hodgkinson 1969). The exposure and the development are such as to produce a transmission characteristic that is almost linearly proportional to the surface defects. Because of the limits on the precision obtainable from photographic work, the measurement of surface defects by this method has a precision of around $\lambda/500$.

It is possible to dispense with the intermediate record of an integrated interferogram if the transmitted wavefront is sensed by a spatially scanning photodetector (or a high-resolution vidicon camera) and the information is stored for detailed analysis (Fig. 6.13). The reference mirror can be mounted with piezoelectrical scanning devices, and the separation between the mirrors adjusted precisely so that the detector records half the peak transmission, say, from a small central area. Then the transmission from this central spot can be used as the reference signal to measure the deviation from planeness of the other points. For a direct graphical computation the transmission curve along any diameter of the test surface is compared against the "ideal" Airy curve, recorded through the reference central spot as an oscilloscopic trace while scanning one of the

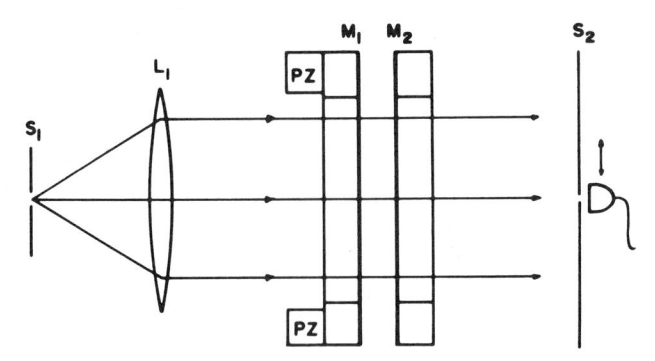

Figure 6.13. A Fabry–Perot interferometric arrangement for evaluating the surface defects of all the points of a pair of mirrors. The entire interferometer is illuminated by a collimated beam. Defects are determined from the change in transmission. One of the mirrors is mounted on piezoelectric material (PZ) to choose the desired value of transmission.

mirrors. This is illustrated in Fig. 6.14. The half-width of the "ideal" Airy curve is given by

$$pp' = \frac{2\pi}{N_R} \equiv \frac{\lambda}{2N_R} \tag{6.11}$$

Also, the abscissa (Fig. 6.14a) is linear in phase, and the finesse N is known either by measurement from the actual Airy curve obtained from a small central

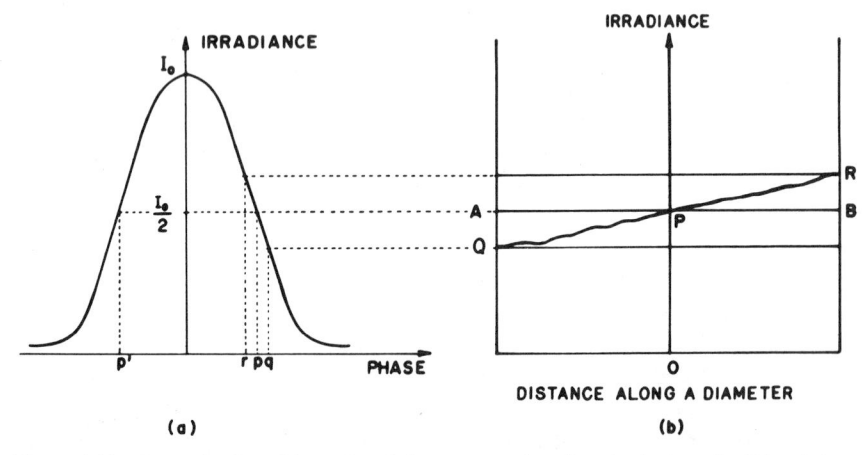

Figure 6.14. Determination of the surface defects every point of a pair of mirrors by Fabry–Perot interferometry. (a) Reference (or ideal) Airy transmission curve for a Fabry–Perot with calibrated axes. (b) Transmission curve along a particular diameter of the Fabry–Perot when the transmission from the small central reference region was adjusted (using the PZ) to exactly half the peak transmission.

6.6. PLANE PARALLEL FABRY-PEROT INTERFEROMETER

reference region or from the value of the reflection coating (Fig. 6.15). Then any fluctuation in the transmission, AQ or BR, that corresponds to an optical path change of pq or pr, respectively, can be directly transformed into fractions of a wavelength using Eq. (6.11). The precision of the method depends on the minimum detectable signal like pq that is controlled by the slope of the Airy curve (Polster 1969). This precision is limited by the noise inherent in the photodetection. Nevertheless, a precision of $\lambda/1000$ or better can be obtained with an oscilloscope and a photodetector of moderate quality when the reflective finesse N_R of the interferometer is about 100 (reflectivity $\sim 97\%$).

We conclude this section by mentioning another advantage of Fabry-Perot interferometry. So far, we have discussed the problem of determining the positions and sizes of local defects of surfaces. However, in most optical testing the approximate overall (average) quality, λ/m, of the entire surface is also of interest. This is most readily obtained by assembling a master plate and the

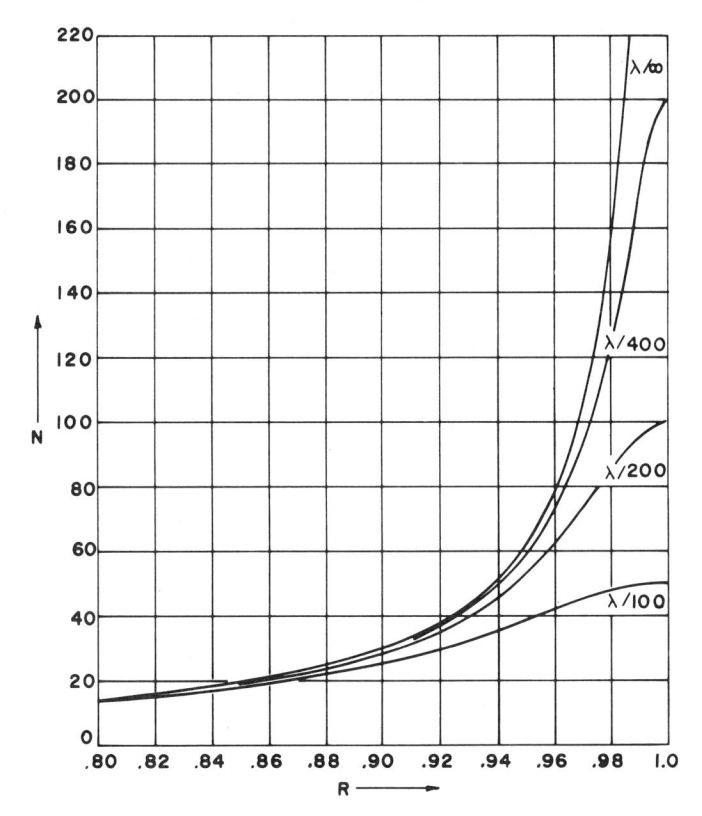

Figure 6.15. Effective surface finesse N plotted against reflectance R for a Fabry-Perot with mirrors having a surface defect of residual spherical curvature. Curves for four different cases of surface flatness are shown.

plate under test to form a Fabry–Perot of known ideal reflective finesse and comparing this quantity with the experimental finesse give by the interferometer. Presented in Figs. 6.15 and 6.16 are two sets of curves for effective finesse against surface reflectivity; the first one is for pure residual spherical curvature, and the second one is for Gaussian random deviation from perfect flatness. These curves were computed (Roychoudhuri 1973b) using Chabbal's analysis (1958). The usefulness of the curves is illustrated by a simple example. Suppose that the reference flat and the flat under test have 98% reflectivity, giving an ideal reflectivity finesse close to 155, but the experimental finesse is, say, 80. Then, with a careful look at the curves of Figs. 6.15 and 6.16, one can conclude that, if the surface under test has a pure residual spherical curvature, the magnitude of the mean deviation from flatness is somewhat below $\lambda/200$; or if the

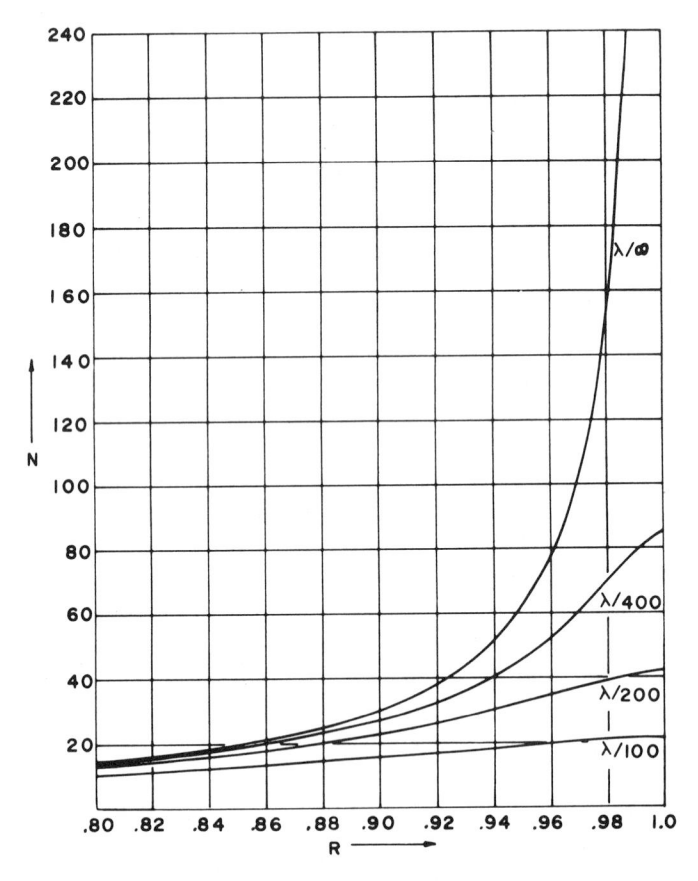

Figure 6.16. Effective surface finesse N plotted against reflectance R for a Fabry–Perot with mirrors having a surface defect of the Gaussian type. Curves for four different cases of surface flatness are shown.

deviation is of Gaussian nature or is a combination of both regular and Gaussian, the mean deviation must certainly be less than $\lambda/400$. Furthermore, one can use the value of the peak transmission (Chabbal 1958; Jacquinot 1960; Hodgkinson 1969) or the nature of the broadened Fabry–Perot fringes (Hill 1963; Bhatnagar et al. 1974) to discern and characterize the surface deviation more precisely.

6.7. TOLANSKY FRINGES WITH FABRY–PEROT INTERFEROMETER

When a pair of highly reflecting plane parallel surfaces is illuminated by a point source, a characteristic series of composite circular fringes (with decaying secondary maxima) can be observed at any distance from the surfaces. Such nonlocalized fringes appearing on diverging conical surfaces were first noted and used for the study of thin crystal plates by Tolansky (1943, 1946). Therefore we shall call these nonlocalized multiple-beam fringes Tolansky fringes.

Tolansky fringes can be used for localized precision testing by imaging a point source between the mirror surfaces. For this method to be effective the mirrors must be fairly close. The observation can be carried out without the use of any microscope or similar device by simply intercepting the fringes with a distant wall (Fig. 6.17). Tolansky (1948a) gave a simplified analysis. The an-

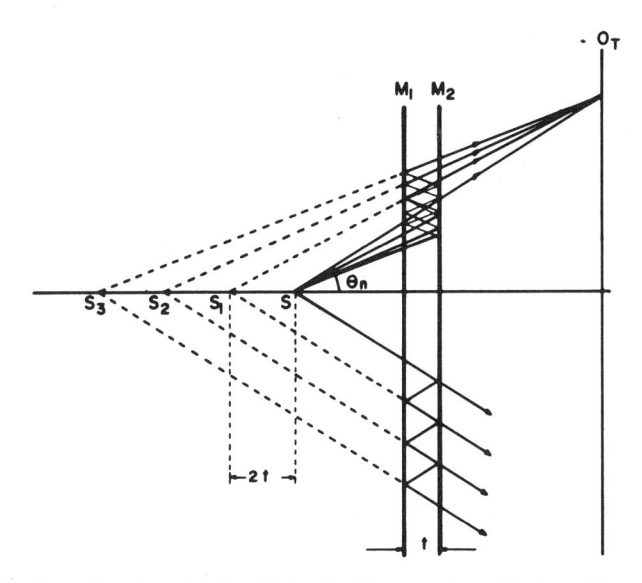

Figure 6.17. Formation of nonlocalized Tolansky fringes at plane (O_T) by a Fabry–Perot $(M_1$–$M_2)$ when it is illuminated by a point source S.

gular diameters of the fringes are approximately given by $2\theta_n$, where θ_n follows the Fabry–Perot formula,

$$2t \cos \theta_n = n\lambda. \qquad (6.12)$$

A detailed computation of the characteristics of these fringes was made by Aebischer (1971).

Tolansky fringes also finds useful application in the quick alignment of a laser cavity (Bergman and Thompson 1968) and the Fabry–Perot interferometer (Ford and Shaw 1969; Roychoudhuri 1973b). For this purpose, if transmission fringes are used, the secondary maxima should be aligned to be perfectly concentric with the primary maxima; a clear residual tilt in the alignment is indicated when the secondary maxima cross the primary maxima (Figs. 6.18a and 6.18b). However, when the fringes in reflection are used for alignment (the weak secondary, dark fringes are almost invisible), the fringe system should be made concentric with the axis of the cone of fringes (or the illuminating point source).

Another application of Tolansky fringes involves measurements of the long-term stability of high-quality Fabry–Perot interferometers (or a laser cavity). As mentioned before, the fringes formed by the Fabry–Perot under test can be intercepted with a distant wall, and the contraction (expansion) of the fringes measured in millimeters to obtain the longitudinal expansion (contraction) of the Fabry–Perot separation in fractions of a wavelength. The tilt of the plates can also be measured by observing the crossing of the secondary fringes over the primary ones. We have used this technque to measure the long-term stability of a commercial Fabry–Perot (all-Invar structure) and one of our own construction (all-Cervit structure with thermally compensated mounts) (Roychoudhuri 1973a, 1973b). The latter Fabry–Perot shows better stability.

6.8. MULTIPLE-BEAM INTERFEROMETER FOR CURVED SURFACES

As mentioned in the introduction, multiple-beam interferometers to test curved surfaces were developed independently by Herriott (1966), by Heintze et al. (1967), and by Biddles (1969). Such interferometers are very useful for precision-testing of various surfaces of different curvatures against a suitable master surface when they are arranged as a concentric system. A similar interferometric setup can also be adapted to other possible precision measurements, such as thermal, pressure, or composition gradients in wind tunnels, shock tubes, and so on (Herriott 1966).

The essential elements of the interferometer are portrayed in Fig. 6.19, where a master aplanatic M_1 and a surface under test M_2 form a concentric system. The point source S is imaged at C, the common center, by the lens L_1. With

6.8. MULTIPLE-BEAM INTERFEROMETER FOR CURVED SURFACES 233

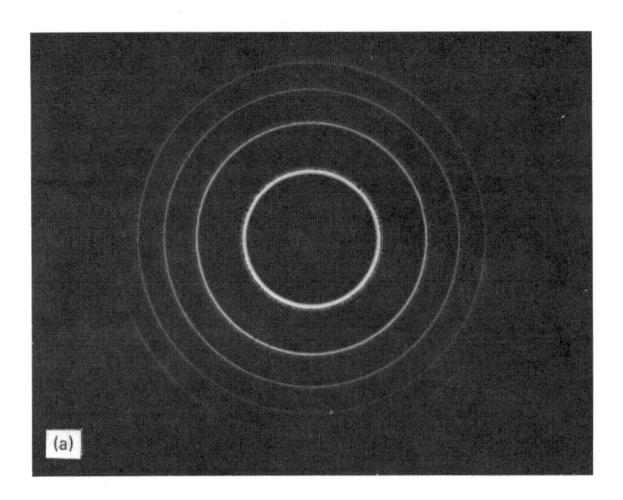

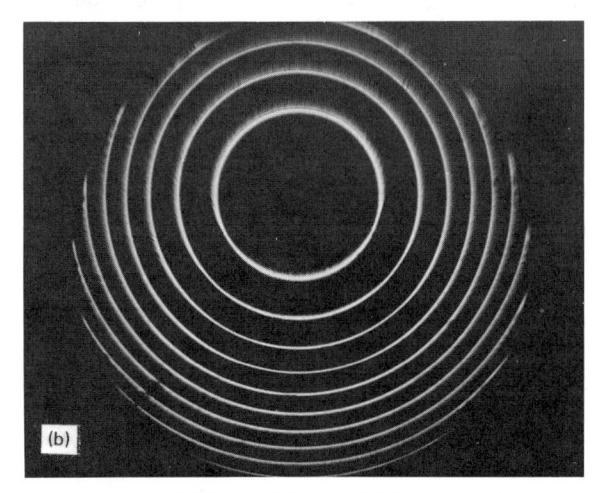

Figure 6.18. Some nonlocalized Tolansky fringes. (*a*) Fabry–Perot plates were "perfectly" parallel, and (*b*) the plates were tilted.

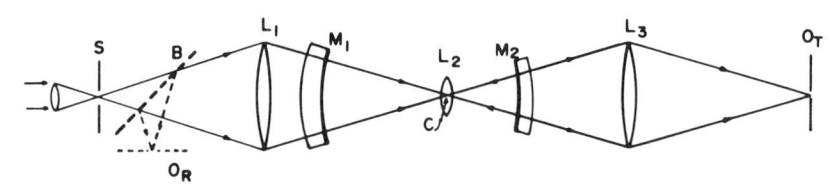

Figure 6.19. Multiple-beam interferometric arrangement for testing curved surfaces. Mirrors M_1 and M_2 are set almost concentric with point C. Lens L_1 images point source S at common center C. Observation can be carried out both in transmission (O_T) and reflection (O_R). Lens L_2 images surface M_1 on M_2 and vice versa to correct the walk-off defect.

the precise concentric arrangement, one finds at the observation plane (O_T or O_R) a uniform wavefront with perfectly spherical surfaces. To produce contour fringes, one displaces one of the mirrors by a small amount either laterally (to obtain straight fringes) or longitudinally (to obtain circular fringes). [The latter technique is also used with a confocal spherical Fabry–Perot to produce high-dispersion spectral fringes (Persin and Vukicevic 1973).] Since such displacement introduces walk-off of the beams, Herriott (1966; see his Fig. 5) introduced a compensating lens at the common center to image one mirror to the other in such a way that the reflected rays are deviated back to the same points on the mirrors, thus preserving the localized surface testing capability. For various modifications of this basic interferometer see the original references (Herriott 1966, Heintze et al. 1967; Biddles 1969). See Rafalowski (1988, 1990) for testing coma of decentration and asymmetric wavefronts with confocal Fabry–Perot interferometers.

Such spherical interferometers can also be used with multiple (say, n) wavelength illumination (Herriott 1966) to reduce the fringe interval from $\lambda/2$ to some $\lambda/2n$ and thus increase the precision of testing, discussed in Section 6.5. Since such interferometers normally have large separation, one can use regular multilongitudinal mode lasers to advantage by appropriately matching the lengths of the laser and the interferometer cavities. In matching the cavities, one should carefully take into account the separation of the laser modes (laser cavity length), the free spectral range of the interferometer (interferometric cavity length), and the finesse of the interferometer. Here matching does not imply equality of the cavity lengths; they can be integral or fractional multiples of each other as the situation demands (Herriott 1966). In using a regular multilongitudinal laser for long-path interferometers, one must be aware of the difficulties in obtaining stable high-contrast fringes (Batishko and Shannon 1972).

6.9. COUPLED AND SERIES INTERFEROMETERS

In this section we describe two interesting and useful modifications of multiple-beam Fizeau interferometers. The first one consists in coupling a Fizeau interferometer into a Twyman–Green interferometer as the dual interferometric arrangement shown in Fig. 6.20 (Pastor and Lee 1968: Langenbeck 1968; Aebischer 1970). For more details on such interferometers see Cagnet (1954) and Candler (1951). The second modification uses three plates in series instead of two (see Fig. 6.22), as in conventional Fizeau interferometers (Saunders 1954; Post 1954; Roberts and Langenbeck 1969).

6.9.1. Coupled Interferometer

The coupled interferometer, which consists of a Twyman–Green interferometer with one of its mirrors replaced by a Fizeau interferometer, combines several

6.9. COUPLED AND SERIES INTERFEROMETERS

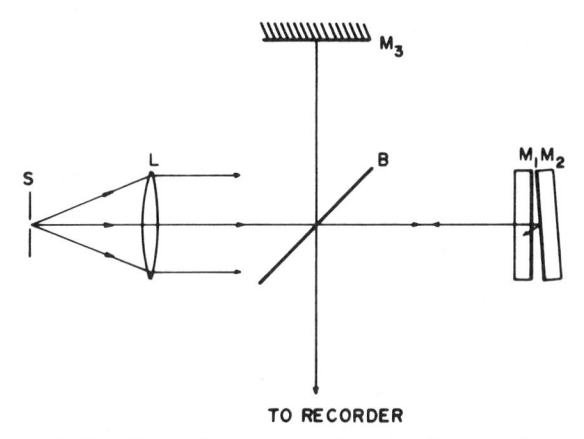

Figure 6.20. A coupled interferometric arrangement between a Twyman–Green and a Fizeau to control the contrast and exploit the dynamic moiré technique in Fizeau interferometry.

advantages in a single setup. Studying opaque optical flats in a Fizeau setup poses a general drawback because the fringes in reflection are sharp black lines against a bright background. This in itself does not create the main problem, although the contrast is poor with absorbing reflection coatings like silver film (Tolansky 1948a; Schulz 1951b). The Tolansky condition [Eq. (6.8)] for good-quality Fizeau fringes forces one to have low-frequency fringes in the field of view, and since the fringes are very sharp, most of the area does not produce any information, as has already been mentioned. (Several solutions to this problem were discussed in Sections 6.4 and 6.5.) The use of white-light (Section 6.4) or multiple-wavelength illumination (Section 6.5) is ruled out for reflection fringes since the large bright background due to one wavelength will wash out the sharp dark fringes due to another wavelength. This problem, which is inherent in studying opaque surfaces by reflection Fizeau, can be eliminated by the use of this dual interferometer (Fig. 6.20) to reverse the contrast and obtain transmission-like fringes in reflection Fizeau interferometry (Pastor and Lee 1968). However, in such a coupled interferometer the reference surface (of the Fizeau arm), the Twyman mirror surface, the beam splitter, and the light collimation must be of very good optical quality.

The other advantages of this coupled interferometer have been described in detail by Langenbeck (1968), and we shall mention only the major points. It is well known that the Moiré interferometry increases the sensitivity of the test under proper conditions (see Chapter 12 for detailed background), where one superposes the test interferogram against a suitable master reference pattern. A Moiré pattern displaying only the absolute error will be presented, provided that the master reference pattern has been made to include the inherent aberrations of the interferometer itself (e.g., the errors introduced by the beam splitter in a Twyman–Green). Such a master reference can be obtained "live" from the

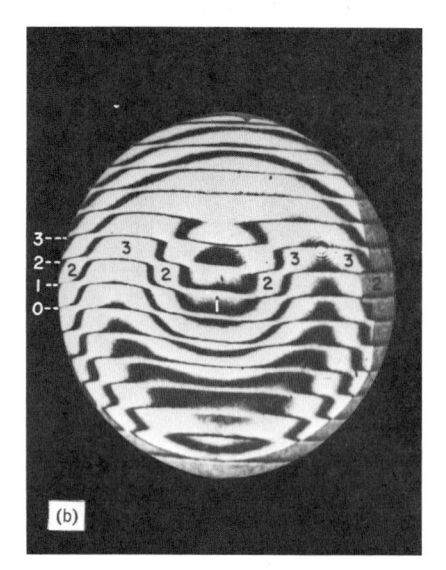

Figure 6.21. (*a*) Twyman-Green interferogram of a deformed mirror. Concentric fringes may be high or low areas. Dynamic observation is needed to determine the direction of order or interference. (*b*) Same, but with superimposed wedge field, permitting determination of the direction of the order of interference. (From Langenbeck 1968).

Fizeau wedge in one arm of the dual interferometer. Such "live" fringes also have another very useful advantage: The absolute direction of a surface deviation ("hills" and "valleys") can be read directly from the resultant interferogram, especially if the Fizeau fringes are used as the reference Moiré grid (because of the direct knowledge of the order of the reference fringes). This is illustrated by the photograph in Fig. 6.21, taken from Langenbeck (1968).

6.9.2. Series Interferometer

In this second modification of a Fizeau interferometer, introduced to measure the homogeneity of optical plates, the plate under test is inserted between two reference Fizeau mirrors (Fig. 6.22). Such a device is necessary for testing high-quality beam splitters whose function is not only to reflect but also to transmit wavefronts without distortion. Hence a simple surface flatness test of the reflecting surface of the beam splitter is not sufficient. The Fizeau fringes alone in reflection, with the plate under test on the side of the observer, cannot directly map the index variation when there is a simultaneous surface variation. Since the interfering light beams in such a three-plate interferometer (Fig. 6.22) pass through the plate under test many times, the sensitivity of the measurement of the index variation or homogeneity increases by a large factor. Such interferometers have been named by their independent discoverers as the "in-line interferometer" (Saunders 1954) and the "series interferometer" (Post 1954).

6.11. FINAL COMMENTS

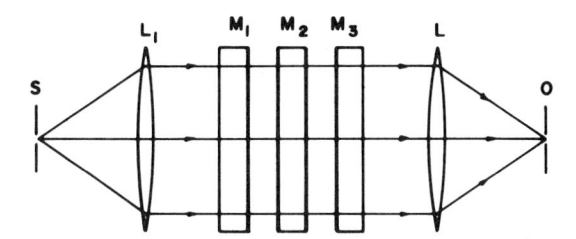

Figure 6.22. A series interferometer with three mirrors for precision testing of the homogeneity of a flat (M_2).

The various different conditions and reflectances for the plates of such interferometers have been described in the references cited. Of the more recent papers along similar lines by Ashton and Marchant (1967) and by Roberts and Langenbeck (1969), the latter describes how to evaluate and obtain a contour map of fractional index variation better than 10^{-6}.

6.10. HOLOGRAPHIC MULTIPLE-BEAM INTERFEROMETERS

The advantages of holographic interferometric tests of optical components are described in Chapter 12. When the choice has been made in favor of holographic interferometry, one can introduce the precision associated with multiple-beam interferometry by employing the hologram to reconstruct many wavefronts that can be combined to produce sharp multiple-beam fringes. This is achieved by the hologram, nonlinearly recorded, that produces several higher-order reconstructed beams. Matsumoto (1969) and Bryngdahl (1969) described such techniques, by which they recorded single-exposure holograms with single-wavelength illumination. An obvious advantage of this method is that one can obtain multiple-beam Fizeau type fringes from diffuse surfaces under test.

Burch et al. (1966) described a different method that uses linear holographic recording, but of multiple exposures made while bending the object. Multiple exposure can also be obtained by changing the direction of illumination or by using multiple-wavelength illumination, instead of bending the object. But before undertaking multiple-beam holographic testing with its attendant difficulties, one should make reasonably certain that simple two-beam interferometry with a good-quality fringe detecting device is insufficient for the precision necessary.

6.11. FINAL COMMENTS

Multiple-beam interferometry provides very high precision surface measurements compared to two-beam interferometry under comparable data processing environments. However, the requirements on source coherence and the basic experimental setup is more expensive and time consuming, so multiple-beam interferometry is less useful in routine shop floor testing where $\lambda/10$ precision is adequate. Measuring surface quality to much better than $\lambda/50$ is more efficient using various forms of multiple-beam interferometry exploiting principles of Fabry–Perot interferometry and multiple-beam Fizeau interferometers including fringes of equal chromatic order (FECO). There are many unique niches where multiple-beam interferometers are ideally suited. The first historic example is the characterization of surface microtopography using multiple-beam Fizeau interferometer (Bennett and Bennett 1967). However, readers interested in measuring surface roughness to a root mean square (rms) accuracy of 20 Å using two-beam Fizeau interferometer may see Eastman (1980) where he uses a piezoelectrically scanned interferometer. A natural application of Fabry–Perot interferometry is to test the component Fabry–Perot plates or solid etalons requiring a precision of $\lambda/200$ or better (Hernandez 1988; Hariharan et al. 1984). Measurements of parallelism of optically contacted etalons have been described by Killeen et al. (1981). Modern electronic industry is requiring bonding optically flat, large silicon crystal wafers with precision micrometric spacing. Rhee et al. (1990) have successfully exploited the principle of Fabry–Perot interferometry.

Multiple-beam interferometry using a grating as one of the two conventional mirrors has been developed for special applications. An example is noncontact testing of optical thin films developed by Pedraza-Contreras et al. (1984) and Bates and Li (1986). A combination of a Fabry–Perot interferometer and a two-beam interferometer has been cleverly used to very accurately measure the thickness of air layers by Schwider (1979). Precision associated with multiple-beam fringes can also be generated using holography that is suitable for real-time testing of deformation of microobjects (Shaalan and Jonathan 1978).

Sometimes, four-beam and three-beam interferometry are also useful. Makosch and Jaerisch (1978) have described four-beam Fizeau interferometry for contactless surface testing using a grating as the reference flat. Lin and Cowley (1986) have described the advantages of using three-beam Ronchigram in measuring primary aberrations. Strains of curved surfaces that are coated with photosensitive material have been measured by Chiang and Kim (1984) by three-beam holographic interferometeric techniques.

Finally, we should note that further contribution has been made in the 1980s on the understanding of the process of multiple-beam fringe formation. Fringe shifts in multiple-beam Fizeau interferometry are described by Rogers (1982) and the fringe characteristic formed by point source illumination has been analyzed and verified by Zerbino et al. (1984).

REFERENCES

Aebischer, N., "Etudes d'Interférences en Ondes Multiples par Diagramme Coplexe Visualiser les Franges en Réflexion," *Nouv. Rev. Opte. Appl.*, **1**, 233 (1970).

Aebischer, N., "Calculs de Profils Dissymétriques Observables sur des Figures d'Interférences en Ondes Multiples Sphériques," *Nouv. Rev. Opte. Appl.*, **2**, 351 (1971).

Airy, G., *Mathematical Transactions*, 2nd ed., 1836, p. 301. (obtainable from Library of British Museum, London).

Ashton, A. and A. C. Marchant, "Note on the Testing of Large Glass Panels," *Opt. Acta*, **14**, 203 (1967).

Baird, K. M., "Interferometry: Some Modern Techniques," in *Advances in Optical Techniques*, A. C. S. Van Heel, Ed., North-Holland, Amsterdam, 1967, Chap. 4.

Baird, K. M. and G. R. Hanes, "Interferometers," in *Applied Optics and Optical Engineering*, Vol. IV, R. Kingslake, Ed., Academic Press, New York, 1967, Chap. 9.

Barakat, N., A. S. Farghaly, and A. Abd-El-Azim, "Studies on Multiple-Beam Interference Fringes Formed on High Order Planes of Localization: Intensity Distribution and Fringe Shift between Successive Planes of Localization," *Opt. Acta*, **12**, 205 (1965).

Bates, B. and Y. Li, "Multiple-Beam Grating Interferometry: A Comparison of Techniques Applied to the Measurement of Coating Thickness," *Appl. Opt.*, **25** (6), 835–36 (1986).

Batishko, C. R. and R. R. Shannon, "Problem in Large-Path-Difference Laser Interferometry," *Appl. Opt.*, **11**, 195 (1972).

Benedetti-Michelangeli, G., "A New Technique for the Evaluation of the Quality of Plane Surfaces," *Appl. Opt.*, **7**, 712 (1968).

Bennett, H. E. and J. M. Bennett, "Precision Measurements in Thin Film Optics," in *Physics of Thin Films*, Vol. IV, G. Hass and R. F. Thun, Eds., Academic Press, New York, 1967, Chap. 1.

Bennett, J. M., "Measurement of the RMS Roughness, Autocovariance Function and Other Statistical Properties of Optical Surfaces Using a FECO Scanning Interferometer," *Appl. Opt.*, **15**, 2705 (1976).

Bergman, T. G. and J. L. Thompson, "An Interference Method for Determining the Degree of Parallelism of Laser Surfaces," *Appl. Opt.*, **7**, 923 (1968).

Bhatnagar, G. S., K. Singh, and B. N. Gupta, "Transmission Profile of a Fabry–Perot Interferometer Suffering from Asymmetric Surface Defects," *Nouv. Rev. Opt.*, **5**, 237 (1974).

Biddles, B. J., "A Non-contacting Interferometer for Testing Steeply Curved Surfaces," *Opt. Acta*, **16**, 137 (1969).

Boersch, H., H. J. Eichler, M. Pfundstein, and W. Wisemann, "Measurement of Length Shifts Down to 10^{-5} Å with a Three-Mode Laser," *IEEE J. Quantum Electron*, **QE-10**, 501 (1974).

Born, M. and E. Wolf, *Principles of Optics*, 5th ed., Pergamon, Oxford, 1975.

Boulouch, M. R., "Dédoblement des Franges d'Interférence en Lumiére Naturelle," *J. Phys.*, 3rd series, **2**, 316 (1893).

Boyd, G. D. and J. P. Gordon, "Confocal Multimode Resonator for Millimeter through Optical Wavelength Masers," *Bell Syst. Tech. J.*, **40**, 489 (1961).

Briers, J. D. "Interferometric Testing of Optical Systems and Components: a Review," *Opt. Laser Technol.*, **4**, 28 (1972).

Brossel, J., "Multiple-Beam Localized Fringes. I: Intensity Distribution and Localization; II: Conditions of Observation and Formation of Ghosts," *Proc. Phys. Soc.*, **59**, 224, 234 (1947).

Bruce, C. F. and F. P. Sharpless, "Relative Flatness Measurement of Uncoated Optical Flats," *Appl. Opt.*, **14**, 3082 (1975).

Bryngdahl, O., "Multiple-Beam Interferometry by Wavefront Reconstruction," *J. Opt. Soc. Am.*, **59**, 1171 (1969).

Bünnagel, R., H. A. Oehring, and K. Steiner, "Fizeau Interferometer for Measuring the Flatness of Optical Surfaces," *Appl. Opt.*, **7**, 331 (1968).

Burch, J. M., A. E. Ennos, and R. J. Wilton, "Dual- and Multiple-Beam Interferometry by Wavefront Reconstruction," *Nature*, **209**, 1015 (1966).

Cagnet, M., "Méthodes Interférométriques Utilisant les Franges de Superposition," *Rev. Opt.*, **33**, 1, 113, 229 (1954).

Candler, C., *Modern Interferometers*, Hilger and Watts, London, 1951.

Chabbal, R., "Recherche des Meilleures Conditions d'Utilisation d'un Spectromètre Photoelectrique Fabry–Perot," *J. Rech. Cent. Nat. Rech. Sci.*, No. 24, p. 138, 1953.

Chabbal, R., "Finese Limite d'un Fabry–Perot Forme de Lames Imparfaites," *J. Phys. Radium.*, **19**, 295 (1958).

Chiang, F. P. and C. C. Kin, "Three-Beam Interferometric Technique for Determination of Strain of Curved Surfaces," *Opt. Eng.*, **23**(6), 766–68 (1984).

Clapham, P. B. and G. D. Dew, "Surface Coated Reference Flats for Testing Fully Aluminized Surfaces by Means of a Fizeau Interferometer," *J. Sci. Instrum.*, **44**, 899 (1967).

Connes, P., "L'Etalon de Fabry–Perot Sphérique," *J. Phys. Radium*, **19**, 262 (1958).

Cook, A. H., *Interference of Electromagnetic Waves*, Clarendon Press, Oxford, 1971.

Dew, G. D., "A Method for Precise Evaluation of Interferograms," *J. Sci. Instrum.*, **41**, 160 (1964).

Dew, G. D. "The Measurement of Optical Flatness," *J. Sci. Instrum.*, **43**, 409 (1966).

Dyson, J., "Optics in a Hostile Environment," *Appl. Opt.*, **7**, 569, (1968).

Eastman, J. M., "Effects and Measurement of Scattering and Absorption of Thin Films," *Proc. Soc. Photo-Opt. Instrum. Eng.*, **50**, 43 (1975).

Eastman, J. M., "The Scanning Fizeau Interferometer: An Automated Instrument for Characterizing Optical Surfaces," *Opt. Eng.*, **19**(6), 810–14 (1980).

Eastman, J. M. and P. W. Baumeister, "Measurement of the Microtopography of Optical Surfaces Using a Scanning Fizeau Interferometer," *J. Opt. Soc. Am.*, **64**, 1369 A (1974).

Fabry, C. and A. Perot, "Sur les Franges des Lames Minces Argentées et Leur Application a la Mesure de Petites Epaisseurs d'Air," *Ann. Chim. Phys.*, **12**, 459 (1897).

Fizeau, M. H. "Recherches sur les Modifications que Subit la Vitesse de la Lumiére dans le Verre et Plusieurs," *Ann. Chim. Phys.*, **66**, 429 (1862a).

Fizeau, M. H., "Recherches sur les Modifications que Subit la Vitesse de la Lumiére dans le Verre et Plusieurs Autres Corps Solides sous L'Influence de la Chaleur," *C. R. Acad. Sci. Paris*, **54**, 1237 (1862b).

Ford, D. L. and J. H. Shaw, "Rapid Method of Aligning Fabry-Perot Etalons," *Appl. Opt.*, **8,** 2555 (1969).

Fox, A. G. and T. Li, "Resonant Modes in a Master Interferometer," *Bell Syst. Tech. J.*, **40,** 453 (1961).

Hanes, G. R., "Limiting Precision in Optical Interferometry," *Can. J. Phys.*, **37,** 1283 (1959).

Hanes, G. R., "Quantum Limit to Precision of Wavelength Determination," *Appl. Opt.*, **2,** 465 (1963).

Hariharan, P. B., F. Oreb, and A. J. Leistner, "High Precision Digital Interferometry: Its Application to the Production of An Ultrathin Solid Fabry-Perot Etalon," *Opt. Eng.*, **23**(3), 294–97, (1984).

Heintze, L. R., H. D. Polster, and J. Vrabel, "A Multiple-Beam Interferometer for Use with Spherical Wavefront," *Appl. Opt.*, **6,** 1924 (1967).

Heisenberg, W., *The Physical Principles of the Quantum Theory*, Dover, New York, 1949.

Hernandez, G., *Fabry-Perot Interferometers*, Cambridge University Press, Cambridge, 1988.

Herriott, D. R., "Multiple-Wavelength Multiple-Beam Interferometric Observations of Flat Surfaces," *J. Opt. Soc. Am.*, **51,** 1142 (1961).

Herriot, D. R., "Spherical-Mirror Oscillating Interferometer," *Appl. Opt.*, **2,** 865 (1963).

Herriot, D. R., "Long-Path Multiple-Wavelength Multiple-Beam Interference Fringes," *J. Opt. Soc. Am.*, **56,** 719 (1966).

Hill, R. M., "Some Fringe-Broadening Defects in a Fabry-Perot Etalon," *Opt. Acta*, **10,** 141 (1963).

Hill, R. M. and C. F. Bruce, "Limiting Precision in a Scanning Optical Interferometer," *Aust. J. Phys.*, **15,** 194 (1962).

Hodgkinson, I. J., "A Method for Mapping and Determining the Surface Defects Function of Pairs of Coated Optical Flats," *Appl. Opt.*, **8,** 1373 (1969).

Hodgkinson, I. J., "The Application of Fringes of Equal Chromatic Order to the Assessment of the Surface Roughness of Polished Fused Silica," *J. Phys.*, **E3,** 300 (1970).

Jacquinot, M. P., "New Developments in Interference Spectroscopy," *Rep. Prog. Phys.*, **23,** 267 (1960).

Jenkins, F. A. and H. E. White, *Fundamentals of Optics*, 3rd ed. McGraw-Hill, New York, 1957.

Killeen, T. L., P. B. Hays, and J. DeVos, "Parallelism Maps for Optically Contacted Etalons," *Appl. Opt.*, **20**(15), 26116–119 (1981).

Kinosita, K., "Numerical Evaluation of the Intensity Curve of a Multiple-Beam Fizeau Fringe," *J. Phys. Soc. Jap.*, **8,** 219 (1953).

Koehler, W. F., "Multiple-Beam Fringes of Equal Chromatic Order. I: Phase Change Considerations. II: Mechanism of Polishing Glasses," *J. Opt. Soc. Am.*, **43,** 738, 743 (1953).

Koehler, W. F., "Multiple-Beam Fringes of Equal Chromatic Order. IV: Use of Multilayer Film," *J. Opt. Soc. Am.*, **45,** 934 (1955a).

Koehler, W. F., "Multiple-Beam Fringes of Equal Chromatic Order. VII: Mechanism of Polishing Glass," *J. Opt. Soc. Am.*, **45,** 1015 (1955b).

Koehler, W. F. and A. Eberstein, "Multiple-Beam Fringes of Equal Chromatic Order. III: The Cleavage of Topaz," *J. Opt. Soc. Am.*, **43,** 747 (1953).

Koehler, W. F. and W. C. White, "Multiple-Beam Fringes of Equal Chromatic Order. V: Fringe Fine Structure; VI: Method of Measuring Roughness," *J. Opt. Soc. Am.*, **45,** 940, 1011 (1955).

Koppelmann, G., "Eine beugungsbedingte Auflösungsgrenze in der Mehrstrahl-Interferometrie," *Opt. Acta*, **13,** 211 (1966).

Koppelmann, G., "Multiple-Beam Interference and Natural Modes in Open Resonators," in *Progress in Optics*, Vol. VII, E. Wolf, Ed., Wiley-Interscience, New York, 1969, Chap. 1.

Koppelmann, G. and K. Krebs, "Eine Registriermethode zur Vermessung der Reliefs höchsterbener Oberflächen," *Optik*, **18,** 349 (1961).

Kuhn, H., "New Technique in Optical Interferometry," *Rep. Phys. Soc. Prog. Phys.*, **14,** 80 (1951).

Lang, J. and G. Scott, "Resolution Limits in Multiple-Beam Interferometry," *J. Opt. Soc. Am.*, **58,** 81 (1968).

Langenbeck, P., "Optical Wavefront Mapping by Dual Interferometry," *J. Opt. Soc. Am.*, **58,** 499 (1968).

Langenbeck, P., "Fizeau Interferometer—Fringe Sharpening," *Appl. Opt.*, **9,** 2053 (1970).

Laurent, L., "Sur Phisieurs Appareils d'Optique, Destinés à Controler les Surfaces Planes: Parallèles, Perpendiculaires et Obliques," *C. R. Acad. Sci. Paris*, **96,** 1035 (1883).

Logan, J. L., "Gravitational Waves—a Progess Report," *Phys. Today*, March, p. 44 (1973).

Lin, J.-A. and J. M. Cowley, "Aberration Analysis by Three-Beam Interferograms," *Appl. Opt.*, **25**(14), 2245–46 (1986).

Makosch, G. and W. Jaerich, "Mapping of Optical Surfaces with Quarter Wavelength Fringes," **17**(5), 744–747, **17**(13), 1990–91 (1978).

Malacara, D., A. Cornejo, and M. V. R. K. Murty, "Bibliography of Various Optical Testing Methods," *Appl. Opt.*, **14,** 1065 (1975).

Matsumoto, K., "Holographic Multiple-Beam Interferometry," *J. Opt. Soc. Am.*, **59,** 777 (1969).

Moos, H. W., G. F. Imbusch, L. F. Mollenauer, and A. L. Schawlow, "Tilted-Plate Interferometry with Large Plate Separations," *Appl. Opt.*, **2,** 817 (1963).

Moss, G. E., L. R. Miller, and R. L. Forward, "Photon-Noise-Limited Laser Transducer for Gravitational Antenna," *Appl. Opt.*, **10,** 2495 (1971).

Murty, M. V. R. K., "Multiple-Pinhole Multiple-Beam Interferometric Observation of Flat Surfaces," *Appl. Opt.*, **1,** 364 (1962).

Pastor, J. and P. H. Lee, "Transmission Fringes in Reflection Multiple-Beam Interferometry," *J. Opt. Soc. Am.*, **58,** 149 (1968).

Persin, A. and D. Vukicevic, "Block Defocused Spherical Fabry–Perot Interferometer," *Appl. Opt.*, **12,** 275 (1973).

Pilston, R. G. and G. N. Steinberg, "Multiple-Wavelength Interferometry with Tunable Source," *Appl. Opt.*, **8,** 553 (1969).

Polster, H. D., "Multiple-Beam Interferometry," *Appl. Opt.*, **8,** 522 (1969).

Post, D., "Characteristics of the Series Interferometer," *J. Opt. Soc. Am.*, **44,** 243 (1954).

Rafalowski, M., "Testing the Coma of Decentration With a Confocal Fabry–Perot Interferometer," *Appl. Opt.*, **27**(14), 3046–50 (1988).

Rafalowski, M., "Precise Testing of Asymmetric Wavefronts With An Inverting Interferometer," *Appl. Opt.*, **29**(10), 1477–81 (1990).

Raymond, O. J., "Limiting Fringe Pointing Precision in a Scanning Two-Beam Interferometry," *Appl. Opt.*, **9,** 1140 (1970).

Rhee, I., F. M. Gasparini, A. Petrou, and D. J. Bishop: "Si Wafers Uniformly Spaced; Bonding and Diagnostics," *Rev. Sci. Instrum.*, **61**(5), 1528–36 (1990).

Roberts, F. E. and P. Langenbeck, "Homogeneity Evaluation of Very Large Disks," *Appl. Opt.*, **8,** 2311 (1969).

Rosier, F. L., "Mapping of High Quality Optical Flats without Reflection Coating," *J. Opt. Soc. Am.*, **52,** 471 (1962).

Roesler, F. L. and W. Traub, "Precision Mapping of Pairs of Uncoated Optical Flats," *Appl. Opt.*, **5,** 463 (1966).

Rogers, J. R.: "Fringe Shifts in Multiple Beam Fizeau Interferometry," *J. Opt. Soc. Am.*, **72**(5), 638–43 (1982).

Roychoudhuri, C., "Brillouin Spectra of CaF_2 Microcrystals Using a Stable 3-Pass Fabry–Perot Interferometer," *Appl. Phys. Lett.*, **23,** 543 (1937a).

Roychoudhuri, C., *Multi-pass Fabry–Perot Interferometer for Brillouin Scatter Measurements*, Ph.D. Thesis, University of Rochester, New York, 1973b; University Microfilms No. 74-14413.

Roychoudhuri, C., "Dynamic and Multiplex Holography with Scanning Fabry–Perot Fringes," *Opt. Commun.*, **10,** 160 (1974).

Roychoudhuri, C., "Response of Fabry–Perot Interferometers to Light Pulses of Very Short Duration," *J. Opt. Soc. Am.*, **65,** 1418 (1975).

Saunders, J. B., "Precise Topography of Optical Surfaces," *J. Res. Nat. Bur. Stand.*, **47,** 148 (1951).

Saunders, J. B., "In-Line Interferometer," *J. Opt. Soc. Am.*, **44,** 241 (1954).

Schulz, G., "Ein Interferenzverfahren zur absoluten Ebenheitsprufung langs beliebiger Zentralschnitte," *Opt. Acta*, **14,** 375 (1967).

Schulz, G. and J. Schwider, "Precise Measurement of Planeness," *Appl. Opt.*, **6,** 1077 (1967).

Schulz, G. and J. Schwider, "Interferometric Testing of Smooth Surfaces," in *Progess in Optics*, Vol. XIII, E. Wolf, Ed., American Elsevier, New York, 1976.

Schulz, G., J. Schwider, C. Hiller, and B. Kicker, "Establishing an Optical Flatness Standard," *Appl. Opt.*, **10**, 929 (1971).

Schulz, L. G., "Accurate Thickness Measurements with a Fabry–Perot Interferometer," *J. Opt. Soc. Am.*, **40**, 177 (1950a).

Schulz, L. G., "An Interferometric Method for Accurate Thickness Measurements of Thin Evaporated Films," *J. Opt. Soc. Am.*, **40**, 690 (1950b).

Schulz, L. G., "The Effect of Phase Changes in White Light Interferometry," *J. Opt. Soc. Am.*, **41**, 261 (1951a).

Schulz, L. G., "An Interferometric Method for the Determination of the Absorption Coefficients of Metals, with Results for Silver and Aluminum," *J. Opt. Soc. Am.*, **41**, 1047 (1951b).

Schulz, L. G. and E. J. Scheibner, "An Experimental Study of the Change in Phase Accompanying Reflection of Light from Thin Evaprorated Films," *J. Opt. Soc. Am.*, **40**, 761 (1950).

Schwider, J., "Informationssteigerung in der Vielstrahlinterferometrie," *Opt. Acta*, **15**, 351 (1968).

Schwider, J., "Superposition Fringers As A Measuring Tool in Optical Testing," *Appl. Opt.*, **18**(14), 2364–67 (1979).

Shaalan, M. S. and J. M. Jonathan, "Multiple Beam Holographic Interferometry," *Opt. Acta*, **25**(11), 1025–34 (1978).

Shaalan, M. S. and V. I. Little, "The Application of Multiple-Beam White-Light Fringes to the Study of Surfaces," *J. Phys.*, **D8**, 1003 (1975).

SIRA, "New Aid for the Optical Industry: an Interferometer for Testing Deeply Curved Lens Surfaces," *SIRA News-Sheet* ("Impact"), No. 5, 1967.

Thorton, B. S., "An Uncertainty Relation in Interferometry," *Opt. Acta*, **4**, 41 (1957).

For a complete list of Tolansky's work see the bibliography in Tolansky (1960).

Tolansky, S., "New Non-localized Interference Fringes," *Philos. Mag.*, **34**, 555 (1943).

Tolansky, S., "New Contributions to Interferometry. II: New Interference Phenomena with Newton's Rings; III: The Differential Polarization Phase Change on Reflection at a Thin Silver Film," *Philos. Mag.*, Ser. 7, **35**, 120, 179 (1944).

Tolansky, S., "New Contributions to Interferometry. V: New Multiple-Beam White Light Interference Fringes and Their Applications," *Philos. Mag.*, Ser. 7, **36**, 225 (1945a).

Tolansky, S., "Topography of Crystal Faces. I: The Topography of a (100) Face of a Left-Handed Quartz Crystal; II: The Topography of Cleavage Faces of Mica and Selenite," *Proc. Roy. Soc.*, **A184**, 41, 51 (1945b).

Tolansky, S., "Further Interferometric Studies with Mica: New Multiple-Beam Fringes and Their Applications," *Proc. Roy. Soc.*, **A186**, 261 (1946).

Tolansky, S., *Multiple-Beam Interferometry of Surfaces and Films*, Oxford University Press, Oxford, 1948a, Dover, New York, 1970.

Tolansky, S., *Multiple-Beam Interferometry*, Oxford University Press, Oxford, 1948b.

Tolansky, S., *Multiple-Beam Fringes*, Clarendon Press, Oxford, 1948c.

REFERENCES

Tolansky, S., *Surface Microtopography*, Interscience, New York, 1960.

Tolansky, S., *Introduction to Interferometry*, Longmans, London, 1966.

Tolansky, S., *Microstructure of Surfaces Using Interferometry*, Edward Arnold, London, 1968.

Tolansky, S. and S. H. Emara, "Precision Multiple-Beam Interference Fringes with High Lateral Microcopic Resolution," *J. Opt. Soc. Am.*, **45**, 792 (1955).

Vinokurov, V. M., A. L. Ardamatski, and L. V. Popov, "The Surface of the Disrupted Layer," in *Generation of Optical Surfaces*, K. G. Kumanin, Ed., Focal Library, London, 1962, Chap. I.

Vrabel, J. and E. B. Brown, "The Practice of Interferometry," *Opt. Eng.*, **14**, 124 (1975).

Williams, W. E., *Applications of Interferometry*, Methuen, London, 1950.

Zerbino, K. M., R. Torroba, N. Rodriguez, and M. Garavaglia, "Interference Profiles with Multiple Spherical Waves: General Case," *J. Opt. Soc. Am.*, **A1**(5), 495–501 (1984).

7
Multiple-Pass Interferometers

P. Hariharan

This chapter discusses some variants of the conventional interferometers used for optical testing in which one (or more) of the wavefronts is sent back and makes two or more traverses of either the whole system or a part of it. Such double- or multiple-pass interferometers offer definite advantages for some testing applications.

7.1. DOUBLE-PASS INTERFEROMETERS

7.1.1. Separation of Aberrations

The interference pattern obtained with a lens in a Twyman–Green interferometer gives a contour map of the wavefront leaving the lens aperture; however, when many aberrations are present, estimation of the individual aberrations becomes difficult. This problem can be simplified if the Twyman–Green interferometer is used in a double-pass configuration (Hariharan and Sen 1961d) so that the symmetrical and antisymmetrical parts of the wave aberration (see Chapter 13) are displayed in separate interferograms.

As shown in Fig. 7.1, the beams emerging from the interferometer through the lens L_2 are reflected back through it by the plane mirror M_3 placed at its focus, and the double-pass beams emerging from L_1 are brought to a focus at the eye stop by the auxiliary beam divider S_2. If the source is shifted very slightly sideways, the two images formed at the eye stop move off the axis in opposite directions, and it is possible to view either the fringes produced by the double-pass beams or the normal interference pattern.

The four double-pass rays derived from a ray incident on the beam divider S_1 at O can be identified as the AA' ray ($SOAOM_3O'A'O'S$), the AB' ray ($SOAOM_3O'B'O'S$), the BA' ray ($SOBOM_3O'A'O'S$), and the BB' ray ($SOBOM_3O'B'O'S$), corresponding to the paths they follow on the outward and

---

Optical Shop Testing, Second Edition, Edited by Daniel Malacara.
ISBN 0-471-52232-5 © 1992, John Wiley & Sons, Inc.

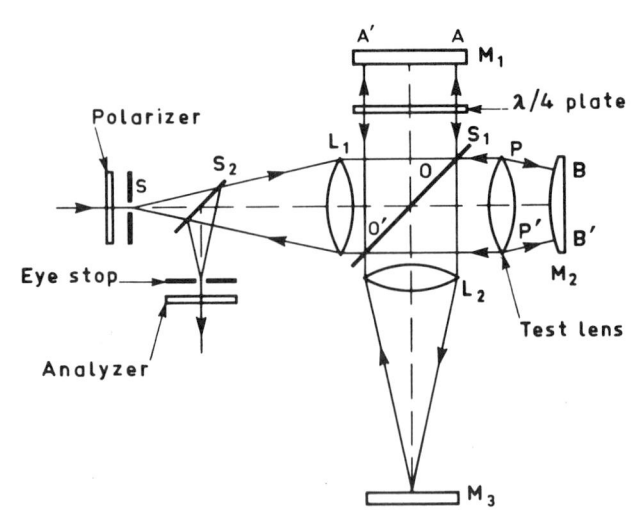

Figure 7.1. Double-pass Twyman–Green interferometer for separation of symmetrical and antisymmetrical wavefront aberrations. (From Hariharan and Sen 1961d.)

return journeys. Since the wavefronts emerging from the interferometer are inverted before they are sent back, it is easily seen that, if on the first pass the BB' ray traverses the pupil of the lens under test at a point $P(r, \phi)$, it traverses the pupil on the second pass at the diametrically opposite point $P'(r, \pi + \phi)$. In the same manner, while the BA' ray traverses the pupil at P, the AB' ray traverses it at P'.

At these two points the terms in the expressions for the total wavefront aberration (Hopkins 1950) involving even powers of $\cos \phi$ (i.e., the defocusing, spherical aberration, and astigmatism terms) have the same value, whereas the terms involving odd powers of $\cos \phi$ (i.e., the distortion and coma terms) are of equal magnitude but have opposite signs. The total paths of the four double-pass rays can therefore be written as follows:

$$D_{AA'} = D \text{ (say)},$$
$$D_{BA'} = D + 2W_{\text{even}} + 2W_{\text{odd}} + \Delta D,$$
$$D_{AB'} = D + 2W_{\text{even}} - 2W_{\text{odd}} + \Delta D, \quad (7.1)$$
$$D_{BB'} = D + 2W_{\text{even}} + 2W_{\text{odd}} + \Delta D + 2W_{\text{even}} - 2W_{\text{odd}} + \Delta D$$
$$= D + 4W_{\text{even}} + 2\Delta D,$$

where W_{even} is the sum of the terms involving even powers of $\cos \phi$, W_{odd} is the sum of the terms involving odd powers of $\cos \phi$ (see Appendix 3), and ΔD is the difference in the lengths of the A and B paths for the principal ray.

To select the required combinations of double-pass beams, the beam from

7.1. DOUBLE-PASS INTERFEROMETERS

the collimator is polarized in the vertical plane and a quarter-wave plate is introduced in the A path. The plane of polarization of the AB' and BA' beams is then rotated through 90°, while the plane of polarization of the AA' and BB' beams remains unchanged. Hence, when the analyzer is set with its axis vertical, the AB' and BA' beams are extinguished, and interference takes place between the AA' and BB' beams. The path difference between these is

$$D_{BB'} - D_{AA'} = 4W_{\text{even}} + 2\Delta D. \qquad (7.2)$$

Accordingly, symmetric errors such as change of focus, spherical aberration, and astigmatism are shown in this interferogram with doubled sensitivity, whereas features that depend on odd powers of cos ϕ are eliminated.

When the analyzer is rotated so that its axis is horizontal, the AB' and BA' rays are isolated and made to interfere. The path difference between these rays is

$$D_{BA'} - D_{AB'} = 4W_{\text{odd}}. \qquad (7.3)$$

Only antisymmetric aberrations such as distortion and coma are shown in this interferogram, along with any tilt of the reference mirror (which is equivalent to the introduction of an additional "odd" term).

A typical set of interferograms obtained with an uncorrected lens is shown in Fig. 7.2. Figure 7.2a is a normal Twyman–Green interferogram, and Figs. 7.2b and 7.2c are double-pass interferograms, which show the "even" and "odd" components of the wavefront aberration. The patterns obtained by plotting the expressions for the most important aberration terms are shown below each interferogram.

7.1.2. Reduction of Coherence Requirements

Residual wedge errors in plane parallel plates are commonly measured with the Fizeau interferometer. This has the disadvantage that the path difference between the interfering wavefronts is twice the optical thickness of the plate under test. It is therefore essential to use a well-collimated beam of monochromatic light when testing thick plates. Even with a laser source, a well-corrected objective must be used. This problem can be eliminated, however, if the interferometer is double passed, so that superposition fringes are formed between the plate under test and its inverted image.

One method (Cagnet 1954) is to use the wavefronts transmitted through the plate under test and reflect them back by means of an auxiliary afocal system. In this case it is necessary to coat the surfaces of the plate so that their reflectivity R is fairly high. With a broadband source such as a high-pressure mercury vapor lamp, the fringes then have the same intensity distribution as those obtained with strictly monochromatic light in a conventional Fizeau interferometer with surfaces having reflectivity R^2.

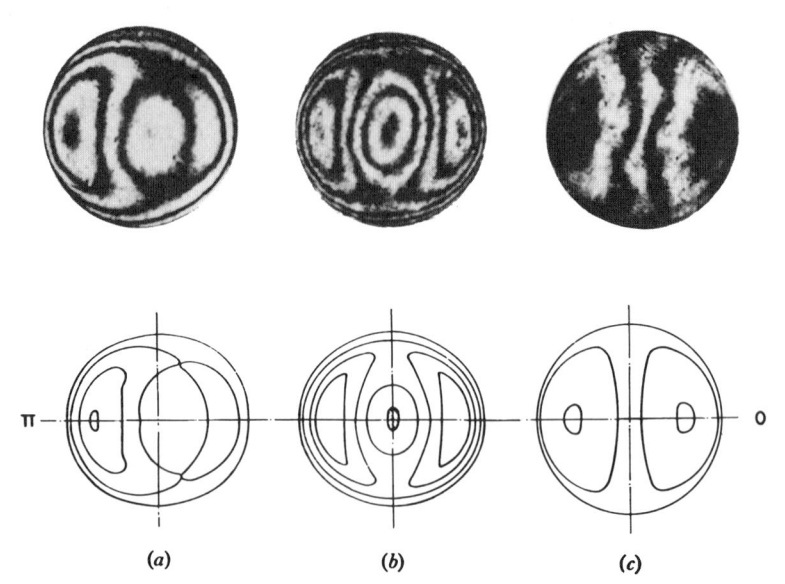

Figure 7.2. Interferograms obtained with the double-pass Twyman interferometer. The patterns obtained by plotting the expressions for the most important aberration terms are shown below each interferogram. (a) Normal interferogram: $2W = 0.25 + 1.85r \cos \phi - 3.99r^2 - 2.75r^2 \cos^2 \phi - 1.57r^3 \cos \phi + 4.33r^4 + 1.72r^4 \cos^2 \phi$. (b) Double-pass interferogram showing only the "even" terms: $4W_{even} = 0.50 - 7.98r^2 - 5.50r^2 \cos^2 \phi + 8.66r^4 + 3.44r^4 \cos^2 \phi$. (c) Double-pass interferogram showing only the "odd" terms: $4W_{odd} = 3.70r \cos \phi - 3.14r^3 \cos \phi$. (From Hariharan and Sen 1961d.)

An alternative that does not require the surfaces to be coated and involves very little in the way of additional optics is to use a doubly reflected system of fringes (Sen and Puntambekar 1965). The optical arrangement shown in Fig. 7.3 is the same as that in a conventional Fizeau interferometer, except that the pinhole source is on the reflecting surface of a plane mirror. If this pinhole is shifted slightly off axis, the rays reflected back from the plate under test are brought to a focus at a point on this mirror on the other side of the axis, so that they are reflected back once more through the system. The double-pass rays are finally brought to a focus at the eye stop by a beam divider.

A ray traversing the plate under test at a point $P(x, y)$ on the first pass traverses it at $P'(-x, -y)$ on the second pass. Let $t(x, y)$ be the optical thickness of the plate at P, $\zeta = (2\pi/\lambda)[t(x, y) + t(-x, -y)]$, and $\psi = (2\pi/\lambda)[t(x, y) - t(-x, -y)]$, and assume that the spectral bandwidth of the source is such that no interference can take place for a phase difference of ζ, while the variation of ψ over this bandwidth can be neglected. With an uncoated glass plate whose reflectivity R is small, the fringe irradiance at any point is then

$$I(x, y) = 2I_0 R^2 (2 + \cos 2\psi) \quad (7.4)$$

7.1. DOUBLE-PASS INTERFEROMETERS

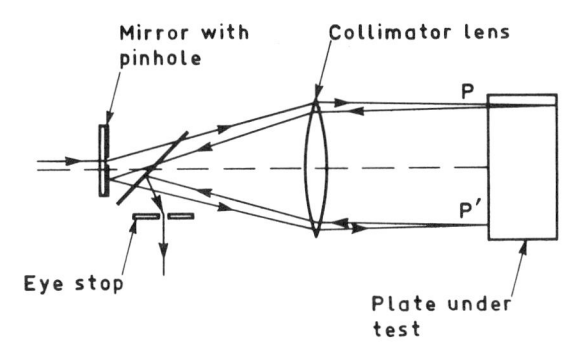

Figure 7.3. Double-pass Fizeau interferometer for testing plane parallel plates. (From Sen and Puntambekar 1965.)

where I_0 is the irradiance of the original incident ray. This corresponds to the irradiance distribution in two-beam fringes with a visibility of 0.5, which is adequate for most measurements. The interferogram shows only the antisymmetrical (wedge) errors in the plate under test, for which its sensitivity is twice that of the normal Fizeau interferometer.

The double-pass Fizeau interferometer can also be used for testing right-angle prisms and cube corners (Sen and Puntambekar 1966). The setup for this is shown in Fig. 7.4. One-half of the diagonal face of the prism is covered to isolate two reflected wavefronts, one formed by reflection from the uncovered half of the diagonal face, and the other transmitted through the prism and reflected back by the covered half of the diagonal face. The angle between these wavefronts is doubled on the second pass, so that the sensitivity to angular errors is twice that obtainable with a Twyman–Green interferometer.

Another version of this interferometer, shown schematically in Fig. 7.5, can be used to test concave surfaces (Puntambekar and Sen 1971). With a thermal

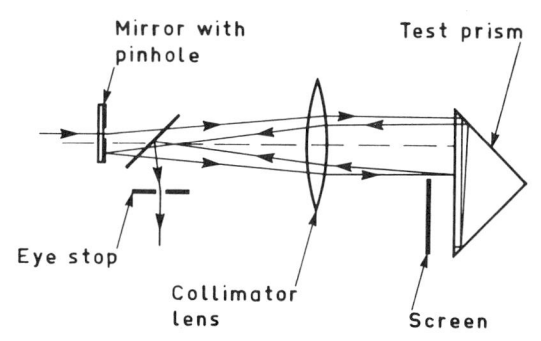

Figure 7.4. Double-pass Fizeau interferometer for testing reflecting prisms. (From Sen and Puntambekar 1966.)

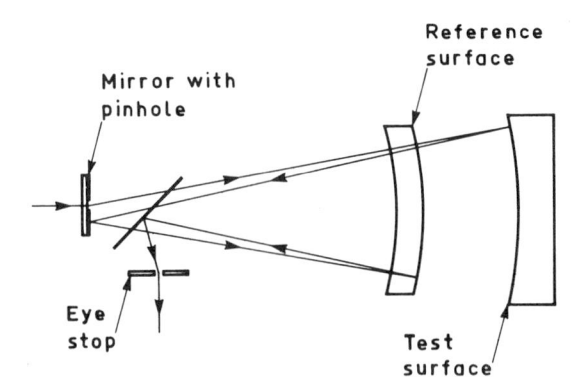

Figure 7.5. Double-pass Fizeau interferometer for testing concave surfaces. (From Puntambekar and Sen 1971.)

source the distance between the reference surface and the concave mirror in a conventional Fizeau interferometer cannot be very large, so that only a limited range of curvatures can be tested with a given reference surface. This problem can be overcome if a laser source is used, but unwanted fringes due to reflection at other surfaces then appear. Double passing permits a wide range of curvatures to be tested with a single reference surface, using a thermal source.

In this case also, when the ray paths during the first and second passes are symmetrical about the optical axis of the system under test, only antisymmetrical errors appear in the interferogram. However, an additional shear can be introduced between the inverted wavefronts if the pinhole mirror is rotated about an axis on its surface passing through the common center of curvature; the symmetrical errors can then be evaluated from the resulting interferograms. Different parts of the reference surface are used for the sheared and unsheared interference patterns, but if the reference surface is accurately spherical this does not introduce any error.

7.1.3. Double Passing for Increased Accuracy

The accuracy with which measurements can be made with a conventional two-beam interferometer is determined by the precision with which a local fringe displacement ΔQ can be estimated as a fraction of Q, the average interfringe spacing. This precision is limited by the sinusoidal irradiance distribution in two-beam fringes. One method to obtain greater accuracy, which was discussed in Chapter 6, is multiple-beam interferometry. Double-pass interferometry (Hariharan and Sen 1960a, 1961b) can also be used to advantage in some cases.

Consider the double-pass Twyman–Green interferometer shown in Fig. 7.6, and assume that the end mirrors M_1 and M_2 are normal to the original incident beams in the vertical plane but make an angle equal to $(\pi/2) + \alpha$ with them

7.1. DOUBLE-PASS INTERFEROMETERS

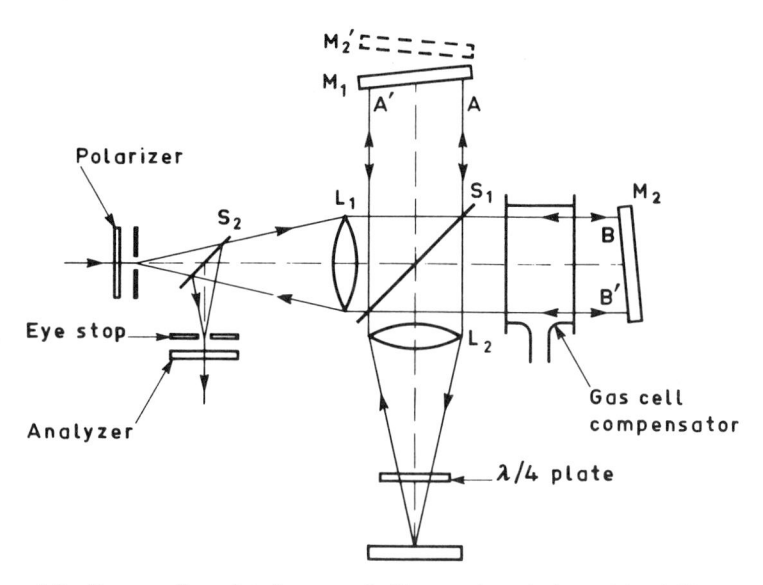

Figure 7.6. Twyman–Green interferometer double passed to obtain modulated fringes. (From Hariharan and Sen 1960a.)

in the horizontal plane, so that M_1 and M_2', the virtual image of M_2 in the beam divider, make an angle 2α with each other. In this case the only aberration term present is the "odd" tilt term, and from Eq. (7.1) the total optical paths of the four double-pass rays emerging from the interferometer at a distance x from the center of the field in the horizontal plane are as follows:

$$D_{AA'} = D \quad \text{(say)},$$
$$D_{AB'} = D + 4x\alpha + 2d, \quad (7.5)$$
$$D_{BA'} = D - 4x\alpha + 2d,$$
$$D_{BB'} = D + 4d,$$

where d is the distance between M_1 and M_2' at the center of the field. If all four beams are made to interfere, the complex amplitude A_p at this point can be written as

$$A_p = A \exp - \left(\frac{2\pi i D}{\lambda}\right) [1 + \exp - 2i(\eta - \xi)$$
$$+ \exp - 2i(\eta + \xi) + \exp - 4i\eta] \quad (7.6)$$

where $\xi = (2\pi/\lambda)2x\alpha$, $\eta = (2\pi/\lambda)d$, and A is the amplitude corresponding to a single beam. The irradiance I_p at this point, obtained by multiplying A_p by its complex conjugate, is then

$$I_P = 4I_0(\cos 2\xi + \cos 2\eta)^2 \tag{7.7}$$

where I_0 is the irradiance due to a single beam.

The irradiance distribution in the fringe system given by Eq. (7.7) is shown in Fig. 7.7 for three different values of the phase difference η between the two paths at the center of the field. A change in η does not result in a displacement of the fringes but causes only a change in the irradiance distribution. When η

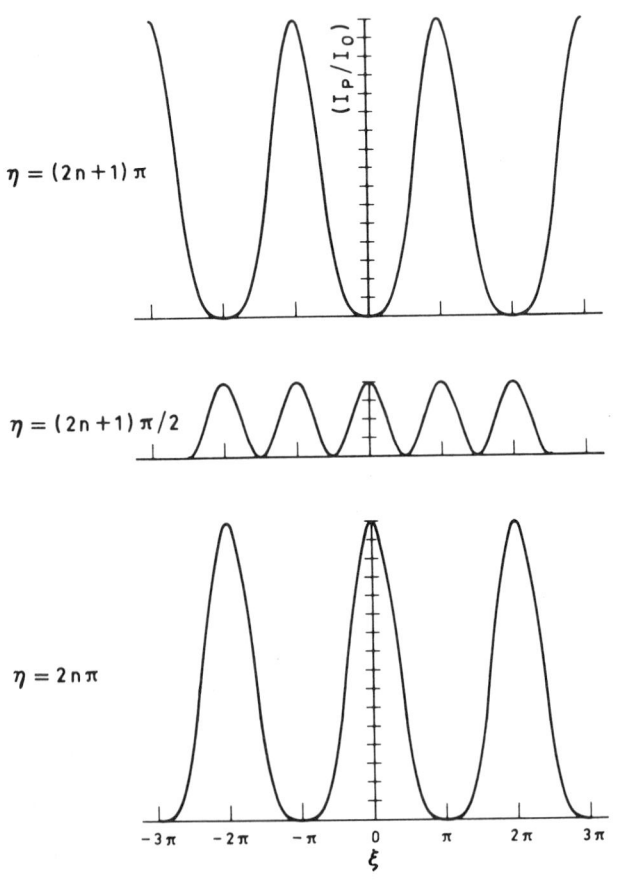

Figure 7.7. Irradiance distribution in the fringe pattern obtained with the double-pass Twyman-Green interferometer for various values of η, the phase difference between the two paths. (From Hariharan and Sen 1960a.)

= 0 or $n\pi$, where n is an integer, alternate fringes are suppressed, whereas when $\eta = (2n + 1)\pi/2$ all the fringes have the same irradiance. At the latter point a very small change in the value of η results in an appreciable difference in irradiance between adjacent fringes. This effect can be used with a suitable compensator to achieve an accurate null setting, permitting measurements of the changes in the optical path of $\lambda/500$.

This technique is well adapted to measurement of the refractive index and thickness of thin films (Hariharan and Sen 1961c). Even higher accuracy is possible by double passing a Fizeau interferometer (Hariharan and Sen 1960b, 1961a).

7.2. MULTIPASS INTERFEROMETRY

Direct visual presentation of wavefront errors with increased accuracy is obtained in multipass interferometry. In this technique the fact that the total deformation of the wavefront is proportional to the number of times it is reflected from or transmitted through the optical system under test is utilized to increase the ratio of the fringe displacement ΔQ to the average interfringe spacing Q.

The optical system of the multipass Twyman–Green interferometer (Langenbeck 1967) is shown schematically in Fig. 7.8. An additional beam divider is inserted in one arm of the interferometer. If this makes a slight wedge angle with respect to the mirror under test, the multiply reflected beams formed be-

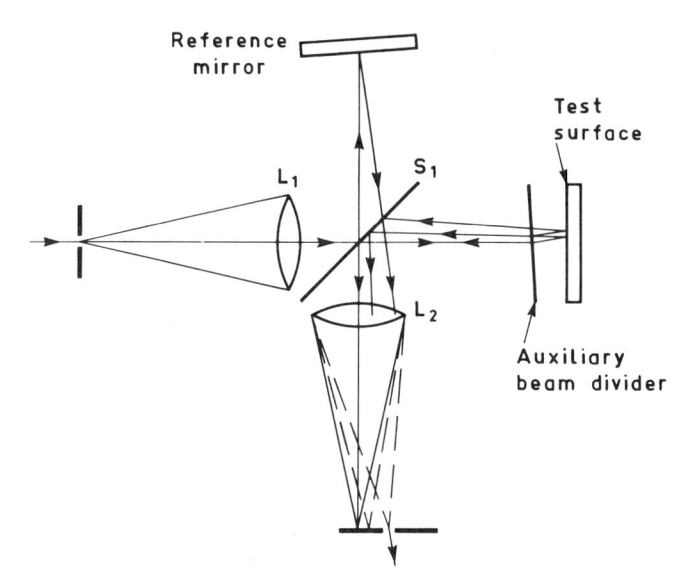

Figure 7.8. Multipass Twyman–Green interferometer. (From Langenbeck 1969a.)

tween them give rise to a series of laterally separated images of the light source in the rear focal plane of lens L_2. This permits selecting a beam that has undergone any desired number of reflections at the mirror under test by an aperture in the focal plane of L_2. If the reference mirror is tilted so that the light reflected by it also passes through the same aperture, two-beam interference fringes are obtained between the reference wavefront and a wavefront that has undergone n reflections at the mirror under test and therefore exhibits a deformation of $2n\,\Delta t$, where Δt is the deviation of its surface from flatness.

In the resulting interferogram $\Delta Q/Q = 2n\,\Delta t/\lambda$. Hence the sensitivity is n times greater than that obtained with a normal Twyman–Green interferometer.

The irradiances of the two interfering beams can be made approximately equal and the visibility of the fringes optimized, if the auxiliary beam divider has a high transmittance and an uncoated glass surface is used for the reference mirror. A better solution (Langenbeck 1969a) is to use a modified optical system in which the reference beam is reflected off an uncoated glass surface at Brewster's angle. Rotation of the plane of polarization of the illuminating beam by a $\lambda/2$ plate makes it possible to control the attenuation of the reference beam.

A typical series of multipass interferograms of a flat surface with a square, $\lambda/8$ deep step at its center is shown in Fig. 7.9. Figure 7.9a is the normal Twyman–Green interferogram of the first-order reflection, and Figs. 7.9b to 7.9e are the second-, third-, fourth-, and fifth-order multipass interferograms, showing the increase in the relative fringe displacement $\Delta Q/Q$ with the order of the reflection.

A modified multipass interferometer for testing large concave surfaces has been described by Bubis (1972). This is essentially an unequal-path interferometer using a laser source in which an auxiliary concave mirror is introduced in the measuring arm.

As can be seen from Fig. 7.10, the converging beam from the laser is split at the beam divider into two beams, one of which is reflected back from a small concave mirror to form the reference wavefront. The other beam undergoes a predetermined number of reflections in the system comprising the mirror under

Figure 7.9. Interferograms obtained with the multipass Twyman–Green interferometer from a flat surface with a square, $\lambda/8$ deep step at its center. (*a*) Normal interferogram (first order). *b*) Second order. (*c*) Third order. (*d*) Fourth order. (*e*) Fifth order. (From Langenbeck 1967.)

7.2. MULTIPASS INTERFEROMETRY

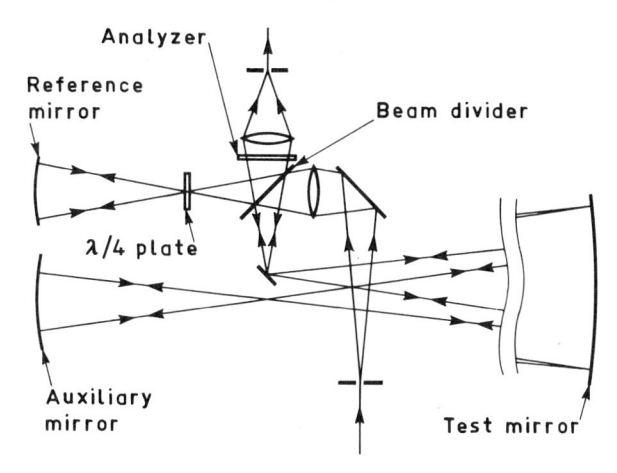

Figure 7.10. Multipass interferometer for testing large concave surfaces. (From Bubis 1972).

test and the auxiliary concave mirror before it is reflected back to the beam divider to interfere with the reference wavefront. (The number of reflections that the beam undergoes can be controlled by the relative offsets of the centers of curvature of the mirror under test and the auxiliary mirror.) A mica $\lambda/4$ plate in the comparison arm and a Polaroid analyzer are used to equalize the amplitudes of the interfering beams. A fivefold increase in sensitivity has been obtained.

The Fizeau interferometer can also be operated in a multipass configuration (Langenbeck 1969a, 1969b) if the zero-order reflected beam and one of the higher-order reflected beams are made to interfere. For this purpose the wedge angle between the surfaces is made large enough that the individual orders of reflection can be separated in the focal plane. As shown in Fig. 7.11, a stop with two openings is then used to select the zero-order and a suitable higher-order beam. An auxiliary lens whose front focal plane coincides with the stop gives interference fringes between nominally plane wavefronts.

The relative amplitudes of the interfering beams are equalized by a simple polarizing system. The zero-order beam is reflected from the uncoated part of a partially coated mirror at the Brewster angle, while the higher-order beams are reflected from the adjacent metal-coated part. Rotation of the plane of polarization of the illuminating beam with a $\lambda/2$ plate then permits attenuating the zero-order beam to any required degree.

Because of the relatively large angle between the zero-order beam and the nth-order beam, the resulting interference fringes are too narrow to be viewed directly, and the inteference pattern is best observed by a moiré technique. For this, a grating with a line spacing equal to the average fringe spacing is introduced into the image plane. The moiré fringes seen have the same spacing;

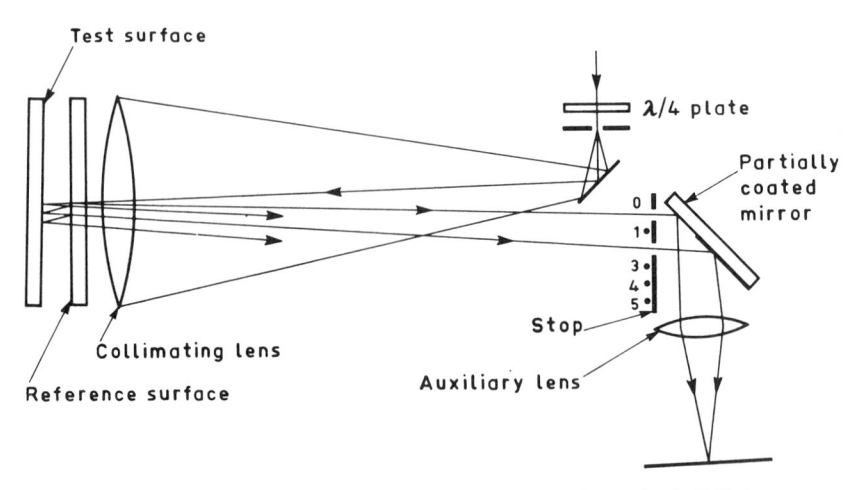

Figure 7.11. Multipass Fizeau interferometer. (From Langenbeck 1969a.)

their spacing can be controlled by rotating the grating with respect to the interferogram.

Higher-order moiré fringes can also be obtained by using a grating with a line spacing equal to $1/m$ the average fringe spacing. This gives m times as many moiré fringes across the field and a relative sensitivity m times as high, though at the expense of fringe contrast.

The most serious problem in multipass interferometry is beam "walk-off." This integrates the errors over a finite region of the surface under test and can lead to fringe shifts even in areas where there is no deformation of the test surface. The extent of this walk-off is determined by the distance separating the multipass mirrors and by the wedge angle. It can be minimized by tilting the interferometer with respect to the incoming beam so that the beam, which is initially reflected toward the apex of the wedge in the test path, is finally reflected at normal incidence by the auxiliary beam divider and returns along the same path. However, in most cases the order of reflection must be chosen so that the walk-off is small compared to the lateral dimensions of the defects under examination.[†]

Misinterpretation of multipass interferograms can also result from misalignment of the stop in the focal plane. If this stop accidentally cuts off low spatial frequencies in the image, strong diffraction effects are observed, as in the Foucault test.

Applications where multipass interferometry is useful include measurements of the homogeneity of glass plates and the precise determination of 90° or 45°

[†]This limitation does not apply in some cases, such as thickness measurements of uniform thin films, for which specialized optical systems using a narrow beam and permitting as many as 50 passes have been described by Dupoisot and Lostis (1973).

7.2. MULTIPASS INTERFEROMETRY

angles on roof prisms and cube corners, where high accuracy is required (Langenbeck 1969b). Another interesting application is in grazing incidence interferometry. Grazing incidence interferometry (Linnik 1941; Saunders and Gross 1959) makes it possible to test large surfaces with a system of relatively small aperture. In addition it is possible to test many nonoptical surfaces, such as fine-ground glass and metal, from which a specular reflection of adequate intensity can be obtained at low angles of incidence (Birch 1973; Hariharan 1975, Murty 1976).

With incidence at an angle α, a beam of diameter d covers a strip of length

$$L = \frac{d}{\cos \alpha} \tag{7.8}$$

However, a deviation Δt of the test surface from flatness results in a deviation of the fringes from straightness given by the relation

$$\frac{\Delta Q}{Q} = \frac{4\Delta t \cos \alpha}{\lambda} \tag{7.9}$$

where ΔQ is the local fringe displacement and Q is the average spacing of the fringes. As a result, the sensitivity of the interferometer is decreased by a factor proportional to the cosine of the angle of incidence. The sensitivity can be increased by a factor of 2, even with ground surfaces having a low reflectance, by using a double-pass system in which the reflectance of the beam splitter is adjusted to equalize the intensity of the two beams (Wilson 1983). If the reflectivity of the test surface is adequate, it is even possible to use the multiple-pass arrangement shown in Fig. 7.12 (Langenbeck 1969a). In this case, for an angle of incidence α, the deviation of the fringes from straightness in the fringe pattern obtained with the nth-order reflected beam is

$$\left(\frac{\Delta Q}{Q}\right)_n = \frac{4n \Delta t \cos \alpha}{\lambda}, \tag{7.10}$$

so that the loss in sensitivity due to the high angle of incidence can be regained by choosing a suitable higher-order beam.

Another interesting application of multiple-pass interferometry is accurate measurement of the radius of curvature of a concave surface with a very long radius of curvature, where the working distance available is limited (Gerchman and Hunter 1979).

In this procedure a reference plane surface and the test concave surface are set up to form a confocal cavity. Collimated light entering such a cavity through the plane surface is brought to a focus, as shown in Fig. 7.13, on one or the

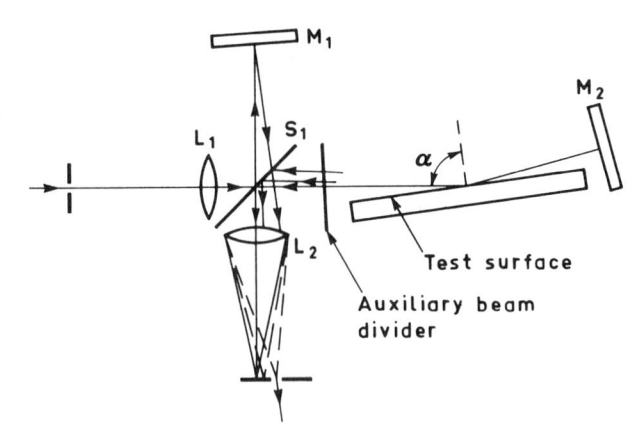

Figure 7.12. Grazing incidence multipass interferometer used for testing large surfaces with an optical system of small aperture. (From Langenbeck 1969a.)

other of the surfaces. In such a confocal configuration a collimated beam that has undergone the appropriate number of reflections is reflected back as a collimated beam. The focus setting can therefore be determined accurately by observing the interference fringes formed by the reflected beam from the plane surface and the reflected beam from the confocal cavity, using a setup similar to a Fizeau interferometer.

Configurations in which the order number n is even bring the light to a focus on the plane surface, while configurations in which n is odd focus the light on

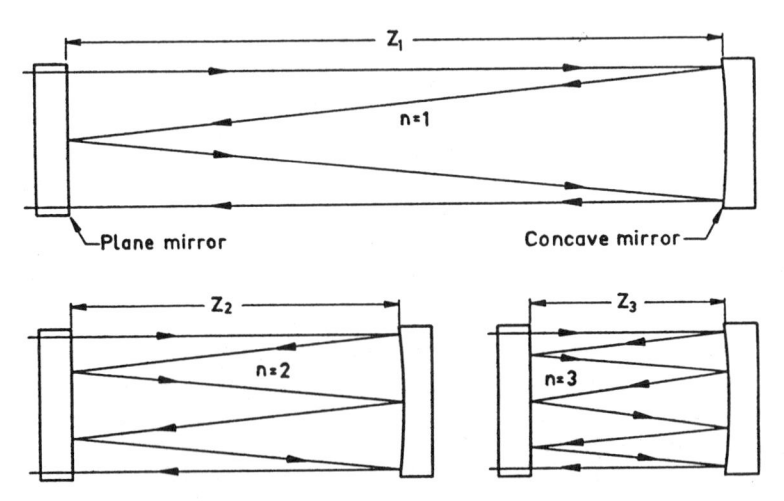

Figure 7.13. Ray paths in confocal cavities of orders 1, 2, and 3. (From Gerchman and Hunter 1979.)

7.2. MULTIPASS INTERFEROMETRY

Table 7.1 Cavity Length and Radius of Curvature $z_n = C_n R$

n	C_n
1	0.5
2	0.25
3	0.1464466
4	0.0954915
5	0.0669873
6	0.0495156
7	0.0380603
8	0.0301537
9	0.0244717

the concave surface. The equations relating z_n, the separation of the surfaces in the nth-order configuration, and R, the radius of curvature of the concave surface, can be obtained from a paraxial analysis. This analysis shows that the expressions for the odd configurations are similar to the even-order Chebyshev polynomials of the first kind, $T_{n+1}(x)$, ($n = 1, 3, 5, \ldots$), where $x = z_n/R$. Similarly, the expressions for the even configuration are similar to the even-order Chebyshev polynomials of the second kind, $U_n(x)$, ($n = 2, 4, 6, \ldots$). In both cases the solutions of interest are the first positive root.

Accordingly, if the change in the separation of the mirrors between two successive confocal configurations is measured, it is possible to calculate the radius of the concave surface. Table 7.1 lists the values of z_n, the cavity length for successive confocal configurations, as a function of R, the radius of curvature of the concave surface, while Table 7.2 lists $z_n - z_{n+1}$, the corresponding differences in the cavity length between successive configurations.

As can be seen, the separation of the surfaces in the higher-order configurations is much smaller than the radius of curvature of the concave surface, while the distance through which it has to be moved between successive con-

Table 7.2 Difference in Cavity Lengths and Radius of Curvature

$$R = 4(z_1 - z_2)$$
$$R = 9.65685(z_2 - z_3)$$
$$R = 19.62512(z_3 - z_4)$$
$$R = 35.08255(z_4 - z_5)$$
$$R = 57.23525(z_5 - z_6)$$
$$R = 87.29584(z_6 - z_7)$$
$$R = 126.47741(z_7 - z_8)$$
$$R = 175.99437(z_8 - z_9)$$

figurations is even smaller. As a result, surfaces with very long radii can be measured in a limited working space using a short measuring slide.

The accuracy with which measurements can be made is limited by two factors. One is the accuracy with which the position of best focus can be estimated; the other is the accuracy with which the movement of the mirror can be measured. A limit is also imposed on the highest order test configuration by the drop in the visibility of the fringes. To optimize the visibility for higher-order configurations, the plane surface should have a reflectivity of about 0.6, while the concave surface should have a highly reflecting coating. Acceptable fringe visibility can then be obtained even with an $n = 9$ configuration, and measurements of the radius of curvature can be made with an accuracy better than 0.01%.

REFERENCES

Bubis, I. Ya., "Multipass Interferometer for Surface Shape Inspection," *Sov. J. Opt. Technol.*, **39**, 411 (1972).

Birch, K. G., "Oblique Incidence Interferometry Applied to Non-Optical Surfaces," *J. Phys. E.*, **6**, 1045 (1973).

Cagnet, M., "Méthodes Interférometriques Utilisant les Franges de Superposition (voir Erratum, p. 552)," *Rev. Opt.*, **33**(1), 113 (1954).

Dupoisot, H. and P. Lostis, "Interférométrie à Passages Multiples et Mesure des Très Faibles Epaisseurs," *Nouv. Rev. Opt.*, **4**, 227 (1973).

Gerchman, M. C. and G. C. Hunter, "Differential Technique for Accurately Measuring the Radius of Curvature of Long Radius Concave Optical Surfaces," *Proc. SPIE*, **192**, 75 (1979).

Hariharan, P. and D. Sen, "Double-Passed Two-Beam Interferometers," *J. Opt. Soc. Am.*, **50**, 357 (1960a).

Hariharan, P. and D. Sen, "The Double-Passed Fizeau Interferometer," *J. Opt. Soc. Am.*, **50**, 999 (1960b).

Hariharan, P. and D. Sen, "Double-Passed Fizeau Interferometer. II: Fringe Systems Formed by the Reflected Beams," *J. Opt. Soc. Am.*, **51**, 400 (1961a).

Hariharan, P. and D. Sen, "Fringes of Equal Inclination in the Double-Passed Michelson Interferometer," *J. Opt. Soc. Am.*, **51**, 617 (1961b).

Hariharan, P. and D. Sen, "Double-Passed Two-Beam Interferometers, II: Effects of Specimen Absorption and Finite Path Difference," *J. Opt. Soc. Am.*, **51**, 1212 (1961c).

Hariharan, P. and D. Sen, "The Separation of Symmetrical and Asymmetrical Wave-Front Aberrations in the Twyman Interferometer," *Proc. Phys. Soc.*, **77**, 328 (1961d).

Hariharan, P., "Improved Oblique-Incidence Interferometer," *Opt. Eng.*, **14**, 257 (1975).

Hopkins, H. H., *Wave Theory of Aberrations*, Clarendon Press, Oxford, 1950, p. 48.

Langenbeck, P., "Multipass Twyman–Green Interferometer," *Appl. Opt.*, **6,** 1425 (1967).

Langenbeck, P., "Multipass Interferometry," *Appl. Opt.*, **8,** 543 (1969a).

Langenbeck, P., in *Optical Instruments and Techniques*, J. Home Dickson, Ed., Oriel Press, Newcastle-Upon-Tyne, 1966b.

Linnik, V. P., "An Interferometer for the Investigation of Large Plane Surfaces," *Computes Rendus (Doklady) Acad. Sci. URSS*, **32,** 189 (1941).

Murty, M. V. R. K. and Shukla, R. P., "An Oblique Incidence Interferometer," *Opt. Eng.*, **15,** 461 (1976).

Puntambekar, P. N. and D. Sen, "A Simple Inverting Interferometer," *Opt. Acta*, **18,** 719 (1971).

Saunders, J. B. and Gross, F. L., "Interferometer for Large Surfaces," *J. Res. Natl. Bur. Stand.*, **62,** 137 (1959).

Sen, D. and P. N. Puntambekar, "An Inverting Fizeau Interferometer," *Opt. Acta*, **12,** 137 (1965).

Sen, D. and P. N. Puntambekar, "Shearing Interferometers for Testing Corner Cubes and Right Angle Prisms," *Appl. Opt.*, **5,** 1009 (1966).

Wilson, I. J., "Double-Pass Oblique Incidence Interferometer for the Inspection of Nonoptical Surfaces," *Appl. Opt.*, **22,** 1144 (1983).

8

Foucault, Wire, and Phase Modulation Tests

J. Ojeda-Castañeda

8.1 INTRODUCTION

The knife-edge method, introduced by Foucault (1858, 1859), and, in general, all the Schlieren techniques—Töpler (1864, 1866, 1868), Ritchey (1904), Hartmann (1908), Platzeck and Gaviola (1939), Wolter (1949 and 1956)—have proved to be extremely useful for testing optical surfaces. Rays may be moved from their expected trajectories (or wavefronts may be deformed) by optical aberrations, by diffraction, or by a deformed unfinished surface. The basic idea behind the Schlieren techniques is to detect lateral displacements of rays by blocking out or modifying these displaced rays. The blocking or modification can be accomplished by placing screens in any of the planes of convergence of the light passing through, or being reflected from, the optical surface under test.

The main advantages of the Schlieren techniques are their high sensitivity and their simplicity, both in apparatus and in qualitative interpretation (at least from the geometrical point of view). Of course, to appreciate the relative merits of one method over others in this class, it is necessary to study its characteristics when it is used to detect the presence of (a) aberrations greater than the wavelength of the illuminating radiation (geometrical theory of image formation) and (b) aberrations smaller than the wavelength (physical theory of image formation).

The choice of test depends on the circumstances.

8.2. FOUCAULT OR KNIFE-EDGE TEST

8.2.1. Description

The knife-edge test may be considered, in general, as a method for detecting the presence of transversal aberrations. This is done by blocking out one part

266   FOUCAULT, WIRE, AND PHASE MODULATION TESTS

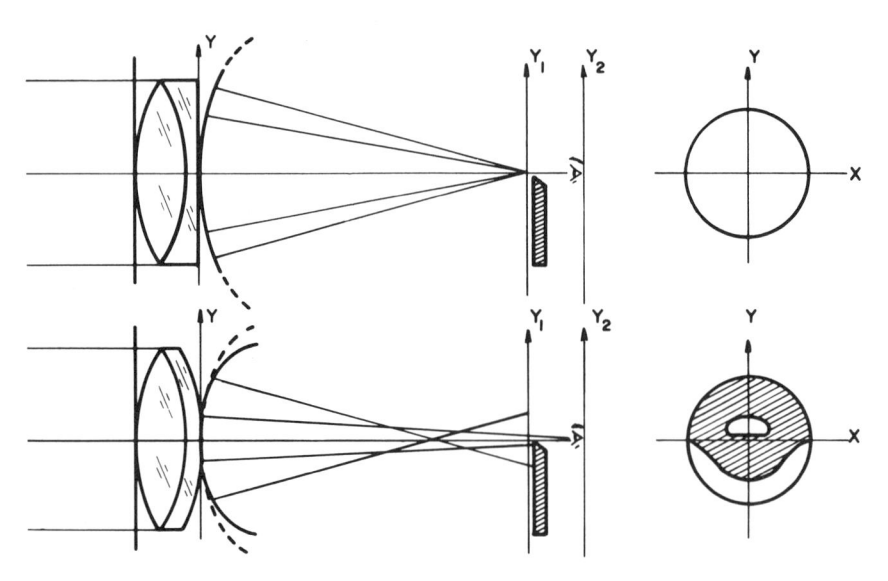

Figure 8.1. Foucault method for testing a lens.

of a plane traversed by rays or diffracted light so that a shadow appears over the aberrated region, as shown in Fig. 8.1. This high simplicity of operation and interpretation makes the Foucault test unique among optical testing methods. This test may be well considered as the first optical shop test, from which many other optical tests were developed.

The Foucault test consists in placing an illuminated pinhole size source at one side of the center of curvature of a spherical concave mirror, as shown in Fig. 8.2. The image of this illuminated pinhole size source has the shape of a point source, localized on the other side of the center of curvature of the spherical mirror.

When a knife edge is introduced, cutting the illuminating beam that forms

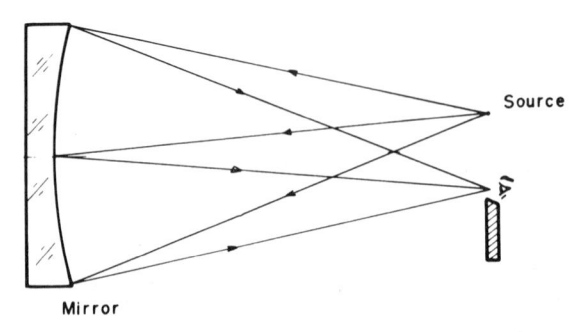

Figure 8.2. Knife-edge setup when testing a concave mirror.

8.2. FOUCAULT OR KNIFE-EDGE TEST

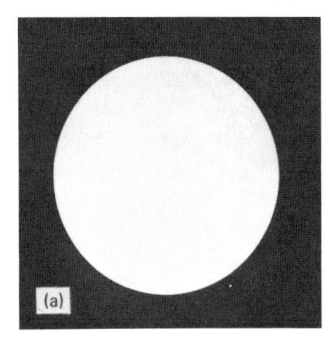

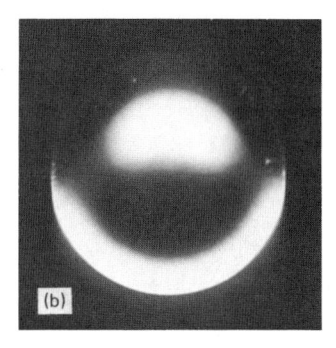

Figure 8.3. Foucault graphs of a parabolic mirror. (*a*) Before introducing the knife edge. (*b*) After introducing the knife edge.

the image, an observer (placed behind the image) sees a shadow pattern appearing over the otherwise uniformly illuminated surface of the mirror (see Figs. 8.1 and 8.3). In the case of a "perfect" spherical mirror, when the knife edge is introduced *inside the focus* (toward the mirror), the shadow pattern consists of a dark region and a bright region, sharply separated, as in Figs. 8.4 and 8.5. As the knife is moved, the dark region appears to move across the mirror in the *same direction* in which the knife edge moves. On the other hand, when the knife edge is introduced *outside the focus* (away from the mirror), the dark region of the shadow pattern moves across the mirror in a direction *opposite* to that in which the knife edge moves. When the knife edge is introduced precisely

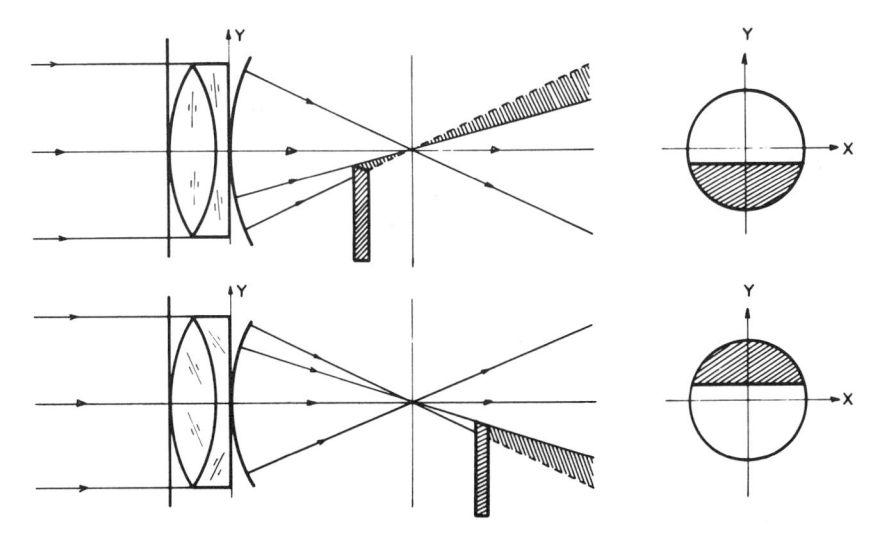

Figure 8.4. Knife-edge testing of an aberration-free lens.

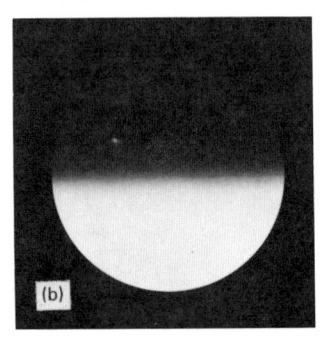

Figure 8.5. Foucault graphs of an aberration-free mirror with focus errors. (*a*) Inside of focus. (*b*) Outside of focus. Knife introduced from below.

at the focus, the mirror appears to *darken suddenly*, without any apparent motion of the shadow pattern. This is a very accurate way of finding the center of curvature.

Moreover, if the concave mirror is not spherical, each zone over the mirror has a different radius of curvature. In this case, when the knife edge is introduced, the corresponding zone darkens for different positions of the knife edge along the optical axis.

For an irregular concave mirror there are many zones with different radii of curvature and different centers of curvature. In this case, when the knife edge is introduced, many different zones all over the mirror are darkened. An intuitive way to visualize how this shadow pattern is formed (given by R. W. Porter 1953) is to imagine an illuminating source at one side of the mirror, at the opposite side of the knife edge, as shown in Fig. 8.6. With this kind of illumination it follows that the regions with positive slope are illuminated and those with negative slope are not illuminated, or vice versa, as in Fig. 8.7.

To summarize, the Foucault test can be used (1) to measure the radius of curvature of each zone over the mirror and (2) as a *null test*, for checking that all the zones of a supposedly spherical wavefront have the same radius of curvature.

It has been found in practice that the Foucault test may be done perfectly by employing an illuminating slit instead of a point source, as shown in Figs. 8.8*b* and 8.8*c*. The reason is that each point of the slit source produces an image with the same shape and at the same distance from the knife edge. Therefore all of the points produce identical Foucault patterns, resulting in a tremendous gain in the brightness of the observed pattern. The patterns, however, must be exactly identical if the accuracy is to be preserved. This means that the slit and the knife edge must be exactly parallel. To avoid the problem of setting the knife edge exactly parallel to the slit, the arrangement in Fig. 8.8*d* was suggested by Dakin (1967).

8.2. FOUCAULT OR KNIFE-EDGE TEST

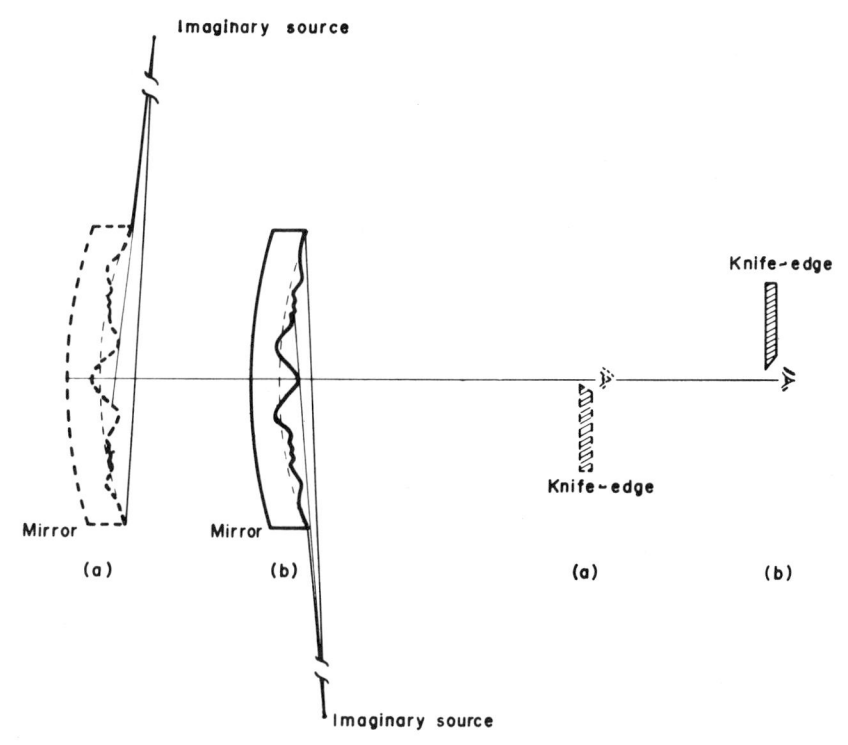

Figure 8.6. Visualization of the shadow patterns formed when using the knife-edge test. Notice that the imaginary light source and the knife edge are at opposite sides of the optical axis.

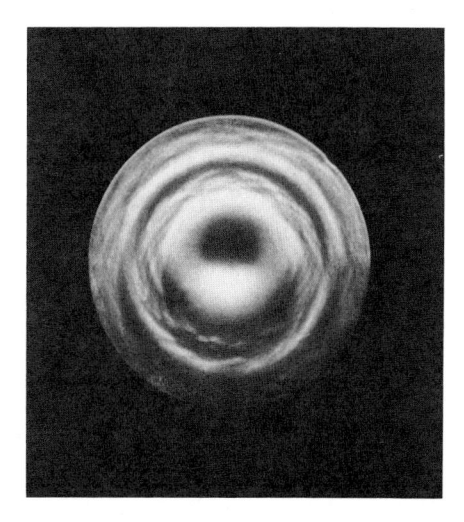

Figure 8.7. Foucault graph of an irregular mirror.

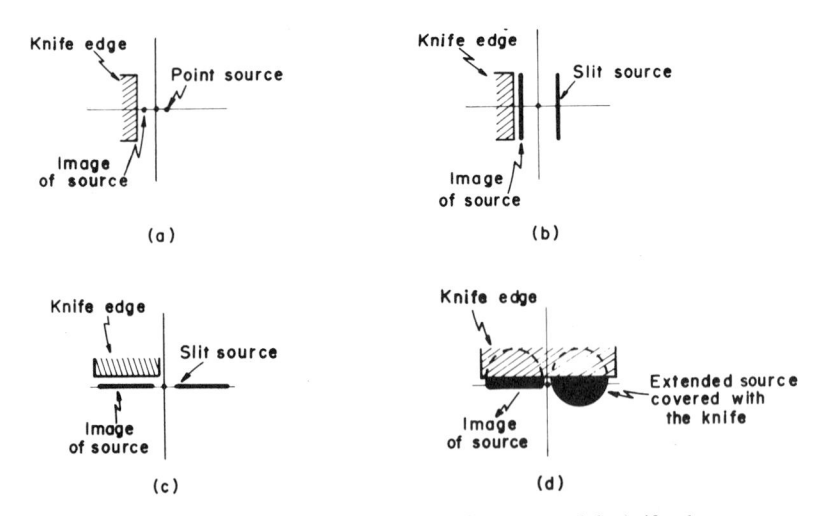

Figure 8.8. Different arrangements for the source and the knife edge.

The use of an extended source, in the form of a slit, was previously used in flow visualization (Wolter 1956, Stolzenburg 1965) rather than in optical testing. In the previous field it is also common to employ white-light sources for encoding optical path differences in terms of color variations; see for example Schardin (1942a, 1942b), Wolter (1956), North (1954), O'Hare (1969), Banerji (1918), and Weinberg (1961). The Foucault test in white light has been analyzed by Bescos and Berriel-Valdos (1986).

Figures 8.9 and 8.10 show an instrument to perform the Foucault test. The instrument is mounted in a carriage capable of translating it, along the optical axis or perpendicularly to the optical axis, laterally or up and down.

8.2.2. Geometrical Theory

The border of the knife in the Foucault test is placed at a distance r_1 from the chief ray intersection (the origin of the X_1-Y_1 plane), and an angle ϕ_1 is subtended between the Y_1 axis and the knife edge, as shown in Fig. 8.11. The angle ϕ_1 will be defined as positive if the slope of the knife edge is positive. The following equation defines the border:

$$x_1 \cos \phi_1 - y_1 \sin \phi_1 = r_1. \tag{8.1}$$

The transmittance over this plane may be expressed as

$$T(x_1, y_1) = \begin{cases} 1 & \text{if } x_1 \cos \phi_1 - y_1 \sin \phi_1 < r_1, \\ 0 & \text{if } x_1 \cos \phi_1 - y_1 \sin \phi_1 \geq r_1. \end{cases} \tag{8.2}$$

8.2. FOUCAULT OR KNIFE-EDGE TEST

Figure 8.9. Photograph of the Foucault apparatus.

But since the X_1-Y_1 plane defines the paraxial plane of convergence of the ideal wavefront W, any point (x_1, y_1) over this plane satisfies, approximately, the following property (Rayces 1964):

$$(x_1, y_1) = \left(-R\frac{\partial W}{\partial x}, -R\frac{\partial W}{\partial y}\right); \qquad (8.3)$$

where R is the distance between the X-Y plane and the X_1-Y_1 plane.

Using this result in Eq. (8.2), we have, then, that the transmittance function becomes

$$T\left(\frac{\partial W}{\partial x}, \frac{\partial W}{\partial y}\right) = \begin{cases} 1 & \text{if } -\frac{\partial W}{\partial x}\cos\phi_1 + \frac{\partial W}{\partial y}\sin\phi_1 < \frac{r_1}{R}, \\ 0 & \text{if } -\frac{\partial W}{\partial x}\cos\phi_1 + \frac{\partial W}{\partial y}\sin\phi_1 \geq \frac{r_1}{R}. \end{cases} \qquad (8.4)$$

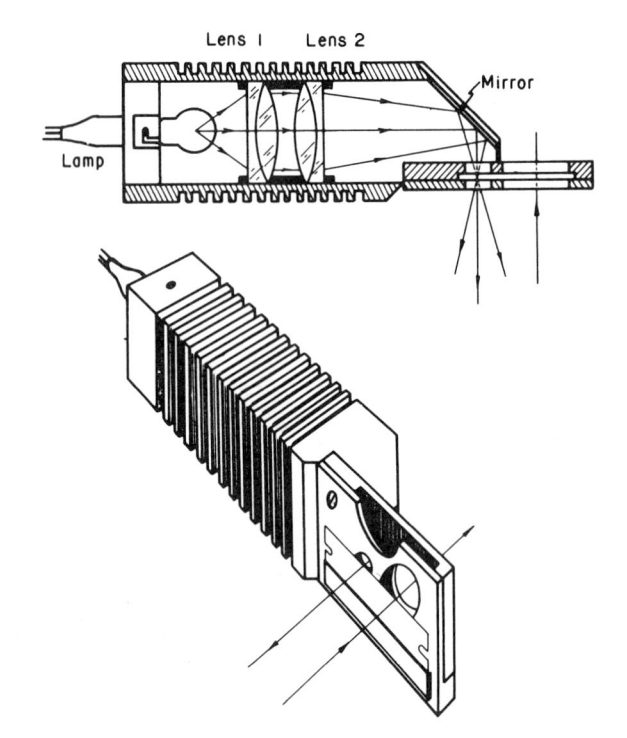

Figure 8.10. Diagrams of the Foucault apparatus.

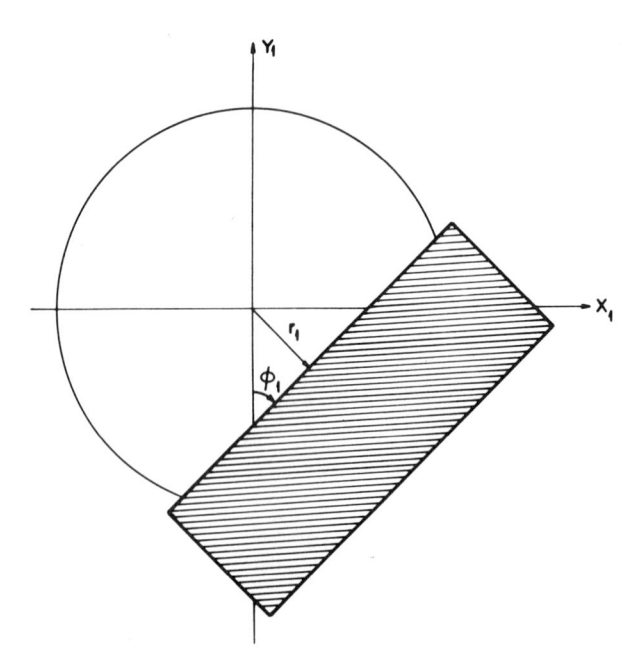

Figure 8.11. Knife-edge position projected over the entrance pupil plane of the viewing system.

8.2. FOUCAULT OR KNIFE-EDGE TEST

For the particular case of aberrations with rotational symmetry, any axis over the X_1-Y_1 plane may be used as reference to specify the position of the knife edge. It proves convenient, for our purposes, to choose the X axis, so that $\phi_1 = 90°$. Thus Eq. (8.4) is reduced to

$$T\left(\frac{\partial W}{\partial y}\right) = \begin{cases} 1 & \text{if } \frac{\partial W}{\partial y} < \frac{r_1}{R}, \\ 0 & \text{if } \frac{\partial W}{\partial y} \geq \frac{r_1}{R}. \end{cases} \quad (8.5)$$

Focus Error. Let us consider our first example, where we have an aberration-free optical surface but the knife is placed at some distance from the converging plane X_1-Y_1. In other words only focus error is present in the wavefront aberration function, that is,

$$W(x, y) = D(x^2 + y^2). \quad (8.6)$$

From Eq. (8.5) it follows that the boundary between dark and bright regions (bright region, $T = 1$ if $y_1 < r_1$) may be expressed as

$$y_1 = \frac{r_1}{2DR}, \quad (8.7)$$

when $D \neq 0$. Hence, depending on whether the knife is placed inside the focus (D negative) or outside the focus (D positive), the shadow along the Y axis (i.e., parallel to the X axis) will change sides, as shown in Fig. 8.5, from the negative side of the Y axis to the positive side of the Y axis, respectively.

This situation is precisely the one described before when indicating on which side of a "perfect" spherical mirror the dark region appears when the knife edge is introduced.

A particular case occurs when the knife edge touches the optical axis. The shadow pattern will be a dark half-circle with the other half-circle bright. In other words the boundary between the bright and the dark regions is at $y_1 = 0$ when $r_1 = 0$ for any $D \neq 0$.

Primary Spherical Aberration. When both primary spherical aberration and focus errors are present, the aberration function is given by

$$W(x, y) = A(x^2 + y^2)^2 + D(x^2 + y^2).$$

Thus, using Eq. (8.5), the equation of the borders of the shadow pattern may be written as

$$y^3 + \left(\frac{D}{2A} + x^2\right) y - \frac{r_1}{4AR} = 0. \tag{8.8}$$

It is to be noted that, since the equation has dependence on x, the boundary of the shadow will not, in general, be limited by straight lines.

The boundaries of the shadow pattern along the Y axis for this case may be found by letting $x = 0$ in Eq. (8.8), which may then be written as

$$y^3 + \frac{D}{2A} y - \frac{r_1}{4AR} = 0. \tag{8.9}$$

Being a cubic equation, it has three roots. Of course, we are interested only in the roots that are real numbers. Defining the parameter Δ as

$$\Delta = \left(\frac{r_1}{8AR}\right)^2 + \left(\frac{D}{6A}\right)^3, \tag{8.10}$$

we have, from a known result from algebra, that:

1. If $\Delta < 0$, there will be three real and unequal roots.
2. If $\Delta = 0$, there will be three real roots of which at least two are equal.
3. If $\Delta > 0$, there will be only one real root and two conjugate imaginary roots.

Consequently, the shadow pattern will show more than one dark region if conditions 1 and 2 are satisfied. This happens when the defocusing coefficient and the primary aberration coefficient have different signs. In other words, by letting $\Delta \leq 0$, we can find from Eq. (8.10) that

$$\left(\frac{r_1}{R}\right)^2 \leq \frac{(-\frac{2}{3}D)^3}{A}. \tag{8.11}$$

Now, since the left-hand term is always positive, it follows that when the knife is inside the focus (D negative) and there is positive primary aberration, or, equivalently, when the knife is outside the focus and there is negative primary aberration, the shadow pattern will show two dark regions.

Again, a particular case of this result occurs when the knife edge is touching the optical axis. In this case $r_1 = 0$ and Eq. (8.9) becomes

$$y\left(y^2 + \frac{D}{2A}\right) = 0, \tag{8.12}$$

8.2. FOUCAULT OR KNIFE-EDGE TEST

solutions of which are

$$y = 0 \tag{8.13a}$$

and

$$y = \pm\sqrt{-\frac{D}{2A}}. \tag{8.13b}$$

It is, indeed, clear that D and A must have different signs in order to obtain real numbers as solutions of Eq. (8.13b). In Figs. 8.1, 8.12, and 8.13, Foucault graphs for spherical aberration are shown. For different treatments for spherical aberration the reader is referred to Conrady (1924) and Kingslake (1937).

Primary Coma. In the case of primary coma and focus error, the aberration function is given by

$$W(x, y) = By(x^2 + y^2) + D(x^2 + y^2). \tag{8.14}$$

Since primary coma does not have radial symmetry, we will consider the two cases in which the knife edge is being displaced along the X_1 axis and the Y_1

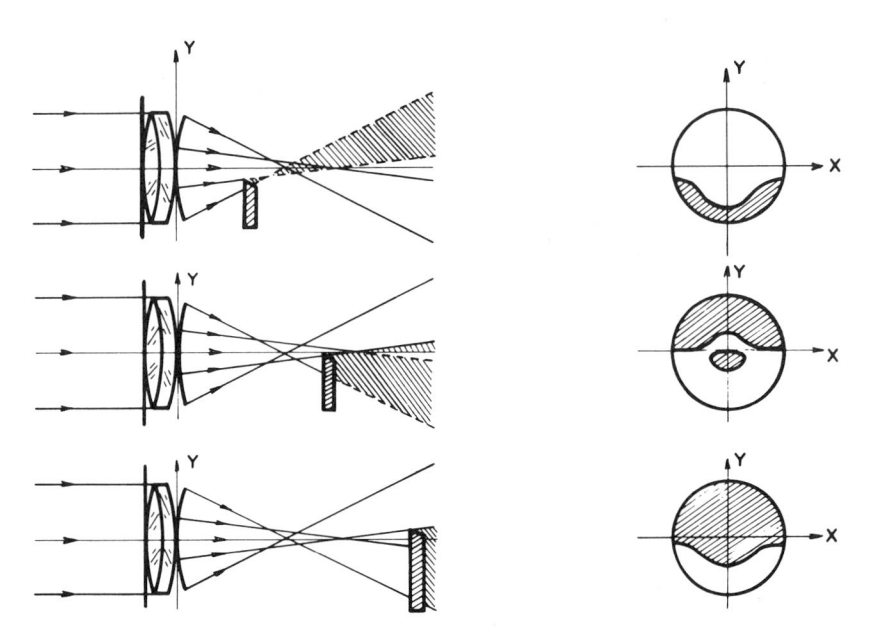

Figure 8.12. Knife-edge testing of a lens with spherical aberration.

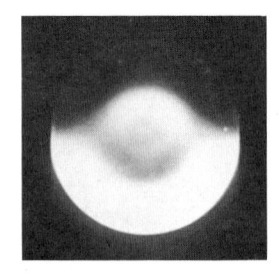

Figure 8.13. Foucault graphs of an aspherical mirror.

axis, respectively. Using the same procedure employed in the preceding examples, we have, in the case in which the knife edge is placed at a point on the X_1 axis at a distance r_1 from the origin (parallel to the Y_1 axis, $\phi_1 = 0$), that the shadow pattern satisfies the following equation:

$$2Bxy + 2Dx = \frac{-r_1}{R}, \tag{8.15}$$

which may usefully be written as

$$\left(y + \frac{D}{B}\right)x = \frac{-r_1}{2RB}. \tag{8.16}$$

It is clear, then, that the shadow pattern consists of rectangular hyperbolas, centered at the point $[O, -(D/B)]$. See Fig. 8.14a.

On the other hand, when the knife edge is introduced at a point on the Y_1 axis (parallel to the X_1 axis, $\phi_1 = 90°$), the partial derivative of the wavefront aberration function with respect to y is

$$\frac{\partial W}{\partial y} = B(x^2 + 3y^2) + 2yD. \tag{8.17}$$

Hence the borders of the shadow pattern may be obtained from the following equation:

$$x^2 + 3\left(y + \frac{D}{3B}\right) = \frac{r_1}{RB} + \frac{1}{3}\left(\frac{D}{B}\right)^2, \tag{8.18}$$

which is the equation for an ellipse centered at point $[O, -(D/3B)]$. The major axis of the ellipse will be parallel to the X axis, and the minor axis will be along the Y axis. See Fig. 8.14b.

8.2. FOUCAULT OR KNIFE-EDGE TEST

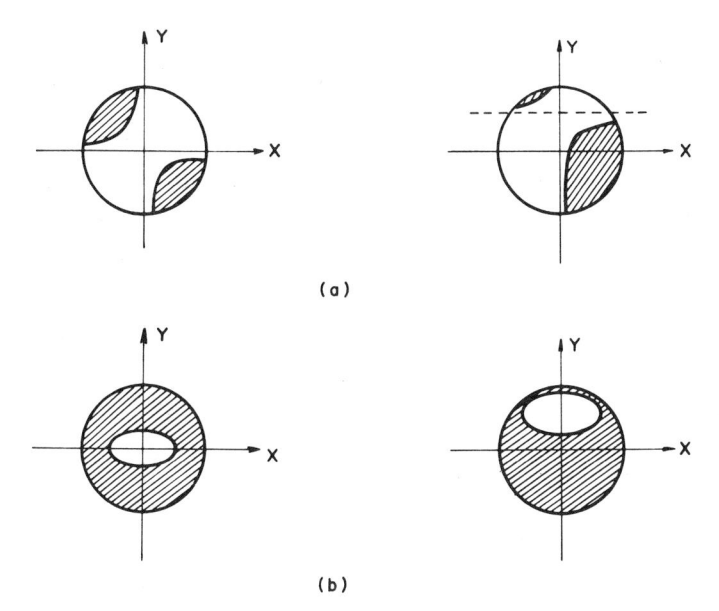

Figure 8.14. Diagrams showing the patterns associated with comatic aberrations. (*a*) Knife edge parallel to the y_1 axis. (*b*) Knife edge parallel to the x_1 axis.

The common procedure employed in the knife-edge test is to place the knife in the paraxial plane. Thus the shadow patterns will be centered at the origin of the optical surface under test. Under this condition a scale is placed over the shadow pattern, and either the parameters of the rectangular hyperbola or the ellipse may be readily obtained. In other words the knife edge is located at the paraxial plane, $D = 0$. Hence, when the knife is oriented parallel to the Y_1 axis, in the Gaussian plane, we know from Eq. (8.16) that the shadow pattern will be described by

$$yx = \frac{-r_1}{2RB}; \tag{8.19}$$

and since r_1, the knife-edge position along the X_1 axis, and the distance R between the optical surface under test and the paraxial plane are known parameters, the value of B may be immediately obtained by measuring the ratio $r_1/2RB$.

In a similar manner, when the knife edge is oriented parallel to the X_1 axis, in the Gaussian plane, the shadow pattern in Eq. (8.18) may be written as

$$x^2 + 3y^2 = \frac{r_1}{RB}. \tag{8.20}$$

Here, again, by knowing the parameters r_1 and R, and measuring the ratio r_1/RB, the amount of coma present may be determined.

Astigmatism. If the optical surface under test has astigmatism, and (in order to incorporate the position of the knife edge in any plane) we add focus error to the wavefront aberration function, we have

$$W(x, y) = C(x^2 + 3y^2) + D(x^2 + y^2). \tag{8.21}$$

As in the case of coma, since astigmatism does not have radial symmetry, we need to consider the effect of introducing the knife edge in a particular direction. In what follows we consider the cases in which the knife edge is introduced along an axis that subtends an angle ϕ_1 with the Y_1 axis. Since the partial derivative of the wavefront aberration function with respect to x is

$$\frac{\partial W}{\partial x} = 2Cx + 2Dx, \tag{8.22a}$$

and with respect to y is

$$\frac{\partial W}{\partial y} = 6Cy + 2Dy, \tag{8.22b}$$

the border of the shadow pattern can be obtained from Eq. (8.4) to give

$$-(C + D)x \cos \phi_1 + (3C + D) y \sin \phi_1 = \frac{r_1}{2R}. \tag{8.23}$$

It is clear, then, that an optical surface with astigmatism will have a straight line with slope ψ, given by

$$\tan \psi = \frac{3C + D}{C + D} \tan \phi_1. \tag{8.24}$$

From Eq. (8.24) it follows that, if the knife edge (oriented parallel to the Y_1 axis, i.e., $\phi_1 = 0°$) is introduced along the X_1 axis, the boundary of the shadow pattern consists of a straight line parallel to the Y axis, as seen in Fig. 8.15a, namely,

$$x = \frac{-r}{2R(C + D)}, \tag{8.25}$$

8.2. FOUCAULT OR KNIFE-EDGE TEST

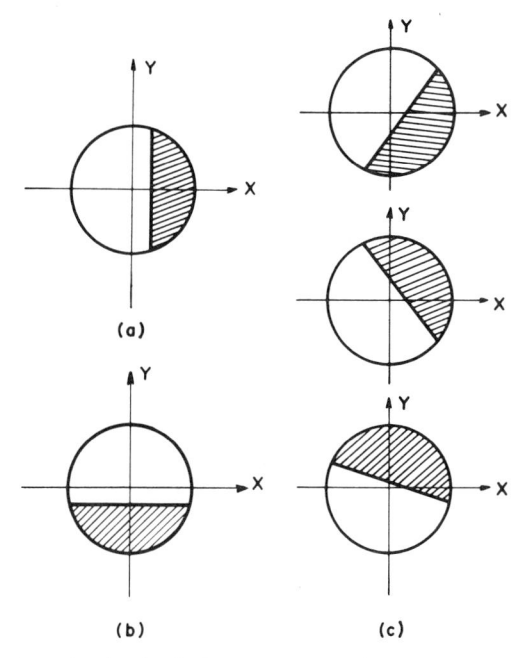

Figure 8.15. Diagrams showing the shadow patterns associated with astigmatism. (*a*) Knife edge parallel to the y_1 axis. (*b*) Knife edge parallel to the x_1 axis. (*c*) Knife edge along an axis making an angle ϕ with the y_1 axis.

and $\psi = 0°$. On the other hand, when the knife edge (oriented parallel to the X_1 axis, i.e., $\phi_1 = 90°$) is introduced along the Y_1 axis, the boundary of the shadow pattern consists also of a straight line parallel to the X_1 axis, as seen in Fig. 8.15*b*, namely,

$$y = \frac{r_1}{2R(3C + D)}, \qquad (8.26)$$

and $\psi = 90°$. Employing Eq. (8.24), it proves convenient to write the slope associated with the angle subtended between the knife edge and the boundary of the shadow pattern as in the Ronchi test (see Chapter 9), that is,

$$\tan(\psi - \phi_1) = \frac{C \sin 2\phi_1}{D + C(2 - \cos 2\phi_1)}. \qquad (8.27)$$

From this it follows that in the paraxial plane ($D = 0$) the angle between the knife edge and the boundary of the shadow pattern is independent of the amount of astigmatism present in the wavefront leaving the optical surface under test.

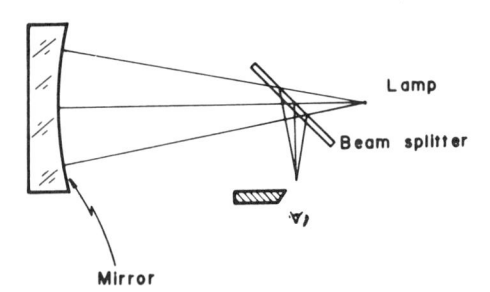

Figure 8.16. Foucault setup for testing on the optical axis.

Moreover, as the knife is displaced along the optical axis (i.e., a variation in D), the slope of the boundary of the shadow pattern varies when ϕ_1 is different from $0°$ or $90°$. Consequently, a rotation of the shadow pattern, as shown in Fig. 8.15, is observed when the optical surface under test suffers from astigmatism.

It is worth noting that, when a concave mirror with a small f number is tested with a pinhole size source placed off axis (as shown in Fig. 8.2), the mirror, even when perfect, appears to give rise to an astigmatic wavefront (see Appendix 1). To avoid this effect of false astigmatic error, the mirror can be tested on axis, using a beam splitter, as shown in Fig. 8.16.

For the purpose of training students and technicians, it is convenient to generate the previous shadow patterns of the Foucault test, or indeed of other tests, by computer simulation using a personal computer and the concept of spot diagrams (Herzberger, 1947; Miyamoto 1963) as follows (Bermudez et al. 1985):

From the set of the aberration coefficients (Chapter 13), one can find the spot diagram associated with the intersection of rays coming from different points (inside a grid) of the surface under test; as in the Hartmann test (Chapter 10). Then, a computer procedure sets to zero value all the ray intersection points that were covered by the knife edge or other related test. If we visualize the missing points as dark spots, on the surface under test, then we obtain patterns as those shown in Fig. 8.17.

8.2.3. Physical Theory

As indicated in the introduction of this chapter, we restrict the analysis of the knife-edge test in terms of diffraction theory to the case in which the aberrations present are smaller than the wavelength of the illuminating radiation. A reader interested in a more general treatment is referred to the following publications: Barakat (1969), Banerji (1918), Gascoigne (1944), Linfoot (1945a, 1945b, 1946a, 1946b, 1946c, 1955), Rayleigh (1917), and Zernike (1934a, 1934b, 1942).

In particular, a Fourier optics treatment of the Schlieren techniques, under coherent illumination or partially coherent illumination is considered by Menzel

8.2. FOUCAULT OR KNIFE-EDGE TEST

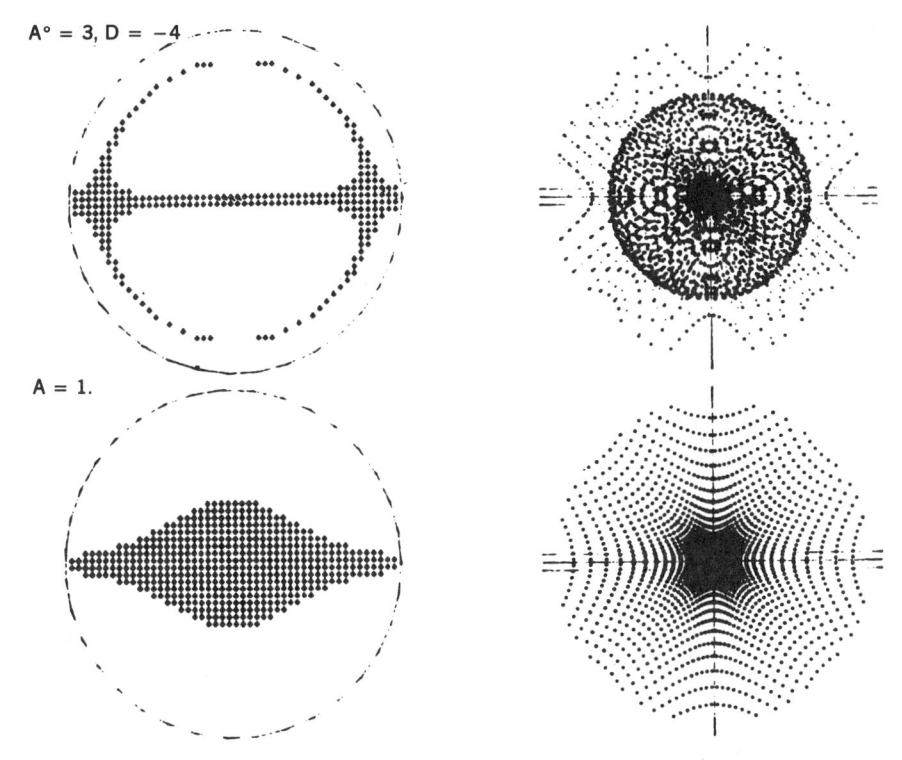

Figure 8.17. Computer-simulated aberration patterns. On the left side we show the wire test patterns for spherical aberration and defocusing. On the right side we show the corresponding spot diagrams.

(1960), Lowenthal et al. (1967), Ojeda-Castañeda (1978, 1979, 1980a, 1980b), and Soroko (1981).

An optical surface under test may be conveniently thought of as being composed of the required surface and an unwanted surface. When the shift from the required to the unwanted surface produces aberrations smaller than the wavelength of the illuminating radiation, the complex amplitude leaving the optical surface under test may be usefully simplified as the sum of a uniform background (due to the required surface) plus a complex amplitude describing the aberrations (due to the unwanted surface), with a phase difference of $\pi/2$ between the terms, that is,

$$A(x, y) = \begin{cases} \exp\left[i\frac{2\pi}{\lambda}W(x, y)\right] \simeq 1 + i\frac{2\pi}{\lambda}W(x, y), & x^2 + y^2 \leq S_{max}^2, \\ 0, & x^2 + y^2 > S_{max}^2, \end{cases}$$

(8.28)

where $W(x, y)$ describes the aberrations present over the optical surface (see Chapter 13) and λ is the wavelength of the illuminating radiation. In the present treatment we examine the case in which

$$|W(x, y)| < \lambda. \tag{8.29}$$

Of course, if the optical surface is free from aberrations, Eq. (8.28) reduces to

$$A(x, y) = \begin{cases} 1, & x^2 + y^2 \leq S_{max}^2, \\ 0, & x^2 + y^2 < S_{max}^2. \end{cases} \tag{8.30}$$

A "perfect" image of the optical surface under test (in the sense that the image is formed by an optical system that has neither aberrations nor focus error, and has an infinite aperture) will have an irradiance distribution equal to the irradiance distribution on the surface. In other words the image irradiance is given by

$$I(x, y) = A(x, y) A^*(x, y)$$
$$= 1 \tag{8.31}$$

where the asterisk denotes a complex conjugate, and this gives no information about the presence of aberrations. Therefore, to obtain information about the presence of aberrations, it is necessary to produce a modified image of the optical surface under test. In our case this is done by introducing the knife edge across the paraxial plane of the converging wavefront coming from the optical surface under test.

Since the complex amplitude $a(x_1, y_1)$ in the paraxial plane is proportional to the Fourier transform of Eq. (8.28), then we have

$$a(x_1, y_1) = \iint_{x^2 + y^2 \leq S^2_{max}} A(x, y) \exp\left[-i\frac{2\pi}{\lambda R}(x_1 x + y_1 y)\right] dx \, dy$$

$$= \frac{2J_1(\rho)}{\rho} + iw(x_1, y_1); \tag{8.32}$$

where $\rho = (2\pi/\lambda R) [x_1^2 + y_1^2]^{1/2}$, $J_1(\rho)$ denotes the Bessel function of the first kind and order 1, and

$$w(x_1, y_1) = \iint_{x^2 + y^2 \leq S^2_{max}} W(x, y) \exp\left[-i\frac{2\pi}{\lambda R}(x_1 x + y_1 y)\right] dx \, dy \tag{8.33}$$

8.2. FOUCAULT OR KNIFE-EDGE TEST

is the Fourier transform of $W(x, y)$ and represents the complex amplitude over the paraxial plane produced by the presence of aberrations. The complex amplitude distribution over the paraxial plane is blocked out by the presence of the knife edge, which has the following amplitude transmittance:

$$M(x_1, y_1) = \begin{cases} 1, & -r_1 < y_1, \\ 0, & -r_1 > y_1, \end{cases} \quad (8.34)$$

where r_1 is a real number indicating the position of the knife edge (oriented parallel to the X_1 axis) along the Y_1 axis as in Eq. (8.1).

It follows from Eqs. (8.33) and (8.34) that, if $(2\pi/\lambda R) r_1 \gg 3.83$, the radius of the Airy disk at the paraxial plane due to the uniform background will pass "practically" unmodified, and a uniform irradiance distribution will be present over the image of the optical surface under test. If the knife edge is placed at distances of the order of the radius of the Airy disk ($\rho = 3.83$), the complex amplitude associated with the direct light in the image will be also modulated. Hence, the complex amplitude of the image due to an aberration-free optical surface is given by

$$A'(x_2, y_2) = \int\!\!\int_{-\infty}^{+\infty} M(x_1, y_1) \frac{2J_1(\rho)}{\rho} \exp\left[-\frac{i2\pi}{\lambda R_1}(x_2 x_1 + y_2 y_1)\right] dx_1\, dy_1, \quad (8.35)$$

which can conveniently be written, employing the convolution theorem, as the Hilbert transform of the aberration-free part of the complex amplitude of the object, that is,

$$A'(x_2, y_2) = \int\!\!\int_{-\infty}^{+\infty} \frac{1}{y_2/\lambda R_1 - y/\lambda R} \operatorname{circ}(x, y)\, dx\, dy, \quad (8.36)$$

where $\operatorname{circ}(x, y)$ is equal to unity inside the optical surface under test, $x^2 + y^2 \leq S_{\max}^2$, and is zero otherwise. Equation (8.36) can be integrated to give

$$A'(x_2, y_2) = \operatorname{constant} \ln \left| \frac{y_2 + (S_{2\max}^2 - x_2^2)^{1/2}}{y_2 - (S_{2\max}^2 - x_2^2)^{1/2}} \right|, \quad (8.37)$$

where ln indicates the natural logarithmic function. From this function [Eq. (8.37)] it follows that over the edge of the image of an aberration-free surface there is a bright ring, known as the Rayleigh diffraction ring, which is not to be interpreted as a turned edge or to be confused with any other error on the surface under test. Of course, in the present formulation the image irradiance over the rim goes to infinity, which does not happen in practice. This is so because, as was pointed out by Rayleigh (1917), the modulating plane in which the knife edge is placed is considered as infinitely extended, for purely mathematical convenience, neglecting the fact that any viewing system (forming the image of the optical surface under test) has a finite aperture. Welford (1970) showed that the singularity is replaced by a finite maximum value of the image irradiance, when the finite aperture of the viewing system is considered.

Berriel-Valdos (1978, 1981, 1982) has performed several numerical calculations that show the influence of the finite size of the viewing system, which is used to observe the Foucault test and other related tests.

To summarize, the complex amplitude leaving the paraxial plane and going toward the image plane is given by

$$a'(x_1, y_1) = M(x_1, y_1) a(x_1, y_1)$$
$$= 2 \frac{J_1(\rho)}{\rho} + i \frac{2\pi}{\lambda} M(x_1, y_1) w(x_1, y_1). \quad (8.38)$$

Thus the complex amplitude over the image plane is obtained by calculating the inverse Fourier transform of Eq. (8.38), that is,

$$A'(x_2, y_2) = \iint_{-\infty}^{+\infty} a'(x_1, y_1) \exp\left[i \frac{2\pi}{\lambda R_1} (x_2 x_1 + y_2 y_1)\right] dx_1 \, dy_1$$
$$= 1 + i \frac{2\pi}{\lambda} W'(x_2, y_2) \quad (8.39)$$

where

$$W'(x_2, y_2) = \iint_{-\infty}^{+\infty} M(x_1, y_1) w(x_1, y_1) \exp\left[i \frac{2\pi}{\lambda R_1} (x_2 x_1 + y_2 y_1)\right] dx_1 \, dy_1 \quad (8.40)$$

is the inverse Fourier transform of the diffracted light, produced by the presence of aberrations that are not blocked out by the knife. It is clear that, if the amplitude transmittance function $M(x_1, y_1)$ is equal to unity over the paraxial plane, we have

$$W'(x_2, y_2) = W(x_2, y_2),$$

8.2. FOUCAULT OR KNIFE-EDGE TEST

and the result in Eq. (8.31) holds again. That is, the information about the presence of aberrations is lost. Consequently, we need to modify the function $w(x_1, y_1)$ in such a manner that $W'(x_2, y_2)$ will be a purely imaginary function or a complex function with its imaginary part different from zero, that is,

$$W'(x_2, y_2) = W'_r(x_2, y_2) + iW'_i(x_2, y_2) \tag{8.41}$$

where the subindices r and i denote the real and imaginary parts, respectively.

By substituting Eq. (8.41) in Eq. (8.39), we can calculate the image irradiance distribution as

$$I'(x_2, y_2) = A'(x_2, y_2) A'^*(x_2, y_2)$$

$$= 1 - \frac{4\pi}{\lambda} W'_i(x_2, y_2). \tag{8.42}$$

It is clear, then, that the irradiance variations in the image plane show the presence of aberrations, provided that a modification is done over the Fourier spectrum of the object in the paraxial plane, in order to obtain $W'_i(x_2, y_2) \neq 0$. The contrast of the image irradiance variations is over the uniform background,

$$\gamma = \frac{4\pi}{\lambda} W'_i(x_2, y_2). \tag{8.43}$$

To interpret clearly how the image irradiance variations are related to the aberration over the optical surface, let us look closely at the term $W'_i(x_2, y_2)$. Since $W(x_1, y_1)$ is the Fourier transform of a purely real function $W(x, y)$, from Eq. (8.40) it follows that $W'_i(x_2, y_2)$ is different from zero only if the modulating function $M_1(x_1, y_1)$ has a term that is a purely real-odd function and/or a purely imaginary-even function (Ojeda-Castañeda 1978, 1979, 1980a, 1980b). Now, since any function can be decomposed in terms of its even and odd parts, it is always possible to find out which term in the modulating function $M(x_1, y_1)$ gives rise to $W'_i(x_2, y_2) \neq 0$. In this manner, by identifying and studying the properties of such a term, one can relate the image irradiance variations in Eq. (8.42) to the aberration function $W(x, y)$.

In the case of the knife edge, the term in the modulating function responsible for the image irradiance variation is the purely real-odd function

$$g(x_1, y_1) = \begin{cases} -0.5, & y_1 < -r_1, \\ 0, & |y_1| < r_1, \\ 0.5, & y_1 > +r_1, \end{cases} \tag{8.44}$$

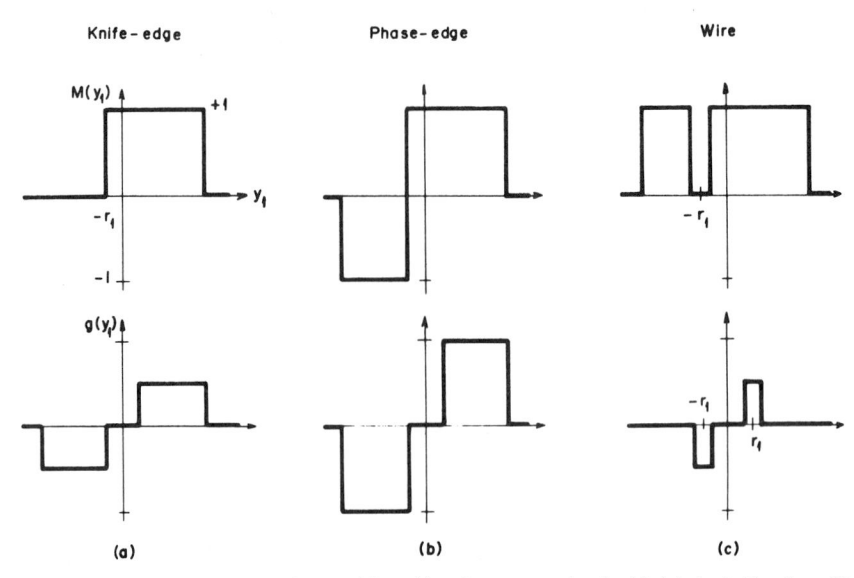

Figure 8.18. Modulating function, and its odd-real part, associated with (a) the knife-edge, (b) the phase-edge, and (c) the wire tests.

as shown in Fig. 8.18a. Consequently, the image irradiance variation term is obtained from

$$W'_i(x_2, y_2) = \frac{1}{i} \int_{x_1 = -\infty}^{+\infty} \left\{ -0.5 \int_{y_1 = -\infty}^{-r_1} w(x_1, y_1) \left[\exp i \frac{2\pi}{\lambda R_1} (x_2 x_1 + y_2 y_1) \right] dy_1 \right.$$
$$\left. + 0.5 \int_{y_1 = r_1}^{+\infty} w(x_1, y_1) \left[\exp i \frac{2\pi}{\lambda R_1} (x_2 x_1 + y_2 y_1) \right] dx_1 \right\} dy_1,$$

which may conveniently be written as

$$W'_i(x_2, y_2) = \int_{x_1 = -\infty}^{+\infty} \int_{y_1 = r_1}^{-\infty} \left[u(x_1, y_1) \sin \frac{2\pi}{\lambda R_1} y_2 y_1 + v(x_1, y_1) \right.$$
$$\left. \cdot \cos \frac{2\pi}{\lambda R_1} y_2 y_1 \right] \times dy_1 \exp \left(i \frac{2\pi}{\lambda R_1} x_2 x_1 \right) dx_1,$$

(8.45)

where $u(x_1, y_1)$ and $v(x_1, y_1)$ denote the even and odd parts, respectively, in y_1 of the function $w(x_1, y_1)$. That is, $w(x_1, y_1) = u(x_1, y_1) + iv(x_1, y_1)$.

As Zernike (1942) pointed out, the aberration function is obtained from

8.2. FOUCAULT OR KNIFE-EDGE TEST

$$W(x_2, y_2) = 2 \int_{x_1=-\infty}^{+\infty} \int_{y_1=0}^{+\infty} \left[u(x_1, y_1) \cos \frac{2\pi}{\lambda R_1} y_2 y_1 - v(x_1, y_1) \sin \frac{2\pi}{\lambda R_1} y_2 y_1 \right]$$

$$\times dy_1 \left(\exp i \frac{2\pi}{\lambda R_1} x_2 x_1 \right) dx_1, \tag{8.46}$$

which is to be compared with Eq. (8.45). It can be now appreciated that the irradiance variations in the image plane are different from the aberration function $W(x_0, y_0)$ in the following respects:

1. The magnitude of the former is attenuated by a factor of 2 as compared to the latter.
2. The Fourier spectrum of the aberrations inside the interval $(0, r_1)$ does not participate in forming the image.
3. An interchange in the coefficients of the sine and cosine function has taken place. This change may be expressed as

$$W'_i(x_2, y_2)$$

$$= -\frac{\lambda R_1}{2\pi y_2} \int_{x_1=-\infty}^{+\infty} \int_{y_1=r_1}^{+\infty} \left[u(x_1, y_1) \frac{d}{dy_1} \left(\cos \frac{2\pi}{\lambda R_1} y_2 y_1 \right) \right.$$

$$\left. - v(x_1, y_1) \frac{d}{dy_1} \left(\sin \frac{2\pi}{\lambda R_1} y_2 y_1 \right) \right] dy_1 \exp \left(i \frac{2\pi}{\lambda R_1} x_2 x_1 \right) dx_1,$$

or, equivalently, as

$$W'_i(x_2, y_2) = \frac{\lambda R_1}{2\pi y_2} \int_{x_1=-\infty}^{+\infty} \left[u(x_1, r_1) \cos \frac{2\pi}{\lambda R_1} y_2 r_1 - v(x_1, r_1) \right.$$

$$\left. \cdot \sin \frac{2\pi}{\lambda R_1} y_2 r_1 \right]$$

$$\times \exp \left(i \frac{2\pi}{\lambda R_1} x_2 x_1 \right) dx_1 + \frac{\lambda R_1}{2\pi y_2} \int_{x_1=-\infty}^{+\infty} \int_{y_1=r_1}^{+\infty}$$

$$\times \left(\frac{\partial u}{\partial y_1} \cos \frac{2\pi}{\lambda R_1} x_2 x_1 - \frac{\partial v}{\partial y_1} \sin \frac{2\pi}{\lambda R_1} y_2 y_1 \right) dy_1$$

$$\cdot \exp \left(i \frac{2\pi}{\lambda R_1} x_2 x_1 \right) dx_1, \tag{8.47}$$

assuming that $u(x_1, \infty) = v(x_1, \infty) = 0$. The image irradiance variations are related to the inverse Fourier transform of the derivative of the Fourier spectrum of the aberration function.

Finally, if the minimum contrast that the naked eye can detect ($\gamma = 2\%$) is employed in Eq. (8.40), it is possible to obtain, in the present approximation, the theoretical limit of sensitivity of the method, which is

$$\frac{\lambda}{200\pi} = W_i'(x_2, y_2). \tag{8.48}$$

This is in good agreement with $\lambda/600$, the value reported by Texereau (1957) when he checked the sensitivity of the method by comparing it with the Lyot (1946) phase contrast technique.

8.3. WIRE TEST

From Section 8.2 it is clear that the knife-edge method is very sensitive in detecting zonal errors; however, the Foucault test is less sensitive in measuring the errors. Moreover, Foucault graphs are not very useful in testing nonspherical mirrors, since small details are lost, immersed on the mirror asphericity.

To measure the mirror asphericity, Ritchey (1904) suggested using a screen with zonal slots over any aspherical surface, when testing mirrors by the Foucault method. In this way it is possible to calculate the amount of transversal ray aberration of each zone; when it just becomes dark for a given position of the knife edge along the optical axis; see Fig. 8.18. It is also possible to compare the illumination of the two slots of the zone, which are localized at opposite sides of the optical axis, to check whether the regions behind the slots are regular, and to match between them.

Because parabolic surfaces were the most commonly used mirrors, a screen for testing these optical surfaces was designed to allow illumination to only the zones over the mirror: the central (paraxial) region, the rim (marginal region), and an intermediate region between them, which are associated with the two ends and the center of the caustic (see Appendix 1), respectively (Fig. 8.19a). Once the screen has been placed over the mirror, the knife edge can be used to find the places along the optical axis in which the rays, coming from two opposite ends of a given zone, converge (Gaviola, 1936). In other words, by finding each of the places where the three zones darken, we can localize the two extremes of the interval of the caustic as well as its center (Simon, 1976).

Since the amount of transverse aberration of the normals (associated with the presence of spherical aberration in a paraboloid tested at the center of curvature) is given by cS^2, assuming that the light source and the knife move together, where again $S^2 = x^2 + y^2$ is the radius of any zone over the mirror and $c = 1/$radius of curvature of the paraxial region of the mirror, the radius of the intermediate region in the mask has to be $S = 0.707\, S_{max}$ if we want the

8.3. WIRE TEST

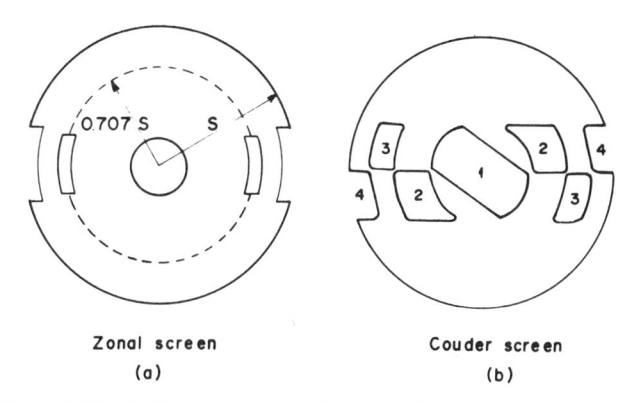

Figure 8.19. (a) Zonal and (b) Couder screens for zonal Foucault testing.

ideal transverse aberration of the zone to be half the ideal transverse aberration at the edge, in order to intersect the optical axis at the center of the caustic.

The width of the slots over the screen must be found experimentally. The slots must be narrow enough to increase local precision but wide enough to avoid diffraction effects.

Couder (1932) designed a more versatile screen (see Fig. 8.19b) in which only the width of the exterior zone (marginal region) has to be found experimentally; the radii and the width of the inner zones can be obtained from the facts that the zones immediately adjoin each other, and that the difference between the square of the radii of successive zones is a constant. In other words, if the experimental value of the radial width for the rim is δS, the difference between the outside and inside radii of this n-fold zone is

$$S_n^2 - S_{n-1}^2 = (\delta S)^2, \tag{8.49}$$

since the inside radius of the n-fold slot is the outside radius of the $(n-1)$-fold region. Consequently, the inside radius of the $(n-1)$-fold region, or the outside radius of the $(n-2)$-fold region, is given by

$$S_{n-1}^2 - S_{n-2}^2 = (\delta S)^2, \tag{8.50}$$

and in this way the radii of all the zones can be determined.

The advantage of this screen is that the whole mirror is observed and the overall shape of a surface can be tested. On the other hand, the disadvantage of any screen test is that the error in the measurement is of the order of magnitude of the tolerance to which the optical surface is to be polished (Schroader 1953; Simon 1976). Therefore a finer test for aspherical surfaces is required, and it is in these cases that the wire test and the Platzeck–Gaviola test (Section 8.4) can be usefully employed. The wire test is applicable only to aspherical

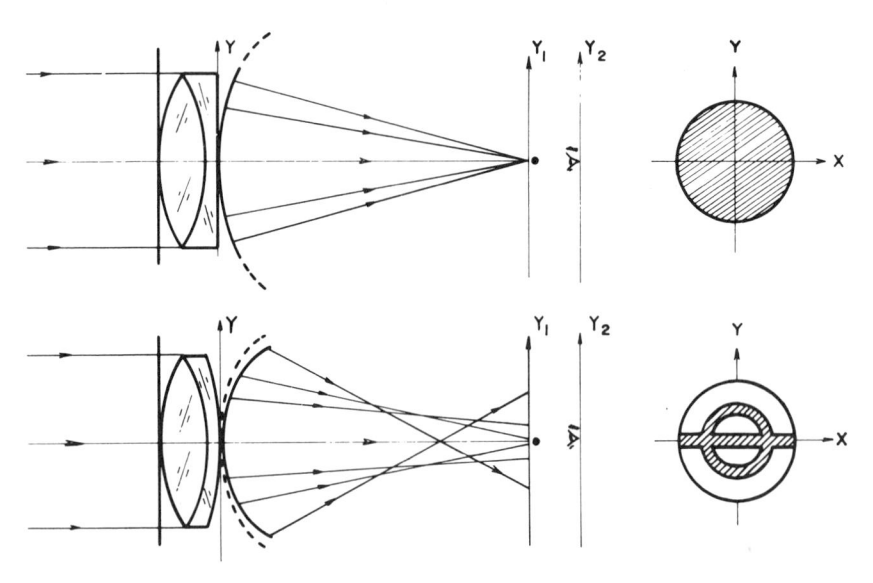

Figure 8.20. Wire setup for testing a lens.

surfaces, conical or not, and works in a manner similar to that of the zonal knife-edge test, by blocking out the deviated rays (or diffracted light) across a very narrow region. For this purpose a thin opaque wire is placed sequentially near the intersections of the normals of various annular zones with the optical axis of the mirror or surface under test (see Fig. 8.20). In this way a particular intersection of the normals to a particular annulus on the optical surface is found when the annulus in question appears dark against a uniformly bright background. In practice this is done by displacing the wire together with the pinhole size source (i.e., displacing the plane in which both the wire and the pinhole size source are contained). The pinhole size source does not need to be monochromatic.

The main advantage of the wire test over the zonal knife-edge method, used when masking the optical surface under test, is the ability of the former to restrict the blocking-out action to a very narrow region at the annulus being tested. In addition, diffraction effects are symmetrical with respect to the annulus. As a result measurement of the intersection of the normals with the optical axis is more precise, and in this way the deviations between the experimental values of the intersections and the theoretical values can be measured. In other words the aberrations of the normals are determined (see Fig. 8.21).

Moreover, the wire test has the advantage that the measurements of the intersections of the normals with the optical axis are referred to the theoretical value of the intersection of the first or paraxial annulus. This feature can prove to be extremely useful when testing an optical blank with a central hole.

8.3. WIRE TEST

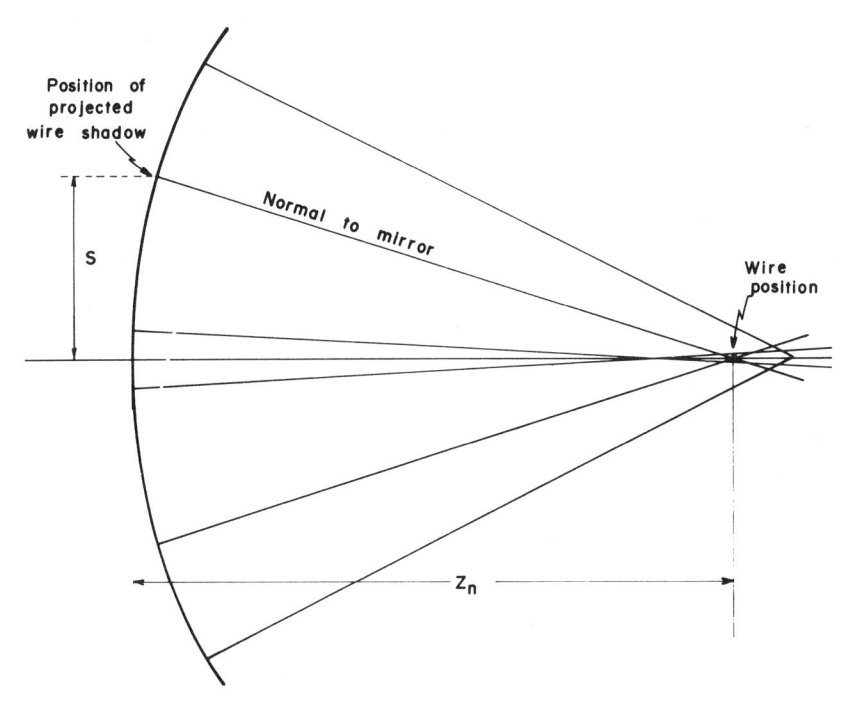

Figure 8.21. Diagram showing the wire method.

Since the region covered by the wire corresponds only to the edge of the knife in the Foucault test, the shadow patterns obtained with the wire test consist only of thin, dark contours, as shown in Fig. 8.22. Compare these wire patterns with the corresponding Foucault patterns in Fig. 8.23.

The wire test proves to be a good technique for testing an aspherical optical surface that is being figured; that is, it is an effective method of checking the longitudinal ray aberration, from which it is possible to obtain the wavefront aberration by using a numerical integration technique (Buchroader et al. 1972a, 1972b; Fell 1968; Loomis 1968; Smith 1966). The main disadvantage of the wire test is that the width of the shadow pattern sometimes masks small displacement of the whole shadow while the wire is displaced along the optical axis.

Meinel (1968) indicated that the wire test can also be used for testing astigmatism by determining whether or not the annular shadows are circular.

8.3.1. Geometrical Theory

If the position of the central part of the wire is denoted by r_1, and the thickness of the wire by $2\delta r$, the amplitude transmittance in the paraxial plane is given

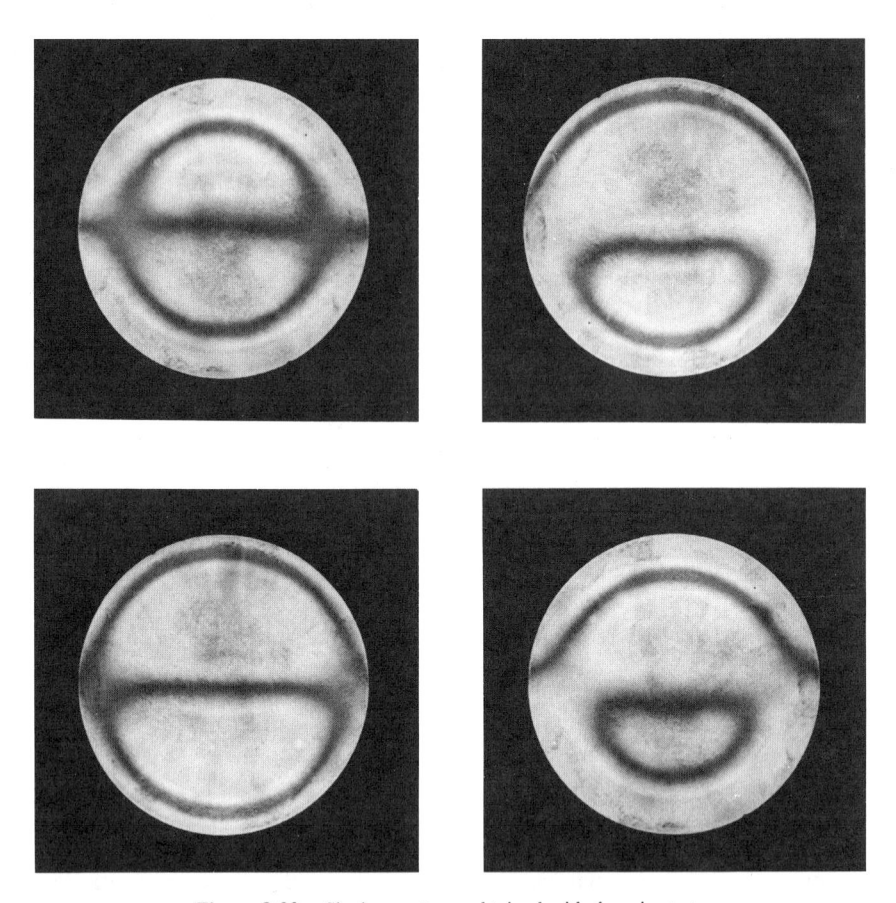

Figure 8.22. Shadow patterns obtained with the wire test.

by

$$T(x_1, y_1) = \begin{cases} 1 & \text{if } |x_1 \cos \phi_1 - y_1 \sin \phi_1| > \delta r, \\ 0 & \text{if } |x_1 \cos \phi_1 - y_1 \sin \phi_1| < \delta r. \end{cases} \quad (8.51)$$

The center of the dark contour obtained with the wire test coincides with the boundary between the dark and bright regions of the shadow pattern obtained with the Foucault test. Of course, since the wire does not have the covering extension of the knife, the dark areas are reduced to dark lines along the boundary positions.

In this section we describe the procedure used for obtaining the wavefront aberration of the optical surface under test from the experimental measurement

8.3. WIRE TEST

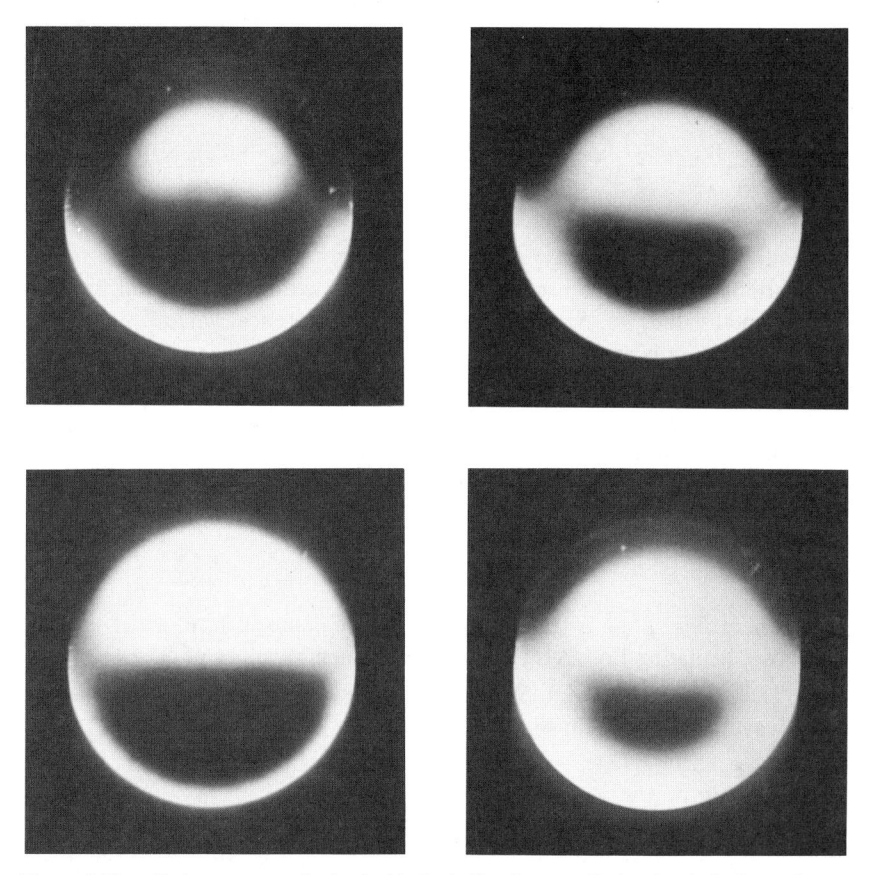

Figure 8.23. Shadow patterns obtained with the knife-edge test. Notice the similarity to the patterns in Fig. 8.22.

of the intersection of the normals with the optical axis. Let us denote the experimental and theoretical values of the intersection of the normal of the optical surface under test with the optical axis by Z_n' and Z_n, respectively. The transverse ray aberrations are then given by

$$TA = (Z_n' - Z_n)Sc \qquad (8.52)$$

where c is the inverse of the radius of curvature of the central zone and $S^2 = x^2 + y^2$. The wavefront aberration is obtained from

$$W = c^2 \int_{S=0}^{S} (Z_n' - Z_n) S \, dS. \qquad (8.53)$$

This integral may be solved numerically by employing the trapezoidal rule (dividing the region of integration into M subintervals) to give

$$W_M = \frac{c^2}{2} \sum_{m=1}^{M-1} [(Z'_n - Z_n)_{m+1} S_{m+1} + (Z'_n - Z_n)_m S_m] \times (S_{m+1} - S_m). \tag{8.54}$$

This formula can then be used to calculate the wavefront aberration from the experimental values Z'_n, S_m, S_{m+1} and from the theoretical value Z_n.

The procedure is as follows:

1. The theoretical value for the intersection of the normal with the optical axis is obtained from Eq. (A1.23) (Appendix 1):

$$Z_n = \frac{1}{c} - Kz \tag{8.55}$$

where $K = e^2$ (e is the eccentricity of the conic surface) and z is found by calculating Eq. (A1.1) for the experimental values of S_m.

2. The terms $(Z'_n - Z_n)_m S_m$ can then be evaluated, to obtain finally the wavefront aberration W_M over the optical surface under test.

8.3.2. Physical Theory

In the same way that the diffraction theory was used for describing the knife-edge method, the diffraction theory for the wire test is given here. The modulating function for the wire test is

$$M(x_1, y_1) = \begin{cases} 1 & \text{if } |x_1 + r_1| > \delta r, \\ 0 & \text{if } |x_1 + r_1| < \delta r, \end{cases} \tag{8.56}$$

where r_1 denotes the position of the wire and $2\delta r$ its width. After crossing the paraxial plane, the complex amplitude is obtained as before by $M(x_1, y_1) a(x_1, y_1)$. The complex amplitude distribution over the image plane is found by taking the Fourier transform of the complex amplitude leaving the paraxial plane, to give

$$A'(x_2, y_2) = 1 + i \frac{2\pi}{\lambda} W'(x_2, y_2) \tag{8.57}$$

where again

$$W'(x_2, y_2) = \int \int_{-\infty}^{+\infty} M(x_1, y_1) w(x_1, y_1)$$
$$\cdot \exp\left[-i \frac{2\pi}{\lambda R_1} (x_1 x_2 + y_1 y_2)\right] dx_1\, dx_2 \quad (8.58)$$

denotes the inverse Fourier transform of $M(x_1, y_1) w(x_1, y_1)$. If we employ the same arguments as those used in calculating the image irradiance in the knife-edge test, that is, to find the component in the modulating function that gives the rise to the visible term in the image irradiance variations, we obtain,

$$g(x_1, y_1) = \begin{cases} -0.5, & |x_1 + r_1| < \delta r, \\ 0, & |x_1 \pm r_1| > \delta r, \\ 0.5, & |x_1 - r_1| < \delta r, \end{cases} \quad (8.59)$$

which is to be compared with Eq. (8.44), as well as curves a and c in Fig. 8.17. It follows, then, from Eq. (8.59) that the only difference between employing the amplitude knife-edge method and the wire test for checking the presence of small aberration is that in the latter the image irradiance variations are due only to the light being diffracted, by the optical surface under test, around the interval between $(r_1 - \delta r)$ and $(r_1 + \delta r)$.

Moreover, the contrast of the image irradiance variations is the same for both methods. Consequently, the only advantage in using the wire test over the amplitude knife-edge procedure is that the light diffracted may be selected by choosing the position and width of the wire used. In other words a selective method for detecting small zones over the optical surface under test is provided by the wire test.

8.4. PLATZECK-GAVIOLA TEST

The so-called caustic test, introduced by Platzeck and Gaviola (1939), is based on the fact that the center of curvature of any off-axis segment over a "perfect" parabolic mirror lies, not on the optical axis, but off center on a curve termed the caustic, as was pointed out by Wadsworth (1902). To prove this point Platzeck and Gaviola performed a beautiful experiment, masking a parabolic mirror except in two symmetrical zones of elliptical shape. The mirror was then illuminated by a slit source, and the images formed with this optical system were recorded for different positions of the illuminating slit.

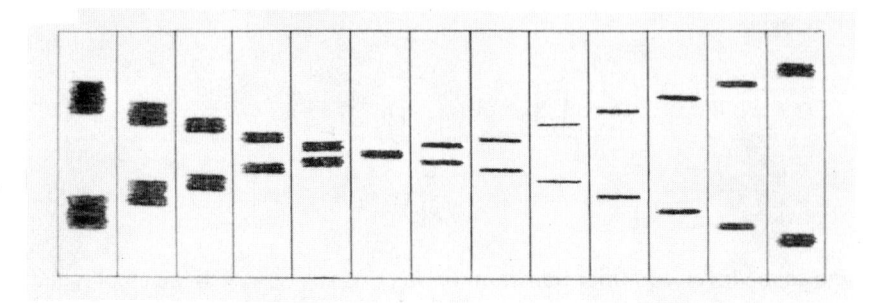

Figure 8.24. Sketch showing the appearance of the photographs taken by Platzeck–Gaviola using two slits over the mirror.

A series of photographs (showing the images of the illuminating slit) taken from different positions inside and outside of the focus of the mirror clearly indicated (see Fig. 8.24) which is a sketch of the photographs taken by Platzeck and Gaviola) that, when both beams cross the optical axis, the image is not sharp. But it shows the diffraction structure associated with the images obtained when the optical system has focus errors. Furthermore, in the same series of photographs it appears that, behind the region where the two images cross the optical axis, the focal plane for these images is found.

Hence Platzeck and Gaviola concluded that, to obtain accurate measurements, it was necessary to use this focal plane with the images outside the optical axis as the modifying plane. Also, a thin opaque wire would be more suitable as the modifying screen, since the wire would coincide with the image of the illuminating slit.

The method itself has proved to be useful and accurate. As Schroader (1953) has pointed out, it has the following main advantages:

1. It is possible to observe one hole of the mask at a time.
2. Holes considerably wider than those used in testing zonal errors with the amplitude knife-edge test may be perforated in the mask.
3. The criteria for knowing the center of curvature of the portion of a mirror exposed by one hole in the mask, that is, the minimum of illumination that passes an opaque wire or strip, are more reliable than the criteria for matching the irradiance distribution within the two holes, when using the knife edge on axis.

8.4.1. Geometrical Theory

In this section we first show how to calculate the theoretical center of curvature for a section of any conic of revolution (Cornejo and Malacara 1978), and then indicate the procedure employed in measuring the shape of the mirror from the

8.4. PLATZECK-GAVIOLA TEST

experimental and theoretical data for the centers of curvature of the several sections over the optical surface under test.

We define a new set of coordinates (η, ξ), also called caustic coordinates, such that the origin of the plane is at the center of curvature of the central zone, as shown in Fig. 8.25. The center of curvature of any zone will be referred to this origin.

The coordinates (η, ξ) for the center of curvature of any conic of revolution can be found from the following formula:

$$\frac{\xi}{2(\eta + Kz)} = \frac{S}{(1/c) - (K + 1)z} \qquad (8.60)$$

for the triangles in dashed lines in Fig. 8.25. From the same figure, the distance R between a point on the mirror with coordinates (z, S) and the center of curvature at $(1/c + \eta, \xi/2)$ can be obtained from

$$R^2 = \left(S + \frac{\xi}{2}\right)^2 + \left(\frac{1}{c} + \eta - z\right)^2. \qquad (8.61)$$

Of course, R is the radius of curvature, which can also be found by using a well-known formula of calculus for the radius of curvature of any function S, that is:

$$R = \frac{[1 + (dz/dS)^2]^{3/2}}{d^2z/dS^2}. \qquad (8.62)$$

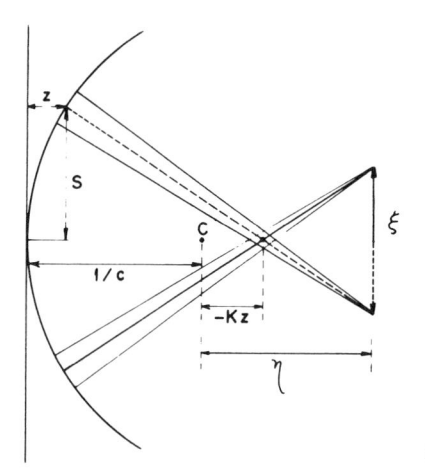

Figure 8.25. Geometry of the caustic.

Now, the first derivative of z with respect to S is given by Eq. (A1.21), namely,

$$\frac{dz}{dS} = cS[1 - (K + 1)c^2 S^2]^{-1/2} \tag{8.63}$$

from which the second derivative can be calculated to give

$$\frac{d^2 z}{dS^2} = c[1 - (K + 1)c^2 S^2]^{-3/2}. \tag{8.64}$$

If these two results are employed in Eq. (8.62), we have

$$R = \frac{1}{c}(1 - Kc^2 S^2)^{3/2}. \tag{8.65}$$

If, however, the value of $\xi/2$ in Eq. (8.60) is used in Eq. (8.61), we have

$$R^2 = \left(\frac{1}{c} + \eta - z\right)^2 \left\{1 + \frac{S^2}{[(1/c) - (K + 1)z]^2}\right\}. \tag{8.66}$$

Now, from formulas (8.65) and (8.66) we can obtain the value of η, namely,

$$\eta = \frac{(1/c)(1 - Kc^2 S^2)^{3/2}}{\left\{1 + \dfrac{S^2}{[(1/c) - (K + 1)z]^2}\right\}^{1/2}} + z - \frac{1}{c} \tag{8.67}$$

which, to eliminate the dependence on S, can be rewritten by using the expression for z, as

$$\eta = -Kz\{3 + cz(K + 1)[cz(K + 1) - 3]\}. \tag{8.68}$$

This value can be used in Eq. (8.60) to obtain the value of ξ, that is,

$$\xi = -2ScKz\left\{\frac{2 + cz(K + 1)[cz(K + 1) - 3]}{1 - cz(K + 1)}\right\}. \tag{8.69}$$

From Eq. (8.69) it is possible to calculate the theoretical position of the center of curvature $(\eta, \xi/2)$ of a particular zone of the mirror with coordinates (z, S).

The usual procedure when employing the caustic method is as follows (Schroader 1953):

1. Measure c from the radius of curvature of the central zone of the mirror.

8.5. PHASE MODULATION TESTS

2. Cut a mask with an odd number of zones in order to have a zone at the center.
3. Determine the center of curvature of the central zone, with the shadow pattern over the mirror while the wire is displaced along the optical axis.
4. Calculate the theoretical value of (η, ξ) and the experimental values of (η', ξ') for the center of curvature of a particular zone, as the wire blocks out the light coming from each slot.

   It is to be noted that only in this method is each slot, for a particular zone, tested separately. Consequently, the method may be particularly useful for testing errors that are not symmetrically distributed.
5. Calculate the theoretical minus the experimental values. If the difference is positive, the particular zone is overcorrected, and vice versa.

The diffraction theory of this test has been considered by Simon (1971), Platzeck and Simon (1974), and Simon et al. (1979). An interesting variation of this method is considered by Teocaris et al. (1976).

8.5. PHASE MODULATION TESTS

Zernike (1934a) analyzed the Foucault test in terms of diffraction theory. At the same time he suggested an improved form of the knife-edge test, which he named "phase contrast" (Linfoot 1946b). Essentially the same method was discovered quite independently by Lyot (1946). A further application of the phase modulation technique was later developed by Wolter (1956).

8.5.1. Zernike Test and Its Relation to the Smartt Interferometer

The phase contrast technique consists in employing as the modulating screen a disk of radius r, with an optical height of $\lambda/4$. In other words there is an optical path difference of $\lambda/4$ between a ray passing through the disk, or slit, and a ray passing outside it (λ is the wavelength of the illuminating radiation).

If the wavefront aberrations are small with respect to the wavelength, the direct light (associated with the "perfect" part of the optical surface under test) and the deviated light (associated with the "aberrated" part of the optical surface) have a 90° phase difference between them, as shown by Eq. (8.28). Since the aberrations are small, the resultant amplitude over the image of the surface under test is such that its associated irradiance is the same as that of the direct light. Thus the aberrations are invisible on the image, as shown in a general manner by Eq. (8.31).

When the modulating screen described earlier is employed, the direct light undergoes a phase delay of 90° (see Fig. 8.26) after crossing this screen. Now the direct and the deviated light have the same phase, producing an image of

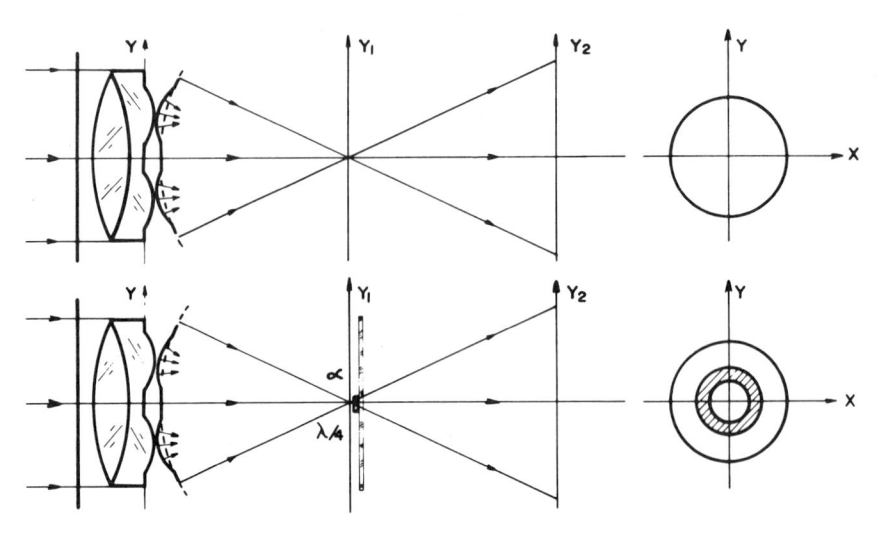

Figure 8.26. The phase contrast method.

the surface under test with the wavefront aberrations visible as variations in the irradiance, which indicate their magnitude and sign.

Moreover, by suitably decreasing the background irradiance, it is possible to enhance the visibility of very weak deviated light (which otherwise would remain undetectable). In other words it is possible to increase the contrast of the irradiance term that contains the presence of the aberrations, that is, the superimposed pattern, with respect to the background irradiance in the image plane. In this way a highly sensitive test for the presence of small aberrations is obtained.

In summary, the phase contrast method has the following advantages for testing an optical surface with aberrations smaller than the wavelength of the illuminating radiation:

1. The aberrations appear as a linear superposition term over a uniform background irradiance.
2. The contrast of the aberrations, over this uniform background irradiance can be increased by decreasing the irradiance in the background illumination, that is, by decreasing the optical transmittance over the $\lambda/4$ region. The mathematical formulation associated with this method is given in the following discussion.

The modifying function employed in the phase contrast test may be described as an amplitude-absorbing disk with a real amplitude transmittance equal to α and a radius equal to the radius of the Airy disk, $\rho = 3.83$. Outside this disk

8.5. PHASE MODULATION TESTS

the complex amplitude transmittance is equal to i, that is,

$$M(x_1, y_1) = \begin{cases} \alpha, & x_1^2 + y_1^2 < r_1^2, \\ i, & x_1^2 + y_1^2 > r_1^2. \end{cases} \tag{8.70}$$

Hence the complex amplitude leaving toward the image plane is given by (Eq. 8.38):

$$a'(x_1, y_1) = M(x_1, y_1)\left[2\frac{J_1(\rho)}{\rho} + i\frac{2\pi}{\lambda}w(x_1, y_1)\right]. \tag{8.71}$$

Consequently, the complex amplitude over the image plane can be calculated from the inverse Fourier transform of Eq. (8.71) to give

$$A'(x_2, y_2) = \alpha - \frac{2\pi}{\lambda}W'(x_2, y_2) \tag{8.72}$$

where

$$W'(x_2, y_2) = \int\int_{x_1^2+y_1^2>r_1^2} w(x_1, y_1)\exp\left[-i\frac{2\pi}{\lambda R_1}(x_2 x_1 + y_2 y_1)\right]dx_1\,dy_1. \tag{8.73}$$

The image irradiance can be calculated from

$$I'(x_2, y_2) = A'(x_2, y_2)A'^*(x_2, y_2)$$

to give

$$I'(x_2, y_2) = \alpha^2 - \alpha\frac{4\pi}{\lambda}W'(x_2, y_2). \tag{8.74}$$

From Eq. (8.72) it follows that the image irradiance variations are proportional to the aberration function. Of course, if the aberration is a smoothly varying function fairly distributed over the object, its diffraction pattern may be passing by the central disk. Therefore it will remain invisible in the image irradiance distribution when the phase contrast method is used.

The contrast of the image irradiance variations for the phase contrast test is

$$\gamma = \frac{4\pi}{\alpha\lambda}W'(x, y_2). \tag{8.75}$$

It is clear that the contrast γ can be increased by decreasing the background of the image irradiance, that is, by increasing the absorption of the central disk, as was pointed out in item 2 when summarizing the advantages of the method.

The phase contrast method is often presented as the result of two-beam or multiple-beam interference (Françon 1952a, 1952b; Golden 1977a). Since this approach may be more familiar for many people in the optical testing field, and since Smartt (1975) used the same basic idea in proposing a point diffraction interferometer, this approach is now examined; anyone who is not interested in it may skip this part.

Let us represent the complex amplitude leaving the surface under test by

$$A(x, y) = \exp\left[i\frac{2\pi}{\lambda} W(x, y)\right] \qquad (8.76)$$

where again $W(x, y)$ is the aberration function of the wavefront. This expression for the complex amplitude may conveniently be written as

$$A(x, y) = \exp\left[i\frac{2\pi}{\lambda} \overline{W}(x, y)\right]$$
$$+ \left\{\exp i\left[\frac{2\pi}{\lambda} W(x, y)\right] - \exp\left[i\frac{2\pi}{\lambda} \overline{W}(x, y)\right]\right\} \qquad (8.77)$$

where $\overline{W}(x, y)$ is the phase variation that a wavefront leaving an "ideal" optical surface would have (see Fig. 8.27). From formula (8.77) is it clear that the complex amplitude may be expressed as a beam associated with a reference wavefront (first term) plus a beam associated with the aberration over the reference wavefront (last term). The complex amplitude over the image plane may also be written as

$$A'(x_2, y_2) = \exp\left[i\frac{2\pi}{\lambda} \overline{W}(x_2, y_2)\right]$$
$$+ \left\{\exp\left[i\frac{2\pi}{\lambda} W(x_2, y_2)\right] - \exp\left[i\frac{2\pi}{\lambda} \overline{W}(x_2, y_2)\right]\right\}. \qquad (8.78)$$

Of course, the mathematical separation in terms of the two beams in formula (8.78) is purely convenient since the image irradiance is given by

$$I(x_2, y_2) = A'(x_2, y_2) A'^*(x_2, y_2)$$
$$= 1,$$

8.5. PHASE MODULATION TESTS

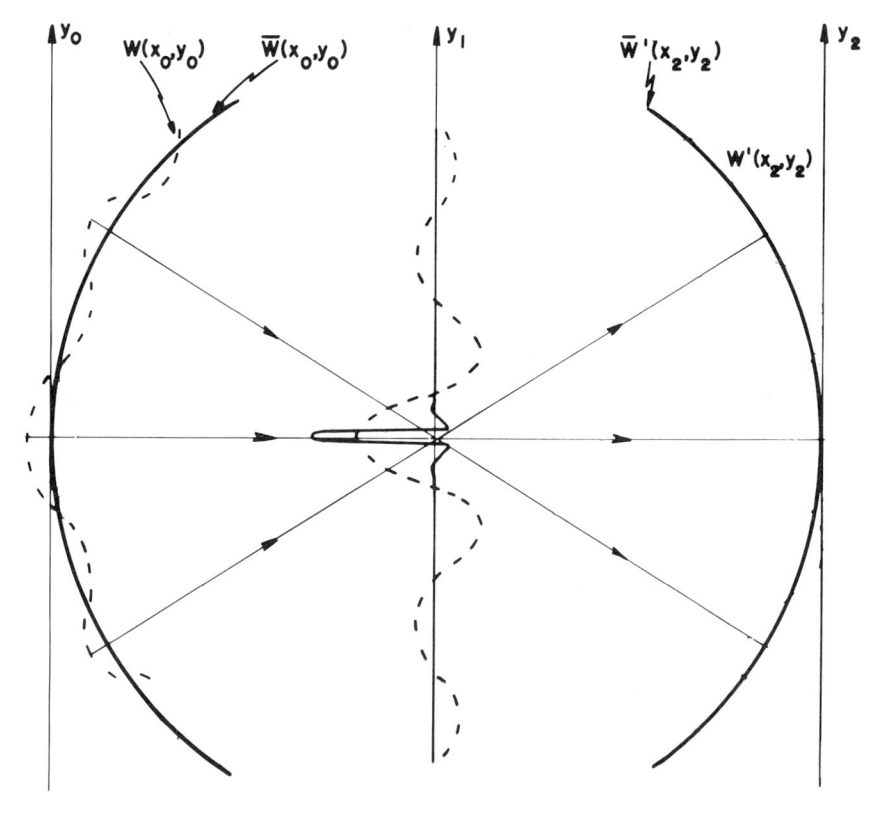

Figure 8.27. Geometry of the phase contrast method and Smartt's interferometer.

as is to be expected. But since in the paraxial plane X_1-Y_1 both components are differently distributed, they may, to a certain extent, be separately modified. Therefore the image may usefully be thought of as the interference of two different beams.

It is precisely this technique that is employed in the phase contrast method and in the point diffraction interferometer. In other words the idea is to alter the amplitude and phase of the first beam when it converges at the paraxial plane, in order to obtain an image irradiance variation different from unity. The complex amplitude in the image is then given by

$$A'(x_2, y_2) = \alpha \exp(i\delta) \exp\left[i\frac{2\pi}{\lambda} \overline{W}(x, y)\right] + \exp\left[i\frac{2\pi}{\lambda} W(x, y)\right]$$
$$- \exp\left[i\frac{2\pi}{\lambda} \overline{W}(x, y)\right], \qquad (8.79)$$

where α is the real amplitude transmittance and δ is the phase difference introduced in the reference beam. The image irradiance can be calculated from Eq. (8.79) to give

$$I(x_2, y_2) = \alpha^2 + 4 \sin^2\left\{\frac{\pi}{\lambda}[\overline{W}(x_2, y_2) - W(x_2, y_2)]\right\}$$
$$- 4 \sin\left\{\frac{\pi}{\lambda}[\overline{W}(x_2, y_2) - W(x_2, y_2)]\right\}$$
$$\times \alpha \sin\left\{\frac{\pi}{\lambda}[\overline{W}(x_2, y_2) - W(x_2, y_2)] + \delta\right\}. \quad (8.80)$$

Of course, the trivial case in which the image irradiance is equal to unity over the whole image plane occurs when $\alpha = 1$ and $\delta = 0$.

From general formula (8.80) we obtain the phase contrast method as a particular case by setting $\delta = \pi/2$. Hence the image irradiance is given by

$$I(x_2, y_2) = \alpha^2 + 4 \sin^2\left\{\frac{\pi}{\lambda}[\overline{W}(x_2, y_2) - W(x_2, y_2)]\right\}$$
$$- 2\alpha \sin\left\{\frac{2\pi}{\lambda}[\overline{W}(x_2, y_2) - W(x_2, y_2)]\right\}. \quad (8.81)$$

It is clear that, when $|\overline{W}(x_2, y_2) - W(x_2, y_2)| \ll 1$, the second term may be neglected, and formula (8.74) is again obtained if we set $W' = \overline{W} - W$.

The point diffraction interferometric technique can be easily obtained from general formula (8.81) by letting $\delta = 0$. Hence the image irradiance variations are given by

$$I(x_2, y_2) = \alpha^2 + (1 - \alpha)4 \sin^2 \frac{\pi}{\lambda}[\overline{W}(x_2, y_2) - W(x_2, y_2)]. \quad (8.82)$$

Since no restriction has so far been imposed on the form of the function $\overline{W}(x_2, y_2)$ or, equivalently, on $\overline{W}(x, y)$, this function may indicate tilt or focus error (as in the point diffraction interferometer) placing the point containing the change in amplitude α along the X_1-Y_1 plane, inside or outside the focus, respectively.

It is also clear that, in the point diffraction interferometer, the contrast γ in the image intensity variations is given by

$$\gamma = 4\left(\frac{1}{\alpha} - 1\right)\sin^2 \frac{\pi}{\lambda}[\overline{W}(x_2, y_2) - W(x_2, y_2)]. \quad (8.83)$$

8.5. PHASE MODULATION TESTS

The contrast can be increased by decreasing α, and the method will indicate the presence of small aberrations around the function $\overline{W}(x_2, y_2)$.

Any other values of δ in formula (8.80) will correspond to different optical thicknesses or, equivalently, to displacements of one of the arms of a two-beam interferometer. But since the most sensitive region for small aberrations corresponds to $\delta = \pi/2$, they will not be considered here.

8.5.2. Lyot Test

Finally, as Texereau (1957) pointed out, although the phase contrast technique has proved to be extremely useful in microscopy, its value in optical testing seems to be limited. The problem is that the test is photometric, that is, is based on irradiance variations, and the interpretation and application of the results obtained are not straightforward tasks. This is especially true when a clear understanding, in terms of physical optics, of the properties of the test is lacking. The most common application in optical testing of the phase contrast method involves the detection of surface defects that are small in area and amplitude.

As mentioned before, the Zernike test usually is performed using a point source (white or monochromatic) and a phase plate with a small circular disk, smaller than or equal to the Airy disk. Another alternative, used with modern phase contrast microscopes, is to employ a circular, or annular, extended light source and a phase plate whose retarding zone exactly coincides with the geometrical image of the source. The advantages of this procedure are that more light is used and that it is easier to manufacture a larger phase plate. Using this principle, Lyot (1946), independently of Zernike, developed a technique in which he used a relatively wide slit (100 to 200 μm) as an illuminating light source. This slit is imaged by the optical surface under test upon a phase-retarding slit with an optical thickness of $\lambda/4$. To increase the contrast, the transmittance of the retarding slit is reduced to about $T = 10^{-D}$, with D being the optical density, which ranges from 1.5 to 3.0. See for example Françon (1952b).

Since the light source is a slit and not circular, the light diffracted in the direction of the slit will not alter its phase relative to the undiffracted light and thus the corresponding errors will not be detected. This means that only errors with variations in the direction perpendicular to the slit will be detected. This preferential orientation of the errors is shown in Fig. 8.28, taken with a vertical slit. This detected pseudo-orientation of the error can be ignored, assuming that the real errors are randomly oriented.

The Lyot version of the phase contrast method is so sensitive that it is possible to see defects that deviate from the mean surface by 1 Å with a contrast equal to 15%.

When a polychromatic light source is employed, the phase-modulating screen can have different optical thicknesses for the different wavelengths present, and the image irradiance variations can appear in color.

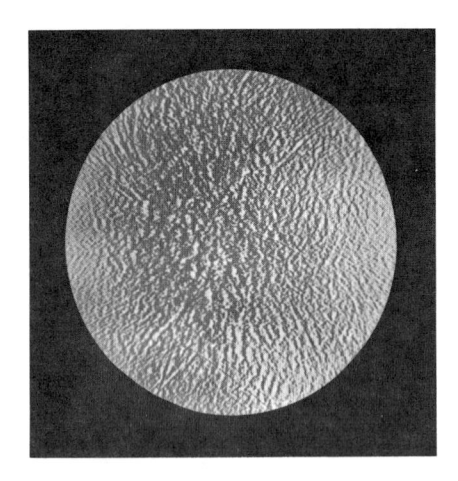

Figure 8.28. Lyotgraph taken with a vertical phase slit. (From M. Françon, *Modern Applications of Physical Optics*, Interscience, New York, 1963.)

Information about the experimental procedures for making phase disks or phase slits can be found in the articles of Burch (1934), Zernike (1942), Françon and Nomarski (1952), and Golden (1977b). This test can also be implemented in polarized light (Kastler and Momtarnal 1948).

8.5.3. Wolter Test

The $\lambda/2$ phase-edge or phase-knife test, proposed by Wolter (1949), and independently by Kastler (1950), can be regarded as an improvement of the wire test or the zonal Foucault method. The idea is to employ usefully the diffracted light at one side, instead of blocking it out as when the knife edge is used.

The method is based on reversing the sign of the complex amplitude across a half-plane, while the other half remains unaltered. This is achieved by introducing a transparent half-plane with an optical thickness equal to $\lambda/2$. This method is also known as the Hilbert transform method (Belvaux and Vareille 1970, 1971; Lowenthal and Belvaux 1967) because mathematically the image is obtained as the Hilbert transform of the object.

Since in this test more diffracted light is collected by the eye of the observer, this method would be expected to produce a shadow pattern of higher contrast. Therefore the $\lambda/2$ phase-edge test is expected to be more sensitive and accurate, in locating on the aspherical surface under test the projected shadow of the phase edge, than the normal knife-edge test. Here, again, as in the phase contrast method, it is not possible to explain in terms of geometrical optics the irradiance variations over the image plane when a $\lambda/2$ phase edge is introduced near the paraxial focal plane. However, Landgrave (1974) made the additional assumption that only the rays passing through the border of the $\lambda/2$ phase edge interfere destructively.

8.5. PHASE MODULATION TESTS 307

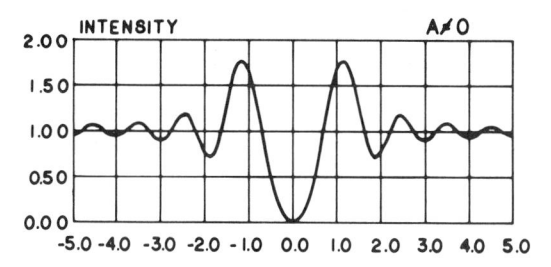

Figure 8.29. Fresnel diffraction image of a $\lambda/2$ phase edge, observed over the surface under test.

Although the later approach may appear primitive, it is in accordance with the fact that the Fresnel diffraction images of a $\lambda/2$ phase edge have zero irradiance values at the image of the border (Kastler 1950; Hopkins 1952; Ojeda-Castañeda 1976), as shown in Fig. 8.29. Indeed, the Wolter test proves to be extremely useful since the $\lambda/2$ phase edge is considered to be equivalent to the wire test. In other words the border of the $\lambda/2$ phase edge may be thought of as the border of the opaque knife edge, and the "reversal" side of the $\lambda/2$ phase edge may be considered to have a transmittance equal to unity, as on its "clear" side. Therefore the shadow patterns associated with the $\lambda/2$ phase edge are, from the point of view of geometrical optics, considered to be identical with the shadow patterns associated with the wire test. Upon analysis the boundaries of the shadow patterns are, then, identical to those presented in Section 8.2.2. Notice the similarity between the Foucault test drawings in Fig. 8.9 and the phase-edge photographs of a comatic lens (kindly provided by Landgrave) in Fig. 8.30.

It has been reported (Vasil'ev 1971; Landgrave 1974) that the dark lines of the shadow pattern associated with the $\lambda/2$ phase edge are better contrasted against the background irradiance intensity than the dark lines associated with both the knife-edge test and the wire test. Thus the sensitivity of the $\lambda/2$ phase-edge test is greater than the sensitivity obtained when using an opaque knife

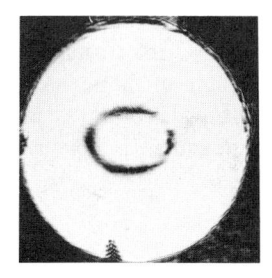

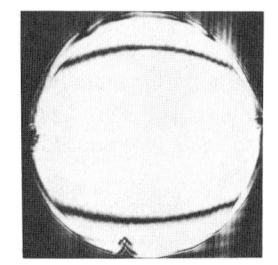

Figure 8.30. Shadow patterns obtained when using a $\lambda/2$ phase edge for testing a lens with comatic aberrations.

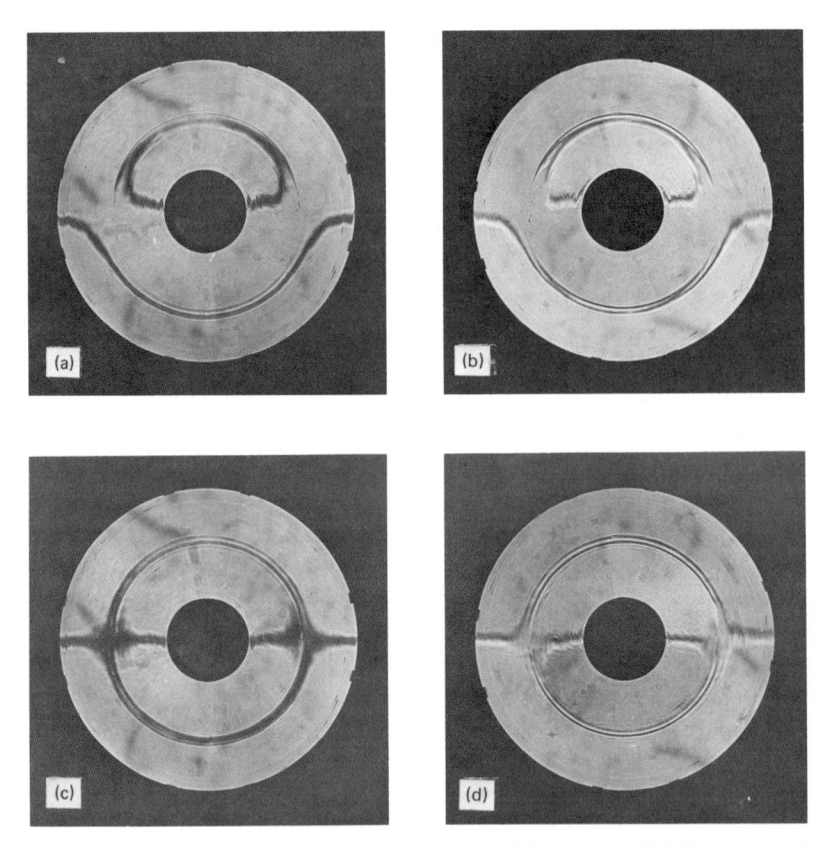

Figure 8.31. Shadow patterns of a hyperbolic mirror. (*a*) Off-axis wire. (*b*) Off-axis phase edge. (*c*) On-axis wire. (*d*) On-axis phase edge.

edge or the wire test. The wire test photographs in Figs. 8.31*a* and 8.31*b* should be compared with the phase-edge photographs, Figs. 8.31*c* and 8.31*d*, respectively, taken with the same aspherical mirror.

Obviously, a formal analysis of the sensitivity is, indeed, more complicated, and it is not considered here. However, it is possible to show (Ojeda-Castañeda 1976, 1979) that, for aberrations smaller than the wavelength of the incident radiation, the $\lambda/2$ phase edge will produce images with higher contrast than the images produced with the opaque knife edge. The treatment is here repeated.

The modulating function for the phase-edge method is

$$M(x_1, y_1) = \begin{cases} -1, & x_1 < -r_1, \\ 1, & x_1 > -r_1. \end{cases} \tag{8.84}$$

8.6. RITCHEY–COMMON TEST

The complex amplitude leaving the plane of modulation and traveling toward the image plane, after being modulated by function (8.84), is again obtained by $M(x_1, y_1)a(x_1, y_1)$. The complex amplitude distribution over the image plane can be calculated by taking the inverse Fourier transform, giving

$$A'(x_2, y_2) = 1 + i\frac{2\pi}{\lambda} W'(x_2, y_2) \quad (8.85)$$

where again $W'(x_2, y_2)$ denotes the inverse Fourier transform of $M(x_1, y_1)w_1(x_1, y_1)$. When calculating the image intensity, the same approximation, neglecting second-order terms in λ^{-1}, may be made. Under this approximation, as was pointed out in the sections dealing with the diffraction theory of the knife-edge and wire tests, only the real-odd component of the modulating function gives rise to the visible term in the image irradiance variations. Since in this case the function is

$$g(x_1, y_1) = \begin{cases} -1, & x_1 < -r_1, \\ 0, & |x_1| < r_1, \\ 1, & x_1 > r_1, \end{cases} \quad (8.86)$$

we do not have to calculate the image irradiance, for we know that the result will be the same as that obtained when the image irradiance is calculated using the knife edge. The only exception to this statement that should be noted is that function (8.86) has double amplitude transmittance (see curves a and b in Fig. 8.17). Therefore the contrast of the image irradiance variations when the phase edge is used is twice that obtained with the knife edge, that is

$$\gamma_{pe} = 2\gamma_{ke} = \frac{8\pi}{\lambda} W_i(x_2, y_2). \quad (8.87)$$

The other characteristics associated with the Wolter test are the same as those listed for the knife-edge and the wire tests.

8.6. RITCHEY–COMMON TEST

According to Ritchey (1904), this test was suggested by a Mr. Common as a way of testing large optical flats (Shu 1983). For this purpose a good spherical mirror is used, with the flat acting also as a mirror, to obtain an apparatus similar to the one used for the knife-edge test. See Fig. 8.32.

Any small spherical concavity or convexity of the surface under test appears as an astigmatic aberration in the image of the illuminating point source. The

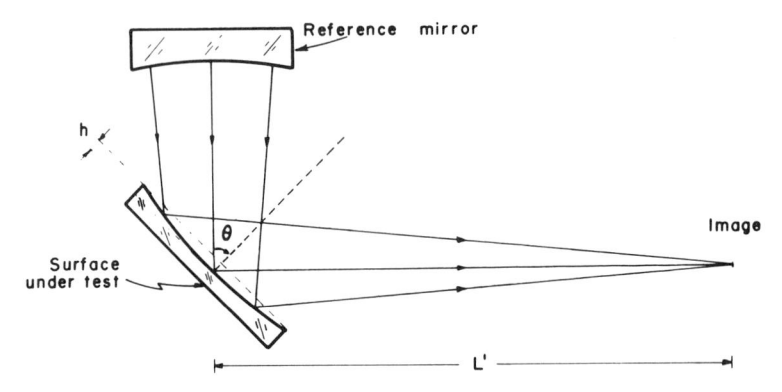

Figure 8.32. Geometry of the Ritchey–Common test.

two focal planes associated with the sagittal and tangential foci can be accurately found (by employing the knife-edge test as indicated in Section 8.1), and with this information the radius of curvature of the surface under test can be calculated. Of course, any localized deformation or deviation from flatness can also be observed.

The advantage of the Ritchey–Common test over other methods is that a reference flat is not needed as in the Newton or Fizeau interferometer. The autocollimation method requires a paraboloid as large as the largest dimension of the flat under test. On the other hand, the auxiliary mirror used in the Ritchey–Common test has a spherical shape, which is the easiest to make and test. An additional advantage is that the spherical mirror needs to be only as large as the minor diameter, if it is elliptical, like many diagonal mirrors used in telescopes or other instruments.

If we measure the astigmatism, we can obtain the magnitude of the concavity or convexity of the "flat" mirror, but it is assumed that this mirror is spherical, not toroidal, in shape. If the mirror is not circular but is elliptical, the possibility that it will develop a toroidal shape while being polished is very high. This toroidal shape introduces an additional astigmatism that complicates the analysis. Interested readers can consult the work of Tatian (1967), Silvernail (1973), and Shu (1983).

The mathematical description given here assumes that the mirror under test is not toroidal and follows the approach used by Couder (1932) and Strong (1938), that is, finding the curvature from the Coddington equations for astigmatism. In what follows it is assumed that the pinhole size source and the wire travel together, in the same plane. Hence, when evaluating the transversal ray aberrations, it is necessary to multiply the value obtained in this way by a factor of 0.5 to get the real value of the transverse ray aberrations. However, since the aberrations of the wavefront are duplicated in the Ritchey–Common test because of the double reflection on the optical flat under test, we can conve-

8.6. RITCHEY–COMMON TEST

niently omit the double reflection factor of 2 and the factor 0.5 when calculating the value of the transversal ray aberrations.

Applying the well-known Coddington equations for astigmatism to a reflecting surface, we find that the sagittal focus of a converging beam incident on an optical reflecting surface is given (see Fig. 8.33) by

$$\frac{1}{s'} = \frac{1}{s} + \frac{2\cos\theta}{r} \qquad (8.88)$$

and the tangential focus by

$$\frac{\cos^2\theta}{t'} = \frac{\cos^2\theta}{t} + \frac{2\cos\theta}{r} \qquad (8.89)$$

where θ is the angle of incidence at the surface under test and r is the radius of curvature of the optical surface under test.

If we assume that the incident beam is free from astigmatism, that is, $s = t$, Eqs. (8.88) and (8.89) can be used to obtain

$$\frac{s' - t'}{s't'} = \frac{2\sin^2\theta}{r\cos\theta}. \qquad (8.90)$$

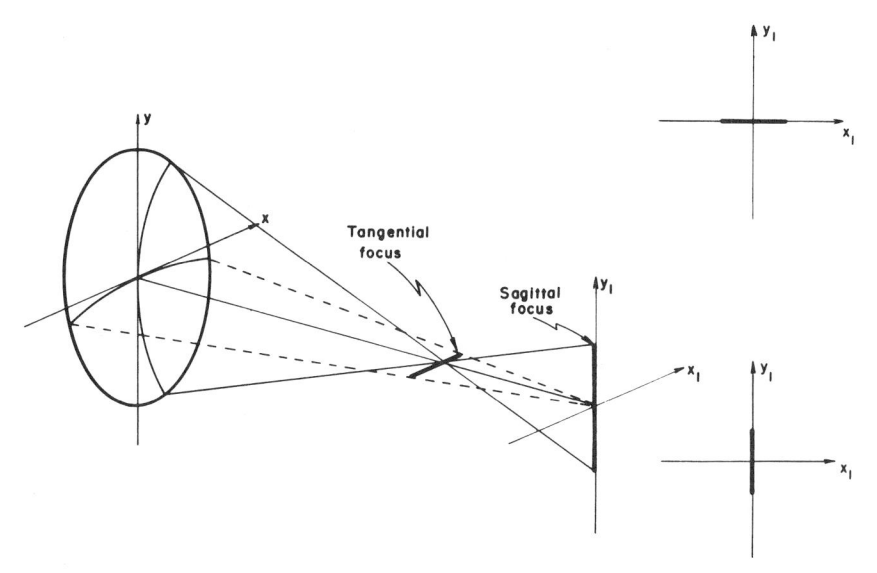

Figure 8.33. An astigmatic wavefront and its associated focal images. It is to be noted that the orientations of the images are changed when the setup in Fig. 8.32 is used. That is, the sagittal image is a horizontal line and the tangential image is a vertical one.

Now, by denoting $s' = L' + \delta L'$ and $t' = L'$, and substituting these values in Eq. (8.90), we have

$$r = 2L'\left(1 + \frac{L'}{\delta L'}\right)\frac{\sin^2 \theta}{\cos \theta}. \tag{8.91}$$

Since $|L'/\delta L'| > 1$, it follows from Eq. (8.91) that, when the sagittal focus is longer than the tangential focus ($\delta L' > 0$), the surface under test has a positive radius of curvature, that is, the surface is convex. On the other hand, when the sagittal focus is shorter than the tangential focus ($\delta L' < 0$), the surface has a negative radius of curvature and is concave. The sagitta h of a surface with a diameter D and a radius of curvature r can be approximated by

$$h = \frac{D^2}{8r}. \tag{8.92}$$

Using Eq. (8.91) in (8.92) we have

$$h = \frac{D^2 \cos \theta}{16L'(1 + L'/\delta L') \sin^2 \theta}. \tag{8.93}$$

For $L' \gg \delta L' (|L'/\delta L'| \gg 1)$, Eq. (8.93) reduces to

$$h = \frac{D^2 \delta L' \cos \theta}{16 L'^2 \sin^2 \theta}, \tag{8.94}$$

which gives the value for the sagitta of the optical surface under test for any angle of incidence θ. For the particular case of $\theta = 45°$, the sagitta is given by

$$h = \frac{D^2 \delta L'}{11.3 L'^2}. \tag{8.95}$$

Couder (1932 in Texereau 1957) has indicated that an optimum angle for testing optical flats is $\theta = 54°45'$. Of course, in practice it is more convenient to set $\theta = 60°$. It is also common practice to use any eyepiece, instead of the knife edge, to determine visually the positions of the sagittal and tangential foci, by locating the positions of the patterns shown to the right of Fig. 8.33.

8.7 CONCLUSIONS

As mentioned in the introduction of this chapter, the examples examined demonstrate the versatility and usefulness of the Schlieren techniques and the related phase modulation tests. Of course, the choice of any particular test depends on the circumstances.

Ackowledgments. I am indeted to D. Malacara, A. Cornejo, R. Noble, N. Noble, and E. Landgrave for useful discussions. I thank also J. J. Pedraza, who deposited the $\lambda/2$ phase edge; E. Jara and S. Jara for the drawings and the photographic work; and Rocio Arce for her patience in typing the manuscript.

REFERENCES

Belvaux, Y. and J. C. Vareille, "Controle de L'etat de Surface ou d'Homogeneite de Materiaux Optiques par Contrast de Phase a Dephasage Quelconque," *Optics Comm.*, **2**, 101 (1970).

Belvaux, Y. and J. C. Vareille, "Visualisation d'Objects de Phase par Transformation de Hilbert," *Nouv. Rev. Opt.*, **2**, 149 (1971).

Banerji, S., "On Some Phenomena Observed in the Foucault Test," *Astrophys. J.*, **48**, 50 (1918).

Barakat, R., "General Diffraction Theory of Optical Aberration Tests, from the Point of View of Spatial Filtering," *J. Opt. Soc. Am.*, **59**, 1432 (1969).

Bermúdez, B., J. Ojeda-Castañeda, and L. R. Berriel-Valdos, "Shadow Diagrams," (in Spanish), *Technical Report No. 41* INAOE, Puebla Mexico (1985).

Berriel-Valdos, L. R., "A Quantitative Study of the Foucault, Phase Edge and Wire Optical Tests," *Proc. ICO-11*, J. Bescós, A. Hidalgo, L. Plaza, and J. Santamaría, Ed., Madrid, 1978, p. 291.

Berriel-Valdos, L. R., "Numerical Simulation of Optical Workshop Tests Based on Spatial Filtering," (in Spanish), *Technical Report No, 33 A*, INAOE, Puebla, Mexico (1981).

Berriel-Valdos, L. R., "Influence of the Finite Size of the Viewing System in the Phase Edge Test," *Bol. Inst. Tonantzintla*, **2**, 117 (1982).

Bescós, J. and L. R. Berriel-Valdos, "Foucault Test in White Light Imagery," *Opt. Commun.*, **57**, 10 (1986).

Buchroader, R. A., L. H. Elmore, R. V. Shack, and P. N. Slater, "The Design Construction and Testing of the Optics for a 147 cm Aperture Telescope," Technical Report N. 79, Univ. of Arizona, Tucson, AZ, 1972a.

Buchroader, R. A., L. H. Elmore, R. V. Shack, and P. N. Slater, *Optical Sciences Center Technical Report No. 79*, 1972b.

Burch, C. R., "On the Phase-Contrast Test of F. Zernike," *Mon. Not. R. Astron. Soc.*, **94**, 384 (1934).

Conrady, H. G., "Study of the Significance of the Foucault Knife-edge Test When Applied to Refracting Systems," *Trans. Opt. Soc.*, **25**, 219 (1924).

Cornejo-Rodriguez, A. and D. Malacara, "Caustic Coordinates in Platzeck-Gaviola Test of Conic Mirrors," *Appl. Opt.*, **17**, 18 (1978).

Couder, A., *Recherches sur les Deformations des Grands Miroirs Employes aux Observations Astronomiques*, Thesis Faculte des Sciences Paris, *Bull. Astron.*, **7** (1932).

Dakin, R. K., "An Improved Focault Testing Device," *Sky Telesc.*, **33**, 45 (1967).

Fell, B., "Optical Surfaces: Computer Program Facilitates Work in Optical Shop," *Opt. Sci. Newslett.*, **2**, 5 (1968).

Foucault, L. M., "Description des Procedees Employes pour Reconnaitre la Configuration des Surfaces Optiques," *C. R. Acad. Sci. (Paris)*, **47**, 958 (1858).

Foucault, L. M. "Memoire sur la Construction des Telescopes en Verre Argente," *Ann. Obs. Imp. Paris.*, **5**, 197 (1859).

Françon, M. Ed., *Etude Theorique Comparative du Contraste de Phase et de Methods Interferentielles, Contraste de Phazse et Contraste par Interferometrie*, Edition de la Revue d'Optique Théorique et Instrumentale, Paris, 1952a, p. 48.

Françon, M. and G. Nomarski, Eds., *Les l'Ames de Phase par Reflexion Contraste de Phase et Contraste par Interferometrie*, Edition de la Revue d'Optique Theorique et Instrumentale, Paris, 1952b, p. 136.

Gascoign, S. C. B., "The Theory of the Foucault," *Test. Mon. Not. R. Astron. Soc.*, **104**, 326 (1944).

Gaviola, E., "On the Quantitative Use of the Foucault Knife-Edge Test," *J. Opt. Soc. Am.*, **26**, 163 (1936).

Golden, L. J., "Zernike Test. 1: Analytical Aspects," *Appl. Opt.*, **16**, 205 (1977a).

Golden, L. J., "Zernike Test. 2: Experimental Aspects," *Appl. Opt.*, **16**, 214 (1977b).

Hartmann, J., "An Improvement of the Foucault Knife-Edge Test in the Investigation of Telescope Objectives," *Astrophys. J.*, **27**, 254 (1908).

Herzberger, M. J., "Light Distribution in the Optical IMAGE," *Opt. Soc. Am.*, **37**, 485 (1947).

Hopkins, H. H., "Phase Structures Seen in the Ordinary Microscope," in *Contraste de Phase et Contraste par Interferometrie*, M. Françon, Ed., Rev. d'Opti. Theor. et Inst, Paris, 1952, p. 142.

Kastler, A. "Un Systeme de Franges de Diffraction a Grand Contraste," *Rev. Opt.*, **29**, 307 (1950).

Kastler, A. and Momtarnal, R., "Phase Contrast in Polarized Light," *Nature*, **161**, 357 (1948).

Kingslake, R., "The Knife-Edge Test for Spherical Aberration," *Proc. Phys. Soc.*, **49**, 376 (1937).

Landgrave, J. E. A., *Phase Knife Edge Testing*, M.Sc. Report Imp. Coll. Sc. and Tech., London, 1974.

Linfoot, E. H., "Astigmatism Under the Foucault Test," *Mon. Not. R. Astron. Soc.*, **105**, 193 (1945a).

Linfoot, E. H., "A Contribution to the Theory of the Foucault Test," *Mon. Not. R. Astron. Soc.*, **105**, 193 (1945b).

Linfoot, E. H., "A Contribution to the Theory of the Foucault Test," *Proc. Roy. Soc. Ser. A*, **186**, 72 (1946a).

Linfoot, E. H., "On the Zernike Phase-Contrast Test," *R. Astron. Soc.*, **58**, 759 (1946b).

Linfoot, E. H., "On the Interpretation of the Foucault Test," *Proc. Roy. Soc. Ser. A*, **193**, 248 (1946c).

Linfoot, E. H., "The Foucault Test," in *Recent Advances in Optics*, Oxford University Press, Dover, 1955, Chap. II.

Loomis, D. A., "Testing Techniques Arouse Interest," *Opt. Sci. Newslett.*, **2**(3), 73 (1968).

Lowenthal, S. and Y. Belvaux, "Observation of Phase Objects by Optically Processed Hilbert Transform," *Appl. Phys. Lett.*, No. 2, 49 (1967).

Lyot, B., "Procedes Perme Hand d'Etudier les Irregularities d'une Surface Optique Bien Polie," *C. R. Acad. Sci. (Paris)*, **222**, 765 (1946).

Meinel, A. B., "Quantitative Reduction of a Wire Test (Azimuthal is Explained)," *Opt. Sci. Newlett.*, **2**(5), 134 (1968).

Menzel, E. "Die Abbildung von Phasen objekten in der optischen übertragungstheorie," in *Optics in Metrology*, Ed. Pol Mollet, Pergamon Press, New York, 1960, p. 283.

Miyamoto, K. "Image Evaluation by Spot Diagram Using a Computer," *Appl. Opt.*, **2**, 1247 (1963).

North, R. J., "A Colour Schlieren System Using Multicolor Filters of Simple Construction," National Physics Laboratory, England, Aero 26 (1954).

O'Hare, J. E. and J. D. Trolinger, "Holographic Colour Schlieren," *Appl. Opt.*, **8**, 2047 (1969).

Ojeda-Castañeda, J., *Images of Certain Type of Phase Objects*, Ph.D. Thesis, University of Reading, England, 1976.

Ojeda-Castañeda, J., "Nota Sobre el Microscopio de Contraste de Modulación," *Bol. Inst. Tonantzintla*, **2**, 293 (1978).

Ojeda-Castañeda, J. and L. R. Barriel-Valdos, "Classification Scheme and Properties of Schlieren Techniques," *Appl. Opt.*, **18**, 3338 (1979).

Ojeda-Castañeda, J., "Necessary and Sufficient Conditions for Thin Phase Imagery," *Opt. Acta*, **27**, 905 (1980a).

Ojeda-Castañeda, J., "A Proposal to Classify Methods Employed to Detect Thin Phase Structures under Coherent Illumination," *Opt. Acta*, **27**, 917 (1980b).

Platzeck, R. P. and E. Gaviola, "On the Errors of Testing a New Method for Surveying Optical Surfaces and Systems," *J. Opt. Soc. Am.*, **29**, 484 (1939).

Platzeck, R. P. and J. M. Simon, "The Method of the Caustic for Measuring Optical Surfaces," *Opt. Acta*, **21**, 267 (1974).

Porter, R. W., "Mirror Making for Reflecting Telescopes," in *Amateur Telescope Making*, Vol. 1, A. G. Ingalls, Ed., Scientific American, New York, 1953, p. 1.

Rayces, J. L., "Exact Relation Between Wave Aberration and Ray Aberration," *Opt. Acta*, **11**, 85 (1964).

Rayleigh, Lord, "On Methods for Detecting Small Optical Retardations, and on the Theory of Foucault's Test," *Philos. Mag.*, **33**, 161 (1917).

Ritchey, G. W., "On the Modern Reflecting Telescope and the Making and Testing of Optical Mirrors," *Smithson. Contrib. Knowl.*, **34**, 3 (1904).

Schardin, H., "Die Schlierenverfahren und ihre Anwendungen," *Ergebn Exakt Naturw.*, **20**, 303 (1942a).

Schardin, H. and G. Stamm, "Prüfung von Flachglass mit Hilfe Eines Farbigen Schlierenverfahrens," *Glastechn. Ber.*, **20**, 249 (1942b).

Schroader, I. H., "The Caustic Test," in *Amateur Telescope Making*, Vol. 3, A. G. Ingalls, Ed., Scientific American, New York, 1953, p. 429.

Shu, K. L., "Ray-Trace Analysis and Data Reduction Methods for the Ritchey-Common Test," *Appl. Opt.*, **22**, 1879 (1983).

Silvernail, C. J., "Extension of the Ritchey Test," *Appl. Opt.*, **12**, 445 (1973).

Simon, J. M., "Diffraction Theory of the Method of the Caustic for Measurement of Optical Surfaces," *Opt. Acta*, **18**, 369 (1971).

Simon, J. M., "Difracción en el Método de Foucault por Zonas," Physics 01/76, University of Buenos Aires, Argentina, 1976.

Simon, M. C. and J. M. Simon, "Testing of Flat Optical Surfaces by the Quantitative Foucault Method," *Appl. Opt.*, **17**, 132 (1978).

Simon, M. C., J. M. Simon, and E. L. Zenobi, "Testing Optical Surfaces by the Method of the Caustic Using a Phase-Stripe as Spatial Filter," *Appl. Opt.*, **18**, 1463 (1979).

Smartt, R. N. and W. H. Steel, "Theory and Application of Point Diffraction Interferometers," Proceedings of the ICO Conference on Optical Methods in Scientific and Industrial Measurements, Tokyo, 1974; *Jap. J. Appl. Phys.*, **14**, Suppl. I, 351 (1975).

Smith, W. J., *Modern Optical Engineering*, Mc-Graw Hill, New York, 1966, p. 439.

Sokoro, L. M., *Hilbert Optics*, Nauka Editorial, Moscow, 1981.

Stolzenburg, W. A., "The Double Knife Edge Technique for Improved Schlieren Sensitivity in Low Density Hypersonic Aerodynamic Testing," *J. SMPTE*, **74**, 654 (1965).

Strong, J., *Procedures in Experimental Physics*, Prentice-Hall, New York, 1938.

Tatian, B., An Analysis of the Ritchey-Common Test for Large Plane Mirrors, Internal Report, ITEK Corporation, Massachusetts, 1967.

Texereau, J., *How to Make a Telescope* (Translation and Adaptation from *La Construction du Telescope d'Amateur*), American Museum of Natural History, Anchor Books Doubleday, New York, 1957.

Teocaris, P. S. and E. E. Gdoutos, "Surface Topography by Caustics," *Appl. Opt.*, **15**, 1629 (1976).

Töpler, A., *Beobachtungen Nach Einer Neuen Optischen Methode*, Max. Cohen. und Sohn, Bonn, 1864.

Töpler, A., "Beobachtungen Nach Einer Neuen Optischen Methode," *Pogg. Ann.*, **127**, 556 (1866).

Töpler, A., "Beobachtungen Singender Flammenmit clem Schlierenapparat," *Pogg. Ann.*, **128**, 126 (1866); also in **131**, 33 and 180 (1867) and **134**, 194 (1868).

Vasil'ev, L. A., *Schlieren Methods*, Israel Prog. for Scient. Transl., New York, 1971.

Wadsworth, F. L. O., "Some Notes on the Correction and Testing of Parabolic Mirrors," *Pop. Astron.*, **10**, 337 (1902).

Weinberg, F. J., *Optics of Flames*, Butterworth, London, 1961.

Welford, W. T., "A Note on the Theory of the Foucault Knife-Edge Test," *Opt. Commun.*, **1,** 443 (1970).

Wolter, H., Minimum Strahl Kenn Zeichnung, German Patent No. 819925, assigned to E. Leitz, Wetzlar, 1949.

Wolter, H., "Schlieren-Phase Kontrast und Lichtschnittverfahren," in *Handbuch der Physik*, Vol. 24, Springer-Verlag, Berlin, 1956, p. 582.

Zernike, F., "Diffraction Theory of Knife-Edge Test and Its Improved Form, the Phase Contrast," *Mon. Not. R. Astron. Soc.*, **94,** 371 (1934a).

Zernike, F., "Begungstheorie des Schneidenver-Fahrans und Seiner Verbesserten Form, der Phasenkontrastmethode," *Physica*, **1,** 44 (1934b).

Zernike, F., "Phase Contrast, a New Method for the Microscopic Observation of Transparent Objects," *Physica*, **9,** 686 (1942).

ADDITIONAL REFERENCES

Barakat, R. and R. F. van Ligten, "Determination of the Transfer Function of an Optical System from Experimental Measurements of the Wavefront," *Appl. Opt.*, **4,** 749 (1965).

Barnes, N. F. and S. L. Bellinger, "Schlieren and Shadowgraph Equipment for Air Flow Analysis," *J. Opt. Soc. Am.*, **35,** 497 (1945).

Beskind, G. M., E. A. Bogudlov, O. A. Vitrichenko, O. A. Euseev, and S. M. Soldatov, "An Improved Foucault Philibert Method," *Izv. Spetz. Astrofiz. Obs.*, **7,** 182 (1975).

Bowen, I. S., "The Final Adjustments and Test of the Hale Telescope," *Publ. Astron. Soc. Pac.*, **62,** 91 (1950).

Boyd, W. R., "Fiber Optic Point Sources," *Appl. Opt.*, **8,** 73 (1969).

Bubis, I. Ya., "Determination of the Surface Quality of Large Astronomical Optical Components by the Slit and the Wire Method," *Sov. J. Opt. Technol.*, **38,** 4 (1971).

Bubis, I. Ya, A. I. Kusnetsov, and N. V. Konstantinovskaya, "Large-Aperture Adapter for the Schlieren Instrument for Evaluating the Quality of Large Optical Surfaces," *Sov. J. Opt. Technol.*, **45,** 19 (1978).

Burner, E. C., "Sensitive Visual Test for Concave Diffraction Gratings," *Appl. Opt.*, **11,** 1357 (1972).

Burner, A. W. and J. M. Franke, "Schlieren with a Laser Diode Source," *Opt. Eng.*, **20,** 801 (1981).

Cox, R. E. "The Hot Wire Foucault Test," *Sky Telesc.*, **25,** 114 (1953).

Dakin, R. K., "A Single-Edge Slit Foucault Tester," in *Advanced Telescope Techniques, Vol. 1: Optics*, A. Mackintosh, Ed., Willmann-Bell, Richmond, VA, 1987, p. 30.

Dakin, R. K., "Use of an Auxiliary Telescope in Foucault Testing," in *Advanced Telescope Techniques, Vol. 1: Optics*, A. Mackintosh, Ed., Willmann-Bell, Richmond, VA, 1987, p. 32.

Davies, T. P., "Schlieren Photography—Short Bibliography and Review," *Opt. Laser Technol.*, **13**, 37 (1981).

De Vany, Arthur S., "Spherical Aberration Analysis by Double Wire Testing," *Appl. Opt.*, **6**, 1073 (1967).

De Vany, Arthur S., Supplement to: "Aberration Analysis by Double Wire Testing," *Appl. Opt.*, **9**, 1720 (1970).

Ellison, W. F. A., "Figuring," in *Amateur Telescope Making, Vol. 1*, A. G. Ingalls, Ed. Scientific American, New York, 1953, p. 89.

Ellison, W. F. A., "Testing and Figuring," in *Amateur Telescope Making, Vol. 1*, A. G. Ingalls, Ed., Scientific American, New York, 1953, p. 120.

Ellison, W. F. A., "Testing; Foucault Shadow Test," in *Amateur Telescope Making, Vol. 1*, A. G. Ingalls, Ed., Scientific American, New York, 1953, p. 82.

Eastman Kodak Company, "Schlieren Photography," KODAK Publication P-11, Rochester, 1977.

Françon, M. and D. Wagner, "Etude des Default d'Homogeneite par la Methode l'Ombre Portee," *C. R. Acad. Sci. (Paris)*, **230**, 1850 (1950).

Françon M., Ed., *Le Contraste de Phase en Optique et Microscopie*, Edition de la Revue d'Optique Théorique et Instrumentale, Paris, 1950.

Gaviola, E., "A New Method for Testing Cassegrain Mirrors," *J. Opt. Soc. Am.*, **29**, 480 (1939).

Granger, E. M., "Wavefront Measurements from a Knife Edge Test," *Proc. SPIE*, **429**, 174 (1983).

Hansler, R. L., "A Holographic Foucault Knife-Edge for Optical Elements of Arbitrary Design," *Appl. Opt.*, **7**, 1863 (1968).

Haven, A. C., Jr., "A Simple Recording Foucault Test," *Sky Telesc.*, **38**, 51 (1969).

Hayslett, C. R., J. L. Rosson, and W. H. Swantner, "Testing and Analysis of Optical Components and Systems at White Sands Missile Range," *Proc. SPIE*, **134**, 12 (1978).

Holder, D. W., R. J. North, *Schlieren Methods Notes on Applied Science Series*, National Physical Lab., London, 1963.

Hung, Y. Y., J. L. Turner, M. Tafalian, J. D. Hovanesian, and C. E. Taylor, "Optical Method for Measuring Contour Slopes of an Object," *Appl. Opt.*, **17**, 128 (1978).

Ingalls, A. G., "Miscellany" in *Advanced Telescope Making, Vol. 1*, A. G. Ingalls, Ed., Scientific American, New York, 1953, p. 278.

Knöös, S. "A Quantitative Schlieren Technique for Measuring One Dimensional Density Gradients in Transparent Media," *Proc. 8th International Cong. on High-Speed Photography*, Almquist & Wiksell, Stockholm, 1968.

Kuchel, F. M., "Absolute Measurement of Flat Mirrors in the Ritchey-Common Test," in *Workshop on Optical Fabrication and Testing, Vol. 1*, Pag 114, Optical Society of America, Washington, 1986, p. 114.

Kutter, A., "Testing Long Focus Convex Secondary Mirrors," *Sky Telesc.*, **18**, 348 (1958).

Leonard, A. S., "High Intensity Light Source," in *Advanced Telescope Techniques, Vol. 1: Optics*, A. Mackintosh, Ed., Willmann-Bell, Richmond, VA 1987, p. 35.

Luhe, O. von der, "Wavefront Error Measurement Technique Using Extended, Incoherent Light Sources," *Opt. Eng.*, **27**, 1078 (1988).

Lyot, B. and M. Françon, "Etude des Défauts d'Homogénéité de Grands Disques de Verre," *Rev. Opt.*, **29**, 499 (1950).

Mackintosh, A., "Caustic Testers," in *Advanced Telescope Techniques, Vol. 1: Optics*, A. Mackintosh, Ed., Willmann-Bell, Richmond, VA 1987, p. 33.

Mackintosh, A., "A Precision Foucault Tester," in *Advanced Telescope Techniques, Vol. 1: Optics*, A. Mackintosh, Ed., Willmann-Bell, Richmond, VA, 1987, p. 25.

Maksutov, D. D., *Trudi GOI*, No. 6 (1932).

Maréchal, A., *Imagerie Géométrique-Aberrations*, Editions de la Revue d'Optique Théorique et Instrumentale, Paris, 1967, p. 162.

Martin, L. C. and W. T. Welford, *Technical Optics, Vols. 1, 2*, 2nd ed., Pitman, London, 1966.

Merzkirch, W., *Flow Visualization*, Academic Press, New York, 1974.

Meyer-Arendt, Jr., J., "Testing of Glass Surfaces by an Incident Light Schlieren Method," *J. Opt. Soc. Am.*, **46**, 1090 (1956).

Ojeda-Castañeda, J., "Edge Filters for Phase Contrast Imagery. 1. Theoretical Design," *Appl. Opt.*, **18**, 2355 (1979).

Ojeda-Castañeda, J., J. Moya, and G. Rodriguez, "Función de Transferencia Efectiva para Objetos Delgados en Fase, en Iliminación Parcialmente Coherente," *Bol. Inst. Tonantzintla*, **2**, 209 (1977).

Ojeda-Castañeda, J. and E. Jara, "Isotropic Hilbert Transform by Anisotropic Spatial Filtering," *Appl. Opt.*, **25**, 4035 (1986).

Obreimov, I. V., *Trudi GOI*, No. 23 (1924).

Onions, E. G., "The Hot-Wire Light Source," in *Advanced Telescope Techniques, Vol. 1: Optics*, A. Mackintosh, Ed., Willmann-Bell, Richmond, VA, 1987, p. 42.

Oppenheim, A. K., P. A. Urtiew, and F. J. Weinberg, "On the Use of Laser Light Sources in Schlieren Interferometer Systems," *Proc. Roy. Soc.*, **A291**, 279 (1966).

Orlov, P. V. and V. A. Bystrov, "General-Purpose Instrument for Testing Astro-Optical Surfaces," *Sov. J. Opt. Technol.*, **49**, 162 (1982).

Parks, R. E. and R. E. Sumner, "A Bright Inexpensive Pinhole Source," *Appl. Opt.*, **17**, 2469 (1978).

Philbert, M., "Procede Analogique Associe a la Methode de Focault pour la Determination Rapide du Profil d'une Surface d'Onda para Voie Electr.," *Opt. Acta*, **14**, 169 (1967).

Phillips, F. W., "Aspherizing and Other Problems in Making Maksutov Telescopes," *Sky Telesc.*, **25**, 110 (1963).

Plaskett, J. S., "83-Inch Mirror of McDonald Observatory," *Astrophys. J.*, **89**, 84 (1939).

Porter, R. W., "Knife-Edge Shadows. Photography as an Aid in Testing Mirrors," *Astrophys. J.*, **47**, 324 (1918).

Raiskii, S. M., "One Way of Realizing the Shadow Method," *J. Exp. Theor. Phys.*, **20**, 378 (1950).

Rodriguez-Zurita, G. and R. Uribe, "Schlieren Effects in Amici Prisms," *Appl. Opt.*, **29,** 531 (1990).

Schulz, L. Gunter, "Quantitative Test for Off-Axis Parabolic Mirrors," *J. Opt. Soc. Am.*, **36,** 588 (1946).

Sigler, R. D., "Extending the Caustic Test to Mirrors Other than Paraboloids," in *Advanced Telescope Techniques, Vol. 1: Optics*, A. Mackintosh, Ed., Willmann-Bell, Richmond, VA, 1987, p. 440.

Stetson, H. I., "Optical Tests of the 69-Inch Perkins Observatory Reflector," *J. Opt. Soc. Am.*, **23,** 293 (1933).

Strong, C. L., "Foucault Test Gear," in *Scient. Am.*, **193,** 124 (1955).

Yoder, P. R., Jr., F. B. Patrick, and A. E. Gee, "Permitted Tolerance on Percentage Correction of Paraboloidal Mirrors," *J. Opt. Soc. Am.*, **43,** 702 (1953).

9
Ronchi Test

A. Cornejo-Rodriguez

9.1. INTRODUCTION

Since its discovery and application to the testing of optical surfaces, the Ronchi test has been used widely but in a qualitative rather than a quantitative way. Also, because it is simple to accomplish and easy to interpret the experimental observations, the Ronchi test has almost always been conceptually interpreted from the point of view of geometrical optics. Sections 9.2 and 9.5 reflect this point of view. However, the author believed that a more comprehensive presentation of two aspects of the Ronchi test would provide a fuller view of the subject. These two aspects are (a) the need for a quantitative analysis of the data and (b) the development of the test from the point of view of the principles of physical optics, and a comparison to the geometrical approach (Toraldo di Francia 1941a). Sections 9.3 and 9.4 are devoted to these areas of study, respectively, and the need for mathematical treatment in these two sections was unavoidable. Perhaps some readers would like to skip these two sections (as can be done without losing continuity); others seeking a more comprehensive treatment of the test will find this material of interest. Section 9.6 is a brief review of some tests that can be related to the classical Ronchi test in various ways.

With respect to the previous edition of this book, in the present one the author has added some corrections in the original equations and figures and new references have been included.

9.1.1. Historical Introduction

The Ronchi test is one of the simplest and most powerful methods to evaluate and measure the aberrations of an optical system. The Italian physicist Vasco Ronchi (1923a) discovered that, when a ruling was placed near the center of curvature of a mirror, the image of the grating was superimposed on the grating

Optical Shop Testing, Second Edition, Edited by Daniel Malacara.
ISBN 0-471-52232-5 © 1992, John Wiley & Sons, Inc.

itself, producing a kind of Moiré pattern that he called combination fringes. Since the shape of these combination fringes depended on the aberrations of the mirror, he immediately thought of applying the phenomenon to the quality testing of mirrors. However, the combination fringes proved extremely difficult to interpret. The Ronchi test in the form that we now know appeared when Ronchi (1923b) published his thesis in order to obtain his final diploma at the Scuola Normale Superiore di Pisa.

One of the first applications of the method was to measure the aberrations of the telescope made by Galileo (Ronchi 1923c) and a lens made by Toricelli Ronchi 1923d). The first serious applications of this test to astronomical telescopes were made by Anderson and Porter (1929), and since then it has been very widely used by professional and amateur astronomers (King 1934; Kirkham 1953; Phillips 1963; Porter 1953; Strong 1938).

A few months after Ronchi's invention, Lenouvel (1924a, 1924b, 1925a, 1925b) in France published an extensive study of this test, along the same lines as those followed by Ronchi.

General descriptions of the Ronchi method can be found in many review articles (Adachi 1962a; Briers 1972; Morais 1958; Murty 1967; Ronchi 1925, 1954, 1958; Wehn 1962; Cornejo-Rodriguez 1983; Rosenbruch 1985; Briers 1979; and Briers and Cochrane 1979). The history of the test has been wonderfully described by Ronchi (1962, 1964) himself.

The name Ronchigrams for the patterns observed with a Ronchi ruling was coined by Schulz (1948). It is interesting that an attempt has been made to use this method for the measurement of the optical transfer function of lenses (Adachi 1962b) and even their chromatic aberration (Malacara and Cornejo 1971; and Toraldo di Francia 1942b). Salzmann (1970) used it to evaluate the quality of laser rods. Stoltsmann (1978) has applied this test to evaluate the quality of large-aperture flat mirrors, Assa et al. (1977) measured slope and curvature contours of flexed plates, Brookman et al. (1983) measured Gaussian beam diameters, and Kasana et al. (1984) have measured glass constants. Ronchi rulings have also been used in nontraditional configurations to evaluate optical components, for example, Patorski (1979) measured the wavefront curvature of small-diameter laser beams, using the Fourier imaging phenomena, and Kessler and Shack (1981) performed dynamic optical tests of a high-speed polygon.

New analysis and proposals to improve the sensitivity and range of applications of the Ronchi test have recently been done by several authors, for example, Patorski (1984) described the reversed path Ronchi test, Patorski (1986) described a method using spatial filtering techniques, and Lin et al. (1990) described a quantitative three-beam Ronchi test.

9.2. GEOMETRICAL THEORY

The Ronchi test has two equivalent models; one is geometrical, interpreting the fringes as shadows of the ruling bands, and the other is physical, interpreting

9.2. GEOMETRICAL THEORY 323

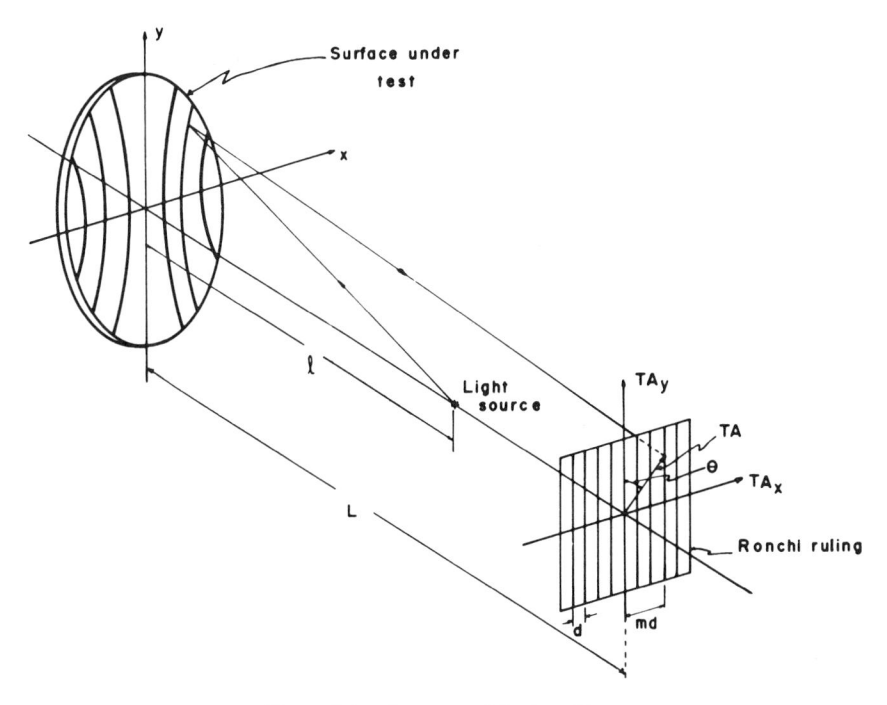

Figure 9.1. Geometry of the Ronchi test.

the fringes as due to diffraction and interference. It will be shown in this chapter that, when the frequency of the ruling is not very high, the two models arrive at the same result. Both the geometrical and physical models were described by Ronchi in his original paper. A good treatment of the geometrical model was developed by Jentzsch (1928), and it is briefly explained by Martin (1960).

As explained by Malacara (1965c), the Ronchi test really measures the transverse aberration TA in a direct way, as shown in Fig. 9.1. In this figure both the object and the image are on the optic axis, so the transverse aberration is measured from the axis and can be seen to include defocusing as well as other aberrations.

The wave aberration is defined in this exit pupil of the optical system under test, using a formula given by Rayces (1964), as

$$\frac{\partial W}{\partial x} = -\frac{TA_x}{r - W}; \quad \frac{\partial W}{\partial y} = -\frac{TA_y}{r - W}. \tag{9.1}$$

For all practical purposes very accurate results can be obtained if we write formula (9.1) as equal to

$$\frac{\partial W}{\partial x} = -\frac{TA_x}{r}; \quad \frac{\partial W}{\partial y} = -\frac{TA_y}{r}, \tag{9.2}$$

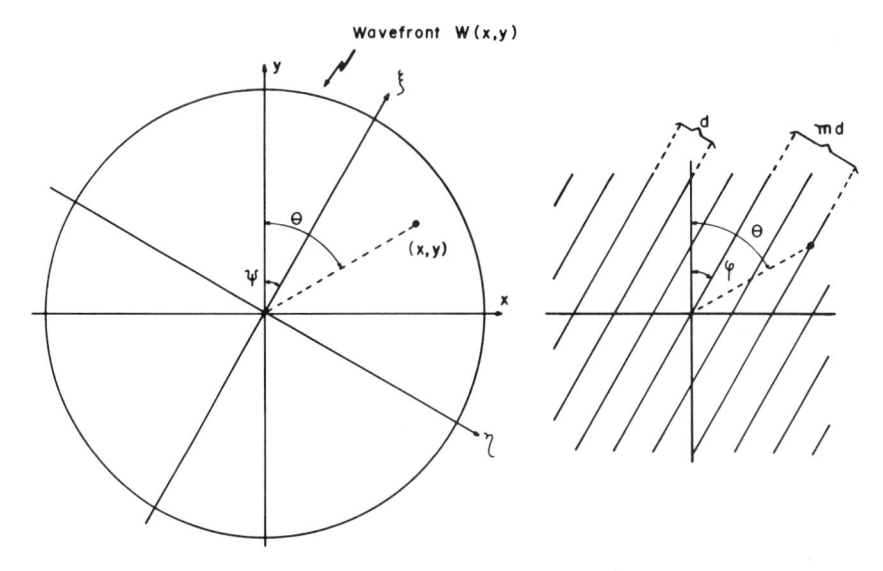

Figure 9.2. Wavefront and ruling orientation.

where r is the radius of curvature of the wavefront. Thus, if we assume a Ronchi ruling with spacing d between the slits, for a point (x, y) on the mth fringe we may write, in general,

$$\frac{\partial W}{\partial x} \cos \varphi - \frac{\partial W}{\partial y} \sin \varphi = -\frac{md}{r}, \tag{9.3}$$

where it is assumed that the ruling lines are inclined at an angle φ with respect to the y axis, as shown in Fig. 9.2. This is the basic formula for the geometrical model of the Ronchi test.

9.2.1. Ronchi Patterns for Primary Aberrations

Ronchi patterns obtained with primary aberrations have been studied by many authors, as we will see later on. General treatments of them have been published by Adachi (1960a, 1960b), Crino (1933), Schulz (1928), and Toraldo di Francia (1947, 1954).

The wavefront of a system with primary aberrations can be written (see Appendix 3) as

$$W = A(x^2 + y^2)^2 + By(x^2 + y^2) + C(x^2 + 3y^2) + D(x^2 + y^2), \tag{9.4}$$

where A, B, and C are the spherical aberration, coma, and astigmatism coefficients, respectively. The last coefficient D is the defocusing, given by the dis-

9.2. GEOMETRICAL THEORY

tance l' from the Ronchi ruling to the paraxial focus, as

$$D = \frac{l'}{2r^2}. \qquad (9.5)$$

No tilt terms are included since the Ronchi test is insensitive to them.

If we substitute Eq. (9.4) in Eq. (9.3), we obtain

$$4A(x^2 + y^2)(x \cos \varphi - y \sin \varphi) + B[2xy \cos \varphi - (3y^2 + x^2) \sin \varphi]$$
$$+ 2C(x \cos \varphi - 3y \sin \varphi) + 2D(x \cos \varphi - y \sin \varphi) = -\frac{md}{r}. \qquad (9.6)$$

In the study of each of the aberrations, it will often be convenient to apply a rotation ψ to this expression by means of the relations

$$x = \eta \cos \psi + \xi \sin \psi,$$
$$y = -\eta \sin \psi + \xi \cos \psi, \qquad (9.7)$$

where η and ξ are the new coordinate axes.

Defocusing. By applying the rotation $\psi = \varphi$ to the defocusing term, we obtain

$$2D\eta = -\frac{md}{r}. \qquad (9.8)$$

Hence we obtain straight, equidistant bands, which are parallel to each other and to the ruling slits. The separation S between these bands on the wavefront under study is

$$S = \frac{d}{2Dr}. \qquad (9.9)$$

These bands are illustrated in Fig. 9.3.

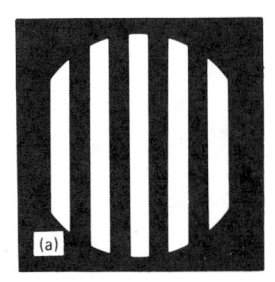

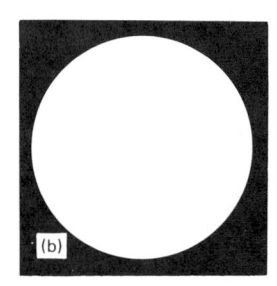

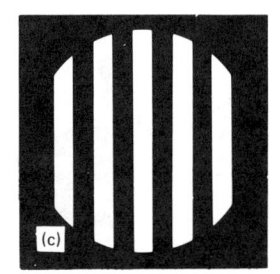

Figure 9.3. Ronchigrams with defocusing. (*a*) Outside of focus. (*b*) In focus. (*c*) Inside of focus.

Spherical Aberration with Defocusing. Spherical aberration patterns were first studied by Bocchino (1943) and Scandone (1933). By applying the rotation $\psi = \varphi$, we obtain for the spherical aberration and defocusing terms

$$4A(\eta^2 + \xi^2)\eta + 2D\eta = -\frac{md}{r}, \qquad (9.10)$$

where we can see that the axis of symmetry of the pattern is parallel to the ruling slits, as shown in Fig. 9.4. The lines are cubics in η. This Ronchi pattern without defocusing is identical to the Twyman–Green interferogram for coma. Adding defocusing in this Ronchigram is equivalent to adding tilt in the Twyman–Green interferogram.

In the absence of defocusing, the central fringe is very broad and for this reason the paraxial focus is called the "fusiform" or uniform focus (Crino 1939; Di Jorio 1939a; Ricci 1939).

Fifth-order spherical aberration Ronchigrams were studied by Bocchino (1940), Erdös (1959), and Scandone (1930).

Coma. Ronchigrams for coma were studied by Bruscaglioni (1932b), Villani (1930), and Villani and Bruscaglioni (1932). Applying a rotation $\psi = \varphi/2 +$

Figure 9.4. Ronchigrams with spherical aberration ($A = -20$). (*a*) Paraxial focus. (*b*) Medium focus. (*c*) Marginal focus. (*d*) Outside of focus.

9.2. GEOMETRICAL THEORY

$\pi/4$ to the coma term in Eq. (9.6), we obtain

$$B[-\eta^2 (1 + 2 \sin \varphi) + \xi^2(1 - 2 \sin \varphi)] = -\frac{md}{r}. \quad (9.11)$$

Different figures are obtained, depending on the value of φ (ruling inclination with respect to the meridional plane), as follows (see Fig. 9.5):

$\varphi = 0°$	Hyperbolas
$\varphi = 90°$	Ellipses with semiaxes in the ratio $\sqrt{3}$ to 1
$0° < \varphi < 30°$	Hyperbolas inclined at an angle ψ
$\varphi = 30°$	Straight bands
$30° < \varphi < 90°$	Ellipses inclined at an angle ψ

If we rewrite Eq. (9.11) as

$$B[-2(1 + \sin \varphi)(\eta^2 + \xi^2) + (\eta^2 + 3\xi^2)] = -\frac{md}{r}, \quad (9.12)$$

we can see that this pattern is identical to that of a Twyman–Green interferogram for astigmatism with defocusing, where the magnitude of this apparent defocusing is given by the angle φ.

Astigmatism with Defocusing. Astigmatism Ronchigrams were extensively studied by Bruscaglioni (1932b), Calamai (1938), Scandone (1931a), Villani (1930), and Villani and Bruscaglioni (1932). By applying a rotation $\psi = \varphi$ to the corresponding terms in Eq. (9.6), we obtain

$$2C[\eta(2 - \cos 2\varphi) - \xi \sin 2\varphi] + 2D\eta = -\frac{md}{r}. \quad (9.13)$$

We can see that the Ronchigram is formed by straight, equidistant, parallel bands whose inclination α with respect to the ruling slits is given (see Fig. 9.6) by

$$\tan \alpha = \frac{C \sin 2\varphi}{D + C(2 - \cos 2\varphi)}. \quad (9.14)$$

The intersections of the bands with the ξ axis are fixed, independently of the focusing term D, at equally separated points whose separation $\Delta\xi$ is

$$\Delta\xi = \frac{d}{2rC \sin 2\varphi}. \quad (9.15)$$

Figure 9.5. Ronchigrams with coma ($B = -30$).

9.2. GEOMETRICAL THEORY

Figure 9.6. Ronchigrams with astigmatism ($C = -20$).

The value of α changes with the focus term D (ruling position along the optical axis), making the bands rotate as the ruling is moved along the optical axis. The effect, called "capriola," was studied in detail by Bruscaglioni (1932a) and Scandone (1931b). The bands become perpendicular to the ruling slits when

$$D + C(2 - \cos 2\varphi) = 0. \tag{9.16}$$

a condition that occurs for a value of D/C between -1 (at the sagittal focus, when $\varphi = 0°$) and -3 (at the tangential focus, when $\varphi = 90°$). The precise values of D/C equal to -1 and -3 are excluded because at those points $\sin 2\varphi = 0$ and hence $\Delta\xi \to \infty$, rendering the test insensitive to astigmatism. The maximum sensitivity is obtained when $\Delta\xi$ is as small as possible with respect to the wavefront diameter. Thus the optimum angle for measuring the astigmatism is $\varphi = 45°$, and then the bands become perpendicular to the ruling slits ($\alpha = 90°$) when $D/C = -2$.

Spherical Aberration with Astigmatism and Defocusing. The patterns obtained with this combination of aberrations were studied by Scandone (1931a, 1931b). Applying a rotation $\psi = \varphi$ to the corresponding terms in Eq. (9.6), we obtain

$$4A\eta(\eta^2 + \xi^2) + 2\eta[D + C(2 - \cos 2\varphi)] - 2C\xi \sin 2\varphi = -\frac{md}{r}. \quad (9.17)$$

This combination of aberrations produces the serpentine fringes described by Scandone and shown in Fig. 9.7. The first term comes from the spherical aberration and is identical to the Twyman–Green coma term. The second term comes from the defocusing D (ruling position) and the astigmatism C and is identical to a tilt term about the ξ axis in a Twyman–Green interferometer. The third term comes from the astigmatism and is identical to a tilt term about the η axis in a Twyman–Green interferometer.

Using this analogy between astigmatism in the Ronchi test and tilt in the Twyman–Green, Murty (1971) suggested adding an apparent tilt about the η axis to the Ronchi test by introducing astigmatism by means of a cylindrical lens in the beam, with its axis at 45° with respect to the ruling slits. To have apparent tilt only about the η axis, we need

$$D + C(2 - \cos 2\varphi) = 0,$$

$$C \sin 2\varphi \neq 0, \quad (9.18)$$

for the particular case of $\varphi = 45°$,

$$D = -2C. \quad (9.19)$$

Expression (9.19) assumes that the defocusing term is just enough to cancel the effect of the astigmatism in producing zero tilt about the ξ axis. To produce an apparent tilt about the η axis of the Ronchigram at the marginal focus, we must introduce an additional defocusing equal to $-2A$, thus obtaining

$$D = -2(C + A). \quad (9.20)$$

9.2. GEOMETRICAL THEORY

$\varphi = 45°$ $\varphi = 45°$

Spherical aberration (Paraxial focus)

Spherical aberration (Marginal focus)

Spherical aberration with astigmatism ($D = -2C$)

Spherical aberration with astigmatism ($D = -2(C + A)$)

Figure 9.7. Ronchigrams with sperical aberration ($A = -10$) alone (*top*) and combined with astigmatism ($C = -10$) (*bottom*).

9.2.2. Ronchi Patterns for Aspherical Surfaces

The Ronchi test is very useful for testing aspherical surfaces, including large mirrors for astronomical telescopes (Popov 1972). Using Eq. (9.3) with $\varphi = 0°$, we can compute the ideal Ronchigram for any aspherical surface, as defined in Appendix 1, by assuming valid the approximate relation

$$z(x, y) - z_0(x, y) = 2W(x, y) \tag{9.21}$$

where z is for the aspheric surface and z_0 is for the osculating sphere. This method is, however, only approximate, not exact. An alternative and more accurate procedure is to trace rays by using the law of reflection. It is interesting that two early attempts to use the Ronchi test for aspherical surfaces were made by Waland (1938) and Schulz (1948).

Sherwood (1958) and Malacara (1965a, 1965b) independently showed by two different methods that, when the configuration in Fig. 9.1 is used, the transverse aberration *TA* at the Ronchi ruling plane is given by

$$TA(S) = \frac{(l + L - 2z)\left[1 - \left(\frac{dz}{dS}\right)^2\right] + 2\frac{dz}{dS}\left[S - \frac{(l-z)(L-z)}{S}\right]}{\frac{l-z}{S}\left[1 - \left(\frac{dz}{dS}\right)^2\right] + 2\frac{dz}{dS}}, \quad (9.22)$$

where S is the distance from the optical axis to the point on the mirror. In Fig. 9.1, we can also see that

$$TA = \frac{md}{\sin \theta}. \quad (9.23)$$

The Ronchigram is obtained by assigning many values of S and then calculating θ for different values of m. In general, one wishes to obtain the Ronchigram over a flat surface, parallel to and near the mirror, since this is what is obtained when a photograph is taken. In most cases the error introduced by considering the fringes over the mirror surface is very small; but when the radius of curvature of the mirror is small with respect to its diameter, the error becomes important.

The error is compensated for if, when plotting the Ronchigram, S_p, which is given by

$$S_p = S\left\{1 - \left[\frac{z(S_{\max}) - z(S)}{l - z(S)}\right]\left(1 - \frac{TA}{S}\right)\right\} \quad (9.24)$$

is used instead of the calculated S (Malacara 1965b), as shown in Fig. 9.8.

Ideal Ronchigrams for paraboloids have been computed for the use of amateur astronomers by Lumley (1960, 1961) and Sherwood (1960).

For easy interpretation of the Ronchigram, it is desirable to avoid any closed-loop fringes. This is possible only outside the caustic limits described in Appendix 1, as shown in Fig. 9.9, where it is evident that, when the ruling is at the marginal focus, the only closed-loop fringes are inside the mirror and the outer one closes near the edge of the mirror.

It is also interesting to see that much information about the center of the mirror is lost when the fringes are curved, as at the paraxial focus. Therefore, to obtain the maximum information, the Ronchigram should be taken at the paraxial focus if $K > 0$ or at the end of the caustic if $K < 0$.

9.2.3. Null Ronchi Rulings

We have shown that, when the surface under test is aspherical, the Ronchi fringes are not straight but curved. The surface errors can then be computed from the deviations of the fringes from the desired shape. This procedure can

9.2. GEOMETRICAL THEORY

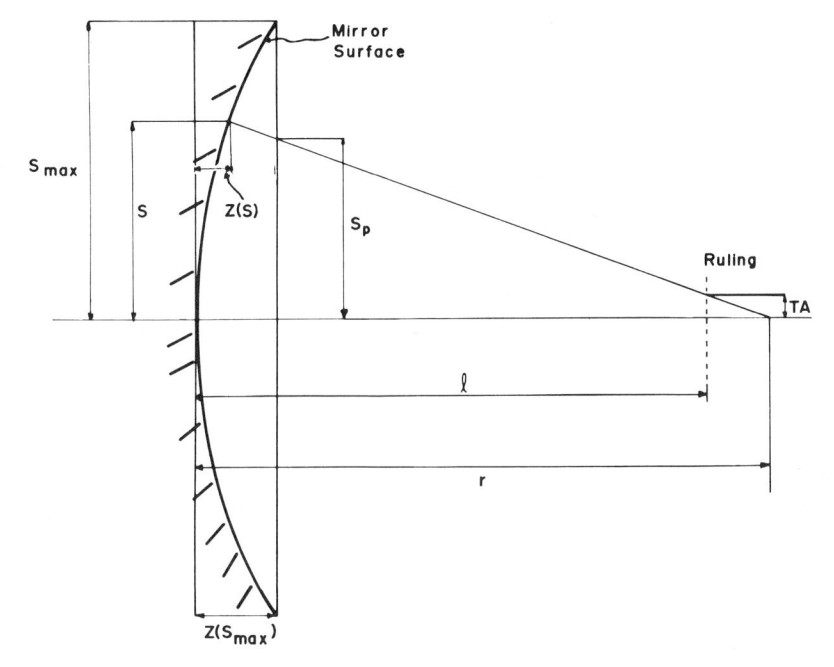

Figure 9.8. Projection of fringes over a plane.

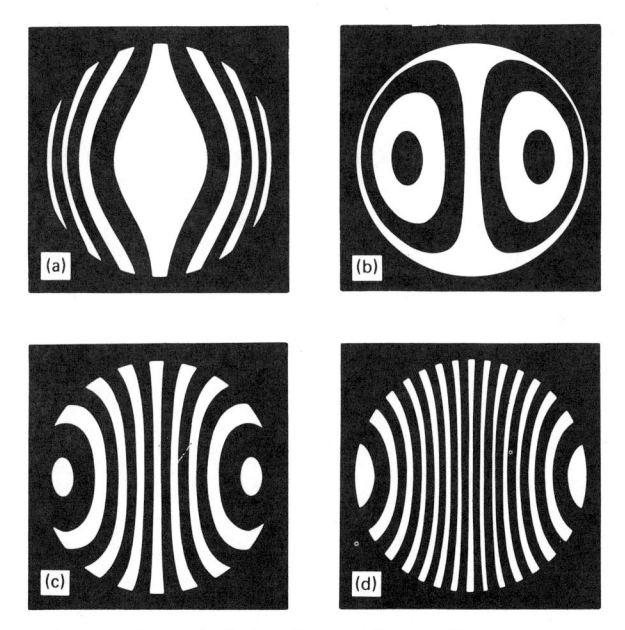

Figure 9.9. Ronchigram of an aspherical surface at different ruling positions on the caustic. (*a*) Paraxial focus. (*b*) Marginal focus. (*c*) Between marginal focus and end of caustic. (*d*) End of caustic.

give very good results when the data are processed in a computer, as will be shown in Section 9.3. If a computer is not used, it is easier for a technician to detect deviations from a straight line than from a curved line. Also, when the fringes are curved, the diffraction effects tend to diffuse the fringes, making the measurements even more difficult.

These disadvantages can be overcome by means of a special kind of Ronchi ruling with curved lines, the curvature of which compensates for the asphericity of the surface in order to produce straight fringes of constant widths. This idea was first qualitatively suggested by Pastor (1969) and later was developed quantitatively by Popov (1972) and independently by Malacara and Cornejo (1974c), using a ray-tracing program and then solving five linear equations. An approximate method to produce these null Ronchi rulings for paraboloids was devised by Mobsby (1973, 1974). More recently Hopkins and Shagan (1977) used spot diagrams for the design of the null gratings.

An alternative exact procedure uses Eq. (9.22) to compute $TA(S)$ for different values of S from the center to the edge of the mirror. Then we take the intersections of the desired straight fringe on the aspheric mirror with the circles on which the $TA(S)$ was computed. For each of those intersections there is an angle θ, which is the same on the ruling to be computed. The radial coordinate on the ruling is the computed $TA(S)$ for that circle with radius S on the mirror.

Still another alternative and approximate method is based on the computation of the third-order transverse spherical aberration (Malacara and Cornejo 1976b). If $\bar{x}$ and $\bar{\rho}$ are the x and radial coordinates of a point over a fringe in the null Ronchi ruling, respectively, the ruling is computed with

$$\bar{\rho}^2 = -\frac{r^2 \bar{x}^3}{Km^3(\Delta x)^3} - \frac{r\Delta r \bar{x}^2}{Km^2(\Delta x)^2} \qquad (9.25)$$

where r is the radius of curvature of the mirror, K is the conic constant, m is the fringe order number, Δx is the fringe separation on the mirror, and Δr is the displacement of the ruling from the paraxial focus, being positive if it is displaced outside the focus.

The ruling is computed by assigning values of $\bar{x}$ and then computing $\bar{\rho}$, starting in $\bar{x}_{min}$, given by

$$\bar{x}_{min} = -\frac{Km^3(\Delta x)^3}{r^2} - \frac{m(\Delta x)(\Delta r)}{r}. \qquad (9.26)$$

Here it is assumed that the light source is at the center of curvature. This approximate method has proved to be very accurate for most practical purposes.

Examples of computed null Ronchi rulings are given in Fig. 9.10. Figure 9.11 shows a normal and a null Ronchigram. It is evident that the fringes, in addition to being straight, are much better defined in the null test (Fig. 9.11b).

9.3. WAVEFRONT SHAPE DETERMINATION

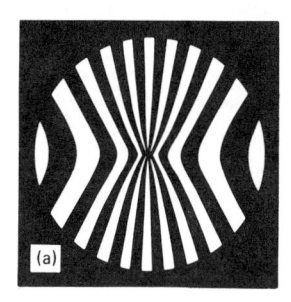

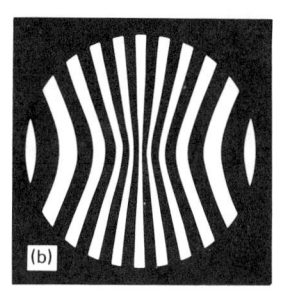

Figure 9.10. Two null Ronchi rulings. (*a*) At paraxial focus. (*b*) Inside paraxial focus.

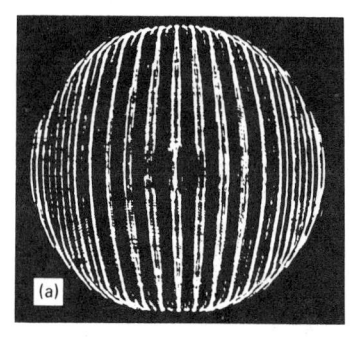

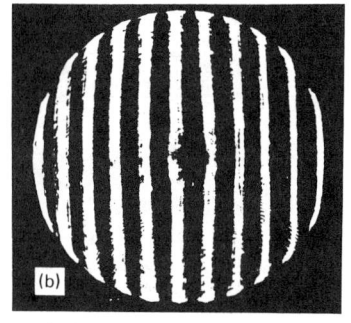

Figure 9.11. Normal and null Ronchigrams of an aspheric surface. (From Malacara and Conejo 1974c.)

A disadvantage of the null test, as compared to the normal test, is that the ruling must be positioned very precisely on the calculated place along the optical axis when the test is made. The reason is that the asphericity compensation depends very critically on that position. It is a great help to draw a circle on the ruling so that its projection on the mirror coincides with the outer edge of that surface.

Another important restriction is that the light source cannot be extended with a ruling covering it, as in the normal Ronchi test, but must be a point source.

9.3. WAVEFRONT SHAPE DETERMINATION

The wavefront or mirror deformations can be determined from the Ronchigram. Probably Pacella (1927) was the first to attempt this and to identify the type of aberration. A qualitative idea about the mirror deformations can very easily be obtained (De Vany 1965, 1970) just by observing the pattern. Figure 9.12 shows some possible patterns, and their corresponding surface deviations are shown in Fig. 9.13. Table 9.1 indicates the relationship between the Ronchigrams and

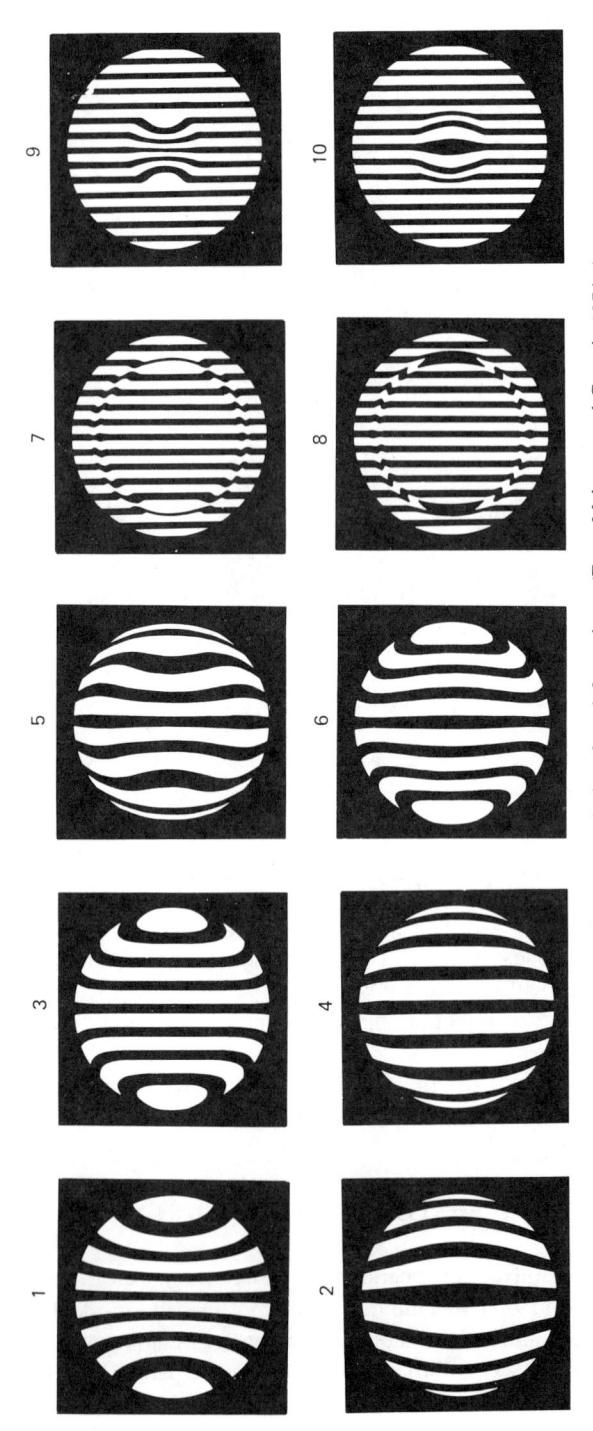

Figure 9.12. Ronchigrams for some typical surface deformations. (From Malacara and Cornejo 1974a.)

9.3. WAVEFRONT SHAPE DETERMINATION

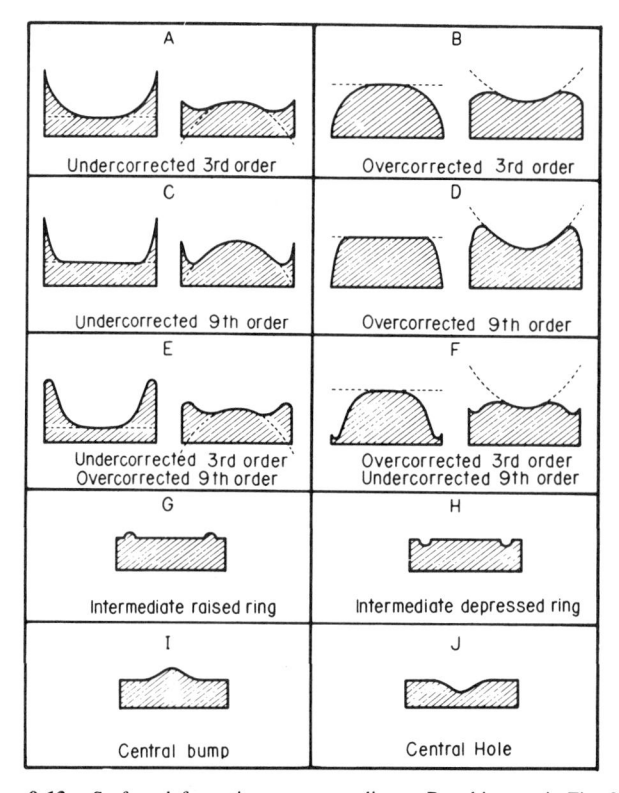

Figure 9.13. Surface deformations corresponding to Ronchigrams in Fig. 9.12.

Table 9.1. Relation Between the Ronchigrams of Figure 9.12 and the Surface Profile Deformations Shown in Fig. 9.13.

	Surface			
	Reflecting		Refracting	
Ronchigram	Ruling Outside of Focus	Ruling Inside of Focus	Ruling Outside of Focus	Ruling Inside of Focus
1	B	A	A	B
2	A	B	B	A
3	D	C	C	D
4	C	D	D	C
5	F	E	E	F
6	E	F	F	E
7	H	G	G	H
8	G	H	H	G
9	J	I	I	J
10	I	J	J	I

deformations. The identification and measurement of the primary aberration was treated in detail by Adachi (1960a, 1960b).

Following a combination of shearing diffraction interferometer theory and some geometrical equations of the Ronchi test, Briers did a thorough study about the theory and experiment of the Ronchi test in order to have semiquantitative results (Briers 1979; Briers and Cochrane 1979). Also De Vany extended his previous work, and established a relation between observed patterns and surface defects (De Vany 1978, 1980, 1981).

A more general treatment for surfaces with any kind of deformations is given in the next section.

9.3.1. General Case

The only assumption we make in considering a surface without any symmetry is that it is smooth enough to be represented by the following two-dimesnional polynomial of kth degree (Cornejo-Rodriguez and Malacara 1976):

$$W(x, y) = \sum_{i=0}^{k} \sum_{j=0}^{i} B_{ij} x^j y^{i-j}. \qquad (9.27)$$

Then it can be shown that the partial derivatives of $W(x, y)$ with respect to x and y are

$$\frac{\partial W}{\partial x} = \sum_{i=0}^{k-1} \sum_{j=0}^{i} (j + 1) B_{i+1, j+1} x^j y^{i-j} \qquad (9.28)$$

and

$$\frac{\partial W}{\partial y} = \sum_{i=0}^{k-1} \sum_{j=0}^{i} (i - j + 1) B_{i+1, j} x^j y^{i-j}. \qquad (9.29)$$

But from the fundamental relation for the Ronchi test in Eq. (9.3) we can write

$$\frac{\partial W}{\partial x} \cos \varphi - \frac{\partial W}{\partial y} \sin \varphi = - \frac{[m(x, y) - m_0(x, y]}{r} d, \qquad (9.30)$$

where $m(x, y)$ is the measured value of m at point (x, y) in the real Ronchigram and $m_0(x, y)$ is the computed value of m at point (x, y) in the perfect Ronchigram, using Eqs. (9.22) and (9.23).

We now define $m_x(x, y)$ and $m_y(x, y)$ as the values of $m(x, y)$ when the ruling orientations $\varphi = 0°$ and $\varphi = 90°$, respectively, are used. These are obtained with two Ronchigrams having mutually perpendicular rulings. Then we may write for these two patterns

9.3. WAVEFRONT SHAPE DETERMINATION

$$\frac{\partial W}{\partial x} = -\frac{[m_x(x, y) - m_0(x, y)]}{r} d \quad (\varphi = 0°) \tag{9.31}$$

and

$$\frac{\partial W}{\partial y} = \frac{[m_y(x, y) - m_0(x, y)]}{r} d \quad (\varphi = 90°). \tag{9.32}$$

The difference function $(m_y - m_0)$ can be fitted to a two-dimensional polynomial of $(k - 1)$th degree by means of a least-squares procedure to find

$$\frac{\partial W}{\partial x} = \sum_{i=0}^{k-1} \sum_{j=0}^{i} C_{ij} x^j y^{i-j} \tag{9.33}$$

and

$$\frac{\partial W}{\partial y} = \sum_{i=0}^{k-1} \sum_{j=0}^{i} D_{ij} x^j y^{i-j}. \tag{9.34}$$

If we compute Eqs. (9.28) and (9.29) with these two relations, we can see that

$$B_{ij} = \frac{C_{i-1,j-1}}{j} \quad \text{for} \quad \begin{cases} i = 1, 2, 3, \ldots, k \\ j = 1, 2, 3, \ldots, i \end{cases} \tag{9.35}$$

and

$$B_{ij} = \frac{D_{i-1,j}}{i-j} \quad \text{for} \quad \begin{cases} i = 1, 2, 3, \ldots, k \\ j = 0, 1, 2, \ldots, (i-1). \end{cases} \tag{9.36}$$

For the value of $m = 0$ only Eq. (9.36) can be used; therefore

$$B_{i0} = \frac{D_{i-1,0}}{i} \quad \text{for } i = 1, 2, 3, \ldots, k. \tag{9.37}$$

For the value $m = n$ only Eq. (9.35) can be used; therefore

$$B_{ii} = \frac{C_{i-1,i-1}}{i} \quad \text{for } i = 1, 2, 3, \ldots, k. \tag{9.38}$$

For all other combinations of n and m, either Eq. (9.35) or Eq. (9.36) can be used. Therefore, to increase the accuracy, we take the average of both values:

$$B_{ij} = \frac{1}{2}\left(\frac{C_{i-1,j-1}}{j} + \frac{D_{i-1,j-1}}{i-j}\right) \quad \text{for} \quad \begin{cases} i = 2, 3, \ldots, k \\ j = 1, 2, 3, \ldots, (i-1). \end{cases} \tag{9.39}$$

Once the coefficients $B_{i,j}$ have been found, the wavefront deviations $W(x, y)$ can be computed with Eq. (9.27) and then the mirror surface deviations found with Eq. (9.21).

9.3.2. Surfaces with Rotational Symmetry

If we assume that the surface has a rotational symmetry, one Ronchigram with $\varphi = 0°$ is enough to completely determine the shape of the surface. One approach is to measure the fringe intersections with the x axis (Malacara 1965a, 1965b) to determine the transverse aberration $TA(S)$ at those points. The value of S is the distance from the center of the pattern to the fringe intersection with the x axis, and TA is equal to md. Then these measurements are fitted to a polynomial with only odd powers of S, using a least-squares procedure, and the wavefront is computed by a simple integration of $TA(S)$.

This method, however, does not provide enough data because no information is obtained for zones between the fringe intersections, and for this reason a polynomial interpolation is used. Unfortunately, with this procedure the surface profile is sometimes smoothed more than is desired.

Using a procedure by Malacara and Cornejo (1975), we may assume that the ideal fringes on the mirror are not necessarily straight but can be curved, as in Fig. 9.14. The solid lines are the actual shape of the fringes on an imperfect mirror, while the dotted lines show the ideal shape for a perfect surface. The residual transverse aberration $TA(S)$ can be defined by

$$TA(S) = TA_A(S) - TA_0(S), \qquad (9.40)$$

where $TA_A(S)$ is the actual total transverse aberration for the imperfect surface and $TA_0(S)$ is the computed ideal transverse aberration. It can be seen in Fig. 9.14 that

$$\frac{S_0}{x_0} = \frac{S}{x} = \frac{TA_A(S)}{TA_x} \qquad (9.41)$$

Since the points on the actual and ideal fringes are considered to be aligned with the center of the surface, they must correspond to the same point on the ruling. Therefore

$$TA_0(S_0) = TA(S). \qquad (9.42)$$

Now, since the lines on the ruling are straight, we can write $TA_x = md$, obtaining from Eq. (9.41)

9.3. WAVEFRONT SHAPE DETERMINATION

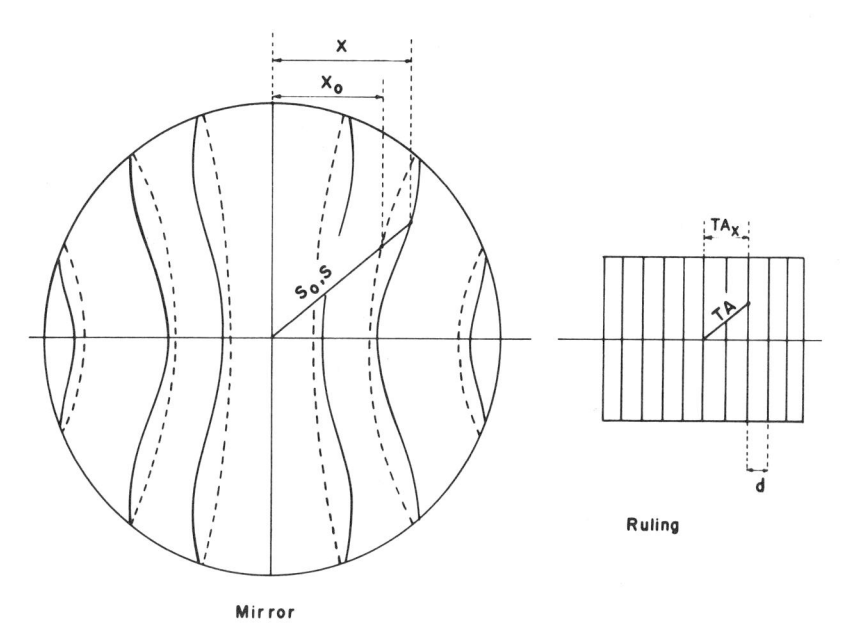

Figure 9.14. Ronchigram formation in a surface with rotational symmetry.

$$TA_A(S) = \frac{md}{x} S, \qquad (9.43)$$

which, substituted in Eq. (9.40), gives

$$TA(S) = \frac{md}{x} S - TA_0(S). \qquad (9.44)$$

To determine $TA(S)$, we need to measure one value of x for each value of S. There are, however, in general, several values of x, one for each fringe, for a given value of S. An average value of x/m can be shown to be

$$\left(\frac{x}{m}\right)_{average} = \frac{\sum_{i=1}^{N} x_i m_i}{\sum_{i=1}^{N} m_i^2}, \qquad (9.45)$$

where N is the number of data points on the circle with radius S. The actual shape of the mirror is then found by numerical integration without the need for a polynomial fitting to the data.

9.4. PHYSICAL THEORY

As Ronchi pointed out in his first paper (1923b), the Ronchi test can be considered from a physical point of view as an interferometer. At the beginning many attempts were made to make a good physical theory model (Ronchi 1924a, 1924b, 1926a, 1926b, 1927, 1928), from which a reasonably good explanation was produced that assumed the Ronchi ruling really acted as a diffraction grating, producing many diffracted orders, each one giving a laterally sheared image of the pupil, as shown in Fig. 9.15. The theory was further developed by Di Jorio (1939a, 1939b, 1939c, 1939d, 1942, 1943), Pallotino (1941), and Toraldo di Francia (1941a, 1941b, 1942a, 1943a, 1943b, 1946), who obtained the exact shape of the fringes with the physical theory and demonstrated their similarity to the shadow fringes. They also investigated many small details related to this theory. A good account of the history of these developments has been given by Ronchi (1962, 1964).

9.4.1. Mathematical Treatment

A mathematical treatment using Fourier theory was originally developed by Adachi (1963) and later extended by Barakat (1969). It is assumed that the wavefront at the exit pupil in the plane x_0-y_0 is represented by a complex function $F_0(x_0, y_0)$, which is zero outside the limits imposed by the aperture. If the

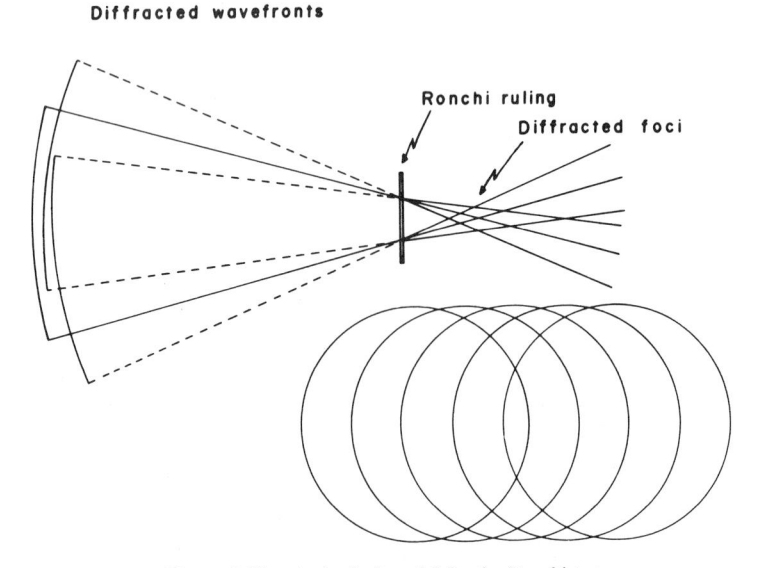

Figure 9.15. A physical model for the Ronchi test.

9.4. PHYSICAL THEORY

system is evenly illuminated, this function $F_0(x_0, y_0)$, inside the free portions of the exit pupil, is given by $\exp[i2\pi W(x_0, y_0)]$, where $W(x_0, y_0)$ is the wavefront deformation function. The phase deviations (wavefront deformations) given by $F_0(x_0, y_0)$ are measured with respect to a sphere with its center at the Ronchi ruling, in the plane x_r-y_r. If r is the radius of curvature of the reference wavefront, the field $U(x_r, y_r)$ at the ruling plane will be

$$U(x_r, y_r) = \int\!\!\int_{-\infty}^{\infty} F_0(x_0, y_0) \exp\left[-i\frac{2\pi}{\lambda r}(x_r x_0 + y_r y_0)\right] dx_0\, dy_0. \quad (9.46)$$

As suggested by Barakat (1969), we can use the concept of spatial filtering to treat the effect of the Ronchi ruling and thus consider it as a filtering mask in the Fourier transform plane x_r-y_r. Then, if the observation plane x_1-y_1 is an image of the pupil plane x_0-y_0, the amplitude in that plane will be given by

$$G(x_1, y_1) = \int\!\!\int_{-\infty}^{\infty} U(x_r, y_r) M(x_r, y_r) \times \left\{\exp\left[i\frac{2\pi}{\lambda r}(x_r x_1 + y_r y_1)\right]\right\} dx_r\, dy_r$$

$$(9.47)$$

where $M(x_r, y_r)$ is the ruling function acting as a filtering or modulating device. If we substitute the value of $U(x_r, y_r)$ from Eq. (9.46) in Eq. (9.47), we obtain

$$G(x_1, y_1) = \int\!\!\int_{-\infty}^{\infty} F_0(x_0, y_0)\, dx_0\, dy_0 \int\!\!\int_{-\infty}^{\infty} M(x_r, y_r)$$

$$\times \left(\exp\left\{i\frac{2\pi}{\lambda r}[(x_1 - x_0)x_r + (y_1 - y_0)y_r]\right\}\right) dx_r\, dy_r. \quad (9.48)$$

This expression is valid for any kind of modulating function. Now, however, we assume that the ruling is formed by straight, parallel, equidistant bands, and so we write

$$M(x_r) = \sum_{n=-\infty}^{\infty} B_n \exp\left(i\frac{2\pi n}{d} x_r\right), \quad (9.49)$$

where d is the ruling period, as usual. Then, substituting this expression in Eq. (9.48), we obtain

$$G(x_1, y_1) = \sum_{n=-\infty}^{\infty} B_n \int\!\!\int_{-\infty}^{\infty} F_0(x_0, y_0)\, dx_0\, dy_0$$

$$\times \left\{ \int_{-\infty}^{\infty} \exp\left[i \frac{2\pi}{\lambda r}\left(x_1 - x_0 + \frac{\lambda rn}{d}\right)x_r\right] dx_r \right\}$$

$$\times \left\{ \int_{-\infty}^{\infty} \exp\left[i \frac{2\pi}{\lambda r}(y_1 - y_0)y_r\right] dy_r \right\}, \qquad (9.50)$$

which, when the definition for the Dirac δ function,

$$\int_{-\infty}^{\infty} \exp\left[i(k - k_0)x\right] dx = \delta(k - k_0), \qquad (9.51)$$

is used, becomes

$$G(x_1, y_1) = \sum_{n=-\infty}^{\infty} B_n \int\!\!\int_{-\infty}^{\infty} F_0(x_0, y_0)\delta\!\left(x_1 - x_0 + \frac{\lambda rn}{d}\right)\delta(y_1 - y_0)\, dx_0\, dy_0. \qquad (9.52)$$

This can be shown to be equal to

$$G(x_1, y_1) = \sum_{n=-\infty}^{\infty} B_n F_0\!\left(x_1 + \frac{\lambda rn}{d}, y_1\right). \qquad (9.53)$$

If the ruling is of the form

$$M(x_r) = 1 + \cos\!\left(\frac{2\pi n x_r}{d}\right), \qquad (9.54)$$

we have $B_0 = B_1 = B_{-1} = 1$, and thus we have three laterally sheared images of the pupil. In general, however, we have many images laterally sheared by an amount $\lambda rn/d$, as in Figs. 9.15 and 9.16. If the ruling has a periodic square-wave profile as in Fig. 9.17, we can show, using Fourier theory, that the coefficients B_n are given by

$$B_n = (-1)^n \frac{\sin n\pi k_0}{n\pi}, \qquad (9.55)$$

9.4. PHYSICAL THEORY

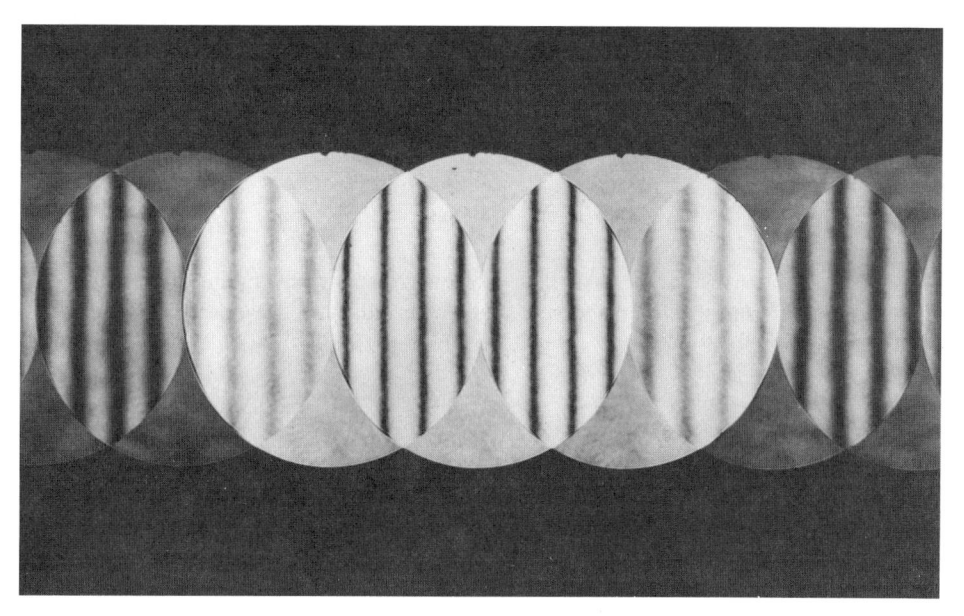

Figure 9.16. Interference fringes between different diffracted orders in a Ronchi ruling.

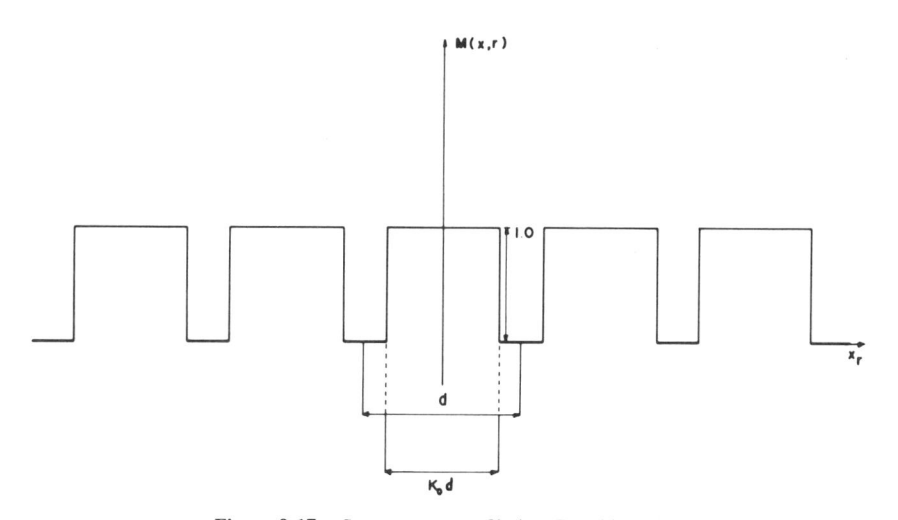

Figure 9.17. Square-wave profile in a Ronchi ruling.

where k_0 is the ratio between the width of the clear slits in the ruling and its period d, as shown in Fig. 9.17. If the width of the clear and dark bands is the same ($k_0 = \frac{1}{2}$), all even orders are missing. Substituting Eq. (9.55) into (9.53) gives

$$G(x_1, y_1) = \sum_{n=-\infty}^{\infty} (-1)^n \frac{\sin n\pi k_0}{n\pi} F_0\left(x_1 + \frac{\lambda rn}{d}, y_1\right). \quad (9.56)$$

Let us now consider the case of a perfect wavefront with a radius of curvature R, different from the reference wavefront with radius r. The defocusing Δr is then equal to $R - r$. Then $F_0(x_1, y_1)$ may be represented by

$$F_0(x_1, y_1) = \begin{cases} \exp\left[i\frac{\pi}{\lambda rR}(x_1^2 + y_1^2)\right] \Delta r; & \text{for } x_1^2 + y_1^2 < S_{max}^2 \\ 0; & \text{for } x_1^2 + y_1^2 > S_{max}^2 \end{cases} \quad (9.57)$$

where S_{max} is the semidiameter of the aperture. Let us assume that the lateral shear $\lambda rn/d$ is very small compared with the semiaperture S_{max}, so that the summation can be extended to large values of n. To find the amplitude profile of the fringes, we also take $y_1 = 0$; then we can show that

$$G(x_1) = \left[k_0 + 2 \sum_{n=-\infty}^{N} (-1)^n \frac{\sin n\pi k_0}{n\pi} \cos\left(\frac{2\pi n \Delta r}{Rd} x_1\right)\right.$$
$$\left. \times \exp\left(i\frac{\pi n^2 \lambda r \Delta r}{Rd^2}\right)\right] - \left[\exp\left(i\frac{\pi \Delta r}{\lambda Rr} x_1^2\right)\right], \quad (9.58)$$

where N is the number of wavefronts that overlap in the region under study.

9.4.2. Fringe Contrast and Sharpness

Expression (9.58) explains quantitatively the lack of sharpness of the Ronchi fringes, which is identical outside and inside the focus. It has been shown by Malacara (1990) that for small defocusings, which are the ones we may find in practical arrangements, the bright fringes become sharper as the defocusing is increased. This means that the closer the fringes are, the sharper they appear. Thus, for the case of a parabolic mirror, the fringes are sharper around the edge of the pattern for inside of focus patterns, and near the center for outside of focus patterns.

The fringes become very sharp when the phase shift between the first and zero orders in Eq. (9.58) is a multiple M of 2π, yielding

$$M \frac{2Rd^2}{\lambda r(R-r)} = 1. \quad (9.59)$$

9.4. PHYSICAL THEORY

This is the Talbot effect (1836) appearing in the Ronchi test, which has been reported by Malacara and Cornejo (1974b). The defocusing Δr given by this condition is extremely large even when M is 1. This condition is equivalent to having a fringe spacing equal to half the lateral shear S, as follows:

$$\alpha = \frac{\theta_s}{2M}, \tag{9.60}$$

where θ_s is the angular lateral shear between any two consecutive orders and α is the angular separation between the fringes. This effect is illustrated in Fig. 9.18.

We have another autoimaginag condition, but with the contrast reversed, when $M = \frac{1}{2}$, that is, at half the Talbot defocusing. When $M = \frac{1}{4}$ and $M = \frac{3}{4}$, the fringe contrast disappears. This fringe blurring occurs when the fringe spacing is equal to twice the lateral shear.

Usually, the Ronchi test is interpreted using the geometrical approximation. This is valid as long as the lateral shear is small compared with the wavefront diameter. On the other hand the number of fringes over the Ronchi pattern cannot be very large. This means that the fringe spacing cannot become extremely small compared with the lateral shear.

Lau (1948) showed that the autoimaging of gratings and rulings, or Talbot effect, also appears if an extended, incoherent, and periodically modulated light source with the proper period is used. This is done in practice by covering the extended light source with a ruling. In the Ronchi test this type of illumination is frequently used, as shown later in this chapter.

Numerous interferometers, based both in the Talbot and the Lau effects using gratings, have been invented. Lohmann and Silva (1971, 1972), Yokoseki and Susuki (1971a, 1971b), Silva (1972), Mallick and Roblin (1972), Hariharan et al. (1974), and Patorski (1982), Patorski (1984). These interferometers have

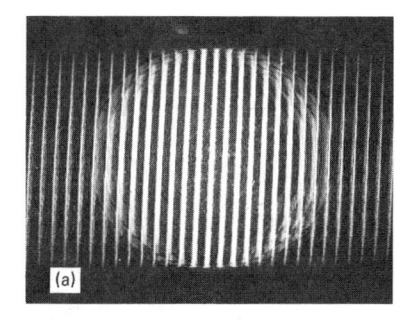

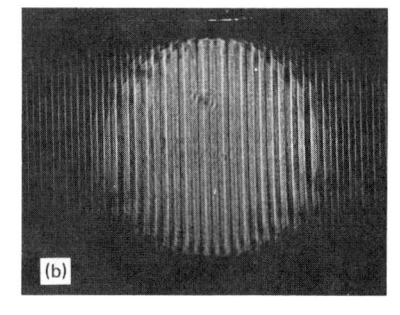

Figure 9.18. Talbot effect in the Ronchi test. (*a*) Ruling at the Talbot effect position. (*b*) Outside it. (From Malacara and Cornejo 1974b.)

been so successful because many applications have been found for them, for example, for measuring focal lengths (Nakano and Murata 1985; Bernardo and Soares 1988). Some interferometers have used a combination of the Talbot effect with Moiré patterns (Glatt and Kafri 1987; Keren et al. 1988).

Experimental attempts have been made to enhance the sharpness of the bands by changing the relative width k_0, shown in Fig. 9.17, of the slits. Some improvements have been obtained (Murty and Cornejo 1973; Cornejo-Rodriguez et al. 1978). Studies about the contrast and sharpness of the observed fringes have been also carried out by Patorski and Cornejo-Rodriguez (1986a), and Malacara (1990).

The phase ruling is one in which, instead of dark and clear bands, there is a periodic change in the optical thickness. Such rulings can be made in many ways, as described by Vogl (1964) and Ronchi (1965) or by any other of the conventional methods to make phase plates. This kind of grating is especially useful when the period is high enough that no more than two wavefronts overlap (Fig. 9.19).

It is interesting that Pallotino (1941) shows experimentally that, with phase gratings, patterns placed symmetrically with the zero order are complementary; in other words, a dark fringe in one pattern corresponds to clear fringe in the other, as in Fig. 9.19. Later, Toraldo di Francia (1941a, 1941b) theoretically proved this effect. The explanation uses the fact that all diffracted orders in a phase ruling have a phase shift of magnitude $m\lambda/4$ with respect to the zero-order beam.

The sensitivity of the Ronchi test has been a subject of research almost since its invention by Ronchi himself (1930, 1940) and Bruscaglioni (1933, 1939). More recently, Cornejo and Malacara (1970) studied the subject.

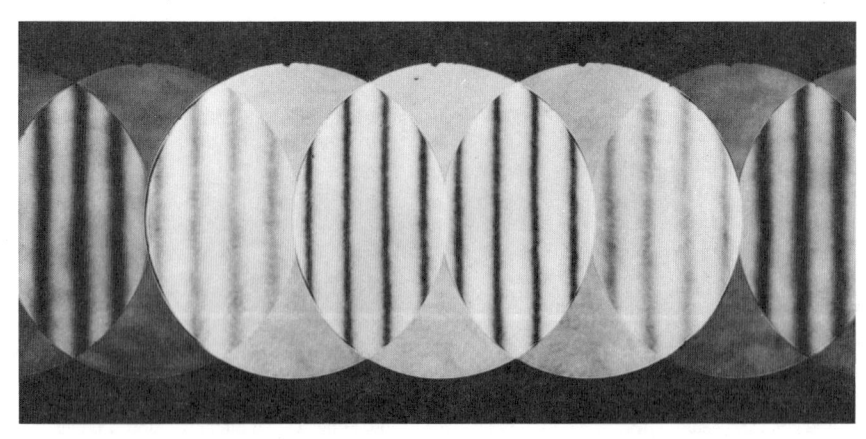

Figure 9.19. Ronchigram with a phase grating.

9.4.3. Physical versus Geometrical Theory

If the irradiance distribution over the exit pupil is a constant, we can write the pupil function $F_0(x_0, y_0)$ as

$$F_0(x_0, y_0) = \exp\left[i\frac{2\pi}{\lambda} W(x_0, y_0)\right], \quad (9.61)$$

where $W(x_0, y_0)$ for the primary aberrations is given by Eq. (9.4).

Let us now assume that the ruling has a period d small enough to produce interference patterns of no more than two overlapping beams, as in Fig. 9.19. Then, from Eq. (9.48), the interference pattern between two consecutive orders n_1 and n_2 can be shown to be given by

$$|G(x_1, y_1)|^2 = B_{n_1}^2 + B_{n_2}^2 + 2B_{n_1}B_{n_2}$$
$$\times \cos\left\{\frac{2\pi}{\lambda}\left[W\left(x_1 + \frac{\lambda r n_1}{d}, y_1\right) - W\left(x_1 + \frac{\lambda r n_2}{d}, y_1\right)\right]\right\}. \quad (9.62)$$

It can be seen here that a bright fringe appears whenever

$$W\left(x_1 + \frac{\lambda r n_2}{d}, y_1\right) - W\left(x_1 + \frac{\lambda r n_1}{d}, y_1\right) = -m\lambda, \quad (9.63)$$

where m is any positive or negative integer, including zero. If we now take the center between the two shears of the wavefront as a new origin, we find

$$W\left[x_1 + \frac{\lambda r(n_2 - n_1)}{2d}, y_1\right] - W\left[x_1 - \frac{\lambda r(n_2 - n_1)}{2d}, y_1\right] = -m\lambda. \quad (9.64)$$

Expanding now in a Taylor series about the origin and taking $n_2 - n_1 = 1$, we have

$$\frac{\partial W(x_1, y_1)}{\partial x_1} + \frac{1}{6}\left(\frac{\lambda r}{2d}\right)^2 \frac{\partial^3 W(x_1, y_1)}{\partial x_1^3} = -\frac{md}{r}. \quad (9.65)$$

When the third derivative term disappears, this expression becomes identical to Eq. (9.3) (with $\varphi = 0°$), which was the basis of the geometrical theory of the Ronchi test. Therefore in such cases the physical and geometrical patterns coincide. This result was derived by Toraldo di Francia (1947, 1954), who pointed out that the third derivative cancels out for focus shifts, and third-order astigmatism and coma when no power of x higher than the second appears [see Eq. (9.4)].

With regard to spherical aberration, we find substituting the expression for spherical aberration with defocusing from Eq. (9.4) in (9.65),

$$4A(x_1^2 + y_1^2)x_1 + 2\left(A\frac{\lambda^2 r^2}{2D^2} + D\right)x_1 = -\frac{md}{r}, \qquad (9.66)$$

which is identical to Eq. (9.10) for the geometrical pattern, except for a difference in the focus shift coefficient. The fusiform of uniform focus is obtained when the apparent focus shift is zero, giving

$$D = -A\frac{\lambda^2 r^2}{2d^2}, \qquad (9.67)$$

but $\lambda r/2d$ is the distance ρ_c from the center of one of the wavefronts to the center of the pattern. Therefore $D = -2A\rho_c^2$, which is the condition necessary to place the focus of the zone with radius ρ_c at the ruling. This effect was studied by Di Jorio (1939d).

9.5. PRACTICAL ASPECTS OF THE RONCHI TEST

The experimental apparatus used to perform the Ronchi test is of many types, as described, for example, by Kirkham (1953). The basic arrangement is shown in Fig. 9.20a. The light source is a white-light tungsten lamp illuminating a pinhole or a slit parallel to the ruling lines. However, De Vany (1974) used a ball bearing as a source, and Patorski and Cornejo-Rodriguez (1986b) used sunlight. The advantage of a slit is that more light is used, producing a brighter pattern. The justification for using a slit is that any point source on the slit produces the same pattern because the corresponding images fall on the ruling displaced only along the grating lines, with no lateral shifts. In the beginning a slit source was used for the test, but a little later Anderson and Porter (1929) suggested allowing the grating to extend over the lamp, as in Fig. 9.20b, instead of employing a slit source. A significant advantage, in addition to the greater luminosity of the pattern, is that there is no need to worry about the parallelism between the ruling lines and the slit. The light source is then a multiple slit, and the images of the slits are separated by a distance equal to the ruling period. This property justifies the use of the ruling over the light source, which may also be justified, however, in a more formal way as follows.

The well-known Van Cittert–Zernike theorem establishes that the coherence function of a source is given by the Fourier transform of the intensity distri-

9.5. PRACTICAL ASPECTS OF THE RONCHI TEST 351

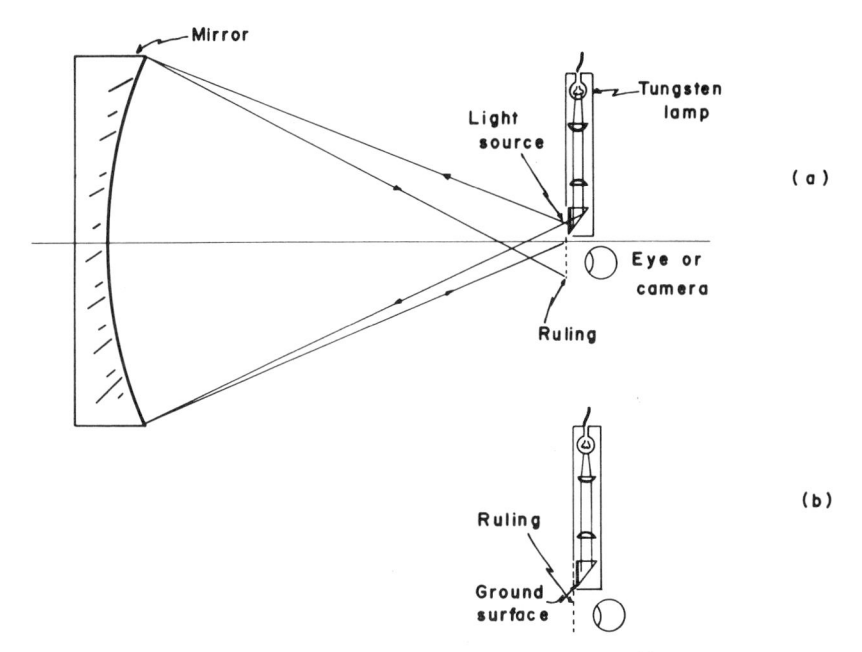

Figure 9.20. Diagram of an instrument to observe Ronchi patterns.

bution at the light source. Since the light source in this case is a ruling, the Fourier transform is as shown in Fig. 9.21. Then, if two points over the wavefront are separated by the distance Δx between any two peaks in the wavefront, they are capable of producing interference fringes with good contrast. If the ruling on the light source and the examining Ronchi ruling are in the same plane, the distance between the peaks on the coherence function is just the lateral shear length between the diffracted wavefronts. This result completely justifies the use of the Ronchi ruling covering the extended light source.

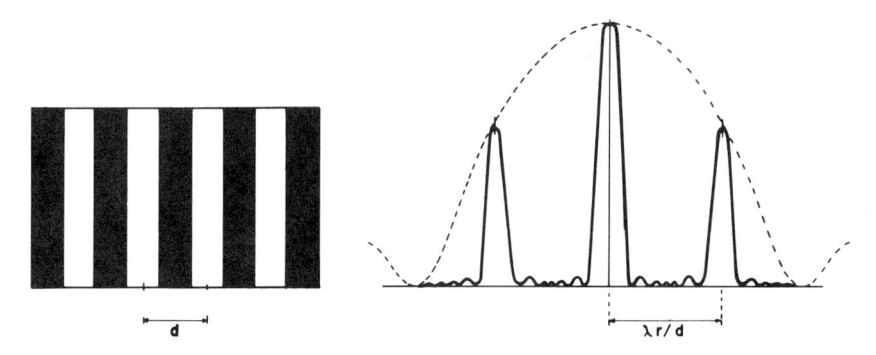

Figure 9.21. Ronchi ruling and its coherence function over the mirror being examined.

It is interesting that, when the dark and clear bands on the ruling have the same width ($k_0 = \frac{1}{2}$), there is no coherence between points on the wavefront separated by a distance Δx equal to $2\lambda r/d$. Therefore both first-order wavefronts interfere with the zero order but do not interfere between themselves. However, when using a point source or slit, all overlapping wavefronts interfere with each other.

The null Ronchi ruling described in Section 9.2.3 does not produce the same lateral shear for all points of the wavefront; therefore the illuminating wavefront must have coherence between any two points at any distance between them. This is possible only if a point source is used. It must be kept in mind that a gas laser is the optimum light source because of its very high coherence.

Some Ronchigrams taken with the instrument in Fig. 9.22 are shown in Fig. 9.23. Although obvious from all past discussions, it should be pointed out once more that the shape of the Ronchi fringes is not directly related to the shape of the surface but rather is related to the transverse aberration function. The photographs were taken with a normal photographic camera focused on the surface plane, because the fringes we want to study are precisely in that plane.

When testing an aspheric mirror, care should be taken that the transverse dimensions of the caustic (see Appendix I) are small enough so that all the light goes through the observing eye or camera. Many times the waist of the caustic is so big (i.e., in large telescope mirrors) that it cannot pass through the pupil of the eye, although it can pass through the lens of a camera.

Another point of great practical interest is that the light passing through each slit on the Ronchi ruling interferes with that passing through the other slits, producing an effect of many laterally sheared pupils. This effect reduces the

Figure 9.22. Instrument to observe Ronchi patterns.

9.6. SOME RELATED TESTS 353

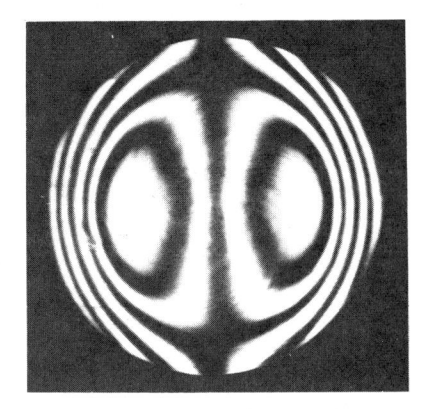

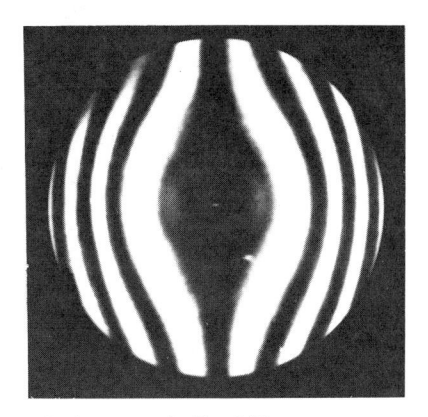

Figure 9.23. Ronchigrams taken with the instrument in Fig. 9.22.

accuracy of the test since the Ronchi pattern cannot be precisely located over the pupil, simply because there are many interfering images of it. An alternative method for improving the accuracy is to use a wire instead of the Ronchi ruling. The wire position is laterally displaced from the optical axis, and such displacement is analogous to the ruling period. The distance of the wire from the optical axis should be accurately measured.

If the Ronchi test is done with care and the patterns are correctly interpreted, the test is a very powerful tool to measure quantitatively or qualitatively the degree of corrections needed in an optical surface or lens.

New instruments to perform the Ronchi test and gratings have been continuously developed. For example, an instrument based on the Ronchi test was reported by Kamalov et al. (1980). The manufacturing of Ronchi gratings has been treated by Kuindzhi et al. (1980), Patorski (1980), Thompson (1987), and Steig (1987).

9.6. SOME RELATED TESTS

9.6.1. Concentric Circular Grid

Instead of straight lines on the ruling, it is also possible to use a grid with circular lines. This modification was studied originally by Scandone (1931c, 1932) and more recently by Murty and Shoemaker (1966), whose treatment we present here. Patterns with this kind of ruling differ from the normal Ronchi patterns. In this case the equation of the fringes can be obtained in the same manner as in Eq. (9.3), giving

$$\left(R\frac{\partial W}{\partial x} - \bar{x}_0\right)^2 + \left(R\frac{\partial W}{\partial y} - \bar{y}_0\right)^2 = M\rho_1^2, \qquad (9.68)$$

where $(\bar{x}_0, \bar{y}_0)$ is the center of the set of concentric circles ρ_1 is the radius of the innermost circle, and M is an integral number defined by

$$M = \begin{cases} n^2 & \text{for equally spaced circles,} \\ n & \text{for a Fresnel zone plate,} \end{cases}$$

n being a positive integer. If we use the wavefront W for primary aberrations as in Eq. (9.4), we find

$$\{R[4A(x^2 + y^2)x + 2Bx + 2Cx + 2Dx + F] - \bar{x}_0\}^2$$
$$+ \{R[4A(x^2 + y^2)y + B(3y^2 + x^2)$$
$$+ 6Cy + 2Dy + E] - \bar{y}_0\}^2 = M\rho_1^2. \tag{9.69}$$

As pointed out by Murty and Shoemaker, this expression becomes extremely complicated, except for a few trivial cases. Some typical patterns are shown in the paper by the same authors.

It is important to point out that this test must be performed with a point source, and that a slit cannot be used.

9.6.2. Phase Shifting Ronchi Test

This class of test was proposed by Thompson (1973), under the name AC grating interferometer, to test laser wavefronts. It is basically a dynamic Ronchi test in which a rapid treatment of the data can be made by means of a minicomputer.

The fringe pattern is not stationary because the ruling is uniformly translated over its own plane, in the direction perpendicular to the ruling slits. Then, instead of photographing a stationary pattern, the moving fringes are recorded by a photoelectric detector such as a silicon diode array of a vidicon tube.

Each individual detector of the array gives an electrical periodical output, all with the same frequency, depending on the velocity of the moving Ronchi ruling. The relative phase of the signals from the detectors depends on the position of the detector over the wavefront and also on the aberrations W. A precise measurement of these phases is enough to reconstruct the fringe pattern and from it the wavefront.

To have a moving Ronchi ruling, Thompson (1973) suggested using a drum ruling with a circumference of about 40 cm. The slight curvature of the ruling is not very important.

Recently, phase shifting Ronchi tests have been described by Koliopoulos (1980), Yatagai (1984), Omura and Yatagai (1988), and Wan and Lin (1990).

9.6. SOME RELATED TESTS 355

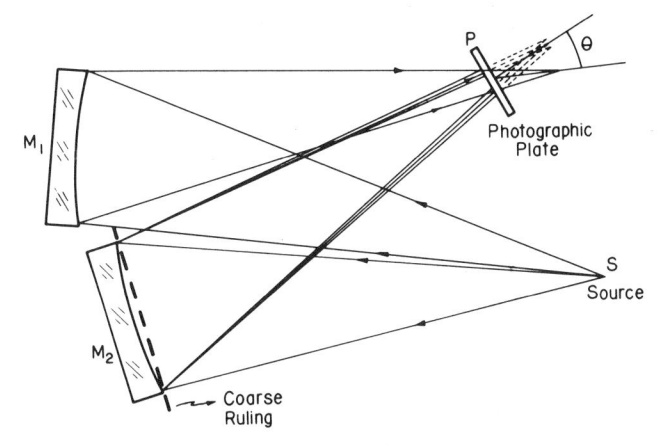

Figure 9.24. Setup to make grating for side band Ronchi test.

9.6.3. Sideband Ronchi Test

This is essentially a holographic test, as devised by Malacara and Cornejo (1976a). The method requires a special ruling (or hologram) that is fabricated as illustrated in Fig. 9.24, where two concave mirrors reflect the light of a point source S over a common area on a photographic plate P. In front of mirror M_2 is placed a very coarse ruling with straight bands. The aim is to make a hologram in order to reconstruct the image of M_2 with the coarse ruling in front of it, using the light from M_1 as a reference beam.

Once the hologram is developed, it is replaced in the beam from M_1, but this time without mirror M_2. If the mirror M_1 used during the reconstruction is identical to the mirror M_1 employed in forming the hologram, a set of straight fringes will be seen where mirror M_2 was. If the mirror is different, the observed fringes will not be straight, indicating a wave aberration given by twice the difference between the two mirror surfaces M_1.

As shown in Fig. 9.25, the main advantage of this method is that the bands

Figure 9.25. Fringes in sideband Ronchi test. (From Malacara and Cornejo 1976a.)

are very sharp and well defined, unlike those in the normal Ronchi test. Further studies and modifications on the sideband Ronchi test have been reported by Malacara and Josse (1978), Schwider (1981), and Patorski and Salbut (1984).

9.6.4. Lower Test

This test was devised by Lower (1937), who pointed out that it slightly resembles an inverted Ronchi test. The normal Ronchi test cannot be used to test a parabolic mirror with a small f ratio, and for obvious reasons it would be even more difficult to test a fast Schmidt camera. Originally Lower used his test to measure the degree of correction of a fast Schmidt camera. The test consists in placing a slit at the focus of the system under test and viewing image (at infinity) from a relatively large distance. This image may be considered as being localized on the aperture of the system, and it is a straight line when there is good correction. The straightness of the line remains as the eye traverses the aperture.

This test is very rapid and simple, but for many practical reasons, like the finite observing distance, the accuracy is very limited. The factors that limit the sensitivity and accuracy of the Lower test have been analyzed by Rank et al. (1949) and Yoder (1959). A related test for paraboloidal reflectors, using a similar principle but a screen line, was developed by Hamsher (1946).

9.6.5. Ronchi-Hartmann and Null Hartmann Tests

Even though the Ronchi and Hartmann tests have been studied separately, Cordero et al. (1990) have shown the existence of a close relation between both tests. In order to prove that relation, it was shown that for the two tests the screen or filter plane can be exchanged with the plane where the pattern is observed or recorded; thus one can go from one to the other test. The physical

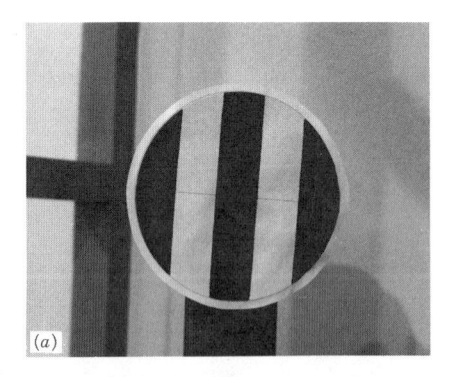

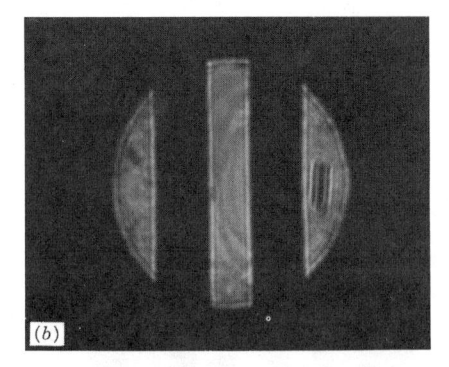

Figure 9.26. Ronchi Hartmann null test. (*a*) Ronchi-Hartmann screen in front of the surface under test. (*b*) Observed null fringe pattern. (From Cordero, et. al. 1990.)

9.6. SOME RELATED TESTS

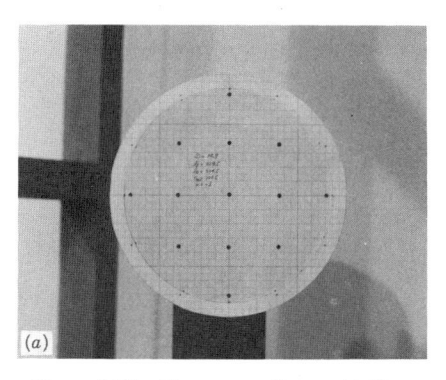

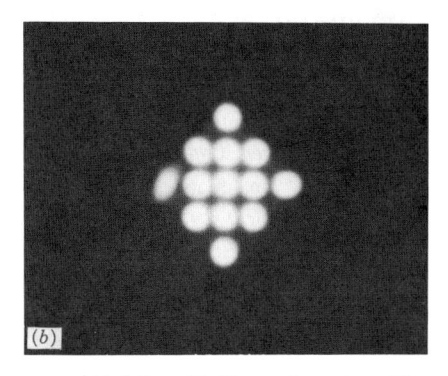

Figure 9.27. Hartmann null test. (*a*) Hartmann screen with holes. (*b*) Observed pattern, with spots along straight lines (null pattern) (From Cordero, et. al. 1990.)

basis for this result rests in the geometrical optics concept of propagation of light between both planes. Using the mathematical development of Section 9.2.2, an ideal Ronchigram (Fig. 9.26*a*) was calculated and used as a one-dimensional Hartmann screen in front of the surface. The observed pattern is a null one because the new Ronchi–Hartmann screen compensates the asphericity of the surface under test. In Fig. 9.26*b* the observed null pattern is shown.

Having as a main aim to develop a null Hartmann test, and after the relation between the Ronchi and Hartmann test was established, the new Hartmann screen for a null Hartmann test was obtained by locating the holes on the screen, as the crossing points of two perpendicular ideal Ronchigrams. In Figs. 9.27*a* and 9.27*b* the screen and observed Hartmanngram are shown for the so-called Null Hartmann test. Recently a common mathematical theory for the Ronchi and Hartmann tests was developed by Cordero, et. al., (1991) which consider the results described in this section.

I do not want to close this chapter without mentioning the scarcity of citations to works translated from languages other than English, French, German, or Italian. Unfortunately, publications in other languages have not been very accessible to me.

Acknowledgments. I am indebted to Dr. Daniel Malacara for many stimulating and enlightening conversations, and to Dr. Robert H. Noble for valuable criticism. I am grateful also to the Instituto Nacional de Astrofísica, Optica y Electrónica for the support given during the preparation of this work, especially to Dr. G. Haro, and the following staff members: Eliezer Jara, José Castro, Ellen Noble, Selma Campos, and Luz Maria Olmos. At the time of this second edition, the collaboration of J. Pedraza C., O. Cardona N., A. Cordero D., and R. Díaz U. has been very valuable.

REFERENCES

Adachi, I., "Quantitative Measurement of Aberration by Ronchi Test," *Atti. Fond. Giorgio Ronchi Contrib. Ist. Naz. Ottica*, **15**, 461-483 (1960a).

Adachi, I., "Quantitative Measurement of Aberration by Ronchi Test," *Atti. Fond. Giorgio Contrib. Ist. Naz. Ottica*, **15**, 550-585 (1960b).

Adachi, I., "The Recent History of the Grating Interferometer and Its Applications," *Atti. Fond. Giorgio Ronchi Contrib. Ist. Naz. Ottica*, **17**, 252-259 (1962a).

Adachi, I., "Measurement of Transfer Function of the Ronchi Test," *Atti. Fond. Giorgio Ronchi Contrib. Ist. Naz. Ottica*, **17**, 523-534 (1962b).

Adachi, I., "The Diffraction Theory of the Ronchi Test," *Atti. Fond. Giorgio Ronchi Contrib. Ist. Naz. Ottica*, **18**, 344-349 (1963).

Anderson, J. A. and R. W. Porter, "Ronchi's Method of Optical Testing," *Astrophys. J.*, **70**, 175-181 (1929).

Assa, A., A. A. Betseer, and J. Politch, "Recording Slope and Curvature Contours of Flexed Plates Using a Grating Shearing Interferometer," *Appl. Opt.*, **16**, 2504-2513 (1977).

Barakat, R., "General Diffraction Theory of Optical Aberration Tests from the Point of View of Spatial Filtering," *J. Opt. Soc. Am.*, **59**, 1432-1439 (1969).

Bernardo, L. M. and O. D. Soares, "Evaluation of the Focal Distance of a Lens by Talbot Interferometry," *Appl. Opt.*, **27**, 296-301 (1988).

Bocchino, G., "L'Aberrazione Sferica Zonale Esaminata con i Reticoli a Bassa Frequenza," *Ottica*, **5**, 286 (1940).

Bocchino, G., "Un Metodo per la Determinazione Rapida e Precise dell' Aberrazione Sferica Semplice, Mediante la Frange d'Ombra," *Ottica*, **8**, 310 (1943).

Briers, J. D., "Interferometric Testing of Optical Systems and Components: A Review," *Opt. Laser Technol.*, **4**, 28-41 (1972).

Briers, J. D., "Ronchi Test Formulae 1: Theory," *Opt. Laser Technol.*, **11**, 189-196 (1979).

Briers, J. D. and M. J. Cochrane, "Ronchi Test Formulae 2: Practical Formulae and Experimental Verification," *Opt. Laser Technol.*, **11**, 245-257, (1979).

Brookman, E. C., L. D. Dickson, and R. S. Fortenberry, "Generalization of the Ronchi Ruling Method for Measuring Gaussian Beam Diameter," *Opt. Eng.*, **22**, 643-647 (1983).

Bruscaglioni, R., "Sulla Forma delle Fringe d'Interferenza Ottenute da Onde Affete da Astigmatismo Puro con Reticoli ad Orientamento Qualunque," *Rend. Accad. Naz. Lincei.*, **15**, 70 (1932a).

Bruscaglioni, R., "Sulla Misura dell'Astigmatismo e del Coma Mediante le Frange d'Ombra," *Boll. Assoc. Atti. Fond. Giorgio Ronchi Contrib. Ist. Naz. Ottica Ital.*, **6**, 46 (1932b).

Bruscaglioni, R., "Sulla Sensibilita della Rivelazione e Sulla Misura dell'Astigmatismo con Metodi Interferenziali," *Boll. Assoc. Atti. Fond. Giorgio Ronchi Contrib. Ist. Naz. Ottica Ital.*, **7**, 78 (1933).

Bruscaglioni, R., "Controllo della Afocalita di una Parte e Controllo di un Piano Campione con l'Interferometro Ronchi a Reticolo," *Ottica*, **4**, 203 (1939).

Calamai, G., "Su di una Formula per la Misura dell'Astigmatismo Mediante i Reticoli," *Ottica*, **3**, 41 (1938).

Cordero-Davila, A., A. Cornejo Rodriguez, and O. Cardona-Nuñez, "Null Hartmann and Ronchi Hartmann Tests," *Appl. Opt.*, **29**, 4618 (1990).

Cordero-Davila, A., A. Cornejo-Rodriguez, O. Cardona-Nuñez, "Common Mathematical Theory for the Ronchi and Hartmann Test," *Appl. Opt.*, to be published.

Cornejo-Rodriguez, A., "The Ronchi Test for Aspherical Optical Surfaces," *Kogaku*, **12**, 278 (1983).

Cornejo, A., and D. Malacara, "Ronchi Test of Aspherical Surfaces: Analysis and Accuracy," *Appl. Opt.*, **9**, 1897-1901 (1970).

Cornejo-Rodriguez, A. and D. Malacara, "Wavefront Determination Using Ronchi and Hartmann Tests," *Bol. Int. Tonantzintla*, **2**, 127-129 (1976).

Cornejo-Rodriguez, A., H. Altamirano, and M. V. R. K. Murty, "Experimental Results in the Sharpening of the Fringes in the Ronchi Test," *Bol. Inst. Tonantzintla*, **2**, 313-315 (1978).

Crino, B., "Sulla Misura dell'Aberrazione Sferica, Coma e Astigmatismo Mediante la Frange d'Ombra Estrassiale Ottenute con Reticoli Rettilinei," *Boll. Assoc. Ottica Ital.*, **7**, 113 (1933).

Crino, B., "Nuvoi Risultati Nello Studio Analitico delle Frange d'Ombra Ottenute per Interferenza di Onde Aberranti," d'Ombra Ottenute per Interferenza di Onde Aberranti," *Ottica*, **4**, 114 (1939).

De Vany, A. S., "Some Aspects of Interferometric Testing and Optical Figuring," *Appl. Opt.*, **4**, 831-833 (1965).

De Vany, A. S., "Quasi-Ronchigrams as Mirror Transitive Images of Interferograms," *Appl. Opt.*, **9**, 1844-1945 (1970).

De Vany, A. S., "Laser Illuminated Divergent Ball Bearing Sources," *Appl. Opt.*, **13**, 457-459 (1974).

De Vany, A. S., "Profiling pitch polishers" *Appl. Opt.*, **17**, 3022-3024 (1978).

De Vany, A. S., "Interpreting Wave-Front and Glass-Error Slopes in an Interferogram," *Appl. Opt.*, **19**, 173-173 (1980).

De Vany, A. S., "Patterns of Correlation of Interferograms and Ronchigrams," *Appl. Opt.*, **20**, A40-A41 (1981).

Di Jorio, M., "Ulteriore Approssimazione dello Studio delle Aberrazioni con l'Interferometro Ronchi a Reticolo," *Ottica*, **4**, 31 (1939a).

Di Jorio, M., "Una Formula Più Precisa delle Frange d'Ombra dell'Interferometro Ronchi a Reticolo," *Ottica*, **4**, 83 (1939b).

Di Jorio, M., "L'Aberrazione Sferica Esaminata con i Reticoli di Alta Frequenza," *Ottica*, **4**, 184 (1939c).

Di Jorio, M., "Estensione del Concetto del Fuoco Uniforme: il Fuoco Uniforme Zonale," *Ottica*, **4**, 254 (1939d).

Di Jorio, M., "Similitudine Degli Interferogrammi dell'Interferometro Ronchi al Variare x_p e Verifica Sperimentale della Costansa del numero b," *Ottica*, **7**, 243 (1942).

Di Iorio, M., "Equazione dell'Interferometro Ronchi per le Onde Sferiche Aperte Fino al Quarto Ordine, e sua Discusione," *Ottica*, **8**, 288 (1943).

Erdös, P., "Ronchi Test of Fifth Order Aberrations," *J. Opt. Soc. Am.*, **49**, 865–868 (1959).

Glatt, I. and O. Kafri, "Determination of the Focal Length of Nonparaxial Lenses by Moire Deflectometry," *Appl. Opt.*, **26**, 2507–2508 (1987).

Hamsher, D. H., "Screen Line Tests of Paraboloidal Reflectors," *J. Opt. Soc. Am.*, **36**, 291–295 (1946).

Hariharan, P., W. H. Steel, and J. C. Wyant, "Doubling Grating Interferometer with Variable Lateral Shear," *Opt. Comm.*, **11**, 317 (1974).

Hopkins, G. H. and R. H. Shagan, "Null Ronchi Gratings from Spot Diagram," *Appl. Opt.*, **16**, 2602–2603 (1977).

Jentzsch, F., "Die Rastermethode: Ein Verfahren zur Demonstration und Messung der Spharischen Aberration," *Phys. Z.*, **24**, 66 (1928).

Kamalov, J. A., V. A. Komissaruk, and N. P. Mende, "Preparation of the IAB-451 Instrument for Operation as a Diffraction Interferometer," *Sov. J. Opt. Technol.*, **47**, 249–250 (1980).

Kasana, R. S., S. Boseck, and K. J. Rosenbrich, "Non-Destructive Collimation Technique for Measuring Glass Constants Using a Ronchi Grating Shearing Interferometer," *Opt. Laser Technol.*, **16**, 101–105 (1984).

Keren, E., K. M. Kreske, and O. Kafri, "Universal Method for Determining the Focal Length of Optical Systems by Moirè Deflectometry," *Appl. Opt.*, **27**, 1383–1385 (1988).

Kessler, D. and R. V. Shack, "Dynamic Optical Tests of a High-Speed Polygon," *Appl. Opt.*, **20**, 1015–1019 (1981).

King, J. H., "Quantitative Optical Test for Telescope Mirrors and Lenses," *J. Opt. Soc. Am.*, **24**, 250 (1934). Reprinted in *Amateur Telescope Making*, Vol. 2, A. G. Ingalls, Ed., Scientific American, New York, 1953, p. 104.

Kirkham, A. R., "The Ronchi Test for Mirrors," in *Amateur Telescope Making*, Vol. 1, A. G. Ingalls, Ed., Scientific American, New York, 1953, p. 264.

Koliopoulus, C. L., "Radial Grating Lateral Shear Heterodyne Interferometer," *Appl. Opt.*, **19**, 1523–1528 (1980).

Kuindzhi, V. V., S. A. Strezhnev, and M. T. Popov, "Gratings for Diffraction Interferometer," *Sov. J. Opt. Technol.*, **47**, 240–241 (1980).

Lau, E., "Bengungserscheinungen an Deppeltrastern," *Ann. Phys.*, **6**, 417 (1948).

Lenouvel, M. L., "Methode de Determination et de Mesure des Aberrations des Systemes Optiques," *Rev. Opt.*, **3**, 211 (1924a).

Lenouvel, M. L., "Methode de Determination et de Mesure des Aberrations des Systemes Optiques," *Rev. Opt.*, **3**, 315 (1924b).

Lenouvel, M. L., "Etude des Objectifs de Reproduction," *Rev. Opt.*, **4**, 294 (1925a).

Lenouvel, M. L., "Essai d'Objectifs par le Coin d'Air," *Rev. Opt.*, **4**, 299 (1925b).

Lin, J. A., J. Hsu, and S. A. Shire, "Quantitative Three Beam Ronchi Test," *Appl. Opt.*, **29**, 1912 (1990).

Lohmann, A. W. and D. E. Silva, "An Interferometer Based on the Talbot Effect," *Opt. Comm.*, **2,** 413 (1971).

Lohmann, A. W. and D. E. Silva, "A Talbot Interferometer with Circular Gratings," *Opt. Comm.*, **4,** 926 (1972).

Lower, H. A., "Notes on the Construction of an F/1 Schmidt Camera," in *Amateur Telescope Making*, Vol. 2, A. G. Ingalls, Ed., Scientific American, New York, 1954, p. 410; formerly published by Munn and Co., 1937.

Lumley, E., "A Method of Making a Ronchi Test on an Aspheric Mirror," in *Amateur Astronomers*, Sidney, 1959; reprinted in *Atti. Fond. Giorgio Ronchi Contrib. Ist. Naz. Ottica*, **15,** 457 (1960).

Lumley, E., "Figuring a Paraboloid with the Ronchi Test," *Sky Telesc.*, **22,** 298 (1961).

Malacara, D., Testing of Optical Surfaces, Ph.D. Thesis, University of Rochester, New York, University Microfilms, Ann Arbor, Mich., Order No. 65-12,013, 1965a.

Malacara, D., "Geometrical Ronchi Test of Aspherical Mirrors," *Appl. Opt.*, **4,** 1371–1374 (1965b).

Malacara, D., "Ronchi Test and Transversal Spherical Aberration," *Bol. Obs. Tonanzintla Tacubaya.*, **4,** 73 (1965c).

Malacara, D., "Analysis of the Interferometric Ronchi Test," *Appl. Opt.*, **29,** 3633 (1990).

Malacara, D. and A. Cornejo, "Modified Ronchi Test to Measure the Axial Chromatic Aberration in Lenses," *Appl. Opt.*, **10,** 679–680 (1971).

Malacara, D. and A. Cornejo, "Relating the Ronchi and Lateral Shearing Interferometer Tests," *Opt. Spectra.*, **8,** (9), 54–55 (1974a).

Malacara, D. and A. Cornejo, "The Talbot Effect in the Ronchi Test," *Bol. Inst. Tonantzintla*, **1,** 193–196 (1974b).

Malacara, D. and A. Cornejo, "Null Ronchi Test for Aspherical Surfaces," *Appl. Opt.*, **13,** 1778–1780 (1974c).

Malacara, D. and A. Cornejo, "Shape Measurement of Optical Systems with Rotational Symmetry Using Ronchigrams," *Bol. Inst. Tonantzintla*, **1,** 277–283 (1975).

Malacara, D. and A. Cornejo, "Side Band Ronchi Test," *Appl. Opt.*, **15,** 2220–2222 (1976a).

Malacara, D. and A. Cornejo, "Third Order Computation of Null Ronchi Rulings," *Bol. Inst. Tonantzintla*, **2,** 91–92, (1976b).

Malacara, D. and M. Josse, "Testing of Aspherical Lenses Using Side Band Ronchi Test," *Appl. Opt.*, **17,** 17–18 (1978).

Mallick, S. and M. L. Roblin, "Coherent Imaging in Presence of Defect of Focus," *Trans. Opt. Soc. Am.*, **64,** 1944 (1972).

Martin, L. C., *Technical Optics*, Vol. 2, Pitman, London, 1960, Chap. VII.

Mobsby, E., "Testing Parabolic Mirrors with Inverse Parabolic Grating," *Astronomy: J. Wessex Astron. Soc.*, **1,** 13 (1973).

Mobsby, E., "A Ronchi Null Test for Parabolids," *Sky Telesc.*, **48,** 325–330 (1974).

Morais, C., "Riassunto delle Applicazioni dei Reticoli allo Studio delle Aberrazione dei Sistemi Ottici," *Atti. Fond. Giorgio Ronchi Contrib. Ist. Naz. Ottica*, **13,** 546–602 (1958).

Murty, M. V. R. K., "Interferometry Applied to Testing of Optics," *Bull. Opt. Soc. India*, **1**, 29-37 (1967).

Murty, M. V. R. K., "A Simple Method of Introducing Tilt in the Ronchi and Cube Type of Shearing Interferometers," *Bull. Opt. Soc. India*, **5**, 1-5 (1971).

Murty, M. V. R. K. and A. Cornejo, "Sharpening the Fringes in the Ronchi Test," *Appl. Opt.*, **12**, 2230-2231 (1973).

Murty, M. V. R. K. and A. H. Shoemaker, "Theory of Concentric Circular Grid," *Appl. Opt.*, **5**, 323-326 (1966).

Nakano, Y. and K. Murata, "Talbot Interferometry for Measuring the Focal Length of a Lens," *Appl. Opt.*, **24**, 3162-3166 (1985).

Omura, K. and T. Yatagai, "Phase Measuring Ronchi Test," *Appl. Opt.*, **27**, 523-528 (1988).

Pacella, G. B., "Sulla Ricerca della Forma delle Onde Luminose dall'Esame delle Frange d'Ombra," *Rend. Accad. Naz. Lincei.*, **5**, 752 (1927).

Pallotino, P., "Sulla Dissimmetria delle Frange Dell'Interferometro Ronchi a Reticolo," *Ottica*, **6**, 26 (1941).

Pastor, J., "Hologram Interferometry and Optical Technology," *Appl. Opt.*, **8**, 525-531 (1969).

Patorski, K., "Measurements of the Wavefront Curvature of Small Diameter Laser Beams Using the Fourier Imaging Phenomenon," *Opt. Laser Technol.*, **11**, 91 (1979).

Pastorski, K., "Production of Binary Amplitude Gratings with Arbitrary Opening," *Opt. Laser Technol.*, **12**, 267-270 (1980).

Patorski, K., "Periodic Source Ronchi-Talbot Shearing Interferometer," *Optik*, **62**, 207-210 (1982).

Patorski, K., "Heuristic Explanation of Grating Shearing Interferometry Using Incoherent Illumination," *Opt. Acta*, **31**, 33-38 (1984).

Patorski, K., "Grating Shearing Interferometer with Variable Shear and Fringe Orientation," *Appl. Opt.*, **25**, 4192-4198 (1986).

Patorski, K. and A. Cornejo-Rodriguez, "Fringe Contrast Interpretation for an Extended Source Ronchi Test," *Appl. Opt.*, **25**, 2790-2795 (1986a).

Patorski, K. and A. Cornejo-Rodriguez, "Ronchi Test with Daylight Illumination," *Appl. Opt.*, **25**, 2031-2032 (1986b).

Patorski, K. and L. Salbut, "Reversed Path Ronchi Test," *Optica Appl.*, **15**, 261 (1984).

Phillips, F. W., "Aspherizing and Other Problems in Making Maksutov Telescopes," *Sky Telesc.*, **25**, 110 (1963).

Popov, G. M., "Methods of Calculation and Testing of Ritchey-Chrêtien Systems," *Izv. Krym. Astrofiz, Obs.*, **45**, 188 (1972).

Porter, R. W., "Notes on the Ronchi Band Patterns," in *Amateur Telescope Making*, Vol. 1, A. G. Ingalls, Ed., Scientific American, New York, 1953, p. 268.

Rank, D. H., P. R. Yoder Jr., and J. Vrabel, "Sensitivity of a Rapid Test for High Speed Parabolic Mirrors," *J. Opt. Soc. Am.*, **39**, 36-38 (1949).

Rayces, J. L., "Exact Relation Between Wave Aberration and Ray Abberation," *Opt. Acta*, **11**, 85-88 (1964).

Ricci, E., "Nuovi Criteri per la Misura dell'Aberrazione Sferice di un Sistema Ottico col. Metodo delle Frange d'Ombra," *Ottica,* **4,** 104 (1939).

Ronchi, V., "Le Frange di Combinazioni Nello Studio delle Superficie e dei Sistemi Ottici," *Riv. Ottica Mecc. Precis.*, **2,** 9 (1923a).

Ronchi, V., "Due Nuovi Metodi per lo Studio delle Superficie e dei Sistemi Ottici," *Ann. Sc. Norm. Super Pisa.*, **15,** (1923b).

Ronchi, V., "Sopra la Caratteristiche dei Cannocchiali di Galileo e la Loro Autenticita," *Rend. Accad. Naz. Lincei.*, **2,** 162 (1923c).

Ronchi, V., "Sopra i Cannocchiali di Galileo e Sopra una Lente di Evangelista Torricelli," *L'Universo*, **4,** 10 (1923d).

Ronchi, V., "Sullo Studio dei Sistemi Ottici col Biprisma e Gli Specchi del," *Ren. Accad. Naz. Lincei.*, **3,** 314 (1924a).

Ronchi, V., "Ancora sull'Impiego dei Reticoli Nelo Studio dei Sistemi Ottici," *Nuovo Cimento*, **1,** 209 (1924b).

Ronchi, V., *La Prova dei Sistemi ottici*, Zanichelli, Bologna, 1925.

Ronchi, V., "Sur la Nature Interferentielle des Franges d'Ombres dans l'Essai des Sistemes Optiques," *Rev. Opt.*, **5,** 441 (1926a).

Ronchi, V., "Uber die Schattenstreifen zum Studium der Lichtwellen," *Zt. Instrumentenkd*, **46,** 553 (1926b).

Ronchi, V., "Sul Comportamento e l'Impiengo delle Frange d'Ombra Nella Prova dei Systemes Optiques," *Nuovo Cimento*, **4,** 297 (1927).

Ronchi, V., "Sul Comportamento e l'Impiengo delle Frange d'Ombra nella Prova dei Systemes Optiques," *Rev. Opt.*, **7,** 49 (1928).

Ronchi, V., "Le Frange d'Ombra Nello Studio delle Aberrazioni Spheriche," *Rend. Accad. Naz. Lincei.*, **11,** 998 (1930).

Ronchi, V., "Sulla Sensibilita delle Frange d'Ombra all Aberrazione Sferica," *Ottica*, **5,** 275 (1940).

Ronchi, V., *Corso di Ottica Tecnica*, 2nd ed. Associacione Ottica Italiana, Firenze, 1954.

Ronchi, V., "An Elementary Introduction to the Use of the Grating Interferometer," *Atti. Fond. Giorgio Ronchi Contrib. Ist. Naz. Ottica*, **13,** 368–403 (1958).

Ronchi, V., "Forty Years of Gratings," *Atti. Fond. Giorgio Ronchi Contrib. Ist. Naz. Ottica*, **17,** 93–143 (1962a).

Ronchi, V., "Forty Years of Gratings," *Atti. Fond. Giorgio Ronchi Contrib. Ist. Naz. Ottica*, **17,** 240–251 (1962b).

Ronchi, V., "Forty Years of History of a Grating Interferometer," *Appl. Opt.*, **3,** 437–451 (1964).

Ronchi, V., "On the Phase Grating Interferometer," *Appl. Opt.*, **4,** 1041–1042 (1965).

Rosenbruch, K. J., "Testing of Optical Components and Systems," *J. Optics (India)*, **14,** 23–43 (1985).

Salzmann, H., "A Simple Interferometer Based on the Ronchi Test," *Appl. Opt.*, **9,** 1943–1944 (1970).

Scandone, F., "Sulla Forma delle Frange d'Ombra Dovute ad Onde Luminose Affette da Aberrazione Zonale," *Nuovo Cimento*, **7**, 289 (1930).

Scandone, F., "Sulla Forma delle Frange d'Ombra Dovute ad Onde Luminose Affette de Aberrazione Estrassiale," *Nuovo Cimento*, **8**, 157 (1931a).

Scandone, F., "Sulla Forma delle Frange d'Ombra Estrassiale Ottenute con un Reticolo a Tratti Inclinati sul Piano de Simmetria del Sistema Ottico," *Nuovo Cimento*, **8**, 310 (1931b).

Scandone, F., "Sulla Forma delle Frange d'Ombra Ottenute con Reticolo Circolari a Frequenza Costante," *Nuovo Cimento*, **8**, 378 (1931c).

Scandone, F., "Sulla Forma delle Frange d'Ombra Ottenute con Reticolo Circolari a Frequenza Costante non Centrato sull'Asse Ottico," *Boll. Assoc. Ottica Ital.*, **6**, 35 (1932).

Scandone, F., "Sulla Frange d'Ombra Estrassiali Ottenute con Reticoli in Presenza di Aberrazione Sferica sull'Asse," *Boll. Assoc. Ottica Ital.*, **7**, 100 (1933).

Schulz, L. G., "Uber die Prufung Optischer Systeme mit Rastern," *Ann. Phys.*, **35**, 189 (1928).

Schulz, L. G., "Preparation of Aspherical Refracting Optical Surfaces by an Evaporation Technique," *J. Opt. Soc. Am.*, **38**, 432–441 (1948).

Schwider, J., "Single Sideband Ronchi Test," *Appl. Opt.*, **20**, 2635–2642 (1981).

Sherwood, A. A., "A Quantitative Analysis of the Ronchi Test in Terms of Ray Optics," *J. Br. Astron. Assoc.*, **68**, 180 (1958).

Sherwood, A. A., "Ronchi Test Charts for Parabolic Mirrors," *J. Proc. R. Soc. New South Wales*, **43**, 19 (1959); reprinted in *Atti. Fond. Giorgio Ronchi Contrib. Ist. Naz. Ottica*, **15**, 340–346 (1960).

Silva, D. E., "Talbot Interferometer for Radial and Lateral Derivatives," *Appl. Opt.*, **11**, 2613–2624 (1972).

Steig, H., "More on Ronchi Gratings," in *Advanced Telescope Techniques, Vol. 1: Optics*, A. Mackintosh, Ed., Willmann-Bell, Inc., Richmond, VA, 1987, p. 46.

Stoltzmann, D. E., "Application of Ronchi Interferometry to Testing Large Aperture Flat Mirrors," *Proc. SPIE*, **153**, 68–75 (1978).

Strong, J., *Procedure in Experimental Physics*, Prentice-Hall, Englewood Cliffs, NJ, 1938, Chap. 11.

Talbot, F., "Facts Relating to Optical Science No. IV," *Philos. Mag.*, **9**, 401 (1836).

Thompson, B. J., Studies in Optics, Technical Report, U.S. AFAL-TR-73-112, U.S. Government, Washington, D.C. 1973.

Thompson, E. T., "Making Ronchi Gratings," in *Advanced Telescope Techniques, Vol. 1: Optics*, A. Mackintosh, Ed., Willmann-Bell, Inc., Richmond, VA, 1987, p. 45.

Toraldo di Francia, G., "Sulla Frange d'Interferenza delle Onde Aberranti," *Ottica*, **6**, 151 (1941a).

Toraldo di Francia, G., "Saggio di una Teoria Generale dei Reticoli," *Ottica*, **6**, 258 (1941b).

Toraldo di Francia, G., "Limiti di Validita dell'Ipotesi della Rotazione Rigida per il Reticolo Rettilineo," *Ottica*, **7**, 282 (1942a).

Toraldo di Francia, G., "La Prova dell'Aberrazione Cromatica con l'Interferometro a Reticolo," *Ottica*, **7**, 302 (1942b).

Toraldo di Francia, G., "Ancora su le Aberrazioni delle Onde Diffratte dal Reticolo Rettilineo," *Ottica*, **8**, 1 (1943a).

Toraldo di Francia, G., "La Formula Esatta per le Frange del'Interferometro Ronchi," *Ottica*, **8**, 225 (1943b).

Toraldo di Francia, G., "Introduzione alla Teoria Geometrica e Interferenziale delle Onde Aberranti," *Atti. Fond. Giorgio Ronchi Contrib. Ist. Naz. Ottica*, **1**, 122–138 (1946).

Toraldo di Francia, G., "Introduzione alla Teoria Geometrica e Interferenziale delle Onde Aberranti," *Atti. Fond. Giorgio Ronchi Contrib. Ist. Naz. Ottica*, **2**, 25–42 (1947).

Toraldo di Francia, G., "Geometrical and Interferential Aspects of the Ronchi Test," in *Optical Image Evaluation*, Nat. Bur. Stand. U.S. Circ. No. 256, U.S. Government Printing Office, Washington, D.C., 1954, Chap. 11, p. 161.

Villani, F., "Sulla Misurazione dell'Astigmatismo e del Coma Mediante le Frange d'Ombra," *Nuovo Cimento*, **7**, 248 (1930).

Villani, F. and R. Bruscaglioni, "Sulla Forma delle Frange d'Ombra ottenute de Onde Affette da Astigmatismo e Coma," *Nuovo Cimento*, **9**, 1 (1932).

Vogl, G., "A Phase Grating Interferometer," *Appl. Opt.*, **3**, 1089 (1964).

Waland, R. L., "Note on Figuring Schmidt Correcting Lenses," *J. Sci. Instrum.*, **15**, 339 (1938).

Wan, D. S. and D. T. Lin, "Ronchi Test and a New Phase Reduction Algorithm," *Appl. Opt.*, **29**, 3255–3265 (1990).

Wehn, R., "Die Methode der Ronchi-Gitter in der Praxis," *Atti. Fond. Giorgio Ronchi Contrib. Ist. Naz. Ottica*, **17**, 39–96 (1962).

Yatagai, T., "Fringe Scanning Ronchi Test for Aspherical Surfaces," *Appl. Opt.*, **23**, 3676–3679 (1984).

Yokoseki, S. and T. Susuki, "Shearing Interferometer Using the Grating as the Beam Splitter," *Appl. Opt.*, **10**, 1575–1580 (1971a).

Yokoseki, S. and T. Susuki, "Shearing Interferometer Using the Grating as a Beam Splitter. Part 2," *Appl. Opt.*, **10**, 1690–1693 (1971b).

Yoder, P. R., Jr., "Further Analysis of the 'Lower' Test for High-Speed Parabolic Mirrors," *J. Opt. Soc. Am.*, **49**, 439–440 (1959).

10

Hartmann and Other Screen Tests

I. Ghozeil

10.1. INTRODUCTION

In optical testing ideally one would like to use methods that give accurate quantitative results rapidly for an entire wavefront of interest. Despite recent advances, to be discussed later in this chapter, such quantitative tests usually are laborious and time consuming. Because of this situation, there often is a choice between a quick qualitative or quasiquantitative method and a not-so-quick quantitative wavefront sampling test.

The optician making a lens generally will opt for the first type of method, usually by employing the Foucault knife edge for the daily tests he or she performs. However, in spite of its convenience, this method has some drawbacks, discussed at length in Chapter 8. A major limitation is that this test is insensitive to small slope deviations of the sampled wavefront when compared to the desired wavefront. In other words it is insensitive to slope deviations that are changing slightly in magnitude or direction. This is equivalent to saying that, when the first and second derivatives of the wavefront errors are small, the knife-edge test is insensitive. This drawback is not objectionable when one is examining a small surface or a system of small aperture. But when one is testing a large aperture, for example, one having a diameter of 2 m or more, a continuous small slope deviation can result in a large net departure at the edge of the aperture. For example, in a 4-m (158-in.) diameter $f/2.7$ concave mirror an astigmatism value of 2λ ($\lambda = 500$ nm) on the surface, or 4λ on the wavefront, will be barely noticeable when the surface is tested with a knife edge near the center of the paraxial radius of curvature. Thus one cannot rely on the Foucault knife-edge test exclusively; occasionally a more lengthy quantitative test of the wavefront is needed.

Another test, often used by opticians, that gives quantitative results rapidly is the wire test, also discussed in Chapter 8. However, the measurements ob-

Optical Shop Testing, Second Edition, Edited by Daniel Malacara.
ISBN 0-471-52232-5 © 1992, John Wiley & Sons, Inc.

tained by this method provide information only on the radial component of surface deviations for one diameter at a time. With some care several tests can be made for several diameters, and an approximation of what the surface looks like can be obtained. How the several tests interrelate becomes somewhat uncertain, however, if the surface tested has a central hole or obscuration, as is often the case. This central obstruction makes it necessary that some assumption be made concerning the surface near its vertex, so that the various profiles can be shifted relative to one another to give a consistent expression for the surface.

The quantitative test that has the potential for giving the most accurate information about a wavefront is the interferometric method. This is so because a wavefront can be sampled theoretically with spacing of λ/n, where λ is the wavelength of light used for testing and n is the number of times the system is traversed by the test beam. Examples of this are given in Figs. 10.1 and 10.2. In Fig. 10.1 a concave paraboloidal mirror is tested at focus with the aid of a previously tested flat mirror whose wavefront characteristics are well known. In this situation the interference between reference and test wavefronts is recorded by placing an interferometer at the focus of the paraboloid. Because the test beam has been reflected twice from the paraboloid, the final wavefront can be sampled with a spacing of $\lambda/2$. This is equivalent to sampling the surface every $\lambda/4$, since the wavefront deviations are twice those of the surface. In Fig. 10.2 a flat mirror is under test with the aid of a concave spherical mirror whose surface characteristics have already been determined. Again, the test beam is reflected twice from the surface under test, and the interferogram has fringes spaced every $\lambda/2$ for the wavefront, or $\lambda/4$ for the surface.

Accurate as the interferometric method can be, its sensitivity at times pre-

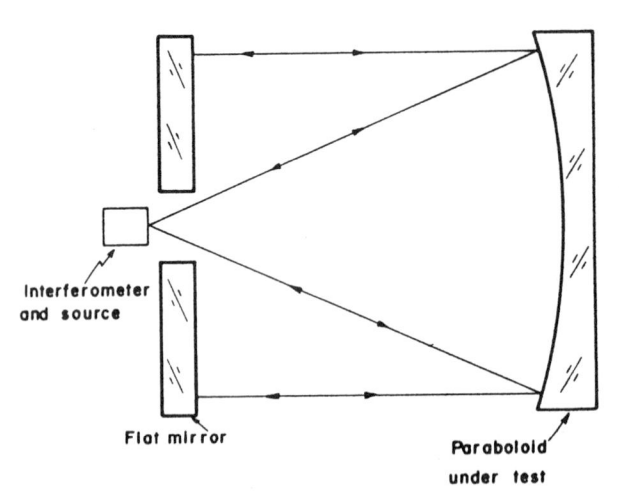

Figure 10.1. Autocollimating arrangement to test a paraboloidal mirror with the aid of a previously tested mirror.

10.1. INTRODUCTION

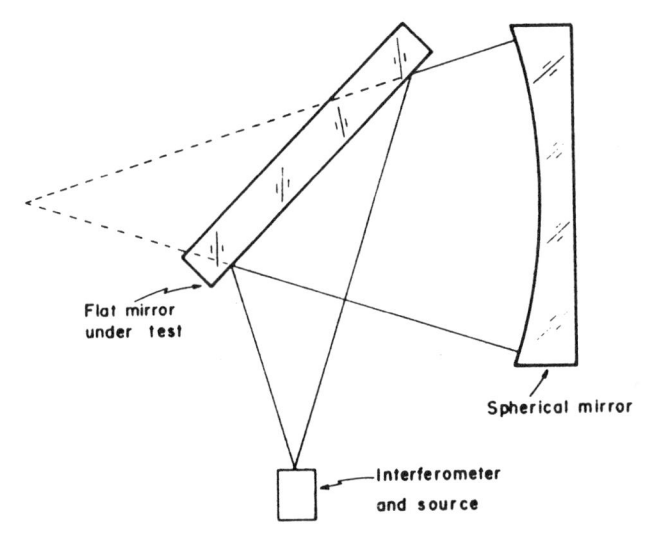

Figure 10.2. Ritchey-Common arrangement to test a flat with the aid of a previously tested sphere.

cludes its use. It usually fails, for example, when the medium between the system under test and the testing station is turbulent or changing rapidly. Another cause of failure is rapid vibration of the system under test in relation to the test station. Although in both situations the wavefront information usually can be obtained from multiple recordings, such retrieval can be laborious and difficult. With the increasing availability of electrooptical detector arrays, such as CCDs (coupled charge devices) and with the advent of the desk-top microcomputer, it is much easier to obtain a large series of interferometric "snap shots" of the wavefronts and to average these in order to average out the effects of random wavefront errors introduced by a randomly varying turbulent medium in the test path.

Air turbulence can be eliminated if the system is tested in a partially evacuated chamber, and vibrations can be reduced to a tolerable level through the use of vibration isolation devices. However, these solutions are not always economically feasible when large systems or those having very long focal lengths are under test. Under those circumstances one is forced to use a theoretically less accurate method. This consists of sampling the wavefront with a perforated screen or mask.

This chapter is concerned mainly with the methods of sampling a wavefront or mirror surface through the use of such screens. Included in the chapter are sections on applications of these methods and on a comparison of the various screen tests among themselves. Although the methods described are applicable to most lens systems, the presentation in the rest of this chapter will be made for the case of large concave mirrors, which are most commonly tested by these methods.

10.2. THEORY

The basic concept behind all screen tests is that a wavefront can be sampled in a number of locations across it in a predetermined fashion, and that the wavefront can then be reconstructed when the sample points are related to each other. These methods are based on a purely geometrical optics approach.

The premise is that a portion of a wavefront, when tilted relative to the ideal wavefront in that region, causes light to come to focus at a distance different from the ideal one, in other words the light does not focus at the ideal focus position. A recording plane located at a some convenient position, such as near the system focus or near the paraxial center of curvature of a mirror, is intercepted by the light arriving from this tilted wavefront region. This interception takes place at a "height," or location, on the recording plane other than at the height that would be obtained with light coming from the ideal wavefront. With this in mind we can see that the converse can be used to determine the tilt error in a portion of a wavefront: By determining where the light from that region intercepts the recording plane, what the difference is between that intersection and the one expected from the perfect wavefront of interest, and knowing the location of the recording plane we can determine the wavefront slope.

If the wavefront is sampled by a number of rays, or beams normal to it, ray deviations at some desired recording plane location can be obtained. A ray deviation is taken to be from the location at which the ray should have been recorded had the desired ideal wavefront been under test. The ideal wavefront need not be perfectly spherical, but in principle can have any shape. The only deviations of interest are those from that wavefront, whatever its shape.

In the Michelson and related tests the beam deviations are found from the interference of light from two different regions. In the other screen tests the light from each of the various regions of the wavefront is recorded in such a way that the beams interfere minimally or not at all. The beam deviations are used to obtain a surface slope error at each sampled point. The tilt at each of these points is established from the measured beam deviation (see Fig. 10.3), which is given by two coordinates in a flat reference plane. The usual method has been to locate a perforated screen, having well-known hole locations, in front of the mirror to be tested and the entire aperture illuminated from a convenient location, such as the paraxial center of curvature of the surface. The reflected light returning through the holes in the screen is recorded, classically on a photographic plate because it is a relatively stable recording medium.

Fig. 10.3 shows how the tilt at each sampled point is established. Here Δx is the value of one of the components of the beam deviation, d is the separation between mirror and recording plane, θ_x is the angular tilt of the area, and h is the resultant height above or below the ideal surface. This height h is approximately related to the wavefront deviation W by $W = 2h$.

10.2. THEORY

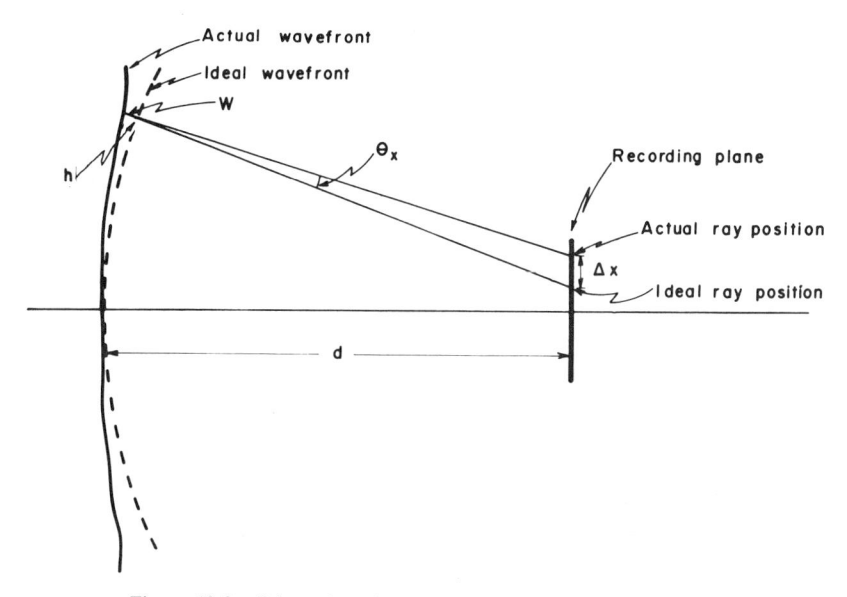

Figure 10.3. Schematics of rays and wavefronts in Hartmann test.

The relation between the wavefront aberration W and the x component Δx of the ray deviation in the testing plane can be obtained from the exact formulas given by Rayces (1964), although the following approximate value is accurate enough for most practical purposes:

$$\frac{\partial W}{\partial x} = \frac{\Delta x}{d}. \tag{10.1}$$

Integrating this expression, we obtain

$$W = \frac{1}{d} \int_0^x \Delta x \, dx; \tag{10.2}$$

if now the surface deviation h is written in wavelengths by defining $H = h/\lambda$, we have

$$H = \frac{1}{2d\lambda} \int_0^x \Delta x \, dx. \tag{10.3}$$

Since the function Δx is sampled only at discrete points, such as the holes in a sampling screen, the integration is usually performed by using the trapezoidal rule, yielding

$$H_N = \frac{1}{2d\lambda} \sum_{n=2}^{N} \left(\frac{\Delta x_{n-1} + \Delta x_n}{2} \right) \delta x_{n-1} \tag{10.4}$$

and similarly,

$$H_M = \frac{1}{2d\lambda} \sum_{m=1}^{M} \left(\frac{\Delta y_{m-1} + \Delta y_m}{2} \right) \delta y_{m-1}, \tag{10.5}$$

where δx_{n-1} is the separation between points n and $n-1$ in the x direction, and δy_{m-1} is the separation between points m and $m-1$ in the y direction. With this procedure the deviations H_N are computed at the same places where the screen holes are located.

Alternative expressions for H_N, derived from Eqs. (10.4) and (10.5), are sometimes quoted (Vitrichenko et al. 1975) for the case in which δx_n and δy_n are constants, as follows:

$$H_N = \frac{1}{d\lambda} \left(\frac{1}{2} \Delta x_1 + \sum_{n=2}^{N-1} \Delta x_n + \frac{1}{2} \Delta x_N \right) \delta x \tag{10.6}$$

and similarly,

$$H_N = \frac{1}{d\lambda} \left(\frac{1}{2} \Delta y_1 + \sum_{m=2}^{M-1} \Delta y_m + \frac{1}{2} \Delta y_M \right) \delta y. \tag{10.7}$$

These expressions give the surface deviation at any Nth point from some reference point $N = 1$.

Other methods of integration can sometimes be used with somewhat more reliable results. In most cases the trapezoidal rule will suffice. There is a temptation to use a polynomial fit to express the changes in slope deviations present in the mirror. The fit is made to agree within some criterion governing how closely the fit and the sample points should agree. Thus, rather than using tilted flat planes to perform the integration to obtain the surface departures, one integrates over a polynomial (Cornejo and Malacara 1976). The procedure employed for the Ronchi test described in Section 9.3.1 can be used. The wavefront can then be written in terms of Zernike polynomials (see Chapter 13). Theoretically, this is ideal because errors due to numerical integration methods are removed. The difficulty lies in obtaining a good polynomial fit. It is known that such fitting can introduce errors by smoothing out relevant sharp features or by introducing oscillations when none exists. The first effect usually is due to low-order polynomial fits and the second effect to high-order fits. Needless to say, this leads to a reduction in the reliability of the test procedure, especially near the edge of the mirror.

10.2. THEORY

Given a choice, it is preferable to use the simple trapezoidal rule, or an equivalent method, rather than one that may mask important features that should not be overlooked in testing a given surface. If any fitting is to be done, a good procedure would be to use two-dimensional cubic spline functions (Ahlberg et al. 1967). Such a fit is merely a point-to-point interpolation method; it uses the slope changes at neighboring points to obtain the best interpolation between the points of interest. This method of fitting would not alter the measured slope deviation values at the sampling points, but presumably would give a better approximation of the events between sampling points than is obtained by the trapezoidal rule, which assumes flat planes between these points.

As can be seen, one of the major difficulties in screen testing is the introduction of errors through the method used to reduce the data in order to obtain the surface deviations. The errors can be reduced if the test procedure used affords ways of obtaining a given result by several independent means. This is so because, in the face of uncertainty, the best approach to reliability is through replication, that is, by averaging the results of many tests. This is expressed by the student's t distribution, which is applicable when fewer than 30 independent observations are made. By using the t test one finds that the range of the likely deviation of the mean of a set of observations from the true mean (i.e., the confidence interval) becomes smaller as the number of independent observations increases. Thus a screen test pattern that allows for several independent ways of obtaining a surface deviation at a sample point is likely to give a closer approximation to the actual value of that deviation than is a pattern that lacks such a feature.

The main assumption inherent in screen testing is that between samples the wavefront changes are gradual rather than abrupt. This is a safe assumption because abrupt changes can be readily detected by other means, such as a Foucault knife-edge test, and in principle their presence can be taken into account or allowance can be made for them. As a consequence the screen test gives results that are more representative of the acutal surface features as the mirror surface becomes smoother. The need for this assumption is due to the fact that each sample point is taken to give the average tilt for a certain area on the mirror surface. For the average-tilt approach to hold, one must assume surface smoothness. This assumption also must be made if a polynomial fit is used, as described earlier.

Another assumption made is that the air turbulence is random in magnitude and direction between the mirror and the recording plane. In fact, all that is needed is that a reflected beam, made to oscillate by air turbulence, traverses most frequently the true position it would register under no turbulence. This criterion allows for turbulence conditions that are not random in amplitude, although random in direction. These conditions can give rise to an asymmetric spot density distribution at the recording plane, unlike the two-dimensional Gaussian distribution one would obtain with ideally random turbulence. However, the distributions will still give accurate spot locations if the location of maximum density is determined. The main turbulence condition to beware of is the case of a laminar sheet of air rising or falling, usually near the edge or through the center hole of a mirror under test. Such a sheet will uniformly displace the light from a number of holes and systematically bias the test results. Occasionally this condition is blatant and causes surprisingly gross errors. Most often, however, if this condition is present, it is on a small scale and is difficult to detect. The blatant condition is often due to a gross oversight, such as a window left open or an air-conditioning unit left on. The small-scale effects are due to one's inability to account for all possible causes of turbulence under normal manufacturing and testing conditions.

A common source for this sort of turbulence is a thermal imbalance across the width of the test beam. The typical situation arises when one is testing with the optical axis horizontal in the presence of a thermal gradient across the optical path, as can be found between the floor and the ceiling of a test area when their temperature is not controlled adequately. If the air reaches a stable condition then a layering results, with the coolest and densest layer at the bottom, and with the warmest and lightest layer at the top. This air density gradient is also an index of refraction gradient, with the higher index at the bottom and the lowest at the top. Although this is not a turbulence in the usual sense, nevertheless it results in an astigmatic wavefront when the system under test may not have such an aberration. This astigmatism is aligned with the vertical axis and can cause further confusion when astigmatism due to a drooping of the optics is expected because of self-weight deflection of those optics under the local gravity gradient.

The other tyical thermal gradient situation arises when the test is performed with the optical axis vertical near an outside wall of a building. In such a case, if the wall is warm, a chimney effect could take place and obviously this would be undesirable. Similarly, if the wall is cold, a cascade of cold air would also cause problems with the test. Proper insulation and care in thermal control are the necessary design criteria in setting up the test under such circumstances.

10.3. TYPES OF SCREENS

Over the years, sampling screens with a number of different hole patterns have been used in testing mirror surfaces. The various types reported in the literature, along with their advantages and disadvantages, are discussed in this section. Although recently there has been an increase in the use of lenticular screens, we will continue to use "hole" as a short-hand instead of the more descriptive term "sampling aperture" in the following discussion. It will be made obvious to the reader, in later parts of the text, when a lenticular screen is being discussed and that the term "hole" is not applicable.

10.3. TYPES OF SCREENS

For convenience in use, it is worthwhile that a few holes be added that do not fall on the hole pattern of interest. These extra holes serve to aid in orienting the screen and subsequently the recorded pattern of dots. Classically, photographic plates were used to record the location of dots obtained from the screen tests and for subsequent measuring of dot locations recorded on these plates.

10.3.1. Hartmann Radial Pattern

By far the most common screen in use, until recently, was the radial pattern shown in Fig. 10.6. This type of screen was used first by Hartmann (1900, 1904) and subsequently, without substantial change in the original concept, by a number of other experimenters. A bibliography of work published on the radial screen test is given at the end of this chapter.

The basic concept of the radial pattern test is that a sample of points on the surface of a concave mirror that has a circular aperture can be taken using an opaque cover in which a number of small holes have been made. The holes are spaced evenly along a number of diameters of the circular mirror aperture, as shown in Fig. 10.4. The choice of such a pattern has several advantages. First, a circular aperture is easily analyzed in a polar coordinate system. Second, the grid pattern will detect the most common flaws of mirror surfaces of any size—zonal errors and concentric "hills" and "valleys" fractions of a micrometer in magnitude—that result from classical grinding and polishing techniques. Thus in principle the radial screen could afford the luxury of an easily analyzed test that detects the major noticeable surface defects.

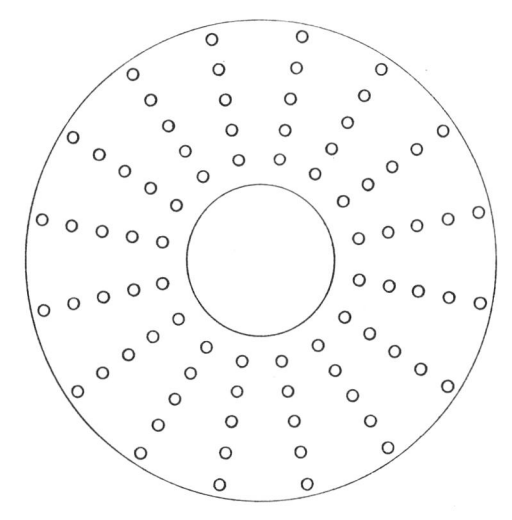

Figure 10.4. Classical Hartmann radial pattern.

In fact, the radial screen test has many shortcomings. A major one is that if one considers each hole in the screen to be sampling an area on an annulus having as mean radius the radial location of the hole, it follows that the area sampled is considerably larger for holes far from the mirror center than for those near that center. This means that the area of the surface contributing the major portion of light gathering is known with least certainty, with the result that asymmetric defects not detectable by other means are likely to go undetected by this method as well. Another major shortcoming is due to the circular symmetry of some of the defects, which go undetected if the hole spacing is not sufficiently small. Intertwined with these procedural disadvantages is the assumption, often made in the data reduction to deduce the surface, that the slope deviations detected are part of a circularly symmetric system. The foregoing objections can be reduced somewhat by using a radial screen with holes spaced moderately close together, with many diameters sampled and consequently with some of the diametral arms not having holes near the mirror center, to prevent overcrowding in that region. The assumption on symmetry could also be abandoned. Incorporating these changes into the classical test, however, would lower its ease of reduction.

Also present in this form of testing is a buildup of integration errors along the circular paths of integration, as discussed in the next section on application. This error accumulation could be eliminated by a method similar to the replication developed for the square array screen (Ghozeil and Simmons 1974), but applied to circular integration.

Application of the Radial Screen. The radial screen can be used for testing a concave mirror in collimated light, or with a point source placed at or near the paraxial center of curvature. The first test is suitable for systems such as telescopes. It requires that the screen be placed either flush with the surface or sufficiently far from the concave mirror so that the beams, after reflection, do not strike the screen before they come to focus. When the mirror is under test with light from a point source, the screen suffers from the same location constraint already described and should be placed as close to the mirror as possible.

In the radial screen method it has been customary to record, on two photographic plates, the light reflected from the mirror, with the mirror illuminated through the screen. One plate can be placed inside and the other outside the theoretical point to which the light should be converging, or both plates can be on the same side of that location. Either two-plate technique allows one to compensate for plate tilt and for errors in hole spacings in the screen. The mirror-to-recording plane distance need not be known accurately, but the plate separation is important in the computation unless a least-squares refocusing is to be done, as described in Section 10.3.3.

In the classical test reduction, where the photographic plates are placed with one inside and the other outside of focus, a mean weighted focus position, F_0,

10.3. TYPES OF SCREENS

can be obtained as follows:

$$F_0 = \frac{\Sigma R_i F_i}{\Sigma R_i} \quad (10.8)$$

where F_i is the focus obtained with the light from a pair of screen holes diametrically located to either side of the vertex, and R_i is the radial distance of the "zone" that contains that pair of points. These values of R_i act as weight factors. To obtain an indication of the size of a point imaged by the lens, the Hartmann T criterion is used, as expressed by

$$T = \frac{200{,}000}{F_0^2} \left(\frac{\Sigma R_i^2 |F_i - F_0|}{\Sigma R_i} \right). \quad (10.9)$$

This gives the point image size in seconds of arc.

The astigmatism of the mirror can be estimated by comparing the average focus obtained from one arm of the radial screen with the average focus obtained from an arm perpendicular to the first one. However, this utilizes a relatively small number of sampling points and consequently does not give a very reliable estimate of that astigmatism.

Only one plate at a time need be reduced if the screen is well known, as was the case for the 3-m (120-in.) Lick Observatory primary mirror (Mayall and Vasilevskis 1960). This modification is a great time saver if there are many points to measure on the photographic plates.

The measured spot locations are reduced to give surface deviations by the general method described in Section 10.2. In the classical form of the test, only radial integration profiles are obtained. More recently, radial and tangential integrations have been used (Mayall and Vasilevskis 1960; Schulte 1968; Vitrichenko et al. 1975) to obtain more reliable results than the radial profiles alone. The two integration modes should differ only by a constant and should be made to agree. In fact, circular error buildup due to the tangential integration does not readily permit a simple shift to bring about the agreement. This method would be improved if some ways were established to introduce replication in the determination of surface deviations.

10.3.2. Helical Screen Test

In an attempt to overcome the disadvantages of the radial screen, a helically distributed set of holes in a screen was used to test the Lick Observatory 3-m (120-in.) primary mirror (Mayall and Vasilevskis 1960). A portion of this pattern is shown schematically in Fig. 10.5. Although basically a radial screen pattern, it differs from the classical pattern insofar as holes along any radius are

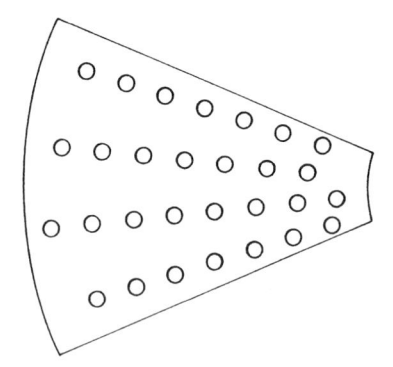

Figure 10.5. A portion of a helical screen, showing the staggering of the holes in the radial arms of the screen.

shifted radially relative to adjacent radii so as to form a helix, as shown in Fig. 10.6.

This choice of pattern affords an opportunity for the pattern to intercept zonal errors that might go undetected with a classical screen having the same hole spacing. Thus it overcomes one of the disadvantages of the classical screen. However, it still suffers from unequal area sampling so that the most poorly sampled area of the mirror is the one that will gather a great percentage of the light that will be used in forming an image with the instrument in actual use.

Application of the Helical Screen. This screen test makes extensive use of both radial and tangential integration paths in its application. This takes into account the reality that mirror deformations result, not merely in radial tilts, but in tilts having two components (Kingslake 1927–1928) that are due to asymmetric surface departures from the ideal mirror surface. Difficulties arise in

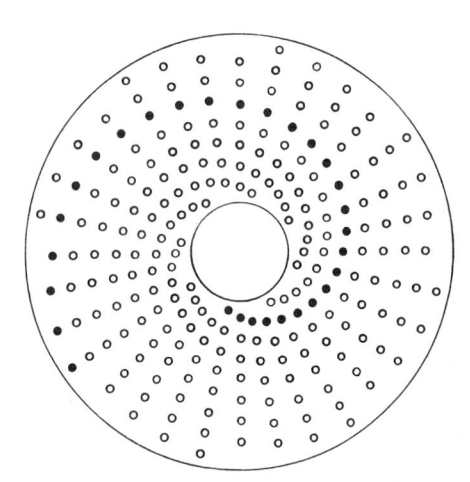

Figure 10.6. Complete helical screen, showing the spiral distribution of holes.

10.3. TYPES OF SCREENS

relating the slopes by sampling holes that are radially displaced relative to neighboring radial arms of the screen. The methods needed to reduce the data are thoroughly reported in the only known occurrence of the use of this screen (Mayall and Vasilevskis 1960).

10.3.3. Square-Array Tests

To overcome the shortcomings of the Hartmann radial screen test, a surface should be sampled in such a way that the sample pattern used cuts across the commonly occurring circular zonal features and samples equal areas on the surface. This is accomplished by setting the sampling points in a so-called square array. Such an array consists of points placed equidistantly at the intersection of lines parallel to the orthogonal axes of a Cartesian coordinate system, as shown in Fig. 10.7.

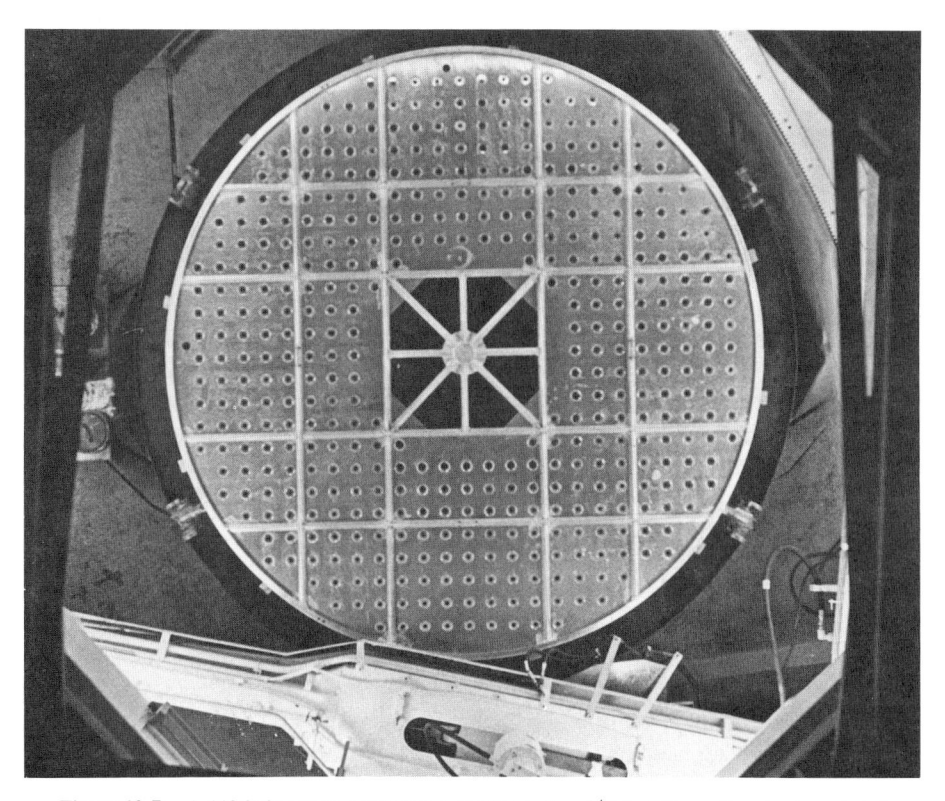

Figure 10.7. A 440-hole square-array screen used to test an $f/2.7$, 400-cm-diameter hyperboloidal mirror made at the Kitt Peak National Observatory (U.S.A.). The hole size is 2.54 cm, and the hole interval is 15.24 cm. (Photo courtesy of Kitt Peak National Observatory; from Ghozeil and Simmons 1974.)

After this pattern had been put into use for obtaining spot diagrams in the computerized analysis of designed systems, it was proposed to obtain spot diagrams for existing systems by means of screens placed in front of the lenses that had been built (Stavroudis and Sutton 1965). Although this proposal had been made so that the imaging characteristics of systems could be studied, and not their wavefront characteristics, it showed a deviation from the radial test pattern that had been prevalent. With the advent of the widespread use of electronic computers in complex massive data analysis, and with the availability of densitometers capable of rapid and accurate measurement of the position of features on photographic plates, it has become possible to test large surfaces more thoroughly than was previously possible.

The use of the square screen has been suggested by R. V. Shack in private communications, and it has been employed to test a number of large telescope primary mirrors, including the mirrors for the following: The first 4-m (158-in.) telescope, made at Kitt Peak National Observatory (Simmons and Ghozeil 1971; Ghozeil and Simmons 1974); the 3.8-m (1950-in.) Anglo-Australian telescope, which was tested at Grubb Parsons (Gascoigne, private communication, 1972); the second 4-m (158-in.) telescope, made at Kitt Peak for the Cerro Tololo Interamerican Observatory (Ghozeil 1974); the 2.6-m (101-in.) telescope for the Irenée Dupont Observatory, made at the Optical Sciences Center of the University of Arizona (Ruda, private communication, 1975); and the 3.6-m (144-in.) telescope for the Canada–France–Hawaii Observatory, made at the Dominion Astrophysical Observatory (Dancey, private communication, 1975).

The square array not only gives uniform sampling of the surface but also does so in a manner that has no circular symmetry. Thus there is no need to make assumptions concerning the disposition and symmetry of the errors to be detected, and the possibility of an artificial circular error buildup has been removed. Furthermore, a much higher surface-sampling frequency can be obtained with this method than is possible with radial or helical screens. The tests performed in this way are independent of random air turbulence and can be done with or without intervening optics, such as null lenses (discussed in Chapter 14). The second option of no intervening matter other than air is preferred, so that the introduction of confounding alignment and surface errors caused by intervening materials can be prevented.

Another advantage of this screen is obtained if the screen is made as a rigid solid unit. Under these conditions the hole spacing can be controlled accurately, and a screen whose characteristics are well known is obtained. The fact that the screen then becomes a well-known fixed reference means, in turn, that only one photographic plate need be taken to record the reflected light from the mirror. Not only is this a time- and labor-saving feature but also it results in a more reliable data analysis through minimization of the uncertainties in obtaining the raw data from which the surface features are derived.

With this test method, as with other screens, it is not possible to detect small-

10.3. TYPES OF SCREENS

scale surface changes taking place between holes in the screen. However, these changes are detected easily with the knife-edge test. Consequently, the square-array test must be used in conjunction with a method capable of small-scale detection. However, this test, and potentially the helical screen method, can be made to show zones that are small in extent through a composite of several tests of the same mirror under the same conditions, but with the screen rotated a known amount relative to the surface (Ghozeil 1974). In this fashion a high rate of replication and complementariness in sampling is achieved. This leads to a more accurate analysis capable of providing a more detailed view of the surface under test than can be obtained by a single screen placement.

The use of a square screen implies almost automatically the use of a high-speed electronic computer if thorough sampling and analysis are to be achieved. This is due to the large number of computations involved in the data reduction for this test. This aspect may be considered a drawback if fast computers are not readily accessible, but with now prevalent desk-top microcomputer this should be of diminishing concern.

Application of the Square Screen Test. The square screen has been applied in three forms: with a projected lenticular screen (Shack and Platt 1971), with a scanning pentaprism (Hochgraf 1971; Hooker 1971), and with a physical screen having holes in it (Simmons and Ghozeil 1971; Ghozeil and Simmons 1974). The treatment that follows is for a physical screen placed on a mirror that is tested at its center of curvature without intervening optical elements. The application of the method is generally valid for mirrors tested at focus and for the other two methods just mentioned.

The screen is placed near the mirror as shown in Fig. 10.8, which shows the

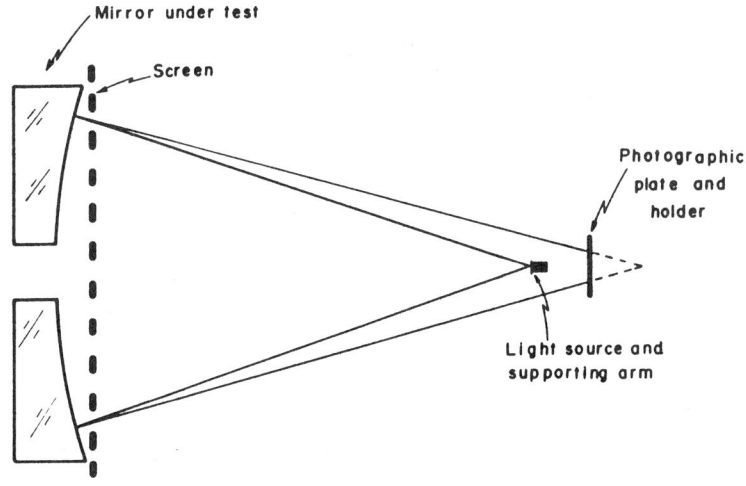

Figure 10.8. Hartmann test setup.

screen used with the two 4-m (158-in.) primary mirrors made at Kitt Peak National Observatory. The view is from the test station in the tower directly above the mirror, and shows the 440 holes in the screen, the extra reference holes, and the ribbed structure incorporated in the screen to make it rigid. The optical polishing was performed under the direction of N. C. Cole and H. J. Wirth, and the mirrors were tested on a support system resembling that used in the telescope. The supports were designed by E. T. Pearson (Pearson 1968).[†]

10.4. HARTMANN TEST IMPLEMENTATION

In this section first we describe the classical approach of testing in which a photographic plate is used to record the light returned from the mirror under test, and through the Hartmann screen. At the end of the section we consider the more recent developments in data collection and analysis.

10.4.1. Classical Test

The holes on the Hartmann screen should be placed very accurately to avoid errors in computing the surface deviations. The hole diameter should be small but not so small that their diffraction images on the photographic plate overlap each other and of sufficiently small size as to permit an accurate measurement of the wavefront aberrations present (Golden 1975; Vitrichenko 1976; Morales and Malacara 1983).

The screen must be centered accurately on the aperture of the mirror under test. This is particularly true in the presence of spherical aberration, which is the case for aspheric mirrors tested at their center of curvature. A decentering of the screen in such a situation leads to an apparent presence of coma, as described by Landgrave and Moya (1986). They present results for parabolic mirrors of various F numbers.

After the screen is placed and centered on the mirror, but before any plates are taken, the point light source used to illuminate the mirror through the screen must be centered properly relative to the mirror to prevent the introduction of off-axis aberrations. This can be done conveniently if the housing of the point source is of such a size that it is narrowly missed by the beams returned by the region near the center of the mirror. Without elaborate equipment one can then center the source by judging and equalizing visually the gaps between the housing and the light beams. With the Kitt Peak and Cerro Tololo 4-m (158-in.) Ritchey–Chrétien primaries, the light could be centered easily to less than 0.2 cm (0.08 in.) to produce no noticeable off-axis effects.

[†]It is important that a well-thought-out support system be used with a mirror under test: only deformations that will be present in actual use of the mirror are allowable. Some of the many effects of support systems have been described by Malvick (1972).

10.4. HARTMANN TEST IMPLEMENTATION

For convenience the point source is placed at a position closer to the mirror than the center of curvature. This position places the conjugate image beyond the center of curvature and permits the interception of the converging light at inside- and outside-of-focus modes without obstruction of the illuminating beam; the so-called focus referred to is the conjugate image location of the point source. In general, the image is highly aberrated for mirrors other than a sphere, and for classical telescope primaries spherical aberration should be the only one present. To facilitate the identification of the dark spots on the Hartmann plate, it is necessary that the photographic plate be placed outside the caustic limits (see Appendix 1). The source can be any light that is dim, or dimmed, enough to permit long exposure times, which do not result in overexposure of the photographic plate. Exposure times in the range of 90 to 180 s, depending on source brightness, should be sufficient to average random air turbulence.

The photographic plate should be very close to being perpendicular to the optical axis; otherwise a fictitious astigmatism will be obtained from the data reduction. There are several ways of orienting the plate holder correctly. A straightforward approach is to load the holder with a clear glass plate that has a cross hair scribed at its center. Then, with an alignment telescope, a zone on the mirror is viewed through the center of the plate. The alignment scope is first brought into alignment with the mirror, and then the fainter return from the glass plate is centered by tilting its holder.

With the screen and source properly located, one or more photographic plates are recorded, usually at a position between the source and its conjugate image. The inside-of-focus pattern obtained in this manner with the 4-m (158-in.) primaries is shown in Fig. 10.9. The location of the dots obtained on the photographic plate can then be determined to a high degree of accuracy with a microdensitometer having an x–y traveling stage. Typically, a position of uncertainty of less than 0.003 mm (0.0012 in.) can be expected. This error represents a surface tilt uncertainty of 0.015 s of arc (1.41×10^{-7} rad) for a mirror having a radius of curvature of 21.3 m (850 in.), as in the case of the 4-m (158-in.) primaries. The measuring stage is one of the few critical aspects of the test. When the photographic plate is placed in the microdensitometer, this instrument is made to have its two perpendicular cross hairs present on a viewing screen through which the plate also can be seen. These cross hairs are aligned with the two rows of dots that correspond to the two perpendicular rows of holes in the square array, and which intersect at the center of the screen. The rows of dots do not form straight lines unless a very good mirror is under test. The plate and cross hairs are aligned through a succession of small plate rotations and displacements until some minimal departure is obtained. The task is easy with a mirror nearing completion because the dots line up nicely.

In general, if the pattern is physically large, say of the order of 60 mm (2.4 in.), alignment can be made with a possible rotation error of less than 30 s of arc for the plate relative to the cross hairs. This error has little effect on the test

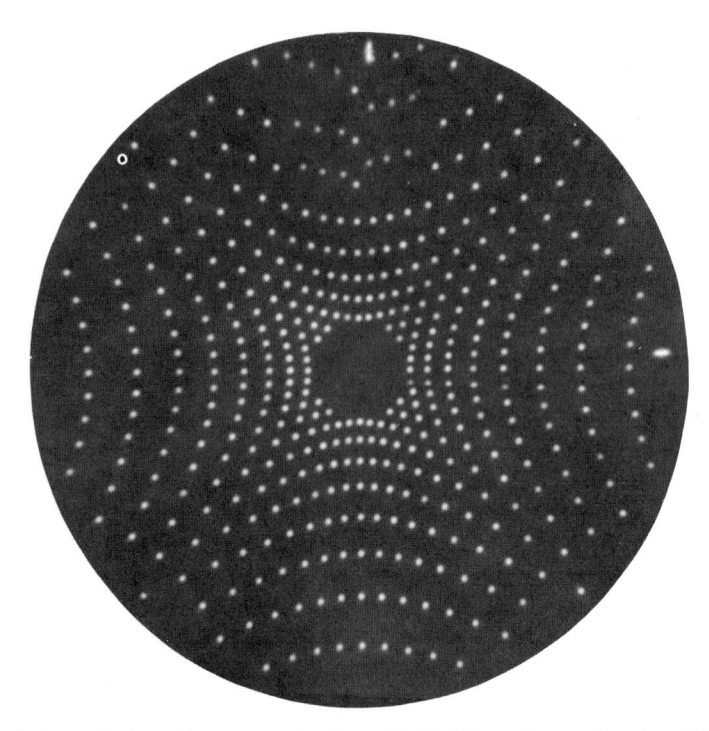

Figure 10.9. Inside focus Hartmann pattern for an $f/2.7$, 400-cm-diameter hyperboloidal mirror. The two oblong bright spots are used to give position referencing for the screen. (Photo courtesy of Kitt Peak National Observatory.)

results, as can be shown through repeated measurements of the same plate with slightly differing alignment criteria. If a gross error of the order of 0.5° to 1° is introduced, the resultant data reduction gives a somewhat astigmatic-looking, saddle-shaped aspect for the surface figure.

10.4.2. Recent Developments

Because of its compactness, ease of calibration, high surface-sampling frequency, and other advantages, the method of reimaging the aperture onto a small screen is finding increasing applications (Shack and Platt 1971; Loibl 1980). The screen can be a lenticular one, as proposed by Shack and Platt (1971).

One of the recent developments has been modification of the classical test to incorporate an electrooptical detector array instead of the customary photographic plate (Hausler and Schneider 1988; Pearson 1990). This permits rapid data collection and analysis of several spot patterns, and with the aid of a microcomputer these can be averaged. The averaging can be done either with the raw data consisting of spot centroids or with the wavefront maps that result from the analysis of the individual patterns. This approach also permits the dynamical sampling of the aperture (Golden 1975; Hausler and Schneider 1988).

An improvement on the accuracy in measuring the spot centroids has been obtained by the intentional overlap of these spots, thus causing an interference effect to occur (Korhonen 1983; Korhonen et al. 1986). Such an approach allows a closer packing of the spots and a possible higher surface-sampling frequency. This would make the use of detector arrays more attractive, especially since the detector could be biased to ignore the low-intensity noise present between adjacent spots.

10.5. DATA REDUCTION

The next step after the determination of the dot positions is the computer-aided data reduction. First, the centroid of the measured dot positions is found and established as the origin of the coordinate system to be used in the reduction. This is done by averaging all the x values and all the y values of the measured dot locations, and then subtracting these average values from each measurement. The reason for this step is that the intersection of the plate with the optical axis of the mirror is unknown, and the centroid is the best method of establishing this intersection. Surface deformations that occur randomly or symmetrically will result in a set of spots whose centroid corresponds with the location of the intersection of the optical axis and the plate. Furthermore, a number of surface deformations resulting in a systematic displacement of the dots would be equivalent to a tilt of the optical axis, which would be represented by a centroid displacement. Since the entire surface is sampled uniformly, there is no need for the area-weighted compensation used in reducing the data from a radial screen.

After the centroid has been established as the origin of the coordinate system to be used, the difference between the newly established x and y coordinates of each dot and the corresponding coordinates expected from a perfect mirror is found. The expected coordinates of the spot can be obtained from geometrical ray tracing or from closed-form equations such as Eq. (9.22) of this book (Malacara 1965). For the 4-m (158-in.) primaries, rays were traced through the location of the centers of the holes in the screen and reflected by the mirror to a collecting plane, not necessarily located where the photographic plate was placed. The net beam deviations obtained from these differences in x and y coordinates can be considered as being the result of an out-of-focus situation. This is especially true for the case where the photographic plate is not placed at the same distance from the mirror as was the collecting plane for which ray intercepts were calculated. A defocus situation results in a linear departure of the ray intercepts, with the rays from the edge of the aperture displaced most and the ones near the center displaced least. This means that the residual beam deviations can be brought to optimum focus through the subtraction of a linear term, which is found for the x-coordinate residuals and y-coordinate residuals independently by means of two linear least-square fits. For the x values the general equation to be fit is

$$\Delta x_0 = A + Bx, \tag{10.10}$$

where Δx_0 is the x component of the residual beam deviation, A and B are constants, and x is the x coordinate of the area on the mirror from which the beam deviation has been detected. The area on the mirror is found by projecting the hole in the screen onto the mirror, as viewed from a point source.

The Δx_0 values are the result of measurement, the x values are also known because the screen is known, and the only unknowns are the values of the constants A and B. These can be found by means of the least-squares equations

$$\sum_{m=1}^{N} \Delta x_0 = AN + B \sum_{m=1}^{N} x, \tag{10.11}$$

$$\sum_{m=1}^{N} x(\Delta x_0) = A\Sigma x + B \sum_{m=1}^{N} x^2, \tag{10.12}$$

where N is the number of holes over which the summations are to take place. These equations can be solved easily, but they are simplified if a symmetrical screen is used. This has the result that for every x coordinate of a hole there is a negative counterpart. As a consequence, summations over x reduce to zero, and Eqs. (10.11) and (10.12) reduce to

$$\sum_{m=1}^{N} \Delta x_0 = AN, \tag{10.13}$$

$$\sum_{m=1}^{N} x(\Delta x_0) = B \sum_{m=1}^{N} x^2. \tag{10.14}$$

From these the values of the constants A and B are found. The value of A should equal zero if the x coordinate of the centroid was found properly, since Eq. (10.13) is the formula for finding the centroid and that was taken as the system's origin.

Once the constants have been found, the values of the residuals at the best focus can be determined from Eqs. (10.13), (10.14), and (10.10). These residuals are found by subtracting the linear term and the independent constant. This

10.5. DATA REDUCTIONS

gives

$$\Delta x = \Delta x_0 - A - Bx, \qquad (10.15)$$

where Δx is the residual, as defined for Eq. (10.3). The Δy's, which are the y coordinates of the residuals, are found by using Eqs. (10.13) through (10.15), with y's substituted for x's.

The residuals can be used to obtain an estimate of the energy distribution of the light returned by the mirror under test. This is done by counting the number of points having a distance $r = \sqrt{(\Delta x)^2 + (\Delta y)^2}$ from the origin of the coordinate system. If the test is performed at the center of curvature and the energy distribution at the focus is desired, the residuals can be divided by a factor of 2. This division is a good approximation that introduces a small error, equivalent to a surface deformation of less than 0.05 wavelength, at the edge of an $f/2.7$ mirror. It is worth remembering that this energy distribution is the result of geometrical optics considerations and that the real image obtained with the mirror will be affected by diffraction. In general, that effect will tend to give a higher energy concentration in the core of the image.

The residuals, Δx and Δy, can also be integrated by some means, such as is given in Eqs. (10.6) and (10.7), for example, to obtain surface departures. Because numerical integration methods tend to have an accumulation of errors inherent in them, some care must be taken to reduce that error. This is best done through the use of paths of integration that intersect at only one point, which means that a surface height at any location is obtained by independent means. It is also worthwhile to integrate along any path in one direction and in its reverse direction and then average the results obtained.

The scheme used in the analysis of the two 4-m (158-in.) primaries follows that philosophy and is shown schematically in Fig. 10.10. First, the summations originate on the x and y axes and have as starting value of each integral the value obtained from the other coordinates integral going through the point in question. Since the surface departure values obtained through the x and y integrations should be the same at each point, the results of the two integrations are averaged at each point. Then the reverse path summation is done and averaged with its corresponding integral.

The next step is to rotate the coordinate axes by 45° about the origin and to repeat the entire integration process, using a different hole interval and different integration paths. Since these integrations should give the same results as those obtained with the first scheme, the values obtained by both processes are averaged at each point. This means that each surface departure value is obtained in at least four ways, with a majority obtained in eight ways. This replication tends to reduce the systematic accumulation of errors, as well as the introduction of spurious ones.

To further reduce spurious errors, two photographic plates may be measured

SQUARE SCREEN HOLE ARRAY

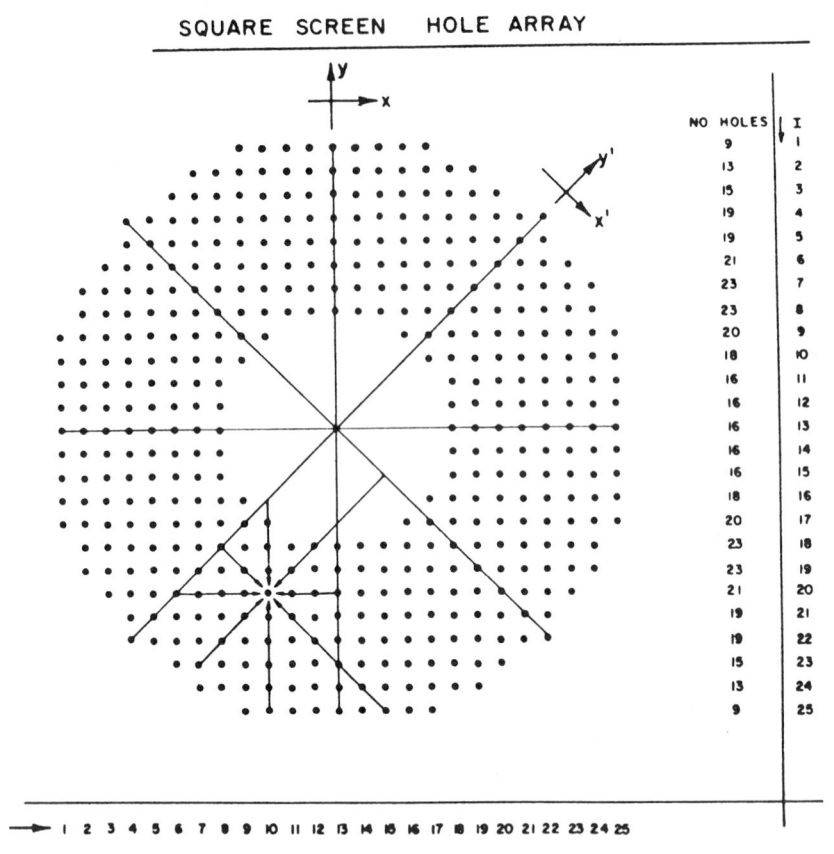

Figure 10.10. Integration paths used to obtain a value for the surface departure at a given sample point. (From Ghozeil and Simmons 1974.)

and reduced in this fashion, and the results of the two tests averaged together. This reduces errors due to imperfect photographic emulsions and to nonrandom air turbulence between mirror and plate, as well as measuring errors.

To eliminate any residual errors that may have been introduced by the integration scheme, or that may be due to errors in finding the centroid and in determining the best focus, a two-dimensional least-squares fit is done to refocus and reorient the final results. If such errors are present, they will have the following form:

$$H = A + Bx + Cy + Dx^2 + Ey^2 + Fxy, \qquad (10.16)$$

where H is the surface deviation at some point, $A, B, \ldots, F$ are unknown constants, and x and y are projected screen coordinates on the mirror. The least-

10.5. DATA REDUCTIONS

squares fit equations become, with N equal to the number of holes over which the summations are to take place,

$$\Sigma H = AN + B\Sigma x + C\Sigma y + D\Sigma x^2 + E\Sigma y^2 + F\Sigma xy, \qquad (10.17)$$

$$\Sigma xH = A\Sigma x + B\Sigma x^2 + C\Sigma xy + D\Sigma x^3 + E\Sigma xy^2 + F\Sigma x^2 y, \qquad (10.18)$$

$$\Sigma yH = A\Sigma y + B\Sigma xy + C\Sigma y^2 + D\Sigma x^2 y + E\Sigma y^3 + F\Sigma xy^2, \qquad (10.19)$$

$$\Sigma x^2 H = A\Sigma x^2 + B\Sigma x^3 + C\Sigma x^2 y + D\Sigma x^4 + E\Sigma x^2 y^2 + F\Sigma x^3 y, \qquad (10.20)$$

$$\Sigma y^2 H = A\Sigma y^2 + B\Sigma xy^2 + C\Sigma y^3 + D\Sigma x^2 y^2 + E\Sigma y^4 + F\Sigma xy^3, \qquad (10.21)$$

$$\Sigma xyH = A\Sigma xy + B\Sigma x^2 y + C\Sigma xy^2 + D\Sigma x^3 y + E\Sigma xy^3 + F\Sigma x^2 y^2. \qquad (10.22)$$

If a symmetrical screen is used, all the sums containing odd powers of either x or y, or both, reduce to zero. The simplified set of equations can be solved for the constants $A, B, \ldots, F$, the resultant fit subtracted from the values of H in the manner of Eq. (10.15). The coefficient B and C represent wavefront tilts due to errors in centering the plate. If $D = E$ and $F = 0$, these coefficients represent a focus error. If $D \neq E$ and $F = 0$, the error is due to astigmatism with a horizontal or vertical axis; if $F \neq 0$, the error is due to astigmatism with an arbitrary axis orientation. It is important to print out the values of the D, E, and F coefficients in order to obtain knowledge of any possible astigmatism on the mirror. If there is any astigmatism, it is better not to subtract it but to use, instead of Eq. (10.16),

$$H = A + Bx + Cy + Dx^2 + Dy^2. \qquad (10.23)$$

The resultant values for the surface departures can be printed out in an array. They are most usefully displayed on contour maps such as the one shown in Fig. 10.11. The accuracy that can be expected from this form of testing is better than 0.1 wavelength.

As the mirror nears completion, there is a need for a more detailed and reliable representation of the surface under test. This can be achieved through the superposition and averaging of several tests obtained with the screen rotated some known amount about its center, relative to the mirror (Ghozeil 1974). The superposition can be done numerically with an electronic computer, or photographically by expressing different surface deviations by corresponding densities on a photographic plate. The deviations obtained from separate tests for each region are used to obtain new average values for all the regions. This composite method is capable of giving surface deviations with an uncertainty of less than 0.05 wavelength. Results obtained in this way have shown faint small features observable with a knife-edge test, while simultaneously revealing overall slowly varying asymmetrical features.

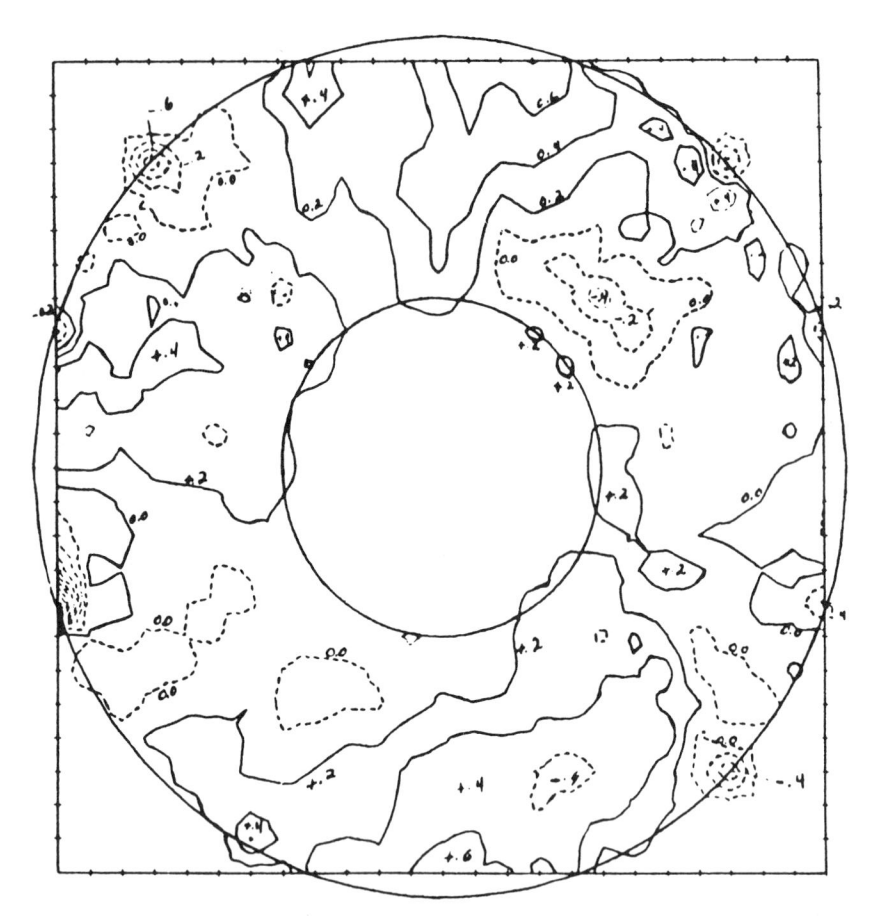

Figure 10.11. Contour map for a mirror early in the figuring process. High areas are shown by solid lines, and shallow areas by dashed lines. Surface deviations shown are in wavelengths of light at 550 nm. (Photo courtesy of Kitt Peak National Observatory.)

A different approach that is more computer intensive because of its use of a two-dimensional fast Fourier transform (FFT) algorithm has been reported (Freischlad and Koliopoulos 1985). Its main advantage is that it is less sensitive to noise in the data, which has been used to obtain two-dimensional wavefront errors.

A polynomial fit can be made through the surface departure values obtained by these or other methods. From a polynomial fit one can then extract classical aberration terms such as coma and astigmatism (see Chapter 13). Also, the FFT has been used with results of Hartmann tests to obtain point spread functions (PSF) and modulation transfer functions (MTF) (Anderson 1982; Dame and Vakili 1984).

10.6. THE MICHELSON AND GARDNER–BENNETT TESTS

A test was devised by Michelson (1918) to determine mirror surface deformations by the interference of light reflected from the surface illuminated through two holes in a screen. The Michelson setup is shown schematically in Fig. 10.12. The light source is a slit illuminated by a monochromatic lamp. The image of the slit is observed by means of a microscope. A series of screens with two apertures or, equivalently, a double-slit mechanism is placed in front of the surface under test. One aperture is fixed at the center, and the other changes its radial distance. Interference bands are observed in the microscope, with the central fringe exactly at the position of the slit image if the wavefront is spherical. When the wavefront is not spherical, the distance between the central fringe and the image of the slit gives the error. The fringe displacement, in fractions of the separation between fringes, is the wavefront error. By placing the movable hole over many places on the aperture, a complete mapping of the wavefront errors can be obtained.

Michelson's method was later modified to a diametral screen with several holes (Merland 1924; Gardner and Bennett 1921) for testing refractive elements and is also applicable to mirrors. The Gardner–Bennett setup is illustrated in Fig. 10.13, which shows that the interference fringes between adjacent holes

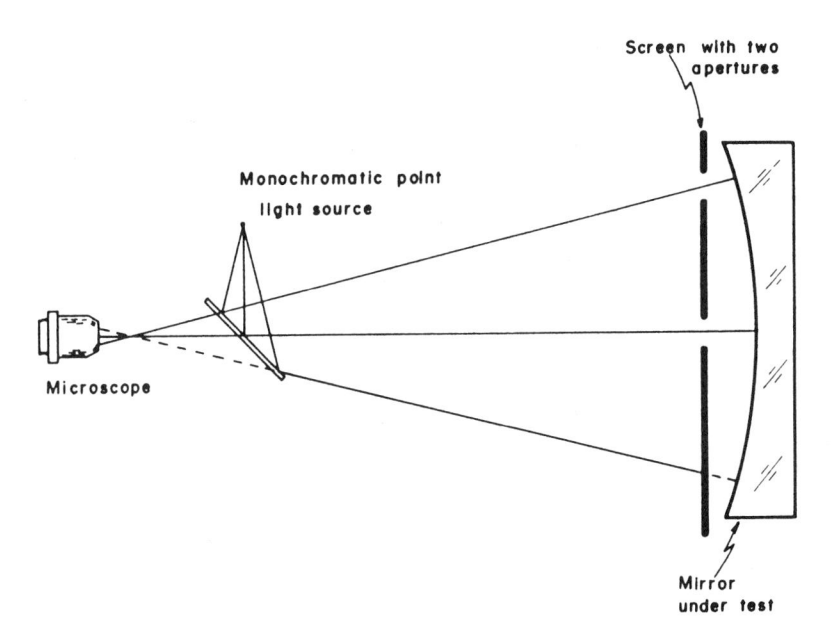

Figure 10.12. Michelson test setup. The central screen hole is not moved, whereas the other hole is moved in the survey of the entire spherical surface.

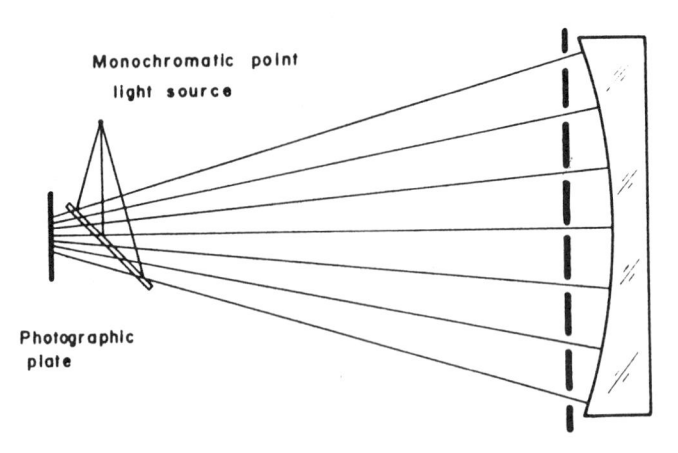

Figure 10.13. Gardner-Bennett test setup.

on the screen are recorded, not at the focus, as in the Michelson test, but outside or inside the focus, as in the Hartmann test. The defocusing must be sufficiently small, however, so that the light from adjacent holes does interfere. The deviation of the central fringe from its ideal position gives an indication of the phase difference between the wavefront regions covered by two holes, as in the Michelson test.

Although these forms of testing have not gained much popularity, they may be worth a thorough investigation for potential use in uncommon testing situations. A bibliography of publications on applications of various forms of this method is given at the end of this chapter.

10.7. HARTMANN TESTS OF THE FUTURE

A number of innovative modifications to the classical test have been proposed. The suggested variations cover possible masks that differ from the usual ones, and the application of Fourier transform methods to the data analysis. These are very promising approaches that need to be pursued further (Roddier 1990).

10.8. SUMMARY

In this chapter it has been shown that surface sampling through a screen is best suited for testing mirrors that cannot be placed in an environment free of vibrations or air turbulence. The most uniform and densest sampling is obtained with a square-array screen; the classical radial screens do not permit uniformity or high density in their surface sampling. Furthermore, the square array permits the damping of errors introduced by the data reduction used in obtaining the surface deviations of the mirror under test. This is done through the use of multiple integration paths in obtaining these surface deviations. The square-array method of testing requires the use of a precision centroid-finding method, and of an electronic computer for the data reduction because of the large numbers of sample points and computations needed in this reduction. With the use of electrooptical detector arrays and fast microcomputers, a rapid turn-around in testing can be achieved. Also this combination makes it possible to test the same surface several times, without changing it, and thus to reduce test errors by averaging the several test results. Finally, we have seen that there is continuing work being done in the application and analysis of the Hartmann test, with very promising innovations being considered.

REFERENCES

Ahlberg, J. H., E. N. Nilson, and J. L. Brown, *The Theory of Splines and Their Applications*, Academic Press, New York, 1967.

Anderson, D., "FRINGE Manual, Version 3," University of Arizona, Tucson, Arizona, 1982.

Cornejo, A. and D. Malacara, "Wavefront Determination Using Ronchi and Hartmann Tests," *Bol. Inst. Tonantzintla*, **2**, 127 (1976).

Dame, L. and F. Vakili, "Ultraviolet Resolution of Large Mirrors via Hartmann Tests and Two-Dimensional Fast Fourier Transform Analysis," *Opt. Eng.*, **23**, 759, (1984).

Freischlad, K. and C. Koliopoulos, "Wavefront Reconstruction from Noisy Slope or Difference Data Using the Discrete Fourier Transform," *Soc. Photo-Opt. Eng.*, **551**, 74, (1985).

Gardner, I. C. and A. H. Bennett, "A modified Hartmann Test Based on Interference," *J. Opt. Soc. Am.*, **11**, 441 (1921).

Ghozeil, I., "Use of Screen Rotation in Testing Large Mirrors," *Soc. Photo-Opt. Inst. Eng.*, **44**, 247 (1974).

Ghozeil, I. and J. E. Simmons, "Screen Test for Large Mirrors," *Appl. Opt.*, **13**, 1773, (1974).

Golden, L. J., "Dynamic Hartmann Test," *Appl. Opt.*, **14**, 2391, (1975).

Hartmann, J., "Bemerkungen uber den Bau und die Justirung von Spektrographen," *Zt. Instrumentenkd.*, **20**, 47 (1900).

Hartmann, J., "Objektuvuntersuchungen," *Zt. Instrumentenkd.*, **24**, 1 (1904).

Hausler, G. and G. Schneider, "Testing Optics by Experimental Ray Tracing with a Lateral Effect Photodiode," *Appl. Opt.*, **27**, 5160, (1988).

Hochgraf, N. A., "Angular Surface Measurements by Scanning Penta-Prism Test Improved and Extended," *J. Opt. Soc. Am.*, **61**, 655 (1971).

Hooker, R. B., "Automated, Non-interferometric Device for Testing Large Optical Surfaces," *J. Opt. Soc. Am.*, **61**, 655 (1971).

Kingslake, R., "The Absolute Hartmann Test," *Trans. Opt. Soc.*, **29**, 133 (1927-1928).

Korhonen, T. K., "Interferometric Method for Optical Testing and Wavefront Error Sensing," *Soc. Photo-Opt. Eng.*, **444**, 249, (1983).

Korhonen, T. K. et al., "Interferometric Optical Test and Diffraction Based Image Analysis," *Soc. Photo-Opt. Eng.*, **628**, 486, (1986).

Landgrave, J. E. A. and J. R. Moya, "Effect of a Small Centering Error of the Hartmann Screen on the Computed Wavefront Aberration," *Appl. Opt.*, **25**, 533, (1986).

Loibl, B., "Hartmann Tests on Large Telescopes Carried out with a Small Screen in a Pupil Image," *Astron. Astrophys. (Germany)*, **91**, 265, (1980).

Malacara, D., "Geometrical Ronchi Test of Aspherical Mirrors," *Appl. Opt.*, **4**, 1371 (1965).

Malvick, A. J., "Theoretical Elastic Deformation of the Steward Observatory 230 cm and the Optical Sciences Center 154-cm Mirrors," *Appl. Opt.*, **11**, 575 (1972).

Mayall, N. U. and S. Vasilevskis, "Quantitative Tests of the Lick Observatory 120-Inch Mirror," *Astron. J.*, **65**, 304 (1960); reprinted in *Lick Obs. Bull.* No. 567, 1960.

Merland, M. A., "Sur la Methode de MM. Michelson et Cotton pour l'Etude des Systemes Optiques," *Rev. Opt.*, **3**, 401 (1924).

Michelson, A. A., "On the Correction of Optical Surfaces," *Astrophys. J.*, **47**, 283 (1918).

Morales, A. and D. Malacara, "Geometrical Parameters in the Hartmann Test of Aspherical Mirrors," *Appl. Opt.*, **22**, 3957, (1983).

Pearson, E. T., "Design Philosophy of the Primary Mirror Supports for the KPNO 150-inch Telescope," Kitt Peak National Observatory, *AURA Engineering Technical Report* No. 5, Tucson, Ariz., 1968.

Pearson, E. T., "Hartmann Test Data Reduction," *Proc. SPIE.* **1236**, 628 (1990).

Rayces, J. L., "Exact Relation between Wave Aberration and Ray Aberration," *Opt. Acta*, **11**, 85 (1964).

Roddier, F., "Variations on a Hartmann Theme" *Soc. Photo-Opt. Eng.*, **1237**, 70 (1990).

Schulte, D. H., "A Hartmann Test Reduction Program," *Appl. Opt.*, **7**, 119 (1968).

Shack, R. V. and B. C. Platt, "Production and Use of a Lenticular Hartmann Screen" (abstract only), *J. Opt. Soc. Am.*, **61**, 656 (1971).

Simmons, J. E. and I. Ghozeil, "Double-Option Technique for Testing Large Astronomical Mirrors" (abstract only), *J. Opt. Soc. Am.*, **61**, 1586 (1971).

Stavroudis, O. N. and L. E. Sutton, *Spot Diagrams for the Prediction of Lens Performance from Design Data*, U.S. Department of Commerce, *National Bureau of Standards Monograph* No. 93, Washington, D.C., 1965.

Vitrichenko, E. A., "Methods of Studying Astronomical Optics Limitations of the Hartmann Method," *Soc. Astron.*, **20**, 373 (1976).

Vitrichenko, E. A., F. K. Katagarov, and B. G. Lipovetskaya, "Methods of Investigation of Astronomical Optics. II: Hartmann Method," *Izv. Spetz, Astrofiz. Obs.*, **7**, 167 (1975).

ADDITIONAL REFERENCES

Hartmann Test

Andersen, J. and J. V. Clausen, "Adjustment and Testing of Schmidt Telescopes," *Astron. Astrophys.*, **34**, 423, (1974).

Bowen, I. S., "Final Adjustments and Tests of the Hale Telescope," *Publ. Astron. Soc. Pac.*, **62**, 91 (1950).

Fox, P., "An Investigation of the Forty-Inch Objective of the Yerkes Obsevatory," *Astrophys. J.*, **27**, 237 (1908).

Golden, L. J., R. V. Shack, and P. N. Slater, "Study of an Instrument for Sensing Errors in a Telescope Wavefront," NASA Report No. NASA-CR-120353, University of Arizona, Tucson, Arizona, 1974.

Hartmann, J., "Objektivuntersuchungen," *Zt. Instrumentenkd.*, **24**, 33 (1904).

Hartmann, J., "Objektivuntersuchungen," *Zt. Instrumentenkd.*, **24**, 97 (1904).

Hoag, A. A. et al., *Installation, Tests and Initial Performance of the 61-Inch Astrometric Reflector*, Publication of the United States Naval Observatory, Vol. 20, Part 2, Washington, D.C., 1967.

Kamper, K. W., "Empirical Measures of the Effect of Coma on Photographic Astrometric Measurements" (abstract only), *Bull. Am. Astron. Soc.*, **3**, 372, (1971).

Kingslake, R., "Application of the Hartmann Test to the Measurement of Oblique Aberrations," *Trans. Opt. Soc.*, **27**, 221 (1925-1926).

Kingslake, R., "The Measurement of the Aberrations of a Microscope Objective," *J. Opt. Soc. Am.*, **26**, 251 (1936).

Lehmann, H., "Angewendung der Hartmann'schen Methode der Zonenprufung auf astronomische Objective, I," *Zt. Instrumentenkd*, **22**, 103 (1902).

Lehmann, H., "Angewendung der Hartmann'schen Methode der Zonenprufung auf astronomische Objective, II," *Zt. Instrumentenkd*, **22**, 325 (1902).

Malacara, D., "Hartmann Test of Aspherical Mirrors," *Appl. Opt.*, **11**, 99 (1972).

Malacara, D. and J. Castro, "A Simple Hartmann Test Interpretation," *Proc. SPIE*, **1164**, 2 (1989).

Martin, L. C., *Technical Optics*, Vol. 2, 2nd ed., Pitman, London, 1954, p. 280.

Plaskett, H. H., "The Oxford Solar Telescope and Hartmann Tests of Its Performance," *Mon. Not. R. Astron. Soc.*, **99**, 219 (1938).

Plaskett, J. S., "82-Inch Mirror of McDonald Observatory," *Astrophys. J.*, **89**, 84 (1939).

Richardson, M. F., "Implementation of the Shack-Hartmann Test for Astronomical Telescope Testing," Ph.D. Dissertation, University of Arizona, Tucson, Arizona, 1983.

Snezhko, L. I., "On the Accuracy of the Hartmann Method for Testing Aspherical Wavefronts," *Sov. J. Opt. Technol.*, **47**, 505, (1980).

Snezhko, L. I., "An Analysis of the Foundations of the Hartmann Test Method," *Bull. Spec. Astrophys. Obs.*, **14**, 1 (1981).

Southwell, W. H., "Wave-Front Estimation from Wave-Front Slope Measurements," *J. Opt. Soc. Am.*, **70,** 998 (1980).

Statton, Charles M., Mauri Donn Bauer, and Jurgen R. Meyer-Arendt, "Evaluation of Ophthalmic Spectacle Lenses Using the Hartmann Test," *Am. J. Optom. and Physiol. Opt.*, **58,** 766 (1981).

Stetson, H. T., "Optical Tests of the 69-Inch Perkins Observatory Reflector," *J. Opt. Soc. Am.*, **23,** 293 (1933).

Suzuki, K., I. Ogura and T. Ose, "Measurement of Spherical Aberration Using a Solid-State Image Sensor," *Appl. Opt.*, **18,** 3866 (1979).

Tull, R. G., "Shop-Testing a 107-Inch Telescope Mirror," *Sky Telesc.*, **36,** 213 (1968).

Van Breda, I. G., "The Adjustment of Telescopes Using the Hartmann Test," *Mon. Not. R. Astron Soc.*, **144,** 73 (1969).

Van Breda, I. G., "Adjustment of Telescopes," *Phys. Bull. (England)*, **21,** 207 (1970).

Van Zuylen, L., "Zur qualitative Untersuchung der spharischen abweichung optischer Systema," *Physica.*, **3,** 243 (1936).

Williams, T. L., "Dynamic Hartmann Test," *Appl. Opt.*, **15,** 599, (1976).

Washer, F. E., "An Instrument for Measuring Longitudinal Spherical Aberration of Lenses," *J. Res. Nat. Bur. Stand.*, **43,** 137 (1949).

Zverev, V. A. et al., "Mathematical Principles of Hartmann Test of the Primary Mirror of the Large Azimuthal Telescope," *Sov. J. Opt. Technol.*, **44,** 78, (1977).

Zverev, V. A. et al., "Use of Hartmann Diaphragm in a Converging Beam to Test Telescopes in an Observatory," *Soc. J. Opt. Technol.*, **47,** 110, (1980).

Michelson and Gardner-Bennett Tests

Cassasent, D. and T. Luu, "Performance Measurement Techniques for Simple Fourier Transform Lenses," *Appl. Opt.*, **17,** 2973 (1978).

De, M and M. K. Sen Gupta, "Measurement of Wave Aberrations of Microscope and Other Objectives," *J. Opt. Soc. Am.*, **51,** 158 (1961).

Vaidya, W. M. and M. K. Sen Gupta, "Measurement of Axial and Off-Axis Geometrical Aberrations of Microscope Objectives," *J. Opt. Soc. Am.*, **50,** 467 (1960).

11

Star Tests

W. T. Welford

The star test is conceptually perhaps the most basic and simplest of all methods of testing image-forming optical systems: We examine the image of a point source formed by the system and judge the image quality according to the departure from the ideal image form. In principle the test can be made quantitative by, for example, photoelectric measurement techniques, but in practice the star test in the workshop is almost always carried out visually and semiquantitatively. We shall discuss mainly visual techniques. We can divide star testing methods into two groups: (a) those in which very small aberrations, near or below the Strehl tolerance limit,[†] are examined, and (b) those in which relatively large aberrations are studied. Group (a) is typified by tests on microscope and telescope objectives, and (b) by tests on camera lenses in which, for example, the star test is used to plot the astigmatic field surfaces or to estimate transverse chromatic aberration.

The monochromatic image of a point source, or point spread function as it is often called, has a very complex structure, particularly in the presence of aberrations. The structure depends in a complicated way on the geometrical aberrations, but it would be out of place in a book on practical methods to go deeply into this aspect. Moreover, it is always possible in principle to calculate the point spread function from the aberrations, although the calculation may in practice be costly in computer time, but for the present purposes we should like to be able to estimate the aberrations from the form of the point spread function; this, however, is generally impossible in principle. If the aberration is axially symmetrical, this calculation could be done from very careful measurements of the light intensity in the star image. We have to make estimates based on experience and on the many examples of point spread functions that have been computed and photographed from known aberrations. Thus the star test is semi-

---

[†]When the intensity of the central maximum is 80% of that of the ideal central maximum, the Strehl tolerance limit is said to have been reached.

Optical Shop Testing, Second Edition, Edited by Daniel Malacara.
ISBN 0-471-52232-5 © 1992, John Wiley & Sons, Inc.

quantitative, and considerable experience is needed to get the best results from it. Nevertheless it is an important testing technique because it is rapid and reliable in experienced hands and because it is very sensitive. For example, the test is used in the final adjustment of a critical air space in microscopic objectives of high numerical aperture, since in this process it is desirable to try a new spacing and retest rapidly.

There are not many publications about the star test, probably because it depends so much on experience. The basic reference is still the work of Taylor (1891). Useful descriptions of the method have also been given by Martin (1961) and by Twyman (1942); the latter reproduces much of Taylor's descriptive material.

In discussing the star test we shall mostly assume that the pupil of the system under test is filled with light of uniform intensity, and we shall also assume that the star image is formed at a reasonably large convergence angle, say, not less than 0.01. Modest variations of intensity across the pupil generally do not affect the star image much, but we make an exception in the case of Gaussian beams from lasers. (A true Gaussian beam has an indefinitely large aperture, while at the same time the theory by which the Gaussian beam profile is derived uses the paraxial approximation! This apparently blatant contradiction is resolved by the fact that in practice the intensity in the Gaussian profile is negligibly small at distances from the central maximum greater than, say, three times the $1/e^2$ radius.) Gaussian beams will be discussed in Section 11.1.6, and the effects of very small convergence angles will be mentioned in Section 11.1.7. At very large convergence angles, say, numerical aperture greater than 0.5, the shape of the point spread function is again different in detail from that to be described here, and an excellent study of such effects was recently published by Stamnes (1986); for reasons of technical convenience it is very unlikely that a practical star test would be carried out at such large numerical aperture (NA); the usual procedure is to transform the NA by a reliably aberration-free system such as a well-corrected microscope objective, itself having been star tested. Thus for our main descriptions of the unaberrated point spread function (Figs. 11.1 to 11.9) we have used the usual scalar wave approximation as the basis of the calculations.

In Section 11.1 of this chapter we discuss principles and results of calculations, in Section 11.2 we explain the techniques as applied to systems with small aberrations, and in Section 11.3 we deal with large aberrations.

11.1. PRINCIPLES OF THE STAR TEST FOR SMALL ABERRATIONS

In star testing systems that are nearly diffraction limited (possibly apart from chromatic aberration), we need a background of information about the appear-

11.1. PRINCIPLES OF THE STAR TEST FOR SMALL ABERRATIONS

ance of the point spread function with small aberrations. This is given in Sections 11.1.1 to 11.1.5, but without mathematical details.

Photographs and computations of point spread functions are to be found throughout the literature on applied optics, although often not in a form particularly applicable to practical star testing. One of the most useful sets of photographs is that used as a frontispiece by Taylor (1891) and subsequently reproduced in many other publications, for example, Martin (1961). Among other photographs we may mention the work of Nienhuis (1948), reproduced by, for example, Born and Wolf (1975); however, we should note Nienhuis took some of his very beautiful spread function photographs with a coherent background in order to enhance the secondary rings and fringes, and this rather falsifies them for the purposes of star testing. In fact all photographs are to be mistrusted for star testing: The nonlinearity of the photographic emulsion is compounded with that of the halftone process, and the result cannot be relied on over any reasonable dynamic range. We therefore give only graphical presentations of spread functions in this chapter.

11.1.1. The Unaberrated Airy Pattern

Figure 11.1 shows the Airy pattern, the monochromatic aberration-free point spread function for a system with a circular aperture of uniform transmission, and Fig. 11.2 shows the same with logarithmic vertical scale. What is plotted in these two figures is the light intensity (vertical) against the radial distance from the center of the image (horizontal). The light intensity is scaled to unity at the center of the pattern. The radial coordinate z, which lies in the image plane, is given by

$$z = \frac{2\pi}{\lambda} \sin \alpha \eta \qquad (11.1)$$

where λ is the wavelength of the light, α is the convergence angle (cone semiangle) of the image-forming beam, and η is the actual radial distance. Thus the radial coordinate as shown is dimensionless and is often said to be in z units or diffraction units. These are such that the radius of the first dark ring in the Airy pattern is 3.83 z units, so that the size of a z unit in a point spread function under inspection can be estimated immediately. Since the Airy pattern is of basic importance in star testing, we give in Table 11.1 some of its numerical properties. Table 11.1 includes the encircled flux function, which gives the proportion of the total flux within a given radius from the center. The encircled flux (sometimes called the encircled energy) has occasionally been proposed for use as a development of the star test (see, e.g., Barakat and Newman 1963), but probably the practical difficulties are too great for it to be generally adopted. However, the function is useful as a guide to the general properties of the Airy pattern. This function is plotted in Fig. 11.3.

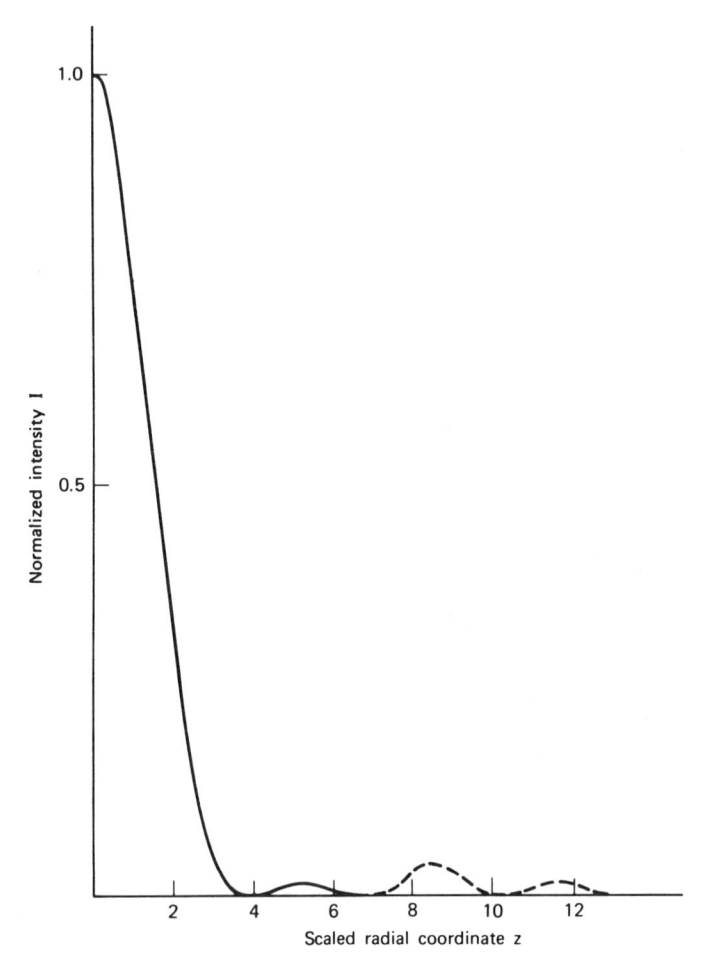

Figure 11.1. The Airy pattern, the aberration-free image of a bright point formed in monochromatic light by a system with a circular aperture. The function plotted is

$$I = \left[\frac{2J_1(z)}{z}\right]^2$$

where $z = (2\pi/\lambda) \sin \alpha \eta$, $\sin \alpha$ is the convergence angle of the image-forming beam, and η is the radial distance from the center of the aperture. The second and third rings are plotted at 10 times the actual ordinate.

Figure 11.4 shows an alternative way of presenting the Airy pattern; the vertical height of the figure represents the relative intensity. Such a figure is sometimes called a diffraction solid. Figure 11.5 represents the same solid with increased vertical scale, to show the details of the outer rings.

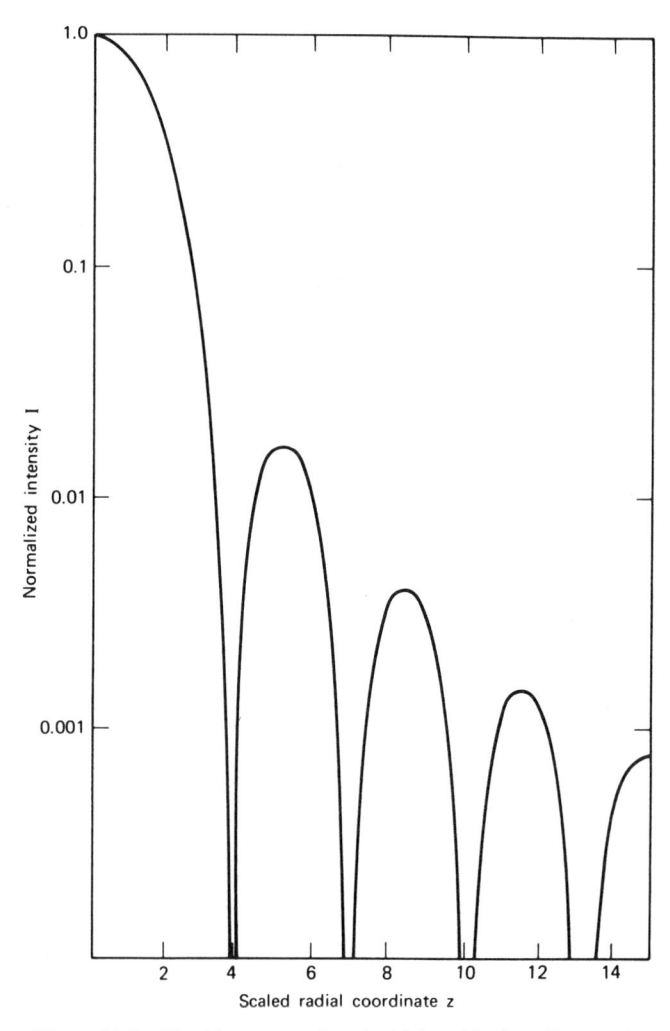

11.1. PRINCIPLES OF THE STAR TEST FOR SMALL ABERRATIONS

Figure 11.2. The Airy pattern plotted with logarithmic ordinate scale.

11.1.2. The Defocused Airy Pattern

One of the earliest findings of practical experience in star testing is that it is very useful to exmaine the defocused point spread function on both sides of the best focal plane. This may seem paradoxical, but it is found to be much easier to diagnose the aberrations from the defocused image than from the image at best focus. We therefore show in Fig. 11.6 contours of constant intensity (isophots) in a plane containing the optical axis. The axes correspond to a radial coordinate in the image plane (vertical axis) and to the defocus distance along

Table 11.1. The Airy Pattern, $I = \left(\dfrac{2J_1(z)}{z}\right)^2$

1. Radii of dark rings (zeros of first-order Bessel function, J_1)

Number of ring:	1	2	3	4	5	6
Radius:	3.83	7.02	10.17	13.32	16.43	19.62

2. Radii and intensities of bright rings (zeros of second-order Bessel function, J_2)

Number of ring:	1	2	3	4
Radius:	5.14	8.42	11.62	14.80
Intensity:	0.0175	0.00416	0.00160	0.000781

3. Halfwidth: the radius at which the intensity = 0.5 is 1.615. (*Note*: This is appreciably less than half the radius of the first dark ring.)

4. Encircled flux: the proportion of the total light flux inside a circle centered on the central maximum; it is given by $1 - J_0^2(z) - J_1^2(z)$.

Radius:	0.5	1	2	3	4	5	6	7	8
Flux:	0.0605	0.221	0.617	0.817	0.838	0.861	0.901	0.910	0.916

See also Fig. 11.3.

---

the principal way. The image plane coordinate is the same dimensionless coordinate as was used in Figs. 11.1 and 11.2, and the defocus coordinate, explained in the caption, is also dimensionless. The scales are such that the geometrical cone of rays would have a semiangle of 45°, but by using appropriate values of the convergence angle α and the wavelength λ, the diagram can represent any cone.[†] The complete light distribution is obtained by revolving the figure about the principal ray.

Several features of this distribution are worth noting. It is symmetrical about the true focal plane.[‡] There are zeros ("dark spots") along the axis, equally spaced (apart from one missing at the focal plane) and alternating with maxima. The intensity along the axis is shown in Fig. 11.7. In Fig. 11.6 the out-of-focus patterns consist of bright and dark rings, spreading out with increasing defocus. This cannot be seen very clearly in Fig. 11.6 but is more evident in Fig. 11.8, which is again a set of isophots, but taken over a greater range of both coor-

---

[†] It should be noted, however, that this applies only to the case in which the diameter of the first dark ring is much less than the diameter of the exit pupil of the optical system. This situation would, of course, almost invariably exist in most workshop testing situations, but occasionally it may happen that the convergence angle is very small and then Fig. 11.6 would not be a true picture. This happens more frequently in beams with Gaussian profiles, as produced by a single-mode TEM$_{00}$) laser, since these are often used nearly collimated.

[‡] Again, this is not true unless the central maximum is much smaller than the exit pupil.

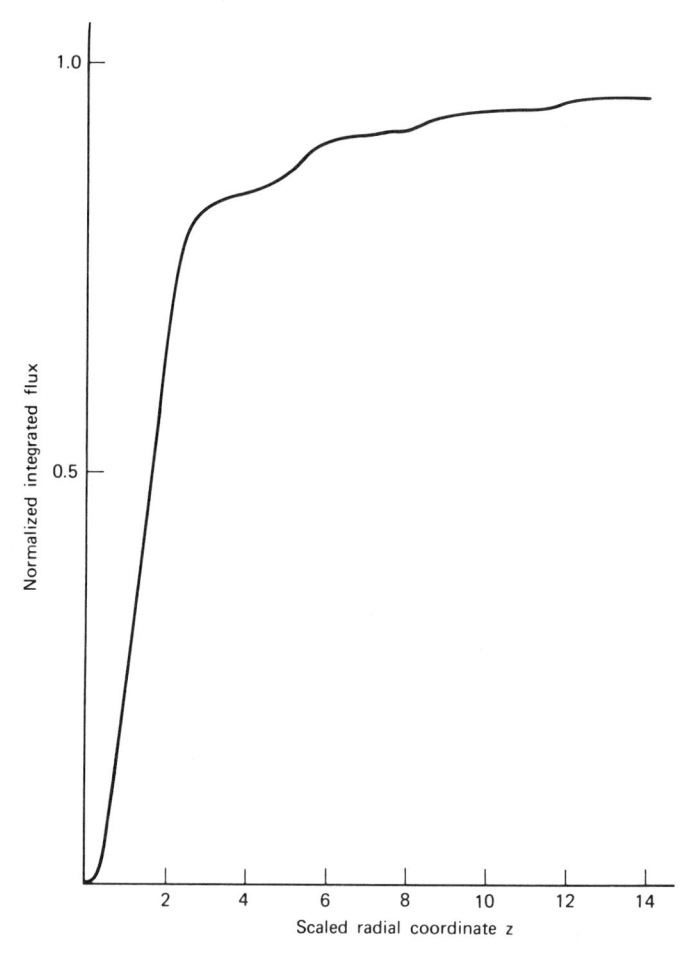

11.1. PRINCIPLES OF THE STAR TEST FOR SMALL ABERRATIONS

Figure 11.3. The "encircled energy," that is, the proportion of total light flux inside a circle of radius z in the image plane. See Table 11.1.

dinates and showing only one quadrant of the symmetrical pattern. The isophots are labeled with the logarithm, to base 10, of the relative intensity, so as to give a greater range of intensities in the diagram. From Fig. 11.8 it can be seen that a typical defocused pattern, easily recognizable by anyone who has looked at point spread functions, will show a series of roughly equally spaced rings, increasing in intensity toward the outside and having the outermost ring brighter, wider, and larger in diameter than would be expected from the general progression of the rings. In fact the rings are not so regularly graded as they appear at first to be; this can be seen from a very careful inspection of the monochro-

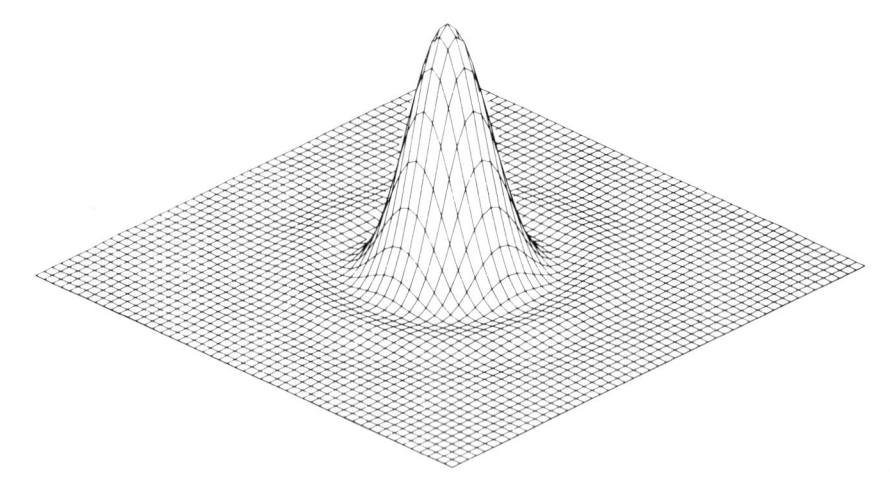

Figure 11.4. The diffraction solid for the Airy pattern; 2.5 squares in the horizontal plane correspond to 1 z unit. (Computed and plotted by M. W. L. Wheeler.)

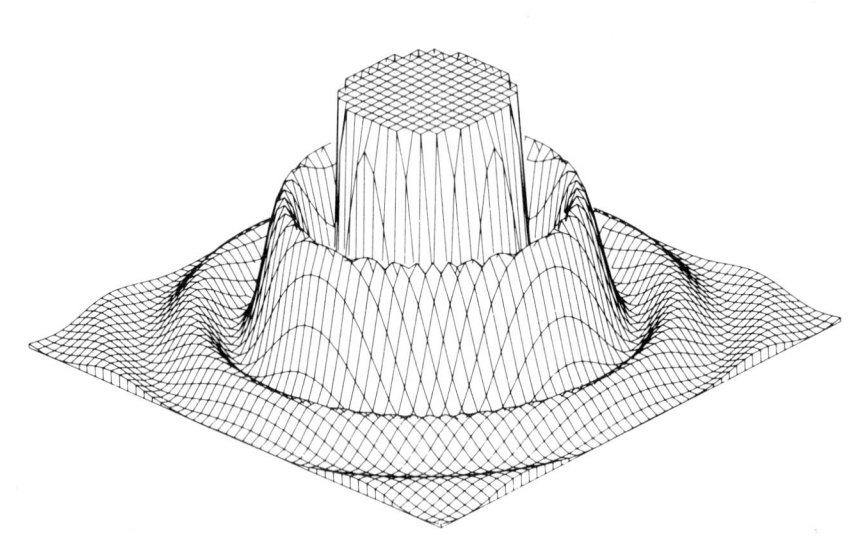

Figure 11.5. Similar to Fig. 11.4, but with increased vertical scale truncated to relative intensity 0.03. (Computed and plotted by M. W. L. Wheeler.)

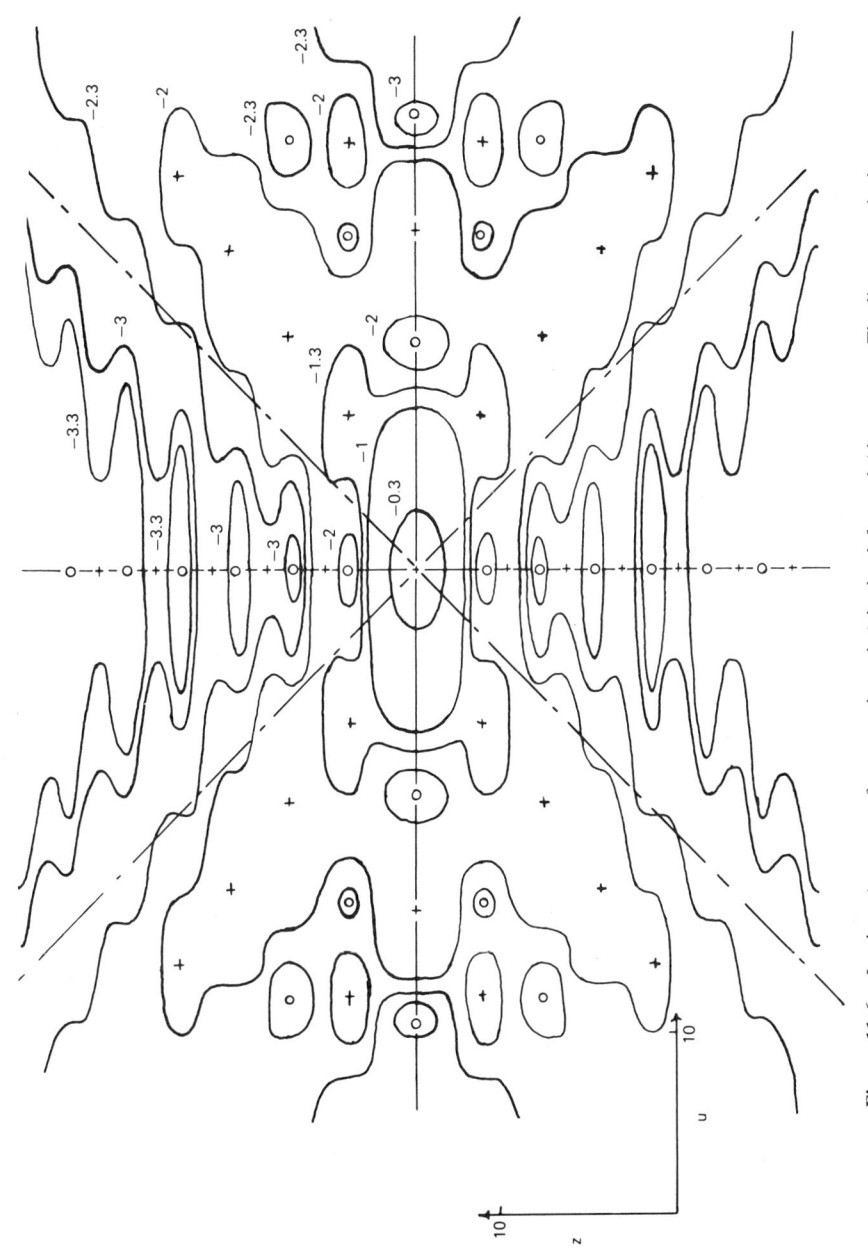

Figure 11.6. Isophots (contours of constant intensity) in the defocused Airy pattern. The line $z = u$ is the boundary of the ray cone. The circles denote minima (zeros on the axes); the crosses, maxima. The contours are labeled with the logarithm of the intensity. (From values computed by J. C. Dainty.)

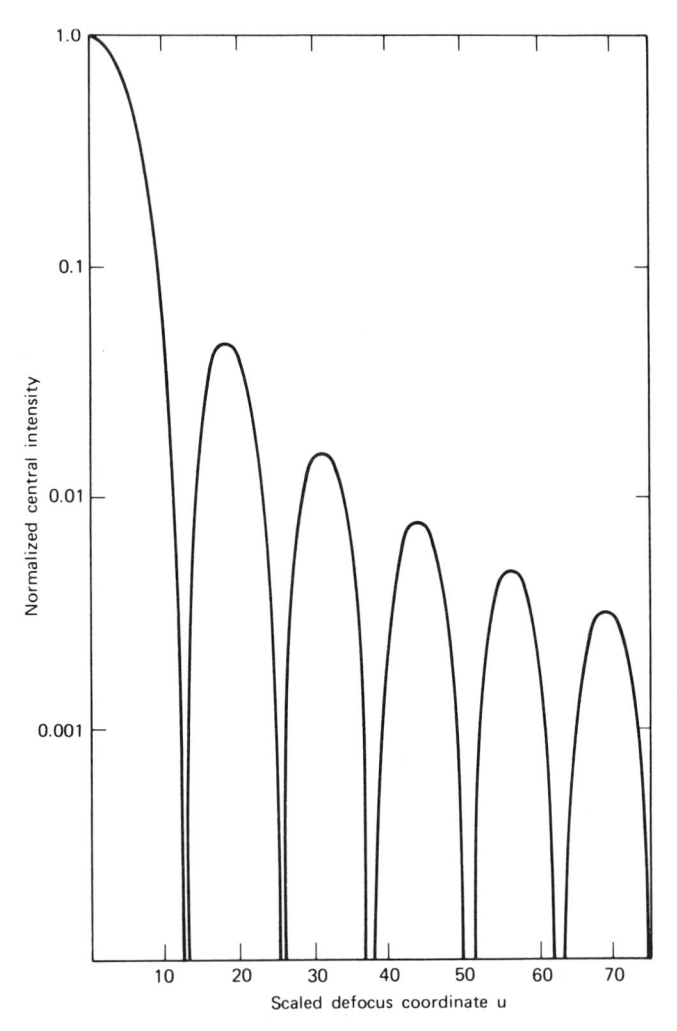

Figure 11.7. The intensity at the center of the defocused Airy pattern. The quantity plotted is $(\sin\frac{1}{4}u/\frac{1}{4}u)^2$, where $u = (2\pi/\lambda)\zeta\sin^2\alpha$, α is the convergence angle of the beam, and ζ is the defocus distance.

matic point spread function or from a study of Fig. 11.8. Nevertheless this apparent regularity is easily upset by small aberrations, and this is a very useful property of the star test. As the defocus is steadily increased, the appearance is as if new rings appear from the center and spread out, taking the rest of the pattern with them like the ripples on a pond.

Some of these effects and those described in Section 11.1.5 are illustrated by photographs taken by Cagnet et al. (1962). Beiser (1966) gave a perspective

Figure 11.8. Isophots in the defocused Airy pattern; this diagram covers a greater range of defocus than Fig. 11.6. The chain line ($z = u$) denotes the boundary of the cone of rays forming the pattern. The circles denote minima (zeros on the axes); the crosses, maxima. The contours are labeled with the logarithm of the intensity, scaled to unity at the origin. (From values computed by J. C. Dainty.)

sketch of the intensity distribution through focus, and Taylor and Thompson (1958) reported careful measurements to verify the predicted distributions.

11.1.3. Polychromatic Light

The descriptions in Sections 11.1.2 and 11.1.3 apply, of course, to monochromatic light. All features of the point spread function (diameters of dark rings, distances between axial zeros, etc.) scale in direct proportion to the wavelength. Thus, even if an optical system has no chromatic aberration according to geometrical optics (i.e., a purely reflecting system), we would still expect to see color effects when star testing; however, in practice these are not very noticeable. We can, therefore, use white light for testing most reflecting systems without danger of confusion from the weak chromatic effects.

In refracting systems, however, the star test easily reveals "secondary spectrum," that is, the residual uncorrectable chromatic aberration. We are not always concerned to actually test chromatic correction by star testing, because achromatism depends on overall choices of glass types and powers of components and is therefore affected only by relatively gross errors in construction that would affect other aberrations more markedly. Thus in testing refracting systems we should use either a fairly narrow wavelength band, selected by means of a filter (Section 11.3.1), or a laser (but see Section 11.2.2). In this way the monochromatic aberrations can be more clearly seen; this is usually the object in workshop tests, since it is these aberrations that are susceptible to correction by refiguring or by tuning air spaces.

Linfoot and Wolf (1952) calculated the total flux in polychromatic star images from refracting telescopes, weighted according to the visibility curve of the human eye. These results are of great interest for optical designers, particularly since the calculations were done for existing telescope objectives, but they are probably not very useful as background to the star test, since in actual operation the varying colors seen probably make at least as much impression as the total brightness. However, they do show how the interesting color changes that are observed arise from secondary spectrum effects.

11.1.4. Systems with Central Obstructions

Most large telescopes and many other optical systems have a central "hole" in the pupil, due to a secondary mirror. Some of the effects of this central obstruction on the point spread function are rather surprising. There are three main effects: (a) The central maximum becomes narrower; (b) the outer rings become irregular in brightness—some brighter, some dimmer; and (c) the spacing along the axis of the dark spots increases. A detailed theoretical study made by Linfoot and Wolf (1953) showed that, if the obstruction ratio (diameter, not area) is ϵ, the axial spreading of the features is according to the factor $(1 - \epsilon^2)^{-1}$. Figure 11.9 shows the in-focus pattern for an obstruction ratio $\epsilon = 0.25$, which

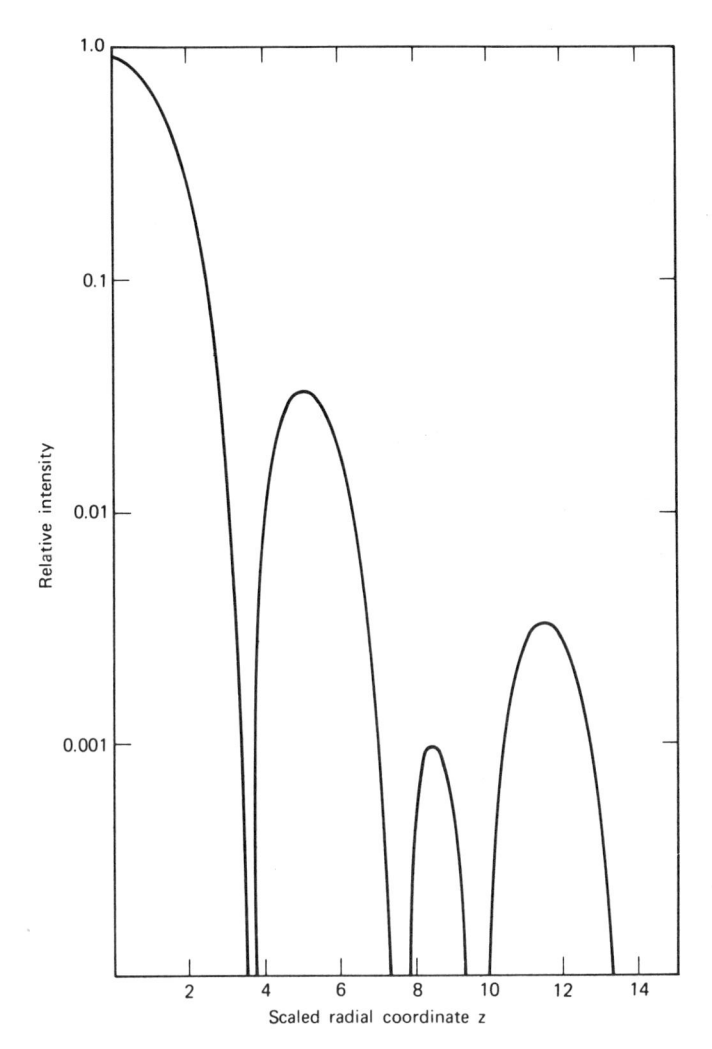

11.1. PRINCIPLES OF THE STAR TEST FOR SMALL ABERRATIONS

Figure 11.9. The in-focus point spread function for a circular aperture with a central obstruction one quarter of the full diameter. This should be compared with Fig. 11.2 to see the effect on the central intensity and the outer rings.

is typical for an astronomical telescope. Features (a) and (b) may be seen on comparison with Fig. 11.2. Linfoot and Wolf (1953) give isophot diagrams showing the out-of-focus effects.

11.1.5. Effects of Small Aberrations

In an optical design context "small aberration" usually means a distortion of the wavefront from the ideal spherical shape of order of magnitude $\lambda/4$. This is so because the system of aberration tolerances based on the Strehl criterion leads to aberration tolerances of this order. It is generally accepted that, when the aberrations are at this tolerance limit, as described later, the effect on the formation of images of extended objects is negligible, and the optical system is said to be diffraction limited.[†] However, even if images of extended objects are barely affected by a quarter-wavelength of aberration, this amount is very easily detected by the star test, and one could probably see $\lambda/20$ or even less, depending on the type of aberration.

The Strehl aberration tolerance system is explained by, for example, Born and Wolf (1975). It depends on the concept that the initial effect of introducing a small amount of aberration of any kind into the wavefront forming the point spread function is to reduce slightly the maximum intensity: The width at half maximum of the central maximum stays unchanged, and the flux taken from the center is redistributed to the outer rings. A 20% drop in the maximum intensity is generally taken as the tolerance limit, and simple formulas are given (e.g., by Born and Wolf 1975) for the amounts of different aberrations that will produce this tolerance limit.

However, the eye is not very good at judging changes in absolute intensity levels, so that the Strehl criterion is not very useful in workshop practice, although it is invaluable in optical design. Also it is found that, as mentioned earlier, one can see much less than the Strehl limit in many cases. This can be understood by looking at the photographs of aberrated point spread functions in some of the references already cited; to choose an example at random. Born and Wolf (1975) reproduce in their Figure 9.8 a photograph of coma at about the Strehl tolerance limit, taken by Nienhuis (1948), and it is quite easy to see that this is an aberrated point spread function. The same point is made by Fig. 11.10, which shows the diffraction solid for coma at the Strehl limit. There seems no doubt, on the basis of experience, that the semiqualitative methods beautifully described by Taylor (1891) are extremely sensitive; however, it is difficult to set a limit to the sensitivity since this depends on the kind of aberration, for example, whether the wavefront shape varies rapidly or slowly across the aperture. In a study of spherical aberration, Welford (1960) suggested that it is possible to detect $\lambda/20$ of slowly varying aberration and $\lambda/60$ of rapidly varying aberration.

11.1.6. Gaussian Beams

By convention the equivalent of a focus for a Gaussian beam is called the beam waist; the beam profile is the same at all points along the beam, simply scaling in intensity and width depending on the distance from the beam waist. There are well-known formulas relating the width of the beam to the distance from

---

[†]Sometimes the term "Rayleigh limit" is used, since Lord Rayleigh proposed the quarter-wavelength criterion.

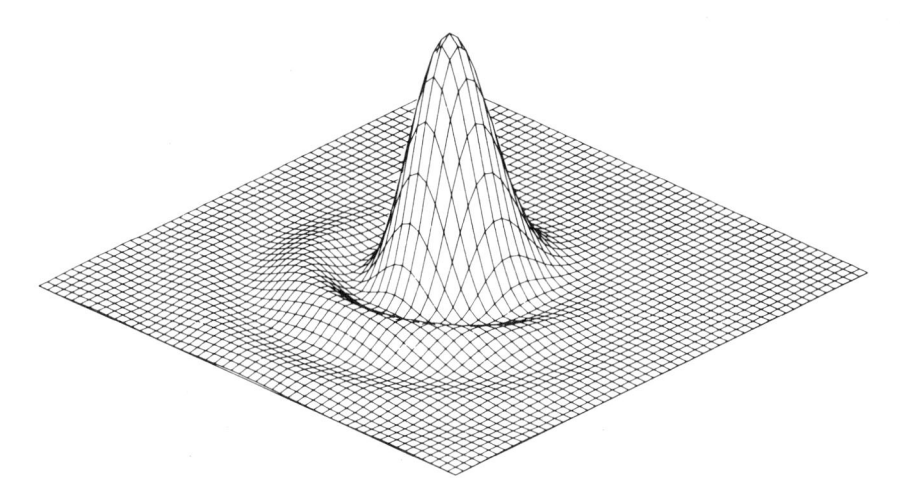

11.1. PRINCIPLES OF THE STAR TEST FOR SMALL ABERRATIONS

Figure 11.10. The diffraction solid for 0.6 of a wavelength of coma (the Strehl tolerance limit). (Computed and plotted by M. W. L. Wheeler.)

the waist and giving the curvature of the phase front (e.g., Kogelnik and Li 1966). If a system is to be used with Gaussian beams, then by implication the lenses and other components must have enough spare aperture to ensure negligible truncation of the Gaussian profile; the equivalent of the star test would then be to check that the final beam waist has a true Gaussian profile; but since the defocused waist does not change qualitatively in a Gaussian beam, this is not very easy to assess. In practice, it is probably better to expand the input beam and test the system with fully filled aperture (so-called hard-edge beam) and with more or less uniform amplitude; this can be an overrigorous test, since the outer parts of the aperture have less amplitude with a Gaussian beam. However, unless the Gaussian beam very much underfills the aperture, the difference is negligible. If a Gaussian beam *is* used in the test, there is the danger that any slight truncation of the beam at some point in the system will introduce ringing or fringing in the phase front; it is then not easy to tell if this is due to truncation or to aberrations in the optical system.

11.1.7. Very Small Convergence Angles (Low Fresnel Numbers)

When a hard-edged beam is almost collimated, so that the focus is formed with a beam of very small convergence angle, the diffraction pattern in the focal region is qualitatively different from that described in Section 11.1. This is discussed in some detail by Stamnes (1986) and by Li and Wolf (1984). From the point of view of star testing the most marked effect is that for an aberration-free beam the intensity distribution is no longer symmetrical on either side of the focus; this would considerably complicate the task of assessing the quality of the system, so that it is advisable to avoid testing in a nearly collimated space.

This raises the question, how nearly collimated may we be without upsetting the simpler descriptions of the star image in Section 11.1? This is best answered in terms of the *Fresnel number* of the beam: Let a be the radius of the exit pupil and let R be the radius of curvature of the emerging phase front, that is, R is the distance from the pupil to the focus then the Fresnel number N is defined by

$$N = a^2/\lambda R \qquad (11.2)$$

The significance of the Fresnel number is that $N/2$ is the number of whole wavelengths depth of curvature of the phase front at the edge of the pupil, and it is shown in the references quoted earlier that the asymmetric effects in the focal region occur when N is not very large, say, less than 10. Thus star testing should always be done in a space where the Fresnel number is not too small according to this criterion.

11.2. PRACTICAL ASPECTS WITH SMALL ABERRATIONS

We now discuss in more detail the effects to be looked for in visual star testing for small aberrations. In what follows we assume that approximately monochromatic light is used.

11.2.1. Effects of Visual Star Testing

Asymmetry of the in-focus pattern denotes a comalike aberration. This may be either genuine coma due to working off-axis or a "manufacturer's aberration" due to poor centering of components (often quite strong asymmetric color effects can be seen because of bad centering), to nonsymmetrical polishing of surfaces, or even to nonuniform distribution of the refractive index in a lens component. A rare source of comalike effects is an asymmetrical phase change on reflection at a mirror surface. This can be caused by uneven coating of multilayer dielectrics; but if a dielectric-coated mirror is used to fold a beam at a large angle, and if the beam has a large convergence angle, enough asymmetrical phase change may result to give comalike effects, due to varying angles of incidence across the beam.

In practice, asymmetries always show up more clearly in the defocused image: With coma, in particular, there is a very marked effect of the out-of-focus asymmetry, which is useful in finding the axis or center of the field of a system; this is what is meant by the term "squaring-on," applied to telescope objectives for astronomical use.

Small amounts of astigmatism show up as a "Maltese cross" effect in focus, and again there is a considerable gain in defocusing. In particular, on moving

11.2. PRACTICAL ASPECTS WITH SMALL ABERRATIONS

steadily through focus from one side to the other, the effect of a switch in the direction of the astigmatic focal lines is easily detected for much less than the Strehl limit.

Of course, neither astigmatism nor coma should be present in the axial point spread function of a well-constructed system, and the most important thing to be able to judge is the state of spherical aberration correction. Here almost nothing can be obtained from studying only the in-focus image (but this is not so for large aberrations; see Section 11.3.1), and it is essential to use defocus. In describing the effects we use the generally accepted terms "undercorrection" and "overcorrection"; undercorrection means that for primary spherical aberration the rays from the edge of the pupil focus nearer the optical system than the paraxial rays, if a real image is formed, and vice versa for overcorrection. In terms of wavefront shapes a zone of the aperture has undercorrection if the wavefront in that zone is in advance of the reference sphere (ideal or unaberrated wavefront), and again, vice versa for overcorrection.

With primary spherical aberration or with spherical aberration of any single higher order, we find that undercorrection produces sharper, well-defined rings on the side of focus nearer the optical system ("inside focus") with a particularly bright outermost ring, whereas with "outside focus" all rings are blurred and show less contrast and the outer ring, in particular, fades to a vague smudge with poor definition. For overcorrection the same appearances are seen, but with "inside focus" and "outside focus" interchanged. Figure 11.11 shows diagramatically the case of undercorrection in relation to the ray diagram; it can

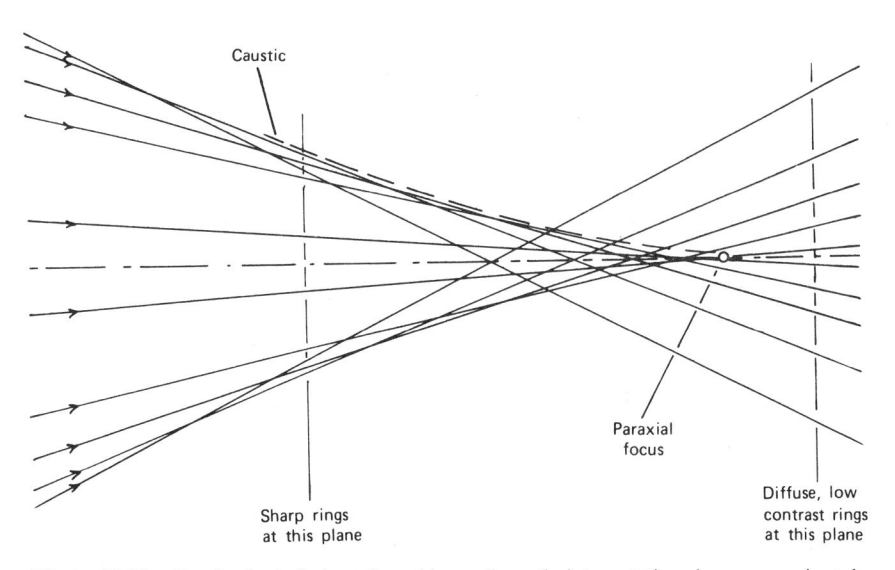

Figure 11.11. Zonal spherical aberration: this ray-theoretic interpretation shows approximately how effects due to the zone would appear.

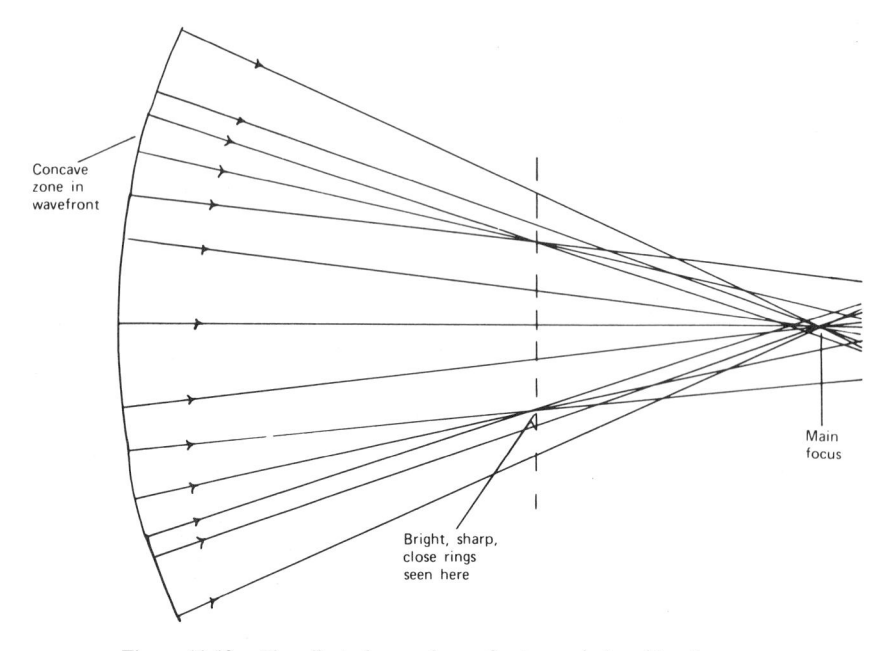

Figure 11.12. The effect of a zonal wavefront error in bunching the rays.

be seen that the much brighter outermost ring occurs on the side where the ray caustic is formed. This is a convenient way to remember how to relate these effects, since we would naturally expect a great concentration of light flux near the caustic. In other words we see more light intensity near a greater concentration of rays. This rule is generally true, provided that we are far enough from the main focus; it offers a simple way of interpreting zonal aberration effects. Thus suppose that we go a considerable distance inside focus and we see that at, say, two thirds of the diameter of the spread-out pattern the rings are brighter and sharper; this indicates that the rays from the corresponding part of the aperture are bunching together here or, alternatively, that the corresponding zone of the wavefront is more concave than it should be. The opposite effect of fainter and more diffuse rings would be seen outside focus. Figure 11.12 shows the effect of a zonal wavefront error in bunching the rays.

11.2.2. The Light Source for Star Testing

In practice we never use a real star as a test object for star testing because atmospheric turbulence causes the star image to vary in intensity, position, and aberrations to the extent that a critical appraisal is impossible.

The laboratory artificial star is a pinhole with a lamp of the right spectral composition. The pinhole must be small enough to be quite unresolved, that is,

11.2. PRACTICAL ASPECTS WITH SMALL ABERRATIONS

its angular subtense at the objective must be much less than λ/D, where D is the diameter of the objective aperture. In practice it is easy to check that the pinhole is the correct size by noticing whether any trace of the pinhole edges can be seen in the star image, since most pinholes are irregular enough in outline to show an image; if only diffraction structure can be seen, the pinhole is small enough.

Very costly pinholes are available in special mounts for use as spatial filters in laser beams. For workshop use it is cheaper and simpler to employ electron microscope aperture stops. These come as thin, electroformed copper disks about 3 mm in diameter with holes from about 5 to 50 μm in diameter, as ordered; the holes are not always very accurately circular in shape, but this does not matter for testing purposes.

Before electron microscope apertures became generally available, several other devices were used, as described in the references cited at the beginning of this chapter. Among these was the reflection of a distant light source in a small, clean globule of mercury, obtained by condensing mercury vapor onto a microscope slide. (In particular this has been suggested for star-testing microscope objectives, but see Section 11.2.4 for modern alternatives.) It is worth mentioning in connection with workshop practice that mercury is a cumulative poison with that appreciable amounts can be absorbed from mercury vapor at room temperature and pressure. This method should therefore never be used.

An intense light source, such as a quartz–halogen lamp or a high-pressure mercury lamp focused onto the pinhole, is desirable in order to be able to pick up the star image easily. A filter can be placed immediately after the pinhole, as in Fig. 11.13. It is usually not advisable to use a laser as a source for star testing. The main objection is that the coherence length of a helium–neon laser is long enough to produce interference effects between many beams multiple reflected off the surfaces of a refracting system. These effects can be bright enough to obscure the details of the star test, particularly since the workshop testing would probably be done before any antireflection coating was put on the surfaces. Another objection to the use of a laser, for refracting systems at any rate, is that the system may be corrected for a different wavelength and the spherical aberration correction vary appreciably with wavelength. However, for testing large mirror systems a laser is probably the best source, since it provides enough intensity to work in daylight.

Figure 11.13. Lamp and pinhole assembly. The drawing shows a tungsten-halogen lamp, which should be on a centering and focusing mount, followed by the condenser, pinhole, and filter. The condenser should focus the filament onto the pinhole.

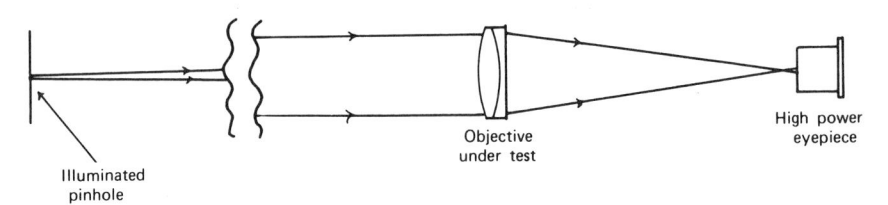

Figure 11.14. Arrangement for testing a small refracting objective; the light may or may not be filtered, according to the detailed requirements of the test.

11.2.3. The Arrangement of the Optical System for Star Testing

For testing small refracting objectives the simplest arrangement is the one shown in Fig. 11.14, with the star at a suitable distance. What this distance needs to be is, of course, arguable, and it is only possible to know for certain by computing the effect of using a finite conjugate from the design specification. For most practical purposes the rule "infinity is a distance greater than 20 times the focal length of the system under test" is good enough.

The eyepiece for viewing the star must have enough magnification to show all the detail in the point spread function. The facts that the scale of detail is of the order of the radius of the first dark ring, and that the angular resolution limit of the eye under favorable conditions is about 1 arc-min, lead to the handy rule that the focal length of the eyepiece in millimeters should be equal to or less than the f number of the system under test;[†] thus to test an $f/10$ beam we need a 10-mm eyepiece, that is, magnification × 25. With some experience it is easy to tell when the eyepiece has enough magnification, since only then can all the diffraction structure in the star image be seen.

Needless to say, we must have an eyepiece that itself has negligible aberrations to star-test a good objective. It is rather unlikely that this would not be so, since our rule implies that the maximum diameter of the ray pencils in the eyepiece will be 1 mm and most eyepiece designs will be substantially perfect at this aperture. However, there may be accidental defects such as scratches or slight decentering, and these can easily be identified by rotating the eyepiece.

For looking at beams with very short focal ratios, it may be necessary to use a low-power microscope instead of an eyepiece. The rule then takes the form that the overall magnification of the microscope should not be less than 250 divided by the f number.

Sometimes a collimator, either refracting or reflecting, is used to place the star exactly at infinity. This is undesirable, partly because added cost is incurred and partly because the aberrations of the collimator are then added to those of

---

[†]Strictly speaking, these figures give twice the f number, but it is as well to have a factor of 2 at hand in magnification.

11.2. PRACTICAL ASPECTS WITH SMALL ABERRATIONS

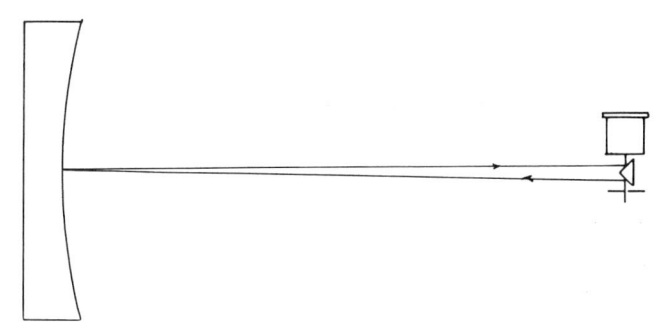

Figure 11.15. Testing a concave spherical mirror. The right-angle prism is aluminized on the two faces used as mirrors. For a mirror with a short focal ratio the eyepiece would have to be replaced by a low-power microscope. Care must be taken that the pinhole is illuminated so as to fill the whole aperture of the mirror; likewise, the microscope objective must have a numerical aperture large enough to collect the light from all the aperture.

the system under test. However, under some workshop conditions the long air path needed for the arrangement of Fig. 11.14 may be subject to unavoidable thermal turbulence, and it may be necessary to use a collimator instead. Usually, under average conditions in air path exceeding 1 or 2 m is enough to show turbulence effects with a critical optical system, so we must either screen a longer path, put it in a separate test bay, or wait for a particular time of day or night when the disturbances are least. Occasionally the whole system has been put in a helium atmosphere or even in a vacuum tank for test purposes, but this is not practicable in most workshops.

A large concave spherical mirror can be tested in Fig. 11.15 since it has no spherical aberration for equal conjugates. However, it is necessary to keep the separation of the two conjugates small to avoid astigmatism. This separation should be less than about

$$\frac{R\sqrt{\lambda R}}{D} \tag{11.3}$$

where R is the radius of curvature of the mirror and D its diameter. For example, if $R = 1$ m and $D = 200$ mm, the separation must be less than 7 mm.

A paraboloidal mirror needs one infinite conjugate. There seems little doubt that star testing is not the best method for short focal ratio paraboloids; on the contrary, however, paraboloidal mirrors are very useful as collimators for star testing other systems. Figure 11.16 illustrates how a paraboloid can be star tested in double pass, using an auxiliary flat, and Fig. 11.17 shows a convenient arrangement of a paraboloid as a collimator. Most large concave mirrors need central holes for various purposes, so a paraboloid would naturally have a hole in the center, as in Fig. 11.17.

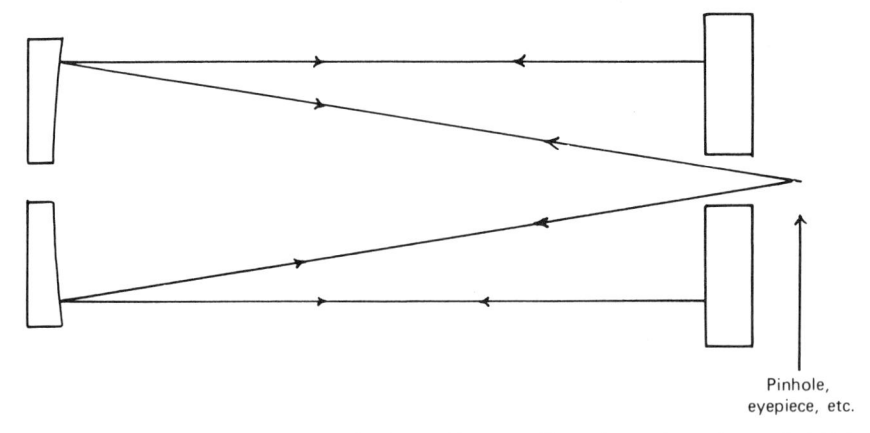

Figure 11.16. Testing a paraboloidal mirror with an auxiliary plane mirror; the paraboloid is double passed, so that the effect of any aberrations is doubled.

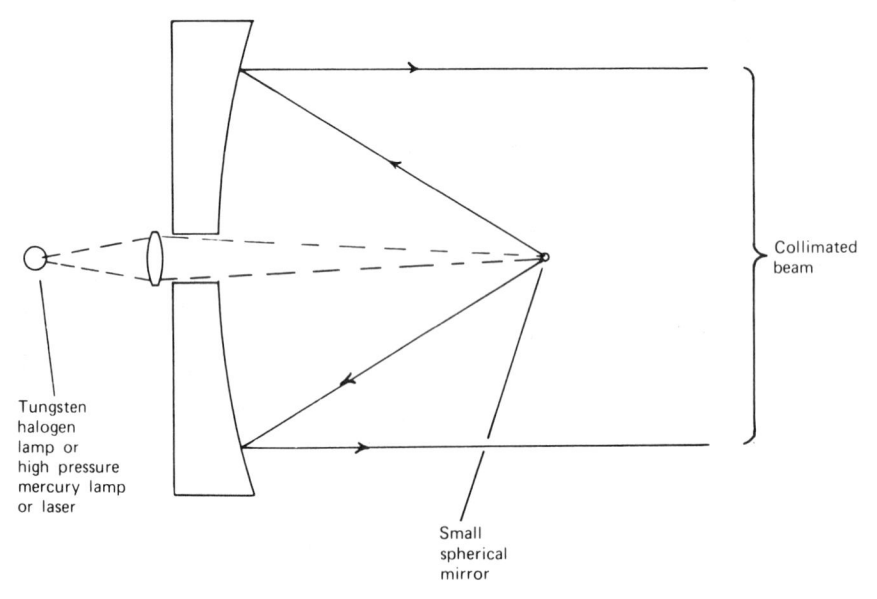

Figure 11.17. A paraboloid as a collimator for the star test. The small spherical mirror can be a 1-mm-precision tungsten carbide sphere or a small sized globule of glass melted on the end of a drawn-out glass rod.

A large, general-purpose optical flat is almost certain to be available for general testing purposes, and it can be used to double-pass large aperture systems, as in Fig. 11.18; this arrangement avoids long optical paths for getting the infinite conjugate and also doubles the sensitivity of the test. The flat must, of course, be as good as the rest of the system is intended to be. Again it is

11.2. PRACTICAL ASPECTS WITH SMALL ABERRATIONS

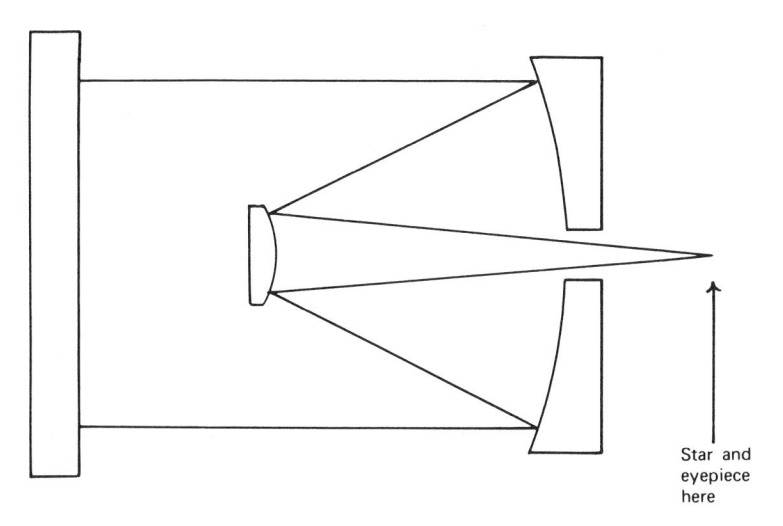

Figure 11.18. Testing a Cassegrain in double pass (autocollimation or autostigmatism).

necessary to keep as close to the axis as possible to avoid astigmatism. In a system such as the Cassegrain in Fig. 11.18, the test can be used to establish correct centering between the two elements and to get the optimum axial spacing by minimizing the spherical aberration. We note, however, that when a system is double passed, as in Fig. 11.18 (or, as it is sometimes called, autocollimated or autostigmatized), any coma, transverse color, and distortion inherent in the optical design are canceled out.

11.2.4. Microscope Objectives

The star test is particularly useful for microscope objectives (Martin and Welford 1971), and the techniques are rather different from those so far described. The artificial star is made by vacuum-alumizing or silvering a microscope slide and gently wiping it with a clean cloth before cementing a coverslip to it. Unless the slide was cleaned very carefully before coating, the wiping will produce several pinholes, among which will be some suitable as artificial stars. An alternative technique described by Slater (1960) will produce pinholes of precisely determined size down to about 0.1 μm diameter. The test is then carried out with a bright lamp and an efficient substage condenser.

As mentioned at the beginning of this chapter, the star test is used in the rapid final adjustment of a critical air space to balance the spherical aberration correction for objectives of high numerical aperture. The spherical aberration of these objectives is also sensitive to both the thickness of the slide cover glass (except for oil immersion objectives) and the magnification; the latter is set by the length of the microscope draw tube, if there is one. Thus for critical work it is necessary to determine the cover glass thickness for which a given objective has least spherical aberration (usually a combination of high orders); alternatively, we may be able to find the magnification at which the objective performs best with a given cover glass thickness.

The star test is also convenient for checking the flatness of allegedly flat-field objectives and for checking the chromatic correction of apochromatic objectives. Finally, it is known that some specimens of fluorite scatter light strongly, and the star test shows whether the fluorite in a given apochromat or semiapochromat has this undesirable property: the scattered light shows up as a haze, easily distinguished from the diffraction rings, around the star image.

11.2.5. Can the Star Test Be Made Quantitative?

The qualitative nature of the star test for small aberrations is a considerable disadvantage, and it must be admitted that one needs experience to obtain the test results from this test. However, it is possible to speed up the process of getting experience by using sample optical systems with known aberrations to train novice workers. In the Applied Optics Group at Imperial College this has been done by taking ordinary well-corrected doublet objectives and modifying them to produce small known amounts of aberrations; these can be set up on, for example, a Twyman–Green interferometer to show, say, half a fringe of astigmatism and then star tested. This procedure serves to give the worker an order-of-magnitude feeling for the amounts of aberrations in the test.

A 300-mm $f/12$ achromatic doublet will produce several wavelengths of primary spherical aberration is simply reversed, but this is too much for the present purposes. Smaller amounts can be produced by simply changing the spacing between the two components. It is not possible to introduce higher order zonal aberrations in this way if they are not already present; the best approach is either to have them polished in or to find a suitable microscope objective and use it. Astigmatism can be demonstrated by simply tilting a coma-corrected doublet so that it is working 1° or 2° off axis. A suitable amount of coma can usually be produced by slightly decentering one component with respect to the other, by 1 or 2% of the diameter. If this is carefully done without introducing any overall tilt of the objective, a small amount of axial coma can be produced. The set of pathological objectives can then be calibrated interferometrically and used as a training set for star testing.

11.3. THE STAR TEST WITH LARGE ABERRATIONS

Very simple and useful quantitative tests can be made on systems such as photographic objectives, projection objectives, and television camera objectives that do not have diffraction-limited aberration correction. The system will normally

11.3. THE STAR TEST WITH LARGE ABERRATIONS

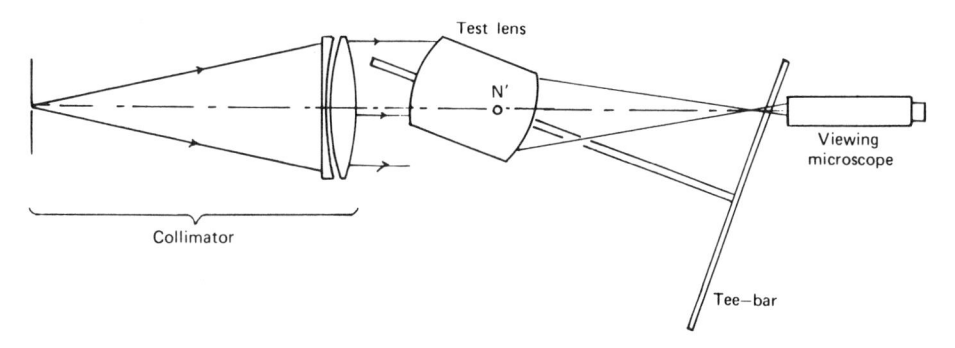

Figure 11.19. A nodal slide optical bench as used for star testing photographic objectives and similar systems. The Tee-bar represents the optical axis and the focal plane of the lens under test; as it turns off axis it pushes the viewing microscope away so that it is always focused on the correct focal plane. (But in some versions the microscope does not move, and the lens is moved forward by the linkage instead; in yet other versions there is no Tee-bar, and the microscope must be moved back a calculated amount.)

be set up on a nodal slide optical bench with a collimator, as in Fig. 11.19, so that off-axis aberrations can be measured. Many detailed descriptions of such benches are available; see, for example, Leistner et al. (1953). The principles in star testing for large aberrations are quite different from those for small aberrations; for the latter we examine the diffraction structure of the point spread function and try to estimate the types and amounts of aberrations on the basis of experience, but for large aberrations we use essentially concepts of geometrical optics and determine ray aberrations. Wandersleb (1952) gave some good photographs of heavily aberrated point spread functions.

11.3.1. Spherical Aberration

Here we set the objective under test on axis, and use a filter for the appropriate wavelength. Then we use a series of annular diaphragms in the aperture of the system and determine the focus of the hollow pencil of rays from each, as in Fig. 11.20. In this way the longitudinal spherical aberration curve is obtained, and this can be treated in the usual ways to obtain any other form of description

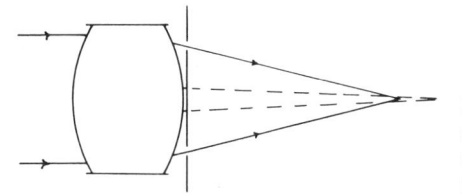

Figure 11.20. Measuring longitudinal spherical aberrations by the star test with a zonal diaphragm. The broken lines indicate a paraxial pencil of rays.

of the aberration; that is to say, when we know the longitudinal spherical aberration, we can calculate the transverse spherical aberration.

In this procedure, as in the others described in Section 11.3, there is some advantage in arranging that the object pinhole is not quite diffraction limited; this not only gives a little more light flux, a useful bonus in most cases, but also makes it rather easier to judge the ray intersections.

A quicker procedure, which gives only a qualitative picture of the aberrations, is simply to put an opaque straightedge across half the aperture as in Fig. 11.21, and look at the image in various focal planes. As indicated, the general nature of the aberration appears as semicircular strips on either side of the axis, and with a little practice this image can easily be interpreted.

11.3.2. Longitudinal Chromatic Aberration

Suitable filters are used, and the aperture is stopped down until the spherical aberration is negligible; the longitudinal chromatic aberration can then be plotted directly from the measured focal positions for the different wavelengths. If we use annular diaphragms, we can also measure the chromatic variation of spherical aberration; however, this refinement is rarely of interest in a practical workshop situation.

Again the straightedge across half the aperture can be used as for spherical aberration, without filters in the light source. The distribution of colors then gives a rapid indication of the balancing of the chromatic correction.

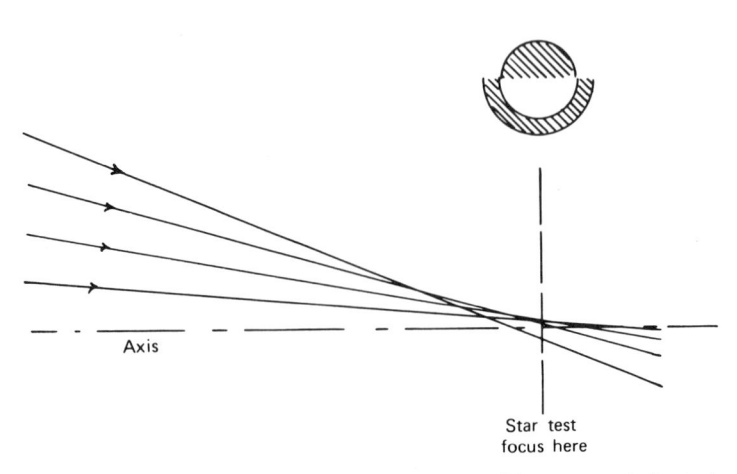

Figure 11.21. Testing for spherical aberration by masking half of the aperture; the hatched pattern will appear bright at the focal plane indicated.

11.3.3. Axial Symmetry

In the systems with which Section 11.3 is concerned it is usual to find some asymmetry on axis due to the building up of constructional tolerances. This is particularly so with zoom lenses, where the decentering may vary over the zoom range. The axial star image provides an instant check on the centering and may be used as a control in improving it, if this is required. To aid in this it is useful to have an accurate rotation mechanism on the nodal slide mount for the lens under test.

11.3.4. Astigmatism

We can plot the positions of the saggital and tangential focal lines by turning the lens off axis and focusing in turn on each line. Of course, in this process we also see all the other off-axis aberrations (coma, transverse chromatic aberration, etc.). We therefore close down the lens iris enough to eliminate or nearly eliminate spherical aberration and coma, but not, of course, to the extent that the astigmatism disappears as well.

In a large system it may be as well to check the symmetry of the lens field by rotating the lens and watching the astigmatic focal lines at a chosen field angle. Sometimes more asymmetry is found off axis than on axis. A star test measurement of the astigmatic field curvature, together with some assessment of coma, is often used to tune the air spacings of large photographic objectives.

11.3.5. Distortion

The accurate calibration of, for example, photogrammetric objectives for distortion is a very large topic that is not considered in this book. However, it is worth noting that the nodal slide optical bench used for star testing enables us to make a simple measurement of distortion that is useful in many cases.

When the lens under test is being set up, we set the axis rotation of the nodal slide under the second nodal point by the following procedure. We find the star image in the microscope, and turn the slide a small amount off axis. If the star image moves, say, to the left, we move the lens along the nodal slide in the direction that brings the image back to the center of the microscope field of view, at the same time keeping it in focus by following it with the microscope. This procedure is continued until the image does not move sideways at all as the nodal slide is rotated; the lens is then correctly positioned. However, it is almost invariably found that there is no position of the lens along the nodal slide that will give no movement at all of the star image; the best we can do is, rather, as shown in Fig. 11.22, where the image returns to the axis after a short excursion. The residual displacement is, of course, a manifestation of the distor-

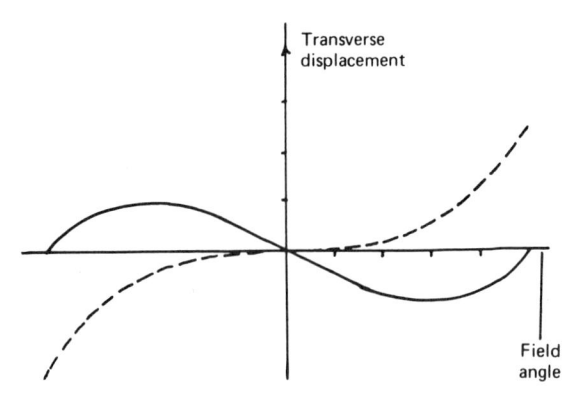

Figure 11.22. Distortion measurement. The full-line graph represents transverse displacement of the star image as a function of field angle, assuming that a compromise position for the nodal axis has been found. The broken line shows the effect of adding a linear term; this amounts to a shift of the nodal point to the paraxial position.

tion of the lens. The full-line graph indicates what might be measured. It would be more usual to subtract from this a suitable linear term to give the broken-line graph for distortion, which depends on cubic and higher powers of the field. The subtraction of a linear term merely implies a slightly different choice of nominal focal length for the system under test.

A detailed description of this procedure, with a discussion of errors, has been given by Washer and Darling (1959).

11.3.6. Nonnull Tests

By implication we have so far considered only null tests, that is, tests of systems that ought to have no aberrations. Sometimes, however, we need to test a system to see whether it has prescribed nonzero aberrations. The classical example is, of course, the paraboloid mirror with equal conjugates, which is, geometrically more convenient than one conjugate at infinity. Another example might be an aspheric singlet lens for use in the infrared, since its aberrations would be different if tested in visible light.

Since the star test is not very accurately quantitative, even when used as described in Section 11.3, for actually measuring large aberrations it is best to convert such a test into a null test by adding an auxiliary optical system with the required aberrations (see Chapter 14). This is a designed optical system that should be easy to make accurately; Dall (1947) described the widely used system of an auxiliary lens for null-testing paraboloids,[†] and many other examples have been published.

[†]The paper cited refers to knife-edge testing, but the principle of the auxiliary lens is, of course, the same.

A more recent alternative, which can be used if it is permissible to test the system with a laser as light source, is to employ a computer-generated hologram as the auxiliary system. A useful description of this method is given by Birch and Green (1972). The computer-generated hologram can synthesize a wavefront of any desired shape, provided that the available computer graphics system has enough resolution to draw it. As usually described, the computer-generated hologram corrector is part of an interferometric testing scheme, but it could equally well be used to synthesize an aberrated wavefront for transmission through the system under test; the result, if the system had the desired aberrations, would be an unaberrated star image.

REFERENCES

Barakat, R. and A. Newman, "Measurement of Total Illuminance in a Diffraction Image. I: Point Sources," *J. Opt. Soc. Am.*, **53**, 1965 (1963).

Beiser, L., "Perspective Rendering of the Field Intensity Diffracted at a Circular Aperture," *Appl. Opt.*, **5**, 869 (1966).

Birch, K. G. and F. J. Green, "The Application of Computer-Generated Holograms to Testing Optical Elements," *J. Phys.*, D: *Appl. Phys.*, **5**, 1982 (1972).

Born, M. and E. Wolf, *Principles of Optics*, 5th ed., Pergamon Press, Oxford and New York, 1975.

Cagnet, M., M. Francon, and J. C. Thrierr, *Atlas of Optical Phenomena*, Springer-Verlag, Heidelberg and New York, 1962.

Dall, H. E., "A Null Test for Paraboloids," *J. Br. Astron. Assoc.*, **57**, (1947); reprinted in *Amateur Telescope Making*, Vol. 3, A. E. Ingalls, Ed., Scientific American, New York, 1953.

Kogelnik, H. and T. Li, "Laser Beams and Resonators," *Appl. Opt.*, **5**, 1550–1567 (1966).

Leistner, K., B. Marcus, and B. W. Wheeler, "Lens Testing Bench," *J. Opt. Soc. Am.*, **43**, 44 (1953).

Li, Y. and E. Wolf, "Three-Dimensional Intensity Distribution Near Focus in Systems of Different Fresnel Numbers," *J. Opt. Soc. Am.*, **1a**, 801–808 (1984).

Linfoot, E. H. and E. Wolf, "On Telescopic Star Images," *Mon. Not. Ry. Astron. Soc.*, **112**, 452 (1952).

Linfoot, E. H. and E. Wolf, "Diffraction Images in Systems with an Annular Aperture," *Proc. Phys. Soc.*, **B66**, 145 (1953).

Martin, L. C., *Technical Optics*, Vol. 2, 2nd ed., Pitman, London, 1961.

Martin, L. C. and W. T. Welford, in *Physical Techniques in Biological Research*, Vol. 1, Part A, 2nd ed., G. Oster, Ed., Academic Press, New York and London, 1971.

Nienhuis, K., *On the Influence of Diffraction on Image Formation in the Presence of Aberrations*, Thesis, J. B. Wolters, Groningen, 1948.

Slater, P. N., in *Optics and Metrology*, P. Mollet, Ed., Pergamon Press, Oxford and New York, 1960.

Stamnes, J. J., *Waves in Focal Regions*, Adam Hilger, Bristol and Boston, 1986.

Taylor, C. A. and B. J. Thompson, "Attempt to Investigate Experimentally the Intensity Distribution near the Focus in the Error-free Diffraction Patterns of Circular and Annular Apertures," *J. Opt. Soc. Am.*, **48,** 844 (1958).

Taylor, H. Dennis, *The Adjustment and Testing of Telescope Objectives*, Sir Howard Grubb, Parsons & Co., Newcastle-upon-Tyne, 1891; 4th ed., 1946, 5th ed. 1983, Adam Hilger Ltd, Bristol, UK.

Twyman, F., *Prism and Lens Making*, Adam Hilger, London, 1942. 2nd rev. ed. 1988, Adam Hilger, Bristol, UK.

Wandersleb, E., *Die Lichtverteilung in der axialen Kaustik eines mil sphoärischer Aberration behafteten Objektivs*, Academie-Verlag, Berlin, 1952.

Washer, F. E. and W. R. Darling, "Factors Affecting the Accuracy of Distortion Measurements Made on the Nodal Slide Optical Bench," *J. Opt. Soc. Am.*, **49,** 517 (1959).

Welford, W. T., "On the Limiting Sensitivity of the Star Test for Optical Instruments," *J. Opt. Soc. Am.*, **50,** 21 (1960).

12
Null Tests Using Compensators

A. Offner and D. Malacara

12.1. INTRODUCTION AND HISTORICAL BACKGROUND

A number of methods for testing the quality of aspheric surfaces have been developed. In one that is particularly useful during the fabrication of the surface, an auxiliary optical system is designed so that, in combination with the aspheric surface, it forms a stigmatic image of a point source. The auxiliary optical system is called a null corrector or null compensator.

The ability of opticians to manufacture accurate astronomical mirrors was enormously enhanced with the development of the knife-edge test (Foucault 1859). The test is very sensitive and easy to interpret when it is used to assess the astigmatism of a pencil because it is then a null test. This is the case when a spherical surface is tested with the knife edge and light source near its center of curvature. Any departure of the cutoff shadows from smoothness over the face of the mirror indicates a figure error at the location.

The corresponding test of a paraboloidal mirror is a lengthy and less accurate process. In this case the figure errors must be deduced from measurements of position made for a large number of zones, since a concave aspheric or conic surface tested at the center of curvature has a spherical aberration wavefront deformation given in third-order approximation by

$$W(\rho) = \frac{(8A_1 + Kc^3)\rho^4}{4}, \qquad (12.1)$$

where A_1 is the first aspheric deformation term, K is the conic constant, and $c = 1/R$ is the curvature. For a conic surface we may write

$$W = 1/4KR(\rho/R)^4 \qquad (12.2)$$

Optical Shop Testing, Second Edition, Edited by Daniel Malacara.
ISBN 0-471-52232-5 © 1992, John Wiley & Sons, Inc.

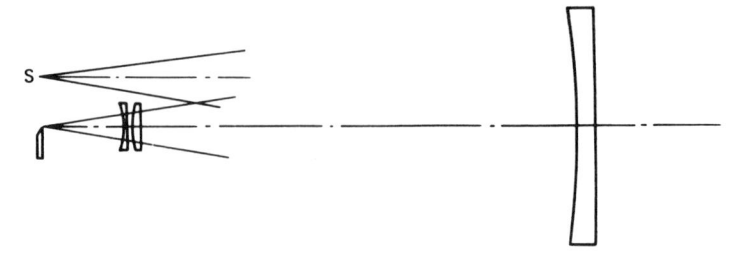

Figure 12.1. Couder two-element compensator, 1927.

For many years the only alternative to the method of using the knife-edge test during the manufacture of a paraboloidal mirror was to test the mirror by autocollimation with the aid of an optical flat, which had to be as large and as accurately figured as the mirror being manufactured.

Couder (1927) pointed out that departure from stigmatism of the image of a point source at the center of curvature of a paraboloidal mirror can be removed by interposing a small compensating lens between the image and the mirror. He used a two-element compensator in the arrangement shown in Fig. 12.1. Two elements were necessary because he required a null corrector of zero net power to conveniently carry out the manufacturing process desired in his paper. To manufacture a 30-cm $f/5$ paraboloidal mirror, he used a null corrector whose aperture (scaled from his drawing) was about 4 cm.

The use of a spherical mirror beyond the center of curvature of a paraboloid to compensate for the aberrations of a paraboloid used with source and knife edge near its center of curvature was described by Burch (1936). He derived the fifth-order aberration of null systems of this type and showed that with the two-mirror arrangement of Fig. 12.2 the residual aberration of the paraboloid

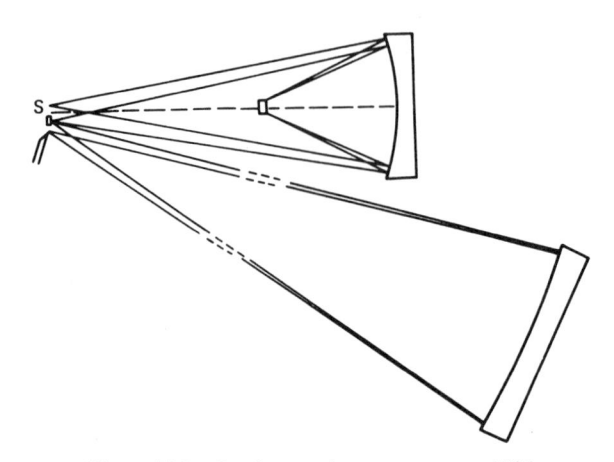

Figure 12.2. Burch two-mirror compensator, 1936.

12.1. INTRODUCTION AND HISTORICAL BACKGROUND

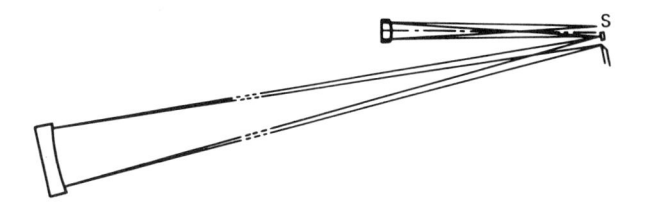

Figure 12.3. Burch plano-convex compensator, 1938.

is less than one-fortieth of a wave for paraboloids as fast as $f/5$ and with apertures up to 80 cm, when the compensating mirror aperture is one-fourth that of the paraboloid. For larger apertures and/or faster paraboloids he suggested figuring the convex mirror with a down edge that departed from the base sphere by an eighth power law so that seventh-order spherical aberration could be balanced. He calculated that an asphericity of about 2.8λ would be required to compensate for the aberrations of the 5-m $f/3.33$ Mount Palomar mirror.

A simple third-order solution for a refracting compensator was also published by Burch (1938). The refractor was a plano-convex lens of focal length f and refractive index used in the arrangement shown in Fig. 12.3, in which its plane surface is reflecting. For a paraboloid of base radius R, the third-order aberration of the image at the center of curvature is balanced when $f = RN^2/(N-1)^2$, so that for $N = 1.52$ the lens has an aperture about one-eighth that of the paraboloid. Burch expected the residual aberration with this null corrector to be negligible for paraboloids of aperture ratio not exceeding $f/8$. He added, "Anyone with an aptitude for analytical optics or for ray-tracing could earn the gratitude of practical opticians by calculating the secondary aberrations for this and other kinds of compensating lens systems." This plea was answered 30 years later (Holleran 1968).

During the manufacture of the 5-m $f/3.33$ Mount Palomar mirror, a 25-cm-diameter compensator was used to form a stigmatic retroreflected image near the center of curvature of the mirror (Ross 1943). The arrangement used is shown in Fig. 12.4.

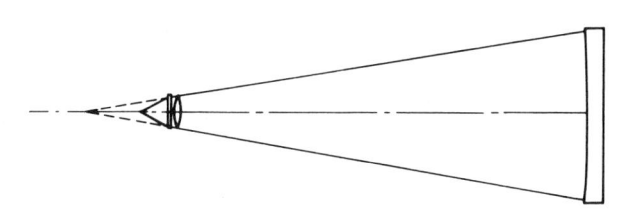

Figure 12.4. Ross aspheric compensator used during the fabrication of the 5-meter Mount Palomar mirror.

To obtain a degree of compensation such that the residual zonal aberration resulted in a disk of confusion that was small compared to that caused by the atmosphere, Ross found it necessary to add an aspheric corrector plate to a refractive element with spherical surfaces, which by itself balanced the spherical aberration of the paraboloidal mirror. The retroreflective arrangement used by Ross has the advantage that it is coma-free and therefore insensitive to departures of the source and knife edge from the axis of the system. Moreover, since the compensator is used twice, it has to contribute only one-half as much aberration as is required in Couder's arrangement (Fig. 12.1).

12.2. THE DALL COMPENSATOR

The plano-convex lens of Burch shown in the arrangement of Fig. 12.3 is a convenient and easily used solution to the problem of making a null corrector for a paraboloidal mirror of moderate aperture. However, with this method a plano-convex lens used in the manufacture of a paraboloidal mirror can serve as a compensator only for other paraboloidal mirrors of the same focal length.

Dall (1947, 1953) noted that, since the spherical aberration of a lens is a function of its conjugates, the same plano-convex lens can be used as a compensator for many paraboloids. Dall employed the arrangement of Fig. 12.5, which is quite similar to the arrangement used by Couder (Fig. 12.1). The main difference is that the Dall compensator is placed in front of the light source and not in front of the observer's eye. The reason for this is that if it is used on the convergent beam, the convergence of the beam may become so large that an observation of the whole surface under test may not be possible.

The Dall compensator is normally used in an off-axis arrangement, with the light passing only once through the compensator. Two double-pass configurations, with the divergent as well as the convergent beams passing through the compensator, may also be used: (a) exactly on axis, using a beam splitter and (b) slightly off axis. Then, the compensating lens needs to compensate only half the total aberration on each pass, making it less strong. Then, an additional factor of 2 has to be used in front of Eq. (12.1). Another advantage of the double-pass symmetrical arrangement is that it is coma free. Thus any coma introduced by the off-axis lateral displacement of the light source and the ob-

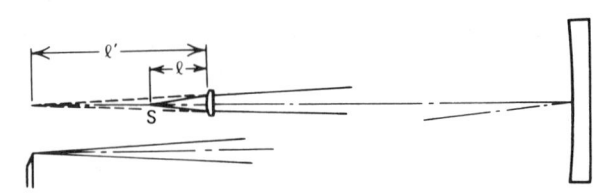

Figure 12.5. Dall plano-convex compensator.

12.2. THE DALL COMPENSATOR

server or any misalignment of the lens compensator does not introduce any coma. Unfortunately, the double-pass arrangement may be used only if the radius of curvature of the surface under test is very large compared with its diameter.

Dall found that proper choice of the short conjugate of the lens provides adequate compensation if the ratio of the radius of curvature R of the paraboloid to the focal length f of the lens is between 10 and 40. The relation required to balance the third-order aberration of the parabola at its center of curvature is

$$\frac{R}{f} = \frac{1}{2}(m-1)^2 \left[\frac{N^2(m-1)^2}{(N-1)^2} + \frac{(3N+1)(m-1)}{N-1} + \frac{3N+2}{N} \right] \quad (12.3)$$

where m is the ratio of the long conjugate distance l' to the short conjugate distance l, and N is the index of refraction of the plano-convex lens. (The sign convention is such that in the Dall arrangement $m > 1$.)

The Dall compensator has been widely used, especially by amateur telescope makers. The degree of compensation that can be attained with this extremely simple null corrector is illustrated by the following example.

A Dall compensator is desired for a 0.6-m $f/5$ paraboloidal mirror. Taking $m = 2$, $N = 1.52$, and $F = 3$ m in Eq. (12.3), we find $R/f = 11.776$, which is within the bounds prescribed by Dall. The compensator specifications are then $f = 50.950$ cm, $l = -25.475$ cm, and $l' = -50.950$ cm. With this null corrector the computed root-mean-square (RMS) departure from the closest sphere [RMS optical path difference (OPD)] of the wavefront that converges to form an image of the light source is 0.048λ at $\lambda = 632.8$ nm. A paraboloid fabricated to give a null test with this compensator would have an RMS figure error of 0.024λ. The Strehl intensity resulting from this figure error is 0.91. The diameter of the plano-convex lens required to achieve this compensation is about one-twelfth that of the paraboloidal mirror.

The paraboloid of the examples is about the largest for which a Dall compensator is adequate. Since the arrangement suggested by Dall is not coma-free, the light source must be located accurately on the axis of the convex lens, and this axis should be directed to pass through the pole of the paraboloid.

Practical instructions for making and using a Dall null tester are given in two papers by Schlauch (1959) and by Stoltzmann and Hatch (1976). By restricting the refractive index of the plano-convex lens to 1.52, the computation of Eq. (12.3) can be avoided with the help of a curve in Schlauch's paper, which is adapted from one published by Dall (1953).

A Dall lens made with BK-7 glass and using red light can also be calculated with the curve in Fig. 12.6. The radius of curvature is 25 mm and the thickness is 5 mm. As in Fig. 12.7, it has been assumed that the flat surface of the lens and the testing point (knife edge) are in the same plane. Assuming only the presence of third-order spherical aberration, from Eq. (12.2) we may say that

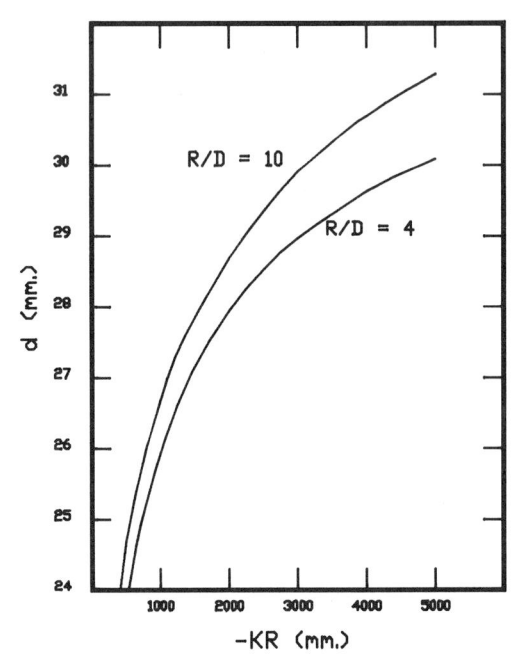

Figure 12.6. Distance d from the light source to the Dall compensator lens in Fig. 12.7, for different values of the product of the radius of curvature times the conic constant.

the distance d from the light source to the vertex of the convex surface of the lens has to be a function only of the product KR. This is true only for small apertures (large ratio R/D). If the aperture is large (ratio R/D small), the fifth-order spherical aberration is present in the Dall lens, and it has to be partially compensated with some overcompensation of the third-order aberration. Then, a zonal spherical aberration remains uncorrected. This remaining aberration is

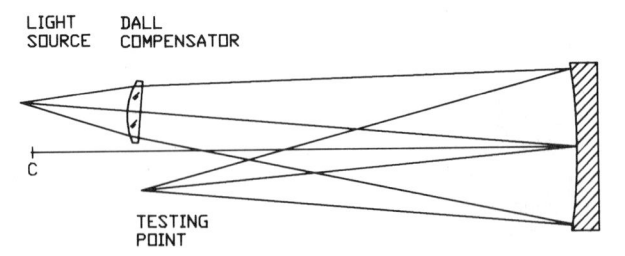

Figure 12.7. A Dall compensator. The testing point is assumed to be in the same plane as the flat surface of the Dall lens.

12.3. THE SHAFER COMPENSATOR

not a severe problem if the aperture is small or if there is a large central obscuration, as pointed out by Rodgers (1986).

A Dall compensator can be modified to be coma-free by using the plano-convex lens in the retroreflective arrangement of Ross (Fig. 12.4). The relation to be satisfied with this arrangement is a modification of Eq. (12.3) in which the factor $\frac{1}{2}$ is eliminated since the lens is traversed twice. With this arrangement, for example, the compensator for the 0.6-m $f/5$ paraboloid has $R/f = 23.552$, $f = 25.475$ cm, $l = -12.737$ cm, and $l' = -25.475$ cm. The diameter of the plano-convex lens is then $\frac{1}{24}$ that of the paraboloid. The residual figure error of the paraboloid that gives a null test with this arrangement is exactly the same as that computed in the preceding example. The retroreflective arrangement has the advantage, however, that since it is coma-free, it is not affected by small departures of the light source from the axis of the plano-convex lens.

In an interesting variation of the Dall compensator proposed by Puryayev (1973), an afocal meniscus whose concave surface is conicoidal is substituted for Dall's plano-convex lens in the autostigmatic arrangement used by Ross (Fig. 12.4). For an afocal meniscus,

$$r_1 - r_2 = \frac{d(N-1)}{N}, \tag{12.4}$$

where r_1 is the radius of the concave surface of the meniscus, r_2 is the radius of its convex surface, d is its thickness, and N is its index of refraction.

The third-order value of the conic constant K of the concave surface required to compensate for the aberration of a paraboloid of radius R is

$$K = \frac{R}{(N-1)(r_2/r_1)^4 l}, \tag{12.5}$$

where l is the distance from the light source to the meniscus. (The sign convention is such that K is negative.)

With the same 20-cm-diameter meniscus compensator, Puryayev achieved compensation for any paraboloidal or near-paraboloidal surface whose focal length does not exceed 24 m and whose aperture ratio does not exceed 1:4. The maximum residual wave aberration of the retroreflected wave for any paraboloid in this range is about $\lambda/2$ at 632.8 nm. This residual can be computed and taken into account when the figure of the test mirror is being determined.

12.3. THE SHAFER COMPENSATOR

This compensator is a triplet, designed by Shafer (1979), so that the following three conditions are satisfied:

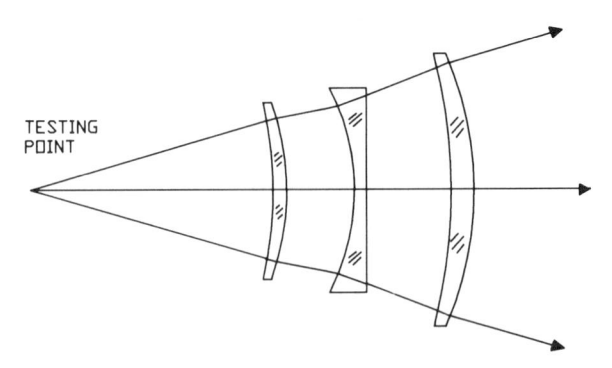

Figure 12.8. A compensator designed by Shafer, with all lenses made out of BK-7 glass.

1. For a certain distance from the light source to the compensator, the spherical wavefront from the light source preserves its spherical shape after passing the compensator. Then, positive or negative compensations may be achieved by displacing the system along the optical axis.
2. The system is afocal (effective focal length infinite), so that the angles with respect to the optical axis of the light rays entering the system are preserved after passing through the compensator.
3. The angular magnification of the afocal system, (lateral magnification for a near object) must be equal to +1. Thus the apparent position of the light source does not change when moving the compensator along the optical axis.

A system with a negative lens between two positive lenses, as in Fig. 12.8, is appropriate for positive conic constants (oblate spheroids) of any magnitude, and negative conic constants (paraboloids or hyperboloids) of moderate magnitude, as shown in the graph in Fig. 12.9. This graph is inverted when a positive lens is placed between two negative lenses.

12.4. THE OFFNER COMPENSATOR

As described earlier (Burch 1936; Ross 1943), compensation for the spherical aberration of a paraboloid or other aspheric concave mirror can be achieved to any desired degree of accuracy by the incorporation of an aspheric element in the null corrector. This method is limited to cases in which the figure of the aspheric element can be ascertained with an accuracy better than that desired for the aspheric mirror. Primary mirrors that are to be incorporated into diffraction-limited space-borne optical systems are now required to have RMS figure errors as small as one-hundredth of the wavelength of visible light. It is there-

12.4. THE OFFNER COMPENSATOR

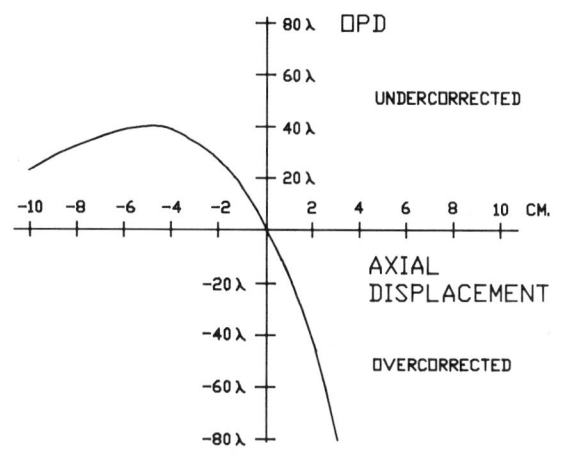

Figure 12.9. Wavefront aberration due to third-order spherical aberration vs. lens motion for Shafer's compensator and an $f/2$ system and a wavelength of 0.6328 μm.

fore desirable that the components of a null corrector for these mirrors be spherical or flat so that their figure errors can be measured to the required accuracy.

In the design of his null corrector, Ross found that the farther he put the lens from the center of curvature of the mirror, the less residual aberration there was when the compensation was exact at the center and the edge. This comes about because, although the longitudinal spherical aberration S of the normals to the paraboloid follows the simple law $S = y^2/2R$, where y is the distance of the normal from the axis of the paraboloid and R is its radius, additional terms of a power series would be required to describe the same spherical aberration distribution in a coordinate system with its origin at the null corrector. For a null corrector in contact with the paraboloid, the compensating spherical aberration would be described by the same simple law. Unfortunately, this null corrector would be as large as the paraboloid.

12.4.1. Refractive Offner Compensator

Offner (1963) pointed out that a small lens that forms a real image of a point source at the center of curvature of a paraboloid, in combination with a field lens at the center of curvature that images the small lens at the paraboloid, is optically equivalent to a large lens at the paraboloid. [The use of a field lens in this way was first suggested by Schupmann to control secondary spectrum (Schupmann 1899; Offner 1969).]

With a field lens that images the compensating lens c at the paraboloid in Offner's arrangement (Fig. 12.10), the spherical aberration of the compensating lens must follow the same law for aperture as do the normals to the paraboloid.

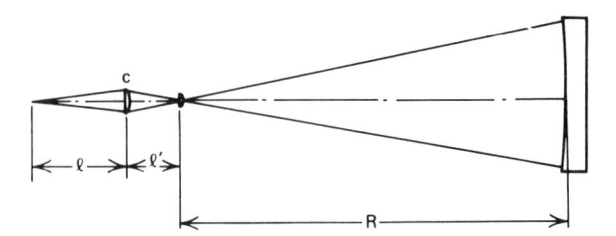

Figure 12.10. Refracting compensator with field lens.

This restriction on the compensating lens is not necessary. All that is required is that lens c provide sufficient third-order spherical aberration to compensate for that of the normals of the paraboloid. The power of the field lens (and thus the location of the image of lens c) is then varied to minimize the high-order aberration.

To balance the third-order aberration of the normals to a conicoidal mirror with conic constant K and base radius R, a plano-convex lens of focal length f and index of refraction N must satisfy the relation

$$-\frac{KR}{f} = (1-m)^2 \left[\frac{N^2(1-m)^2}{(N-1)^2} + \frac{(3N+1)m(1-m)}{N-1} + \frac{(3N+2)m^2}{N} \right],$$

(12.6)

where m is the ratio l'/l (Fig. 12.10). (The conicoid must have $K < 0$ if the aberration of its normals are to be balanced by the spherical aberration of a plano-convex lens. The sign convention is such that $m < 0$.)

Since the mirror under test in general has a large spherical aberration, the field lens does not have an ideal position. It may be placed between the mirror and its caustic, inside the caustic region, or outside the caustic. Any of these possible locations produce good compensations, with slight variations, as explained by Sasian (1989). Different techniques for designing this compensator have been explored by several authors, among others, by Landgrave and Moya (1987) and by Moya and Landgrave (1987).

The main purpose of the field lens is to avoid fifth-order spherical aberration, but another equally important consequence of its presence is that the wavefront at the mirror under test is imaged on the plane where the aberrations are observed, that is, at the compensating lens. The need for these imaging lenses in optical testing systems has been described in Chapter 2.

The importance of the field lens in the Offner arrangement is evident from the example of the design of a null corrector for a 1-m $f/4$ paraboloid using plano-convex lenses with refractive index 1.52. The quantity m that results in the desired convergence angle of the retroreflected wavefront is first chosen.

12.4. THE OFFNER COMPENSATOR

Choosing a convergence of $f/12$ leads to $m = -0.6667$. The value of the focal length of the compensating lens required to balance the third-order aberration of the paraboloid normals is then seen from Eq. (12.6) to be 20.9115 cm, since $K = -1$ and $R = 800.0$ cm. The conjugates for $m = -0.6667$ are $l = -52.2772$ cm, $l' = -34.8532$ cm. The retroreflective system formed by placing the source at the long conjugate of this lens and the paraboloidal mirror center of curvature at its short conjugate is corrected for third-order spherical aberration but has fifth-order lateral spherical aberration of -0.0205 mm. The RMS OPD of the retroreflected wavefront is 0.23λ at $\lambda = 632.8$ nm.

A field lens of focal length 33.3976 cm at the center of curvature of the paraboloid forms an image of the compensating lens on the paraboloidal mirror. With this addition the sign of the fifth-order spherical aberration is reversed, its value being $+0.0207$ mm. The RMS OPD of the retroreflected wave is increased slightly to 0.26λ.

The focal length of the field lens that minimizes the high-order spherical aberration is found to be 66.8900 cm. With this field lens the computed RMS OPD of the retroreflected wavefront is reduced to 0.0003λ, a value well below what can be measured. The diameter of the compensating lens required for this degree of correction is one-twentieth that of the $f/4$ paraboloid.

In Eq. (12.6) it is assumed that the curved surface of the plano-convex field lens is at the center of curvature of the paraboloid. It is sometimes convenient to move the field lens to a position close to, but not at the center of curvature. In this case the field lens introduces an additional magnification m_f. The condition for compensation of third-order aberration then becomes

$$-\frac{KR}{fm_f^2} = (1 - \overline{m})^2 \left[\frac{N^2(1 - \overline{m})^2}{(N - 1)^2} + \frac{(3N + 1)\overline{m}(1 - m)}{N - 1} + \frac{(3N + 2)\overline{m}^2}{N} \right],$$

(12.7)

where $\overline{m} = m_f^2 m$.

Like Ross's arrangement, the retroreflective arrangement of Offner is inherently coma-free so that the correction of the retroreflected wavefront is maintained when the source is near the axis but not exactly on it.

The high degree of stigmatism that can be achieved by the use of the Offner corrector has led to its application for the quantitative assessment of the figures of large aperture concave conicoidal mirrors. For this purpose the retroreflected wavefront can be compared with a reference sphere in a spherical wave interferometer (Houston 1967). A multipass version of the spherical wave interferometer in which the retroreflected wavefront and the reference sphere are optically conjugate (Heintze et al. 1967) is particularly useful for making this measurement with the greatest accuracy. With this interferometer, which has been given the acronym SWIM (spherical wave interferometer multibeam), individual points of a wavefront have been measured with an accuracy of 0.003λ (private communication).

12.4.2. Reflecting Offner Compensators

The weak point in making measurements with the Offner null corrector of Fig. 12.10 is the difficulty in measuring the index variations of the nulling element to the required degree of accuracy. In the example described (Offner 1963), the thickness of the 4.5-cm-diameter compensating lens was 1.05 cm. An average index difference of 3×10^{-7} along the paths of two rays that traverse this lens twice results in an optical path difference of $\lambda/100$ at $\lambda = 632.8$ nm. Larger aperture, faster aspheric mirrors require larger compensating lens diameters and thicknesses. For these, even smaller average index differences result in optical path errors of this magnitude. Fabrication and qualification of large glass elements to this degree of homogeneity is not feasible at present.

These difficulties can be avoided by substituting spherical mirrors for the plano-convex refracting compensating element of Fig. 12.10. The figure errors of such elements can be determined with great accuracy. A small field lens can be retained since it is possible to select small pieces of glass with satisfactorily small index variations.

Single-Mirror Compensator. It is well known (Burch 1936) that the axial aberration of a spherical mirror used at a magnification other than -1 can be used to compensate for the aberration of the normals of a concave conicoid with negative conic constant. The high degree of compensation achieved with the Offner refractive null corrector can also be obtained by a reflecting compensator used with a field at the center of curvature of the conicoid, as shown in Fig. 12.11. As in the refractive version, the radius of the nulling mirror R_N and its conjugates l and l' are chosen to balance the third-order aberration of the normals to the conicoid of radius R_c and the conic constant K. The power of the field lens is then varied to minimize the high-order aberration. The relations to be satisfied are as follows:

$$R_N = -\frac{8KR_c}{(m^2 - 1)^2}, \tag{12.8}$$

$$l = \frac{(1 - m)R_n}{2}, \tag{12.9}$$

$$m = \frac{l}{l'} = -\frac{2\eta_c}{\eta_N}, \tag{12.10}$$

where $2\eta_c$ and η_N are the f numbers of the beam at the center of curvature of the conicoid and at the retroreflected image, respectively.

12.4. THE OFFNER COMPENSATOR

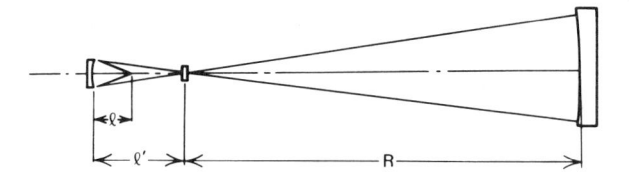

Figure 12.11. Single-mirror compensator with field lens.

The ratio of the diameter of the conicoid A_c to that of the null mirror A_N can be computed from the relation

$$\frac{A_c}{A_N} = \frac{(m^2 - 1)^2}{4K(m - 1)}. \qquad (12.11)$$

Some values of the ratio of the diameter of a paraboloid to that of a single-mirror compensator, computed from Eq. (12.11), are listed in Table 12.1. Diameter ratios of more than 10 require values of $-m$ greater than 4. A practical limit on the value of m is set by the resultant value of h_N, the f number at the retroreflected image, which is inversely proportional to m [Eq. (12.10)]. If the compensated wavefront is to be examined interferometrically without transfer optics, the interferometer must be capable of handling an f/η_N beam. The single-mirror compensator thus requires large compensating mirrors if η_c, the f number of the conicoidal mirror, is small. The permissible residual aberration of the compensated image also must be taken into account, and in some cases this results in a value of A_c greater than that set by the lower limit on η_N.

For example, a one-mirror compensator was designed for a 3-m $f/2.45$ hyperboloidal primary mirror of a proposed Ritchey–Chrétien system. The conic constant of the mirror was -1.003313. A value of -4.9 was chosen for m, resulting in an $f/1$ beam at the retroreflected image. The third-order mirror specifications computed from Eqs. (12.4) to (12.7) and the specifications of the optimized design are shown in Table 12.2. The focal length of the field lens that minimizes the high-order aberration is 55.4849 cm. The computed RMS

Table 12.1. Diameter Ratio and Magnification, Single-Mirror Compensator

Magnification	Diameter ratio
-3.0	4.0
-3.5	7.0
-4.0	11.25
-4.5	16.8
-5.0	24.0

Table 12.2. One-Mirror Null Corrector for 3-M $f/2.45$ Hyperboloid

Type of design	m	R_N	l	A_N	η_N
Third order	−4.9	22.2849	65.7404	13.42	1.000
Optimized	−4.7	22.2849	65.7518	13.50	1.044

OPD of the retroreflected wavefront is 0.009λ at $\lambda = 632.8$ nm. If a smaller residual had been required, a smaller value of $-m$ would have been chosen for the compensator. The resulting compensator mirror would then have been larger, and the convergence angle of the retroreflected beam would have been smaller.

Two-Mirror Compensator. Although the single-mirror compensator of Fig. 12.11 is the least complicated optically of the reflecting compensators, practical implementations require an additional element, such as the folding flat of Fig. 12.12, to make the retroreflected image accessible. The quality of the flat must, of course, be comparable to that of the spherical mirror.

The same number of accurately fabricated optical components is required in the two-mirror compensator shown in Fig. 12.13. The in-line arrangement facilitates accurate alignment and provides an accessible retroreflected image.

With the two-mirror compensator the central portion of the aspheric mirror cannot be observed because of the holes in the nulling mirrors. These null correctors should be designed so that the obscured portion of the aspheric mirror in the null test is no larger than the obscured portion of the aspheric mirror in actual use.

The third-order design of a two-mirror compensator is affected by the value of the obscuration ratio at each of the mirrors. The following equations apply when the obscuration ratio r is the same at the two mirrors. As in the case of the single-mirror null corrector, the parameters and apertures are functions of a magnification. For the two-mirror compensator the magnification is that from the intermediate image (Fig. 12.13) to the image at the center of curvature of the aspheric mirror, defined by the relation

$$m_1 = -\frac{2\eta_c}{\eta_1}, \qquad (12.12)$$

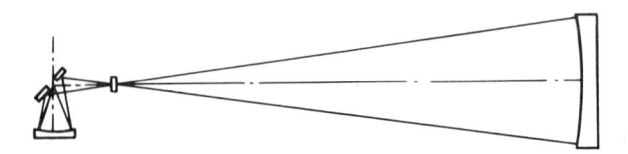

Figure 12.12. Practical implementation of single-mirror compensator.

12.4. THE OFFNER COMPENSATOR

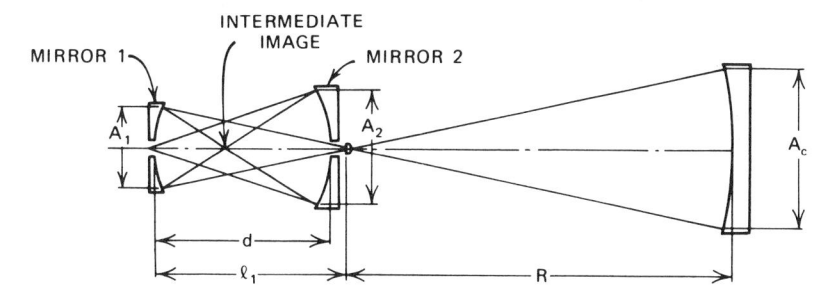

Figure 12.13. Two-mirror compensator with field lens.

where η_1 is the f number at the intermediate image. The ratios of the apertures of the two nulling mirrors A_1 and A_2 to the aperture of the conicoid A_c, when the third-order aberration of the conicoid normals is compensated for by that of the null corrector, can be computed from the following relations:

$$\frac{A_c}{A_1} = \frac{1}{4K}(m_1 + 1)[m_1^2(1 + 2r - r^2) - 2m_1(1 - r) - 2]. \quad (12.13)$$

$$\frac{A_c}{A_2} = \frac{A_c}{A_1}\frac{m_1 r - 1}{m_1 + 1}. \quad (12.14)$$

The other relations required for the third-order design are as follows:

$$R_1 = \frac{4A_1\eta_c}{1 - m_1}, \quad (12.15)$$

$$R_2 = \frac{4A_2\eta_c}{m_1(2 - r) + 1}, \quad (12.16)$$

$$l_1 = 2\eta_c A_1, \quad (12.17)$$

$$d = \eta_1(A_1 + A_2), \quad (12.18)$$

$$\eta_N = -\frac{2\eta_c}{m_1(1 - r) + 1}, \quad (12.19)$$

where l_1 is the distance from the center of curvature of the conicoid to null mirror 1, d is the distance between the two null mirrors, and η_N is the f number at the retroreflected image.

Table 12.3. Diameter Ratios and Magnification, Two-Mirror Compensator

	$r = 0.2$			$r = 0.3$		
m_1	A_c/A_1	A_c/A_2	$2\eta_c/\eta_N$	A_c/A_1	A_c/A_2	$2\eta_c/\eta_N$
−3.0	7.52	6.02	1.4	7.90	7.50	1.10
−3.5	12.66	8.61	1.8	13.37	10.97	1.45
−4.0	19.62	11.77	2.2	20.82	15.27	1.80
−4.5	28.65	15.56	2.6	30.52	20.49	2.15
−5.0	40.00	20.00	3.0	42.75	26.72	2.50

The way in which the apertures of the two mirrors and the f number of the retroreflected image vary as a function of the magnification m_1 for two values of the obscuration ratio can be seen in Table 12.3. Comparison with Table 12.1 shows that for a given magnification the larger mirror of the two-mirror compensator is approximately the same size as the single-mirror compensator. However, the difference between the magnitudes of m_1 and $2\eta_c/\eta_N$ in Table 12.3 indicates that, for a given maximum size compensator element, the convergence angle at the retroreflected image with the two-mirror compensator is approximately one-half that with a single-mirror compensator.

The degree of compensation attainable with the two-mirror compensator is extremely high, as shown by the following example. A compensator was required for a 3-m $f/1.5$ paraboloid that was to be used with an obscuration ratio of 0.3. A spherical wave interferometer that could accommodate convergence angles up to $f/1.2$ was available. The values $r = 0.25$ and $m_1 = -4$ were chosen to give the safe value $\eta_N = 1.5$. Equations (14.9) and (14.10) led to the acceptable values $A_1 = 14.82$ cm, $A_2 = 27.22$ cm. The parameters obtained from Eqs. (12.15) to (12.19) are listed as third-order design parameters in Table 12.4. The field lens required to optimize this compensator is a meniscus lens of refractive index 1.519, thickness 0.5 cm, and radii 14.619 cm (convex) and 71.656 cm (concave), with the convex surface facing the paraboloid. With this field lens and the slight modifications of the other parameters shown in Table 12.4, the RMS OPD of the retroreflected wavefront is 0.009λ at $\lambda = 632.8$ nm. Had a smaller residual aberration been required, the value of m_1 would have had to be reduced. This would have resulted in larger values of A_1, A_2, and η_N.

Table 12.4. Two-Mirror Null Corrector for 3-M $f/1.5$ Paraboloid

Type of design	m_1	R_1	R_2	l_1	d	A_2	η_N
Third order	−4.00	17.7778	22.2222	44.4444	27.7778	22.22	1.50
Optimized	−3.99	17.7776	22.2227	44.3868	27.7789	21.34	1.52

12.4.3. General Comments about Refracting Compensators

The main refracting compensators are the Couder, Dall, Shafer, and the Offner compensators. They have the following properties in common:

1. All compensate the spherical aberration but with a different degree of perfection. The best in this respect is the Offner compensator.
2. All can be used in a single or double pass, but the Couder and Dall compensators are typically used in a single pass, generally in front of the light source, while the Shafer and Offner compensators are normally used in double pass.
3. The chromatic aberration is not corrected in any of these compensators, hence monochromatic light has to be used. One possibility is to use laser light and another is to use a color filter. A red filter used close to the eye or image detector (after the compensator) is suggested so that the wavefront shape is not affected by this filter.
4. The amount of spherical aberration correction depends very critically on the axial position of the compensating lens so, unless its position is very accurately measured, we can never be sure about the exact value of the conic constant of the concave surface under test. This value has to be measured by some other test that does not use any compensator, like the Hartmann test. However, only with a compensator can the general smoothness of the surface be easily determined.
5. In a double-pass configuration any lateral displacement of the light source and the observer in opposite direction and by equal amounts, with respect to the optical axis, does not introduce any coma. It is assumed, however, that the optical axis of both the conic surface under test and the compensating lens coincide, otherwise some coma is introduced.

12.5. OTHER NULL TESTS FOR CONCAVE CONICOIDS

The success of the simple small compensators described in the preceding sections results from a fortunate combination of conditions.

1. The concave aspheric mirrors by themselves transform the divergent wave from a point source into a converging though aberrated wave.
2. The greater part of the aberration introduced by the aspheric is low order.
3. The sense of the aberration is opposite to that introduced by a concave mirror or a simple convex lens.

The first condition must be met if the null corrector is to be smaller than the aspheric mirror being tested. The second condition makes it possible to get good compensation with a single element of convenient form. The third condition

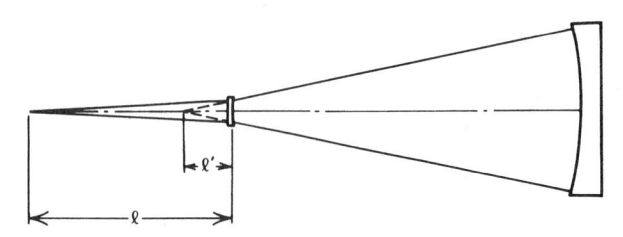

Figure 12.14. Modified Dall-type compensator for oblate spheroid.

makes possible the use of a simple relay lens that provides a position for a field lens. Concave spherical mirrors can be used for compensators only when this condition is met.

A small null corrector of the same general form can be designed for any concave mirror whose surface is generated by rotating a conic section about its major axis.

Concave prolate spheroids do not require compensators since their geometrical foci are accessible and, as is the case with all conicoids that have geometrical foci, their imagery is stigmatic when these are the conjugates. However, when one of the geometrical foci is at a large distance from the mirror, it may be more convenient to perform a null test at the center of curvature with one of the null correctors described in Sections 12.2, 12.3, and 12.4.

An oblate spheroid, such as is used as the primary mirror of a Wright–Schmidt system, does not satisfy condition 3. Nevertheless, a null test of the modified Dall type can be obtained by substituting a plano-concave lens for the Dall plano-convex lens (Fig. 12.14). Since the curved surface of this lens faces the oblate spheroid, the third-order solution is formally the same as that for the Offner plano-convex compensator (Fig. 12.10). The parameters and conjugates of the plano-convex lens required to balance the third-order aberration of the normals to the prolate spheroid can thus be obtained by the use of Eq. (12.6). The quantity m is the ratio l'/l (Fig. 12.14). In this case it is positive and has a value less than 1. The effect of the choice of m on the ratio of the size of the oblate spheroid to that of the null corrector is shown in Table 12.5; the values

Table 12.5. Aperture of Plano-Concave Null Lens Compensator for Oblate Spheroid ($K = 1$, $N = 1.52$)

m	Diameter Asphere/Diameter Compensator
0	8.54
0.1	7.13
0.2	5.88
0.3	4.77
0.4	3.80

12.5. OTHER NULL TESTS FOR CONCAVE CONICOIDS

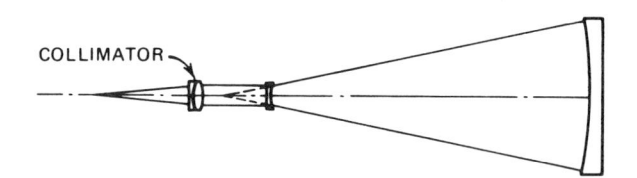

Figure 12.15. Compensator for oblate spheroid used with collimator.

were computed for $N = 1.52$ and $K = 1$. The implementation of the value $m = 0$ requires the addition of a collimator to put the source optically at an infinite conjugate. The resulting arrangement is shown in Fig. 12.15.

A null corrector of this type was designed for a 0.6-m aperture $f/5$ oblate spheroid with conic constant $K = 1$. The departure of this aspheric from the base sphere is equal in magnitude but opposite in sense to that of the paraboloid used as an example in Section 12.2. The values $N = 1.52$ and $m = 0$ were chosen. The focal length of the plano-concave lens, obtained from Eq. (12.2), is 70.2216 cm. The RMS OPD of the retroreflected plane wave is 0.033λ at $\lambda = 632.8$ nm. The diameter of the null lens is 0.7 cm.

Holleran (1963, 1964) described a null test for concave conicoids that has the virture that no auxiliary optical elements need be manufactured. The surface to be tested is made level and is immersed in a liquid that forms a plano-convex lens in contact with the surface under test. In the simplest form of the test a pinhole light source and knife edge are placed at a distance d above the plane surface of the liquid. For liquid lens thickness t and refractive index N,

$$d = \frac{R}{N} - t, \qquad (12.20)$$

where R is the vertex radius of curvature of the conicoid. The retroreflected image is corrected for third-order spherical aberration if

$$N^2 = 1 - \frac{KR}{R - t}, \qquad (12.21)$$

where K is the conic constant of the mirror.

The accuracy of the test is very good for shallow curves of moderate aperture. The peak-to-valley departure ΔW of the surface from the desired conicoid when Eq. (12.21) is satisfied and $t \ll R$ is

$$\Delta W \approx \frac{KR}{41.5 \, \eta^6} \text{ waves.} \qquad (12.22)$$

In Eq. (12.22) η is the f number of the conicoidal mirror, R is its radius in millimeters, and the wavelength is 632.8 nm. The peak-to-valley error of a 0.5-m-diameter $f/3$ paraboloid figured to give a perfect null by this test is 0.10 wave. Decreasing the f number to 2.5 for a mirror having the same aperture increases the figure error to 0.25 wave. The error of an $f/2.5$ mirror reaches 0.10 wave for an aperture of 0.2 m.

The immersion test in this form can also be applied to convex aspherics by observing them through a plane back surface. In this case the optical material replaces the immersion liquid. Puryayev (1971) analyzed an extension of this method in which an immersion fluid is placed above the plane surface. Since the liquid must extend to the retroreflected image, this extension is practical only for small elements. Puryayev's equations reduce to those of Holleran when the immersion fluid is air.

A related test for convex hyperboloids, described by Normal (1957), makes use of the fact that a plano-convex lens forms a stigmatic image of a collimated source on its axis if the convex surface is a conicoid with eccentricity equal to the refractive index of the material of which the lens is made. The autocollimated image of a point source distant one focal length from the convex surface of lens and reflected from its plane surface, or from a flat mirror parallel to its plane surface, can then be examined to determine the figure of the convex surface. As in the Holleran and Puryayev tests, the range of conicoids that can be tested by this method is limited by the range of refractive indices available. The test can be applied to hyperboloids with magnifications between 3.5 and 5 when the range of glass indices is restricted to 1.5 to 1.8. Holleran (1966) showed that a spherical back surface can provide a null test compensator for a concave oblate spheroid when the latter is tested from the back, through the spherical surface.

12.6. COMPENSATORS FOR CONVEX CONICOIDS

The testing of convex hyperboloids is very important for astronomical instruments. The most common test for these surfaces has been implemented by using a Hindle sphere, as described in the following section. The problem with this method is that a very large concave spherical surface, much larger than the surface under test, is required. Various other methods using compensators have been reported for testing convex hyperboloids, as described by Parks and Shao (1988).

Descartes discovered that a convex hyperboloidal surface or a convex ellipsoidal surface produce aplanatic (no spherical aberration and no coma) images, when the object is located as in Fig. 12.16 and the refractive index N is equal to the square of the eccentricity ($N = \sqrt{-K}$. Using this property, a convex hyperboloid may be tested with the arrangement in Fig. 12.17, if the index of

12.6. COMPENSATORS FOR CONVEX CONICOIDS 447

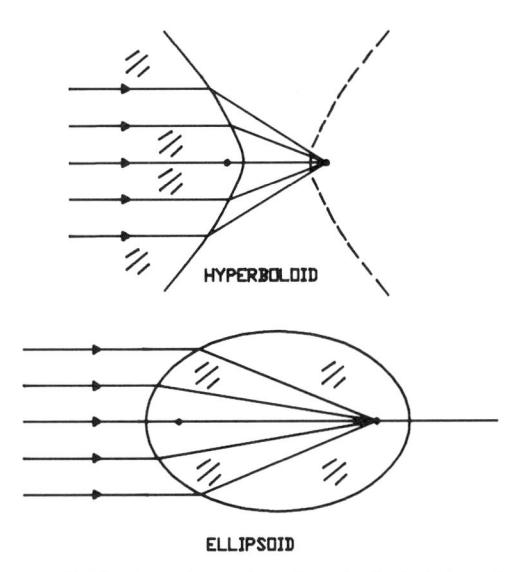

Figure 12.16. Cartesian surfaces free of spherical aberration.

refraction of the glass is of the proper value. The glass has to be clear and homogeneous, so that optical glass or fused silica have to be used. Pyrex or glass-ceramic materials are not adequate. In this test the wavefront deformations OPD and the surface deformations W are related by $W = \text{OPD}/2(N - 1)$, which is about half the sensitivity obtained in a reflective arrangement. If the refractive index is not adequate to produce an aplanatic image, an object distance may be selected so that the image is free of primary spherical aberration. Then, as suggested by James and Waterworth (1965), a lens with a convex hyperboloidal surface may be tested as in Fig. 12.18. In this case, however, the compensation is not perfect, since some small residual aberration may re-

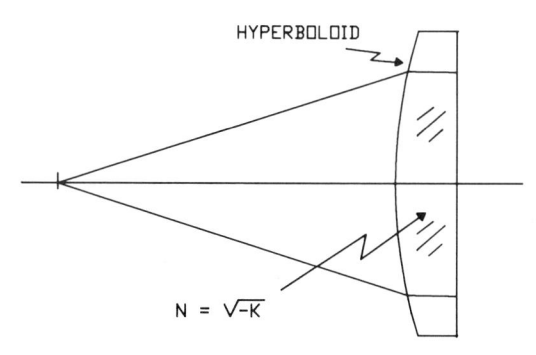

Figure 12.17. Testing of a hyperboloid using a Cartesian configuration.

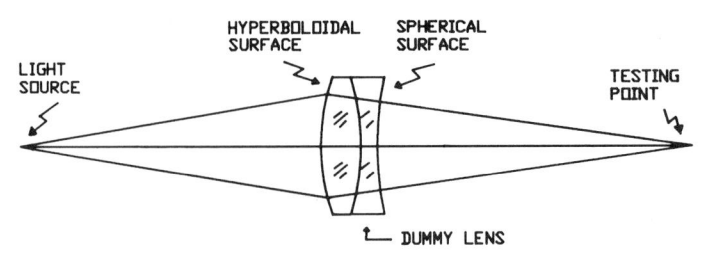

Figure 12.18. Testing an aspheric surface by selecting the conjugate distance that produce the minimum amount of spherical aberration.

main. The sensitivity in about one-fourth that obtained in a reflective arrangement.

Another method to test a convex hyperboloid was suggested by Meinel and Meinel (1983a, 1983b). The surface is tested from the back, through the glass. The back surface is polished with a slightly convex surface, computed by ray tracing so that the object and image are located in the same plane and free from spherical aberration, as in Fig. 12.19. If the back surface is plane, the spherical aberration is eliminated by using different positions for the light source and the image, as in Fig. 12.20. This second solution is even better than with equal conjugates because the residual spherical aberration is smaller.

Still another test for hyperboloidal convex mirrors was proposed by Bruns (1983). A convergent lens is placed in front of the convex surface under test. Unfortunately, no lens with spherical surfaces may correct the spherical aberration of the hyperboloid. So, Bruns uses a lens with a hyperboloidal surface. This hyperboloid has conic constant K_L related to the glass refraction index by $N_L = \sqrt{-K_L}$. This makes the lens very simple to test when a collimated beam enters the lens through the flat face. When this lens is used in reverse, that is,

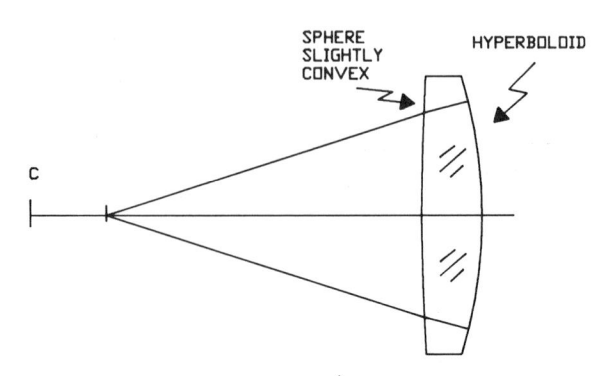

Figure 12.19. Meinel's test of a hyperboloidal surface, using equal conjugates. (Light source and testing point at the same position.)

12.6. COMPENSATORS FOR CONVEX CONICOIDS

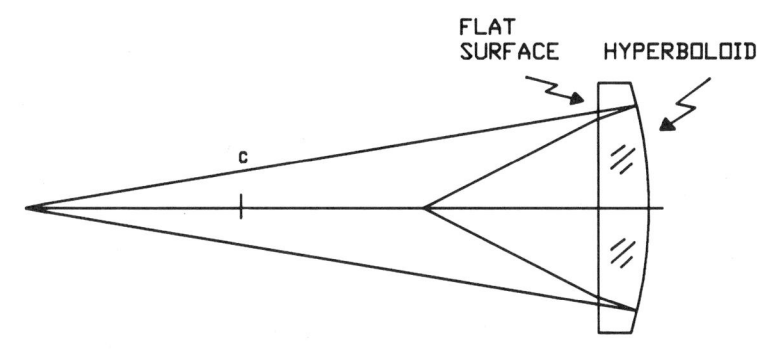

Figure 12.20. Meinel's test of hyperboloidal surface, using unequal conjugates. The light source and the testing point are at different positions, but they can be made to coincide if a double-pass configuration is used, by placing a small flat mirror at one of these points.

with the collimated beam entering the convex surface, it has spherical aberration with the proper sign to cancel that of the convex mirror, as in Fig. 12.21.

If the desired conic constant for the convex mirror is K_M, with radius of curvature R_M, Bruns has shown that the spherical aberration of this system is compensated when the refractive index N_L of the lens is given by the following relation:

$$K_M = -\frac{2(N_L + 1)(R_M + d)}{N_L(N_L - 1)R_M}, \qquad (12.23)$$

where d is the separation between the flat surface of the lens and the mirror, and an infinitely thin lens was assumed. The focal length of the lens is equal to $R_M + d$. The value of R_M is fixed, thus N_L and d must be chosen so that this relation is satisfied for the desired value of K_M.

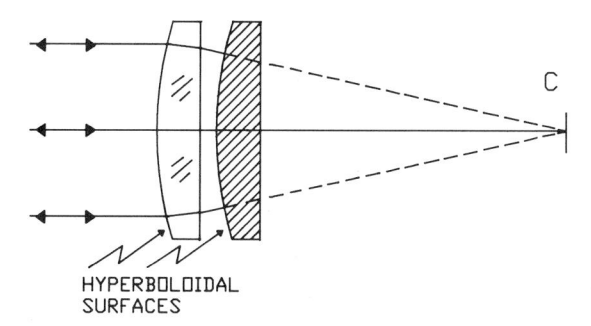

Figure 12.21. Testing of a hyperboloidal surface by using an aspherical lens to compensate the spherical aberration.

12.7. HINDLE-TYPE TESTS

Since convex mirrors that are to be tested in reflection do not fulfill condition 1 of Section 12.4, a compensator for one of these must contain an element that is at least as large as the convex mirror.

When the convex mirror is conicoid, its imagery is stigmatic for conjugates at its geometrical foci. A null test for the imagery with these conjugates does not require a compensator; but, since at least one of the geometrical foci is inaccessible, additional optical elements are required to employ this test.

Hindle (1931) showed how an autostigmatic arrangement for testing a convex hyperboloid can be implemented by retroreflection from a sphere whose center is at the inaccessible focus of the hyperboloid. The arrangement is used as shown in Fig. 12.22.

Figures 12.23 and 12.24 show Hindle arrangements for testing convex paraboloids and convex prolate spheroids. In addition to the Hindle sphere, a collimator is required to test a convex paraboloid. To test a prolate spheroid, an additional optical system is needed to provide a beam that converges to one of the spheroidal foci. Small concave hyperboloids can also be tested in this way (Silvertooth 1940) (Fig. 12.25).

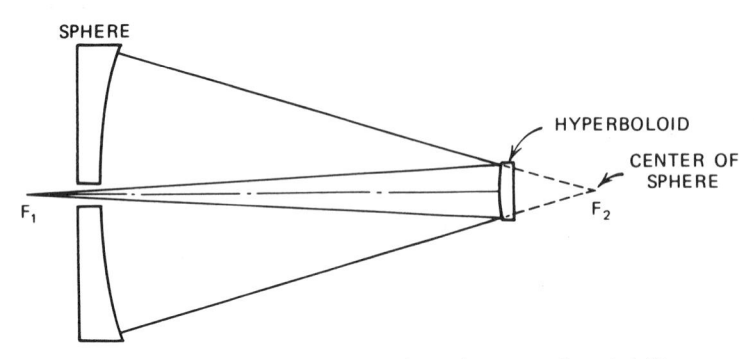

Figure 12.22. Hindle arrangement for testing convex hyperboloid.

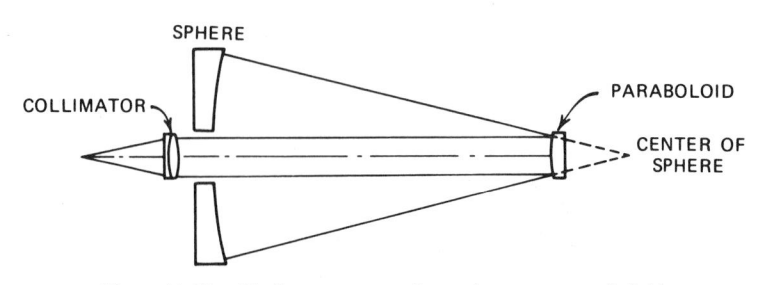

Figure 12.23. Hindle arrangement for testing convex paraboloid.

12.7. HINDLE-TYPE TESTS

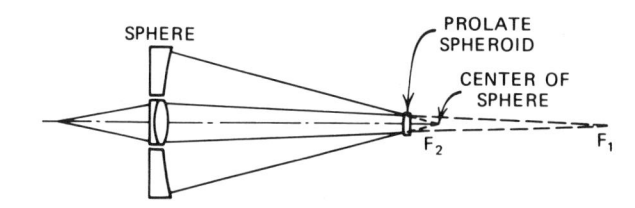

Figure 12.24. Hindle arrangement for testing convex prolate spheroid.

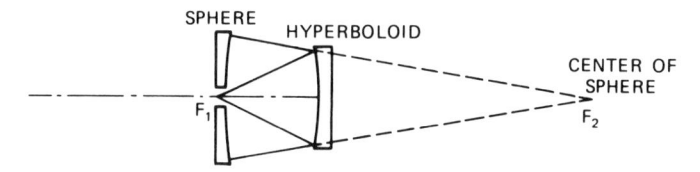

Figure 12.25. Hindle arrangement for testing concave hyperboloid.

Although the Hindle test for convex conicoids affords a stigmatic retroreflected image, its implementation is often impractical because keeping the obscuration inherent in the test within permissible bounds results in a prohibitively large spherical mirror. In the case of a hyperboloid of diameter A, the aperture of the Hindle sphere A_H is given by the relation

$$A_H = \frac{A(m + 1)}{mr + 1}, \qquad (12.24)$$

where r is the permissible obscuration ratio and m is the magnification of the hyperboloid for its stigmatic conjugates. Thus a 0.25-m hyperboloid with $m = 10$ and permissible $r = 0.2$ requires a 0.92-m-Hindle sphere.

A modification of the Hindle test that avoids this difficulty was described by Simpson et al. (1974). Their arrangement is shown in Fig. 12.26. The concave surface of a meniscus element serves as the Hindle sphere. This surface is half-

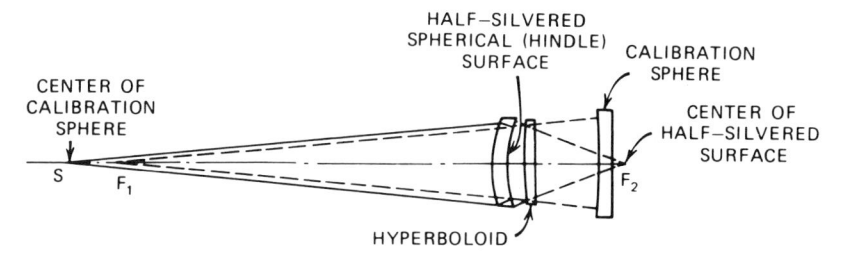

Figure 12.26. Simpson–Oland–Meckel-modified Hindle arrangement.

silvered so that it can be placed close to the conicoid without introducing obscuration. The radius of the convex surface can be chosen to compensate for the spherical aberration introduced by the concave surface. This testing configuration has been studied by Robbert (1979) and by Howard et al. (1983).

To test the effect of the meniscus Hindle element on the retroreflected wave, the hyperboloid is removed and the retroreflected image of S from a calibration sphere with center at F_1 is examined. Any significant aberration introduced by the meniscus can then be subtracted from the measurement of the figure error of the hyperboloid.

A meniscus was designed to test the 0.25-m 10× hyperboloid mentioned earlier. The geometrical foci of the hyperboloid were 0.6 and 6 m. The base radius of the hyperboloid is thus 1.33333 m, and its conic constant is $K = -1.49383$. Glass of index 1.52 was chosen for the meniscus element. For a 5-cm separation of the meniscus from the hyperboloid, the radius of the concave surface is 65 cm. With a meniscus thickness of 5 cm the radius of the convex surface that results in aberration compensation at the edge of the aperture is 66.6637 cm. The required clear aperture of the meniscus is 0.254 m. The RMS OPD of the retroreflected wave is 0.0016λ at $\lambda = 632.8$ nm. The stigmatic quality of the retroreflected image is thus retained in this modified Hindle test. Its use for testing convex conicoids is limited only by the availability of refracting material of approximately the same size as the surface being tested.

REFERENCES

Bruns, D. G., "Null Test for Hyperbolic Convex Mirrors," *Appl. Opt.*, **22**, 12 (1983).

Burch, C. R., "On Reflection Compensators for Testing Paraboloids," *Mon. Not. R. Astron. Soc.*, **96**, 438 (1936).

Burch, C. R., "Report of the General Meeting of the Association," *J. B. Astron. Assoc.*, **48**, 99 (1938).

Couder, A., "Procede d'Examen d'un Miroir Concave Non-spherique," *Rec. Opt. Theor. Instrum.*, **6**, 49 (1927).

Dall, H. E., "A Null Test for Paraboloids," *J. Br. Astron. Assoc.*, **57**, 201 (1947).

Dall, H. E., "A Null Test for Paraboloids," in *Amateur Telescope Making*, Vol. 3, A. E. Ingalls, Ed., Scientific American, New York, 1953, pp. 149–153.

Foucault, L., "Memoire sur la Construction des Telescopes en Verre Argente" (On the Construction of Telescopes in Silvered Glass), *Annal. Obs. Paris*, **5**, 197 (1859).

Heintze, L. R., H. D. Polster, and J. Vrabel, "A Multiple-Beam Interferometer for Use with Spherical Wavefronts," *Appl. Opt.*, **6**, 1924 (1967).

Hindle, J. H., "A New Test for Cassegrainian and Gregorian Secondary Mirrors," *Mon. Not. R. Astron. Soc.*, **91**, 592 (1931).

Holleran, R. T., "Immersion Null Test for Aspherics," *Appl. Opt.*, **2**, 1336 (1963).

Holleran, R. T., "Null Testing Telescope Mirrors by Immersion," *Sky Telesc.*, **28**, 242 (1964).

Holleran, R. T., "Third-Order Wavefronts and Related Null Tests," *Appl. Opt.*, **5**, 1244 (1966).

Holleran, R. T., "An Algebraic Solution for the Small Lens Null Compensator," *Appl. Opt.*, **7**, 137 (1968).

Houston, J. B., Jr., C. J. Buccini, and P. K. O'Neill, "A Laser Unequal Path Interferometer for the Optical Shop," *Appl. Opt.*, **6**, 1237 (1967).

Howard, J. W., M. H. Beaulieu, and R. J. Zielinski, "A Nonobscuring, Easily-Calibrated Method for Testing Ellipsoids on a Fizeau Interferometer," *Proc. SPIE*, **389**, 26 (1983).

Jaems, W. E. and M. D. Waterworth, "A Method for Testing Aspheric Surfaces," *Opt. Acta*, **12**, 223 (1965).

Landgrave, J. E. A. and J. R. Moya, "Dummy Lens for the Computer Optimization of Autostigmatic Null Correctors," *Appl. Opt.*, **26**, 2673 (1987).

Meinel, A. B. and M. P. Meinel, "Self-Null Corrector Test for Telescope Hyperbolic Secondaries," *Appl. Opt.*, **22**, 520 (1983a).

Meinel, A. B. and M. P. Meinel, "Self-Null Corrector Test for Telescope Hyperbolic Secondaries: Comments," *Appl. Opt.*, **22**, 2405 (1983b).

Moya, J. R. and J. E. A. Landgrave, "Third-Order Design of Refractive Offner Compensators," *Appl. Opt.*, **26**, 2667 (1987).

Norman, B. A., "New Test for Cassegrainian Secondaries," *Sky Telesc.*, **17**, 38 (1957).

Offner, A., "A Null Corrector for Paraboloidal Mirrors," *Appl. Opt.*, **2**, 153 (1963).

Offner, A., "Field Lenses and Secondary Axial Aberration," *Appl. Opt.*, **8**, 1735 (1969).

Parks, R. E. and L. Z. Shao, "Testing Large Hyperbolic Secondary Mirrors," *Opt. Eng.*, **27**, 1057 (1988).

Puryayev, D. T., "A Quality Control Technique for Convex Elliptical, Parabolic and Hyperbolic Surfaces of Simple Lenses," *Soc. J. Opt. Technol.*, **38**, 684 (1971).

Puryayev, D. T., "Compensator for Inspecting the Quality of Large-Diameter Parabolic Mirrors," *Sov. J. Opt. Technol.*, **40**, 238 (1973).

Robbert, C. F., "Typical Error Budget for Testing High-Performance Aspheric Telescope Mirror," *Proc. SPIE*, **181**, 56 (1979).

Rodgers, J. M., "A Null-Lens Design Approach for Centrally Oscured Components," *Proc. SPIE*, **679**, 17 (1986).

Ross, F. E., "Parabolizing Mirrors without a Flat," *Astrophys. J.*, **98**, 341 (1943).

Sasian, J. M., "Optimum Configuration of the Offner Null Corrector: Testing an $F/1$ Paraboloid," *Proc. SPIE*, **1164**, 8 (1989).

Schlauch, J., "Construction of a Dall Null Tester," *Sky Telesc.*, **18**, 222 (1959).

Schupmann, L., *Die Medial-Fernrohre: Eine neue Konstruktion für grosse astronomische Instrumente*, Teubner, Leipzig, 1899.

Shafer, D. R., "Zoom Null Lens," *Appl. Opt.*, **18**, 3863 (1979).

Silvertooth, W., "A Modification of the Hindle Test for Cassegrain Secondaries," *J. Opt. Soc Am.*, **30**, 140 (1940).

Simpson, F. A., B. H. Oland, and J. Meckel, "Testing Convex Aspheric Lens Surfaces with a Modified Hindle Arrangement," *Opt. Eng.*, **13,** G101 (1974).

Stoltzmann, D. E. and M. Hatch, "Extensions of the Dall Null Test," *Sky Telesc.*, **52,** 210 (1976).

ADDITIONAL REFERENCES

De Voe, C. E. "Limitations on Aspheric Surface Testing with Simple Null Correctors," Thesis, The University of Arizona, Tucson, AZ, 1989.

Ruda, M. C. "Methods for Null Testing Sections of Aspheric Surfaces," Ph.D. Thesis, The University of Arizona, Tucson, AZ, 1979.

Sasian, J. M., "Design of Null Correctors for the Testing of Astronomical Optics," *Opt. Eng.*, **27,** 1051 (1988).

Willstrop, R. V., "Simple Null Test for a Schmidt Camera Aspheric Corrector," *Mon. Not. R. Astr. Soc.*, **192,** 455 (1980).

13

Interferogram Evaluation and Wavefront Fitting

D. Malacara and S. L. DeVore

13.1. INTRODUCTION

Personal computers have greatly simplified the analysis of interferogram fringes. In this chapter we will examine the automated analysis of single interferograms. By this we mean a single image of the fringes, as opposed to the multiple images required for phase shifting interferometry. Fringe analysis methods can be divided into two classes: (1) Those that first reduce the image to a list of digitized fringe centers and (2) those that directly process the entire fringe to obtain the measured optical path difference (OPD). Most of this chapter will be devoted to working with fringe centers, although an example of the latter class will be included in our discussion of Fourier transform analysis.

The fringes can be located either manually, using a digitizing tablet, or automatically, with the computer directly examining a single-frame fringe image that has been captured using a digital frame grabber. After locating the fringe centers, fringe order numbers must be assigned to each point. The wavefront can then be characterized by direct analysis of the ordered fringe centers, or the fringe centers can be used to generate a uniform grid of data representing a map of the wavefront OPD.

In the first case the analysis results take the form of wavefront statistics, such as peak-to-valley (PV) and root-mean-square (RMS) wavefront error. Converting the fringe centers to a uniform grid of wavefront data makes it possible to graphically display the OPD surface and to perform more advanced analysis such as diffraction patterns and modulation transfer function (MTF) calculations. Reducing all measured wavefronts to a common data grid also allows point-by-point operations, such as differencing and averaging, to be performed among multiple wavefronts. Because the fringe centers represent OPD

Optical Shop Testing, Second Edition, Edited by Daniel Malacara.
ISBN 0-471-52232-5 © 1992, John Wiley & Sons, Inc.

measurements at discrete points, some kind of data interpolation between the fringes will be required to produce the uniform grid. The two methods used most often are global polynomial fitting and local interpolation using splines and other interpolants. The former method has been described by Loomis (1978) while the latter has been described by Hayslett and Swantner (1978, 1980). A number of commercially available computer programs implementing these and other methods have appeared since then.

Fringe analysis is generally less precise than phase shifting interferometry, often by more than one order of magnitude. It also requires more care in handling complex aperture shapes and variations in fringe visibility. In some cases these problems prevent the successful analysis of the fringes. However, fringe analysis does have the advantage of working with a single image of the fringes, an image that may have been acquired over a brief time interval. Phase shifting interferometry requires several images, acquired over a longer time span during which the fringes must be stable. In turbulent or noisy environments one can often obtain acceptable accuracy by averaging the results of several single-image fringe analyses, while a phase-shifted measurement may not work at all.

This chapter will start by discussing polynomials that can be used to describe wavefronts. In many cases measured wavefronts can be described by a few coefficients multiplying the terms of a well-chosen polynomial. We next show how the polynomial coefficients can be determined by performing a least-square fit of the polynomial to the measured fringe data. These coefficients can be used to remove low-order terms such as power and tilt, or they can be evaluated to produce an OPD map over a uniform data grid. We will also mention other methods for interpolating the fringe data. This is followed by a discussion of methods for digitizing the fringe centers and an introduction to the Fourier transform method of analyzing interferograms.

13.2. POLYNOMIAL WAVEFRONT REPRESENTATION

In any interferometric optical testing procedure the main objective is to determine the shape of the wavefront measured with respect to a best fit sphere. If the wavefront is continuous and sufficiently smooth, it can be represented by a two-dimensional function, as we will see here.

13.2.1. Aberration Polynomial for Primary Aberrations

The aberration polynomial describes the wavefront deformations when the optical system has aberrations. For the case of primary aberrations this expression was given by Kingslake (1925-1926) (see Figs. 13.1-13.3):

$$W(x, y) = A(x^2 + y^2)^2 + By(x^2 + y^2) + C(x^2 + 3y^2) \\ + D(x^2 + y^2) + Ey + Fx + G, \quad (13.1)$$

13.2. POLYNOMIAL WAVEFRONT REPRESENTATION

where A = spherical aberration coefficient
B = coma coefficient
C = astigmatism coefficient
D = defocusing coefficient
E = tilt about the x axis
F = tilt about the y axis
G = constant or piston term

In a more general form, a wavefront might include high-order aberrations, which can be expressed in terms of monomials as follows:

$$W(x, y) = \sum_{i=0}^{k} \sum_{j=0}^{i} c_{ij} x^j y^{i-j}, \tag{13.2}$$

where k is the degree of this polynomial.

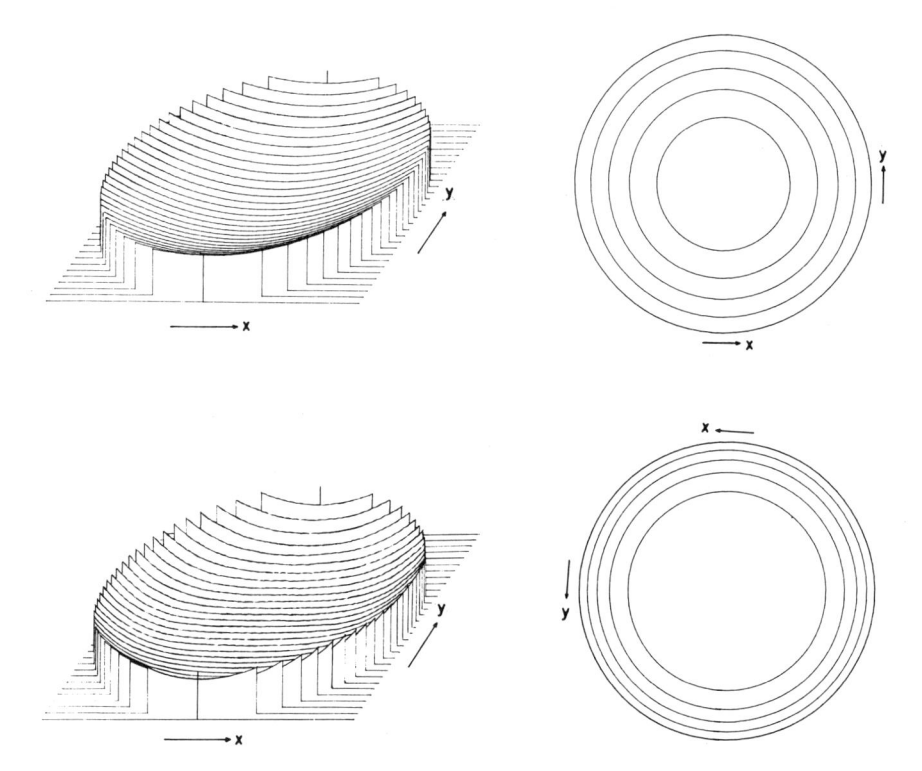

Figure 13.1. Contour and isometric plots for wavefronts with defocusing ($D = 5$) and spherical aberrations ($A = 5$).

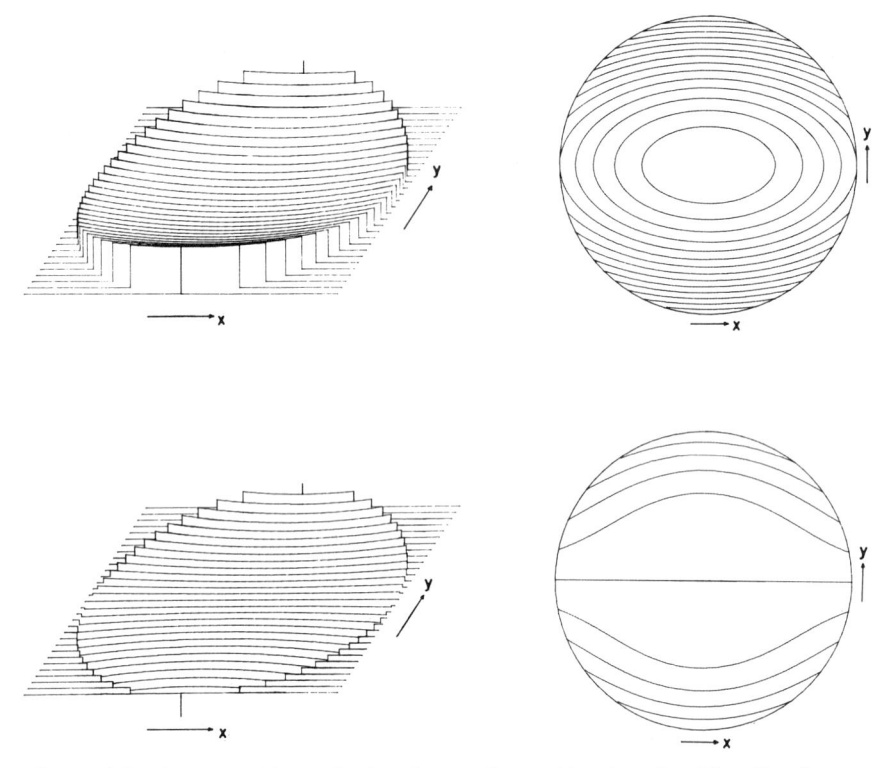

Figure 13.2. Contour and isometric plots for wavefronts with astigmatism ($C = 1$) and coma ($B = 5$).

It is more useful, however, to write W in polar coordinates ρ and θ over a circle with unit radius ($x/S_{\max} = \rho \sin \theta$ and $y/S_{\max} = \rho \cos \theta$), where $S_{\max}$ is the radius of the circle over which the wavefront is defined:

$$W(\rho, \theta) = \sum_{n=0}^{k} \sum_{l=0}^{n} \rho^n (a_{nl} \cos^l \theta + b_{nl} \sin^l \theta) \qquad (13.3)$$

where the $\cos \theta$ and $\sin \theta$ terms describe the symmetrical and antisymmetrical parts of the wavefront, respectively. We must be careful, however, because not all possible values of n and l are permitted. To have a single-valued function, we must require that

$$W(\rho, \theta) = W(-\rho, \theta + \pi) \qquad (13.4)$$

thus obtaining the condition that $(n - l)$ should be even and therefore n and l must be both even or both odd. We also want expression (13.3) to be equivalent

13.2. POLYNOMIAL WAVEFRONT REPRESENTATION

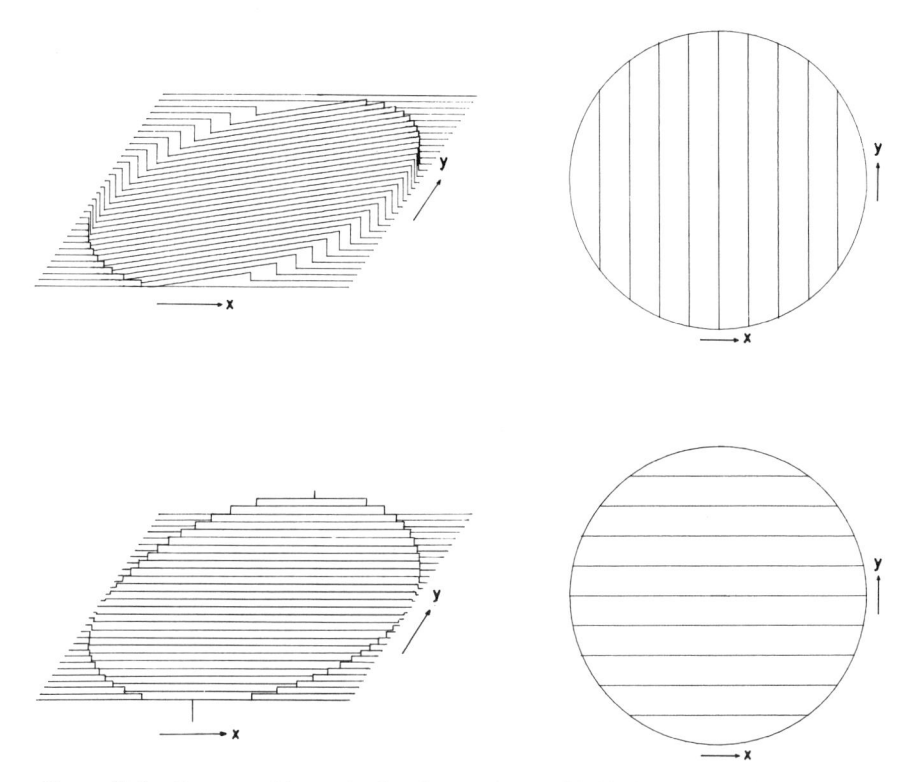

Figure 13.3. Contour and isometric plots for wavefronts with tilt about the Y axis ($F = 5$) and tilt about the X axis ($E = 5$).

to the polynomial (13.1) of k degree. It is possible to show that if $l > n$, expression (13.3) when converted to (x, y) coordinates would become an infinite series. Thus it is desirable to impose the additional condition that $l \leq n$, although in some rare cases this is not satisfied, as shown in Eq. (5.26) in Chapter 5.

If the aberrations are those of an axially symmetric system, the $\sin \theta$ terms disappear, giving

$$W(\rho, \theta) = \sum_{n=0}^{k} \sum_{l=0}^{n} a_{nl} \rho^n \cos^l \theta \qquad (13.5)$$

because the wavefront is symmetric about the tangential or meridional plane (y-z plane). The expressions for the main aberrations of the lens are as follows:

Third-order spherical aberration	$a_{40}\rho^4$
Fifth-order spherical aberration	$a_{60}\rho^6$

Third-order coma	$a_{31}\rho^3 \cos\theta$
Third-order astigmatism	$a_{22}\rho^2 \cos^2\theta$
Longitudinal focus displacement	$a_{20}\rho^2$
Tilt about the y axis	$a_{11}\rho \sin\theta$
Tilt about the x axis	$a_{11}\rho \cos\theta$
Constant term or piston error	a_{00}

Later, when we study the Zernike polynomials, we will see that these results can be obtained directly from the Zernike polynomials, considering that n and l must be both even or both odd. It is often more convenient to use $\cos l\theta$ and $\sin l\theta$ instead of $\cos^l \theta$ and $\sin^l \theta$.

When the aberrations are those of an axially symmetric system, the wavefront $W(\rho, \theta)$ can be expressed in terms of the normalized image height σ, using conditions similar to the ones employed by Hopkins (1950). If we consider only cases in which W has the same value for points symmetrically placed with respect to the meridional plane, then

$$W(\rho, \theta, \sigma) = W(\rho, -\theta, \sigma) \tag{13.6}$$

thus removing all $\sin\theta$ terms. Also, if the sign of the normalized image height is changed, the aberration remains the same if the angle θ is replaced by $\pi \pm \theta$; hence

$$W(\rho, \theta, \sigma) = W(\rho, \theta + \pi, -\sigma). \tag{13.7}$$

Hopkins demonstrated that these conditions are fulfilled provided that the variables (ρ, θ, σ) occur in the aberration function W only in the forms

$$\rho^2, \sigma\rho \cos\theta, \sigma^2 \tag{13.8}$$

and their products. Then we can write W as the most general power series in these variables, as follows:

$$W(\sigma, \rho, \theta) = {}_0c_{20}\rho^2 + {}_1c_{11}\sigma\rho \cos\theta + {}_0c_{40}\rho^4 + {}_1c_{31}\sigma\rho^3 \cos\theta \\ + {}_2c_{22}\sigma^2\rho^2 \cos^2\theta + {}_2c_{20}\sigma^2\rho^2 + {}_3c_{11}\sigma^3\rho \cos\theta. \tag{13.9}$$

The subscript to the left of c denotes the power of σ; the two subscripts to the right denote the powers of r and $\cos\theta$. The first two terms represent defocusing and tilt about the x axis, respectively. The next five terms of the series are the so-called Seidel or third-order aberrations, and by an obvious extension of the nomenclature of these, the general aberration terms can be classified into three categories. Spherical aberration terms are those that are independent of θ, comatic aberration terms involve odd powers of $\cos\theta$, and astigmatic aberration terms involve even powers of $\cos\theta$. The "symmetrical" part of the aberration involves the spherical aberration and astigmatic terms, and the "asymmetrical" part of the comatic terms, so that the total wavefront aberration $W(\sigma, \rho, \theta)$ can be written as

$$W(\sigma, \rho, \theta) = W_{\text{even}}(\sigma, \rho, \theta) + W_{\text{odd}}(\sigma, \rho, \theta), \qquad (13.10)$$

where W_{even} represents the sum of the terms involving even powers of $\cos\theta$, and W_{odd} the sum of the terms involving odd powers of $\cos\theta$.

13.2.2. Zernike Polynomials

The Zernike circular polynomials Z_n^l of nth degree (Zernike 1934, 1954) can be obtained from the following two properties (Born and Wolf 1964; Bathia and Wolf 1952, 1954):

1. These polynomials are orthogonal over the circle with unit radius (wavefront boundary):

$$\int_0^1 \int_0^{2\pi} Z_n^l Z_{n'}^{l'*} \rho \, d\rho \, d\theta = \frac{\pi}{n+1} \delta_{nn'} \delta_{ll'}, \qquad (13.11)$$

2. The mathematical form of the polynomial is preserved when a rotation with pivot at the center of the circle is applied to the function (wavefront).

It will be assumed here that the wavefronts are described in a coordinate system where z is the optical axis and the y–z plane is the meridional plane. The polynomials can be expressed as the product of two functions, one depending only on the radial coordinate and the other depending only on the angular coordinate, as follows:

$$Z_n^l = R_n^l(\rho) e^{il\theta}, \qquad (13.12)$$

where n is the degree of the polynomial, and l is the angular dependence parameter. The complex exponential form of the angular function is dictated by the second of the desired properties listed; the radial function will now be discussed. The coordinate ρ is the normalized radial distance, and θ is the angle from the y axis. It can be shown that $|l|$ is the minimum exponent of the polynomials R_n^l. The numbers n and l are either both even or both odd; thus $n - l$ is always even. There are $(\frac{1}{2})(n+1)(n+2)$ linearly independent polynomials Z_n^l of degree $\leq n$, one for each allowed pair of numbers n and l.

The radial polynomials $R_n^l(\rho)$ of degree n and minimum exponent $|l|$ are functions of ρ alone and satisfy the relation

$$R_n^l = R_n^{-l} = R_n^{|l|}. \tag{13.13}$$

There is a radial polynomial $R^{|l|}{}_n$ of degree n for each pair of numbers n and $|l|$; hence the two Zernike polynomials Z_n^l and Z_n^{-l} contain the same radial polynomial $R_n^{|l|}$. If n is even, the polynomial is symmetrical (all exponents are even); if n is odd, the polynomial is antisymmetrical (all exponents are odd). In Table 13.1 the explicit forms of the radial polynomials $R_n^{|l|}(\rho)$ for $|l| \leq 8$, $n \leq 8$ are given.

Because the angular functions $e^{il\theta}$ are already orthogonal, we need only to be concerned with creating radial functions that satisfy the following orthogonality condition:

$$\int_0^1 R_n^l R_{n'}^l \, \rho \, d\rho = \frac{1}{2(n+1)} \delta_{nn'}. \tag{13.14}$$

These radial polynomials for $n - 2m \geq 0$ can be found by means of the expression

$$R_n^{n-2m}(\rho) = \sum_{s=0}^{m} (-1)^s \frac{(n-s)!}{s!(m-s)!(n-m-s)!} \rho^{n-2s}. \tag{13.15}$$

The Zernike polynomials Z_n^l are complex. However, real Zernike polynomials U_n^l can be defined by

$$U_n^l = \begin{cases} \frac{1}{2}[Z_n^l + Z_n^{-l}] = R_n^l(\rho) \cos l\theta & \text{for } l \leq 0, \\ \frac{1}{2i}[Z_n^l - Z_n^{-l}] = R_n^l(\rho) \sin l\theta & \text{for } l > 0, \end{cases} \tag{13.16}$$

which satisfy the orthogonality condition

$$\int_0^1 \int_0^{2\pi} U_n^l U_{n'}^{l'} \rho \, d\rho \, d\theta = \frac{\pi}{2(n+1)} \delta_{nn'} \delta_{ll'}. \tag{13.17}$$

An extensive catalog showing isometric plots of many Zernike polynomials has been published by Kim and Shannon (1987).

Table 13.1. Radial Polynomials $R_n^{|l|}(\rho)$ for $|l| \leq 8, n \leq 8$

$|l|$	0	1	2	3	4	5	6	7	8
0	1		$2\rho^2 - 1$		$6\rho^4 - 6\rho^2 + 1$		$20\rho^6 - 30\rho^4 + 12\rho^2 - 1$		$70\rho^8 - 140\rho^6 + 90\rho^4 - 20\rho^2 + 1$
1		ρ		$3\rho^3 - 3\rho$		$10\rho^5 - 12\rho^3 + 3\rho$		$35\rho^7 - 60\rho^5 + 30\rho^3 - 4\rho$	
2			ρ^2		$4\rho^4 - 3\rho^2$		$15\rho^6 - 20\rho^4 + 6\rho^2$		$56\rho^8 - 105\rho^6 + 60\rho^4 - 10\rho^2$
3				ρ^3		$5\rho^5 - 4\rho^3$		$21\rho^7 - 30\rho^5 + 10\rho^3$	
4					ρ^4		$6\rho^6 - 5\rho^4$		$28\rho^8 - 42\rho^6 + 15\rho^4$
5						ρ^5		$7\rho^7 - 6\rho^5$	
6							ρ^6		$8\rho^8 - 7\rho^6$
7								ρ^7	
8									ρ^8

13.2.3. Wavefront Representation with Zernike Polynomials

The Zernike polynomials were developed as a convenient set for representing wavefront aberrations over a circular pupil. They are easily related to the classical aberrations as shown in Table 13.2. The only small problem with this representation is that nonrotationally symmetric aberrations, like coma and astigmatism, are decomposed into two components, one along the x axis and another along the y axis. These, however, may be easily combined in a single aberration with a certain orientation that depends on the magnitude of the two components (Malacara 1983). Some other useful properties of these polynomials will be further explored in the next section. We will also show how the orthogonality of the polynomials simplifies fitting the polynomials to measured interferogram data, but first, we will show how Zernike polynomials are used as an aberration function.

The Zernike polynomials are complete; this means that any function (or wavefront) $W(\rho, \theta)$ of degree k can be expressed as a linear combination of Zernike circular polynomials as follows:

$$W(\rho, \theta) = \sum_{n=0}^{k} \sum_{l=-n}^{n} C_{nl} R_n^{|l|} e^{il\theta}. \qquad (13.18)$$

$W(\rho, \theta)$ has to be real, and since $R_n^{|l|}$ is also real, the C_{nl} may be real or imaginary, and must satisfy the relation

$$C_{n,l} = C_{n,-l}^*. \qquad (13.19)$$

Alternatively, in order to use only real coefficients, we may use the real Zernike polynomials U_n^l in the manner of Rimmer and Wyant (1975):

$$W(\rho, \theta) = \sum_{n=0}^{k} \sum_{m=0}^{n} A_{nm} U_{nm} = \sum_{n=0}^{k} \sum_{m=0}^{n} A_{nm} R_n^{n-2m} \begin{Bmatrix} \sin \\ \cos \end{Bmatrix} (n-2m)\theta, \qquad (13.20)$$

where the sine function is used for $n - 2m > 0$ and the cosine function for $n - 2m \leq 0$. A positive number m was defined as

$$m = \frac{n-l}{2}, \qquad (13.21)$$

making use of the fact that $n - l$ is always an even number, and that $n \geq l$.

Table 13.2. Zernike Polynomials U_{nm} up to Fifth Degree

n	m	$n-2m$	Zernike Polynomial	Monomial Representation	Meaning
0	0	0	1	1	Piston or constant term
1	0	1	$\rho \sin\theta$	x	Tilt about y axis
1	1	-1	$\rho \cos\theta$	y	Tilt about x axis
2	0	2	$\rho^2 \sin 2\theta$	$2xy$	Astigmatism with axis at $\pm 45°$
2	1	0	$2\rho^2 - 1$	$-1 + 2y^2 + 2x^2$	Focus shift
2	2	-2	$\rho^2 \cos 2\theta$	$y^2 - x^2$	Astigmatism with axis at $0°$ or $90°$
3	0	3	$\rho^3 \sin 3\theta$	$3xy^2 - x^3$	Triangular astigmatism with base on x axis
3	1	1	$(3\rho^3 - 2\rho) \sin\theta$	$-2x + 3xy^2 + 3x^3$	Third-order coma along x axis
3	2	-1	$(3\rho^3 - 2\rho) \cos\theta$	$-2y + 3y^3 + 3x^2y$	Third-order coma along y axis
3	3	-3	$\rho^3 \cos 3\theta$	$y^3 - 3x^2y$	Triangular astigmatism with base on y axis
4	0	4	$\rho^4 \sin 4\theta$	$4y^3x - 4x^3y$	
4	1	2	$(4\rho^4 - 3\rho^2) \sin 2\theta$	$-6xy + 8y^3x + 8x^3y$	
4	2	0	$6\rho^4 - 6\rho^2 + 1$	$1 - 6y^2 - 6x^2 + 6y^4 + 12x^2y^2 + 6x^4$	Third-order spherical aberration
4	3	-2	$(4\rho^4 - 3\rho^2) \cos 2\theta$	$-3y^2 + 3x^2 + 4y^4 - 4x^4$	
4	4	-4	$\rho^4 \cos 4\theta$	$y^4 - 6x^2y^2 + x^4$	
5	0	5	$\rho^5 \sin 5\theta$	$5xy^4 - 10x^3y^2 + x^5$	
5	1	3	$(5\rho^5 - 4\rho^3) \sin 3\theta$	$-12xy^2 + 4x^3 + 15xy^4 + 10x^3y^2 - 5x^5$	
5	2	1	$(10\rho^5 - 12\rho^3 + 3\rho) \sin\theta$	$3x - 12xy^2 - 12x^3 + 10xy^4 + 20x^3y^2 + 10x^5$	
5	3	-1	$(10\rho^5 - 12\rho^3 + 2\rho) \cos\theta$	$3y - 12y^3 - 12x^2y + 10y^5 + 20x^2y^3 + 10x^4y$	
5	4	-3	$(5\rho^5 - 4\rho^3) \cos 3\theta$	$-4y^3 + 12x^2y + 5y^5 - 10x^2y^3 - 15x^4y$	
5	5	-5	$\rho^5 \cos 5\theta$	$y^5 - 10x^2y^3 + 5x^4y$	

It is also useful sometimes to write Eq. (13.20) as

$$W(\rho, \theta) = \sum_{n=0}^{k} \sum_{l=0}^{n} R_n^l (a_{nl} \cos l\theta + b_{nl} \sin l\theta), \qquad (13.22)$$

where l takes only values with the same parity as n, and use was made of the fact that $R_n^l = R_n^{-l}$. The coefficients a_{nl} and b_{nl} are related to the coefficients A_{nl} by the relations

$$a_{n,l} = A_{n,(n+l)/2},$$
$$b_{n,l} = A_{n,(n-l)/2}. \qquad (13.23)$$

The double-summation approach used thus far can be replaced by a simpler, single-sum notation:

$$W(\rho, \theta) = \sum_{n=0}^{k} \sum_{m=0}^{n} A_{nm} U_{nm} = \sum_{r=1}^{L} A_r U_r, \qquad (13.24)$$

where, instead of the coefficients n and m, a single index r is defined by its appearance order in the double sum, and can be expressed by

$$r = n(n+1)/2 + m + 1. \qquad (13.25)$$

The maximum value of r is the total number of Zernike polynomials used, given by $L = (k+1)(k+2)/2$, where k is the degree of the polynomial. This is one possible ordering of the polynomials, but other orderings are also possible.

Equation (13.25) can be inverted by posing the opposite problem of finding the subscripts n and m from the subscript r. This problem can be solved by means of the formulas

$$n = \text{next integer greater than } \left\{ \frac{-3 + [9 + 8(r-1)]^{1/2}}{2} \right\} \qquad (13.26)$$

and

$$m = r - \frac{(n+1)n}{2} - 1. \qquad (13.27)$$

13.2.4. Some Properties of the Zernike Polynomials

The orthogonality of the Zernike polynomials leads to several useful properties when they are used to describe a wavefront within a circular pupil. Orthogonality also simplifies the task of fitting polynomials to the measured data points. To further explore these properties, we will first define several useful statistics.

13.2. POLYNOMIAL WAVEFRONT REPRESENTATION

If we assume that the mean value of the wavefront surface W is zero, then the *wavefront variance* σ_W^2 is just the mean squared value of the wavefront over the pupil:

$$\sigma_W^2 = \frac{\int_0^1 \int_0^{2\pi} W^2 \rho \, d\rho \, d\theta}{\int_0^1 \int_0^{2\pi} \rho \, d\rho \, d\theta} = \frac{1}{\pi} \int_0^1 \int_0^{2\pi} W^2 \rho \, d\rho \, d\theta. \qquad (13.28)$$

The requirement of zero mean value is not restrictive in practice because the average value of the wavefront phase (sometimes referred to as the piston term in a polynomial representation) is usually arbitrary and can be adjusted to any convenient value. The square root of the wavefront variance is often referred to as the RMS (root-mean-square) wavefront error and is a valuable statistic for characterizing the quality of a wavefront surface.

If we attempt to fit an aberration polynomial to a wavefront, the fitted polynomial may not exactly describe the actual wavefront. The fit error at any point in the pupil will be the difference between the actual wavefront W' and the polynomial evaluated at that point W. We can characterize the quality of the fit over the entire pupil by calculating the *fit variance* σ_f^2, which is the mean-squared fit error:

$$\sigma_f^2 = \frac{\int_0^1 \int_0^{2\pi} (W' - W)^2 \rho \, d\rho \, d\theta}{\int_0^1 \int_0^{2\pi} \rho \, d\rho \, d\theta} = \frac{1}{\pi} \int_0^1 \int_0^{2\pi} (W' - W)^2 \rho \, d\rho \, d\theta. \qquad (13.29)$$

When the fit variance is zero, the polynomial represents a perfect fit to the wavefront; larger values indicate poorer quality fits.

Fitting a polynomial to a wavefront involves selecting polynomial coefficients that minimize the fit variance. This solution is found when the partial derivative of the fit variance with respect to each of the coefficients A_p is made equal to zero. When the aberration polynomial is a linear combination of orthogonal polynomial terms, as described by Eq. (13.24), this problem is easily solved using the orthogonality condition in Eq. (13.17), with the following result:

$$A_p = \frac{2(n+1)}{\pi} \int_0^1 \int_0^{2\pi} W' U_p \rho \, d\rho \, d\theta \qquad (13.30)$$

with $p = 1, \ldots, L$.

This result is important because it shows that given W', and a value of p, the coefficient A_p depends only on U_p and not on any other U_j ($j \neq p$). In other words the coefficients A_p can be found by independently fitting each of the Zernike polynomial terms to be measured data. Note that this problem is not so easily solved when the aberration polynomial does not consist of orthogonal terms. This will be the subject of Section 13.3.

The radial polynomials R_n^l from Eq. (13.15) may be written as

$$R_n^l(\rho) = \sum_{j=1}^{n} a_j \rho^j, \qquad (13.31)$$

where l and n are the minimum and maximum power, respectively, and the coefficients $a_{l+1}, a_{l+3}, \ldots, a_{n-1}$ are all equal to zero.

Now, a method to find these coefficients a_j without using Eq. (13.15) will be described. Let us now assume that a wavefront is represented by only one of the Zernike polynomials and that the magnitude of the coefficient in front of the monomial term with the highest power, that is, the amount of aberration, is completely determined and fixed. Let us consider the variance σ_W of the wavefront deformations, for a given Zernike polynomial alone, defined by the integral in Eq. (13.11) with $n = n'$ and $l = l'$. This variance is to be minimized, but this is possible if the integral on the left side of Eq. (13.14) with $n = n'$ and $l = l'$ is also minimized. In order to do so, the coefficients a_j are chosen in such a way that the partial derivatives of the variance with respect to each of the coefficients a_p is zero, for $p = l, l+2, l+4, \ldots, n-2$. ($p = n$ is excluded because a_n is fixed, since it is the aberration to be compensated). Thus:

$$\frac{\partial}{\partial a_p} \int_0^1 [R_n^l]^2 \rho \, d\rho = 2 \int_0^1 R_n^l \frac{\partial R_n^l}{\partial a_p} \rho \, d\rho = 0. \qquad (13.32)$$

In other words the monomial terms with lower power balance the effect of the highest power term. This property may be proved with the following method suggested by Wolf and Forbes (1990). Since the polynomials R_n^l form a complete set, any other polynomial with the same degree or smaller may be represented by a linear combination of these polynomials. Thus, the monomial ρ^p may be written as

$$\rho^p = \sum_{j=1}^{p} C_j R_j^l, \qquad (13.33)$$

where the c_j are constants. Substituting now this expression for ρ in Eq. (13.31), we find

13.2. POLYNOMIAL WAVEFRONT REPRESENTATION

$$\sum_{j=1}^{p} \int_{0}^{1} R_n^l R_j^l \rho \, d\rho = 0. \qquad (13.34)$$

Because of the orthogonality of the polynomials R_n^l, this is true only if the value of p is never equal to n, which was our initial hypothesis.

This result permits us to calculate the coefficients for the radial polynomials, by defining first $a_n = 1$. At the end all coefficients are multiplied by a common constant in order to satisfy the normalization constant in Eq. (13.14).

From the definition of the Zernike polynomials and the variance statistics described, we can observe the following properties of the Zernike polynomials:

1. Except for the lowest order term ($U_1 = 1$), each Zernike polynomial term has a mean value of zero. This is a direct consequence of orthogonality because the average value of any term U_n ($n > 1$) is the same as the average value of $U_n U_1$, which must be zero because U_n is orthogonal to U_1.

2. When a wavefront is expressed as a linear combination of Zernike polynomials, the wavefront variance is equal to the sum of the variances of the individual polynomial terms. From Eq. (13.24), we obtain

$$W^2(\rho, \theta) = \sum_{r=1}^{L} (A_r U_r)^2 + \sum \text{cross terms}. \qquad (13.35)$$

The polynomial orthogonality dictates that the average value of the cross terms is zero, hence

$$\sigma_W^2 = \sum_{r=1}^{L} \sigma_r^2. \qquad (13.36)$$

3. Each Zernike polynomial term already contains the proper amount of each of the lower order terms (piston, defocusing, and tilt) needed to minimize the variance of that term. Thus, the term $U_5 = 2\rho^2 - 1$ describes only the excess defocus of the wavefront, because the amount of defocus (represented by the ρ^2 monomial) needed to balance higher order terms, such as spherical aberration, is already included in the Zernike polynomial terms describing those aberrations.

4. Adding or subtracting Zernike polynomials does not affect the fit coefficients of other terms. We can subtract one or more fitted terms (defocus, e.g.) from a measured wavefront without having to recalculate the fit coefficients for the other terms. Strictly speaking, this property is valid only if the Zernike polynomial coefficients are obtained by a least-squares fitting to a continuous function, or if the number of data points is infinite and uniformly distributed

over the aperture. However, a good approximation is when the number of data points is very large.

13.2.5. Conversion from Monomials to Zernike Polynomials

A wavefront function is often written in terms of monomials, that is, powers of x and y as in Eq. (13.2). This polynomial is of degree k and contains $N = (k + 1)(k + 2)/2$ terms. Wavefront functions written in terms of the Zernike polynomials can be converted into an equivalent monomial form (Sumita 1969) by substituting into each U_{nm} the following expressions for the angular functions and radial functions (Rimmer 1972):

$$\begin{Bmatrix} \cos \\ \sin \end{Bmatrix} (n - 2m)\theta = \rho^{-(n-2m)} \sum_{j=0}^{q} (-1)^j \begin{bmatrix} n - 2m \\ 2i + p \end{bmatrix} x^{2j+p} y^{n-2m-2j-p},$$

(13.37)

which is valid only for $n - 2m \geq 0$. Values for the parameters p and q are given in Table 13.3. The radial function given in Eq. (13.15) can be converted to monomial form by substituting the following expression for ρ:

$$\rho^{2j} = \sum_{k=0}^{j} \begin{bmatrix} j \\ k \end{bmatrix} x^{2k} y^{2(j-k)}.$$

(13.38)

In Eqs. (13.37) and (13.38) the function $\binom{t}{u}$ is the binomial factor, defined by

$$\begin{bmatrix} t \\ u \end{bmatrix} = \frac{t!}{(t-u)!\, u!}$$

(13.39)

Table 13.3. Values of p and q in Eqs. (13.37) and (13.40)

Terms		n even	n odd
sine	p	1	1
	q	$\dfrac{n-2m}{2}$	$\dfrac{n-2m-1}{2}$
cosine	p	0	0
	q	$-\dfrac{n-2m}{2}$	$-\dfrac{n-2m+1}{2}$

13.2. POLYNOMIAL WAVEFRONT REPRESENTATION

The final expression for the Zernike polynomials U_{nm} in terms of powers x and y is

$$U_{nm} = R_n^{n-2m} \begin{Bmatrix} \sin \\ \cos \end{Bmatrix} (n - 2m)\theta$$

$$= \sum_{i=0}^{q} \sum_{j=0}^{m} \sum_{k=0}^{m-j} (-1)^{i+j} \begin{bmatrix} n - 2m \\ 2i + p \end{bmatrix} \begin{bmatrix} m - j \\ k \end{bmatrix}$$

$$= \frac{(n-j)!}{j!(m-j)!(n-m-j)!} x^{2(i+k)+p} y^{n-2(i+j+k)-p}, \quad (13.40)$$

where m must be replaced by $n - m$ when $n - 2m \leq 0$.

The expansion of some Zernike polynomials in terms of monomials is shown in Table 13.2.

The elements U_{nm} can be represented by a column vector U_r, with $N = (k + 1)(k + 2)/2$ elements, where the subscript r is defined by Eq. (13.25).

Expression (13.40) can be thought to be of the form

$$U_{nm} = \sum_{i=0}^{k} \sum_{j=0}^{i} h_{ijnm} x^j y^{i-j}, \quad (13.41)$$

where $H_{ijnm} = 0$ if i is outside the range $|n - 2m| \leq i \leq n$, since the minimum and maximum degrees of the monomial terms are $|n - 2m|$ and n, respectively. This expression can also be written as

$$U_r = \sum_{s=0}^{k} H_{sr} x^j y^{i-j}, \quad (13.42)$$

where the two subscripts i and j are defined by the single subscript s in the manner described in Eqs. (13.26) and (13.27).

Since the wavefront can be written as a linear combination of Zernike polynomials as in Eq. (13.24), we can, substituting U_r into this expression, find the following relation for the B_{ij} coefficients in Eq. (13.2):

$$c_{ij} = \sum_{n=0}^{k} \sum_{m=0}^{n} H_{ijnm} A_{nm} = \sum_{r=1}^{L} H_{sr} A_r, \quad (13.43)$$

which in matrix notation may be written as

$$c = H \circ \mathbf{A} \quad (13.44)$$

By inverting the H matrix, we can compute the A_{nm} coefficients from c_{ij} coefficients.

It is important to point out that Eqs. (13.15) and (13.40) work only for positive values of $n - m$, but this is not a serious problem because relation (13.13) permits us to change the sign of $n - 2m$ to make it positive. Given the values of n and m, the sign of $n - 2m$ is changed by defining a new value of m, given by $n - m$. The sign of $n - 2m$ thus serves us only to determine whether the sin or cos function should be used.

The matrix H has some interesting properties (Malacara et al. 1976). Of these, the most interesting is that the matrix H for a polynomial of power $k - 1$ is a submatrix on the left superior corner of the matrix for a polynomial of power k. Thus a large matrix for polynomials up to the k power includes all smaller matrices as submatrices of it. The inverse matrix H^{-1} has the same property. Then, two large matrices H and H^{-1} may be computed and used for any desired degree of the polynomial.

Table 13.4 shows the matrix H up to the eighth power. For normal computations the values in this table may be used instead of going through all the equations in this section. Table 13.5 shows the inverse of the matrix H.

13.3. WAVEFRONT FITTING AND DATA ANALYSIS

Once a form of wavefront representation has been chosen, it is often necessary to fit that form to measured data points. When aberration polynomials are used to represent the wavefront, the purpose of the fit is to find the polynomial coefficients that best represent the measured data. When Zernike polynomials are used, each of these coefficients can be related to specific properties of the optical system, such as defocus and coma. The method for performing this fit, and applications of the fit coefficients, will be the topic of this section.

13.3.1. Least-Squares Fit of Polynomials

Least-squares fitting is the method most commonly used to fit measured data to a particular functional form. In some cases Zernike polynomials might be directly fitted to the data, while in others, some other polynomial form might be fitted and then transformed into Zernike polynomials (Forsithe 1957; Freniere et al. 1979, 1981). The measured data is usually known at a series of discrete data points, while the functional form usually represents a continuous function. We start by defining the variance S of the discrete wavefront as follows:

$$S = \frac{1}{N} \sum_{i=1}^{N} [W'_i - W(\rho_i, \theta_i)]^2, \qquad (13.45)$$

13.3. WAVEFRONT FITTING AND DATA ANALYSIS

Table 13.4. Matrix X to Transform from Zernike Polynomials to Monomials, up to the Eighth Power Only Elements Different from Zero are Tabulated

Column	Row	Value	Column	Row	Value
1	1	1.0	17	10	4.0
2	3	1.0	17	17	15.0
3	2	1.0	17	19	10.0
4	5	2.0	17	21	−5.0
5	1	−1.0	18	3	3.0
5	4	2.0	18	8	−12.0
5	6	2.0	18	10	−12.0
6	4	1.0	18	17	10.0
6	6	−1.0	18	19	20.0
7	8	3.0	18	21	10.0
7	10	−1.0	19	2	3.0
8	3	−2.0	19	7	−12.0
8	8	3.0	19	9	−12.0
8	10	3.0	19	16	10.0
9	2	−2.0	19	18	20.0
9	7	3.0	19	20	10.0
9	9	3.0	20	7	−4.0
10	7	1.0	20	9	12.0
10	9	−3.0	20	16	5.0
11	12	4.0	20	18	−10.0
11	14	−4.0	20	20	−15.0
12	5	−6.0	21	16	1.0
12	12	8.0	21	18	−10.0
12	14	8.0	21	20	5.0
13	1	1.0	22	23	6.0
13	4	−6.0	22	25	−20.0
13	6	−6.0	22	27	6.0
13	11	6.0	23	12	−20.0
13	13	12.0	23	14	20.0
13	15	6.0	23	23	24.0
14	4	−3.0	23	27	−24.0
14	6	3.0	24	5	12.0
14	11	4.0	24	12	−40.0
14	15	−4.0	24	14	−40.0
15	11	1.0	24	23	30.0
15	13	−6.0	24	25	60.0
15	15	1.0	24	27	30.0
16	17	5.0	25	1	−1.0
16	19	−10.0	25	4	12.0
16	21	1.0	25	6	12.0
17	8	−12.0	25	11	−30.0

Table 13.4. (*Continued*)

Column	Row	Value	Column	Row	Value
25	13	−60.0	31	32	105.0
25	15	−30.0	31	34	21.0
25	22	20.0	31	36	−21.0
25	24	60.0	32	3	−4.0
25	26	60.0	32	8	30.0
25	28	20.0	32	10	30.0
26	4	6.0	32	17	−60.0
26	6	−6.0	32	19	−120.0
26	11	−20.0	32	21	−60.0
26	15	20.0	32	30	35.0
26	22	15.0	32	32	105.0
26	24	15.0	32	34	105.0
26	26	−15.0	32	36	35.0
26	28	−15.0	33	2	−4.0
27	11	−5.0	33	7	30.0
27	13	30.0	33	9	30.0
27	15	−5.0	33	16	−60.0
27	22	6.0	33	18	−120.0
27	24	−30.0	33	20	−60.0
27	26	−30.0	33	29	35.0
27	28	6.0	33	31	105.0
28	22	1.0	33	33	105.0
28	24	−15.0	33	35	35.0
28	26	15.0	34	7	10.0
28	28	−1.0	34	9	−30.0
29	30	7.0	34	16	−30.0
29	32	−35.0	34	18	60.0
29	34	21.0	34	20	90.0
29	36	−1.0	34	29	21.0
30	17	−30.0	34	31	−21.0
30	19	60.0	34	33	−105.0
30	21	−6.0	34	35	−63.0
30	30	35.0	35	16	−6.0
30	32	−35.0	35	18	60.0
30	34	−63.0	35	20	−30.0
30	36	7.0	35	29	7.0
31	8	30.0	35	31	−63.0
31	10	−10.0	35	33	−35.0
31	17	−90.0	35	35	35.0
31	19	−60.0	36	29	1.0
31	21	30.0	36	31	−21.0
31	30	63.0	36	33	35.0

13.3. WAVEFRONT FITTING AND DATA ANALYSIS

Table 13.4. (*Continued*)

Column	Row	Value	Column	Row	Value
36	35	−7.0	41	39	280.0
37	38	8.0	41	41	420.0
37	40	−56.0	41	43	280.0
37	42	56.0	41	45	70.0
37	44	−8.0	42	4	−10.0
38	23	−42.0	42	6	10.0
38	25	140.0	42	11	60.0
38	27	−42.0	42	15	−60.0
38	38	48.0	42	22	−105.0
38	40	−112.0	42	24	−105.0
38	42	−112.0	42	26	105.0
38	44	48.0	42	28	105.0
39	12	60.0	42	37	56.0
39	14	−60.0	42	39	112.0
39	23	−168.0	42	43	−112.0
39	27	168.0	42	45	−56.0
39	38	112.0	43	11	15.0
39	40	112.0	43	13	−90.0
39	42	−112.0	43	15	15.0
39	44	−112.0	43	22	−42.0
40	5	−20.0	43	24	210.0
40	12	120.0	43	26	210.0
40	14	120.0	43	28	−42.0
40	23	−210.0	43	37	28.0
40	25	−420.0	43	39	−112.0
40	27	−210.0	43	41	−280.0
40	38	112.0	43	43	−112.0
40	40	336.0	43	45	28.0
40	42	336.0	44	22	−7.0
40	44	112.0	44	24	105.0
41	1	1.0	44	26	−105.0
41	4	−20.0	44	28	7.0
41	6	−20.0	44	37	8.0
41	11	90.0	44	39	−112.0
41	13	180.0	44	43	112.0
41	15	90.0	44	45	−8.0
41	22	−140.0	45	37	1.0
41	24	−420.0	45	39	−28.0
41	26	−420.0	45	41	70.0
41	28	−140.0	45	43	−28.0
41	37	70.0	45	45	1.0

Table 13.5. Matrix H^{-1} to Transform from Monomials to Zernike Polynomials, up to the Eighth Power Only Elements Different from Zero Are Tabulated

Column	Row	Value	Column	Row	Value
1	1	1.00000	15	15	0.12500
2	3	1.00000	16	3	0.31250
3	2	1.00000	16	9	0.25000
4	1	0.25000	16	10	0.25000
4	5	0.25000	16	19	0.06250
4	6	0.50000	16	20	0.06250
5	4	0.50000	16	21	0.06250
6	1	0.25000	17	2	0.06250
6	5	0.25000	17	7	0.15000
6	6	−0.50000	17	8	0.05000
7	3	0.50000	17	16	0.06250
7	9	0.25000	17	17	0.03750
7	10	0.25000	17	18	0.01250
8	2	0.16667	18	3	0.06250
8	7	0.25000	18	9	0.05000
8	8	0.08333	18	10	−0.05000
9	3	0.16667	18	19	0.01250
9	9	0.08333	18	20	−0.01250
9	10	−0.25000	18	21	−0.06250
10	2	0.50000	19	2	0.06250
10	7	−0.25000	19	7	0.05000
10	8	0.25000	19	8	0.05000
11	1	0.12500	19	16	−0.06250
11	5	0.18750	19	17	0.01250
11	6	0.37500	19	18	0.01250
11	13	0.06250	20	3	0.06250
11	14	0.12500	20	9	0.05000
11	15	0.12500	20	10	−0.15000
12	4	0.18750	20	19	0.01250
12	11	0.12500	20	20	−0.03750
12	12	0.06250	20	21	0.06250
13	1	0.04167	21	2	0.31250
13	5	0.06250	21	7	−0.25000
13	13	0.02083	21	8	0.25000
13	15	−0.12500	21	16	0.06250
14	4	0.18750	21	17	−0.06250
14	11	−0.12500	21	18	0.06250
14	12	0.06250	22	1	0.07813
15	1	0.12500	22	5	0.14063
15	5	0.18750	22	6	0.28125
15	6	−0.37500	22	13	0.07813
15	13	0.06250	22	14	0.15625
15	14	−0.12500	22	15	0.15625

13.3. WAVEFRONT FITTING AND DATA ANALYSIS

Table 13.5. (*Continued*)

Column	Row	Value	Column	Row	Value
22	25	0.01563	28	14	−0.15625
22	26	0.03125	28	15	0.15625
22	27	0.03125	28	25	0.01563
22	28	0.03125	28	26	−0.03125
23	4	0.09375	28	27	0.03125
23	11	0.10417	28	28	−0.03125
23	12	0.05208	29	3	0.21875
23	22	0.03125	29	9	0.21875
23	23	0.02083	29	10	0.21875
23	24	0.01042	29	19	0.09375
24	1	0.01563	29	20	0.09375
24	5	0.02813	29	21	0.09375
24	6	0.01875	29	33	0.01563
24	13	0.01563	29	34	0.01563
24	14	0.01042	29	35	0.01563
24	15	−0.05208	29	36	0.01563
24	25	0.00313	30	2	0.03125
24	26	0.00208	30	7	0.09375
24	27	−0.01042	30	8	0.03125
24	28	−0.03125	30	16	0.06696
25	4	0.05625	30	17	0.04018
25	12	0.03125	30	18	0.01339
25	22	−0.03125	30	29	0.01562
25	24	0.00625	30	30	0.01116
26	1	0.01563	30	31	0.00670
26	5	0.02813	30	32	0.00223
26	6	−0.01875	31	3	0.03125
26	13	0.01563	31	9	0.03125
26	14	−0.01042	31	10	−0.01042
26	15	−0.05208	31	19	0.01339
26	25	0.00313	31	20	−0.00446
26	26	−0.00208	31	21	−0.04018
26	27	−0.01042	31	33	0.00223
26	28	0.03125	31	34	−0.00074
27	4	0.09375	31	35	−0.00670
27	11	−0.10417	31	36	−0.01563
27	12	0.05208	32	2	0.01875
27	22	0.03125	32	7	0.03125
27	23	−0.02083	32	8	0.01875
27	24	0.01042	32	16	−0.01339
28	1	0.07813	32	17	0.01339
28	5	0.14063	32	18	0.00804
28	6	−0.28125	32	29	−0.01563
28	13	0.07813	32	30	−0.00223

Table 13.5. (*Continued*)

Column	Row	Value	Column	Row	Value
32	31	0.00223	37	6	0.21875
32	32	0.00134	37	13	0.07813
33	3	0.01875	37	14	0.15625
33	9	0.01875	37	15	0.15625
33	10	−0.03125	37	25	0.02734
33	19	0.00804	37	26	0.05469
33	20	−0.01339	37	27	0.05469
33	21	−0.01339	37	28	0.05469
33	33	0.00134	37	41	0.00391
33	34	−0.00223	37	42	0.00781
33	35	−0.00223	37	43	0.00781
33	36	0.01563	37	44	0.00781
34	2	0.03125	37	45	0.00781
34	7	0.01042	38	4	0.05469
34	8	0.03125	38	11	0.07813
34	16	−0.04018	38	12	0.03906
34	17	0.00446	38	22	0.04102
34	18	0.01339	38	23	0.02734
34	29	0.01563	38	24	0.01367
34	30	−0.00670	38	37	0.00781
34	31	0.00074	38	38	0.00586
34	32	0.00223	38	39	0.00391
35	3	0.03125	38	40	0.00195
35	9	0.03125	39	1	0.00781
35	10	−0.09375	39	5	0.01562
35	19	0.01339	39	6	0.01563
35	20	−0.04018	39	13	0.01116
35	21	0.06696	39	14	0.01116
35	33	0.00223	39	15	−0.02232
35	34	−0.00670	39	25	0.00391
35	35	0.01116	39	26	0.00391
35	36	−0.01563	39	27	−0.00781
36	2	0.21875	39	28	−0.02734
36	7	−0.21875	39	41	0.00056
36	8	0.21875	39	42	0.00056
36	16	0.09375	39	43	−0.00112
36	17	−0.09375	39	44	−0.00391
36	18	0.09375	39	45	−0.00781
36	29	−0.01563	40	4	0.02344
36	30	0.01563	40	11	0.01116
36	31	−0.01563	40	12	0.01674
36	32	0.01563	40	22	−0.01367
37	1	0.05469	40	23	0.00391
37	5	0.10938	40	24	0.00586

13.3. WAVEFRONT FITTING AND DATA ANALYSIS

Table 13.5. (*Continued*)

Column	Row	Value	Column	Row	Value
40	37	−0.00781	43	28	0.02734
40	38	−0.00195	43	41	0.00056
40	39	0.00056	43	42	−0.00056
40	40	0.00084	43	43	−0.00112
41	1	0.00469	43	44	0.00391
41	5	0.00938	43	45	−0.00781
41	13	0.00670	44	4	0.05469
41	15	−0.02232	44	11	−0.07813
41	25	0.00234	44	12	0.03906
41	27	−0.00781	44	22	0.04102
41	41	0.00033	44	23	−0.02734
41	43	−0.00112	44	24	0.01367
41	45	0.00781	44	37	−0.00781
42	4	0.02344	44	38	0.00586
42	11	−0.01116	44	39	−0.00391
42	12	0.01674	44	40	0.00195
42	22	−0.01367	45	1	0.05469
42	23	−0.00391	45	5	0.10937
42	24	0.00586	45	6	−0.21875
42	37	0.00781	45	13	0.07812
42	38	−0.00195	45	14	−0.15625
42	39	−0.00056	45	15	0.15625
42	40	0.00084	45	25	0.02734
43	1	0.00781	45	26	−0.05469
43	5	0.01563	45	27	0.05469
43	6	−0.01563	45	28	−0.05469
43	13	0.01116	45	41	0.00391
43	14	−0.01116	45	42	−0.00781
43	15	−0.02232	45	43	0.00781
43	25	0.00391	45	44	−0.00781
43	26	−0.00391	45	45	0.00781
43	27	−0.00781			

where W'_i is the measured wavefront deviation (OPD) for data point i, and $W(\rho_i, \theta_i)$ is the functional representation of the wavefront. The variance σ_f^2 in Eq. (13.29) and this variance S approach equality, if the number of data points is very large and uniformly distributed over the wavefront. In general, however, they will be different. Assume that the wavefront is represented by a linear combination of some polynomials $V(\rho, \theta)$, as follows:

$$W(\rho_i, \theta_i) = \sum_{r=1}^{L} B_r V_r(\rho_i, \theta_i). \tag{13.46}$$

The best fit occurs when S is minimized, given by imposing the following conditions:

$$\frac{\partial S}{\partial B_p} = 0 \qquad (13.47)$$

for $p = 1, 2, 3, \ldots, L$. Thus the following system of L equations with L unknowns is obtained:

$$\sum_{j=1}^{L} B_r \sum_{i=1}^{N} V_r V_p - \sum_{i=1}^{N} W'_i V_p = 0. \qquad (13.48)$$

The matrix of this system of equations becomes diagonal if the polynomials V_r were chosen such as to satisfy the orthogonality condition on the discrete set of data points

$$\sum_{i=1}^{N} V_r V_p = F_r \delta_{rp}, \qquad (13.49)$$

where $F_r = \sum_{i=1}^{N} V_i^2$.

Then, the coefficients B_p are simply given by

$$B_p = \frac{\sum_{i=1}^{N} W'_i V_p}{\sum_{i=1}^{N} V_p^2}. \qquad (13.50)$$

Except for the normalizing constants and the use of summation rather than integration, this expression would be identical to Eq. (13.34), with $U_p = V_p$. This simple form of Eq. (13.34) took advantage of the orthogonality of the Zernike polynomials over the domain of integration; Eq. (13.50) uses the orthogonality of the V_p over the discrete data points.

In general, the Zernike polynomials are not orthogonal over the discrete data points. It is only when the number of data points is large, and they are uniformly distributed over the unit circle, that the Zernike polynomials approach orthogonality over the data points. When this is true, we can use Eq. (13.50) to estimate the least-squares fit coefficients for Zernike polynomials. In general, there are not enough data points or they are not uniformly distributed. In this case fitting Zernike polynomials to the measured data points becomes more complicated and unnecessary. Instead, we must create a set of orthogonal polynomials on the data points by applying the Gram–Schmidt orthogonalization technique.

13.3.2. Gram–Schmidt Orthogonalization

When we wish to fit polynomials that are not orthogonal over the data points, we cannot use the simplified expression of Eq. (13.50) to calculate the poly-

13.3. WAVEFRONT FITTING AND DATA ANALYSIS

nomial coefficients. Instead, we would have to solve the system of equations given by Eq. (13.48). Besides being more computationally intensive, solving these equations is also prone to large round-off errors, especially when the number of data points becomes large.

It is more accurate to first transform the desired polynomials into a new set that are orthogonal over the data points. The coefficients of the new polynomials can then be easily calculated by applying the inverse of the first transform, to obtain the coefficients for the originally desired polynomials. These two methods, with their relative advantages and disadvantages, have been studied and compared by Wang and Silva (1980) and Prata and Rusch (1989).

The process of orthogonalizing a given set of polynomials is known as the Gram–Schmidt orthogonalization technique. We will use it here to obtain from the Zernike polynomials a set of polynomials that are orthogonal over a fine and discrete set of irregularly distributed data points (Malacara et al. 1987). Let us consider the wavefront expression (13.24) with a single index r. As we have pointed out, the Zernike polynomials $U_r(r, \theta)$ are orthogonal on the continuous unit radius circle, but they are not on the discrete set of data points. Then, we need to find a set of orthogonal polynomials defined in a discrete set of data points, within the circle, represented by $V_r(\rho, \theta)$. We would expect that these polynomials will approach the Zernike polynomials when the number of points are uniformly distributed and tend to infinity. The wavefront can be represented by

$$W(\rho, \theta) = \sum_{r=1}^{L} B_r V_r(\rho, \theta), \qquad (13.51)$$

where these polynomials $V_r(\rho, \theta)$ satisfy the discrete orthogonality condition (13.49) on the set of N data points with coordinates (ρ_i, θ_i):

$$\sum_{i=1}^{N} V_r(\rho_i, \theta_i) V_p(\rho_i, \theta_i) = F_r \delta_{rp}, \qquad (13.52)$$

where F_r is already defined as part of Eq. (13.48).

We will now conduct a Gram–Schmidt orthogonalization in order to find these polynomials. We begin by writing

$$V_1 = U_1,$$

$$V_2 = U_2 + D_{21} V_1,$$

$$V_3 = U_3 + D_{31} V_1 + D_{32} V_2,$$

$$\cdots$$

$$V_j = U_j + D_{j1} V_1 + D_{j2} V_2 + \cdots + D_{j,j-1} V_{j-1}. \qquad (13.53)$$

Here, the functions U could be monomials in ascending order or the Zernike polynomials. In a general and more compact manner, we can also write this expression as

$$V_r = U_r + \sum_{s=1}^{r-1} D_{rs} V_s, \qquad (13.54)$$

where $r = 1, 2, 3, \ldots, L$. Since $V_r(\rho, \theta)$ is defined to be orthogonal with $V_p(\rho, \theta)$, we multiply this expression by V_p and then sum for all data points, from $i = 1$ to N. If the orthogonality condition in Eq. (13.49) is then used, we obtain for values of r different from p:

$$\sum_{i=1}^{N} V_r V_p = \sum_{i=1}^{N} U_r V_p + D_{rp} \sum_{i=1}^{N} V_p^2 = 0. \qquad (13.55)$$

Then, D_{rp} can be written

$$D_{rp} = \frac{\sum_{i=1}^{N} U_r V_p}{\sum_{i=1}^{N} V_p^2} \qquad (13.56)$$

with $r = 2, 3, 4, \ldots, L$ and $p = 1, 2, \ldots, r - 1$.

13.3.3. Calculation of the Zernike Polynomials Combination

The next step is to determine the values of coefficients C_{rs} that define the orthogonal polynomials V_r, as a linear combination of the Zernike polynomials U_r. We begin by writing:

$$V_1 = U_1,$$

$$V_2 = U_2 + C_{21} U_1,$$

$$V_3 = U_3 + C_{31} U_1 + C_{32} U_2,$$

$$\ldots$$

$$V_j = U_j + C_{j1} U_1 + C_{j2} U_2 + \cdots + C_{j,j-1} U_{j-1} \qquad (13.57)$$

or in general,

$$V_r = U_r + \sum_{i=1}^{r-1} C_{ri} U_i, \qquad (13.58)$$

13.3. WAVEFRONT FITTING AND DATA ANALYSIS

where $r = 2, 3, \ldots, L$, $C_{rr} = 1$, and $V_1 = U_1$. We can now find a few coefficients C_{ri} as follows:

$$C_{21} = D_{21},$$
$$C_{31} = D_{32}C_{21} + D_{31},$$
$$C_{32} = D_{32},$$
$$C_{41} = D_{43}C_{31} + D_{42}C_{21} + D_{41},$$
$$C_{42} = D_{43}C_{32} + D_{42},$$
$$C_{43} = D_{43},$$
$$\ldots \tag{13.59}$$

These results can be written in a general form as

$$C_{ri} = \sum_{s=1}^{r-i} D_{r,r-s} C_{r-s,i}, \tag{13.60}$$

where $i = 1, 2, 3, 4, \ldots, r - 1$, and $C_{rr} = 1$.

Since we now know the coefficients B_r and C_{ri}, the Zernike coefficients A_r, from Eq. (13.24), can be found by substituting Eq. (13.58) into Eq. (13.51):

$$W(\rho, \theta) = B_1 U_1 + \sum_{r=2}^{L} B_r \left(U_r + \sum_{i=1}^{r-1} C_{ri} U_i \right), \tag{13.61}$$

where the coefficients C_{ri} are given in Eq. (13.60). Then, by rearranging the order of the terms in the sums, we find

$$W(\rho, \theta) = \sum_{r=1}^{L-1} \left(B_r + \sum_{i=r+1}^{L} B_i C_{ir} \right) U_r + B_L U_L. \tag{13.62}$$

Comparing this expression with Eq. (13.24) we can see that the coefficients A_r are given by

$$A_r = B_r + \sum_{i=r+1}^{L} B_i C_{ir} \tag{13.63}$$

with $r = 1, 2, 3, \ldots, (L - 1)$ and $A_L = B_L$.

The procedure for fitting measured wavefront deviations to a linear combination of orthogonal polynomials on the discrete set of data points can be summarized as follows:

1. Choose a polynomial degree k and then arrange in order the first $L = (k + 1)(k + 2)/2$ Zernike polynomials, assigning them the index $r = n(n + 1)/2 + m + 1$.
2. Compute the Zernike polynomials U_r at each of the N measured data points on the wavefront.
3. Making $r = 2, 3, \ldots, L$ and $p = 1, 2, 3, \ldots, (r - 1)$, use Eqs. (13.58) and (13.56) alternately, until all coefficients D_{rp} and the polynomials V_r are computed.
4. By means of Eq. (13.50), with $p = 1, 2, 3, \ldots, L$ and the measured values W_i' of the wavefront deviations, calculate all coefficients B_p.
5. Apply Eq. (13.60), with $i = 1, 2, 3, 4, \ldots, (r - 1)$ and $C_{rr} = 1$, in order to calculate coefficients C_{ri}.
6. Use Eq. (13.63) with $r = 1, 2, 3, \ldots, (L - 1)$ and $A_L = B_L$ to calculate all aberration coefficients A_r.
7. Finally, use Eq. (13.24) to compute the polynomial fitted wavefront deviations.

13.3.4. Analysis and Aberration Removal

We have already mentioned how, by fitting Zernike polynomials to measured wavefronts OPDs, we can relate the polynomial coefficients to specific optical properties of the system under test. We can also use the polynomial fit to remove unwanted artifacts from the data prior to further analysis. For example, many interferometric tests require that some tilt be introduced between the test and the reference wavefronts. It is this tilt that produces a sufficient number of fringes to adequately sample the test aperture. However, the tilt is usually not an intrinsic property of the wavefront being tested. To remove this artifact we need to subtract the Zernike polynomial terms U_2 and U_3, which represent the best fit plane to the measured data. In a similar manner we can also subtract the U_5 term to determine what the surface or wavefront will be when a best fit sphere is subtracted from it, that is, at the best focus.

Polynomial subtraction is a very useful tool for removing systematic errors in the testing system (Parks 1978). This is especially important when the measured errors are of the same order of magnitude as those in the interferometer. Parks has described a technique taking several interferograms of the same optics under test, with different orientations, in order to calibrate and remove the systematic errors.

Optical surfaces and wavefronts are often characterized by their peak-to-valley (PV) and root-mean-square (RMS) errors. The first quality is simply found by calculating the difference between the maximum and minimum values of the sampled data, after the desired terms have been removed. The RMS error is found by calculating the following expression:

13.3. WAVEFRONT FITTING AND DATA ANALYSIS

$$\text{RMS} = \sqrt{\overline{W^2} - \overline{W}^2}, \tag{13.64}$$

where $\overline{W^2}$ is the mean-squared wavefront, $\overline{W}$ is the mean wavefront, and

$$W = W_{\text{measured}} - \text{terms removed.} \tag{13.65}$$

The PV error must be regarded with some skepticism, particularly when it is derived from a large number of measured data points, as is the case with phase shifting interferometry. Even relatively large wavefront errors often have little effect on the optical performance if the error involves only a very small part of the aperture. Because the PV error is calculated from just two data points out of possible thousands, it might make the system under test appear worse than it actually is. The RMS error is a statistic that is calculated from all of the measured data, and it gives a better indication of the overall system performance.

One must be extremely cautious in regarding Zernike polynomials fitted to noncircular apertures. While it is true that by following the Gram–Schmidt process, we can accurately fit Zernike polynomials to any aperture, we must remember that the Zernike polynomials U_r approach orthogonality over the data points only if the wavefront fitting was made with a very large number of data points with uniform density over the wavefront, assuming the aperture is circular with unit radius. Thus, even with a large number of points, uniformly distributed over the wavefront, Zernike polynomials will never approach orthogonality over a noncircular aperture. This means that the properties in Section 13.2.4 are no longer valid. In particular, property 3, which stated that the coefficient of each term could be determined independently of the other terms in the fit, no longer holds. Indeed, the fit coefficient of a particular term will now depend on which other terms are simultaneously included in the fit. For example, if we want to find the best fit tilt and power for data within an elliptical aperture, we would do it by fitting only Zernike terms U_1, U_2, U_3, and U_5. Adding other terms to the fit will change the fit coefficients for the terms of interest.

In some special cases it is possible to overcome this problem by developing special polynomials that are orthogonal over the aperture of interest. This has been done by a number of authors for annular apertures (Tatian 1974; Wang and Silva 1980; Barakat 1980; Swantner and Lowrey 1980; Mahajan 1981, 1984). The broad subject of polynomial fitting of interferograms has been studied by Kim (1982a, 1982b).

Orthogonal polynomials on the discrete set of data points have been used by some authors (Plight 1980) in order to find the wavefront from ray tracing data in a lens evaluation program. Glenn (1983, 1984) has described a set of orthonormal surface error descriptors over the surface of a cylinder, in whose case the pupil cannot be circular. Day and Beery (1986) and Lawrence and Day (1987) described a method to characterize aberrations in full spheres. Swantner (1982) defined a set of orthonormalized functions to represent the wavefront from misaligned or aberrated axicon reflective optics.

13.3.5. Creating a Uniform Grid of Wavefront Data

Fringe center data consist of lists of irregularly positioned points at which we know the measured OPD. In this form the data can be directly analyzed to yield wavefront statistics such as the PV and RMS wavefront errors. We have also shown how aberration polynomials can be fitted to the fringe data to remove tilt and power from the analysis, and to reduce the fringe data to a small number of aberration coefficients. In many cases these few numerical results are all that is required. When more advanced analysis is desired, the first step is usually to convert the fringe centers into a surface map described on a uniform data grid. Surface maps in this format are easier to plot and can be averaged or differenced with other surface maps. Diffraction calculations and slope measurement are also easier to carry out if the data is known on a uniform grid.

Two methods for performing this conversion are the global polynomial fit and the local interpolating function. A global polynomial fit reduces the fringe data to a surface polynomial described by the coefficients of each of the polynomial's terms. The Zernike polynomial fit described in previous sections is an example of this. Once the polynomial has been fitted to the fringe data, it can be evaluated at each of the uniform grid points to generate the new surface map.

One disadvantage of global fits is that they smooth the measured surface. Depending on the degree of the polynomial, there will only be a few degrees of freedom to fit many data points. It is likely that the fitted surface will pass through none of the measured points; the only requirement is that the residual fit error be minimized. If the surface contains irregular features that are not well described by the chosen polynomial (features such as steps or scratches), the polynomial fit will smooth these features so they are not visible in the fitted surface.

Local interpolation divides the aperture into smaller neighborhoods, each containing or bounded by only a few fringe centers. A low-order interpolating function is fitted to the fringe centers, and then evaluated to obtain the OPD surface values for each grid point in the neighborhood. The application of local interpolation to fringe analysis has been described by Hayslett and Swantner (1978, 1980) and Becker et al. (1982). Most local interpolations follow some variation on the following theme. For each point P_{ij} in the data grid:

1. Search through the fringe centers for the n nearest points.
2. Fit the interpolating function to these n points. If the degree of the inter-

13.4. FRINGE DIGITIZATION

polant is less than n, this fit may be a least-squares fit or a weighted least-squares fit.
3. Evaluate the interpolant at P_{ij}.

Common interpolating functions are:

$$z + A + Bx + Cy \qquad \text{(linear)} \qquad (13.66)$$

$$z + A + Bx + Cy + Dxy \qquad \text{(bilinear)} \qquad (13.67)$$

$$z = A + Bx + Cy + Dxy + Ex^2 + Fy^2 \quad \text{(quadratic)} \qquad (13.68)$$

Hayslett and Swantner used the linear ($n = 5$) and quadratic ($n = 9$) forms. Becker et al. used a weighted least-squares fit that sought to minimize the following expression:

$$S = \sum_{r=1}^{n} [W_{ij}(x_r, y_r) - z_r]^2 w[(x_r - x_i)^2 + (y_r - y_j)^2], \qquad (13.69)$$

where

$$w(d) = \exp(-\alpha d^2)$$

is the weighting function with parameter α selected to reduce the contribution of fringe centers at the edge of any given neighborhood. W_{ij} is the interpolant for the grid point P_{ij} located at (x_i, y_j), and the triplets (x_r, y_r, z_r) are the known fringe centers in the neighborhood. The advantage of the weighted fit is that it forces the interpolated surface to pass closer to the measured data, and it improves the continuity of the interpolated surface along the boundaries separating adjacent neighborhoods.

13.4. FRINGE DIGITIZATION

An interferogram can be analyzed in many ways in order to obtain the wavefront. Even manual graphical methods have been devised (Platt et al. 1978). By scanning in a direction perpendicular to the fringes with a photoelectric detector, the fringe maxima may be located by electronic analog methods with a postdigital calculation of the phase of the signal (Toyooka et al. 1987). With a scanning of the whole interferogram, the complete wavefront may then be found. However, the preferred and more precise method is by digitization of the image.

Before fringes can be analyzed by computer, they must first be digitized. Fringe digitization consists of two steps: locating the fringe centers and assigning fringe order numbers. By locating the fringe centers, we mean producing a list of x and y coordinates where each point is in the center of a fringe. Fringe ordering assigns a z value of each of the listed fringe centers. The fringe order is directly proportional to the OPD at that point.

Fringes have been digitized using scanners (Rosenzweig and Alte 1978) and television cameras (Zanoni 1978; Womack et al. 1979). Viswanathan et al. (1974) used photoelectric scanners and digitizing tablets. Commercially available programs since then have concentrated on digitizing tablets and television cameras for fringe input. Reid's (1986, 1988) surveys of fringe analysis methods includes many useful references for fringe digitization using television cameras.

13.4.1. Digitizing Tablets

The interferogram sampling can be performed by means of a digitizing tablet. There are many types of digitizing tablets, but the most common is the electromagnetic. The tablet has a pointing device or cursor that generates a low-intensity magnetic field. Under the tablet surface there is a wire grid that receives the magnetic field and induces a signal. A microprocessor inside the tablet receives the signal and calculates the position of the pointing cursor. The resolution of a digitizing tablet is of the order of 40 lines per millimeter maximum, with a positioning accuracy of about ± 0.5 mm.

Digitizing tablets have the great advantage that an elaborate software analysis is not necessary in order to determine the fringe positions, since these may be visually determined.

The main problem with the analysis of static interferograms is that the order of interference of each fringe is unknown. Any additive constant in this order number may be ignored, but its sign is more important because it determines the sign of the wavefront deviations. This indetermination has to be later defined with some additional information obtained when adjusting the interferometer. Thus, in the order of interference number one may be assigned to any fringe. Then all others are assigned starting with this fringe, increasing in any direction.

13.4.2. Automatic Video Fringe Analysis

Automating the fringe digitization process further improves the efficiency of interferogram analysis and has been considered by many authors (Wyant 1984; Reid 1986; Choudri 1987). The automatic evaluation of interferograms is very important even out of the field of optical testing (Becker et al. 1982). One convenient method for doing this is to capture the video signal output by many

13.4. FRINGE DIGITIZATION

interferometers that use television cameras to image the fringes. The fringe image can then be transferred to the computer's memory using a video frame grabber. Crescentini (1989) has reduced the effect of vibrations in the interferogram by digitizing several video images. This method is not strictly static interferometry, but resembles a little phase shifting interferometry. Fringes that have already been photographed can also be analyzed by viewing the photograph with a video camera.

In the computer's memory, the image is stored as an array of picture elements (pixels), with each pixel represented as a numerical value corresponding to the image gray level at that location. The resolution of the image and number of gray levels depend on the frame grabber design. Common resolutions are 256 by 240 or 512 by 480 pixels. The gray levels are usually either 6 bits (64 levels) or 8 bits (256 levels).

Locating the fringe centers can be complicated by poor contrast in the fringes, variation in the fringe visibility, and image noise due to laser speckle and dust in the optical system. Sometimes, a low-contrast fringe pattern from other surfaces may be superimposed upon the fringes of interest. It may be necessary to preprocess the image to compensate for these effects prior to locating the fringe centers. High-frequency noise (such as speckle) can be reduced by low-pass filtering, a technique that replaces the intensity value of each pixel with the average value of all pixels in a specified window around the replaced pixel. Increasing the size of the window increases the degree of image smoothing. Very large smoothing windows almost entirely remove the fringes, leaving only the low-frequency background variation, which can be subtracted as part of the image preprocessing.

If the fringes are open, and more or less vertically oriented, the fringe centers can be located by scanning along individual horizontal video lines. If smoothing is required, an averaging window can be passed over the data prior to locating the fringe centers. The fringe centers on this line can be located either by thresholding the image to determine the fringe edges (a fringe center lies between two edges) or by sensing intensity minima. In either case once the centers of a particular scan line have been found, they must be matched to the centers that were located on the previous scan to maintain continuity of the fringes. To do this each newly located center should be matched with the closest center of the previous scan. This method works as long as the fringes do not change direction too abruptly in the interval between scan lines. Mastin and Ghiglia (1985) have used a one-dimensional fast Fourier transform (FFT) to locate the centers of the fringes.

This algorithm will be more robust if logic is included to account for spurious digitization. This occurs when noise causes an extra fringe center to be found between two centers that have already been matched to adjacent fringes of the previous scan. These extra centers should be rejected. The algorithm will also be improved if it allows for the possibility of missing a fringe on a particular scan, a frequent occurrence if the fringe contrast is low. Ordering open fringes is trivial because the fringes appear in a simple ascending or descending sequence. However, the choice of ascending or descending order cannot be automatically inferred by examining a single interferogram. This selection must be entered by an operator who has knowledge of the test setup used to produce the fringe image.

This method will not work if the fringes bend so they travel nearly parallel to the scan line, if they are closed, or if they are reentrant (i.e., a fringe exits the aperture and reenters at a different location). In many cases these more complex patterns can be transformed to open fringes by introducing more tilt between the test and reference beams. When this cannot be done, we must apply more sophisticated methods and accept the possibility of ambiguous results for reasons that will be discussed at the end of this section.

One possible way to locate the fringe centers of complex interferograms is to trace each fringe by tracking the intensity minima of the fringe image. However, this method is not fully automated because it requires an operator to indicate each fringe to be tracked.

More automated approaches examine the entire image and divide it into binary values indicating whether a pixel is part of a fringe or not. Yatagai et al. (1982a, 1982b) describe one method for doing this by examining the properties of a 5×5 pixel cell surrounding the pixel in question. If the pixel was determined to be at a local minimum along at least two of four directions ($0°$, $90°$, and $\pm 45°$), then it was considered to be part of a fringe. Becker and colleagues (1982) used a floating threshold applied to each video scan line to binarize the image, but this method would seem to fail if the fringe was parallel to the scan line. Finally, by smoothing the fringes to the point that they are nearly indistinguishable, one obtains an image containing only the low-frequency background variation, which may be used to set individual threshold values for each pixel. This method works best when the average fringe frequency is somewhat higher than that of the background variation.

In any case the resulting binary image will include fringes that are more than one pixel wide, so that the next step will be to skeletonize, or thin, the fringes (Gillies 1988). These processing steps may leave gaps in the skeletonized fringes, which must be repaired before continuing. Gaps can be easily identified because fringes do not normally terminate except at the edges of the test aperture. Once a break is located, it can be repaired by searching for the end of another nearly broken fringe and connecting the two ends. The object is to end up with sets of connected but unordered fringe points, which will be referred to as fringe segments.

Once the points on the fringe segments have been identified, they must be ordered. The usual method is to identify a search path or paths that cross each fringe segment at least once. Along the search path, the first intersecting fringe segment is assigned an arbitrary order number. The initial numbering sequence (ascending or descending) is also selected. Subsequent unordered segments encountered along the path are assigned order numbers by counting in the desired direction in much the same way as for open fringes, with one exception. If two consecutive intersections are found with the same order number, the order of the fringe count is reversed. If the first search path does not encounter all of the fringe segments, then new paths must be started that contain enough previously ordered segments to reestablish the counting order for the new path.

Becker describes a method for finding a single-fringe ordering path. Another method is to find multiple paths by simply bisecting the image, first horizontally, and then vertically. Subsequent ordering scans can be used to further divide the image until all of the segments have been ordered.

Any attempt to order fringes from a single interferogram is limited by a basic ambiguity in the interferogram data: One cannot tell by inspection whether the fringe order is increasing or decreasing. This is especially a problem when the fringes contain sets of concentric or bulls-eye fringes. The usual solution to this problem requires pushing on either the test or reference optic and observing which way the fringes move, but this is not possible with a single, static interferogram. Automated fringe ordering really amounts to a best guess, which may require operator intervention in some cases.

13.5. FOURIER ANALYSIS OF INTERFEROGRAMS

When we analyzed interferograms by locating fringe centers, we often needed to interpolate the OPD between the fringes to complete the OPD map. This interpolation will be inaccurate when the actual OPD between the fringes is not closely represented by the interpolating function. This problem can be avoided by making use of the OPD information between the fringes, which is contained in the sinusoidal modulation of the interferogram intensity (Schemm and Vest 1983). This is more easily done if the background and visibility vary slowly in comparison with the fringes. Mertz (1983) described a sinusoidal fitting method for doing this. Takeda et al. (1982) described a Fourier transform method that analyzed one-dimensional slices of an interferogram. Macy (1983) extended this method to two dimensions, compared its accuracy to the sinusoidal fitting method, and reported an accuracy of about $\lambda/50$ RMS. This method was further refined and analyzed by Womack (1983, 1984), and Roddier and Roddier (1987). In this section we will outline the basic algorithm. (See also Section 14.14.5.)

The fringe pattern irradiance or density is given by

$$g(x, y) = a(x, y) + b(x, y) \cos [2\pi f_0 x + \phi(x, y)], \qquad (13.70)$$

where $a(x, y)$ describes the background illumination, $b(x, y)$ is the fringe modulation, and $\phi(x, y)$ is the wavefront phase. A large amount of tilt between the reference and test wavefronts produces fringes of frequency f_0, which, in this context, will be treated as a spatial carrier frequency. For clarity, we have assumed that the tilt is directed along the x axis. Equation (13.70) can be rewritten as

$$g(x, y) = a(x, y) + c(x, y) \exp(2\pi i f_0 x) + c^*(x, y) \exp(-2\pi i f_0 x), \tag{13.71}$$

where

$$c(x, y) = b(x, y) \exp[i\phi(x, y)]/2. \tag{13.72}$$

The Fourier transform of the preceding expression is

$$G(f_x, f_y) = A(f_x, f_y) + C(f_x - f_0, f_y) + C^*(f_x + f_0, f_y). \tag{13.73}$$

The transform can be numerically implemented by the FFT performed on a two-dimensional array containing an image of the fringes being processed (Fig. 13.4). If we plot the modulus of the transformed data, $|G(f_x, f_y)|$, as a gray scale image in the frequency domain, the central blur represents the transformed background illumination, $|A(f_x, f_y)|$. The spatial carrier frequency serves to separate the transformed phases, C and C^*, from the background illumination. The physical interpretation of this is similar to an amplitude diffraction grating, except that in this case we have transformed the transmitted intensity rather than the transmitted amplitude.

To recover the wavefront phase, we spatially filter all of the terms except for $C(f_x - f_0, f_y)$. The remaining term is then shifted back to the frequency plane origin, prior to performing an inverse transform. We will later explain why the frequency shift is not needed because deleting it will just result in an additional tilt term in the final OPD map. For now, we will continue in the

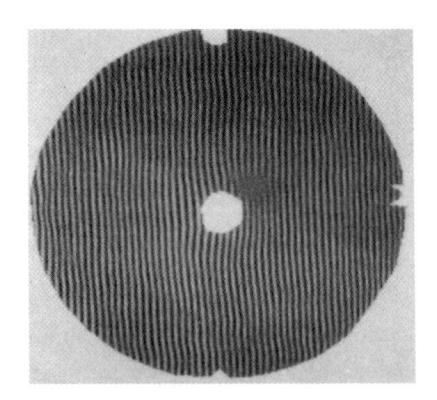

Figure 13.4. Interferogram to be analyzed by Fourier transform. (From Roddier and Roddier, 1987).

13.5. FOURIER ANALYSIS OF INTERFEROGRAMS

manner of Takeda, assuming that the shift has been performed. The result of the inverse transform is simply $c(x, y)$, given in Eq. (13.72). The wavefront phase can be recovered from this using

$$u(x, y) = \text{ATAN2}\ [\text{Im}\{c(x, y)\},\ \text{Re}\{c(x,y)\}], \qquad (13.74)$$

where ATAN2 is a FORTRAN (and C language) function that calculates the arctangent over a full $-\pi$ to π range. Note that the phase given by this equation is modulo 2π and is sometimes referred to as a *wrapped phase*. This is identical to the type of phase map obtained by phase shifting interferometry. Unwrapping this phase map removes the discontinuities that occur when the phase suddenly jumps from $-\pi$ to π. This can be accomplished with the same techniques used in phase shifting interferometry. Unwrapping the phase completes the Fourier transform analysis of the interferogram (see Figs. 13.5 and 13.6).

A physical picture of this method is obtained if we think of the interferogram as a hologram to reconstruct the wavefront. Thus, three wavefronts are generated. One is the flat reference wavefront under test, the second is the wavefront under test, and the third is the conjugate of this wavefront. If the spatial frequency at the interferogram is represented by f, the Fourier transform of the interferogram is formed by a Dirac impulse $\delta(f)$ at the origin (zero order) and two terms shifted from the origin, at frequencies $+f_0$ and $-f_0$. The values of $+f_0$ and $-f_0$ are defined by the tilt between the reference wavefront and the wavefront under test.

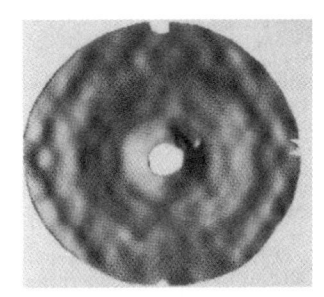

Figure 13.5. Wavefront slopes derived from the interferogram. (From Roddier and Roddier, 1987).

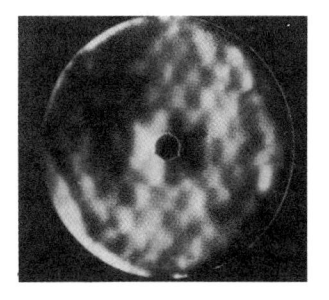

Figure 13.6. Knife-edge test of the same mirror. (From Roddier and Roddier, 1987).

These terms may be found by taking the Fourier transform of the interferogram. The term at $+f_0$ is due to the wavefront under test. So, this wavefront may be obtained by taking the Fourier transform of this term, mathematically isolating it from the others.

The resulting OPD map can be analyzed in the usual manner, including polynomial fitting to determine aberrations and remove the best fit tilt and power. Removing tilt will be necessary if the spatial carrier frequency was not completely removed prior to performing the inverse transform.

Fourier transform analysis is especially useful for analyzing complex or obstructed apertures because it does not require tracing fringes broken by obscurations. Some smoothing of high-frequency features (such as speckle) is obtained by the limited bandwidth of the frequency filter used to isolate the $C(f_x - f_0, f_y)$ term prior to the inverse transform. However, this method is not a solution to the problem of analyzing closed fringes because interferograms of this class generally do not have a large enough spatial carrier frequency to achieve adequate separation of the "diffracted" orders in the frequency domain.

A similar approach to evaluate interferograms has been used by Ru et al. (1988), by using a two-dimensional Fresnel transform. A large amount of defocusing is introduced, instead of the large tilt required in the Fourier analysis. This method is useful in common-path interferometers, where it is easier to introduce defocusing than tilt.

13.6. DIRECT MEASURING INTERFEROMETRY

This is a technique recently developed at Carl Zeiss (Freischlad et al. 1990a, 1990b; Kuchel 1990; Dorband et al. 1990). (See Section 14.14.5.) Like in the Fourier analysis of interferograms, a large amount of tilt has to be introduced, producing a narrow spaced fringe pattern. An obvious advantage of this large tilt is that the wavefront is completely covered with fringes, so that there are no large unsampled zones. This eliminates the need for any kind of interpolation or polynomial fitting, just like in phase shifting interferometry.

A perfect wavefront would produce straight, parallel, and equidistant fringes. If the irradiance in this pattern is scanned with a photometer, along a line perpendicular to the fringes, a perfect sinusoidal signal is obtained. We may think that the spatial frequency along the scanned line is converted to a temporal frequency. If the wavefront is not perfect, the fringes will not be straight and neither will their separation be constant. Then, the signal from the photometer will be phase modulated. This phase modulation is the wavefront deformation.

Following well-known procedures, the phase modulation may be found by heterodyning in the phase-modulated signal with a perfectly sinusoidal signal having the same frequency as the carrier of the phase-modulated signal (Ichioka and Inuiya 1972; Mertz 1983). This heterodyning can be performed optically by superimposing a perfect interferogram with straight fringes on top of the interferogram to be measured, as shown in the chapter on moirè tests, but it is difficult to obtain quantitative information about the phase modulation between the moirè fringes. A better procedure is to do it electronically on the photometer signal.

The interferogram taken by a TV camera is scanned along lines oriented perpendicularly to the fringes. The sine and cosine of the phase modulation of the signal is obtained by an electronic analog multiplication by a sinusoidal and a cosinusoidal signal at the carrier frequency, and a subsequent application of a low-pass filter. Then, the ratio of these two signals is the tangent of the desired phase. Any small error in the fequency of the reference sine and cosine signals will just introduce a tilt in the computed wavefront.

An important advantage of direct measuring interferometry with respect to phase shifting interferometry is its low sensitivity to vibrations. This is because only one interferogram is needed to evaluate the wavefront.

Acknowledgments. The authors would like to thank Walter Hahn for his helpful comments on automated fringe digitization techniques, and Prof. Emil Wolf and Dr. Greg Forbes for very helpful suggestions on the subject of Zernike polynomials.

REFERENCES

Barakat, Richard, "Optimum Balanced Wave-Front Aberrations for Radially Symmetric Amplitude Distributions: Generalizations of Zernike Polynomials," *J. Opt. Soc. Am.*, **70**, 739–742 (1980).

Bathia, A. B. and E. Wolf, "The Zernike Circle Polynomials Occurring in Diffraction Theory," *Proc. Phys. Soc.*, **B65**, 909–910 (1952).

Bathia, A. B. and E. Wolf, "On the Circle Polynomials of Zernike and Related Orthogonal Tests," *Proc. Cambr. Phil. Soc.*, **50**, 40–48 (1954).

Becker, Friedhelm, Gerd E. A. Maier, and Host Wegner, "Automatic Evaluation of Interferograms," *Proc. SPIE*, **359**, 386–393 (1982).

Born, M. and E. Wolf, *Principles of Optics*, Pergamon Press, New York, 1964, p. 464.

Choudri, Amar, "Automated Fringe Reduction Analysis," *Proc. SPIE*, **816**, 49–55 (1987).

Crescentini, Luca, "Fringe Pattern Analysis in Low-Quality Interferograms," *Appl. Opt.*, **28**, 1231–1234 (1988).

Day, Robert D. and Thomas A. Beery, "Sphericity Measurements of Full Spheres Using Subaperture Optical Testing Techniques," *Proc. SPIE*, **661**, 334–341 (1986).

Dörband, D., W. Wiedmann, U. Wegmann, W. Kübler, and K. R. Freischlad, "Software Concept for the New Zeiss Interferometer," *Proc. SPIE*, **1332**, 664–672 (1990).

Forsithe, G. E., "Generation and Use of Orthogonal Polynomials for Data Fitting on a Digital Computer," *J. Soc. Ind. Appl. Math*, **5**, 74 (1957).

Freischlad, K., M. Küchel, K. H. Schuster, U. Wegmann, and W. Kaiser, "Real-time Wavefront Measurement with Lambda/10 Fringe Spacing for the Optical Shop," *Proc. SPIE*, **1332**, 18–24 (1990a).

Freischlad, K., M. Küchel, W. Wiedmann, W. Kaiser, and M. Mayer, "High Precision Interferometric Testing of Spherical Mirrors with Long Radius of Curvature," *Proc. SPIE*, **1332**, 8–17 (1990b).

Freniere, E. R., O. E. Toler, and R. Race, "Interferogram Evaluation Program for the HP-9825A Calculator," *Proc. SPIE*, **171**, 39–42 (1979).

Freniere, E. R., O. E. Toler, and R. Race, "Interferogram Evaluation Program for the HP-9825A Calculator," *Opt. Eng.*, **20**, 253–255 (1981).

Gillies, A. C., "Image Processing Approach to Fringe Patterns," *Opt. Eng.*, **27**, 861–866 (1988).

Glenn, Paul, "Set of Orthonormal Surface Error Descriptors for Near Cylindrical Optics," *Proc. SPIE*, **429**, 178–186 (1983).

Glenn, Paul, "Set of Orthonormal Surface Error Descriptors for Near Cylindrical Optics," *Opt. Eng.*, **23**, 384–390 (1984).

Hayslett, Charles R. and William H. Swantner, "Mathematical Methods for Deriving Wavefronts from Interferograms," in *Optical Interferograms—Reduction and Interpretation, ASTM Symposium*, Am. Soc. for Test. and Mat., Tech. Publ. 666, A. H. Guenther and D. H. Liedbergh, Eds., 1978.

Hayslett, Charles R. and William H. Swantner, "Wave Front Derivation from Interferograms by Three Computer Programs," *Appl. Opt.*, **19**, 3401–3406 (1980).

Hopkins, H. H., *Wave Theory of Aberrations*, Clarendon Press, Oxford, 1950, p. 48.

Ichioka, Y. and M. Inuiya, "Direct Phase Detecting System," *Appl. Opt.*, **11**, 1507–1514 (1972).

Kim, C.-J., Polynomial Fit of Interferograms, Ph.D. Thesis, University of Arizona, Tucson, AZ, 1982a.

Kim, C.-J., "Polynomial Fit of Interferograms," *Appl. Opt.*, **21**, 4521–4525 (1982b).

Kim, C.-J. and Robert Shannon, "Catalog of Zernike Polynomials," in *Applied Optics and Optical Engineering*, Vol. 10, R. Shannon and J. C. Wyant, Eds., Academic Press, New York, 1987, Chap. 4.

Kingslake, Rudolf, "The Interferometer Patterns Due to the Primary Aberrations," *Trans. Opt. Soc.*, **27**, 94 (1925–1926).

Küchel, F. M., Th. Schmieder, Hans J. Tiziani, and E. Mensel, "Beitrag zur Verwendung von Zernike-Polynomen bei der Automatischen Interferrenzstreifenauswertung," *Optik*, **65**, 123–142 (1983).

Küchel, M., "The New Zeiss Interferometer," *Proc. SPIE*, **1332**, 655–663 (1990).

Lawrence, N. George, and R. D. Day, "Interferometric Characterization of Full Spheres: Data Reduction Techniques," *Appl. Opt.*, **26**, 4875–4882 (1987).

Loomis, John S., "A Computer Program for Analysis of Interferometric Data," in *Optical Interferograms—Reduction and Interpretation, ASTM Symposium*, Am. Soc. for Test. and Mat., Tech. Publ. 666, A. H. Guenther and D. H. Liedbergh, Eds., 1978.

Macy, William W., Jr., "Two Dimensional Fringe Pattern Analysis," *Appl. Opt.*, **22**, 3898–3901 (1983).

Mahajan, Virendra N., "Zernike Annular Polynomials for Imaging Systems with Annular Pupils," *J. Opt. Soc. Am.*, **71**, 75–85 (1981).

Mahajan, Virendra N., "Zernike Annular Polynomials for Imaging Systems with Annular Pupils," *J. Opt. Soc. Am.*, **1a**, 685–685 (1984).

Malacara, Daniel, Alejandro Cornejo, and Arquimedes Morales, "Computation of Zernike Polynomials in Optical Testing," *Bol. Inst. Tonantzintla, 1*, **2**, 121–126 (1976).

Malacara, Daniel, "Set of Orthogonal Aberration Coefficients," *Appl. Opt.*, **22**, 1273–1274 (1983).

Malacara, Daniel, J. Martín Carpio-Valadéz, and J. Javier Sánchez-Mondrangón, "Interferometric Data Fitting on Zernike Like Orthogonal Basis," *Proc. SPIE*, **813**, 35–36 (1987).

Mastin, Gary A. and Dennis C. Ghiglia, "Digital Extraction of Interference Fringe Contours," *Appl. Opt.*, **24**, 1727–1728 (1985).

Mertz, L., "Real Time Fringe-Pattern Analysis," *Appl. Opt.*, **22**, 1535–1539 (1983).

Parks, Robert E., "Removal of Test Optics Errors," *Proc. SPIE*, **153**, 56–63, (1978).

Platt, B. C., S. G. Reynolds, and T. R. Holt, "Determining Image Quality and Wavefront Profiles from Interferograms," in *Optical Interferograms—Reduction and Interpretation, ASTM Symposium*, Am. Soc. for Test. and Mat., Tech. Publ. 666, A. H. Guenther, D. H. Liedbergh, Eds., 1978.

Plight, A. M., "The Calculation of the Wavefront Aberration Polynomial," *Opt. Acta*, **27**, 717–721 (1980).

Prata, Aluizio Jr., and W. V. T. Rusch, "Algorithm for Computation of Zernike Polynomials Expansion Coefficients," *Appl. Opt.*, **28**, 749–754 (1989).

Reid, G. T., "Automatic Fringe Pattern Analysis: A Review," *Opt. Lasers Eng.*, **7**, 37–68 (1986).

Reid, G. T., "Image Processing Techniques for Fringe Pattern Analysis," *Proc. SPIE*, **954**, 468–477 (1988).

Rimmer, M. P., Polynomial Fitting of Interferograms, Internal Report, ITEK Corporation, 1972.

Rimmer, M. P. and J. C. Wyant, "Evaluation of Large Aberrations Using a Lateral-Shear Interferometer Having Variable Shear," *Appl. Opt.*, **14**, 142–150 (1975).

Roddier, Claude and François Roddier, "Interferogram Analysis Using Fourier Transform Techniques," *Appl. Opt.*, **26**, 1668–1673 (1987).

Rosenzweig, David and Betsy Alte, "A Facility for the Analysis of Interferograms," in *Optical Interferograms—Reduction and Interpretation, ASTM Symposium*, Am. Soc. for Test. and Mat., Tech. Publ. 666, A. H. Guenther, D. H. Liedbergh, Eds., 1978.

Ru, Qing-Shin, Toshio Honda, Jumpei Tsujiuchi, and Nagaaki Ohyama, "Fringe Analysis By Using 2-D Fresnel Transform," *Opt. Commun.*, **66**, 21–24, (1988).

Sánchez-Mondragón, J. Javier, Daniel Malacara, and Abundio Dávila, "Digital Data Reduction of Rotationally Invariant Interferograms," *Proc. SPIE*, **813**, 259–260 (1987).

Schemm, John B. and Charles M. Vest, "Fringe Pattern Recognition and Interpolation Using Nonlinear Regression Analysis," *Appl. Opt.*, **22**, 2850–2853 (1983).

Sumita, H., "Orthonormal Expansion of the Aberration Difference Function and its Applications to Image Evaluation," *Jpn. J. Appl. Phys.*, **8**, 1027 (1969).

Swantner, W. H. and W. H. Lowrey, "Zernike-Tatian Polynomials for Interferogram Reduction," *Appl. Opt.*, **19**, 161–163 (1980).

Swantner, William H., "Wavefronts of Axicon Systems," *Opt. Eng.*, **21**, 333–339 (1982).

Takeda, Mitsuo, Hideki Ina, and Seiji Kobayashi, "Fourier-Transform Method of Fringe-Pattern Analysis for Computer-Based Topography and Interferometry," *J. Opt. Soc. Am.*, **72**, 156–160 (1982).

Tatian, Berge, "Aberration Balancing in Rotationally Symmetric Lenses," *J. Opt. Soc. Am.*, **64**, 1083–1091 (1974).

Toyooka, S., K., K. Ohashi, K. Yamada, and K. Kobayashi, "Real-Time Fringe Processing by Hibrid Analog-Digital System," *Proc. SPIE*, **813**, 33–35 (1987).

Viswanathan, V. K., J. E. Sollid, W. S. Hall, I. Liberman, and G. Lawrence, "Interferogram Reduction and Interpretation as Applied to the Analysis of a Laser Fusion System," in *Optical Interferograms—Reduction and Interpretation, ASTM Symposium*, Am. Soc. for Test. and Mat., Tech. Publ. 666, A. H. Guenther, D. H. Liedbergh, Eds., 1978.

Wang, J. Y. and D. E. Silva, "Wave-Front Interpretation with Zernike Polynomials," *Appl. Opt.*, **19**, 1518 (1980).

Wolf, Emil and Forbes Greg, personal communication, 1990.

Womack, K. H., J. A. Jonas, C. L. Koliopoulos, K. L. Underwood, J. C. Wyant, J. S. Loomis, and C. R. Hayslett, "Microprocessor-Based Instrument for Analysis of Video Interferograms," *Proc. SPIE*, **192**, 134–139 (1979).

Womack, Kenneth H., "A Frequency Domain Description of Interferogram Analysis," *Proc. SPIE*, **429**, 166–173 (1983).

Womack, Kenneth H., "Frequency Domain Description of Interferogram Analysis," *Opt. Eng.*, **23**, 396–400 (1984).

Wyant, James, "Microprocessor Analysis of Interferometric Optical Testing Data," in *ICO-Conference Digest*, Saporo, Japan, 1984.

Yatagay, Toyohiko, Masanori Idesawa, Y. Yamaashi, and Masene Susuki, "Interactive Fringe Analysis System: Applications to Moire Contourgram and Interferogram," *Opt. Eng.*, **21**, 901–906 (1982a).

Yatagai, Toyohiko, Suezou Nakadate, Idesawa Masanori, and Saito Hiroyoshi, "Automatic Fringe Analysis Using Digital Image Processing Techniques," *Opt. Eng.*, **21**, 432 (1982b).

Zanoni, C. A., "A New, Semiautomatic Interferogram Evaluation Technique," in *Optical Interferograms—Reduction and Interpretation, ASTM Symposium*, Am. Soc. for Test. and Mat., Tech. Publ. 666, A. H. Guenther, D. H. Liedbergh, Eds., 1978.

Zernike, Fritz, "The Diffraction Theory of Aberrations," in *Optical Image Evaluation Circular 526*, Nat. Bur. of Stand., Washington, D.C., 1954.

Zernike, Fritz, "Begunstheorie des Schneidenver-Fahrens und Seiner Verbesserten Form. der Phasenkontrastmethode," *Physica*, **1**, 689 (1934).

ADDITIONAL REFERENCES

Augustin, Walter H., Alvine H. Rosenfeld, and Carl A. Zanoni, "An Automatic Interference Pattern Processor with Interactive Capability," *Proc. SPIE*, **153**, 146-155 (1978).

Augustin, Walter H., "Automatic Data Reduction of Both Simple and Complex Interference Patterns," *Proc. SPIE*, **171**, 22-31 (1979).

Augustin, Walter H., "Versatility of a Microprocessor-Based Interferometric Data Reduction System," *Proc. SPIE*, **192**, 128-133 (1979).

Boutellier, R. and R. Zumbunn, "Digital Interferogram Analysis and DIN Norms," *Proc. SPIE*, **656**, 128-134 (1986).

Chetkareva, L. E., V. A. Evseev, and V. A. Efimov, "Computer Calculation of Interference Pattern," *Sov. J. Opt. Technol.*, **45**, 480 (1978).

Dente, Gregory C., "Separating Misalignment from Misfigure in Interferograms in Off-Axis Aspheres," *Proc. SPIE*, **429**, 187-193 (1983).

Freischlad, Klaus and Chris L. Koliopoulos, "Wavefront Reconstruction from Noisy Slope or Difference Data Using the Discrete Fourier Transform," *Proc. SPIE*, **551**, 74-80 (1985).

Kurita, Hiroyuki, Keisuke Saito, and Kato Masahiko, "Influence of System Aberrations on Interferometric Aspheric Surface Testing," *Proc. SPIE*, **680**, 47-52 (1986).

Lewis, Jeff L., William P. Kuhn, and Philip Stahl H., "The Evaluation of a Random Sampling Error on the Polynomial Fit of Subaperture Test Data," *Proc. SPIE*, **954**, 88-94 (1988).

Loomis, John S., "Analysis of Interferograms from Waxicons," *Proc. SPIE*, **171**, 64-69 (1979).

Wong, Steve, "Fringe Analysis for Testing Optical Surfaces," *Proc. SPIE*, **966**, 316-321 (1988).

Young, Eric W., "Optimal Removal of all Mislocation Effects in Interferometric Tests," *Proc. SPIE*, **661**, 116-124 (1986).

14

Phase Shifting Interferometry

J. E. Greivenkamp and J. H. Bruning

14.1. INTRODUCTION

The single biggest change in all types of instrumentation over the past 20 years or so has been the integration of computers into the measurement system. Interferometry is no exception, and the prime manifestation of this situation has been the development of phase shifting interferometry (PSI). Unlike many of the techniques discussed elsewhere in this book, PSI is not a specific optical hardware configuration but rather a data collection and analysis method that can be applied to a great variety of testing situations.

While the analysis of static interferograms has benefited from computerization, it suffers from the need to find fringe centers and the resulting trade-off between precision and number of data points. Data is collected only along the fringe centers, and most analyses need the data on a regular grid, therefore requiring interpolation. With a static interferogram, an additional piece of information is also required to determine the polarity of the wavefront.

PSI electronically records a series of interferograms while the reference phase of the interferometer is changed. The wavefront phase is encoded in the variations in the intensity pattern of the recorded interferograms, and a simple point-by-point calculation recovers the phase. The need to locate the fringe centers is eliminated, as are the associated problems. While the earliest reference to this technique dates back to 1966 (Carré, 1966), the development of PSI techniques for optical testing really began in the early 1970s (Crane 1969; Bruning et al. 1974; Wyant 1975; Johnson and Moore 1977; Hardy et al. 1977; Stumpf 1979; Bruning 1978; Moore 1979; Massie et al. 1979; Creath 1988; Malacara 1990, Schwider 1990). The applications for this early work included optical testing, real-time wavefront sensing for active optics, distance measuring interferometry, and microscopy.

Over the years this general technique has been known by several names in-

---

Optical Shop Testing, Second Edition, Edited by Daniel Malacara.
ISBN 0-471-52232-5 © 1992, John Wiley & Sons, Inc.

cluding phase measuring interferometry, fringe scanning interferometry, realtime interferometry, AC interferometry, and heterodyne interferometry. All describe the same basic technique. In this chapter we will review the fundamentals of PSI, show some of the various algorithms that are used, consider repeatability issues and error sources, examine some of the alternate implementations, and look at future trends for this technology.

14.2. FUNDAMENTAL CONCEPTS

The relatively simple (in retrospect, anyway) concept behind phase shifting interferometry is that a time-varying phase shift is introduced between the reference wavefront and the test or sample wavefront in the interferometer. A time-varying signal is then produced at each measurement point in the interferogram, and the relative phase between the two wavefronts at that location is encoded in these signals.

For simplicity we will use the wavefront phase for our analysis. This choice allows the analysis to be independent of the specific hardware configuration, and the conversion from phase to surface errors or optical path difference (OPD) is straightforward. For example, a surface with height errors $h(x, y)$ tested in reflection will produce a wavefront error $\phi(x, y)$:

$$\phi(x, y) = 4\pi h(x, y)/\lambda, \tag{14.1}$$

where x and y are the spatial coordinates and λ is the wavelength. This expression is for normal incidence, and obliquity factors must be added in other situations.

General expressions for the reference and test wavefronts in the interferometer are, respectively,

$$w_r(x, y, t) = a_r(x, y)e^{i[\phi_r(x, y) - \delta(t)]} \tag{14.2}$$

and

$$w_t(x, y) = a_t(x, y)e^{i\phi_t(x, y)}, \tag{14.3}$$

where $a_r(x, y)$ and $a_t(x, y)$ are the wavefront amplitudes, $\phi_r(x, y)$ and $\phi_t(x, y)$ are the wavefront phases, and $\delta(t)$ is a time-varying phase shift introduced into the reference beam. To be more precise, $\delta(t)$ is the relative phase shift between the two beams; it can physically result from changes in either the reference or test beams. The resulting intensity pattern is

$$I(x, y, t) = |w_r(x, y, t) + w_t(x, y)|^2 \tag{14.4}$$

or

$$I(x, y, t) = I'(x, y) + I''(x, y) \cos[\phi_t(x, y) - \phi_r(x, y) + \delta(t)], \tag{14.5}$$

14.2. FUNDAMENTAL CONCEPTS

where $I'(x, y) = a_r^2(x, y) + a_t^2(x, y)$ is the average intensity, and $I''(x, y) = 2a_r(x, y)a_t(x, y)$ is the fringe or intensity modulation. If we now define $\phi(x, y)$ to be the wavefront phase difference $\phi_t(x, y) - \phi_r(x, y)$, we obtain

$$I(x, y, t) = I'(x, y) + I''(x, y) \cos [\phi(x, y) + \delta(t)], \qquad (14.6)$$

which is the fundamental equation for PSI. The intensity at each point varies as a sinusoidal function of the introduced phase shift $\delta(t)$ with a temporal offset given by the unknown wavefront phase.

This result can be visualized by looking at the intensity as a function of $\delta(t)$; for a linear phase shift this is equivalent to the variation of intensity with time. The intensity of the interferogram at an individual measurement point varies sinusoidally with phase (or time) as shown in Fig. 14.1, but more importantly, the three unknowns in Eq. (14.6) are easy to identify in this signal. The constant term $I'(x, y)$ is the intensity bias, $I''(x, y)$ is half the peak-to-valley intensity modulation, and the unknown phase $\phi(x, y)$ is related to the temporal phase shift of this sinusoidal variation. The wavefront phase at this location can be easily computed from this temporal delay. The entire map of the unknown wavefront phase $\phi(x, y)$ can be measured by monitoring and comparing this temporal delay at all the required measurement points in the interferogram. In this manner the intensity modulation and bias terms are separated from this delay, and the measured wavefront phase is therefore also decoupled from these terms. Note that the period is independent of the wavefront phase; it is the same for all data points.

If the interferogram is viewed during the phase shifting operation, the fringe pattern will appear to move across the field. This is exactly analogous to the fringe motion generated by pushing on the test flat to determine the high and low points when observing Newton rings. It is important to note that this fringe motion is an artifact of the sinusoidal intensity variations occurring at the same

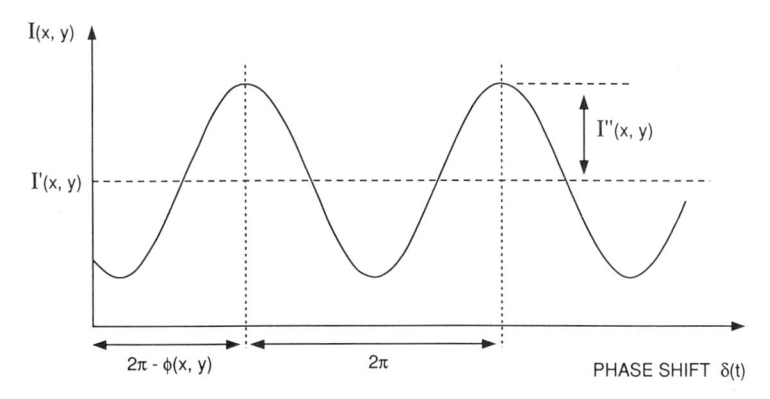

Figure 14.1. The variation of intensity with the reference phase at a point in an interferogram.

temporal frequency but with different phases at each point in the interferogram. PSI does not use this fringe motion to compute the wavefront; it does not find and track fringe centers. PSI uses the intensity variations at each point for this computation.

14.3. ADVANTAGES OF PSI

Traditionally, interferograms have been analyzed by noting the straightness of the fringes or by identifying the fringe centers and assigning a constant surface height along each fringe. Adjacent fringes represent a height change of a half wave. Finding the fringe centers has been the inherent limit to the precision of the technique and has also restricted the amount of data processing that can be done to the results. Visually, the location of the fringe centers can be identified to within a tenth or a twentieth of a fringe spacing under ideal circumstances. It is difficult to precisely locate the maxima or minima of the fringe intensity pattern. With a digitized interferogram produced by electronic camera or an input scanner, the situation can be improved to about a fiftieth of a wave under ideal situations. In a conventional interferometer, the observed fringes represent errors in the test part, the reference surface, and optical errors in the interferometer. Either of these measurement accuracies is well below the inherent resolution of the interferometer system. In addition, the apparent location of the fringe centers can be influenced by intensity variations across the interferogram and spatial sensitivity variations and fixed pattern noise in the detector. The best measurement precision occurs for an interferogram with a few widely spaced fringes, but since the data is collected only along the fringe centers, only a few widely spaced data points are obtained. The number of data points can be increased by adding more tilt fringes to the interferogram pattern, but this is at the cost of reduced repeatability in finding the fringe centers. This trade-off of precision and amount of data is further complicated by the fact that data analysis programs require that the input data be sampled on a regular grid, not along a few curved lines across the part. An interpolation routine is needed to convert the fringe center data to a map of the wavefront on a regular grid.

A further detail needed for analyzing static interferograms is defining the sense of the part. Is it concave or convex? This determination requires another piece of information; for example, the direction the fringes move when the reference surface is pushed. With PSI the phase between the reference and test wavefronts is varied in a known manner and direction during the data collection, and the sense of the part is automatically and unambiguously determined. One way to think of this is that PSI requires that someone "push" on the reference mirror as the data is collected. The use of programs that automatically locate fringe centers sometimes place the requirement on the interferogram that it contain no closed-loop fringes. This is usually accomplished by adding tilt to the

14.3. ADVANTAGES OF PSI

test setup and is needed so that the program can order the fringes from the high point to the low point on the surface.

Since the analysis required for PSI is not dependent on finding fringe centers or following fringes, any type of fringe pattern can be used. This is a more practical situation. Even a fringe pattern with no fringes (one very broad fringe covering the entire field of view) or with a complicated series of closed fringes is analyzed correctly. As we will see later, more accuracy is obtained when the number of fringes in the field is minimized. For the same reason data can be collected and analyzed on any sampling grid that is desired. The most common grid is the regular array of pixels on a solid-state sensor, which is compatible with most data processing programs and packages. Any irregular grid can also be used if the situation demands it.

Another advantage of PSI is that it is insensitive to spatial variations of intensity, detector sensitivity, and fixed pattern noise. This can be seen by modifying Eq. (14.6) to include these effects. Fixed pattern noise is a spatial variation in the bias signal that is unrelated to the incident intensity. The detector sensitivity and fixed pattern noise are denoted by $s(x, y)$ and $n(x, y)$, respectively. The spatial intensity variations are already included in the (x, y) dependence of the terms $I'(x, y)$ and $I''(x, y)$. The resulting output from the detector $v(x, y, t)$ is

$$v(x, y, t) = s(x, y)\{I'(x, y) + I''(x, y) \cos [\phi(x, y) + \delta(t)]\} + n(x, y), \tag{14.7}$$

which can easily be rewritten as

$$v(x, y, t) = [s(x, y)I'(x, y) + n(x, y)] \\ + [s(x, y)I''(x, y)] \cos [\phi(x, y) + \delta(t)]. \tag{14.8}$$

These additional terms have changed the intensity bias and the intensity modulation of the signal recorded at each measurement point, but have little effect on the measured temporal delay. Since the unknown wavefront phase is encoded entirely in this delay, the precision of the interferometric measurement is generally not degraded by these types of spatial variations. The only reason for these last two statements not being expressed as "no effect" is that there are signal-to-noise considerations that may change the repeatability of the measurement. These effects are especially important if the spatial variations are large or if the output signal modulation is small relative to the bias. Since these detector spatial variations show up as terms indistinguishable from the intensity bias and modulation terms already present in Eq. (14.6), the discussion of algorithms and noise considerations in later sections will include these terms implicitly in $I'(x, y)$ and $I''(x, y)$. Time-varying noise sources, such as detector dark current and photon noise, must also be included. These signal-to-noise considerations are discussed at greater length in Section 14.10.

It should not be surprising to learn that the repeatability or precision of PSI is greater than that of static interferogram analysis. Since a series of data is collected over time, some advantage is gained through averaging the sequentially recorded interferograms, but this simple explanation does not give the full reason. The data collection and analysis procedure for PSI is fundamentally different from that used for static interferograms and results in a large improvement in repeatability for PSI. The method used for PSI is much more analogous to that used with lock-in amplifiers or any of the local-oscillator heterodyne detection schemes used for electronic signals. Repeatabilities of a hundredth of a wavelength are routine with PSI, and values of a thousandth of a wavelength have been reported and can be obtained with today's commercial equipment with care. Once again, the particulars of this discussion can be found in Section 14.10.

In Eqs. (14.6) and (14.8) we have assumed that the only term that has any time dependence is $\delta(t)$, and any deviation from this assumption will introduce errors into the PSI analysis. The unknown wavefront phase $\phi(x, y)$ is encoded in the measured time delay at each point, and temporal variations occurring during the measurement interval will degrade the performance of a PSI system. In addition to random noise, a common source of error is vibration and turbulence. This is a significant difference from static interferometry where a snapshot of the interferogram can be obtained. Another possible source of error is temporal laser intensity variations, but this is usually not a large problem as a warmed-up laser is fairly stable over typical measurement times, and stabilized lasers are available for exacting situations. A third source of error unique to PSI is errors or variations in the phase shift $\delta(t)$. The functional form must be correct, and the phase shift should be uniform over the interferogram.

14.4. METHODS OF PHASE SHIFTING

By far the most common method used to introduce the time-varying phase shift in a PSI system is to translate one of the mirrors or optical surfaces in the interferometer with a piezoelectric transducer (Soobitsky 1986; Hayes 1989). Composed of lead–zinc–titanate (PZT) or other ceramic materials, these devices expand or contract with an externally applied voltage. Depending on the configuration, up to a few hundred volts may be needed to obtain the required motion of a wavelength or less. By discretely changing the applied voltage, the induced phase shift varies through a series of steps. If the voltage is programmed to vary smoothly, a phase shift of a desired functional form can be produced. Information from the interferometric test can be used to calibrate the motion of the PZT, and this is further discussed in Section 14.9.

14.4. METHODS OF PHASE SHIFTING

The phase shift configurations for three common interferometer types (the Twyman–Green, the Mach–Zender, and the laser Fizeau) are shown in Fig. 14.2. The first two achieve the phase shift by translating one of the mirrors in the reference arm of the interferometer. At normal incidence in the Twyman–Green, a phase shift of a full wavelength occurs for each half wavelength of translation. Because of the nonnormal angles in the Mach–Zender, the induced phase shift is reduced by a factor of the cosine of the angle of incidence (Fig. 14.3). Since the reference wavefront is plane, the small lateral displacement is ignored. In the laser Fizeau interferometer the transmissive reference objective or the test piece is translated for the phase shift. The reference surface is the final uncoated surface of the objective, and this surface is concentric with the focus. For fast F numbers, errors or nonuniformities in the phase shift result because the ray directions at the edge of the pupil are not parallel to the translation direction (Moore and Slaymaker 1980). It should be noted that phase shift introduced by translating a surface can also be interpreted in terms of a Doppler shift of one of the two beams (Malacara 1969).

An alternative method for producing either a stepped or continuous phase shift is to use a tilted plane-parallel plate in the reference beam of the interferometer (Wyant and Shagam 1978). The optical path within the plate increases as the tilt angle is increased. To avoid introducing aberrations into the reference beam, this method is useful only for a collimated or nearly collimated reference beam.

A continuous phase shift can be produced between the reference and test beams by introducing an optical frequency difference between these two beams. If the two optical frequencies are ν and $\nu + \Delta\nu$, Eqs. (14.2) and (14.3) can be rewritten to include the time dependence:

$$w_r(x, y, t) = a_r(x, y)e^{i[\phi_r(x,y) - 2\pi(\nu + \Delta\nu)t]} \tag{14.9}$$

$$w_t(x, y, t) = a_t(x, y)e^{i[\phi_t(x,y) - 2\pi\nu t]}. \tag{14.10}$$

No other temporal phase shift is included. The resulting interferogram intensity pattern, analogous to Eq. (14.6), is

$$I(x, y, t) = I'(x, y) + I''(x, y) \cos [\phi(x, y) + 2\pi \Delta\nu t]. \tag{14.11}$$

The frequency difference gives rise to a linear phase shift between the test and reference beams:

$$\delta(t) = 2\pi \Delta\nu t, \tag{14.12}$$

and the intensity at a given location varies sinusoidally at the difference frequency $\Delta\nu$. As before, the wavefront phase $\phi(x, y)$ is a spatially varying delay between signals at the various measurement points.

Since even a small optical frequency shift (as a fraction of the optical fre-

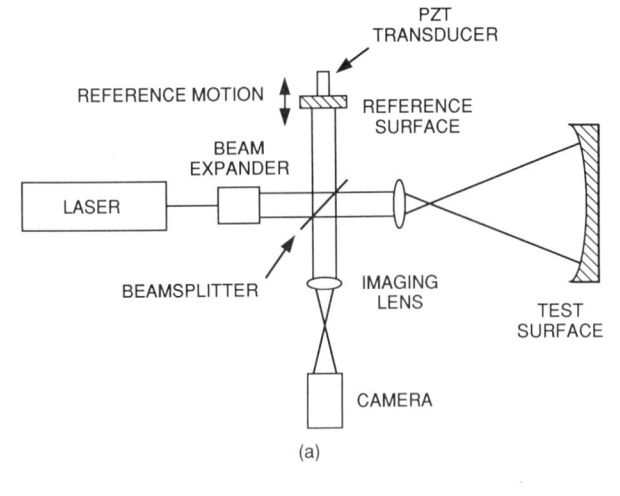

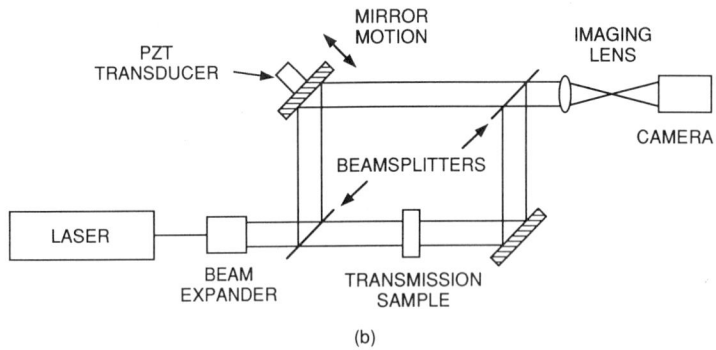

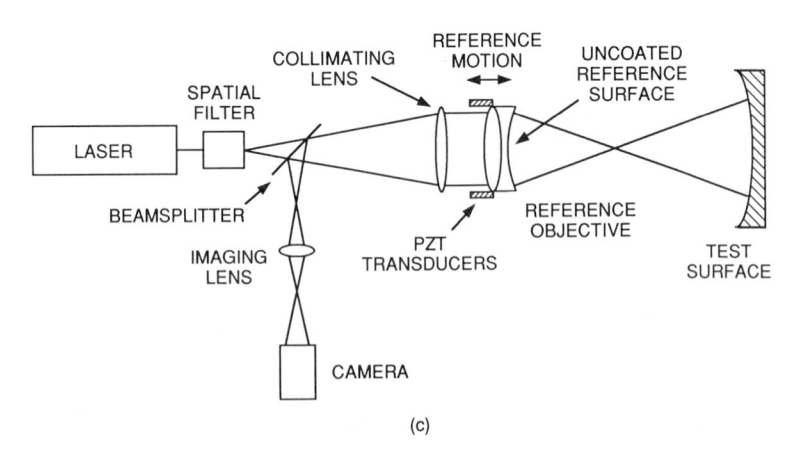

Figure 14.2. Three common interferometer configurations: (*a*) Twyman–Green, (*b*) Mach–Zender, and (*c*) laser Fizeau.

14.4. METHODS OF PHASE SHIFTING

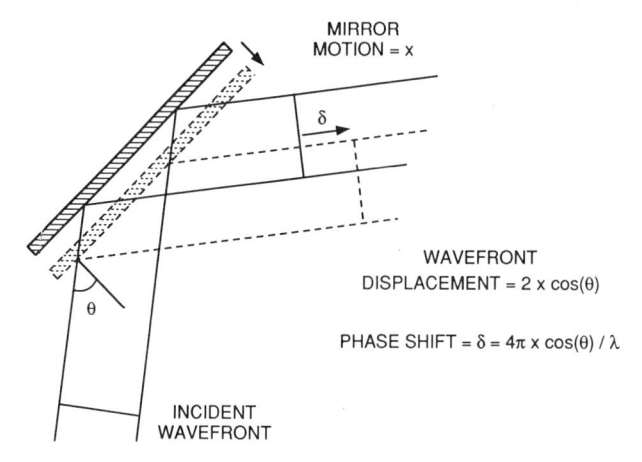

Figure 14.3. The induced phase shift due to a mirror translation at nonnormal incidence.

Labels in figure: MIRROR MOTION = x; WAVEFRONT DISPLACEMENT = 2 x cos(θ); PHASE SHIFT = δ = 4π x cos(θ) / λ; INCIDENT WAVEFRONT.

quency) can result in a large difference frequency, this phase shifting method is very useful for situations where dynamic measurements are required. In these situations the phase shift must occur faster than the changes that are being monitored. Two important applications are the measurement of turbulent flow and distance measurement interferometry. The two primary methods to produce a frequency shift are the Zeeman split laser and Doppler shifts introduced by moving gratings. The application and use of this type of phase shift is discussed in Sections 14.14.3 and 14.16.1.

The output frequency of a laser can be split into two orthogonally polarized output frequencies by the application of a dc magnetic field (Burgwald and Kruger 1970). The frequency separation of a Zeeman split two-frequency laser is controlled by the magnetic field, and values of about 1.8–5 MHz are found for this effect in helium–neon lasers.

When a diffraction grating is translated through a beam of light, a Doppler shift is introduced in the diffracted beams (Suzuki and Hioki 1967; Stevenson 1970; Bryngdahl 1976). The translation direction is perpendicular to the propagation direction (see Fig. 14.4). The frequency shift is proportional to the diffraction order m and the velocity v, and inversely proportional to the grating period d, or

$$\Delta \nu = mv/d. \tag{14.13}$$

The undiffracted beam has no frequency shift; beams diffracted in the same direction as the translation see a positive frequency shift; and beams diffracted in the opposite direction have their frequency decreased. One of the diffracted orders is selected and interfered with the original frequency to produce a phase shifting interferometer. Mechanical translations, including rotating a radial

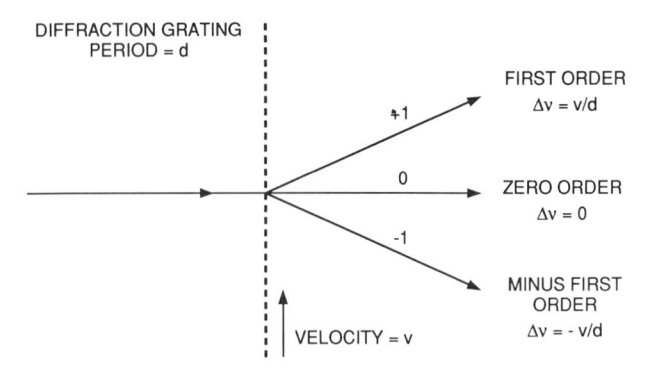

```
                    DIFFRACTION GRATING
                    PERIOD = d
                                                    FIRST ORDER
                                     +1             Δv = v/d

                                      0             ZERO ORDER
                                                    Δv = 0

                                     -1
                                                    MINUS FIRST
                                                    ORDER
                         VELOCITY = v                Δv = - v/d
```

Figure 14.4. The Doppler frequency shifts introduced by a moving grating.

grating about its center, can be used, but relatively low-frequency shifts are obtained. Similarly, a small frequency shift equal to $2v/\lambda$ is produced by the reflection from a moving mirror. Frequency shifts even larger than those of the Zeeman split laser can be produced by an acousto-optic (AO) Bragg cell (Massie and Nelson 1978; Wyant and Shagam 1978; Shagam 1983). Acoustic waves traveling through the cell create periodic index of refraction variations due to the sound pressure, and a moving diffraction grating is set up in the cell. Since the velocity divided by the period of the waves is equal to the acoustic frequency f, the frequency shift for this AO modulator is

$$\Delta \nu = mf. \tag{14.14}$$

Extremely high phase shift rates are achievable with this type of device; some commercially available instruments have operating frequencies of 20–30 MHz.

Frequency shifts can also be obtained through the use of rotating polarization phase retarders, but mechanical limitations generally limit operation to a few kilohertz, and this method is not widely used (Crane 1969; Okoomian 1969; Bryngdahl 1972; Sommargren 1975; Shagam and Wyant 1978; Hu 1983; Kothiyal and Delisle 1984, 1985a). An interesting and potentially significant recent advance is the direct modulation of the output wavelength of a laser diode to produce phase shifts. The operation and implications for this method are discussed in Section 14.16.7.

14.5. DETECTING THE WAVEFRONT PHASE

The major differences between the various PSI detection schemes are the manner in which the reference phase is varied and the number of times and the rate at which the interference pattern is measured. All of the variations we will

14.5. DETECTING THE WAVEFRONT PHASE

discuss aim to find the "best" way to collect and analyze the interferometric data so that Eq. (14.6) may be solved for the unknown wavefront phase. Of course, the best solution for a particular application will depend on a number of factors including computational complexity and speed, sensitivity to phase shift errors and noise, data rates, and compatibility with the detection scheme. To generate a fundamental understanding of the analysis process, we will start with a conceptionally and analytically simple algorithm.

The *four-step algorithm* requires that four separate interferograms of the part under test are recorded and digitized. A 90° optical phase shift is introduced into the reference beam between each of the sequentially recorded interferograms. Since these are now discrete measurements, the time dependence has been changed to the phase step index i. The function $\delta(t)$ now takes on four discrete values:

$$\delta_i = 0, \pi/2, \pi, 3\pi/2; \quad i = 1, 2, 3, 4. \tag{14.15}$$

Substituting each of these four values into Eq. (14.6) results in four equations describing the four measured interferogram intensity patterns:

$$I_1(x, y) = I'(x, y) + I''(x, y) \cos [\phi(x, y)], \tag{14.16}$$

$$I_2(x, y) = I'(x, y) + I''(x, y) \cos [\phi(x, y) + \pi/2], \tag{14.17}$$

$$I_3(x, y) = I'(x, y) + I''(x, y) \cos [\phi(x, y) + \pi], \tag{14.18}$$

and

$$I_4(x, y) = I'(x, y) + I''(x, y) \cos [\phi(x, y) + 3\pi/2]. \tag{14.19}$$

A simple trigonometric identity yields

$$I_1(x, y) = I'(x, y) + I''(x, y) \cos [\phi(x, y)], \tag{14.20}$$

$$I_2(x, y) = I'(x, y) - I''(x, y) \sin [\phi(x, y)], \tag{14.21}$$

$$I_3(x, y) = I'(x, y) - I''(x, y) \cos [\phi(x, y)], \tag{14.22}$$

and

$$I_4(x, y) = I'(x, y) + I''(x, y) \sin [\phi(x, y)]. \tag{14.23}$$

These four equations in three unknowns [$I'(x, y)$, $I''(x, y)$, and $\phi(x, y)$] can now be solved for the value of $\phi(x, y)$ at each point in the interferogram. As will be shown later, three interferograms are all that are required to solve for the wavefront phase, but the fourth is included for computational ease. The intensity bias term $I'(x, y)$ is eliminated by subtracting the equations in pairs:

$$I_4 - I_2 = 2I''(x, y) \sin [\phi(x, y)], \tag{14.24}$$

and

$$I_1 - I_3 = 2I''(x, y) \cos [\phi(x, y)]. \tag{14.25}$$

Taking the ratio of these two equations eliminates the intensity modulation term $I''(x, y)$ to produce a result that contains only the unknown phase $\phi(x, y)$ and the four measured intensities:

$$\frac{I_4 - I_2}{I_1 - I_3} = \frac{\sin [\phi(x, y)]}{\cos [\phi(x, y)]} = \tan [\phi(x, y)]. \tag{14.26}$$

This equation can now be rearranged to produce the result for the four-step PSI algorithm:

$$\phi(x, y) = \tan^{-1} \left[\frac{I_4 - I_2}{I_1 - I_3} \right]. \tag{14.27}$$

This simple equation is evaluated at each measurement point to obtain a map of the measured wavefront. To show the simplicity of the algorithm and to allow easier comparison with other algorithms, the (x, y) dependence of the four measurements is implied. This wavefront can be easily related to the surface height of the part [Eq. (14.1)] or the optical path difference (OPD):

$$\text{OPD}(x, y) = \lambda \phi(x, y)/2\pi. \tag{14.28}$$

Figure 14.5 shows the photographs of five interferograms recorded with a 90° phase shift. The fringes appear to walk across the frame, but more importantly, the intensity at any point varies with phase. Since there is a 360° phase shift between the first and fifth frames, they appear nominally identical.

At this point the differences between PSI and conventional interferometric analysis should be very apparent. PSI calculates the wavefront phase at every measurement location from the time-varying intensity measured at that point. The result at a point is the arctangent of the ratio of the intensities measured at that point, and there is no need to find the fringe centers or to order the fringes. In fact, the fringe pattern is somewhat irrelevant to PSI; there is no requirement on the minimum number of fringes or their shape, and the system precision is relatively independent of the fringe frequency in the interferogram. Since the data can be collected on a regular grid, PSI is well suited for further analysis and processing.

14.5. DETECTING THE WAVEFRONT PHASE

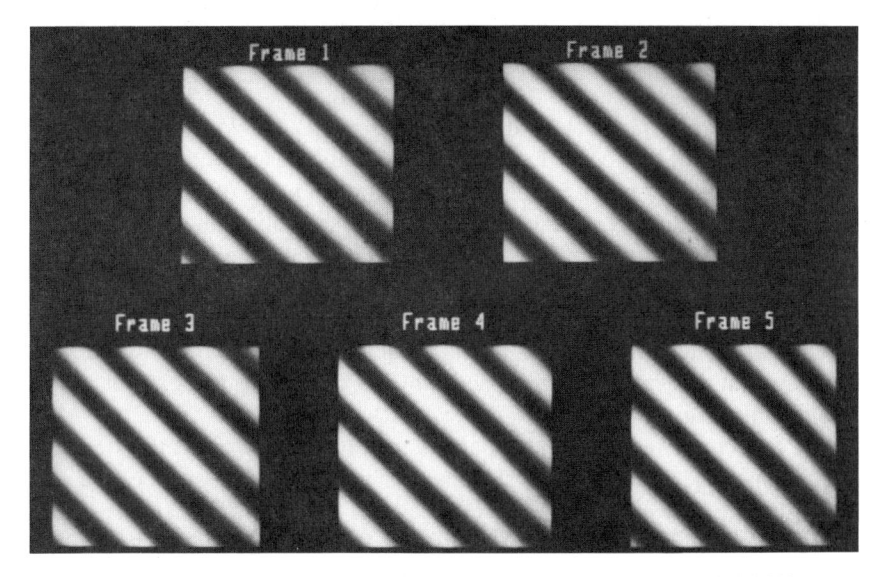

Figure 14.5. Photographs of five interferograms recorded with phase shifts of 90°.

If desired, the intensity data can also be evaluated to determine the data modulation $\gamma(x, y)$ across the interferogram:

$$\gamma(x, y) = \frac{I''(x, y)}{I'(x, y)}. \tag{14.29}$$

Starting with Eqs. (14.24) and (14.25), and also using Eqs. (14.20)–(14.23), we obtain

$$\gamma(x, y) = \frac{2[(I_4 - I_2)^2 + (I_1 - I_3)^2]^{1/2}}{I_1 + I_2 + I_3 + I_4}, \tag{14.30}$$

where the (x, y) dependence of the measurements is implied. The numerator is the intensity modulation, and the denominator is the average intensity or the intensity bias. Once again, all of the detector characteristics such as sensitivity and bias are implicitly included in this analysis as part of $I'(x, y)$ and $I''(x, y)$. This information is useful for evaluating the quality of the data that has been collected. A data modulation near one is good, and a low modulation is bad. Data points with modulations below some threshold will have insufficient signal to noise, and the wavefront phase cannot be reliably calculated at that point. These points are excluded from the analysis, and typical data modulation thresholds are in the range of 5-10%. This value is, of course, application dependent.

14.6. PHASE UNWRAPPING: INTRODUCTION

There is one more operation that must be performed to the calculated phase $\phi(x, y)$ before it is ready to be displayed and evaluated. We must correct for the discontinuities that occur in the phase calculation as a result of the arctangent. The arctangent is defined only over the limited range of angles, $-\pi/2$ to $\pi/2$. Regardless of the actual value of the phase, only values of the phase within this range result from Eq. (14.27), or its equivalent for other algorithms. This limitation would appear to restrict us to measuring OPDs to no more than half a wavelength. Fortunately, there is sufficient information in the calculation to remedy the situation and to provide a usable measurement range. In addition, we know that the wavefront or surface is in fact continuous and extends over a much larger range.

The first correction that can be made to the calculated phase is to extend the calculation range from 0 to 2π. This is possible because the signs of the sine and cosine are known independently of the sign of the tangent. Equations (14.24) and (14.25), for the four-step algorithm, are directly proportional to the sine and the cosine. Similar relationships exist for the other algorithms that are discussed later in this chapter. Table 14.1 gives the formulas needed to convert the arctangent results to values between 0 and 2π as a function of the values of the sine and the cosine, and this process is shown graphically in Fig. 14.6. The result of this correction is to produce the wavefront phase modulo 2π; every time the actual phase equals a multiple of 2π, the calculated value returns to a value of zero. For simplicity, this corrected phase will be referred to as the raw phase or the phase modulo 2π. Note that Table 14.1 is to be used only if the values returned by the arctangent lie between $-\pi/2$ and $\pi/2$.

The final step in the wavefront reconstruction process is to remove the 2π discontinuities that are present in the raw phase data that has been generated. This process is referred to as phase unwrapping, phase integration, or phase continuity, and it converts the modulo 2π phase data into a continuous representation of the wavefront under test. Whenever one of the large discontinuities

Table 14.1. Modulo 2π Phase Correction

Sine	Cosine	Corrected Phase $\phi(x, y)$	Phase Range
0	+	0	0
+	+	$\phi(x, y)$	0 to $\pi/2$
+	0	$\pi/2$	$\pi/2$
+	−	$\phi(x, y) + \pi$	$\pi/2$ to π
0	−	π	π
−	−	$\phi(x, y) + \pi$	π to $3\pi/2$
−	0	$3\pi/2$	$3\pi/2$
−	+	$\phi(x, y) + 2\pi$	$3\pi/2$ to 2π

14.7. INTEGRATING BUCKET DATA COLLECTION

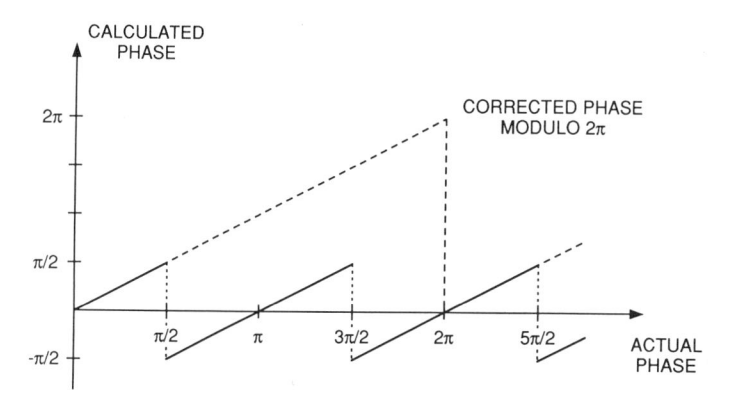

Figure 14.6. The conversion of the phase calculated by the arctangent to the wavefront phase modulo 2π.

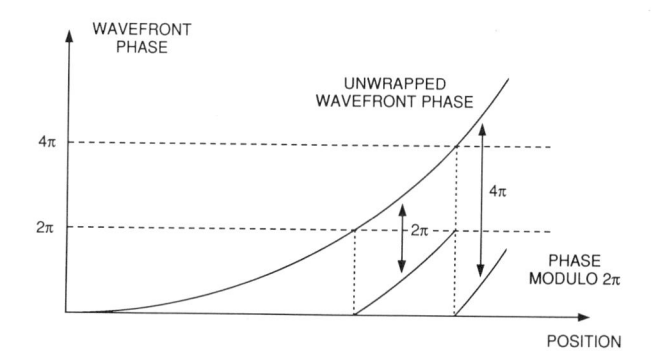

Figure 14.7. The phase unwrapping process in one dimension.

occurs in the reconstruction, 2π or multiples of 2π are added to the adjoining data to remove the discontinuity. This process is diagramed for one-dimensional data in Figure 14.7, and the effects of the phase unwrapping operation in two dimensions are shown in Fig. 14.8 for a spherical wavefront. A continuous surface results. A great deal more about phase unwrapping and its relationship to pixels and spatial sampling will be covered later in this chapter in Section 14.12.

14.7. INTEGRATING BUCKET DATA COLLECTION

In the four-step algorithm used as an introductory example in Section 14.5, the reference phase is stepped through a series of discrete values, and an interferogram is recorded at each step. For practical reasons, it is often desirable to

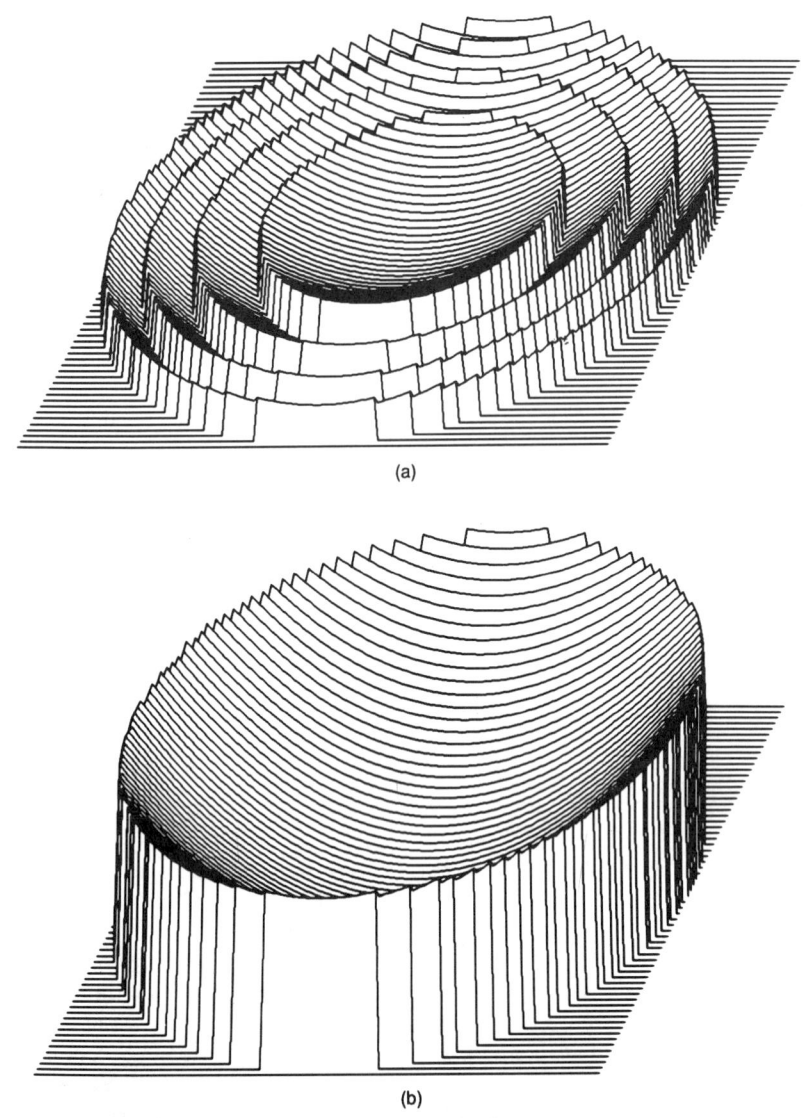

Figure 14.8. The phase unwrapping process for a spherical wavefront: (a) the wavefront phase modulo 2π and (b) the reconstructed wavefront. (Courtesy Zygo Corporation, Middlefield, CT.)

vary the phase smoothly as the interferograms are recorded. One reason to do this is that after a commanded phase step, the PZT transducer may tend to "ring," and the reference mirror position will oscillate until these transients die out (Seligson et al. 1984). This is a likely situation when the reference surface has a large mass or the control electronics are not well damped. It may

14.7. INTEGRATING BUCKET DATA COLLECTION

therefore be necessary to wait after each phase step before collecting the data, and the total data collection time is greatly increased.

The *integrating bucket* data collection scheme, which was first proposed by Wyant (1975), allows the reference phase to vary linearly with time while a series of interferograms is collected. The reference phase changes during the integration time for a single measurement. This integration time may be chosen to equal the frame rate of the sensor, or it could be the time required to collect adequate signal. For this analysis we first rewrite the general expression for an interferogram in Eq. (14.6) as a function of the reference phase δ instead of time:

$$I(x, y, \delta) = I'(x, y) + I''(x, y) \cos [\phi(x, y) + \delta]. \quad (14.31)$$

If the reference phase changes by an amount Δ during the integration or exposure time, the recorded interferogram is found by integrating the intensity over this interval (Greivenkamp 1984):

$$I_i(x, y) = \frac{1}{\Delta} \int_{\delta_i - \Delta/2}^{\delta_i + \Delta/2} \{I'(x, y) + I''(x, y) \cos [\phi(x, y) + \delta]\} \, d\delta, \quad (14.32)$$

where δ_i is the phase shift at the center of the integration period, and $I_i(x, y)$ is the corresponding recorded interferogram. The term $1/\Delta$ in this equation normalizes the results so that the average recorded signal is independent of Δ. Integration yields

$$I_i(x, y) = I'(x, y) + (1/\Delta)I''(x, y) \{\sin [\phi(x, y) + \delta_i + \Delta/2] \\ - \sin [\phi(x, y) + \delta_i - \Delta/2]\}, \quad (14.33)$$

which, after the use of a trigonometric identity, can be rewritten as

$$I_i(x, y) = I'(x, y) + I''(x, y) \operatorname{sinc} (\Delta/2\pi) \cos [\phi(x, y) + \delta_i], \quad (14.34)$$

where

$$\operatorname{sinc} (\beta) = \frac{\sin (\pi \beta)}{\pi \beta}. \quad (14.35)$$

Under this definition the sinc function has a value of one at zero and goes to zero whenever the argument is an integer. Comparing this result to the original expressions in Eqs. (14.31) or (14.6) indicates that the only change due to integrating the intensity while the phase is changing is to reduce the modulation of the recorded intensity at each point in the interferogram by the value of the sinc function. This expression is the most general of those we have discussed.

Note that the value of Δ must be the same for all frames recorded as part of a data set.

The same algorithms can be used for data collected by the phase step method or the integrating bucket method. In fact, we can think of phase stepping as an integrating bucket of zero width ($\Delta = 0$). In this case the integrating bucket result reduces to phase stepping result.

There are some signal-to-noise concerns with using the integrating bucket approach since the PSI analysis depends on having data with good modulation. Some data points may fall below the threshold required for proper analysis as a result of this reduction in modulation. The sinc function serves as a modulation transfer function (MTF) for the measurement. As Δ increases, the MTF decreases. For phase stepping ($\Delta = 0$), it is unity. Small integration periods have minimal effects; for example, the modulation for $\Delta = \pi/4$ (45°) and $\pi/2$ (90°) are 97 and 90%, respectively. The $\Delta = \pi/2$ situation corresponds to collecting four contiguous data frames while the reference phase changes by a full wave, and approximates the condition found in many available interferometers. Integrating the intensity over a full period ($\Delta = 2\pi$) of reference phase shift reduces the signal modulation to zero, and there is no useful information about the wavefront phase in the resulting data. To obtain good data modulation, we are usually restricted to integration periods between 0 and π for each frame. There is little or no advantage to the integrating bucket approach unless there are dynamic or control issues associated with the data collection. A reduction in effective modulation must be weighted against possible improvements in noise reduction by the additional integration time.

14.8. PSI ALGORITHMS

Algorithms using either the phase step or the integrating bucket methods are the most common for PSI. Some other variations of the PSI technique are discussed in Section 14.14. A number of different data collection strategies have been developed, of which the four-step algorithm is just one example. All of these algorithms share common characteristics: They require that a series of interferograms are recorded as the reference phase is varied. The wavefront phase modulo 2π is then calculated at each measurement point as the arctangent of a function of the interferogram intensities measured at that individual point. The final wavefront map is then obtained by unwrapping the phases to remove the 2π phase discontinuities. Differences between the various algorithms relate to the number of recorded interferograms, the phase shift between these interferograms, and the susceptibility of the algorithm to errors in the phase shift or environmental noise such as vibration and turbulence.

Since we have shown that the difference between the phase step and integrating bucket methods relates to the effective modulation of the recorded in-

14.8. PSI ALGORITHMS

tensities, all of the algorithms we will discuss are equally valid for both approaches. To simplify the notation, we will rewrite the general expression in Eq. (14.34) as

$$I_i(x, y) = I'(x, y) + I''(x, y, \Delta) \cos [\phi(x, y) + \delta_i], \quad (14.36)$$

where

$$I''(x, y, \Delta) = I''(x, y) \, \text{sinc} \, (\Delta/2\pi). \quad (14.37)$$

The dependence of the modulation on the integration period Δ is now implicit in $I''(x, y, \Delta)$. In this section we are mainly concerned with algorithm development. The response of these algorithms to various errors and noise sources is more closely examined in Section 14.10.

With these PSI algorithms the starting value of the reference phase is often chosen to produce a simpler mathematical expression for the measured wavefront phase. In practice, we do not know (or need to know) the absolute reference phase; what is important for the algorithms is the phase shift between measurements. We merely define the starting position of the reference mirror to be at the first required phase value, and go from there.

14.8.1. Three-Step Algorithms

Since there are three unknowns in Eq. (14.36), the minimum number of measurements of the interferogram intensity that are required to reconstruct the unknown wavefront phase is three. The general case can be solved using equal phase steps of size α (Gallagher and Herriott 1972; Creath 1988). In this case

$$\delta_i = -\alpha, 0, \alpha; \quad i = 1, 2, 3 \quad (14.38)$$

and

$$I_1(x, y) = I'(x, y) + I''(x, y, \Delta) \cos [\phi(x, y) - \alpha], \quad (14.39)$$

$$I_2(x, y) = I'(x, y) + I''(x, y, \Delta) \cos [\phi(x, y)], \quad (14.40)$$

and

$$I_3(x, y) = I'(x, y) + I''(x, y, \Delta) \cos [\phi(x, y) + \alpha]. \quad (14.41)$$

Using the trigonometric addition identities,

$$I_1(x, y) = I'(x, y) + I''(x, y, \Delta) \{\cos [\phi(x, y)] \cos (\alpha) + \sin [\phi(x, y)] \sin (\alpha)\}, \quad (14.42)$$

$$I_2(x, y) = I'(x, y) + I''(x, y, \Delta) \cos [\phi(x, y)], \quad (14.43)$$

and

$$I_3(x, y) = I'(x, y) + I''(x, y, \Delta) \{\cos [\phi(x, y)] \cos (\alpha) - \sin [\phi(x, y)] \sin (\alpha)\}. \quad (14.44)$$

These three equations can easily be solved for the unknown wavefront phase at each location:

$$\phi(x, y) = \tan^{-1} \left\{ \left[\frac{1 - \cos (\alpha)}{\sin (\alpha)} \right] \frac{I_1 - I_3}{2I_2 - I_1 - I_3} \right\}. \quad (14.45)$$

The signal modulation can also be derived by first solving Eqs. (14.42)–(14.44) for the unknowns $I'(x, y)$ and $I''(x, y, \Delta)$:

$$I'(x, y) = \frac{I_1 + I_3 - 2I_2 \cos (\alpha)}{2[1 - \cos (\alpha)]}, \quad (14.46)$$

and

$$I''(x, y, \Delta) = \frac{\{\{[1 - \cos (\alpha)](I_1 - I_3)\}^2 + [\sin (\alpha)(2I_2 - I_1 - I_3)]^2\}^{1/2}}{2\sin (\alpha)[1 - \cos (\alpha)]}.$$

$$(14.47)$$

The ratio of these two intensities gives the recorded signal modulation at each point in the data set:

$$\gamma(x, y) = \frac{I''(x, y, \Delta)}{I'(x, y)}$$

$$= \frac{\{\{[1 - \cos (\alpha)](I_1 - I_3)\}^2 + [\sin (\alpha)(2I_2 - I_1 - I_3)]^2\}^{1/2}}{[I_1 + I_3 - 2I_2 \cos (\alpha)] \sin (\alpha)}.$$

$$(14.48)$$

It is important to remember that the modulation that is measured here is that of the recorded signal, not the intensity incident on the detector, and it includes the MTF reduction due to using the integrating bucket method and other terms such as fixed pattern noise from the detector. This statement also holds for the modulation previously calculated for the four-step algorithm [Eq. (14.30)]. Calculation of the modulation at each point allows data points with insufficient modulation to be identified and excluded from the analysis.

Two phase step sizes that are commonly used with the three-step algorithm

14.8. PSI ALGORITHMS

are 90° and 120°. For these values of α, the solutions for the wavefront phase are for $\alpha = \pi/2$:

$$\phi(x, y) = \tan^{-1}\left(\frac{I_1 - I_3}{2I_2 - I_1 - I_3}\right), \quad (14.49)$$

where the data modulation is

$$\gamma(x, y) = \frac{[(I_1 - I_3)^2 + (2I_2 - I_1 - I_3)^2]^{1/2}}{I_1 + I_3}, \quad (14.50)$$

and for $\alpha = 2\pi/3$:

$$\phi(x, y) = \tan^{-1}\left(\sqrt{3}\,\frac{I_1 - I_3}{2I_2 - I_1 - I_3}\right), \quad (14.51)$$

and the recorded data modulation for this value can be determined by using Eq. (14.48).

An interesting variation on the three-step algorithm results from using 90° phase steps and a phase offset of 45° (Wyant et al. 1984a; Bhushan et al. 1985). The phase offset is included to take advantage of a trigonometric identity and for computational convenience. It is irrelevant for the operation of the interferometer since we are only interested in the relative phases of the steps. The three chosen values are

$$\delta_i = \pi/4, 3\pi/4, 5\pi/4; \quad i = 1, 2, 3, \quad (14.52)$$

and the resulting wavefront phase is

$$\phi(x, y) = \tan^{-1}\left(\frac{I_3 - I_2}{I_1 - I_2}\right). \quad (14.53)$$

The result for the data modulation is identical to the result we obtained in Eq. (14.50). This should not be surprising since the modulation of the data cannot depend on the starting value of the reference phase ($-\pi/2$ versus $\pi/4$). Similarly, the difference between analyzing the three intensity frames with Eq. (14.49) instead of Eq. (14.53) is a constant wavefront phase offset of $3\pi/4$. These later expressions are useful in that they are analytically less complex and can be used for any set of three measurements with $\pi/2$ phase shifts regardless of the starting phase. An unimportant phase offset or piston error will appear in the result of the calculation.

The three-step algorithm requires the minimum amount of data and is the simplest to use. However, as we will see later, this algorithm is also very sensitive to errors in the phase shift between frames.

14.8.2. Least-Squares Algorithms

As we have noted several times, the measured intensity at a given location in the interferogram varies as a sinusoidal function of the reference phase with a known period and three unknowns: the average signal, the data modulation, and the unknown reference phase. It should not be surprising that the wavefront phase can be determined through a least-squares fit of the measured intensities to a sinusoidal function.

The general least-squares solution for a series of N interferograms recorded at phase shifts δ_i was first discussed by Bruning et al. (1974) and more rigorously by Greivenkamp (1984). Equation (14.36) is first rewritten as

$$I_i(x, y) = I'(x, y) + I''(x, y, \Delta) \cos [\phi(x, y)] \cos (\delta_i)$$
$$- I''(x, y, \Delta) \sin [\phi(x, y)] \sin (\delta_i), \tag{14.54}$$

or

$$I_i(x, y) = a_0(x, y) + a_1(x, y) \cos (\delta_i) + a_2(x, y) \sin (\delta_i), \tag{14.55}$$

where

$$a_0(x, y) = I'(x, y), \tag{14.56}$$

$$a_1(x, y) = I''(x, y, \Delta) \cos [\phi(x, y)], \tag{14.57}$$

and

$$a_2(x, y) = -I''(x, y, \Delta) \sin [\phi(x, y)]. \tag{14.58}$$

Equation (14.55) is in the proper form for a generalized least-squares fit to the measured intensities $I_i(x, y)$ at each location. The squared difference between the measured intensities and the intensities predicted by Eq. (14.55) for a choice of the three unknowns or variables $a_0(x, y)$, $a_1(x, y)$, and $a_2(x, y)$ is

$$E^2 = \sum_{i=1}^{N} [I_i(x, y) - a_0(x, y) - a_1(x, y) \cos (\delta_i) - a_2(x, y) \sin (\delta_i)]^2. \tag{14.59}$$

This error is minimized by differentiating with respect to each of the three unknowns and equating these three results to zero. The simultaneous solution of

14.8. PSI ALGORITHMS

these three equations produces the least-squares result, and it is given by the following matrix equation:

$$\begin{bmatrix} a_0(x, y) \\ a_1(x, y) \\ a_2(x, y) \end{bmatrix} = \mathbf{A}^{-1}(\delta_i)\mathbf{B}(x, y, \delta_i), \quad (14.60)$$

where the component matrices are

$$\mathbf{A}(\delta_i) = \begin{bmatrix} N & \Sigma \cos(\delta_i) & \Sigma \sin(\delta_i) \\ \Sigma \cos(\delta_i) & \Sigma \cos^2(\delta_i) & \Sigma \cos(\delta_i)\sin(\delta_i) \\ \Sigma \sin(\delta_i) & \Sigma \cos(\delta_i)\sin(\delta_i) & \Sigma \sin^2(\delta_i) \end{bmatrix}, \quad (14.61)$$

and

$$\mathbf{B}(x, y, \delta_i) = \begin{bmatrix} \Sigma I_i \\ \Sigma I_i \cos(\delta_i) \\ \Sigma I_i \sin(\delta_i) \end{bmatrix}, \quad (14.62)$$

and all of the summations are from 1 to N. The matrix $\mathbf{A}(\delta_i)$ is a function only of the reference phase shifts and contains the information about the data collection scheme, and therefore it only needs to be calculated and inverted once. The matrix $\mathbf{B}(x, y, \delta_i)$ is composed of weighted sums of the measured interferogram intensities.

Once the values of the three unknowns $a_0(x, y)$, $a_1(x, y)$, and $a_2(x, y)$ are determined at each measurement location, the wavefront phase and data modulation can be easily found from Eqs. (14.56)–(14.58):

$$\phi(x, y) = \tan^{-1}\left(\frac{-a_2(x, y)}{a_1(x, y)}\right), \quad (14.63)$$

and

$$\gamma(x, y) = \frac{I''(x, y, \Delta)}{I'(x, y)} = \frac{[a_1(x, y)^2 + a_2(x, y)^2]^{1/2}}{a_0(x, y)}. \quad (14.64)$$

One feature of PSI that this general algorithm points out is that any combination of three or more reference phase values can be used to measure the wavefront. These values do not need to be evenly spaced and can be spread over a range greater than 2π. However, the choice of the phase shift positions does influence the system signal-to-noise performance and are specific to the $\mathbf{A}(\delta_i)$ matrix.

A specific algorithm that follows from the least-squares analysis is for N evenly spaced phase steps over one period of phase shift (Bruning et al. 1974; Morgan 1982; Greivenkamp 1984), or

$$\delta_i = i2\pi/N; \quad i = 1, \ldots, N. \quad (14.65)$$

In this special case all of the off-diagonal terms in the matrix $\mathbf{A}(\delta_i)$ are zero, and the simple solution for the wavefront phase and the recorded data modulation are

$$\phi(x, y) = \tan^{-1}\left(\frac{-\Sigma I_i \sin(\delta_i)}{\Sigma I_i \cos(\delta_i)}\right), \quad (14.66)$$

and

$$\gamma(x, y) = \frac{2\{[\Sigma I_i \cos(\delta_i)]^2 + [\Sigma I_i \sin(\delta_i)]^2\}^{1/2}}{\Sigma I_i}. \quad (14.67)$$

The similarity between these results and the four-step algorithm in Eqs. (14.27) and (14.30) should be apparent.

All of the three- and four-step algorithms presented in the previous sections of this chapter are derived without the use of the least-squares criterion; they are analytic solutions. However, these algorithms are exactly equivalent to the least-squares solutions and can be derived from Eqs. (14.60)–(14.64).

14.8.3. Carré Algorithm

In all the PSI algorithms that we have discussed so far, it has been assumed that the phase shift between measurements is known. As we will see later, if the actual phase shift differs from this assumed value, errors are introduced into the reconstruction. An error of this sort could result from a change in the slope of the PZT displacement versus voltage curve. All of the phase steps will be the same, but the step size will not be correct (e.g., 88° instead of 90°). Nonuniformities of the phase shift across the pupil that result from applying a linear phase shift in a converging or diverging beam (as discussed in Section 14.4) are another source of this type of error. Nonlinearities in the shape of the PZT displacement versus command curve must also be considered, and this error may result in a series of phase shifts with unequal values. Several PSI algorithms have been developed to minimize the effects of these phase shift variations.

The first of these algorithms is the Carré algorithm (Carré 1966). This algorithm is a variation of the four-step algorithm, but instead of requiring that the data be collected at 90° increments, the reference phase shift between measurements is treated as an unknown and solved for in the analysis. A linear

14.8. PSI ALGORITHMS

phase shift of 2α is assumed between each step, so that the values of the reference phase are

$$\delta_i = -3\alpha, -\alpha, \alpha, 3\alpha; \quad i = 1, 2, 3, 4. \tag{14.68}$$

As noted earlier, these four values are chosen to give simpler expression for the result. The four measured intensity frames are now represented by

$$I_1(x, y) = I'(x, y) + I''(x, y, \Delta) \cos [\phi(x, y) - 3\alpha], \tag{14.69}$$

$$I_2(x, y) = I'(x, y) + I''(x, y, \Delta) \cos [\phi(x, y) - \alpha], \tag{14.70}$$

$$I_3(x, y) = I'(x, y) + I''(x, y, \Delta) \cos [\phi(x, y) + \alpha], \tag{14.71}$$

and

$$I_4(x, y) = I'(x, y) + I''(x, y, \Delta) \cos [\phi(x, y) + 3\alpha], \tag{14.72}$$

This series of four equations contains four unknowns that can be solved for: the three we have been considering up to this point and the now unknown reference phase.

The solution for the reference phase shift can be found by expanding these four equations and applying the trigonometric identity for sine or cosine of 3α:

$$\alpha(x, y) = \tan^{-1} \left[\frac{3(I_2 - I_3) - (I_1 - I_4)}{(I_1 - I_4) + (I_2 - I_3)} \right]^{1/2}. \tag{14.73}$$

Since this equation can be solved at each measurement point, the reference phase shift $2\alpha(x, y)$ can also be determined at each point. This allows spatial variations of the phase shift to be identified. The solution for the wavefront phase at each measurement location is

$$\phi(x, y) = \tan^{-1} \left\{ \tan [\alpha(x, y)] \frac{(I_1 - I_4) + (I_2 - I_3)}{(I_2 + I_3) - (I_1 + I_4)} \right\}, \tag{14.74}$$

or combining these two results

$$\phi(x, y) = \tan^{-1} \left\{ \frac{\{[3(I_2 - I_3) - (I_1 - I_4)][(I_1 - I_4) + (I_2 - I_3)]\}^{1/2}}{(I_2 + I_3) - (I_1 + I_4)} \right\}, \tag{14.75}$$

where the (x, y) dependence of the measured intensities is implied. The value of this algorithm is that it compensates for errors in the amount of phase shift as well as for spatial variations of the phase shift. However, this algorithm requires that the phase shift increments at a given location be equal.

The conversion of the result of the arctangent in the Carré algorithm to the wavefront phase modulo 2π (see Section 14.6) is not as straightforward as for the other algorithms discussed in this chapter. With these other algorithms terms proportional to the sine and cosine of the wavefront phase $\phi(x, y)$ are contained in the numerator and denominator of the arctangent, and these values are used with Table 14.1 to correct the calculated phase. In the Carré algorithm [Eq. (14.75)], the square root in the numerator produces the absolute value of the sin $[\phi(x, y)]$, not the sine. In fact, since the denominator can be either positive or negative, the sign of the phase produced by this equation can be wrong. The conversion to the phase modulo 2π for the Carré algorithm should therefore be based on the absolute value of this calculation, and the appropriate entries in the "Corrected Phase" column of Table 14.1 are 0, $|\phi(x, y)|$, $\pi/2$, $\pi - |\phi(x, y)|$, π, $\pi + |\phi(x, y)|$, $3\pi/2$, and $2\pi - |\phi(x, y)|$. The entries that have changed form are those where the sine and cosine are of opposite sign. In addition, to use this revised table, terms proportional to the sine and cosine of the wavefront phase must also be constructed in order to determine the signs and select the appropriate table entries (Creath 1985). One such set of terms is

$$\sin [\phi(x, y)] \propto (I_2 - I_3)$$
$$= 2I''(x, y, \Delta) \sin [\alpha(x, y)] \sin [\phi(x, y)], \quad (14.76)$$

and

$$\cos [\phi(x, y)] \propto (I_2 + I_3) - (I_1 + I_4)$$
$$= 8I''(x, y, \Delta) \sin^2 [\alpha(x, y)] \cos [\alpha(x, y)] \cos [\phi(x, y)].$$

$$(14.77)$$

With these values the revised Table 14.1 can be used to calculate the wavefront phase modulo 2π for the Carré algorithm. It should also be noted that this revised correction table can be used with any of the other algorithms.

The most commonly used reference phase shift is 90° between measurements ($\alpha = 45°$), and the recorded data modulation can then be calculated with Eq. (14.30), the result for the four-step algorithm. (Once again, the modulation must be independent of the initial value of the phase shift). This expression can be used even in the presence of fairly large variations in the size of the phase step as only small errors in the calculated modulation will result. For a $\pm 10\%$ change in the phase shift, the maximum error in the modulation is approximately $\pm 5\%$. This is quite tolerable as the modulation is generally only used to sort data points with insufficient modulation from the analysis.

14.8.4. Averaging 3 + 3 Algorithm

As we will see in Section 14.10, a linear phase shift error results in a sinusoidal error in the reconstruction of the wavefront phase that has a frequency twice the interferogram fringe frequency. Two calculations of the wavefront phase with a 90° reference phase offset between them can reduce this error (Schwider et al. 1983; Wyant and Prettyjohns 1987). This 90° phase shift will move the fringe pattern by a quarter period, and the reconstruction error will also be shifted by this same physical distance between the two calculations. Since this error occurs at twice the fringe frequency, the periodic error between the two measurements will be offset by half its period. The errors in the two calculations will therefore tend to cancel when they are averaged together.

One implementation of this technique is to record four intensity measurements with a $\pi/2$ phase shift. The first three of these frames (I_1, I_2, and I_3) are analyzed using the three-step algorithm [Eq. (14.53)] to get the first measurement of the wavefront phase. The second three measurements (I_2, I_3, and I_4) are now analyzed with the same equation, substituting I_2 for I_1, I_3 for I_2, and I_4 for I_3, to get the second calculation of the wavefront phase. These two results are then averaged at each measurement location to obtain a result that is significantly less susceptible to phase shift errors that may exist in either of the two individual calculations.

14.8.5. Hariharan Algorithm

Another approach to producing a PSI algorithm that is insensitive to reference phase shift calibration errors is provided by Hariharan et al. (1987a). They use five measurements of the interferogram intensity and initially assume a linear phase shift of α between frames:

$$\delta_i = -2\alpha, -\alpha, 0, \alpha, 2\alpha; \quad i = 1, 2, 3, 4, 5. \tag{14.78}$$

Then,

$$I_1(x, y) = I'(x, y) + I''(x, y, \Delta) \cos [\phi(x, y) - 2\alpha], \tag{14.79}$$

$$I_2(x, y) = I'(x, y) + I''(x, y, \Delta) \cos [\phi(x, y) - \alpha], \tag{14.80}$$

$$I_3(x, y) = I'(x, y) + I''(x, y, \Delta) \cos [\phi(x, y)], \tag{14.81}$$

$$I_4(x, y) = I'(x, y) + I''(x, y, \Delta) \cos [\phi(x, y)], \tag{14.82}$$

and

$$I_5(x, y) = I'(x, y) + I''(x, y, \Delta) \cos [\phi(x, y) + 2\alpha]. \tag{14.83}$$

These five equations are expanded and combined to produce the intermediate result:

$$\frac{\tan [\phi(x, y)]}{2\sin (\alpha)} = \frac{I_2 - I_4}{2I_3 - I_5 - I_1}. \tag{14.84}$$

Since the choice of the phase shift α is open, we can choose it to minimize the variation of this expression to errors in the phase shift. This function is plotted in Fig. 14.9 for $\phi(x, y) = \pi/4$. Differentiating this equation with respect to α, we find

$$\frac{d}{d\alpha} \left\{ \frac{\tan [\phi(x, y)]}{2\sin (\alpha)} \right\} = \frac{-\cos (\alpha) \tan [\phi(x, y)]}{2\sin^2 (\alpha)}, \tag{14.85}$$

which goes to zero when $\alpha = \pi/2$. When using this value for the phase shift, Eq. (14.84) becomes insensitive to phase shift calibration errors and reduces to the final algorithm,

$$\phi(x, y) = \tan^{-1} \left[\frac{2(I_2 - I_4)}{2I_3 - I_5 - I_1} \right]. \tag{14.86}$$

The data modulation at each measurement point can also be determined:

$$\gamma(x, y) = \frac{3[4(I_4 - I_2)^2 + (I_1 + I_5 - 2I_3)^2]^{1/2}}{2(I_1 + I_2 + 2I_3 + I_4 + I_5)}. \tag{14.87}$$

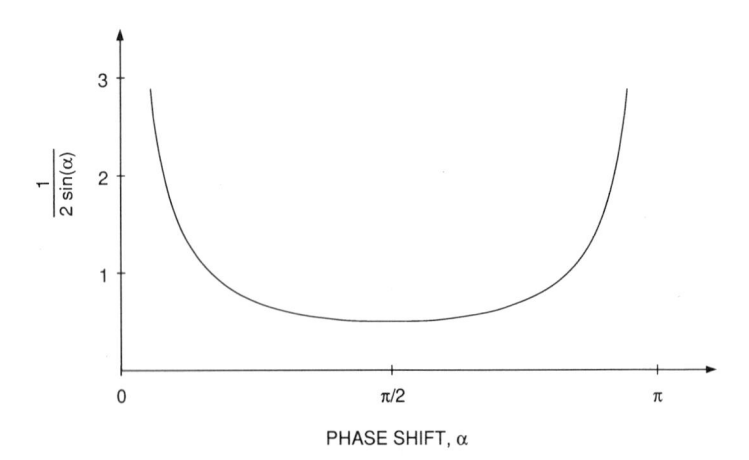

Figure 14.9. A plot of Eq. (14.84) as a function of the value of the phase shift.

14.8. PSI ALGORITHMS

A certain similarity should be noted between the form of these two results and the results for the four-step algorithm [Eqs. (14.27) and (14.30)], which also uses 90° phase shifts. Since I_1 and I_5 are nominally identical, the appearance of both of these terms is offset by an extra I_3, which is the term 180° out of phase.

The minimum in the plot of Eq. (14.84) is centered at 90° and is quite broad (Fig. 14.9). As a result, the Hariharan algorithm can tolerate fairly large errors in the phase shift before significant errors appear in the calculated wavefront phase (Hariharan et al. 1987a). If the actual phase shift between measurements is $\pi/2 + \epsilon$, then there is a corresponding measured phase $\phi'(x, y) = \phi(x, y) + \Delta\phi(x, y)$, which can be approximated using Eq. (14.84) and assuming that ϵ is small:

$$\tan[\phi'(x, y)] \cong [1 + (\epsilon^2/2)] \tan[\phi(x, y)]. \qquad (14.88)$$

The error in the measurement of the phase can now also be easily determined:

$$\Delta\phi(x, y) = \phi'(x, y) - \phi(x, y) \cong (\epsilon^2/4) \sin[2\phi(x, y)]. \qquad (14.89)$$

The error is a function of the wavefront phase, and the maximum error for a 2° change in the linear phase shift between measurements (88° instead of 90°) is about 0.02°. This same phase shift error for one of the standard algorithms such as the three-step algorithm results in a measurement error on the order of 1°.

14.8.6. 2 + 1 Algorithm

PSI algorithms require a series of intensity measurements to calculate the wavefront phase. Since a video camera is normally used to record the intensity patterns, the minimum data acquisition time is typically a few (three to five) video frames or a few tenths of a second. During this time, the optical setup must remain stable and air turbulence minimized. Failure to attain these conditions will result in measurement errors or even a data set that cannot be analyzed. In many situations, such as the testing of large optics, this requirement is difficult or impractical.

The 2 + 1 algorithm (Angel and Wizinowich 1988; Wizinowich 1989, 1990) attacks the problem of PSI testing in the presence of vibration by first rapidly collecting two time-critical interferograms with a 90° phase shift to monitor the intensity modulation, and later recording a third interferogram that gives the average intensity across the field. This third interferogram is the average of two interferograms with a 180° phase shift; the peaks of one set of fringes fall in the troughs of the other set, canceling out the fringe pattern. With this system the phase shifts are

$$\delta_i = 0, -\pi/2, 0 \text{ and } \pi; \quad i = 1, 2, 3, \qquad (14.90)$$

and the three recorded interferograms are

$$I_1(x, y) = I'(x, y) + I''(x, y) \cos [\phi(x, y)], \qquad (14.91)$$

$$I_2(x, y) = I'(x, y) + I''(x, y) \cos [\phi(x, y) - \pi/2],$$
$$= I'(x, y) + I''(x, y) \sin [\phi(x, y)], \qquad (14.92)$$

and

$$I_3(x, y) = \tfrac{1}{2} \{I'(x, y) + I''(x, y) \cos [\phi(x, y)]\}$$
$$+ \tfrac{1}{2} \{I'(x, y) + I''(x, y) \cos [\phi(x, y) + \pi]\}$$
$$= I'(x, y). \qquad (14.93)$$

These three equations can be easily solved for the wavefront phase and recorded data modulation at each measurement point:

$$\phi(x, y) = \tan^{-1} \left(\frac{I_2 - I_3}{I_1 - I_3} \right), \qquad (14.94)$$

and

$$\gamma(x, y) = \frac{[(I_2 - I_3)^2 + (I_1 - I_3)^2]^{1/2}}{I_3}. \qquad (14.95)$$

Perhaps more interesting than the algorithm itself is the hardware needed to implement the 2 + 1 algorithm. This system is described in detail by Wizinowich (1990), and two of the key design features are reviewed here. The first requirement is that two time critical intensities (I_1 and I_2) be recorded with as small a time lag between them as possible. A common solid-state sensor architecture is an interline transfer charge coupled device (see Section 14.11.1). Each photosite on the sensor is accompanied by an adjacent storage pixel. The storage pixels are read out to produce the video signal while the active photosites are integrating the light for the next video field. At the end of this integration period, the charge collected in the active pixels is quickly transferred to the now empty storage sites, and the next video field is collected. This transfer takes place in about a microsecond. It is possible with a synchronized shutter and this sensor to record two exposures of about a millisecond in duration separated by a microsecond; the first is recorded just before the transfer, and the second is recorded just after the transfer. These two recorded interferograms are read out at standard video rates and can be digitized.

14.8. PSI ALGORITHMS

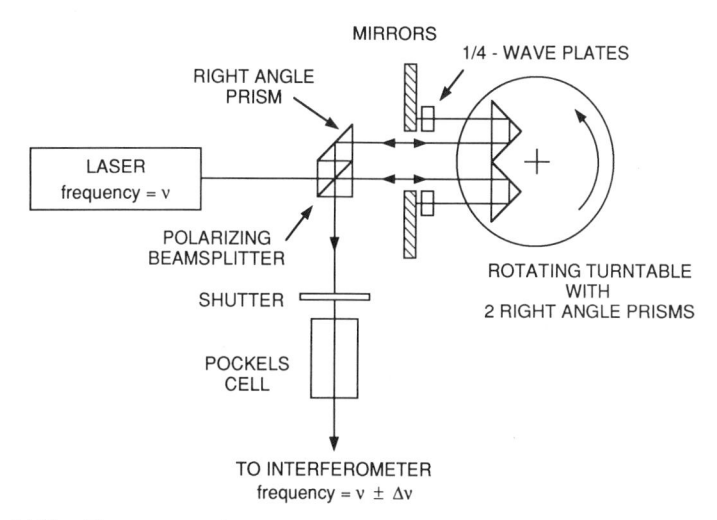

Figure 14.10. The system used to generate the phase shifts required for the 2 + 1 algorithm. (From Wizinowich 1990).

The second system requirement is that the appropriate phase shift is applied to each of these two exposures. The system that has been used is diagramed in Fig. 14.10. The rotating turntable with the two right-angle prisms introduces equal and opposite Doppler frequency shifts of $\pm \Delta \nu$ into the upper and lower beams. Because of the quarterwave plates, the recombined beams are also orthogonally polarized. The Pockels cell selects one or the other or both of the polarizations to enter the interferometer, and the shutter provides the appropriate exposure. All of these devices must be synchronized with the detector array. The interferometer used with the system is nonpolarizing so that any input polarization will enter both the reference and test arms of the interferometer [this is not the situation described in Section 14.4, Eq. (14.12)]. The individual interferograms are therefore produced by just one of the two different optical wavelengths or frequencies, and the relative phase shift α between interferograms recorded with the two frequencies is related to the OPD between the two arms of the interferometer:

$$\alpha = 2\pi \, \text{OPD} \, \Delta\nu/c, \qquad (14.96)$$

where c is the speed of light. This phase shift is equal to the difference in the number of optical cycles that fit into the OPD for the two optical frequencies. The frequency shift of the two beams $\Delta\nu$ is a function of the turntable rotation speed and its size. For path length differences of 10 m, a small frequency shift of about one part in 10^8 is needed. The turntable rotation rate is chosen to produce the desired phase shift.

The system operation is as follows. The phase shift is set for 90°, and the two time-dependent interferograms are recorded on each side of the interline transfer of the charge-coupled device (CCD). The Pockels cell is switched between these exposures, so that only one optical frequency is used for each. After these two interferograms are digitized and stored, the turntable speed is changed for a 180° phase shift. At this point the Pockels cell is set to allow both frequencies into the interferometer, and the third or average interferogram is recorded (both interferograms are imaged on the detector). This system has been shown to freeze out the effects of vibration and allow for PSI analysis.

Another approach to PSI testing in the presence of vibration is to speed up one of the standard algorithms through the use of a high-speed video camera (Greivenkamp 1987a). The speed of the phase shifter must also be increased, but this is not a problem particularly with the integrating bucket approach where stop–start motions are not required. The primary drawback to this system is that the electronic data rates are much higher and high-speed video memory is required. Another approach is to use one of the available diode array cameras that have moderate pixel densities and very high frame rates (10,000 elements at a 1-kHz frame rate). Here the pixel output rate remains roughly the same as standard video, and we are trading frame rate for pixel density. This is not a serious drawback unless a high pixel density is required to resolve a fringe pattern from a highly sloped wavefront. It should be noted, however, that speeding up the frame rate does not guarantee freedom from errors. This is discussed in more detail in Section 14.10.

14.8.7. Summary of Algorithms

In this section we have examined a number of different PSI algorithms. They all require that a series of interferograms are recorded as the reference phase is shifted, and the wavefront phase is then calculated at each measurement site as a function of the intensities measured at that site. The result of these calculations must go through the phase unwrapping process before the final wavefront map is produced. The algorithms are equally valid for either the phase step or integrating bucket data collection methods. In Section 14.10 we will examine characteristics of these algorithms with respect to error sources to aid in the selection of algorithms most appropriate for a particular application. The graphical comparison of many of these algorithms in Fig. 14.11 is obtained by plotting the average reference phase angles required for each of the recorded interferograms. For many applications, the five-step Hariharan algorithm [Eq. (14.86)] has been found to provide a good compromise between computational complexity and susceptibility to errors.

14.9. PHASE SHIFT CALIBRATION

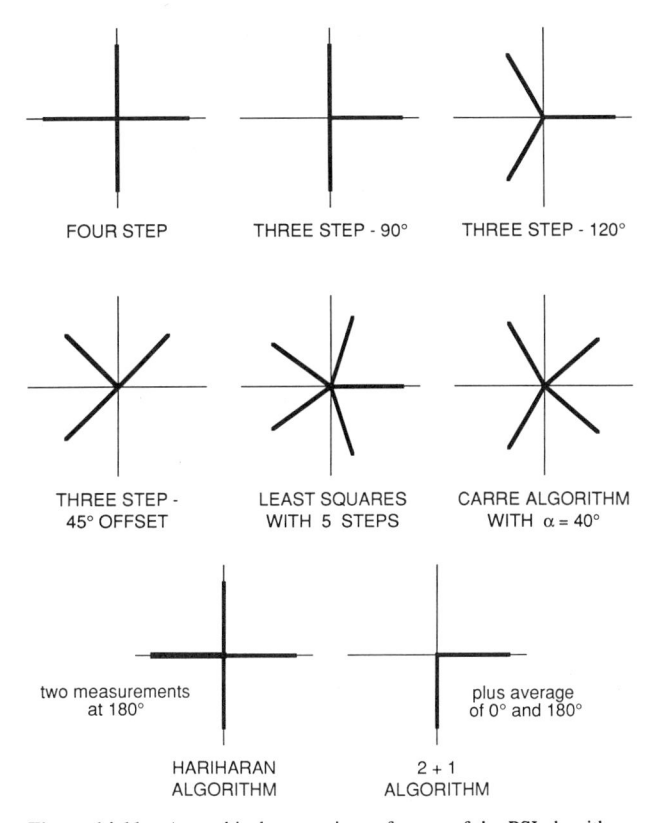

Figure 14.11. A graphical comparison of many of the PSI algorithms.

14.9. PHASE SHIFT CALIBRATION

One step in setting up a PSI system is to calibrate the phase shift α between recorded interferograms. Some of the algorithms are very sensitive to errors in the phase shift, and even the algorithms that are tolerant of phase errors require calibration for best performance. If the integrating bucket approach is used, the phase centroid of each integrating period must be calibrated.

Based on the previous material, the most obvious way to calibrate the phase shift is to use the solution for α from the Carré algorithm [Eq. (14.73), which gives half the step size]. A series of four interferograms are recorded with equal phase steps, and the phase shift can be calculated at each measurement point.

A simpler expression for the phase shift can be found by using five interferograms recorded with equal steps of α (Schwider et al. 1983; Cheng and Wyant 1985b). These are the same interferograms used for the Hariharan algorithm

[Eq. (14.79)–(14.83)], and the solution for the phase shift is

$$\alpha(x, y) = \cos^{-1}\left[\frac{1}{2}\left(\frac{I_5 - I_1}{I_4 - I_2}\right)\right]. \tag{14.97}$$

Note that the third interferogram of the series is not needed for this calculation, and the phase shift can be calculated at each point in the field.

Both of these analytic expressions for the phase shift have singularities for certain values of the wavefront phase $\phi(x, y)$. To avoid errors, a few tilt fringes are introduced into the interferogram, and data points that cause the denominators of the particular equation to fall below some threshold are eliminated from the analysis. A convenient way to display these calculated results is to show a histogram of the measured phase shifts (Creath 1988). A sample histogram for a 90° phase shift is shown in Fig. 14.12, and the controller of the phase shifter should be adjusted to center this curve on the desired value and to minimize its standard deviation. The method using five frames and Eq. (14.97) is the most commonly used calibration procedure. After data acquisition is complete for the particular algorithm, the data is analyzed statistically. The width and shape of the histogram provides an easy way of monitoring the amount of spatial variations of phase shift. The histogram in Fig. 14.12 was generated on a Fizeau interferometer with plano objectives. It is very symmetric and has a half width of 4°. The histogram in Fig. 14.13 was obtained on the same interferometer, but with an $f/0.75$ reference objective. In this case the width of the curve has increased to about 15°, and it is skewed to lower values of phase shift. The skew is due to the high F number rays that make a large angle with the optical axis. These rays see a phase shift that has been reduced by the cosine of this angle, and this situation points out the need for an algorithm that is insensitive to errors in the phase shift.

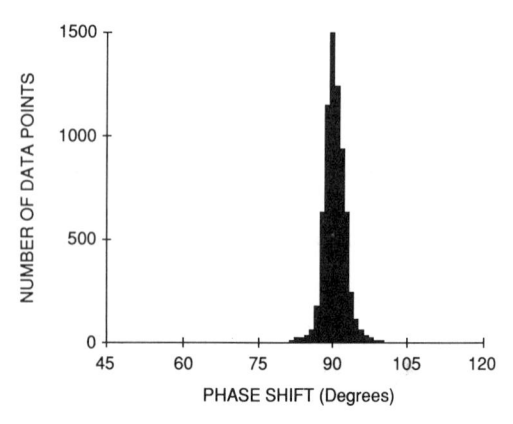

Figure 14.12. A histogram showing the distribution of the measured phase shifts.

14.9. PHASE SHIFT CALIBRATION

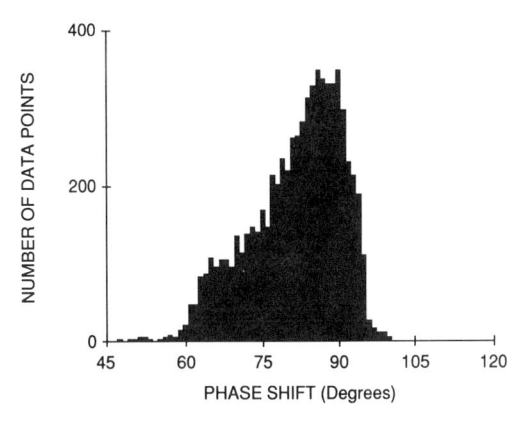

Figure 14.13. A histogram of the measured phase shifts obtained with a fast reference objective on a laser Fizeau interferometer.

A couple of simpler but less accurate methods for calibrating the phase shift involved the visual comparison of a series of interferograms. If five interferograms collected with a 90° phase shift are displayed [as described by Eqs. (14.79)–(14.83) with $\alpha = \pi/2$, and as shown in Fig. 14.5], the first and the fifth frames are 2π out of phase and should appear identical, and complementary intensity patterns should appear in the first and third, second and fourth, and third and fifth because of the π phase shift between each pair. The addition of a few tilt fringes makes the analysis easier. This intensity comparison can also be done by plotting a linear slice across each of these interferograms (Cheng and Wyant 1985b). For these five interferograms the curves from the first and the fifth should lie on top of each other. This is demonstrated in Fig. 14.14, where Fig. 14.14a shows a properly adjusted 90° phase shift and Fig. 14.14b shows the error that results for a 97° phase shift. These techniques can be expanded to other values of phase shift. If $N + 1$ interferograms are collected with a phase shift of $2\pi/N$ [the N-step algorithm in Eq. (14.66)], the first and the last of these interferograms should overlap. With either of these two techniques, the phase shift is calibrated by changing the gain on the PZT driver until the visual match is obtained. An accuracy of plus or minus a few degrees is possible.

Another method for calibrating the phase shift is to use a separate interferometer to monitor the position of the reference mirror as it translates (Hayes 1984). The detected intensity is then used to control the phase shift controller. This reference interferometer can also be a phase shift interferometer. An alternative solution is to use the measured phase positions with the generalized least-squares algorithm (Section 14.8.2). This results in an algorithm that adapts to the actual phase shifts for which the data were collected (Seligson et al. 1984, Lai and Yatagai 1991).

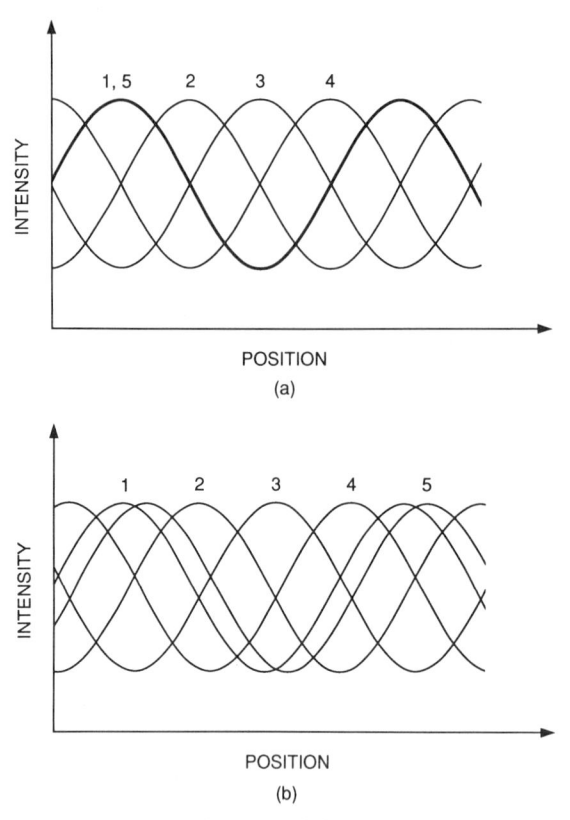

Figure 14.14. Intensity traces across five phase-shifted interferograms: (*a*) 90° phase shift and (*b*) 97° phase shift. (From Cheng and Wyant 1985b).

This type of reference monitoring system is very useful for detecting and correcting nonlinearities in the phase shift. The nonlinearity in even good PZTs can range from less than 1% to a few percent, and they often exhibit hysteresis. The effect of these nonlinearities will be examined in the next section. The other calibration procedures provide a linear correction for the phase shift, but do not correct for nonlinearities. By knowing the actual travel of the reference mirror as a function of drive signal or voltage, the controller signal that will produce a linear phase shift can be generated. In practice, this is usually done through a digital look-up table that is converted to the analog drive signal.

14.10. ERROR SOURCES

There are numerous sources of error that affect the accuracy of phase measurements as determined by the basic PSI algorithms. Some of the PSI algorithms

14.10. ERROR SOURCES

are more sensitive to a particular error source than others, while some sources of error are fundamental and affect the accuracy of all algorithms. Creath (1986) has conducted a thorough simulation of errors for most of the common algorithms. Error contributions specific to each of the algorithms have been conveniently tabulated in a recent reference as linear approximations due to the laser source, the phase shifter, the detector, and the environment (van Wingerden et al. 1991). The PSI algorithms produce a phase measurement that is a relative OPD measurement. Often what is desired is the absolute measurement of an optical surface or OPD of an optical system. There are many factors that get in the way of a direct measurement of the desired quantity. What follows is a discussion of how to identify and minimize the influence of various sources of error.

It is important to emphasize the distinction between precision, repeatability, and accuracy. A system that possesses very small *random* errors has high precision. A system that gives the same result for consecutive measurements has high repeatability. High repeatability, however, does not imply either high accuracy or high precision. A system with very small *systematic* errors is said to have high accuracy because random errors can be reduced by averaging. Generally we try to reduce both types of errors to a minimum, but in all cases the error contributions must be understood if improvements in overall accuracy are to be attained. Some of the largest sources of error in PSI are systematic, quite well understood but often difficult to eliminate. The use of high-performance microprocessors has contributed to the reduction of many of the component errors through the efficient and inexpensive manipulation of large quantities of data. More improvement can be expected.

Error sources generally fall into three categories: (1) those associated with the data acquisition process, (2) environmental effects such as vibration and air turbulence, and (3) those associated with defects in optical and mechanical design and fabrication. The data acquisition process includes errors in the phase shift process, nonlinearities in the detection system, amplitude and frequency stability of the source and quantization errors obtained in the analog-to-digital conversion process. We begin discussion of errors in the first category since these are the most controllable and tend to be quite algorithm dependent.

14.10.1. Phase Shift Errors

All PSI algorithms rely on shifting the phase in a known manner in an interferometer as part of the data acquisition process. The expected variation of intensity at any point in the interference pattern, under ideal conditions, is purely sinusoidal with a linear change in path length of the phase shifting arm [Eq. (14.6)]:

$$I(x, y, t) = I'(x, y) + I''(x, y) \cos [\phi(x, y) + \delta(t)]. \tag{14.98}$$

If the increment and total range of the phase shift $\delta(t)$ during data acquisition is not the predicted subdivision of 2π, then the measured intensity samples do not precisely map to a complete period of a sinusoidal signal. In other words a precisely sinusoidal signal is expected, and the acquired data points are fit to a sinusoidal signal in a least-squares or Fourier series sense, with that expectation. For example, if the reference phase shifter for an N-step algorithm does not shift each step by the expected $2\pi/N$ increment in OPD, but instead is off by some error ϵ_r at each step, the least-squares algorithm will try to fit a string of intensities representing a sinusoidal signal of period $2\pi + \epsilon_r'$ to a sinusoidal function of period 2π, where ϵ_r' represents a deviation from the full 2π period. A sinusoidal signal of period $2\pi + \epsilon_r'$ can be represented by a Fourier series of sinusoids at the fundamental frequency $\nu = 1/2\pi$, together with all higher harmonics, $2m\nu$, where m is an integer. There is a corresponding error that is imparted to the measured phase under these circumstances that shows up most severely in the first harmonic of the measured phase $\phi'(x, y)$. Schwider (1989) has shown that for a small error ϵ_r in the reference phase $\phi_r(x, y)$, there is a corresponding error in the measured phase $\phi'(x, y)$ from the true phase $\phi(x, y)$ of

$$\Delta\phi(x, y) = \phi'(x, y) - \phi(x, y). \tag{14.99}$$

A detailed calculation of the phase error is quite lengthy, but for the general case of an N-step least-squares algorithm, we have

$$\Delta\phi(x, y) \approx a + b \cos[2\phi(x, y)] + c \sin[2\phi(x, y)]. \tag{14.100}$$

The coefficients a, b, and c are summations that can be interpreted as Fourier coefficients for the spatial variation of the interference pattern at *twice* the spatial variation of the true phase $\phi(x, y)$. A higher-order approximation would show higher-order harmonics with correspondingly smaller coefficients.

The error in (14.100) can be shown to be minimized in a number of practical cases. The most notable of these is the Hariharan algorithm (Section 14.8.5) in which five samples are taken with a phase shift of $\pi/2$ between samples. In this case Eq. (14.100) reduces to (14.89) and is repeated here:

$$\Delta\phi(x, y) \approx (\epsilon_r^2/4) \sin[2\phi(x, y)]. \tag{14.101}$$

The Carré algorithm shows no measurement error for a linear error in the phase shift increment since this algorithm analytically determines the phase steps taken during the measurement. This algorithm was developed for long baseline interferometric distance measurements in a coordinate measurement tool (Carré 1966).

The Hariharan algorithm (Hariharan et al. 1987a) is quite insensitive to other

14.10. ERROR SOURCES

error types in addition to the linear calibration error in the phase shift step. Higher-order nonlinearities in the phase shifter are more difficult to predict analytically but quite simple to simulate numerically. Hariharan has shown that deviations in linearity of the phase shifter of 1% from the nominal $\pi/2$ phase shift steps cause a maximum error of less than 0.005° in the measured value of the phase ϕ. Hysteresis errors of the same magnitude give rise to comparable errors.

Linear drift in the phase shifter is minimized with this algorithm but, in fact, can be nearly eliminated by a simple additional operation. The effect of such drifts, if small, can nearly be eliminated by acquiring a second data set in the reverse order and averaging with the first. Hariharan et al. (1987a) have shown in this case that a drift of 4° over two sample periods corresponding to values of $-178°$, $-88.5°$, 1°, 90.5°, and 180° in the first set and 180°, 89.5°, $-1°$, $-91.5°$, and $-182°$ in the second set contributes an error of only 0.001° when the two sets are averaged. Drift in some other part of the interferometer is quite common in practice and may be indistinguishable to the algorithm from drift in the phase shifter since both can appear as uniform changes in OPD over the sample field.

In spite of the relative insensitivity of this algorithm to phase shift errors, it is important to calibrate the phase shift steps as carefully as possible. A simple calculation, which is convenient for the Hariharan algorithm, was proposed by Schwider et al. (1983) in which the phase step α is calculated from the following equation:

$$\cos(\alpha) = \frac{1}{2}\left[\frac{I_5 - I_1}{I_4 - I_2}\right]. \tag{14.102}$$

The use of this equation is restricted because of the singularity when $I_4 - I_2 = 0$ as discussed in Section 14.9. In-depth analyses of the effects of phase shift errors are covered by Kinnstaetter et al. (1988), Ohyama et al. (1988), and by Freischlad and Koliopoulos (1990).

In practice, it is important to choose the algorithm most insensitive to phase shift errors for the given application. This results from the fact that most of the observed phase errors are induced by changes in OPD in the interferometer cavity due to drifts, turbulence, and vibration, not incorrect movements in the reference mirror (i.e., phase shifter).

14.10.2. Detector Nonlinearities

Linearity of the detector represents an error term that could become important when the dynamic range or contrast of the fringes is very high. Correspondingly, when contrast is low, nonlinear detector effects are not significant. Kinnstaetter et al. (1988) have shown that with the four-step algorithm ($\alpha = \pi/2$), phase measurement is free of error not only for a linear detector but even when the detector has second-order nonlinearities. This is not the case for three-step algorithms. Fourth-order detector nonlinearities, however, contribute errors that excite the third harmonic in the spatial variation of the phase $\phi(x, y)$ for the four-step algorithm. More phase steps in the PSI algorithm are effective in reducing higher-order nonlinearities in the detector, but these errors are generally of no consequence. Modern silicon array sensors operate linearly over dynamic ranges of nearly 100:1 at room temperature. The range can be easily extended by at least another order of magnitude with thermoelectric or cryogenic cooling. This is not usually needed except in cases where light levels are very low or pixel integration time, by requirement, is very short.

Detector nonlinearity in CCDs is more likely to be encountered at high intensities where the diode array is operating at or near saturation. This is easily remedied by the insertion of an appropriate attenuator. The integrating CCD-type detector will have the greatest dynamic range if it can be operated just below saturation. This can be adjusted by the selection of the proper laser power, optical attenuation, and integration time (frame rate).

14.10.3. Source Stability

Stability of the interferometer source is important both in frequency as well as amplitude. Frequency instability, particularly in the case of a laser, can give rise to corresponding instabilities in the fringe pattern and reduce the accuracy of the measured wavefront. This is potentially more troublesome for interferometers that operate at large differences in path length in the two arms of the interferometer. The long coherence length of the He–Ne laser has caused many to neglect this factor, but for high accuracy requirements this needs to be considered. For a path length difference in the interferometer of ΔL and a desired measurement accuracy of k wavelengths, the frequency stability $\Delta \nu$ of the source satisfies the inequality

$$\Delta \nu < \frac{ck}{\Delta L}, \qquad (14.103)$$

where c is the velocity of light. This derives from the assumption that we should be able to determine the phase to within one wave if the coherence length of the laser matches the path length difference ΔL. Thus a path difference of 3 m requires a source frequency stability of 1 MHz for $< 0.01 \lambda$ accuracy. While large OPDs present a problem for sources with poor frequency stability, this suggests an alternate method for shifting the fringes if the source frequency or wavelength is an easily controllable parameter in the system. This has been exploited in the case of laser diodes and is covered in greater detail in Section 14.16.7.

14.10. ERROR SOURCES

Intensity fluctuations of the source, if random, can be mitigated by averaging since all points in the field vary together under most circumstances. The effect on the phase measurement of intensity fluctuations depends on the algorithm used. In the case of the N-step least-squares algorithm, it has been shown that the standard deviation in the measured phase σ_ϕ is given by (Bruning 1978; Koliopoulos 1981)

$$\sigma_\phi = \frac{1}{\sqrt{N}\, S}, \qquad (14.104)$$

where S is the signal-to-noise ratio in the detection system. This and other noise sources have been analyzed in great detail for most of the algorithms by Freischlad and Koliopoulos (1990) and Brophy (1990). Intensity fluctuations have the same general behavior as small amplitude vibrations, which are covered in Section 14.10.5. In the presence of systematic and slowly varying intensity fluctuations, a supplementary detector can be used to monitor the variations and either ratio the results at the detector or use the signal in feedback to stabilize the source. The latter is usually preferable.

The ultimate limitation to precision in two-beam interferometry is ascribed to photon shot noise at the detector. Tarbeyev (1986) estimates the precision limit at 6×10^{-4} Å.

14.10.4. Quantization Errors

PSI has become a practical technique because of inexpensive digital computing. The first step is the conversion of analog intensity signals to digital information. A source of error at this stage is the quantization error of the video signal. Since conversion is accomplished with an analog-to-digital converter, the accuracy of this conversion process depends upon the number of bits in the digital word transferred to the computer. Common video converters digitize the analog input signal into an 8- or 10-bit word, meaning that there are $2^8 = 256$ or $2^{10} = 1024$ discrete quantization levels in the digital word. The effect of quantization errors was first discussed by Koliopoulos (1981) for a three-step algorithm and later generalized by Brophy (1990) who rigorously derived the error for most of the common algorithms along with the special case of a 13-step algorithm with phase shift steps of 75°. The simulations are complicated by the fact that quantization error is not a statistical error. Brophy has shown that the standard deviation of the phase due to Q quantization levels, ϕ_4, for a 4-step algorithm ($\alpha = \pi/2$) is approximately

$$\sigma_{\phi_4} \approx \frac{1}{\sqrt{3}\, Q}, \qquad (14.104a)$$

and

$$\sigma_{\phi_N} \approx \frac{2}{\sqrt{3N}\,Q}, \qquad (14.104b)$$

can be inferred for an N-step algorithm. The 13-step algorithm shows about a factor of two improvement in error as a result of reduced correlation between steps. The expressions (14.104a-b) assume that the fringe modulation spans the full dynamic range of quantization levels. For 1024 quantization levels, the root mean square (RMS) phase error of the 4-step algorithm is <0.0001 waves. This is completely negligible under most circumstances, suggesting that an 8-bit converter is sufficient. In practice, when fringe modulation decreases, fewer bits of dynamic range are used, which has the effect of decreasing Q and thus increasing the error. The cost differential between 8- and 10-bit converters is generally less of an issue than the overhead in speed and the cost of moving to a longer word length in the digital section of the video interface electronics.

14.10.5. Vibration Errors

Mechanical stability is paramount in interferometry. The presence of vibration can be detrimental to the point of completely obliterating fringes as seen by the detector. Yet vibration, in some instances, cannot be eliminated. Special techniques have been developed to cope with these situations. Examples are stroboscopic illumination and holographic methods where the data acquisition system can freeze the vibrating fringes so that some of the usual analytical methods can be deployed to analyze the fringes. A condition of large amplitude vibrations could manifest itself by motions in two or three dimensions of the test structure or interferometer while other motions may exist at some other fundamental vibrational mode. This type of vibrational analysis will not be covered here but can be found in other references (Powell and Stetson 1965; Hariharan et al. 1987b; Hariharan 1984).

The usual precautions to minimize vibrations in interferometry are assumed, such as the use of passive or active vibration isolation mounts, acoustical isolation and damping, as well as generally good structural practice in the design of the interferometer and the table to which it is mounted. Location of the interferometer in a mechanically quiet area, far from rotating machinery, is also crucial.

The presence of a *small* amount of vibration usually manifests itself in a manner equivalent to perturbations of the phase shift increment and can thus be analyzed with familiar methods. Equation (14.100) indicates the general behavior of error in the measured phase with error in the phase shift increment. This behavior is characterized by error terms twice the spatial frequency of the fringe pattern. To first order, those algorithms that have the greatest immunity to errors in the phase shift increment have the greatest immunity to vibration and drifts of small amplitude. When the vibration frequency is high compared

14.10. ERROR SOURCES

to the pixel frame rate, the amplitude fluctuations tend to average out and there is little correlation of the vibrational component at each pixel location. As one might expect, the effect of the vibration-induced intensity fluctuations depends on the PSI algorithm used. The standard deviation in the phase introduced by small vibrations, averaged over all possible values of phase ϕ_r, is related to the standard deviation of the intensity fluctuation σ_I through the simple equation (Brophy 1990):

$$\sigma_{\phi_r} = \frac{\sqrt{k}\, \sigma_I}{\gamma I_0}. \tag{14.105}$$

Here γ is the fringe modulation, I_0 is the mean intensity, and k is a constant that depends on the PSI algorithm used. The parameter k decreases roughly in inverse proportion to the number of phase steps in the algorithm. For a four-step algorithm, $k = \frac{1}{2}$, and for an N-step algorithm, $k = 2/N$. The variance in phase (in radians) is always less than the variance in uncorrelated frame-to-frame intensity noise, whether that intensity noise is introduced by small vibrations or intensity variations of the source.

14.10.6. Air Turbulence

Air currents and air turbulence are an onerous source of error, particularly when high accuracy measurements are required. The character at any particular point in the fringe field tends to be a slowly varying phase error with reduced correlation to neighboring points at distances greater than a few millimeters to centimeters (Rosenbluth and Bobroff 1990). The same problems plagued the early holographers until they learned how to properly "mix" the ambient air. Proper mixing of the air is achieved by creating enough airflow to prevent stagnation or layering, while not so much as to generate excessive turbulence.

Air currents and turbulence are not unrelated to thermal gradients, which exist as a result of isolated heat sources and sinks. When a thermal gradient exists in any of the structural parts of the interferometer or the part under test, there are bound to be time-dependent changes in optical path over the wavefront aperture. Sufficient time must be allotted for thermalization of the part and the instrumentation. This can be hours when measurement accuracies at the $\lambda/100$ level are required. The particular measurement strategy, however, must be tailored to take into account the time constants of the potential error sources and the time frames of the data acquisition algorithms.

14.10.7 Extraneous Fringes and Other Coherent Effects

The long coherence length of the laser source allows interference with very large differences in path length in the test and reference arms of a Twyman-Green or Fizeau interferometer. While this may provide mechanical convenience in some cases, reflections from surfaces within the coherence length can interfere with one another and confuse the intended measurement. Extraneous fringes introduce systematic measurement errors that, in some cases, can be eliminated.

Consider first the case of three interfering beams. Let $e_{i\phi_t(x, y)}$ be the test arm field, $e^{i\phi_r(x, y)}$ the reference arm field, and $qe^{i\eta(x, y)}$ be an extraneous reflection field with fractional amplitude q of the primary interfering beams. When these three beams interfere, the resultant intensity is

$$I_r = 2 + q^2 + 2 \cos [\phi(x, y) - \phi_r(x, y)] + 2q \cos [\eta(x, y) - \phi_r(x, y)]$$
$$+ 2q \cos [\phi(x, y) - \eta(x, y)]. \qquad (14.106)$$

The true phase $\phi(x, y)$ is distorted to $\phi'(x, y)$ by the error $\Delta\phi(x, y)$, which can be shown to be (Schwider et al. 1983)

$$\Delta\phi(x, y) = \tan^{-1} \left\{ \frac{q \sin [\eta(x, y) - \phi(x, y)]}{1 + q \cos [\eta(x, y) - \phi(x, y)]} \right\}. \qquad (14.107)$$

The error $\Delta\phi(x, y)$ is seen to be dependent on the true phase $\phi(x, y)$. This reduces to the value reported by Bruning et al. (1974) when $\phi(x, y) = 0$. The incorporation of an additional phase shifter in the *test* arm has been proposed (Schwider et al. 1983). If the test arm phase is shifted by π and another data set is acquired and averaged with the first, a strong decrease in the amplitude of the disturbing wavefront results.

Ai and Wyant (1988) proposed a modification of this procedure that is easier to implement, particularly in the case of the four-step method. In their method four intensities are taken using the four-step bucket, and the individual intensities are stored. Following that, the test beam is blocked just prior to the test surface, and four new intensities are acquired and subtracted from the first set. The phase is calculated from the resultant intensities, which removes the effect of the spurious reflection without error. These analyses apply to cases where the reference and test surfaces have relatively low reflectivity and are matched. The situation becomes much more difficult if the test surface is highly reflective. Hariharan (1987) has shown that the three-step algorithm is much more sensitive to spurious reflections than the four-step algorithm.

Other forms of coherent noise, such as dust and scratches on optical surfaces, inhomogeneities and imperfections in the optical elements and coatings, can cause troublesome interference. A moving diffuser close to the source point can also average out some coherent noise. Scrupulous cleaning of the optics can further reduce the scattered light, improve contrast, and reduce artifacts. Judicious use of quarterwave plates in a polarization interferometer can minimize unwanted surface reflections and allow balancing of reference arm and test arm intensities (Bruning and Herriott 1970).

14.10.8. Interferometer Optical Errors

The overall optical correction of the interferometer impacts the accuracy of the measurements due to wavefront shear. This is a result of the fact that rays from an imperfect wavefront do not retrace themselves even when reflected from a perfectly spherical or flat surface. When the rays do not retrace themselves, they shear. The measurement error introduced by wavefront shear becomes greater with increased wavefront slope errors in the interferometer. Because of inevitable residual aberration, the best results are obtained with fringes that are nulled as completely as possible, by minimizing focus and tilt fringes.

It is difficult to give general formulas that quantify the effect of optical errors, but larger slopes give rise to greater error, in many instances, at quadratic or faster rates. The precise optical design of the interferometer must be ray traced to accurately characterize the effect. Any effect that introduces more slope for a given shear or more shear for a given slope will introduce greater error. Selberg (1987, 1990a) has calculated some of these errors for a Twyman–Green interferometer for several particular configurations. He introduced various amounts of tilt, power and spherical aberration in a simulated test surface and tabulated the calculated measurement error. Table 14.2 gives a summary of these results for 1 wave and 10 waves of peak-to-valley (P-V) error in tilt, plus and minus power errors (without tilt) and spherical aberration of the surface (without tilt and focus errors) for test parts of approximately 100 mm diameter. Note that for a spherical surface, tilt and focus are errors of adjustment whereas spherical aberration cannot be adjusted out. The simulation shows that these errors can seldom be combined analytically due to the nonlinearities. Specific situations must be simulated in their entirety, particularly when aberrations are large. Surface errors that return rays away from the axis introduce larger errors as indicated by the result for convex power error in Table 14.2. It is important that the reference wavefront be reasonably similar in radius to the wavefront under test. The error introduced by an imperfect reference or interferometer is dependent on both the aberrations and the disparity in radius of the test and reference wavefronts.

Optical errors resulting from design, fabrication or alignment imperfections are inevitable. To attain the highest accuracies, we must find ways of minimizing or compensating for these errors. If these errors can be characterized, in

Table 14.2. P-V Measurement Errors Due to Test Conditions

Setup for Test Part Error Type	P-V Ray Mapping Error in Waves for 1 Fringe Error	P-V Ray Mapping Error in Waves for 10 Fringe Error
Tilt	< .0001	.007
Concave power	< .0001	.007
Convex power	< .0001	.017
Spherical aberration	.0004	.089

one form or other, the interferometer can be calibrated. This important process depends on the interferometer configuration.

Geometric distortion in the interferometer imaging or viewing system will introduce errors that are proportional to the slope of the wavefront under test. This error can result from slope errors in the test surface or as a result of defocus. For example, a detector imaging system that has 5% distortion at the edge of field will introduce .01 waves of error for one wave of defocus (Truax, 1988). This emphasizes the need to null out the test fringes as much as possible prior to taking data.

Another source of wavefront measurement error occurs when the aperture or surface being tested is not imaged in sharp focus at the detector. This tends to create strong slope errors at the edge of the test aperture due to Fresnel diffraction.

14.11. DETECTORS AND SPATIAL SAMPLING

In most PSI systems in use today, solid-state detector arrays are used to collect the required intensity frames. They are chosen because the response at each pixel shows good linearity with intensity, no image lag between measured frames, and no geometrical distortion introduced in the recorded interference pattern by the sensor. The intensity measured at each discrete photosite can be digitized and stored in the computer memory for the PSI calculation. Tube-type sensors, such as vidicons, have been successfully used in PSI systems, but the system performance is usually degraded by detector nonlinearities, image lag and distortion.

14.11.1. Solid-State Sensors

Solid-state sensors can be classified by their geometry as either area arrays or linear arrays. Area arrays are the most commonly used for the interferometry and permit the measurement of a two-dimensional section of a surface or wavefront. The system spatial resolution is related to the number of pixels along each dimension of the sensor, and commonly available sensors have resolutions appropriate for television applications; typically about 500 by 500 pixels or less. Some newer sensors designed for machine vision applications or high-definition television have dimensions of about 1000 by 1000 or even 2000 by 2000 pixels, but these sensors are expensive and are difficult to use because of slower frame rates and the amount of computer memory and processing required for this amount of data (Creath 1989). Linear arrays, on the other hand, measure only a one-dimensional trace across the part, but make up for this disadvantage by providing a higher spatial resolution along this line. Linear sensors containing over 7000 pixels are available today, so that measurements with extremely high

14.11. DETECTORS AND SPATIAL SAMPLING

spatial resolution can be obtained. The amount of data from these large linear arrays is small and easily handled when compared to the 100,000 or more pixels on even a low-resolution area array. Because of advances in semiconductor fabrication, we can expect to see the dimensions of available sensors continuing to grow.

There are a variety of different configurations and technologies used for the manufacture of solid-state sensors including frame transfer and interline transfer charge-coupled devices (CCDs), charge injection devices (CIDs), and photodiode arrays. It is beyond the scope of this chapter to fully discuss this technology (for a review, see Hall 1980), but any of these sensors can be used (and have been used) for PSI. For practical considerations the major differences between cameras with these various solid-state sensors are resolution, sensitivity to light, dynamic range, and data output format (Prettyjohns 1984). The resolution is related to the number of pixels on the sensor, and the sensitivity is a function of both the quantum efficiency and geometry (the percentage of the imaging area that is occupied by photosensitive pixels as opposed to storage and transfer registers and control lines). There are also differences in camera performance as measured by the signal-to-noise ratio of the output. Because of all the complexities involved in designing a camera, no single sensor technology or configuration has emerged as best for this application.

For area arrays there are two primary formats for data output: interlaced and line sequential. Sensors designed for television applications produce a video signal where each video frame is composed of two interlaced video fields. The odd-numbered lines of the frame are output during the first field, and even-numbered lines occur in the second field. The video standard requires that there be 242 active lines per field or 484 active lines per frame, and the frame and field rates are fixed at about $\frac{1}{30}$ s and $\frac{1}{60}$ s (U.S. standards). This interlaced signal can be a problem for PSI systems as the two fields are integrated at different times and at different locations. While the use of a series of video frames to record the measured interferograms is possible with phase step data collection, it will not work with the integrating bucket strategy since the average phase shift δ_i will be different for the two fields in each frame. A better approach for either of the data collection approaches is to use only the odd or even fields of the video signal to record the series of interferograms and to ignore the other field. This results in 242 usable lines of data. Some video cameras provide options where only one of the two fields is output in a repeat field mode. Cameras using the line sequential output, such as those produced by EG&G Reticon and others, do not have an interlaced output, and every frame can be used. These cameras usually consist of a square array of pixels (128 by 128 and 256 by 256 are common) and can have an adjustable frame rate. A square array is more appropriate than the 4:3 television format since, for most applications, it is important to retain the geometric integrity of the interferogram and to have equal resolution in x and y.

14.11.2. Spatial Sampling

Regardless of the type of sensor used in the interferometer, there are certain aspects of the data collection that are common and must be considered for their impact on the measured intensity data. All of these sensors consist of an array of pixels with a given active area and spacing. The interferogram is imaged onto the sensor, and the intensity pattern is averaged not only over the integration time of the sensor, but also spatially over this active area. With the integrating bucket approach, the reference phase changes during this integration. As the sensor is read out, the analog voltage in the video signal corresponding to each pixel is digitized and stored in the computer memory. One number is assigned for each pixel. This process is repeated as the phase shift is varied to collect the required number of frames. For a PSI system built around a tube-type sensor without defined pixels, this same digitization occurs, and the effective pixel area is defined by the integration period of the analog-to-digital converter in the horizontal direction and the width of the scan lines in the vertical direction. The digitization process spatially samples the interferogram, and the effects of this spatial sampling on the PSI system will be studied in the remainder of this section and in Section 14.13. In a real sense the operations of PSI sample the interference pattern in three dimensions: time and the two spatial dimensions.

We will model the sensor as a rectangular array of rectangular pixels; see Fig. 14.15. The pixels have dimensions and spacing of a by b and x_s by y_s, respectively. The signal measured at a specific pixel (mx_s, ny_s), m and n are integers, is found by integrating the time-averaged interferogram intensity $I_i(x, y)$ from Eq. (14.36) over the active area of the pixel:

$$\hat{I}_i(mx_s, ny_s) = \int \int_{-\infty}^{\infty} I_i(x, y) \, \text{rect}\left(\frac{\alpha - mx_s}{a}, \frac{\beta - ny_s}{b}\right) d\alpha \, d\beta, \quad (14.108)$$

where this result is for the ith intensity frame, the rect function represents the active area of the pixel, and α and β are variables of integration. The total

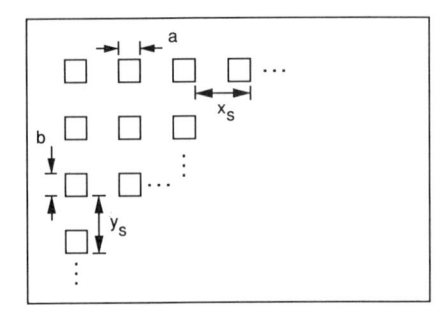

Figure 14.15. Sensor geometry.

14.11. DETECTORS AND SPATIAL SAMPLING

sampled interferogram that is produced by this sensor and stored in memory is given by assigning this result to a specific memory location and summing over all of the pixels:

$$I_i^s(x, y) = \sum_m \sum_n \hat{I}_i(mx_s, ny_s) \, \delta(x - mx_s, y - ny_s), \quad (14.109)$$

where $\delta(x, y)$ is a two-dimensional Kronecker delta function. With some manipulation, Eqs. (14.108) and (14.109) can be combined into the following expression for the ith sampled interferogram:

$$I_i^s(x, y) = \left[I_i(x, y) \ast\ast \text{rect}\left(\frac{x}{a}, \frac{y}{b}\right) \right] \text{comb}\left(\frac{x}{x_s}, \frac{y}{y_s}\right), \quad (14.110)$$

where ** indicates a two-dimensional convolution, and the comb function is an array of δ functions with the same spacings as the pixels. Note that for this discussion all dimensions are measured in the plane of the sensor, so that the incident intensity must be appropriately scaled. Also, constants of proportionality have been dropped. In interpreting this expression, the convolution with the rect function provides for the averaging of the intensity over the pixel, and the comb function generates one value of this average for each pixel.

Since the interferogram intensity pattern is averaged over the pixel active area, it is reasonable to conclude that this process will result in a reduction of the modulation of the digitized interference pattern. This effect is completely analogous to the reduction in modulation that occurs with the temporal averaging occurring in the integrating bucket data collection scheme; see Eq. (14.34). With point detectors there is no loss of modulation because there is no averaging, but as the active areas get larger, the recorded modulation decreases. It should also be clear that the modulation reduction is a function of the spatial frequency of the fringes being recorded. For a given size pixel more loss will occur for high frequencies than for low frequencies; the higher the frequency, the greater the variation in intensity that occurs within the pixel. In fact, when a full fringe period exactly fits within the active area of a pixel (or the fringe frequency equals $1/a$) the recorded modulation is zero; the spatial average over the active area results in a constant that is independent of the position of the fringe relative to the pixel. This situation is exactly the same as the temporal condition of $\Delta = 2\pi$ for the integrating bucket.

This reduction in modulation is a function of the spatial frequency of the fringes. It is useful to examine spatial averaging in the frequency domain. The frequency–space representation of the sampled interferogram is obtained through a Fourier transform of Eq. (14.110):

$$\tilde{I}_i^s(\xi, \eta) = [\tilde{I}_i(\xi, \eta) \, \text{sinc}(a\xi, b\eta)] \ast\ast \text{comb}(x_s\xi, y_s\eta), \quad (14.111)$$

where ξ and η are the spatial frequency coordinates, the $\tilde{}$ indicates the Fourier transform, and the sinc function is the two-dimensional equivalent of the function defined in Eq. (14.35):

$$\text{sinc}(\alpha, \beta) = \frac{\sin(\pi\alpha)}{\pi\alpha} \frac{\sin(\pi\beta)}{\pi\beta}. \tag{14.112}$$

Equation (14.111) states that the sampled spectrum is formed by first multiplying the spectrum of the time-averaged interferogram by the sinc function corresponding to the pixels. This filtered spectrum is then replicated at each multiple of the two-dimensional sampling frequency $(1/x_s, 1/y_s)$. Since the average over the pixel represented by the sinc function serves to low-pass filter the interferogram intensity incident on the sensor, a convenient name for this term is the pixel MTF:

$$\text{Pixel MTF} = \text{sinc}(a\xi, b\eta). \tag{14.113}$$

The reduction in recorded signal modulation due to the nonzero size of the pixels can be found by evaluating this function at the fringe frequency. As mentioned earlier, the first zero of this function occurs at a spatial frequency equal to $1/a$ (along one dimension and scaled to the plane of the sensor). The pixel MTFs for sensors with pixels with width to pitch ratios of 50% and 100% are shown in Fig. 14.16. If nonrectangular pixels are used, these expressions can be modified, and functions other than a rect and a sinc will appear.

The other major consequence of a sampled imaging system is its limitation on resolution. The limiting resolution of a sampled system is the Nyquist frequency, which is defined to be half the sampling frequency:

$$f_N = \text{Nyquist frequency} \equiv \frac{1}{2x_s}, \tag{14.114}$$

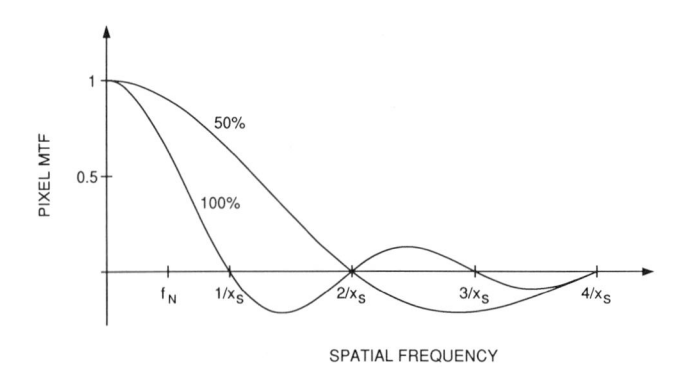

Figure 14.16. The pixel MTFs for sensors with width-to-pitch ratios of 50 and 100%.

where this is measured along the x axis (see, e.g., Gaskill 1978). The vertical Nyquist frequency is often different and is dependent on the vertical pixel pitch. The reasoning behind this limit is that to measure a fringe of a particular frequency or period, at least two samples per period must be measured; we must be capable of measuring the high and low points on a fringe. If the fringe frequency exceeds the Nyquist frequency, aliasing appears in the recorded image of the fringes, and this sampled interferogram is not interpretable by standard PSI techniques. Aliasing, and techniques to get around this limitation, are discussed in Section 14.13. To avoid these problems, it is usually recommended that about four pixels per fringe be used.

14.12. PHASE UNWRAPPING

In Section 14.6 we began our discussion of phase unwrapping, and we will continue it here in the context of spatial sampling. The intensity of an interferogram is a periodic function of the wavefront phase. The arctangent in any of the PSI equations interprets the intensities and returns a value of phase between $-\pi/2$ and $\pi/2$ at each pixel. These values can be easily corrected (Table 14.1) to produce the wavefront phase modulo 2π. The 2π discontinuities in these numbers must be corrected to obtain a usable result.

When the recorded interferograms satisfy the Nyquist criteria (at least two pixels per fringe period), the phase unwrapping process is straightforward. Having at least two pixels per fringe implies that the wavefront phase changes by no more than π per pixel spacing. This criteria is used to reconstruct the wavefront. Starting at some location in the wavefront, normally near the center of the interferogram, we require that the phase between any two adjacent pixels does not change by more than π. If the phase difference calculated for two pixels exceeds π, then 2π or multiples of 2π must be added to or subtracted from the calculated value of the second pixel until this condition is met. The entire wavefront map is calculated by working outward from the starting location.

This phase unwrapping process in one dimension is graphically represented in Fig. 14.17 (Greivenkamp 1987b). Figure 14.17a shows the wavefront phase modulo 2π as calculated by the arctangent at each pixel. The true wavefront phase is found by adding the correct number of 2π's to each of these values, and the correct phase is one of these solutions at each pixel; all of these possible solutions are plotted in Fig. 14.17b. The wavefront reconstruction process can be thought of as a connect-the-dots puzzle where there are many extra dots. The process moves from pixel to pixel and selects the dot that is within π of the preceding dot. Since the dots for each pixel are separated by 2π, this is equivalent to selecting the closest dot Fig. 14.17c, and the wavefront is produced.

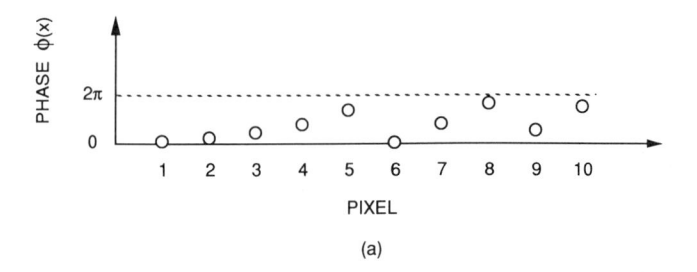

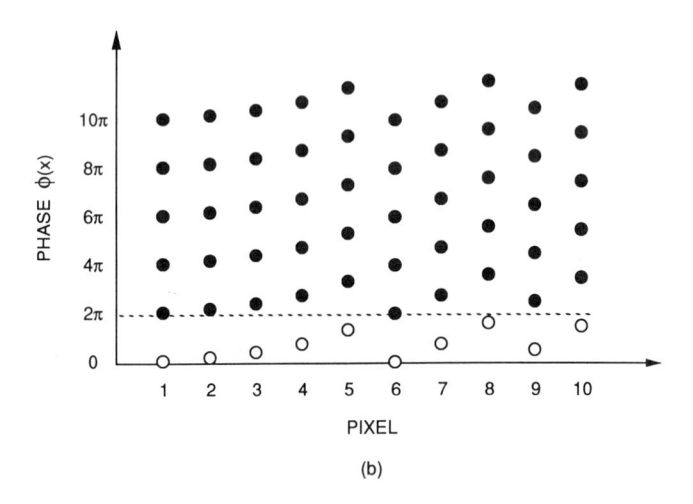

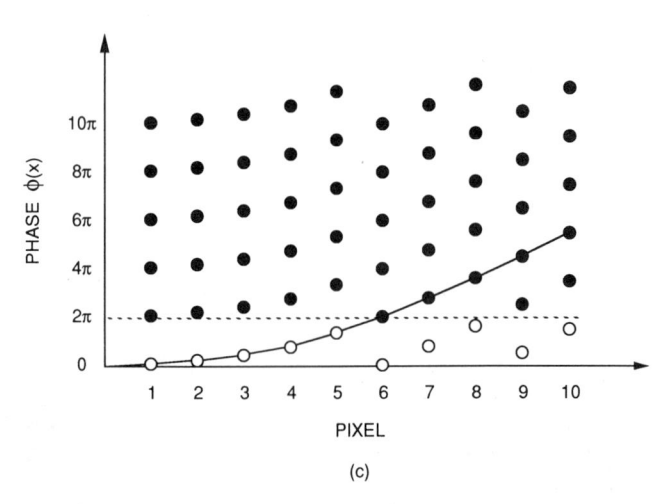

Figure 14.17. The PSI phase unwrapping process: (a) the wavefront data modulo 2π at each pixel, (b) all of the possible solutions for the wavefront phase; and (c) the reconstructed wavefront.

There are many ways to apply this inherently one-dimensional reconstruction process to a two-dimensional wavefront map. The simplest is to start at a single pixel and apply a one-dimensional reconstruction to generate a horizontal line across the part. Each point on this line is then used as the starting pixel for a one-dimensional vertical reconstruction to fill out the two-dimensional data array. Provisions must be made within the reconstruction program to jump over bad pixels with insufficient modulation that have been eliminated from the analysis or to work around obscurations that occur in the wavefront.

14.13. ASPHERES AND EXTENDED RANGE PSI TECHNIQUES

From a testing point of view, the prime characteristic of an aspheric wavefront or surface is that it has a large departure from a best-fit reference sphere, and an interferogram made without some sort of aspheric null contains many, many fringes. The Nyquist condition of having at least two pixels per fringe provides a limit to the amount of asphericity that can be measured with a PSI system. The maximum wavefront slope is limited to π per pixel. The fringe frequency is proportional to the wavefront slope, and even weakly aspheric surfaces will violate this condition when tested against a spherical reference surface with instrumentation available today. Typically, these systems are limited to testing surfaces with no more than 10-20 waves of asphericity. The exact number is not possible to predict since the maximum fringe frequency is related to slope, not peak excursion.

There are currently three approaches available to test aspherics with PSI: (1) Use a null optic (refractive, reflective, or holographic) to reduce the amount of asphericity in the wavefront; (2) use a larger detector array so that the sensor Nyquist frequency is increased; or (3) use a longer wavelength (such as 10.6 μm from a CO_2 laser) to rescale the wavefront departure with this longer wavelength. None of these is entirely satisfying; each involves a trade-off that places long lead times on the design of the test, requires additional fabrication, increases the difficulty of using and calibrating the instrument, decreases its precision, or greatly increases the instrument cost. In order to test aspheric surfaces without resorting to one of these three approaches, a PSI system must be capable of handling fringes that occur above the Nyquist frequency and are therefore aliased. Two such techniques that have been proposed and demonstrated are sub-Nyquist interferometry and two-wavelength PSI. A third technique is to use PSI to test subapertures of the wavefront and this approach is discussed in Section 14.16.5.

14.13.1. Aliasing

Before describing either of these extended range techniques, it is useful to discuss aliasing in more detail. Aliasing is the property of a sampled imaging system that causes high-frequency content in the input to be displayed at lower spatial frequencies. Fringes at frequencies above the Nyquist frequency appear in the display of the sampled image at frequencies less than the Nyquist frequency. This property can be graphically demonstrated in Fig. 14.18, where two different spatial frequency inputs are sampled; the vertical lines indicate the location of the pixels, small pixels are assumed, and the dots represent

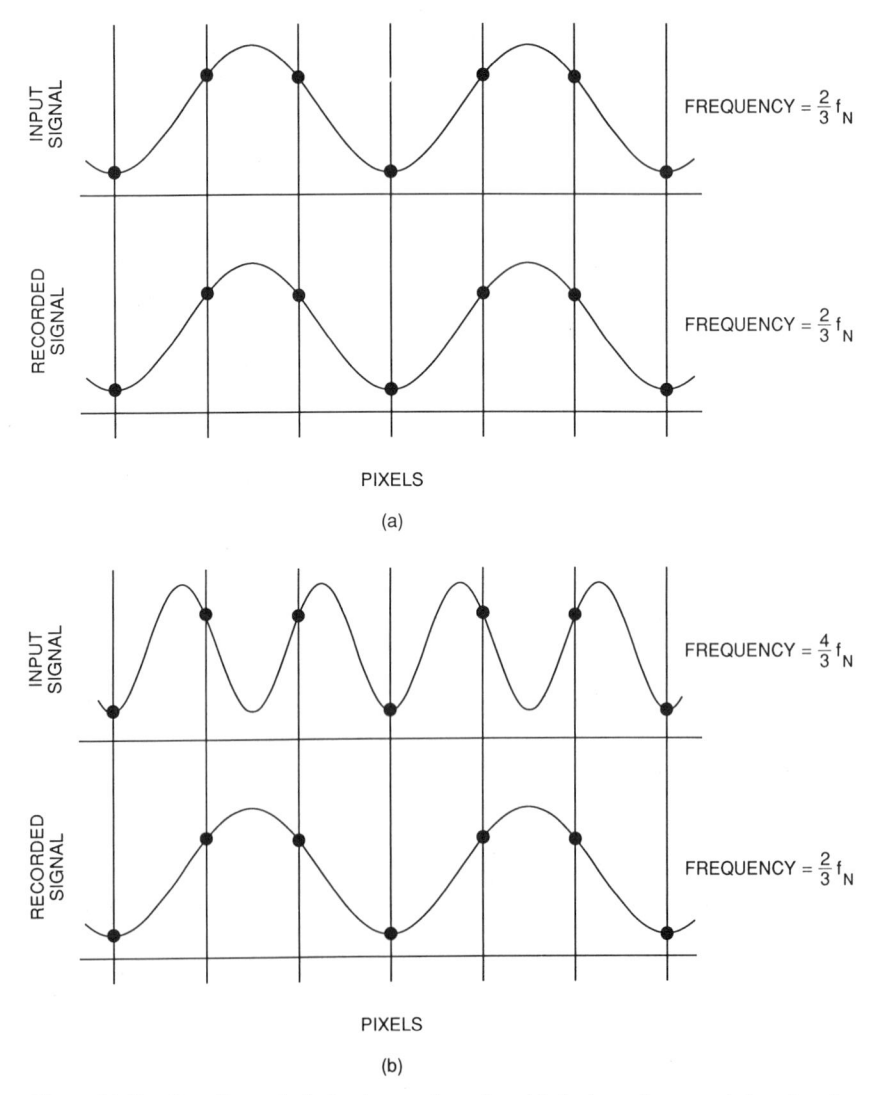

Figure 14.18. Sampling and aliasing in one dimension: (*a*) the input frequency is less than the Nyquist frequency and (*b*) the input frequency exceeds the Nyquist frequency.

14.13. ASPHERES AND EXTENDED RANGE PSI TECHNIQUES

sampled values at the pixels. In Fig. 14.18a, the fringe frequency is two-thirds the Nyquist frequency, or the sampling rate is three pixels per fringe or cycle. The sampled output is clearly at the same frequency as the input. The fringe frequency in Fig. 14.18b is four-thirds the Nyquist frequency, and there are only three pixels for every two fringes. There is an insufficient number of pixels to resolve this frequency, and the recorded samples are exactly the same as for the lower frequency measured in Fig. 14.18a. The displayed output must also be equal and therefore occurs at a frequency of $2f_N/3$ instead of $4f_N/3$. The effect of aliasing on the displayed fringe pattern is to replace the high-frequency fringes with low-frequency fringes, and this greatly impairs our ability to visually interpret this pattern. There is an ambiguity in the meaning of a fringe. The separation between recorded fringes no longer implies a wavefront change of one wave. More importantly for PSI, aliasing is an indication that the wavefront phase is changing at a rate greater than π per pixel. Under this condition the phase unwrapping operation described in Section 14.12 must fail as this process restricts the wavefront change to π per pixel.

A statement of the Whittaker–Shannon sampling theorem is as follows: If a scene is band limited to within the Nyquist frequency of the sensor, the scene can be recovered, without error, from the sampled image. The PSI phase unwrapping operation is based on the assumption that this band-limited requirement is met. It is important to note that this theorem does not state anything about the possibility of image recovery when this condition is not met. If the input scene or fringe pattern is not band limited to the Nyquist frequency, some additional information or a priori knowledge about the wavefront or surface under test is required to properly unwrap the phase.

14.13.2. Sub-Nyquist Interferometry

Sub-Nyquist interferometry (SNI) is a data collection and analysis method that is capable of greatly extending the measurement range of PSI through the use of a priori information (Greivenkamp 1987b). SNI applied to aspheric testing uses the simple assumption that the wavefront or surface under test is smooth and therefore has continuous derivatives. This additional information allows the analysis to interpret fringes that occur at frequencies well in excess of the Nyquist frequency and that are significantly undersampled. The differences between SNI and PSI occur during the phase unwrapping of the modulo 2π data. Instead of using the wavefront height constraint of PSI, SNI requires that derivatives of the reconstructed wavefront not exhibit any large changes from pixel to pixel. This slope continuity constraint limits the change of the wavefront slope to π per pixel per pixel, and large changes of the wavefront height between pixels are permitted. The appropriate number of 2π's are added to each pixel to satisfy this condition, and there is only a single solution at each pixel that produces this result. The slope continuity constraint correctly reconstructs the wavefront from the aliased data until the second derivative of the actual wavefront exceeds the limit imposed by the constraint. When this situation arises, further correction is possible by requiring that the second, or even higher-order, derivative be continuous, and adding more 2π's. This procedure can be continued until a more fundamental limit is reached, as explained later. In practice, however, first derivative or slope continuity is very effective in improving the measurement range of PSI, and higher orders are usually not needed.

The SNI phase unwrapping process for aspherics is graphically demonstrated in Fig. 14.19 using the connect-the-dot pictures similar to those in Section 14.12. Figure 14.19a shows all of the possible solutions to the arctangent at each pixel, and the open circles indicate the phase modulo 2π. The asphere used to generate these points is shown as the dotted line, and it goes through one point at each pixel. The object of the reconstruction is to determine these correct dots. The conventional PSI reconstruction of this data is given in Fig. 14.19b, and this reconstruction fails at pixel 5 where the correct dot is not the closest dot; the wavefront changes by more than π per pixel. The reconstruction shows a large change in wavefront slope. The SNI reconstruction in Fig. 14.19c gives the correct result by applying the slope continuity constraint. The proper dot is chosen by extrapolating a line from the previous two dots into the next pixel, and the dot that is closest to this line gives the solution with the smallest change in wavefront slope. Any other choice would violate the slope continuity

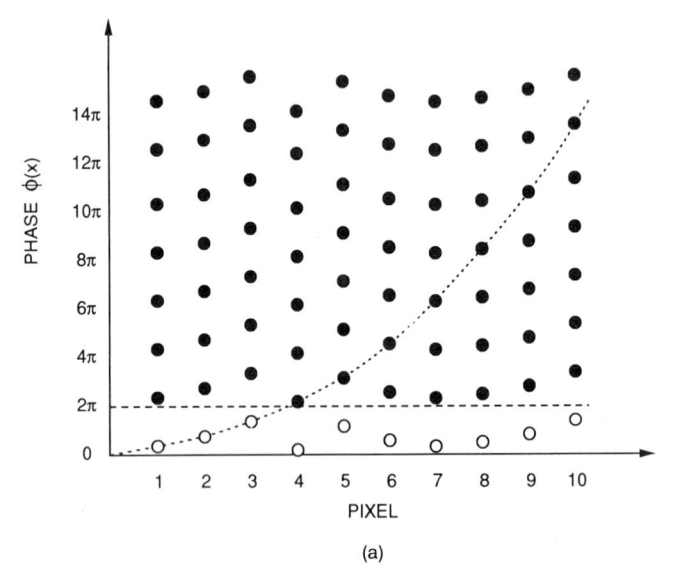

Figure 14.19. The SNI reconstruction process: (a) the possible solutions for the wavefront phase at each pixel, (b) the standard PSI reconstruction of this data; and (c) the SNI reconstruction using slope continuity. (From Greivenkamp 1987b).

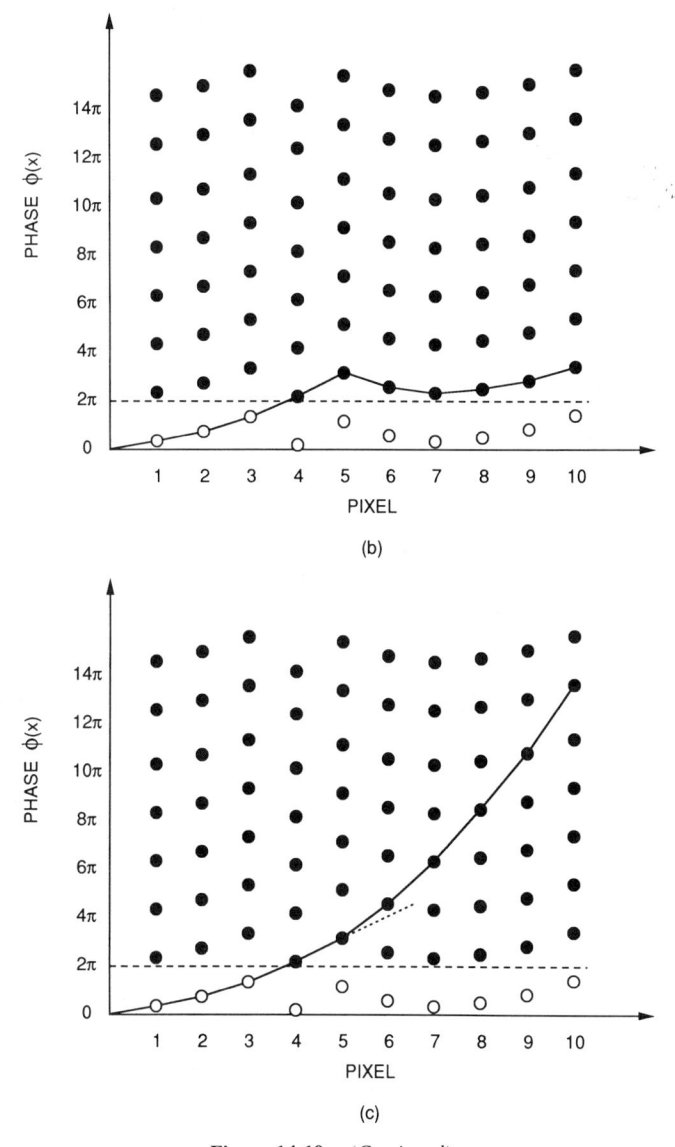

14.13. ASPHERES AND EXTENDED RANGE PSI TECHNIQUES

Figure 14.19. (*Continued*)

condition. Higher-order constraints operate by extrapolating a curve of the appropriate order that has been fit through a number of already selected dots. For example, second-derivative continuity uses a quadratic fit through the previous three points.

The fundamental limit to the measurement range of an SNI system is in the

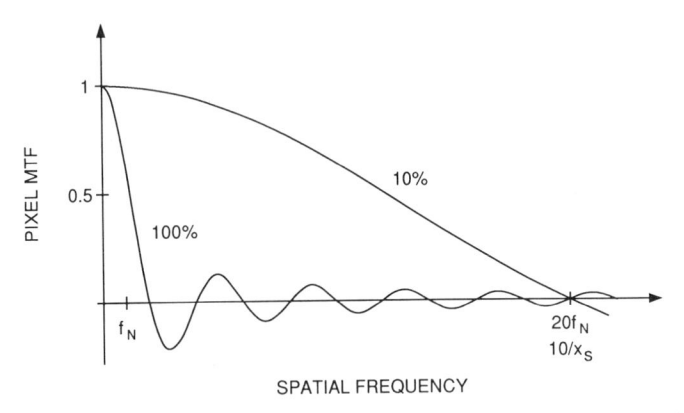

Figure 14.20. The pixel MTF of a sparse array sensor compared to that of a standard sensor.

ability of the sensor to respond to the high-frequency fringes: the pixel MTF. For good results in the arctangent, the measured data modulation must be high, and the sensor must be able to respond to fringes well beyond the Nyquist frequency. From Eq. (14.113), we see that extent of the pixel MTF is increased by using a sensor with a small pixel width-to-pitch ratio. This ratio for standard sensors is usually 50–100%, and the sparse array sensor that is needed for this application should have a smaller ratio. The pixels on this sensor should approximate point detectors, and the first zero of the pixel MTF is at a frequency well beyond the Nyquist frequency. A comparison of the pixel MTF of a sensor with a 10% width-to-pitch ratio to that of a standard sensor is shown in Fig. 14.20. The maximum wavefront slope that can be measured with this sensor and SNI is up to 20 times that which can be measured with PSI and the same number of pixels. The measurement range, which is dependent on the functional form of the wavefront, is increased by an even larger factor.

Some results from a prototype SNI system are displayed in Fig. 14.21. The system is a laser Fizeau interferometer that contains a sparse array sensor with 377 by 242 pixels (Palum and Greivenkamp 1989). The pixels have a width-to-pitch ratio of about 12% in both directions. The wavefront error under test is due to a large defocus; it is a quadratic departure with about 150 waves of error at the edge of the pupil. The sampled interferogram of this wavefront is shown in Fig. 14.21a and is heavily aliased. The ring patterns surrounding the central ring pattern are aliased fringes. The center of each of these additional patterns occur when the actual fringe frequency equals a multiple of the sensor Nyquist frequency. As expected, the PSI reconstruction of this interference pattern in Fig. 14.21b shows a good reconstruction only in the central unaliased region. The aliased fringes are improperly interpreted. The two-dimensional SNI result in Fig. 14.21c smoothly reconstructs the quadratic wavefront. The horizontal streaks in this result are due to the simplicity of the algorithm used;

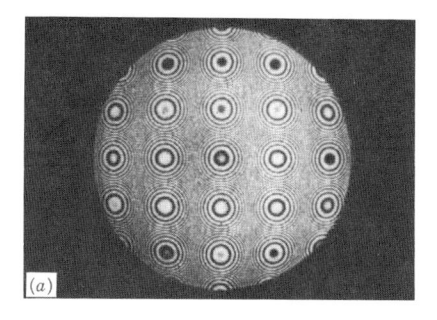

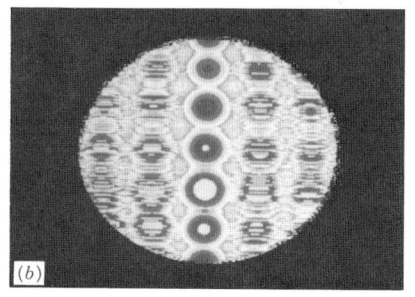

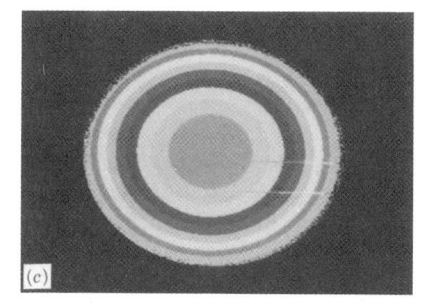

14.13. ASPHERES AND EXTENDED RANGE PSI TECHNIQUES

Figure 14.21. Results from a prototype SNI system: (*a*) the sampled interferogram of a defocused wavefront, (*b*) the standard PSI reconstruction of this data, and (*c*) the SNI reconstruction using slope continuity.

the phase unwrapping algorithm has not been taught to jump over bad data points.

SNI allows the interpretation of very complex fringe patterns, and the primary issue that needs to be addressed before SNI can be practically implemented is calibration. The testing of aspheric surfaces without null optics can lead to errors. In a null configuration, such as testing a sphere at its center of curvature in a laser Fizeau interferometer (Fig. 14.22*a*), the rays reflected by the test surface follow the same path through the converging lens and the interferometer as the rays reflected by the reference surface. Both the reference and test wavefronts see the same amount of additional aberration, and this aberration cancels. This is not the case in testing aspheric surfaces (Fig. 14.22*b*). These two ray paths can be very different, producing different system aberrations in the two interfering beams. A simple way to think about this is that different zones on the aspheric surface use the converging lens at different conjugates and different apertures, and it is only corrected at one conjugate. The wavefront difference measured in the interferogram is not simply the difference between the test and the reference wavefronts. An aspheric measurement system requires that the entire interferometric system, including all of the interferometer optics, be ray traced for the asphere under test.

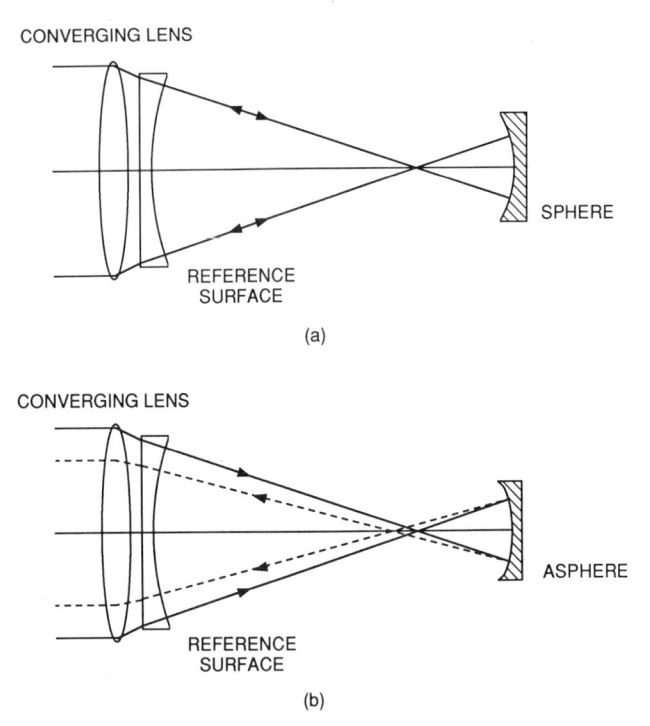

Figure 14.22. The need for calibration and ray tracing: (*a*) testing a spherical surface at its center of curvature and (*b*) the same configuration with an aspheric surface.

14.13.3. Two-Wavelength PSI

The additional information that two-wavelength PSI (TWPSI) uses to extend the PSI measurement range beyond the Nyquist frequency is a separate measurement of the part at a different wavelength (Wyant et al. 1984b; Cheng and Wyant 1984; Fercher et al. 1985; Creath et al. 1985; Creath and Wyant 1986; Wyant and Creath 1989). Two complete sets of interferometric data are collected, and the phase modulo 2π at each wavelength is computed. Since a phase of 2π is a different OPD at each wavelength, only one choice of wavefront deformation will satisfy both data sets. TWPSI can also be interpreted by the connect-the-dot representation (Fig. 14.23). The vertical scale is now OPD instead of phase so that it is not wavelength dependent. Each set of data produces its own column of possible solutions at each pixel, and the spacing of the dots is the wavelength. The locations where the two sets of dots fall on top of each other indicate possible correct solutions for the wavefront; many individual dots are now eliminated. At each pixel, there are multiple locations of coincidence,

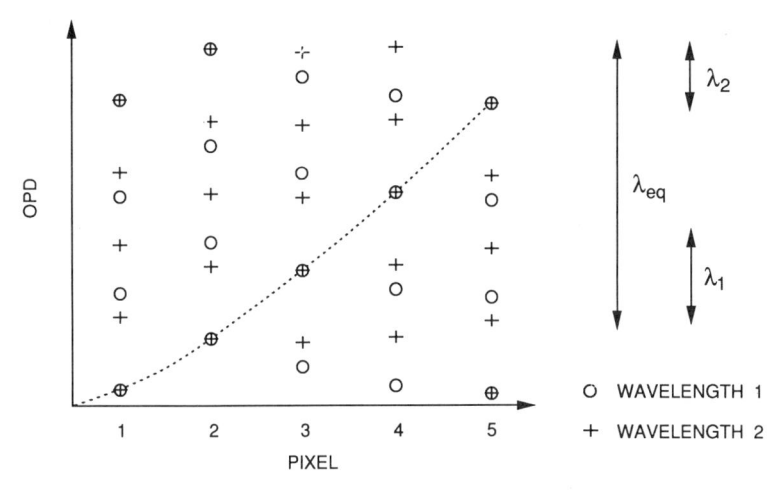

14.13. ASPHERES AND EXTENDED RANGE PSI TECHNIQUES

Figure 14.23. Two wavelength PSI.

and the spacing between these points is the equivalent wavelength:

$$\lambda_{eq} = \frac{\lambda_1 \lambda_2}{|\lambda_1 - \lambda_2|}, \qquad (14.115)$$

where λ_1 and λ_2 are the two measurement wavelengths. Phase unwrapping of the common points is now done using standard PSI techniques at the equivalent wavelength. Since the equivalent wavelength is much longer than the actual wavelengths, large wavefront slopes can be handled before the algorithm breaks down. Two-wavelength operation of an interferometer for testing aspheres requires correction of the optical system at both wavelengths as well as ray tracing at these wavelengths for the asphere under test. It is also possible to extend this technique to more than two wavelengths (Cheng and Wyant 1985a). More will be said about these multiple-wavelength techniques in Chapter 17.

There is an important similarity between SNI and TWPSI for measuring aspheric surfaces: Both depend on being able to measure aliased fringes, and therefore require sparse array sensors with small pixel width-to-pitch ratios. They are both limited by the pixel MTF. A practical problem that has hindered the implementation of TWPSI for this application is chromatic aberration in the reference optics and the interferometer. These optics must be achromatized at both wavelengths.

The application of measuring steps or discontinuities greater than a quarter wave in surface height (π in wavefront phase) requires different solutions than those needed for aspheres. Instead of high-frequency fringes, there is a break in the fringe pattern. Single-wavelength techniques lose track of the fringe order over the step and do not produce the correct result. Multiple-wavelength techniques, on the other hand, allow the fringe order to be determined and can measure step heights well in excess of a quarter wave. TWPSI has been successfully applied to this problem using interference microscopes (Creath 1987a), and this application will also be discussed in Chapter 17. SNI has been used for discontinuous surfaces measured at a single wavelength, but the requirement is added that the step height be known to $\pm\lambda/4$ (Greivenkamp et al. 1989). This additional information allows the correct number of 2π's to be added to the arctangent data. Since neither of these techniques for measuring steps rely on detecting aliased fringes, there are no special sensor considerations.

14.14. OTHER ANALYSIS METHODS

Since it is the most commonly used PSI technique, the bulk of this chapter has concentrated on the phase step and integrating bucket algorithms. A number of other PSI algorithms and systems have been developed, and in this section, we will briefly review each of these. Some are included for historical significance, and others are useful in special situations.

14.14.1. Zero Crossing Analysis

Zero crossing analysis is an electronic technique that directly measures the time delay between the intensity signals at different locations in the interferogram. The relative wavefront phase at the two points can then be calculated (Crane 1969; Moore 1973; Wyant and Shagam 1978). Referring back to Figure 14.1, the intensity at any location in the interferogram varies sinusoidally, and temporal phase at a point is proportional to the wavefront phase. One point in the interferogram is designated as a reference point, and the time delay between the signal at this point and another point is measured. These values can be converted to the wavefront phase modulo 2π by normalizing with the period of the reference signal. The location of the zero crossings is a convenient measure of the time lag. Since the intensity never goes negative, a reference intensity level [such as the average intensity $I'(x, y)$] is established for determining the zero crossing locations. To measure a two-dimensional wavefront, this time delay relative to the reference point must be measured at a large number of points.

14.14.2. Synchronous Detection

Synchronous detection is an early technique that is the clear precursor to the phase stepping techniques in use today. It is based on the well-known methods of radio communication theory, where the phase of an unknown signal is recovered by correlation of this signal with a sinusoidal signal of the same fre-

14.14. OTHER ANALYSIS METHODS

quency (Bruning et al. 1974; Bruning 1978). For a linear reference phase shift, the intensity pattern in the interferogram is [from Eq. (14.6)]

$$I(x, y, t) = I'(x, y) + I''(x, y) \cos [\phi(x, y) + \alpha t], \quad (14.116)$$

where α is the rate of change of the reference phase. Since we are controlling the reference phase, α is well known, and sinusoidal and cosinusoidal signals with this same frequency can be generated. This cosinusoidal signal is correlated with the interferogram intensity at each point:

$$\langle I_1(x, y) \rangle = \lim_{T \to \infty} \frac{1}{T} \int_0^T I(x, y, t) \cos (\alpha t) \, dt, \quad (14.117)$$

which reduces to

$$\langle I_1(x, y) \rangle = I''(x, y) \cos [\phi(x, y)]. \quad (14.118)$$

In a similar manner the correlation of the intensity with a sinusoidal signal produces

$$\langle I_2(x, y) \rangle = -I''(x, y) \sin [\phi(x, y)]. \quad (14.119)$$

As with all of the phase step techniques we have examined, the wavefront phase modulo 2π can be determined through an arctangent of the ratio of these two results:

$$\phi(x, y) = \tan^{-1} \left[\frac{-\langle I_2(x, y) \rangle}{\langle I_1(x, y) \rangle} \right]. \quad (14.120)$$

This analysis also provides a feel for the noise immunity of this technique and for PSI. Additive noise included in the interferogram intensity is not related to the frequency α corresponding to the phase shift, and the correlation operation averages these noise terms to zero.

In a similar manner this correlation can be performed with temporally sampled data. When N measurements of the interferogram intensity are made during a 2π change in the reference phase,

$$\delta_i = i2\pi/N; \quad i = 1, \ldots, N, \quad (14.121)$$

elementary Fourier series analysis can be used to solve for the unknown wavefront phase:

$$\phi(x, y) = \tan^{-1} \left[\frac{-\sum I_i \sin (\delta_i)}{\sum I_i \cos (\delta_i)} \right], \quad (14.122)$$

where the summations run from 1 to N. This result is identical to the result obtained using a least-squares fit of the intensity data to a sinusoidal function of the proper frequency [Eq. (14.66)].

14.14.3. Heterodyne Interferometry

Heterodyne or AC interferometry is the general term used to describe interferometers that produce a continuous phase shift by introducing two different optical frequencies into the two arms of the interferometer. Several methods exist for generating this frequency difference, and the most common are moving gratings and two-frequency lasers (see Section 14.4). With any of these approaches, the interferogram intensity is modulated at the difference frequency [Eq. (14.9)-(14.12)]. The primary applications for heterodyne interferometry are distance measuring interferometry, measuring dynamic systems, and surface profilometry (Crane 1969; Sommargren 1981; Koliopoulos 1980; Barnes 1987).

Digital heterodyne interferometry (DHI) has been developed to address the need to perform high-speed analysis of dynamic systems (Massie 1978, 1980, 1987; Massie et al. 1979; Mottier 1979; Massie et al. 1981; Massie et al. 1983; Evans 1983). Two significant applications of this technology are deformable mirrors and turbulent flow field testing. To obtain meaningful PSI data, the phase shifts must occur at a rate that is faster than the changes that are being observed. Acousto-optic Bragg cells are the most convenient method to obtain the frequency shifts, and a frequency difference of up to 1 MHz has been demonstrated. The interferogram intensity will oscillate at this frequency, and the wavefront phase is encoded in the phase difference of these oscillations at different measuring points. Because of the frequencies involved, a high-speed detector is needed; a standard video camera will not work. The conceptionally simplest method to collect the data is to use an array of discrete detectors. Each will produce an output at 1 MHz with a different phase. One detector is defined as the reference detector, and the phase of the other signals are measured relative to this signal. A phase detection method, such as the zero crossing method described earlier, measures these phases, and a wavefront map modulo 2π is produced. A high frame rate image tube, called an image-dissector tube, can also be used as the detector. This camera functions as a rapidly scannable photomultiplier tube. A repeatability in excess of $\lambda/100$ RMS has been demonstrated with DHI systems.

14.14.4. Phase Lock Interferometry

Phase lock interferometry involves applying a small sinusoidal oscillation to the reference mirror (Johnson and Moore 1977; Johnson et al. 1977, 1979; Moore et al. 1978; Matthews et al. 1986). The average reference phase is also allowed to vary so that

14.14. OTHER ANALYSIS METHODS

$$\delta(t) = \delta + a \sin(\omega t), \qquad (14.123)$$

where a is the amplitude of the oscillation and is much less than a wavelength. The resulting interferogram intensity is

$$I(x, y, t) = I'(x, y) + I''(x, y) \cos[\phi(x, y) + \delta + a \sin(\omega t)]. \qquad (14.124)$$

A series expansion of this intensity produces

$$I(x, y, t) = I'(x, y) + I''(x, y) \{ \cos[\phi(x, y) + \delta][J_0(a) + 2J_2(a) \\ \cdot \cos(2\omega t) + \ldots] - \sin[\phi(x, y) + \delta] \\ \cdot [2J_1(a) \sin(\omega t) + 2J_3(a) \sin(3\omega t) + \ldots] \}, \qquad (14.125)$$

where J_n is a Bessel function of order n. In addition to the fundamental frequency of the oscillation, higher-order terms are also present. This signal is then filtered to allow only the fundamental frequency ω:

$$I_f(x, y, t) = -2I''(x, y)J_1(a) \sin[\phi(x, y) + \delta] \sin(\omega t). \qquad (14.126)$$

There are a couple of interesting features to this signal. First, if the average reference phase δ is varied so that $\phi(x, y) + \delta$ equals a multiple of π, $I_f(x, y, t)$ goes to zero, and there is no intensity modulation due to the reference mirror oscillation or dither. Referring back to Fig. 14.1, this situation corresponds to sitting on either a maximum or a minimum of this curve; the small change in phase due to the dither does not result in any change of intensity. The second thing to notice about $I_f(x, y, t)$ is that its magnitude is proportional to the slope of the interference pattern, and for small values of $\phi(x, y) + \delta$, it is actually proportional to the phase distance from a fringe maximum. The sign of the slope is encoded in the phase of this signal relative to the drive signal; this allows the instrument to find an intensity maximum instead of a minimum.

The interferometer operates by measuring the filtered intensity signal $I_f(x, y, t)$ at a location in the interferogram with a detector. The magnitude and phase of this signal are used to produce a feedback signal that tells the phase shift controller how to change the average phase shift δ to null out this signal. When a null is achieved, the sum $\phi(x, y) + \delta$ is equal to zero modulo 2π; we are sitting on one of the fringe maxima. Since we know δ, the value of wavefront phase $\phi(x, y)$ modulo 2π has been determined. As the wavefront phase changes, the feedback signal is used to vary the average phase shift to maintain the null condition. When the required average phase shift exceeds $\pm 2\pi$, an up/down counter is incremented or decremented, and δ is reset to zero. This reduces the travel requirements on PZT producing the phase shift, and counts the number of fringes. The wavefront phase is then given by the number of 2π's that have been counted plus the current value of δ. To measure an area, the detector is scanned across the wavefront. Since the scan rate can be varied, interferograms with large numbers of fringes can be measured with phase lock interferometry, and repeatabilities on the order of $\lambda/100$ have been demonstrated.

14.14.5. Spatial Synchronous and Fourier Methods

While strictly speaking not PSI techniques, the types of interferogram analysis that are known as spatial synchronous detection and Fourier analysis have some similarity to PSI. Both are fringe analysis methods that look at a single interferogram, and therefore can be used when turbulence and vibration are present. Both of these techniques are more precise than standard interferogram analysis as it is not necessary to determine fringe centers; the data is also collected on a regular grid providing compatibility with computer analysis.

Spatial synchronous detection (see Direct Interferometry in Section 13.6) operates in the spatial domain by multiplying the measured interferogram with a reference pattern (Ichioka and Inuiya 1972; Womak 1984a; Toyooka and Tominaga 1984). Expressions for the unknown interferogram and the reference pattern are

$$I(x, y) = I'(x, y) + I''(x, y) \cos [\phi(x, y) + 2\pi f_0 x], \qquad (14.127)$$

and

$$R(x, y) = \cos (2\pi f_0 x), \qquad (14.128)$$

where a significant amount of wavefront tilt has been added to the interferogram, and f_0 is the fringe frequency due to the tilt. $R(x, y)$ is a uniform fringe pattern (tilt fringes) that closely matches the interferogram and is implemented either physically or computationally. The product of these two patterns is

$$I(x, y)R(x, y) = I'(x, y) \cos (2\pi f_0 x) + [I''(x, y)/2] \cos [\phi(x, y) + 4\pi f_0 x]$$
$$+ [I''(x, y)/2] \cos [\phi(x, y)]. \qquad (14.129)$$

The third term contains low spatial frequencies that can be filtered from the other two terms to produce

$$S_1(x, y) = [I''(x, y)/2] \cos [\phi(x, y)]. \qquad (14.130)$$

Similarly, a second filtered signal results from the multiplication of the interferogram with a reference pattern equal to $\sin (2\pi f_0 x)$:

$$S_2(x, y) = [I''(x, y)/2] \sin [\phi(x, y)]. \qquad (14.131)$$

14.14. OTHER ANALYSIS METHODS

The wavefront phase difference modulo 2π can then be computed from the ratio of these two signals:

$$\phi(x, y) = \tan^{-1}\left[\frac{S_2(x, y)}{S_1(x, y)}\right]. \quad (14.132)$$

The similarity between this algorithm and the many PSI algorithms, especially the synchronous detection algorithm in Eq. (14.120), should be apparent. Similar systems have been built that image the interferogram onto a segmented detector, and the individual outputs are sinusoidally weighted to determine the phase (DeCou 1974; Mertz 1983a, 1983b, 1989).

The Fourier analysis methods (see also Section 13.5) provide a virtually identical analysis, but the computations are done in the spatial frequency domain instead of the spatial domain (Takeda et al. 1982; Macy 1983; Nugent 1985; Bone et al. 1986; Kreis 1986; Roddier and Roddier 1987). An interferogram pattern with tilt as described by Eq. (14.127) can be rewritten as

$$I(x, y) = I'(x, y) + c(x, y)e^{i2\pi f_0 x} + c^*(x, y)e^{-i2\pi f_0 x} \quad (14.133)$$

where

$$c(x, y) = [I''(x, y)/2]e^{i\phi(x, y)}, \quad (14.134)$$

and * indicates the complex conjugate. A one-dimensional Fourier transform of $I(x, y)$ produces

$$\tilde{I}(\xi, y) = \tilde{I}'(\xi, y) + \tilde{c}(\xi - f_0, y) + \tilde{c}^*(\xi + f_0, y), \quad (14.135)$$

where ξ is the spatial frequency coordinate and $\tilde{}$ indicates a Fourier transform. This function is a trimodal function with peaks at $-f_0$, f_0, and the origin as sketched in Fig. 14.24a. The component of this spectrum centered at f_0 can be recovered without the carrier by first bandpass filtering and then shifting the isolated spectrum back to the origin (Fig. 14.24b). This results in the function $\tilde{c}(\xi, y)$. An inverse Fourier transform produces $c(x, y)$ as described in Eq. (14.134). The wavefront phase modulo 2π can then be determined from an arctangent:

$$\phi(x, y) = \tan^{-1}\left\{\frac{\text{Im}[c(x, y)]}{\text{Re}[c(x, y)]}\right\}, \quad (14.136)$$

where Re and Im refer to the real and imaginary part of the function.

Both of these spatial techniques suffer from a problem when the wavefront under test has a large deviation from the reference wavefront. The three components of the interferogram spectrum [Eq. (14.129) or (14.135)] overlap, and

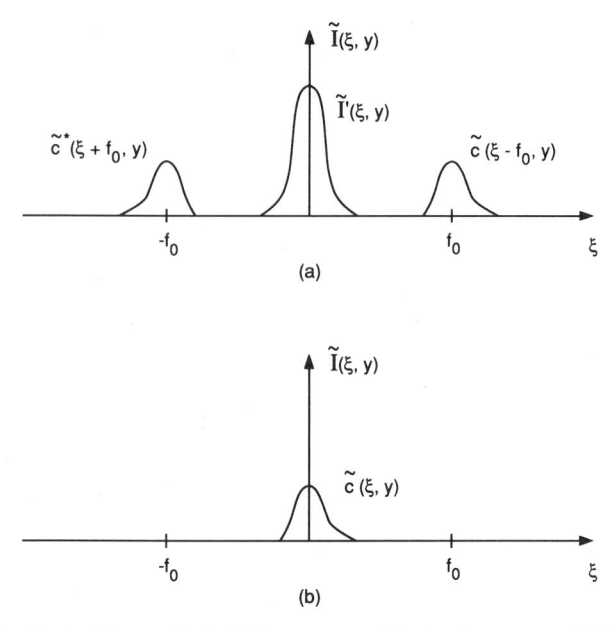

Figure 14.24. Fourier fringe analysis: (*a*) the spectrum of the interferogram and (*b*) the processed spectrum.

it is not possible to completely separate them through filtering. Various filters have been suggested to optimize the output of these techniques (Womak 1984a, 1984b).

14.15. COMPUTER PROCESSING AND OUTPUT

Now that a complete representation of the wavefront phase is stored in the computer memory, we can use the computer for the analysis and display of this data. A number of different analyses can be performed to produce results that are appropriate for the application. The initial analysis often starts by fitting the measured wavefront to a polynomial expansion, usually the Zernike polynomials (Kim and Shannon 1987). The terms in the expansion corresponding to tilt and focus can then be subtracted from the measured wavefront to leave only the wavefront error as measured from a best-fit reference sphere (for a flat, only tilt is subtracted as we are interested in the radius of curvature of the wavefront). Higher-order aberrations, such as spherical, coma, and astigmatism, can also be subtracted.

The point spread function (PSF) and MTF due to the measured wavefront errors and diffraction can be easily computed from the wavefront data. We start

14.15. COMPUTER PROCESSING AND OUTPUT

by expressing the wavefront in the form of Eq. (14.3), which includes an amplitude term:

$$w(x, y) = a(x/d, y/d)e^{i\phi(x, y)}. \tag{14.137}$$

The amplitude term $a(x, y)$ now also defines the exit pupil diameter d. The amplitude PSF is found through a fast Fourier transform (FFT) of this wavefront function, and the intensity PSF results from the squared modulus of this result (Goodman 1968). Since the wavefront represents only the error from a spherical reference surface, this answer must be scaled to image plane with the wavelength and the system $f/\#$. The pupil diameter is already contained in the expression for the wavefront, so that the scaling is done by replacing the spatial frequency variables ξ and η with $x/\lambda f$ and $y/\lambda f$, respectively:

$$\text{PSF} = |\tilde{w}(\xi, \eta)|^2 \big|_{\xi = x/\lambda f; \eta = y/\lambda f}, \tag{14.138}$$

where $\tilde{\ }$ indicates a Fourier transform, and f is the focal distance. The result of this computation for an unaberrated wavefront with a circular pupil is an Airy pattern, and aberrations will distort the PSF. The system MTF contribution of the tested wavefront or surface is found through an FFT of the PSF. This result is properly scaled to the $f/\#$ of the optical system, and the diffraction-limited cutoff frequency of $1/(\lambda f/\#)$ results.

There are times when the system parameters (pupil diameter, focal length, and wavelength) are not known or defined at the time of the test, and it is useful to express the results in arbitrary units. An example might be the test of a flat mirror to be used in a multielement system; we are interested in the influences of the flat mirror on the system performance, and assume that the aberrations of that surface propagate through the system to the final converging wavefront. We may also not be testing the element at the final system wavelength. It is the parameters of this final wavefront that are used to scale the PSF and MTF due to the element. An arbitrary unit that is sometimes used is waves/radius, which can be obtained by setting the exit pupil radius, the focal distance, and the wavelength all equal to one in the preceding calculations. The output dimensions in waves/radius can then be scaled at a later time to a physical distance by multiplying by $\lambda f/r = 2\lambda f/\#$, where r is the pupil radius.

Examples of interferometer output for a test surface are shown in Fig. 14.25, where the OPD and the resulting PSF and MTF are displayed. A number of different plotting options, including contour maps and isometric plots, are available. Some other examples of analyses that can be performed on the wavefront data are one-dimensional traces, surface slope plots, Strehl ratio computations, encircled energy distributions, surface height histograms, and geometrical spot diagrams. The choice of the output can be customized to the requirements of the user (Truax 1986; Creath 1987b). Another feature that can be built into the

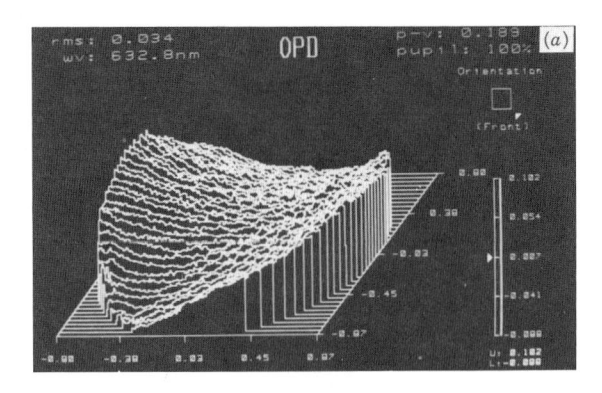

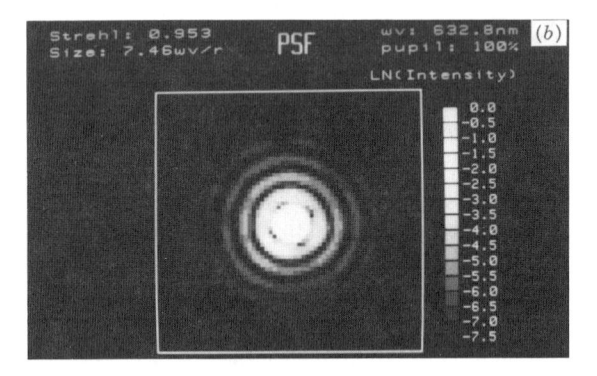

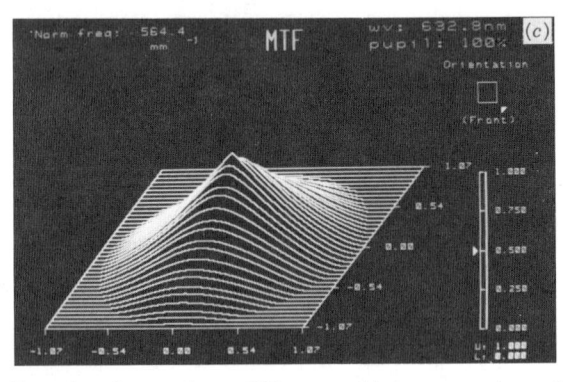

Figure 14.25. Examples of output from a PSI system: (*a*) the measured wavefront, (*b*) the PSF corresponding to this wavefront, and (*c*) the MTF contribution.

software are production menus that will lead the operator through the test, and then provide a pass/fail display based on certain predetermined criteria.

Recently, interfaces between interferometer analysis software and optical design programs have become available. While the exploration of this capability is just beginning, it has been used for several applications (Figoski et al. 1989: Stephenson 1989; Willey and Patchin 1989). The overall system design can be reoptimized based on the actual parts that have been manufactured and tested; this is especially useful for systems that are built around a difficult to manufacture element. In a similar manner the interferometric data can be used to predict the actual system performance for comparison to the design requirements. The alignment of optical systems has also made use of this capability. The wavefront from either the entire optical system or a subsystem is measured and compared to the wavefront predicted by the design program. Based on the differences between these two wavefronts, the design program computes in an iterative fashion the changes needed to bring the system into specification. A fourth, and more general, application of this capability is to create reference aspheric wavefronts in the computer software. The design software calculates the desired wavefront, and the interferometer then subtracts this calculated wavefront from the measurement to determine the wavefront errors. This process is almost exactly analogous to the use of computer-generated holograms, except the "hologram" always resides in computer memory.

14.16. IMPLEMENTATION AND APPLICATIONS

In this section we will look at the practical considerations of PSI and also examine some applications of PSI. A generic PSI system consists of four things: an interferometer, a device for changing the reference phase, a detector or detector array, and a computer. The detector array provides the interface between the wavefront and the computer, and the computer provides control of the interferometer and phase shifter and performs the necessary computations. The advances in detectors and computers have been the enabling technologies that have driven the performance improvements and acceptance of PSI. There is sufficient computing power in personal computers to handle most PSI applications.

14.16.1. Commercial Instrumentation

Commercial instrumentation utilizing PSI technology is available for measuring surface and wavefront shape, surface profile, and distance. PSI-based surface profiling microscopes are widely used for quantitative examination of small surface features and microroughness of both optical and nonoptical compo-

572 PHASE SHIFTING INTERFEROMETRY

Figure 14.26. Commerically available laser Fizeau interferometer. (Photo courtesy of Zygo Corporation, Middlefield, CT.)

nents. Repeatabilities of a few angstroms RMS can be obtained, and these profiling instruments are fully discussed in Chapter 17.

PSI interferometers have been marketed for measuring surfaces and wavefronts based on all three of the optical configurations shown in Fig. 14.2: Twyman–Green, Mach–Zender, and laser Fizeau. However, the vast majority of these instruments are laser Fizeau interferometers, and a photograph of a commercially available system is shown in Fig. 14.26. All of these systems employ video cameras to record the interferograms, the phase shifts are provided by moving the reference objective with PZT transducers, and a variety of reference objectives and flats are available to match the part under test (Smythe et al. 1987; Creath 1987b). A prominent feature of these systems is the flexibility of the user interface provided by the computer. As discussed in Section 14.15, a large number of graphics displays can be produced, and the processing can be customized to meet the needs of the user.

A Mach–Zender interferometer configuration that is available for testing the quality of optical wavefronts with PSI is shown in Fig. 14.27 (Hayes and Lange

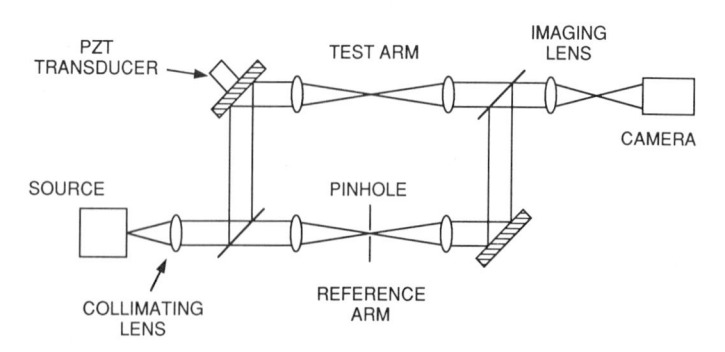

Figure 14.27. Mach–Zender PSI interferometer for measuring the wavefront quality of sources.

14.16. IMPLEMENTATION AND APPLICATIONS

1983; Creath 1987b). The output of a source, such as a laser diode, is collimated as it enters the system. A spherical reference wavefront is provided by the spatial filter in the reference arm. A similar set of optics, without the pinhole, is used in the test arm to match the optical path length in the two arms. The beams are recombined and imaged onto the camera. Both the intensity and phase distributions of the wavefront can be measured.

An interesting system that can be used for testing the shape or flatness of nonoptical parts has been developed based on grazing-incidence interferometry (Synborski 1978). The system is diagramed in Fig. 14.28a. The component under test is placed in close proximity to the hypotenuse of a large glass prism that serves as the reference surface for the interferometer. An air gap is maintained. A closeup of this interface is shown in Fig. 14.28b. Because of refrac-

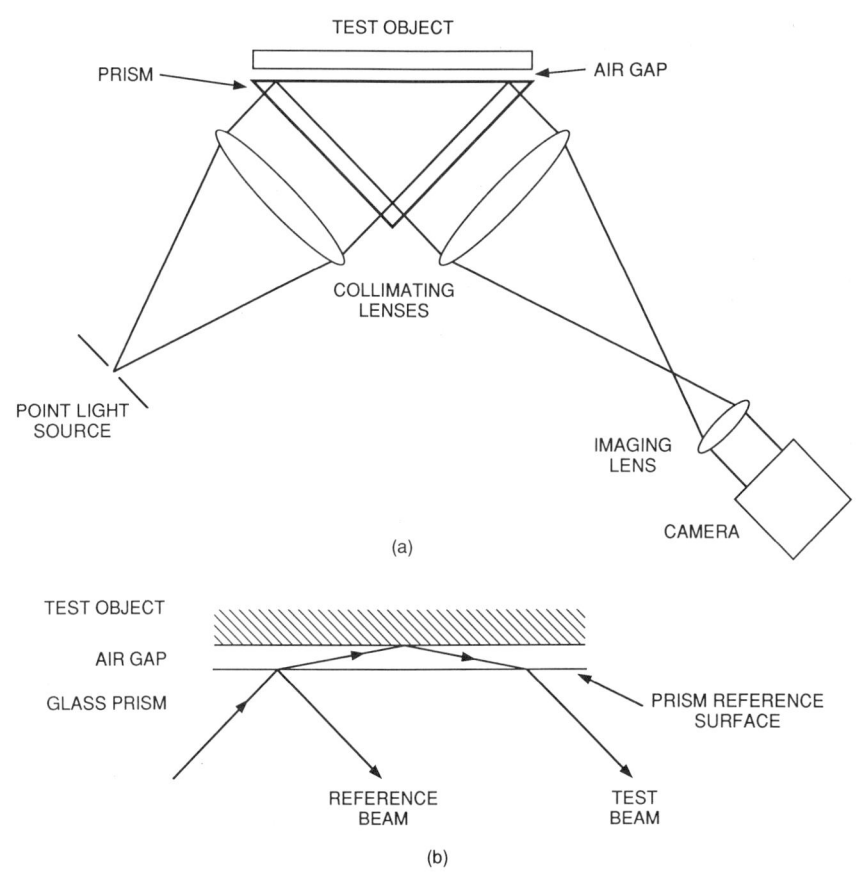

Figure 14.28. Grazing incidence PSI interferometer: (*a*) overall system layout and (*b*) a closeup of the interface at the air gap.

tion, the test beam is incident on the part at a large angle. For a 41.7° angle of incidence and a refractive index of 1.5, the angle of incidence at the test surface is 86°. This is close to the critical angle of the prism. There are two beneficial consequences to this configuration. First, the sensitivity of the test is reduced by the cosine of the angle of incidence. At 86°, the sensitivity is 7% of the normal incidence sensitivity; a fringe represents seven waves of surface height change instead of a half wave. It is as if the surface is tested with a longer wavelength. The second benefit is that almost everything, including a ground surface, is highly reflective at this angle of incidence, and a variety of different materials and samples can be used. If the angle of the light in the prism is changed slightly, the optical path length in the air gap will change much more than the path in the glass. This tilt is used to produce the phase shift between the reference and test beams. Systems are available with up to a 200-mm aperture and have been applied to testing silicon wafers and machined or ground parts. A clear drawback to a system such as this is the need for the large precision glass prism.

Distance measuring interferometers (DMIs) are capable of providing position readout with a resolution of better than a thousandth of a wavelength (about a tenth of a microinch) and have found extensive use in the area of precision machine control (Dukes and Gordon 1970; Steinmetz et al. 1987; Smythe et al. 1987; Steinmetz 1990). Most DMIs use heterodyne techniques (Section 14.14.3), and a basic system is diagramed in Fig. 14.29. The two output frequencies from the source are orthogonally polarized, and the polarizing beam splitter splits the frequencies into the reference and test arms of the interferometer. The two corner cubes return these two beams to the detector, where the polarizer at 45° allows them to interfere. When the movable corner cube is stationary, a beat frequency equal to the difference in the two optical frequencies is obtained; this frequency is identical to a similar reference signal generated in the laser head. A motion of this corner cube introduces a Doppler frequency shift in the light returning from the test arm, and the measured beat frequency changes. The change in the length of the test arm is determined by

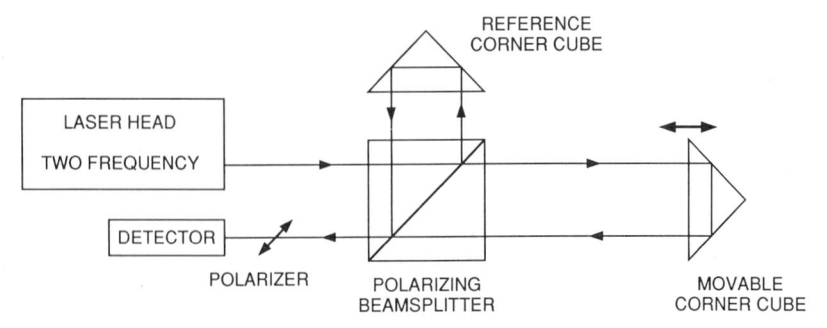

Figure 14.29. Distance measuring interferometer.

comparison of the measured beat frequency with the reference beat frequency. Some specific applications of DMIs are microlithographic step-and-repeat cameras, single-point diamond turning machines, and precision photoplotters. Additional information about DMIs is contained in Chapter 18.

14.16.2. Interferometer Configurations

A number of grating lateral shearing interferometers that use PSI techniques have been demonstrated (Wyant 1975; Hardy et al. 1977; Stumpf 1979; Koliopoulos 1980; Hardy and MacGovern 1987). Translation of the grating produces a frequency shift between the diffracted orders, resulting in a linear phase shift between the interfering beams. The primary application of many of these systems has been wavefront sensing for active control of adaptive optics. More information on grating interferometers can be found in Chapter 4. The wavefront is focused onto the grating, and three or more diffraction orders are produced. The period of the grating is chosen so that there is overlap, and therefore interference, of the diffracted orders. When the grating is translated, the diffracted orders are Doppler shifted (see Section 14.4); the plus one order increases, the minus one order decreases, and the zero order is unchanged. A beat frequency appears in the output that can be analyzed by either heterodyne or integrating bucket techniques to determine the wavefront phase. A convenient grating configuration is to use a radial grating (starburst pattern) that is spun about its center (Stevenson 1970). With this grating the shear distance can be changed by moving the grating center relative to the focus spot.

PSI has also been applied to the double-frequency grating lateral shear interferometer (Wyant 1973). In this system the wavefront is focused onto a grating with two distinct line spacings. Two diffraction patterns are produced, one from each set of grating lines. The shear between the two interfering first orders is determined by the difference in the two grating frequencies, and the average frequency can be chosen to separate these diffracted orders from the undiffracted light. Translation of the grating through the focus will produce different Doppler shifts for each of the two diffracted beams, and a relative phase shift is produced. The resulting interferogram can be evaluated by PSI techniques.

In situations where the test wavefront can change with time, there is a desire to obtain all of the required phase-shifted interferograms simultaneously. There should be no time lag between the measurements. Several of these snapshot interferometers have been demonstrated, and they are also compatible with pulsed light sources.

The first of these instantaneous systems uses polarizing elements to produce the desired PSI signals (Smythe and Moore 1984). A polarization interferometer is used where the two polarizations are each sent to one arm of the interferometer. The returning beams are sent to the decoding module shown in Fig. 14.30. The halfwave plate rotates both polarizations by 45° so that they are at

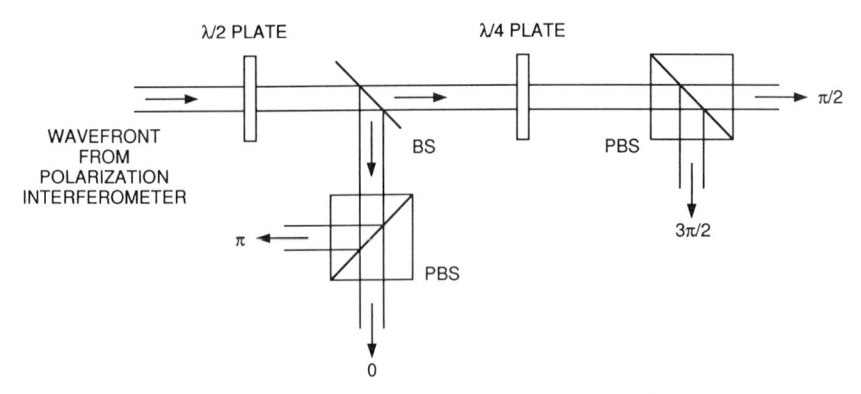

Figure 14.30. Polarization decoding scheme for instantaneous PSI measurements.

45° with the axes of the polarizing components. An intensity beam splitter (BS) creates two beams, and the reflected beam is further split by a polarizing beam splitter (PBS). Two sets of interference fringes are formed with a phase shift of 180° due to conservation of energy. The other beam from the intensity beam splitter goes through a quarterwave plate whose fast axis is aligned with one of the orthogonal polarizations; this introduces a 90° phase shift between the two polarizations. When this beam is analyzed by a second polarization beam splitter, the two interferograms will be 180° out of phase and also shifted by 90° from the first set of fringe patterns. These four interferograms with 90° phase shifts can be recorded and wavefront calculated with the four-step algorithm [Eq. (14.27)]. Similar polarization encoding schemes have been applied to distance measuring interferometers (Dorsey et al. 1983; Crosdale and Palum 1990).

A second stroboscopic interferometer uses a diffraction grating to introduce a phase shift between the reference and multiple test beams (Kwon 1984, 1987; Kwon et al. 1987, Kujawinska et al. 1991). One implementation of the system is shown in Fig. 14.31. The three replicas of the test beam are the diffracted orders from the stationary grating, and the reference beam is provided by the spatial filter. A small lateral shift of the grating will produce a 90° phase shift between the three test beams. A simple way to understand this phase shift is that the grating is in the Fourier transform plane of the wavefront; a shift in this plane will result in a linear phase factor in the plane of the interferograms. A shift of a quarter of the grating period will produce the desired phase shift. The three output interferograms can be evaluated with one of the three step algorithms [Eq. (14.49) or (14.53)].

A third snapshot interferometer (McLaughlin and Horwitz 1986, Horwitz 1990) uses a phase grating to shear the wavefront under test and to create a carrier fringe frequency in the interferogram. The period of these carrier fringes is adjusted to equal four pixels on the sensor, and each pixel sees a relative phase shift of 90°. The four-step algorithm can therefore be used to analyze the

14.16. IMPLEMENTATION AND APPLICATIONS

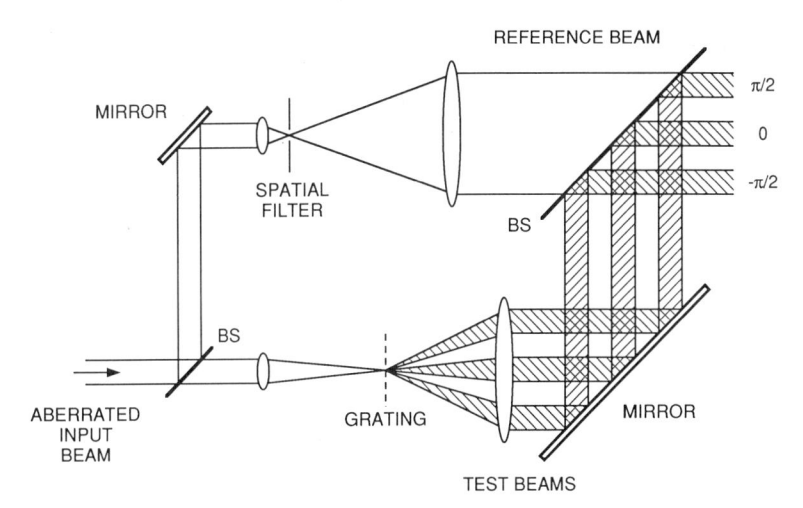

Figure 14.31. Stroboscopic PSI interferometer using a diffraction grating. (From Kwon 1987).

sheared wavefront. There is a great deal of similarity between this method and the spatial synchronous detection methods discussed in Section 14.14.5. A crossed grating is used to simultaneously analyze the shear in orthogonal directions.

14.16.3. Absolute Calibration

Since interference patterns only indicate phase difference, we generally need a reference wavefront or reference surface if we are to measure a surface in any sort of an absolute sense. If a reference surface is to serve as a standard, it must be characterized by some other method. If the reference surface is certified to a high degree, the measurement of that surface on the interferometer will reveal wavefront errors that are characteristic of the interferometer, provided the reference accuracy exceeds the interferometer accuracy. This measured wavefront then closely represents the interferometer error that, if stored in the computer as a *reference*, can be subtracted from future measurements as differences from a calibrated standard. This calibration method is limited by the "goodness" of the reference standard and must be repeated if any adjustments to the interferometer are changed or disturbed, or if there are drifts that would result in significant deviations. The stability of the reference calibration can be assessed by periodic measurement of the reference surface against itself (with the appropriate fiducials to assure that the surface is measured in precisely the same orientation).

In situations where a reference standard does not provide adequate or sufficiently quantified accuracy, a somewhat more involved method is possible with

Twyman–Green and Fizeau interferometers, which have an external and accessible focus. This method is capable of calibrating the interferometer in an absolute sense without a secondary reference and hence is capable of making an absolute measurement of a test surface. This method was first suggested by Jensen (1973) and later demonstrated by Bruning et al. (1974); Truax (1988); Elssner et al. (1989). This method requires the facility to measure and manipulate three wavefronts, W_1, W_2, and W_3. These are shown in Fig. 14.32a. W_1 is the wavefront obtained with the test surface measured at its center of curvature at an azimuthal orientation of $0°$. W_2 is the wavefront obtained similarly to W_1, with the exception that the test surface is rotated $180°$ about its axis from the $0°$ orientation. Finally, W_3 is the position where odd errors in the interferometer cancel and is obtained by placing the vertex of the test surface at the focus of the autocollimator or "cat's-eye" location. By introducing the following notation, the wavefront manipulations and the calibration procedure should become clear:

$$W_1 = W_R + W_T + W_S, \tag{14.139}$$

$$W_2 = W_R + W_T + W_S^{180°} \tag{14.140}$$

$$W_3 = W_R + \tfrac{1}{2}(W_T + W_T^{180°}), \tag{14.141}$$

where

W_R^α = reference arm wavefront;

W_T^α = test arm wavefront;

W_S^α = test surface wavefront;

α = orientation of wavefront (0 or $180°$).

If the interferometer possessed no aberration whatsoever, then wavefronts W_R and W_T would be identically zero, and the surface errors W_S would be precisely described by the wavefront W_1. In the presence of interferometer aberrations, W_1 contains the test surface errors along with others that can be separated using Eqs. (14.139)–(14.141), which yields the desired test surface wavefront:

$$W_S = \tfrac{1}{2}(W_1 + W_2^{180°} - W_3 - W_3^{180°}). \tag{14.142}$$

where all wavefronts have focus and tilts removed. The total error in the interferometer, denoted by W_I, is composed of the sum of W_R and W_T. The interferometer errors can be calculated by substituting (14.142) into (14.139), which

14.16. IMPLEMENTATION AND APPLICATIONS

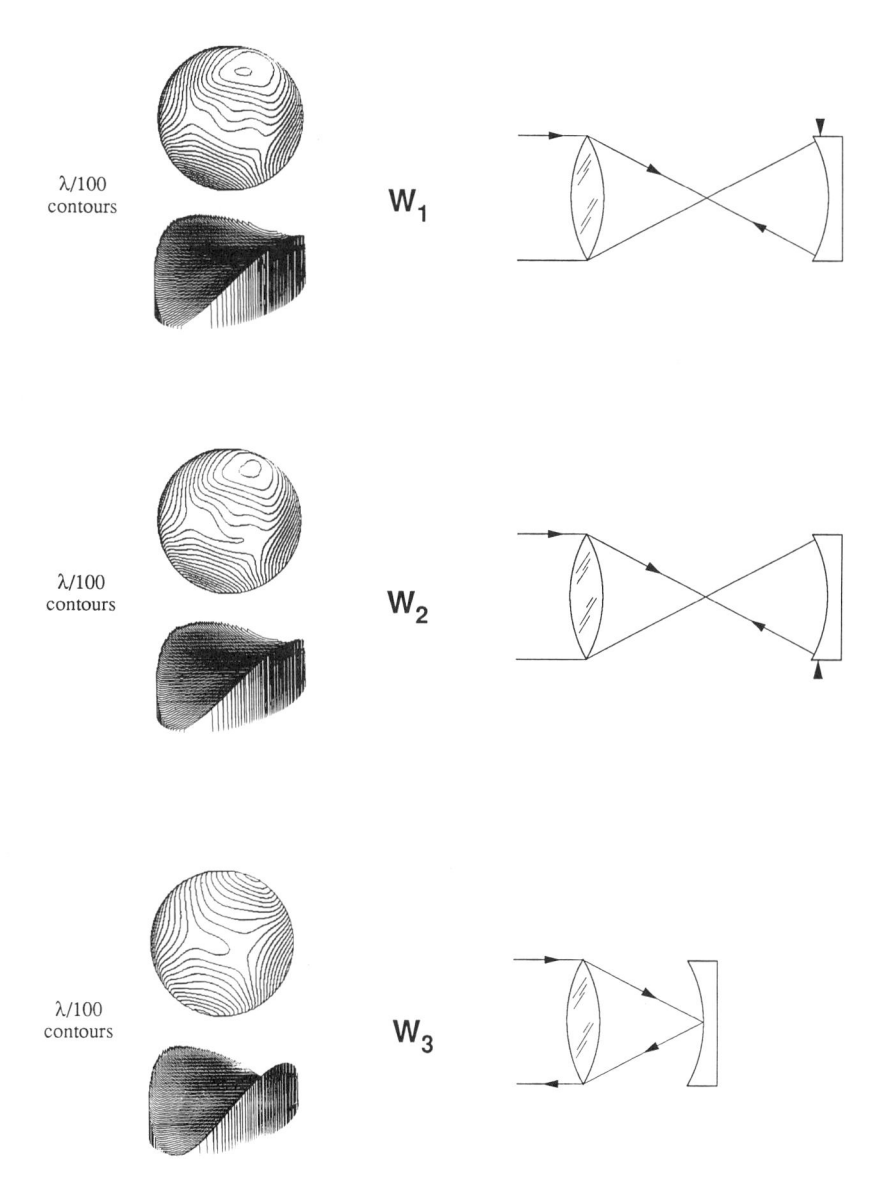

λ/100 contours

W_1

λ/100 contours

W_2

λ/100 contours

W_3

Figure 14.32. Absolute calibration of a spherical surface: (*a*) three raw wavefronts of part with interferometer errors and (*b*) composite errors of the interferometer and test surface errors.

λ/100 contours

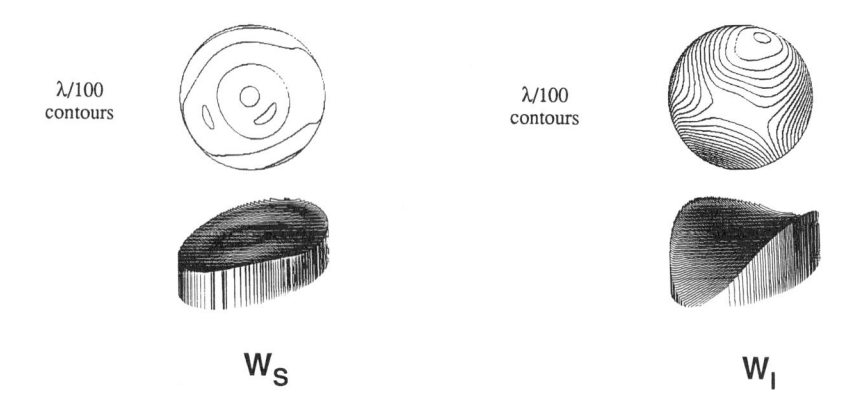

λ/100 contours

W_S

W_I

Figure 14.32. (*Continued*)

gives

$$W_I = \tfrac{1}{2}(W_1 - W_2^{180°} + W_3 + W_3^{180°}). \tag{14.143}$$

Note that both Eqs. (14.142) and (14.143) involve the manipulation of wavefronts that derive from the rotation of the part and the rotation of the wavefront data matrices by 180°. This allows the separation of even and odd errors in the interferometer when we have access to the cat's-eye configuration. Figure 14.32*b* shows the results of such calibration calculations as well as the individual and composite errors in the interferometer and test part. This method is usually employed when it is essential to achieve the very highest accuracies possible. The method is only effective when the interferometer is operated in a stable environment and the 0° and 180° test part configurations of W_1 and W_2 match the pixel coordinates (x', y') and $(-x', -y')$ of the matrix detector. This calibration procedure is conveniently carried out with PSIs that can easily manipulate and store wavefronts in memory. Errors incurred by improper alignment of the various surfaces have been examined by Truax (1988) and Elssner et al. (1989). Generally, lateral alignments are the most critical and need to be held to within one pixel or less.

The special case of flat surfaces can be calibrated by using three surfaces in different orientations. A similar approach has been applied to Fizeau reference calibration (Schulz and Schwider 1976). The use of Zernike polynomials facilitates the calibration process for flats (Fritz 1984).

14.16.4. Calibration of Asymmetric Errors

There are situations when it is not possible to apply this absolute calibration technique, such as when we have a collimated or diverging test space in Twyman–Green or Fizeau interferometers. These configurations cannot directly pro-

14.16. IMPLEMENTATION AND APPLICATIONS

vide data from another position (such as the cat's-eye position) sufficient to determine all errors. It is quite important, however, in many situations to at least be able to accurately quantify the asymmetrical errors in a test piece or optical system. Parks (1978) and later Fritz (1984) showed how this could be achieved through the use of Zernike polynomials. Using the same nomenclature as before, we acquire two or more wavefronts, the only difference being the azimuthal orientations of the test piece. Let W_1 and W_2 be test wavefronts obtained at orientations 0 and α. The angle α will generally not be 90° or 180°. The wavefronts W_1 and W_2 can be broken down into their symmetric and asymmetric parts, which will also include the errors in the interferometer W_I:

$$W_1 = W_{\text{sym}} + W_{\text{asy}} + W_I, \qquad (14.144)$$

$$W_2 = W_{\text{sym}} + W_{\text{asy}}^{\alpha} + W_I. \qquad (14.145)$$

If we subtract these two wavefronts, the difference wavefront

$$\Delta W = W_2 - W_1 = W_{\text{asy}}^{\alpha} - W_{\text{asy}} \qquad (14.146)$$

eliminates the interferometer errors W_I, as well as the symmetric errors of the surface under test, leaving only the asymmetric information of the test surface. We can extract this error using the properties of the Zernike polynomials. Let Z_1 and Z_2 be Zernike polynomial representations of wavefronts W_1 and W_2, which in general form (see Chapter 16) are given by

$$Z(r, \theta) = \sum_{l,k} R_l^k(r)[c_{lk} \cos(k\theta) + d_{lk} \sin(k\theta)], \qquad (14.147)$$

where $R_l^k(r)$ are the radial polynomials (which, by definition, are symmetric), and c_{lk} and d_{lk} are coefficients of the angular terms, which specify the azimuthal or asymmetric variation of the wavefront. It is clear that when we replace θ by $(\theta + \alpha)$, Eq. (14.147) becomes

$$Z(r, \theta + \alpha) = \sum_{l,k} R_l^k(r)[c'_{lk} \cos(k\theta) + d'_{lk} \sin(k\theta)], \qquad (14.148)$$

where the coefficients c'_{lk} and d'_{lk} can be represented in terms of the unrotated coefficients c_{lk} and d_{lk} of Eq. (14.147) and the rotation angle α:

$$c'_{lk} = c_{lk} \cos(k\alpha) + d_{lk} \sin(k\alpha), \qquad (14.149)$$

$$d'_{lk} = d_{lk} \cos(k\alpha) - c_{lk} \sin(k\alpha). \qquad (14.150)$$

Taking the difference of Eqs. (14.147) and (14.148) yields a Zernike polynomial representation ΔZ of the measured wavefront difference ΔW of Eq. (14.146):

$$\Delta Z = \sum_{l,k} R_l^k(r)[(c'_{lk} - c_{lk})\cos(k\theta) + (d'_{lk} - d_{lk})\sin(k\theta)], \quad (14.151)$$

which we now define in terms of difference coefficients Δc_{lk} and Δd_{lk}:

$$\Delta Z = \sum_{l,K} R_l^k(r)[\Delta c_{lk}\cos(k\theta) + \Delta d_{lk}\sin(k\theta)]. \quad (14.152)$$

Using Eq. (14.152), we can fit the Zernike wavefront ΔZ to the measured wavefront difference ΔW of the original and rotated wavefront of Eq. (14.146). The coefficients c_{lk} and d_{lk} of the asymmetrical part of the test surface W_{asy} can now be expressed in terms of the Zernike coefficients Δc_{lk} and Δd_{lk} of Eq. (14.152), independent of the interferometer errors, using Eqs. (14.149) and (14.150):

$$c_{lk} = \frac{1}{2}\left[\Delta c_{lk} + \frac{\Delta d_{lk}\sin(k\alpha)}{(1 - \cos k\alpha)}\right] \quad (14.153)$$

$$d_{lk} = \frac{1}{2}\left[\Delta d_{lk} - \frac{\Delta c_{lk}\sin(k\alpha)}{(1 - \cos k\alpha)}\right]. \quad (14.154)$$

Equation (14.147) can now be evaluated to obtain W_{asy}. These relations may be used to analyze interferograms for any angle of azimuthal shear α, provided $\cos(k\alpha) \neq 1$ for all $k < K$, where K is the desired order of the Zernike fit. In view of the expected decrease in the contributions of the higher-order harmonics, an optimum value of the shear angle α was proposed by Gubin and Sharonov (1990):

$$\alpha \approx \frac{3}{2}\left(\frac{\pi}{K}\right). \quad (14.155)$$

Accuracy can be improved by averaging more interferograms over several azimuthal orientations as described by Parks (1978) and Gubin and Sharonov (1990).

To demonstrate the value of both of these methods, we chose an interferometer that had very large residual errors as can be seen in Fig. 14.32a, and a test surface that we knew to be very good. The results indicate we are able to make meaningful measurements to accuracy levels that are more than 10 times better than the errors in the instrument. Figure 14.33 shows example interferograms W'_1, W'_2 (taken of the same test part as in Fig. 14.32) and the asymmetric contribution W''_{asy} of the test part in the absence of interferometer errors using the method described. The primes are used to distinguish these measurements from the previous measurements. A 37-term Zernike polynomial was fit to the measured test surface in Fig. 14.32b that resulted from the three-wavefront

14.16. IMPLEMENTATION AND APPLICATIONS

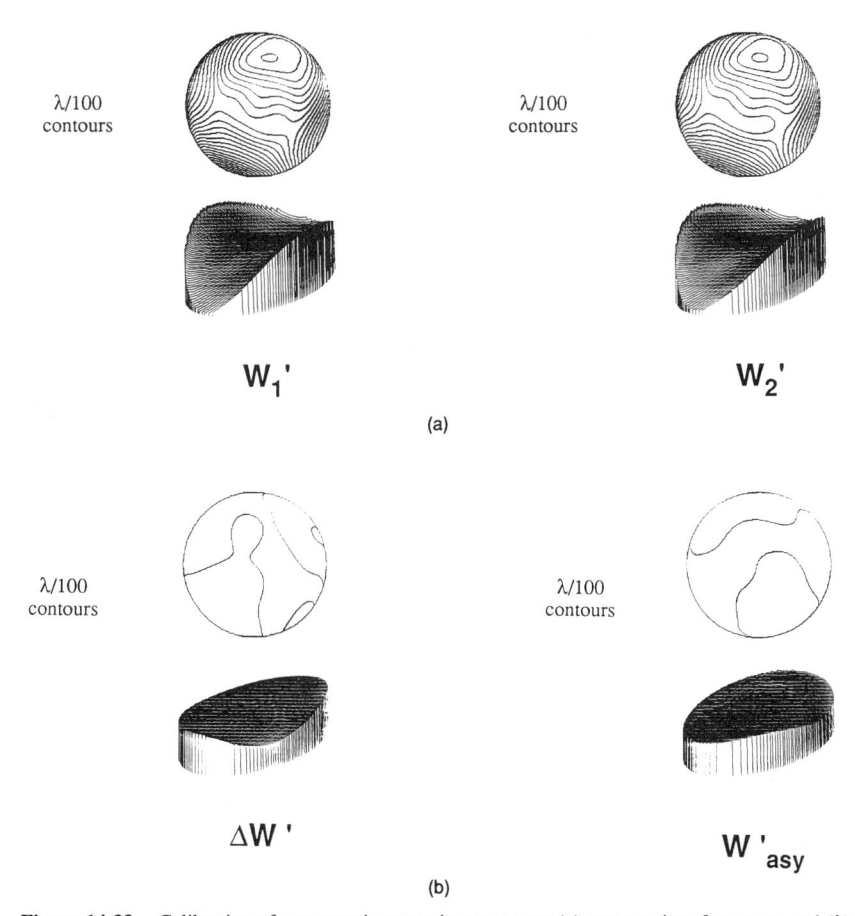

Figure 14.33. Calibration of asymmetric errors in a test part: (*a*) two raw interferograms and (*b*) the difference and asymmetric errors.

method. Figure 14.34 shows the symmetric and asymmetric components of this fit. When the asymmetric parts are compared from the two methods, the agreement is remarkable. The *P-V* difference of the two analyses is < 0.005 waves.

Although a bit more clumsy to implement, the same philosophy could be implemented to find the symmetric component of the surface under test. Here we could shear the test surface with itself in a radial or Cartesian (x, y) direction. A lateral shear by an amount equal to the radius of the part would yield the symmetric component of the part along a meridian. Averaging of results at several azimuthal orientations would improve the result. Accurate translations and rotations of the part are necessary to obtain good results. Radial shearing interferometry could be employed to achieve the same end, but the optical com-

$\lambda/100$ contours

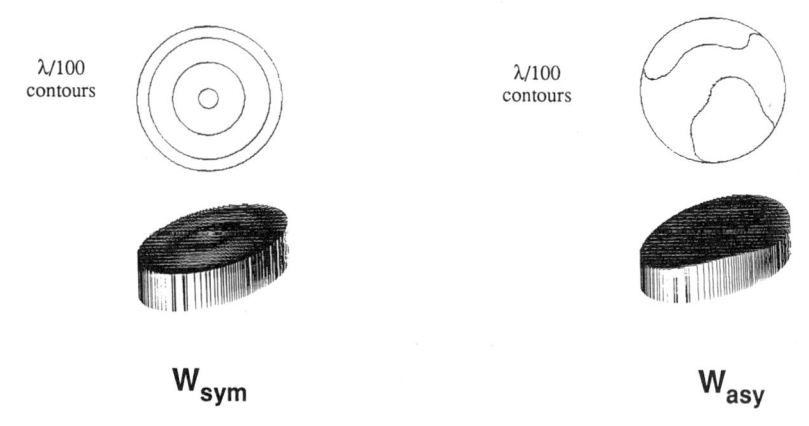

$\lambda/100$ contours

W_{sym} W_{asy}

Figure 14.34. Zernike fit of symmetric and asymmetric errors for test surface wavefront in Fig. 14.32b.

plexity of the interferometer as well as the data analysis are somewhat more troublesome (Kothiyal and Delisle 1985b; Freischlad 1987). See Chapter 4 and further discussions in this chapter.

14.16.5. Aspheric Testing

In addition to the extended-range PSI techniques discussed in Section 14.13, a number of other approaches have been applied to the problem of aspheric testing (Wyant 1987). These tests aim to reduce the number of fringes analyzed in a single interferogram. In particular, two of these methods are shearing interferometry and subaperture testing.

The attractive feature of using shearing interferometry for aspheric testing is that the sensitivity of the test can be varied by changing the shear distance (see Chapter 4). In this way the number of fringes is reduced. However, to measure an asymmetric wavefront, two sets of PSI data with orthogonal shears must be collected. Configurations based on lateral shear, radial shear, and the Ronchi test have been demonstrated (Hariharan et al. 1984; Seligson et al. 1984; Yatagai and Kanou 1984; Yatagai 1984; Kothiyal and Delisle 1985b; Kanoh 1986; Omura and Yatagai 1988). The wavefront slope is measured with shearing interferometry, and the actual wavefront or surface then is found by integration. A practical difficulty is that noise in the measurement tends to spread across the reconstructed wavefront as a result of this integration; a bad pixel can influence large portions of the wavefront map. When the wavefront is measured directly, the effects of a bad pixel are localized at that pixel.

The basic idea behind subaperture testing of aspherics is to divide the wavefront up into small sections; the wavefront departure in each subaperture is within the measurement range of the instrumentation. The maximum fringe fre-

14.16. IMPLEMENTATION AND APPLICATIONS

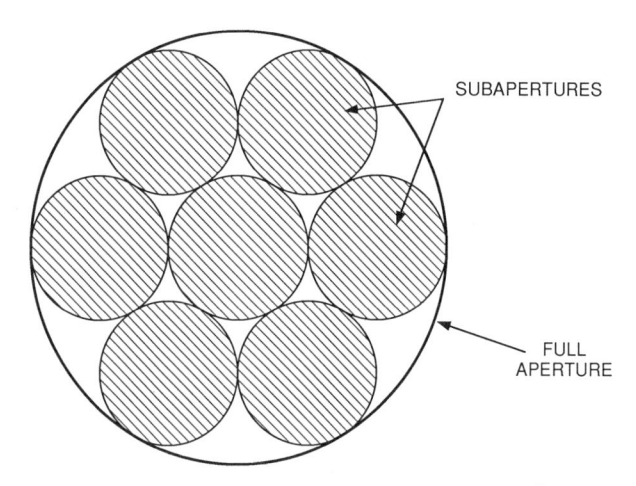

Figure 14.35. One configuration for subaperture testing.

quency is kept below the Nyquist frequency of the sensor. The problem is then reduced to fitting all of these separate measurements, which can contain different amounts of tilt, piston, and sometimes defocus, back into a complete map of the aspheric wavefront or surface (Kim 1982; Thunen and Kwon 1982; Jensen et al. 1984; Negro 1984; Stuhlinger 1986; Liu et al. 1988). An example of a set of subapertures is shown in Fig. 14.35. The overall wavefront is represented by a polynomial expansion, and the subaperture data is analyzed to determine the expansion coefficients. The Zernike polynomials are the usual choice, and a limited number of terms are used. This technique provides the overall aspheric wavefront shape. Small or localized errors will not appear in the final polynomial fit and must be determined from the subaperture data. It is important that the location of each subaperture within the aperture be precisely known.

Another subaperture technique has been developed to expand the field of view of interferometric surface profiling microscopes (Cochran and Wyant 1986; Cochran and Creath 1987, 1988). A series of partial overlapping linear traces of a surface are collected. These measurements are then combined into a single trace by adjusting the tilt, piston, and position to minimize the difference between the traces in the overlap region.

14.16.6. Radius of Curvature Measurement

A classical technique for the measurement of radius of curvature of a spherical surface involves positioning an autocollimator first at the radius of curvature or confocal position of the test part and then at the vertex or cat's-eye position while measuring the distance that the surface moves (Twyman 1952). By measuring wavefronts at these two points and accurately tracking the motion of the surface between points with a distance measuring interferometer or scale, radius measurement can be made in the part per million region. If the wavefront at each position is recorded, and the wavefront focus error at each of these points is calculated from these wavefronts, we can make a correction to the radius measurement determined by distance measuring interferometer. For the purpose of this discussion we will assume that there is no error in the measured travel distance of the part between the two points where the wavefronts are measured. From the calculated focus error at each point, and the knowledge of the numerical aperture for each wavefront, we can calculate the sagittal height correction needed at each point, δ_{confocal} and δ_{vertex}. Thus the correct value of radius of curvature R is given by (Bruning 1978)

$$R = R_{\text{measured}} + \delta_{\text{confocal}} - \delta_{\text{vertex}} \tag{14.156}$$

Selberg and Hunter (1990) have properly indicated that the exact sagittal calculation should be used for high numerical aperture (NA) surfaces, which amounts to

$$\delta_i = \frac{\Delta\phi\lambda}{1 - \sqrt{1 - (\text{NA}_{\text{LIMIT}})^2}}, \tag{14.157}$$

where $\Delta\phi$ is the focus error in waves, λ is the interferometer wavelength, and NA_{LIMIT} is the limiting NA at each of the δ_i positions. At the vertex position the limiting NA is that of the test arm of the interferometer. At the confocal position the limiting NA is that of the interferometer or the part under test. The error introduced can be substantial if the part being tested is very slow (i.e., subtends a small NA), and the focus error is not corrected by a PSI calculation. Equation (14.157) shows that, for He–Ne operation, a 0.01λ focus setting error at the confocal position causes a 1-μm radius measurement error for a surface of speed NA = 0.1. Selberg (1990b) points out that the interferometer errors, if large, can affect the measurement since the errors are different at the confocal position than at the cat's-eye position as seen, for example, in Fig. 14.32. This correction can be made with the absolute calibration procedure discussed in Section 14.16.3. Other error sources, such as distortion in the viewing arm, incorrect numerical aperture calculations, figure errors in the test surface, and alignment errors can contribute additional error. A complete error budget analysis summarizing these effects has been given by Selberg (1990b).

14.16.7. Sources

PSI systems have been built at wavelengths ranging from the ultraviolet to the infrared. The test wavelength is sometimes chosen to test the system at the operating wavelength (Prettyjohns et al. 1985). An improvement in the precision of the test results from using shorter wavelengths; 2π of wavefront phase corresponds to a smaller physical distance. Longer wavelengths reduce the sensitivity of the test and permit the testing of aspheric wavefronts without nulls. A second advantage to these long wavelengths is that high contrast fringes can be obtained even when testing ground surfaces, and testing can commence during the grinding process. A common IR wavelength that has been used for PSI is the 10.6-μm output of a CO_2 laser, and the detector that is used for these systems is a pyroelectric vidicon (PEV) (Kwon et al. 1980; Stahl and Koliopoulos 1987; Creath 1987b; Stahl 1989). PEVs respond only to time-varying imagery, which makes them appropriate for a PSI system. The phase shift provides the intensity variations, but the rate of phase shift must be adjusted to optimize the camera response. A modulation frequency of 7.5 Hz has been found to be an optimum in one system (Stahl and Koliopoulos, 1987). Work in the development of IR solid-state sensors has pushed the cutoff wavelength into the mid-IR, about 6 μm (Zanio 1990; Kozlowski et al. 1990). This work in materials such as HgCdTe may eventually provide an alternative to PEVs.

An area that shows great promise for simplifying the operation of PSI systems is to obtain the phase shifts through modulation of the output wavelength of a laser diode (Tatsuno and Tsunoda 1987; Ishii et al. 1987; Chen and Murata 1988; Chen et al. 1988). The output wavelength of a laser diode varies with both the injection current and temperature. The laser diode output is sent to both arms of the interferometer and the phase shift is proportional to the OPD between the two arms:

$$\alpha = \frac{2\pi \text{OPD} \Delta \lambda}{\lambda_0^2}, \qquad (14.158)$$

where $\Delta \lambda$ is the wavelength change, and λ_0 is the average wavelength. As with the instrumentation for the 2 + 1 algorithm (Section 14.8.6), this phase shift corresponds to the difference in the number of wavelengths that fit into the OPD at the two wavelengths. The laser wavelength can either be stepped to implement the phase step PSI approach or ramped for the integrating bucket approach. This method of phase shift eliminates the need for a mechanical motion of a PZT and the associated high-voltage drive circuit.

A source modulation interferometer has been used to measure both surfaces and the refractive index inhomogeneities of a glass plate from a set of phase-shifted interferograms (Okada et al. 1990). The plate is put in the test arm of a Twyman–Green interferometer, and fringes are formed between the reference surface and the front surface reflection, between the reference and the rear surface reflection, and between the two surface reflections. Since each of these overlapping fringe patterns corresponds to a different OPD, each phase shifts at a different rate, and the analysis is able to separate the three quantities to be measured.

A phase lock laser diode interferometer (see Section 14.14.4) has also been demonstrated by modulating the drive current, and therefore the output wavelength, of the diode (Suzuki et al. 1989). Two potential problems with the use of laser diode sources are changes in the output power as the wavelength is varied and the effects of a small wavelength modulation needed to lock the wavelength to a reference cavity (Hariharan 1989a, 1989b).

14.16.8. Alignment Fiducials

One area that is often overlooked in interferometric testing is the accuracy required in relating the position of a fringe or a point in an interferogram to an actual position in the aperture of the component or system under test. This relationship is accomplished by placing fiducial marks on the test aperture. The image of these marks in the interferogram relates the interferogram coordinates to locations on the part. Precise fiducial location determination allows for the removal of calibration wavefronts, such as those due to the interferometer or reference optics. These alignment aids are important for the absolute calibration and subaperture techniques described earlier. Fiducials also provide a coordinate system for geometric wavefront manipulations, including flips, rotations, and magnification changes, as well as for aberration fits. The importance of the geometrical alignment of a test part with the analysis can be understood by considering the situation of subtracting a stored calibration wavefront from a measured wavefront. Misalignment will shift the error to be subtracted on itself during the analysis. These shifts can result in significant comatic and/or astigmatic errors.

14.17. FUTURE TRENDS FOR PSI

This chapter has concentrated on the theory of phase shifting analysis and its application to the area of interferometry. PSI has provided improved repeatability, precision, and accuracy and allowed for a more complete analysis of the wavefront and surface data. Not surprisingly, these same techniques can be applied to almost any system that uses fringes for measurement. Chapter 15 describes the application of PSI systems to holographic and speckle interferometers (Hariharan et al. 1982a, 1982b, 1983; Creath 1985; Hariharan 1985; Nakadate and Saito 1985; Thalmann and Dändliker 1985), and Chapter 16 discusses its use with Moiré and fringe projection techniques (Indebetouw 1978; Moore and Truax 1979; Shagam 1983; Bell and Koliopoulos 1984). Optical profilers (Chapter 17) and distance measuring interferometers (Chapter 18) are two other areas of instrumentation where phase shifting techniques have been applied with great success.

There are several areas where further improvements to this technology are needed. A partial list includes systems that work in the deep ultraviolet with excimer lasers, systems with improved compatibility with pulsed lasers, improved calibration procedures through ray tracing the aberrations of the interferometer, the integration of alignment fiducials into the analysis programs, and the testing of aspheric surfaces and wavefronts.

Even with the available technology, aspherics remain difficult to test. There exists a need for a device that could be described as a walkup aspheric interferometer. Such a system would allow for the testing of an aspheric surface without the need for the preparation of any special reference or null optics; this is equivalent to the situation that exists for spherical optics today. It remains to be seen if the extended range techniques, such as sub-Nyquist interferometry, will provide a path for reaching this goal. The development of this instrument will almost certainly be coupled with the integration of optical design programs into the interferometer to provide for on-line ray tracing and calibration.

There are some current trends in PSI systems that are certain to remain. The continued influence of the improvements in solid-state electronics on PSI will mean larger detector arrays, detector arrays that are sensitive to a broader range of wavelengths, and ever faster computations. The computer interface will become more flexible and user friendly. The use of interferometers in conjunction with optical design programs will also become important for optimization, alignment, and testing of systems. In manufacturing, the computer analysis of interferograms permits the integration of statistical process control with the manufacturing process to improve quality and yield. Interferometric testing will also be applied to a greater variety of nonoptical parts.

It is clear that over the last 20 years, phase shifting interferometry has become established as an integral part of optical testing. Its success can be measured by its widespread use and the number of applications to which it has been applied.

Acknowledgments. The authors acknowledge the support of Eastman Kodak Company and GCA Tropel during the preparation of this manuscript.

REFERENCES

Ai, C. and J. C. Wyant, "Effect of Spurious Reflection on Phase Shift Interferometry," *Appl. Opt.*, **27**, 3039 (1988).

Angel, J. R. P. and P. L. Wizinowich, "A Method of Phase Shifting in the Presence of Vibration," *European Southern Observatory Conf. Proc.*, **30**, 561 (1988).

Barnes, T. H., "Heterodyne Fizeau Interferometer for Testing Flat Surfaces," *Appl. Opt.*, **26**, 2804 (1987).

Bell, B. W. and C. L. Koliopoulos, "Moire Topography, Sampling Theory, and Charged Coupled Devices," *Opt. Lett.*, **9**, 171 (1984).

Bhushan, B., J. C. Wyant, and C. L. Koliopoulos, "Measurement of Surface Topography of Magnetic Tapes by Mirau Interferometry," *Appl. Opt.*, **24**, 1489 (1985).

Bone, D. A., H. A. Bachor, and R. J. Sandeman, "Fringe Pattern Analysis Using a 2-D Fourier Transform," *Appl. Opt.*, **25**, 1653 (1986).

Brophy, C. P., "Effect of Intensity Error Correlation on the Computed Phase of Phase-Shifting Interferometry," *J. Opt. Soc. Am. A*, **7**, 537 (1990).

Bruning, J. H., "Fringe Scanning Interferometers," in *Optical Shop Testing*, D. Malacara, Ed., Wiley, New York, 1978.

Bruning, J. H. and D. R. Herriott, "A Versatile Laser Interferometer," *Appl. Opt.*, **9**, 2180 (1970).

Bruning, J. H., D. R. Herriott, J. E. Gallagher, D. P. Rosenfeld, A. D. White, and D. J. Brangaccio, "Digital Wavefront Measuring Interferometer for Testing Optical Surfaces and Lenses," *Appl. Opt.*, **13**, 2693 (1974).

Bryngdahl, O., "Polarization-Type Interference Fringe Shifter," *J. Opt. Soc. Am.*, **62**, 462 (1972).

Bryngdahl, O., "Heterodyne Shearing Interferometers Using Diffractive Filters with Rotational Symmetry," *Opt. Comm.*, **17**, 43 (1976).

Burgwald, G. M. and W. P. Kruger, "An Instant-on Laser for Distance Measurement," *Hewlett-Packard J.*, **21**, 14 (1970).

Carré, P. "Installation et Utilisation du Comparateur Photoelectrique et Interferentiel du Bureau International des Poids de Mesures," *Metrologia* **2**, 13 (1966).

Chen, J. and K. Murata, "Digital Phase Measuring Fizeau Interferometer for Testing of Flat and Spherical Surfaces," *Optik*, **81**, 28 (1988).

Chen, J., Y. Ishii, and K. Murata, "Heterodyne Interferometry with a Frequency-Modulated Laser Diode," *Appl. Opt.*, **27**, 124 (1988).

Cheng, Y.-Y. and J. C. Wyant, "Two-Wavelength Phase Shifting Interferometry," *Appl. Opt.*, **23**, 4539 (1984).

Cheng, Y.-Y. and J. C. Wyant, "Multiple-Wavelength Phase-Shifting Interferometry," *Appl. Opt.*, **24**, 804 (1985a).

Cheng, Y.-Y. and J. C. Wyant, "Phase Shifter Calibration in Phase-Shifting Interferometry," *Appl. Opt.*, **24**, 3049 (1985b).

Cochran, E. R. and K. Creath, "A Method for Extending the Measurement Range of a Two-Dimensional Surface Profiling Instrument," *Proc. SPIE*, **818**, 353 (1987).

Cochran, E. R. and K. Creath, "Combining Multiple-Subaperture and Two-Wavelength Techniques to Extend the Measurement Limits of an Optical Surface Profiler," *Appl. Opt.*, **27**, 1960 (1988).

Cochran, E. R. and J. C. Wyant, "Longscan Surface Profile Measurements Using a Phase-Modulated Mirau Interferometer," *Proc. SPIE*, **680**, 112 (1986).

Crane, R., "Interference Phase Measurement," *Appl. Opt.*, **8**, 538 (1969).

Creath, K., "Phase-Shifting Speckle Interferometry," *Appl. Opt.*, **24**, 3053 (1985).

Creath, K., "A Comparison of Phase-Measurement Algorithms," *Proc. SPIE*, **680**, 19 (1986).

Creath, K., "Step Height Measurement Using Two-Wavelength Phase-Shifting Interferometry," *Appl. Opt*, **26**, 2810 (1987a).

REFERENCES

Creath, K., "WYKO Systems for Optical Metrology," *Proc. SPIE*, **816,** 111 (1987b).

Creath, K., "Phase-Measurement Interferometry Techniques," in *Progress in Optics. Vol. XXVI*, E. Wolf, Ed., Elsevier Science Publishers, Amsterdam, 1988, pp. 349–393.

Creath, K., "High Resolution Optical Profiler," *SPIE*, **1164,** 142 (1989).

Creath, K. and J. C. Wyant, "Direct Phase Measurement of Aspheric Surface Contours," *Proc. SPIE*, **645,** 101 (1986).

Creath, K., Y.-Y. Cheng, and J. C. Wyant, "Contouring Aspheric Surfaces Using Two-Wavelength Phase-Shifting Interferometry," *Opt. Acta*, **32,** 1455 (1985).

Crosdale, F. and R. Palum, "Wavelength Control of a Diode Laser for Distance Measuring Interferometry," *Proc. SPIE*, **1219,** 490 (1990).

DeCou, A. B., "Interferometric Star Tracking," *Appl. Opt.*, **13,** 414 (1974).

Dorsey, A., R. J. Hocken, and M. Horowitz, "A Low Cost Laser Interferometer System for Machine Tool Applications," *Prec. Eng.*, **5,** 29 (1983).

Dukes, J. N. and G. B. Gordon, "A Two-hundred Foot Yardstick with Graduations Every Microinch," *Hewlett-Packard J.*, **21,** 2 (1970).

Elssner, K.-E., R. Burow, J. Grzanna, and R. Spolaczyk, "Absolute Sphericity Measurement," *Appl. Opt.*, **28,** 4649 (1989).

Evans, J. T., "Real Time Metrology Using Heterodyne Interferometry," *Proc. SPIE*, **429,** 199 (1983).

Fercher, A. F., H. Z. Hu, and U. Vry, "Rough Surface Interferometry with a Two-Wavelength Heterodyne Speckle Interferometer," *Appl. Opt.*, **24,** 2181 (1985).

Figoski, J. W., T. E. Shrode, and G. F. Moore, "Computer-Aided Alignment of a Wide-Field, Three-Mirror, Unobscured, High-Resolution Sensor," *Proc. SPIE*, **1049,** 166 (1989).

Freischlad, K., "Sensitivity of Heterodyne Shearing Interferometers," *Appl. Opt.*, **26,** 4053 (1987).

Freischlad, K. and C. L. Koliopoulos, "Fourier Description of Digital Phase-Measuring Interferometry," *J. Opt. Soc. Am. A*, **7,** 542 (1990).

Fritz, B. S., "Absolute Calibration of an Optical Flat," *Opt. Eng.*, **23,** 379 (1984).

Gallagher, J. E. and D. R. Herriott, "Wavefront Measurement." U.S. Patent 3,694,088 (1972/1972).

Gaskill, J. D., *Linear Systems, Fourier Transforms and Optics*, Wiley, New York, 1978.

Goodman, J. W. *Introduction to Fourier Optics*, McGraw-Hill, San Francisco, 1968.

Greivenkamp, J. E., "Generalized Data Reduction for Heterodyne Interferometry," *Opt. Eng.*, **23,** 350 (1984).

Greivenkamp, J. E., "Interferometric Measurements at Eastman Kodak Company," *Proc. SPIE*, **816,** 212 (1987a).

Greivenkamp, J. E., "Sub-Nyquist Interferometry," *Appl. Opt.*, **26,** 5245 (1987b).

Greivenkamp, J. E., K. G. Sullivan, and R. J. Palum, "Resolving Interferometric Step Height Measurement Ambiguities Using a Priori Information," *Proc. SPIE*, **1164,** 79 (1989).

Gubin, V. B. and V. N. Sharonov, "Algorithm for Reconstructing the Shape of Optical Surfaces from the Results of Experimental Data," *Sov. J. Opt. Technol.*, **57**(3) (1990).

Hall, J. A., "Arrays and Charge-Coupled Devices," in *Applied Optics and Optical Engineering*, Vol. VIII, R. R. Shannon and J. C. Wyant, Eds., Academic Press, New York, 1980.

Hardy, J. W. and A. J. McGovern, "Shearing Interferometry: A Flexible Technique for Wavefront Measurement," *Proc. SPIE*, **816**, 180 (1987).

Hardy, J. W., J. E. Lefebvre, and C. L. Koliopoulos, "Real-Time Atmospheric Compensation," *J. Opt. Soc. Am.*, **67**, 360 (1977).

Hariharan, P., *Optical Holography*, Cambridge, London, 1984 p. 232.

Hariharan, P., "Quasi-Heterodyne Hologram Interferometry," *Opt. Eng.*, **24**, 632 (1985).

Hariharan, P. "Digital Phase-Stepping Interferometry: Effects of Multiple Reflected Beams," *Appl. Opt.*, **26**, 2506 (1987).

Hariharan, P., "Phase-Stepping Interferometry with Laser Diodes: Effect of Changes in Laser Power with Output Wavelength," *Appl. Opt.*, **28**, 27 (1989a).

Hariharan, P., "Phase-Stepping Interferometry with Laser Diodes. 2: Effects of Laser Wavelength Modulation," *Appl. Opt.*, **28**, 1749 (1989b).

Hariharan, P., B. F. Oreb, and N. Brown, "A Digital Phase-Measurement System for Realtime Holographic Interferometry," *Opt. Comm.*, **41**, 393 (1982a).

Hariharan, P., B. F. Oreb, and N. Brown, "A Digital System for Real-Time Holographic Stress Analysis," *Proc. SPIE*, **370**, 189 (1982b).

Hariharan, P., B. F. Oreb, and N. Brown, "Real-Time Holographic Interferometry: A Microcomputer System for Measurement of Vector Displacements," *Appl. Opt.*, **22**, 876 (1983).

Hariharan, P., B. F. Oreb, and T. Eiju, "Digital Phase-Shifting Interferometry: A Simple Error-Compensating Phase Calculation Algorithm," *Appl. Opt.*, **26**, 2504 (1987a).

Hariharan, P., B. F. Oreb, and C. H. Freund, "Stroboscopic Holographic Interferometry: Measurements of Vector Components of Vibration," *Appl. Opt.*, **26**, 3899 (1987b).

Hariharan, P., B. F. Oreb, and Z. Wanzhi, "Measurement of Aspheric Surfaces Using a Microcomputer-Controlled Digital Radial-Shear Interferometer," *Opt. Acta*, **31**, 989 (1984).

Hayes, J. B., "Linear Methods of Computer Controlled Optical Figuring", Ph.D. Dissertation, Optical Sciences Center, Univ. of Arizona, Tucson, AZ, 1984.

Hayes, J. B., "Compact Micromotion Translator," U.S. Patent 4,884,003 (1989).

Hayes, J. and S. Lange, "A Heterodyne Interferometer for Testing Laser Diodes," *Proc. SPIE*, **429**, 22 (1983).

Horwitz, B. A., "Multiplex Techniques for Real-Time Shearing Interferometry," *Appl. Opt.*, **29**, 1223 (1990).

Hu, H. Z., "Polarization Heterodyne Interferometry Using a Simple Rotating Analyzer. 1: Theory and Error Analysis," *Appl. Opt.*, **22**, 2052 (1983).

Ichioka, Y. and M. Inuiya, "Direct Phase Detecting System," *Appl. Opt.*, **11**, 1507 (1972).

Indebetouw, G., "Profile Measurement Using Projection of Running Fringes," *Appl. Opt.*, **17**, 2930 (1978).

Ishii, Y., J. Chen, and K. Murata, "Digital Phase-Measuring Interferometry with a Tunable Laser Diode," *Opt. Lett.*, **12**, 233 (1987).

Jensen, A. E., "Absolute Calibration Method for Laser Twyman–Green Wavefront Testing Interferometers," (abstract only), *J. Opt. Soc. Am.*, **63**, 1313 (1973).

Jensen, S. C., W. W. Chow, and G. N. Lawrence, "Subaperture Testing Approaches: A Comparison," *Appl. Opt.*, **23**, 740 (1984).

Johnson, G. W. and D. T. Moore, "Design and Construction of a Phase-Locked Interference Microscope," *Proc. SPIE*, **103**, 76 (1977).

Johnson, G. W., D. C. Leiner, and D. T. Moore, "Phase-Locked Interferometry," *Proc. SPIE*, **126**, 152 (1977).

Johnson, G. W., D. C. Leiner, and D. T. Moore, "Phase-Locked Interferometry," *Opt. Eng.*, **18**, 46 (1979).

Kanoh, T., "Automated Interferometric System for Aspheric Surface Testing," *Proc. SPIE*, **680**, 71 (1986).

Kim, C. J., "Polynomial Fit of Interferograms," *Appl. Opt.*, **21**, 4521 (1982).

Kim, C. J. and R. R. Shannon, "Catalog of Zernike Polynomials," in *Applied Optics and Optical Engineering*, Vol. X, R. R. Shannon and J. C. Wyant, Eds., Academic Press, San Diego, 1987.

Kinnstaetter, K., A. W. Lohmann, J. Schwider, and N. Streibl, "Accuracy of Phase-Shifting Interferometry," *Appl. Opt.*, **27**, 5082 (1988).

Koliopoulos, C. L., "Radial Grating Lateral Shear Heterodyne Interferometer," *Appl. Opt.*, **19**, 1523 (1980).

Koliopoulos, C. L., "Interferometric Optical Phase Measurement," Ph.D. Dissertation, Univ. Arizona, Tucson, 1981.

Kothiyal, M. P. and C. Delisle, "Optical Frequency Shifter for Heterodyne Interferometry Using Counterrotating Wave Plates," *Opt. Lett.*, **9**, 319 (1984).

Kothiyal, M. P. and C. Delisle, "Rotating Analyzer Heterodyne Interferometer: Error Analysis," *Appl. Opt.*, **24**, 2288 (1985a).

Kothiyal, M. P. and C. Delisle, "Shearing Interferometer for Phase Shifting Interferometry with Polarization Phase Shifter," *Appl. Opt.*, **24**, 4439 (1985b).

Kozlowski, L. J., K. Vural, V. H. Johnson, J. K. Chen, R. B. Bailey, D. Bui, M. J. Gubala, and J. R. Teague, "256 × 256 PACE-1 PV HgCdTe Focal Plane Arrays for Medium and Short Wavelength Infrared Applications," *Proc. SPIE*, **1308**, 202 (1990).

Kreis, T., "Digital Holographic Interference-Phase Measuring Using the Fourier-Transform Method," *J. Opt. Soc. Am. A*, **3**, 847 (1986).

Kujawinska, M., L. Salbut and K. Patorski, "Three-Channel Phase Stepped System for Moire Interferometry," *Appl. Opt.*, **30**, 1663 (1991).

Kwon, O. Y., "Multichannel Phase-Shifted Interferometer," *Opt. Lett.*, **9**, 59 (1984).

Kwon, O. Y., "Advanced Wavefront Testing at Lockheed," *Proc. SPIE*, **816**, 196 (1987).

Kwon, O. Y., D. M. Shough, and R. A. Williams, "Stroboscopic Phase-Shifting Interferometer," *Opt. Lett.*, **12**, 855 (1987).

Kwon, O. Y., J. C. Wyant, and C. R. Hayslett, "Rough Surface Interferometry at 10.6 μm," *Appl. Opt.*, **19**, 1862 (1980).

Lai, G. and T. Yatagai, "Generalized Phase-Shifting Interferometry," *J. Opt. Soc. Am. A*, **8**, 822 (1991).

Liu, Y.-M., G. N. Lawrence, and C. L. Koliopoulos, "Subaperture Testing of Aspheres with Annular Zones," *Appl. Opt.*, **27**, 4504 (1988).

Macy, W. W., "Two-Dimensional Fringe-Pattern Analysis," *Appl. Opt.*, **22**, 3898 (1983).

Malacara, D., "Phase Shifting Interferometry," *Revista Mexicana Fisica*, **36**, 6 (1990).

Malacara, D., I. Rizo and A. Morales, "Interferometry and the Doppler Effect," *Appl. Opt.*, **8**, 1746 (1969).

Massie, N. A., "Quasi-Real-Time High Precision Interferometric Measurements of Deforming Surfaces," *Proc. SPIE*, **153**, 126 (1978).

Massie, N. A., "Real-Time Digital Heterodyne Interferometry: A System," *Appl. Opt.*, **19**, 154 (1980).

Massie, N. A., "Digital Heterodyne Interferometry," *Proc. SPIE*, **816**, 40 (1987).

Massie, N. A. and R. D. Nelson, "Beam Quality of Acousto-Optic Phase Shifters," *Opt. Lett.*, **3**, 46 (1978).

Massie, N. A., M. Dunn, D. Swain, S. Muenter, and J. Morris, "Measuring Laser Flow Fields with a 64-Channel Heterodyne Interferometer," *Appl. Opt.*, **22**, 2141 (1983).

Massie, N. A., J. Hartlove, D. Jungwirth, and J. Morris, "High Accuracy Interferometric Measurements of Electron-Beam Pumped Transverse-Flow Laser Media with 10 μsec Time Resolution," *Appl. Opt.*, **20**, 2372 (1981).

Massie, N. A., R. D. Nelson, and S. Holly, "High Performance Real-Time Heterodyne Interferometry," *Appl. Opt.*, **18**, 1797 (1979).

Matthews, H. J., D. K. Hamilton, and C. J. R. Sheppard, "Surface Profiling by Phase-Locked Interferometry," *Appl. Opt.*, **25**, 2372 (1986).

McLaughlin, J. L. and B. A. Horwitz, "Real-Time Snapshot Interferometer," *Proc. SPIE*, **680**, 35 (1986).

Mertz, L., "Complex Interferometry," *Appl. Opt.*, **22**, 1530 (1983a).

Mertz, L., "Real-Time Fringe-Pattern Analysis," *Appl. Opt.*, **22**, 1535 (1983b).

Mertz, L., "Optical Homodyne Phase Metrology," *Appl. Opt.*, **28**, 1011 (1989).

Moore, D. T., "Gradient Index Optics Design and Tolerancing," Ph.D. Dissertation, Institute of Optics, Univ. of Rochester, Rochester, NY, (1973).

Moore, D. T., R. Murray, and F. B. Neves, "Large Aperture ac Interferometer for Optical Testing," *Appl. Opt.*, **17**, 3959 (1978).

Moore, D. T. and B. E. Truax, "Phase-locked Moiré Fringe Analysis for Automated Contouring of Diffuse Surfaces," *Appl. Opt.*, **18**, 91 (1979).

REFERENCES

Moore, R. C., "Automatic Method of Real-Time Wavefront Analysis," *Opt. Eng.*, **18**, 461 (1979).

Moore, R. C. and F. H. Slaymaker, "Direct Measurement of Phase in a Spherical-Wave Fizeau Interferometer," *Appl. Opt.*, **19**, 2196 (1980).

Morgan, C. J., "Least-Squares Estimation in Phase-Measurement Interferometry," *Opt. Lett.*, **7**, 368 (1982).

Mottier, F. M., "Microprocessor-Based Automatic Heterodyne Interferometer," *Opt. Eng.*, **18**, 464 (1979).

Nakadate, S. and H. Saito, "Fringe Scanning Speckle-Pattern Interferometry," *Appl. Opt.*, **24**, 2172 (1985).

Negro, J. E., "Subaperture Optical System Testing," *Appl. Opt.*, **23**, 1921 (1984).

Nugent, K. A., "Interferogram Analysis Using an Accurate Fully Automatic Algorithm," *Appl. Opt.*, **24**, 3101 (1985).

Ohyama, N., S. Kinoshita, A. Cornejo-Rodriguez, T. Honda, and T. Tsujiuchi, "Accuracy of Phase Determination with Unequal Reference Phase Shift," *J. Opt. Soc. Am. A*, **5**, 2019 (1988).

Okada, K., H. Sakuta, T. Ose, and J. Tsujiuchi, "Separate Measurements of Surface Shapes and Refractive Index Inhomogeneity of an Optical Element Using Tunable-Source Phase Shifting Interferometry," *Appl. Opt.*, **29**, 3280 (1990).

Okoomian, H. J., "A Two Beam Polarization Technique to Measure Optical Phase," *Appl. Opt.*, **8**,, 2363 (1969).

Omura, K. and T. Yatagai, "Phase Measuring Ronchi Test," *Appl. Opt.*, **27**, 523 (1988).

Palum, R. J. and J. E. Greivenkamp, "Sub-Nyquist Interferometry: Results and Implementation Issues," *Proc. SPIE*, **1162**, 378 (1989).

Parks, R. E., "Removal of Test Optics Errors," *Proc. SPIE*, **153**, 56 (1978).

Powell, R. L. and K. A. Stetson, "Interferometric Vibration Analysis by Wavefront Reconstruction," *J. Opt. Soc. Am.*, **55**, 1593 (1965).

Prettyjohns, K. N., "Charge-Coupled Device Image Acquisition for Digital Phase Measurement Interferometry," *Opt. Eng.*, **23**, 371, (1984).

Prettyjohns, K. N., S. DeVore, E. Dereniak, and J. C. Wyant, "Direct Phase Measurement Interferometer Working at 3.8 μm.," *Appl. Opt.*, **24**, 2211 (1985).

Roddier, C. and F. Roddier, "Interferogram Analysis Using Fourier Transform Techniques," *Appl. Opt.*, **26**, 1668 (1987).

Rosenbluth, A. E. and N. Bobroff, "Optical Sources of Nonlinearity in Heterodyne Interferometers," *Prec. Eng.*, **12**, 7 (1990).

Schulz, G. and J. Schwider, "Interferometric Testing of Smooth Surfaces," in *Progress in Optics, Vol. XIII*, E. Wolf, Ed., North-Holland, Amsterdam, 1976, pp. 93–167.

Schwider, J., "Phase Shifting Interferometry: Reference Phase Error Reduction," *Appl. Opt.*, **28**, 3889 (1989).

Schwider, J., "Advanced Evaluation Techniques in Interferometry," in *Progress in Optics, Vol. XXVIII*, E. Wolf, Ed., Elsevier Science Publishers, Amsterdam, 1990, pp. 271–359.

Schwider, J., R. Burow, K. E. Elssner, J. Grzanna, R. Spolaczyk, and K. Merkel,

"Digital Wavefront Measuring Interferometry: Some Systematic Error Sources," *Appl. Opt.*, **22**, 3421 (1983).

Selberg, L. A., "Interferometer Accuracy and Precision," *Proc. SPIE*, **749**, 8 (1987).

Selberg, L. A., "Interferometer Accuracy and Precision," *Proc. SPIE*, **1400**, 24 (1990a).

Selberg, L. A., "Interferometric Radius Scale System Description and Analysis," Zygo Application Bulletin AB-0050, Zygo Corp., Middlefield, CT (1990b).

Selberg, L. A. and G. C. Hunter, "High Accuracy Interferometric Radius of Curvature Measurement," *OF&T Tech. Dig.*, OSA, **11**, 115 (1990).

Seligson, J. L., C. A. Callari, J. E. Greivenkamp, and J. W. Ward, "Stability of a Lateral-Shearing Heterodyne Twyman-Green Interferometer," *Opt. Eng.*, **23**, 353 (1984).

Shagam, R. N., "AC Phase Measurement Technique for Moiré Interferograms," *Proc. SPIE*, **429**, 35 (1983).

Shagam, R. N. and J. C. Wyant, "Optical Frequency Shifter for Heterodyne Interferometers Using Simple Rotating Polarization Retarders," *Appl. Opt.*, **17**, 3034 (1978).

Smythe, R. A. and R. Moore, "Instantaneous Phase Measuring Interferometry," *Opt. Eng.*, **23**, 361 (1984).

Smythe, R. A., J. A. Soobitsky, and B. E. Truax, "Recent Advances in Interferometry at Zygo," *Proc. SPIE*, **816**, 95 (1987).

Soobitsky, J. A., "Piezoelectric Micromotion Actuator," U.S. Pat. 4,577,131 (1986).

Sommargren, G. E., "Up/Down Frequency Shifter for Optical Heterodyne Interferometry," *J. Opt. Soc. Am.*, **65**, 960 (1975).

Sommargren, G. E., "Optical Heterodyne Profilometry," *Appl. Opt.*, **20**, 610 (1981).

Stahl, H. P., "IR Interferometry Using Pyroelectric Vidicons," *Lasers Optronics*, **57** (Sept. 1989).

Stahl, H. P. and C. L. Koliopoulos, "Interferometric Phase Measurement Using Pyroelectric Vidicons," *Appl. Opt.*, **26**, 1127 (1987).

Steinmetz, C. R., "Sub-Micron Position Measurement and Control on Precision Machine Tools with Laser Interferometry," *Precision Eng.*, **12**, 12 (1990).

Steinmetz, C. R., R. Burgoon, and J. Harris, "Accuracy Analysis and Improvements to the Hewlett-Packard Laser Interferometer System," *Proc. SPIE*, **816**, 79 (1987).

Stephenson, D., "Forget Finding the Problem . . . Just Fix It," *Proc. SPIE*, **1049**, 187 (1989).

Stevenson, W. H., "Optical Frequency Shifting by Means of a Rotating Diffraction Grating," *Appl. Opt.*, **9**, 649 (1970).

Stuhlinger, T. W., "Subaperture Optical Testing: Experimental Verification," *Proc. SPIE*, **656**, 118 (1986).

Stumpf, K. D., "Real-Time Interferometer," *Opt. Eng.*, **18**, 648 (1979).

Suzuki, T. and R. Hioki, "Translation of Light Frequency by a Moving Grating," *J. Opt. Soc. Am.*, **57**, 1551 (1967).

Suzuki, T., O. Sasaki, and T. Maruyama, "Phase Locked Laser Diode Interferometry for Surface Profile Measurement," *Appl. Opt.*, **28**, 4407 (1989).

REFERENCES

Synborski, C. E., "The Interferometric Analysis of Flatness by Eye and Computer," *Proc. SPIE*, **135**, 104 (1978).

Takeda, M., H. Ina, and S. Kobayashi, "Fourier-Transform Method of Fringe-Pattern Analysis for Computer-Based Topography and Interferometry," *J. Opt. Soc. Am.*, **72**, 156 (1982).

Tarbeyev, Y. V., *Proc. 3rd IMEKO TC-8-Symp. Theor. Metrology Berlin GDR*, p. 5 (1986).

Tatsuno, K. and Y. Tsunoda, "Diode Laser Direct Modulation Heterodyne Interferometer," *Appl. Opt.*, **26**, 37 (1987).

Thalmann, R. and R. Dändliker, "Holographic Contouring Using Electronic Phase Measurement," *Opt. Eng.*, **24**, 930 (1985).

Thunen, J. G. and O. Y. Kwon, "Full Aperture Testing with Subaperture Test Optics," *Proc. SPIE*, **351**, 19 (1982).

Toyooka, S. and M. Tominaga, "Spatial Fringe Scanning for Optical Phase Measurement," *Opt. Comm.*, **51**, 68 (1984).

Truax, B. E., "Programmable Interferometry," *Proc. SPIE*, **680**, 10 (1986).

Truax, B. E., "Absolute Interferometric Testing of Spherical Surfaces," *Proc. SPIE*, **966**, 130 (1988).

Twyman, F., *Prism and Lens Making*, Hilger and Watts, London, 1952.

van Wingerden, J., H. J. Frankena, and C. Smorenburg, "Linear Approximation for Measurement Errors in Phase Shifting Interferometry," *Appl. Opt.*, **30**, 2718 (1991).

Willey, G. and R. Patchin, "Optical Design Analysis Incorporating Actual System Interferometric Data," *Proc. SPIE*, **1168**, 176 (1989).

Wizinowich, P. L., "System for Phase Shifting Interferometry in the Presence of Vibration," *Proc. SPIE*, **1164**, 25 (1989).

Wizinowich, P. L., "Phase-Shifting Interferometry in the Presence of Vibration: A New Algorithm and System," *Appl. Opt.*, **29**, 3271 (1990).

Womack, K. H., "Interferometric Phase Measurement Using Spatial Synchronous Detection," *Opt. Eng.*, **23**, 391 (1984a).

Womack, K. H., "Fourier Domain Description of Interferogram Analysis," *Opt. Eng.*, **23**, 396 (1984b).

Wyant, J. C., "Double Frequency Grating Lateral Shear Interferometer," *Appl. Opt.*, **12**, 2057 (1973).

Wyant, J. C., "Use of an ac Heterodyne Lateral Shear Interferometer with Real-Time Wavefront Correction Systems," *Appl. Opt.*, **14**, 2622 (1975).

Wyant, J. C., "Interferometric Testing of Aspheric Surfaces," *Proc. SPIE*, **816**, 19 (1987).

Wyant, J. C. and K. Creath, "Two-Wavelength Phase-Shifting Interferometer and Method," U.S. Pat. 4,832,489 (1989).

Wyant, J. C. and K. N. Prettyjohns, "Optical Profiler Using Improved Phase-Shifting Interferometry," U.S. Pat. 4,639,139 (1987).

Wyant, J. C. and R. N. Shagam, "Use of Electronic Phase Measurement Techniques in Optical Testing," *Proc. ICO, Vol 11, Madrid*, 659 (1978).

Wyant, J. C., C. L. Koliopoulos, B. Bhushan and O. E. George, "An Optical Profilometer for Surface Characterization of Magnetic Media," *ASLE Trans.*, **27,** 101 (1984a).

Wyant, J. C., B. F. Oreb, and P. Hariharan, "Testing Aspherics Using Two-Wavelength Holography: Use of Digital Electronic Techniques," *Appl. Opt.*, **23,** 4020 (1984b).

Yatagai, T., "Fringe Scanning Ronchi Test for Aspherical Surfaces," *Appl. Opt.*, **23,** 3676 (1984).

Yatagai, T. and T. Kanou, "Aspherical Surface Testing with Shearing Interferometer Using Fringe Scanning Detection Method," *Opt. Eng.*, **23,** 357 (1984).

Zanio, K., "HgCdTe on Si for Hybrid and Monolithic FPAs," *Proc. SPIE*, **1308,** 180 (1990).

15

Holographic and Speckle Tests

K. Creath and J. C. Wyant

15.1. INTRODUCTION

One of the most useful applications of holographic and speckle interferometry is the testing of optical components and systems. A wavefront produced by a master optical system can be stored holographically and used to perform null tests of similar optical systems. The holographic test plate is very similar to a test plate used for making optical components. If a master optical system is not available for making a *real* hologram, a *synthetic* or computer-generated hologram (CGH) can be made to provide the reference wavefront. When an aspheric optical element with a large departure from a sphere is tested, a CGH can be combined with null optics to perform a null test. Phase-shifting interferometry techniques as described in Chapter 14 can be applied to all of these tests.

Optical surfaces that may require the use of a long wavelength source to perform the test can be measured using two-wavelength holography (TWH). This enables visible light to perform an interferometric test having a very large effective wavelength. Two-wavelength interferometry can be performed in conjunction with phase-shifting techniques to easily obtain quantitative data of the surface contour. The two-wavelength interferometric measurement at a single wavelength can be corrected using the information contained in a second wavelength. Similarly, multiple wavelengths can be used to correctly determine the height of discontinuous surface features such as steps.

Diffusely reflecting or polished surfaces that are subject to stress can be interferometrically compared with their normal states using holographic interferometry techniques. Both static and dynamic measurements can be made. Speckle-interferometry techniques, which do not require the use of an intermediate holographic recording, can provide interference fringes corresponding to a change in the object shape. These techniques replace the holographic recording medium with a TV camera, electronics (or computer or array proces-

---

Optical Shop Testing, Second Edition, Edited by Daniel Malacara.
ISBN 0-471-52232-5 © 1992, John Wiley & Sons, Inc.

sor), and a monitor. They electronically process fringe-intensity data from speckle patterns and holograms at video frame rates.

This chapter describes holographic and speckle techniques useful in optical testing. Since this subject has such a large amount of information to cover, emphasis is placed on basic descriptions of the techniques and examples showing results of measurements using these tests.

15.2. INTERFEROMETERS USING REAL HOLOGRAMS

Many different experimental setups can be used for the holographic testing of optical elements. Because a hologram is simply an interferogram with a large tilt angle between the reference and object wavefronts, holographic tests can be performed either with standard interferometers or with setups having a larger angle between the object and reference beams. Figures 15.1 and 15.2 show interferometers that can be used for making a hologram of a concave mirror. The hologram is made in a plane conjugate to the test mirror. Once the hologram is made, it can be replaced in the same location and reconstructed by illuminating with a plane wave and imaging onto a viewing screen. When the object beam is blocked and the reference mirror is tilted so that the plane reference wave interferes with the first-order diffraction from the hologram, the wavefront due to the mirror will be reconstructed. Because several diffraction orders are produced by the hologram, it is usually necessary to select one of the diffraction orders using a spatial filter. The imaging lens and spatial filter are only necessary for the reconstruction of the hologram.

Holograms can be recorded on photographic plates, thermoplastic materials, or in photorefractive crystals. Photographic plates provide the highest resolution; however, they require a lot of chemicals for processing and unless they are processed in situ, they are hard to replace in the correct location for real-time techniques (Biedermann 1975; Kurtz and Owen 1975). Thermoplastic materials provide up to 1000 lines/mm resolution and can be erased and reprocessed hundreds of times (Leung et al. 1979; Friesem et al. 1980). They also have a very fast turnaround time. Newport Corporation sells a thermoplastic camera using thermoplastic plates with 800 lines/mm resolution for holo-

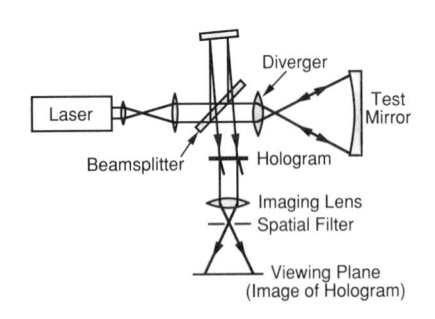

Figure 15.1. Interferometer utilizing a hologram.

15.2. INTERFEROMETERS USING REAL HOLOGRAMS

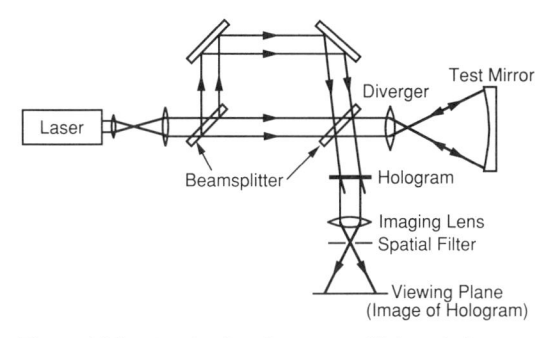

Figure 15.2. Another interferometer utilizing a hologram.

graphic interferometry purposes. Photorefractive crystals have a lot of potential as a high-resolution recording medium (Lam et al. 1984; Uhrich and Hesselink 1988); however, the optical setup is more complex and getting a high-quality crystal is not as easy as getting other recording materials. A very promising recording medium is the use of high-resolution detector arrays or charge-coupled device (CCD) cameras to record a hologram directly. These techniques will be discussed in greater detail in the section on TV holography.

15.2.1. Holographic Wavefront Storage

Sometimes it is convenient to holographically store a wavefront produced by an optical system and analyze the wavefront later without the test system present (Hildebrand et al. 1967; Hansler 1968b).

Care must be taken to ensure that the reconstructed wavefront is identical to the wavefront used to record the hologram. Errors are possible from differences between the reconstructing geometry and the recording geometry, recording-material deformation, and aberrations introduced by the recording-material substrate. Errors introduced because of differences in the recording and reconstructing geometries are greatly reduced if the reference wavefront is collimated and the object wavefront is as collimated as possible. Collimated wavefronts are particularly important if the reconstruction wavelength is different from the recording wavelength. Recording-material deformation can change the shape of the recorded interference fringes and thereby change the shape of the reconstructed wavefront. It is possible to keep the root-mean-square (*rms*) wavefront error less than $\lambda/40$ by using Kodak 649F photographic plates (Wyant and Bennett 1972). Glass substrates used for photographic plates generally show optical-thickness variations of at least one fringe per inch. For wavefront storage, this magnitude of error is not acceptable. This problem can be solved by either putting the hologram in a fluid gate or index matching the two surfaces to good optical flats. Thickness variations can also be minimized by sending both beams through the hologram to cancel the errors.

Stored holograms can be used to test for symmetry in optical components (Greivenkamp 1987). A hologram is made of the test surface, and then the test surface is rotated with respect to the hologram. The fringes from the interference between the stored wavefront and the wavefront produced by the rotated test surface will correspond to symmetry deviations in the test surface. Care must be taken not to translate the test surface as it is rotated because unwanted fringes will affect the test.

15.2.2. Holographic Test Plate

If an ideal optical system is available, the wavefront produced by the system can be stored holographically and used to test other optical systems. This is very similar to the use of a test plate in the testing of optical components (Hansler 1968a; Pastor 1969; Snow and Vandewarker 1970; Lurionov et al. 1972; Broder-Bursztyn and Malacara-Hernández 1975). As an example, the setup shown in Fig. 15.1 is used to make a hologram of a wavefront produced by a concave mirror. After processing the hologram, it is placed in its original position and the master mirror is replaced with a test mirror. The wavefront stored in the hologram is interferometrically compared with the wavefront produced by the mirror under test. The secondary interference between the stored wavefront and the test wavefront should be recorded in a plane conjugate to the exit pupil of the test surface. If the hologram is made in a plane conjugate to the exit pupil of the mirror under test, the amount of tilt in the resulting interferogram can be selected by simply changing the tilt of the wavefront used in the hologram reconstruction process. If the hologram is not made in a plane conjugate to the exit pupil of the test surface, tilting the reference beam not only adds tilt to the interferogram, it also causes a displacement between the image of the exit pupil of the test surface and the exit pupil stored by the hologram.

The holographic test plate interferometer can also be thought of in terms of moiré patterns (Pastor et al. 1970). Interference fringes resulting from the wavefront stored in the hologram and the wavefront coming from the optics under test can be regarded as the moiré pattern between the interference fringes recorded on the hologram plate (formed by the wavefront produced by the master optics and a plane wave), and the real-time interference fringes formed by the wavefront under test and a plane wavefront. The contrast in this moiré pattern is increased with spatial filtering by selecting only the wavefront produced by the mirror under test and the diffraction order from the hologram giving the stored wavefront produced by the master optics.

In addition to the error sources already mentioned, there can be error due to improper positioning of the hologram in the interferometer. Any translation or rotation of the hologram will introduce error. If the hologram is made conjugate to the exit pupil of the master optical system, the exit pupil of the system under test must coincide with the hologram. If the test wavefront in the hologram

15.3. INTERFEROMETERS USING SYNTHETIC HOLOGRAMS

plane is described by the function $\phi(x, y)$, a displacement of the hologram a distance Δx in the x direction produces an error

$$\Delta\phi(x, y) \approx \frac{\partial\phi(x, y)}{\partial x} \Delta x, \qquad (15.1)$$

where $\partial\phi/\partial x$ is the slope of the wavefront in the x direction. Similarly, for a wavefront described by $\phi(r, \theta)$, the rotational error $\Delta\theta$ is given by

$$\Delta\phi(r, \theta) \approx \frac{\partial\phi(r, \theta)}{\partial \theta} \Delta\theta. \qquad (15.2)$$

Phase-shifting techniques can be used to measure the phase of the secondary interference fringes by placing a phase shifter in the reference beam of the interferometer and shifting the phase of the secondary interference fringes (Hariharan et al. 1982; Thalmann and Dändliker 1985a, 1985b; Hariharan 1985). Because the secondary interference fringe spacing corresponds to one wavelength of optical path difference (OPD) between the stored wavefront and the live test surface wavefront, a $\pi/2$ phase shift of the fringes for the test surface will cause a $\pi/2$ phase shift in the secondary interference fringes. The calculated phase surface will correspond to the difference between the master optical component and the test optical component. To ensure that the fringes actually correspond to the test surface, the hologram must be made in the image plane of the test surface, and the hologram plane must be imaged onto the detector array when the phase measurement is performed.

15.3. INTERFEROMETERS USING SYNTHETIC HOLOGRAMS

When master optics are not available to make a real hologram, a computer-generated (or synthetic) hologram (CGH) can be made (Pastor 1969; Lee 1970, 1974, 1978, 1980; MacGovern and Wyant 1971; Wyant and Bennett 1972; Ichioka and Lohmann 1972; Fercher and Kriese 1972; Birch and Green 1972; Faulde et al. 1973; Deever 1975; Takahashi et al. 1975; Fercher 1976; Schwider and Burow 1976; Bartelt and Forster 1978; Yatagai and Saito 1978; Schwider et al. 1979; Leung et al. 1980; Caulfield et al. 1981; Smith 1981; Wyant 1974; Ono and Wyant 1984; Dörband and Tiziani 1985; Arnold 1989). A CGH is a binary representation of the actual interferogram (hologram) that would be obtained if the ideal wavefront from the test system is interfered with a tilted plane wavefront. The test setup is the same as that for a real hologram used as a holographic test plate. CGHs are an alternative to null optics when testing aspheric optical components.

15.3.1. Computer-Generated Holograms (CGHs)

To make a CGH, the test setup must be ray traced to obtain the fringes in the hologram plane that result from the interference of the tilted plane wave and the wavefront that would be obtained if the mirror under test were perfect. Just like a real hologram used as a test plate, the CGH should be made in a plane conjugate to the exit pupil of the system under test. These fringes are then represented as a binary grating, commonly having a 50% duty cycle. Methods of calculating these fringes are outlined by Wyant and Bennett (1972), Lee (1978), and Arnold (1989). A procedure that encodes the fringes as a series of exposure rectangles is discussed by Leung et al. (1980) and Dvore (1983). The process of breaking fringes into rectangles or polygons reduces the amount of computer storage necessary and the time needed to plot the CGH. The procedure used to make a CGH can be employed for any general optical system as long as all the optics in the interferometer are known and can be ray traced. A typical CGH is shown in Fig. 15.3.

An example of an interferometer used to test a steep aspheric optical element is shown in Fig. 15.4. If the deviation of the test surface from a sphere is substantial, the rays will follow a different path back through the diverger lens after they have reflected off the test surface. This means that there will be additional aberration added to the final interferogram due to the ray paths through the diverger lens. However, when the entire system is ray traced, the hologram will automatically correct for these aberrations if a null test is performed. Another important consideration is that the test element may deviate the rays so much that light reflected from the test surface will not get back through the interferometer. In this case, a partial null lens must be used to ensure that light will get back through the system. This is discussed in more detail in the section on the combination of CGHs with null optics.

Once the fringes are calculated, they are either plotted directly on a holographic substrate or plotted and photographically reduced onto a holographic substrate. The techniques of plotting have been substantially improved over the years. Early work utilized pen plotters to make an enlarged version of the hologram that was then photographically reduced to the appropriate size (MacGovern and Wyant 1971; Wyant and Bennett 1972; Wyant et al. 1974; Wyant and O'Neill 1974). The large format enabled a high-resolution CGH to be formed. However, problems due to plotter irregularities such as line thickness, pen quality, plotter distortion, and quantization caused errors in the reconstructed wavefront. Nonlinearities inherent in the photographic process and distortion in the reduction optics caused further degradation. With the advent of laser-beam recorders, resolution improved due to machine speed and an increased number of distortion-free recording points (Wyant et al. 1974). The most recent advances in the recording of CGHs have been made using the electron beam (e-beam) recorders used for producing masks in the semiconductor industry (Emmel and Leung 1979; Leung et al. 1980; Leung et al. 1981; Arnold

15.3. INTERFEROMETERS USING SYNTHETIC HOLOGRAMS 605

Figure 15.3. Sample computer-generated hologram.

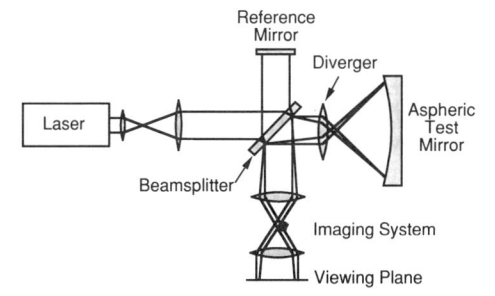

Figure 15.4. Aspheric test surface in interferometer showing rays not following same path after reflection off test surface.

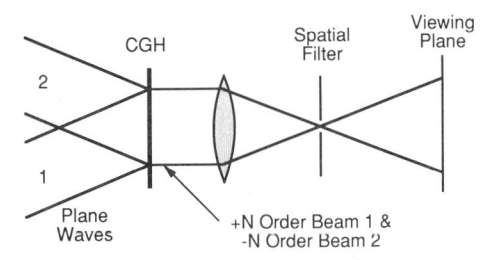

Figure 15.5. Test setup with $\pm N$ orders of hologram interfering to test the quality of a CGH plotter.

1983, 1989). These machines write onto photoresist deposited on an optical-quality glass plate and currently produce the highest quality CGHs. Patterns with as many as 10^8 data points can be produced in a hologram of the desired size. Typical e-beam recorders will write a 1-mm area with a resolution of 0.25 μm. Large patterns are generated by stitching a number of 1-mm scans together. Errors in this technique are due to aberrations in the electron optics, beam drift, instabilities in the controlling electronics, and positioning of the stepper stage. Many of these errors are reproducible and can be compensated for in the software controlling the recorder. Plotter errors can be evaluated by generating a hologram that is composed of straight lines in orthogonal directions forming a grid (Wyant et al. 1974). This test hologram is then illuminated with two plane waves as shown in Fig. 15.5 to interfere the $+N$ and $-N$ orders. Deviations of the fringes from straight lines will correspond to errors in the plotting process. The resulting aberration in the interferogram is $2N$ times that of the first order.

15.3.2. Using a CGH in an Interferometer

A CGH test is performed by interfering the test wavefront with a reference wavefront stored in the hologram. This entails overlapping the zero-order test beam and the first-order reference beam from the hologram in the Fourier plane of the hologram. The test can also be performed by interfering the minus first-order test beam with the zero-order reference beam to compare the two plane waves instead of the two aspheric wavefronts. When the test wavefront departs from the reference wavefront, fringes corresponding to the difference between the wavefronts appear. In the Fourier plane of the hologram, the zero- and first-order diffracted spots of the reference wavefront will overlap the first- and zero-order diffracted spots of the test wavefront when the interferometer is correctly aligned. Both outputs yield the same interferogram. Spatial filtering can be used to improve the fringe contrast if the tilt of the plane reference wavefront used for the hologram is large enough. To ensure no overlapping of the first and second orders in the Fourier plane (where the spatial filter is located), the tilt

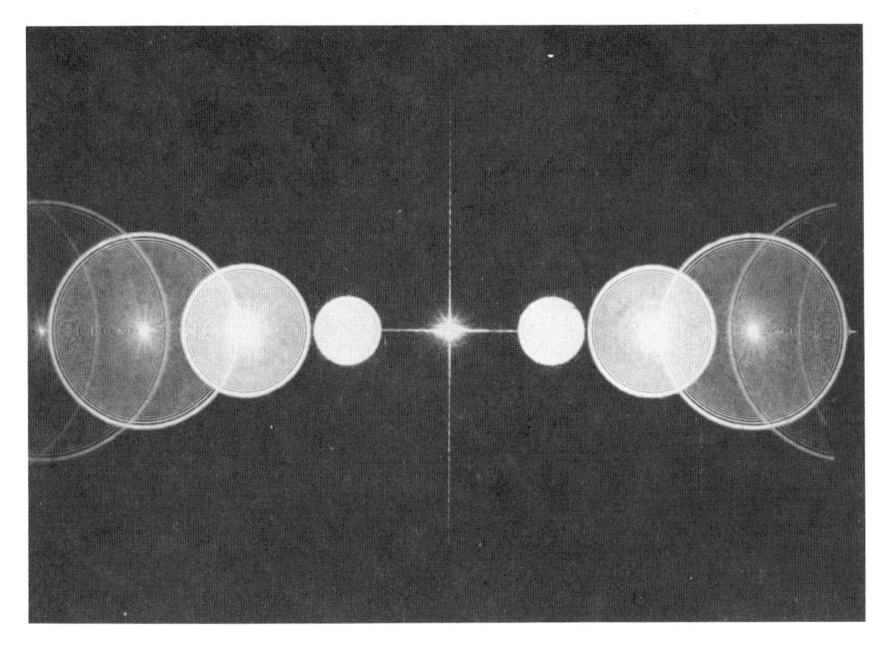

15.3. INTERFEROMETERS USING SYNTHETIC HOLOGRAMS

Figure 15.6. Diffracted orders in Fourier plane of CGH.

angle of the reference plane wave needs to be greater than three times the maximum wavefront slope of the aberrated wave (Goodman 1968, p. 214). (Note that there are no even orders for a grating with a 50% duty cycle.) A photograph of the diffracted orders from the hologram in the Fourier plane is shown in Fig. 15.6, and a diagram detailing the necessary separations of the orders in the Fourier plane is shown in Fig. 15.7. The bandwidth of the Nth order is given by $2NS$, where S is the maximum wavefront slope in waves per radius of the wavefront to be reconstructed. This bandwidth (diameter of the diffracted beam in the Fourier plane) determines the size of the spatial-filtering aperture. By moving the spatial-filtering aperture, the output of the interferometer can either be the interference of two plane waves or two aspheric waves.

A CGH is sensitive to the same errors as real holograms. Because of this, the CGH should be placed in the interferometer, as shown in Fig. 15.1, so that thickness variations in the hologram substrate have no effect on the results. If the hologram is recorded on a very high-quality optical flat or used in reflection, it can be placed in a single beam of the interferometer, as illustrated in Fig. 15.8 using a Fizeau configuration. In addition to the error sources associated with a real hologram, a CGH has additional error sources due to plotter distortion, incorrect hologram size, and photoreduction distortion if the hologram is photographically reduced in size. These errors are proportional to the maximum slope of the departure of the test wavefront from a spherical or plane wavefront.

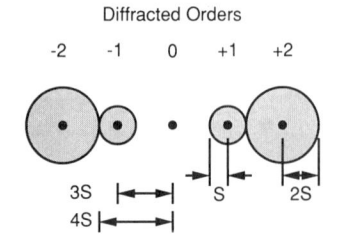

Figure 15.7. Diffracted orders in Fourier plane of CGH.

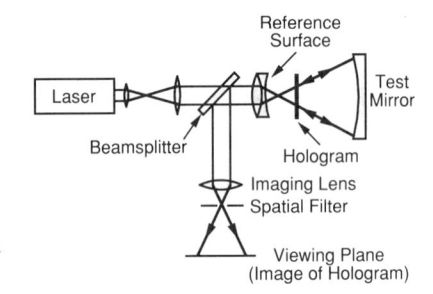

Figure 15.8. Fizeau interferometer utilizing a CGH in one beam.

These errors can be minimized when the test wavefront is calculated relative to the spherical wavefront, which minimizes the slope of the test wavefront departure from a spherical wavefront. Errors due to photographic reduction can be eliminated by writing a hologram of the correct size directly onto a glass substrate using an e-beam recorder.

One source of error is incorrect hologram size. If the aberrated test wavefront in the plane of the hologram is given by $\phi(r, \theta)$, a hologram of incorrect size will be given by $\phi(r/M, \theta)$, where M is a magnification factor. The error due to incorrect hologram size will be given by the difference $\phi(r/M, \theta) - \phi(r, \theta)$, and can be written in terms of a Taylor expansion as

$$\phi\left(\frac{r}{M}, \theta\right) - \phi(r, \theta) = \phi\left[r + \left(\frac{1}{M} - 1\right)r, \theta\right] - \phi(r, \theta)$$

$$= \left[\frac{\partial \phi(r, \theta)}{\partial r}\right]\left(\frac{1}{M} - 1\right)r + \cdots, \quad (15.3)$$

where terms higher than first order can be neglected if M is sufficiently close to 1 and a small region is examined. Note that this error is similar to a radial shear. When the CGH is plotted, alignment aids, which can help in obtaining the proper hologram size, must be drawn on the hologram plot.

The largest source of error is distortion in the hologram plotter (Wyant et al. 1974). The CGH wavefront accuracy depends on the number of plotter resolution points and the maximum slope of the aspheric wavefront being tested.

15.3. INTERFEROMETERS USING SYNTHETIC HOLOGRAMS

Assuming that the plotter has $P \times P$ resolution points, there are $P/2$ resolution points across the radius of the hologram. Since the maximum error in plotting any point is half a resolution unit, any portion of each line making up the hologram can be displaced by a distance equal to $1/P$. If the maximum difference between the slope of the test wavefront and the tilted plane wave is $4S$ waves per hologram radius, the phase of the plane wave at the hologram lines can differ from that of the required wavefront at the same lines by as much as $4S/P$ waves (Loomis 1980a, 1980b). The maximum error in the reconstructed wavefront will be $4S/P$ waves, and since the final interferogram is recorded in the image plane of the hologram, the quantization due to the finite number of resolution points causes a peak error in the final interferogram of $4S/P$ waves. It is important to note that the peak wavefront error of $4S/P$ waves is really a worst-case situation; it occurs only if, in the region of the hologram where the slope difference is maximum, the plotter distortion is also a maximum. This systematic error due to plotter distortion can be calibrated out when the plotter distortion is known (Fercher and Kriese 1972; Wyant et al. 1974). When the maximum plotting error is equal to one-half of the resolution spot size, the sensitivity of the CGH test ΔW is given by $4S/P$, where P is the number of distortion-free plotter points. Using an e-beam recorder with 0.25-μm resolution over a 10-mm-diameter hologram would enable the measurement of an aspheric wavefront with a maximum wavefront slope of 1000 waves per radius to be tested to a sensitivity of $\lambda/10$ (assuming a perfect plotter).

Figure 15.9 shows the results of measuring a 10-cm-diameter $F/2$ parabola using a CGH generated with an e-beam recorder (Leung et al. 1980). The fringes obtained in a Twyman–Green interferometer using a helium-neon source without the CGH present are shown in Fig. 15.9a. After the CGH is placed in the interferometer, a much-less-complicated interferogram is obtained as shown in Fig. 15.9b. The CGH corrects for about 80 fringes of spherical aberration and makes the test much easier to perform.

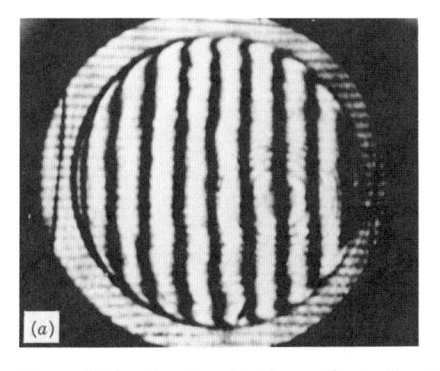

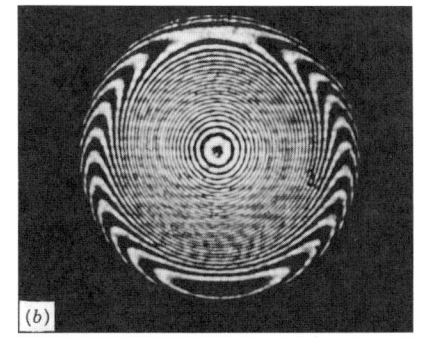

Figure 15.9. Results of testing a 10-cm-diameter F/2 parabola (a) with and (b) without a CGH made using an e-beam recorder.

15.3.3. Combination of CGH with Null Optics

Although a CGH can be designed for any optical system, a point is reached where the time and expense required to make a CGH are unreasonable. Also, given enough time and money, null optics, either reflective or refractive, can be designed and built to test almost any arbitrarily complicated optical system. It is possible to replace the complicated CGH or the complicated null optics required to test complicated optical surfaces (notably aspherical surfaces) with a combination of relatively simple null optics and a relatively simple CGH.

To illustrate the potential of the combined test, results for a CGH-null-lens test of the primary mirror of an eccentric Cassegrain system with a departure of approximately 455 waves (at 514.5 nm) and a maximum slope of approximately 1500 waves per radius are shown in Fig. 15.11 (Wyant and O'Neill 1974). The mirror was a 69-cm-diameter off-axis segment whose center lies 81 cm from the axis of symmetry of the parent aspheric surface. The null optics was a Maksutov sphere (as illustrated in Fig. 15.10), which reduces the departure and slope of the aspheric wavefront from 910 to 45 waves, and 3000 to 70 waves per radius, respectively. A hologram was then used to remove the remaining asphericity.

Figure 15.11a shows interferograms obtained using the CGH-Maksutov test of the mirror under test. Figure 15.11b shows the results when the same test was performed using a rather-expensive refractive null lens. When allowance is made for the fact that the interferogram obtained with the null lens has much more distortion than the CGH-Maksutov interferogram, and for the difference in sensitivity ($\lambda = 632.8$ nm for the null lens test and 514.5 nm for the CGH-Maksutov test), the results for the two tests are seen to be very similar. The "hills" and "valleys" on the mirror surface appear the same for both tests, as expected. The peak-to-peak surface error measured using the null lens was 0.46

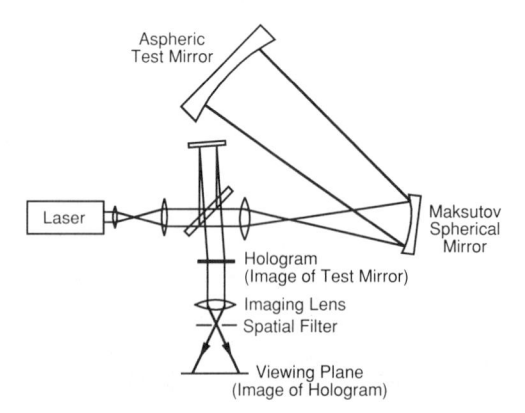

Figure 15.10. Setup to test the primary mirror of a Cassegrain telescope using a Maksutov sphere as a partial null with a CGH.

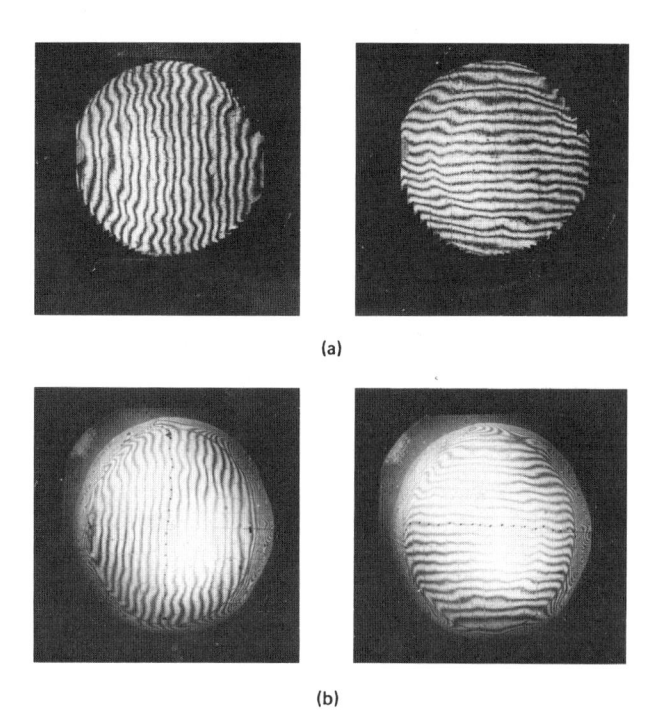

15.3. INTERFEROMETERS USING SYNTHETIC HOLOGRAMS

Figure 15.11. Results of testing using setup of Fig. 15.10: (*a*) CGH–Maksutov test ($\lambda = 514.5$ nm), and (*b*) using null lens ($\lambda = 632.8$ nm). (From Wyant and O'Neill 1974.)

waves (632.8 nm), while for the CGH–Maksutov test it was 0.39 waves (514.5 nm). The *rms* surface error measured was 0.06 waves (632.8 nm) for the null lens, while the CGH–Maksutov test gave 0.07 waves (514.5 nm). These results certainly demonstrate that expensive null optics can be replaced by a combination of relatively inexpensive null optics and a CGH.

The difficult problem of testing aspheric surfaces, which are becoming increasingly popular in optical design, is made easier by the use of CGHs. The main problem with testing aspheric optical elements is reducing the aberration sufficiently to ensure that light gets back through the interferometer. Combinations of simple null optics with a CGH to perform a test enable the measurement of almost any optical surface. The making and use of a CGH are analogous to using an interferometer setup that yields a large number of interference fringes and measuring the interferogram at a large number of data points. Difficulties involved in recording and analyzing a high-density interferogram and making a CGH are very similar. In both cases, a large number of data points are necessary, and the interferometer must be ray traced so that the aberrations due to the interferometer are well known. The advantage of the CGH technique is that once the CGH is made, it can be used for testing a single piece of optics many times or for testing several identical optical components (Greivenkamp 1987). In addition, it is much easier for an optician to work with a null setup.

15.4. TWO-WAVELENGTH AND MULTIPLE-WAVELENGTH TECHNIQUES

The test surface is often not known accurately enough to perform a null test. Even if a null test is attempted, the resulting interferogram may contain too many fringes to analyze. Since high accuracy may not be needed, a longer wavelength light source could be used in the interferometer to reduce the number of fringes. Unfortunately, a long wavelength light source creates problems because film and detector arrays may not be available to record the interferogram directly, and the inability to see the radiation causes considerable experimental difficulty. Two-wavelength and multiple-wavelength techniques provide a means of synthesizing a long effective wavelength using visible light to obtain an interferogram identical to the one that would be obtained if a longer wavelength source were used (Hildebrand and Haines 1967; Zelenka and Varner 1968, 1969; Wyant 1971; Polhemus 1973; Leung et al. 1979; Wyant et al. 1984; Lam et al. 1984; Cheng and Wyant 1984, 1985; Creath et al. 1985; Fercher et al. 1985; Creath 1986a; Creath and Wyant 1986; Wyant and Creath 1989).

15.4.1. Two-Wavelength Holography

Two-wavelength holography (TWH) consists of first photographing the fringe pattern obtained by testing an optical element using a wavelength λ_1 in an interferometer such as the interferometer shown in Fig. 15.1. This photographic recording of the fringe pattern (hologram) is then developed, replaced in the interferometer in the exact position it occupied during exposure, and illuminated with the fringe pattern obtained by testing the optical element using a different wavelength, λ_2. The resulting two-wavelength fringes can either be thought of as the moiré between the interference fringes stored in the hologram (recorded at λ_1 and replayed at λ_2) and the live interference fringes (at λ_2) or as the secondary interference between the test wavefront stored in the hologram and the live test wavefront. These fringes are identical to those that would be obtained if the optical element was tested using a long effective wavelength given by (Wyant 1971)

$$\lambda_e = \frac{\lambda_1 \lambda_2}{|\lambda_1 - \lambda_2|}. \qquad (15.4)$$

Table 15.1 lists the values of λ_e that can be obtained using various pairs of wavelengths from an argon-ion and a helium-neon laser. By use of a dye laser,

15.4. TWO-WAVELENGTH AND MULTIPLE-WAVELENGTH TECHNIQUES

Table 15.1. Possible Effective Wavelengths Obtainable with Argon–Ion and Helium–Neon Lasers

λ (μm)	0.4597	0.4765	0.4880	0.4965	0.5017	0.5145	0.6328
0.4597	—	11.73	9.95	5.89	5.24	4.16	1.66
0.4765	11.73	—	20.22	11.83	9.49	6.45	1.93
0.4880	9.95	20.22	—	28.50	17.87	9.47	2.13
0.4965	5.89	11.83	28.50	—	47.90	14.19	2.30
0.5017	5.24	9.49	17.87	47.90	—	20.16	2.42
0.5145	4.16	6.45	9.47	14.19	20.16	—	2.75
0.6328	1.66	1.93	2.13	2.30	2.42	2.75	—

a large range of equivalent wavelengths can be obtained (Schmidt and Fercher 1971). Tunable helium–neon lasers with four or five distinct wavelengths ranging from green to red are also available.

The contrast in the final interferogram can be increased by spatial filtering. If the filtering is to be effective, the angle between the two interfering beams in the interferometer must be such that only the object beam, not the reference beam, passes through the spatial filter (aperture) shown in Fig. 15.1. The spatially filtered pattern yields the interference between the wavefront produced by illuminating (with wavelength λ_2) the hologram recorded at wavelength λ_1 and the wavefront obtained from the optical element using wavelength λ_2. Since the TWH interferogram provides the difference between the two interfering beams only in the plane of the hologram, the fringe pattern (hologram) must be recorded in the plane conjugate to the test surface. The final photograph of the interferogram should be recorded in the image plane of the hologram.

Figure 15.12a shows an interferogram of an optical element tested using a wavelength of 0.4880 μm. The other interferograms were obtained using TWH to test the same optical element. The interferograms in Figs. 15.12–15.12e were obtained by recording an interferogram (hologram) using a wavelength of 0.5145 μm and illuminating the recording with the fringe pattern obtained using a wavelength of 0.4765 μm for Figs. 15.12b and 15.12c and of 0.4889 μm for Figs. 15.12d and 15.12e. The interferograms were spatially filtered, and the amount of tilt shown was adjusted in real time by changing the angle at which the reference wavefront was incident on the hologram during the reconstruction. The interferograms in Figs. 15.12f and 15.12g were obtained by recording an interferogram using a wavelength of 0.4880 μm and illuminating this recording with the fringe pattern obtained using a wavelength of 0.4765 μm and of 0.4965 μm, respectively.

In TWH, the final interferogram gives the difference between a fringe pattern recorded at one instant of time and a fringe pattern existing at some later instant of time. If the two fringe patterns are different for reasons other than wavelength

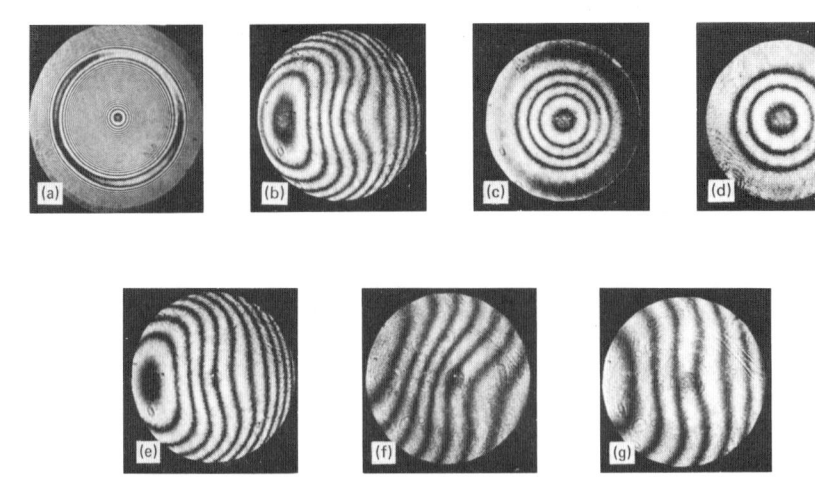

Figure 15.12. Interferograms of optical element at a number of different effective wavelengths: (a) λ = 0.488 μm, (b) λ_e = 6.45 μm, (c) λ_e = 6.45 μm, (d) λ_e = 9.47 μm, (e) λ_e = 9.47 μm, (f) λ_e = 20.22 μm, and (g) λ_e = 28.5 μm. (From Wyant 1971.)

change (e.g., air turbulence), incorrect results are obtained. This means that if air turbulence causes a one-fringe change between the fringe pattern obtained using λ_1 = 0.4880 μm and the fringe pattern obtained using λ_2 = 0.5145 μm, the final interferogram will contain one fringe of error corresponding to 9.47 μm.

The effect of air turbulence can be reduced by simultaneously recording the interferograms resulting from the two wavelengths. If the recording process is sufficiently nonlinear and the interferograms have sufficiently high contrast, the interferogram obtained shows the moiré pattern described earlier. Generally, this moiré pattern is too low in contrast to be useful. However, when this interferogram (hologram) is illuminated with a plane wave, is spatially filtered, and is reimaged in the manner shown in Fig. 15.1, the result is a high-contrast interferogram, identical to that obtained using the TWH method described earlier. Since both fringe patterns are recorded simultaneously, and air dispersion is small, the sensitivity of the interferometer to air turbulence is essentially the same as if a long wavelength light source were used in the interferometer.

15.4.2. Two-Wavelength Phase Measurement

Although two-wavelength holography provides a practical variable-sensitivity interferometric measurement, the major drawback is the creation of an intermediate hologram. As long as the fringes generated at each wavelength can be resolved by the detection system, two measurement wavelengths can be used with phase-shifting interferometry techniques (Wyant et al. 1984; Cheng and

15.4. TWO-WAVELENGTH AND MULTIPLE-WAVELENGTH TECHNIQUES

Wyant 1984; Fercher et al. 1985; Creath et al. 1985; Creath 1986a; Creath and Wyant 1986; Wyant and Creath 1989). By using the information from a second wavelength, the height range of the measurement can be significantly increased. A two-wavelength phase measurement is performed by first taking data at one wavelength while shifting the phase the appropriate amount for that wavelength and calculating the modulo 2π phase for that wavelength. The illumination wavelength is then changed, data are taken at the second wavelength with the appropriate phase shifts, and the modulo 2π phase is calculated for that wavelength. These two-phase measurements can be combined to produce a phase corresponding to a long synthesized wavelength, which is the beat between the two measured wavelengths (see Fig. 15.13). The long effective wavelength for this technique is given by Eq. (15.4), and its corresponding phase is given by

$$\phi_e = \phi_1 - \phi_2, \tag{15.5}$$

where ϕ_1 and ϕ_2 are modulo 2π phases measured at wavelengths λ_1 and λ_2 ($\lambda_2 > \lambda_1$). Once the phase is determined modulo 2π at the effective wavelength, the 2π ambiguities are removed with the same phase-unwrapping techniques used with a single wavelength.

An alternative method for calculating the effective wavelength phase is to take all the frames of data for both wavelengths and then calculate the effective wavelength phase directly from the intensity data. This can be written as

$$\phi_e = \tan^{-1}\left[\frac{\sin(\phi_1 - \phi_2)}{\cos(\phi_1 - \phi_2)}\right] = \tan^{-1}\left[\frac{\sin\phi_1 \cos\phi_2 - \cos\phi_1 \sin\phi_2}{\cos\phi_1 \cos\phi_2 + \sin\phi_1 \sin\phi_2}\right], \tag{15.6}$$

which for the four-frame method becomes

$$\phi_e = \tan^{-1}\left[\frac{(I_{14} - I_{12})(I_{21} - I_{23}) - (I_{11} - I_{13})(I_{24} - I_{22})}{(I_{11} - I_{13})(I_{21} - I_{23}) - (I_{14} - I_{12})(I_{24} - I_{22})}\right], \tag{15.7}$$

where I_{ji} is the ith data frame taken with wavelength λ_j.

Two-wavelength techniques are useful for measuring surfaces with discontinuities such as step heights. Because an intermediate hologram does not need

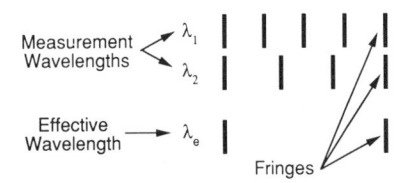

Figure 15.13. Beat wavelength for two-wavelength interferometry.

to be recorded, two-wavelength interferometry techniques can be implemented using either laser sources or a white-light source followed by narrowband spectral filters (Creath 1986a). For the measurement of aspheric surfaces, care must be taken to minimize chromatic aberrations so that the image size and location is the same for both measurement wavelengths.

15.4.3. Correction of Single-Wavelength Measurements

The noise in the two-wavelength measurement will be scaled by the wavelength. If there is an *rms* measurement noise of 0.01 μm at $\lambda = 0.5$ μm, there will be an *rms* noise of 0.1 μm with an effective wavelength of $\lambda_e = 5$ μm. A two-wavelength measurement can be used to correct the phase ambiguities in the modulo 2π single-wavelength phase to provide a measurement with visible-wavelength precision and extended height range (Creath 1986a). This is done by comparing a scaled version of the long effective wavelength phase with the single-wavelength phase. The number of 2π's to add to the single-wavelength data is determined by looking at the height changes in the scaled effective phase. The correction works well for smooth data. If the noise in the scaled effective phase is greater than $\pm\lambda/4$ at the single wavelength, then there will be unwanted 2π jumps in the corrected data.

15.4.4. Multiple-Wavelength Interferometry

Two-wavelength techniques can be extended to multiple wavelengths in order to correct single-wavelength data (Cheng and Wyant 1985). A number of wavelengths are chosen such that a series of effective wavelengths are produced that are proportionally spaced from the single wavelength up to the wavelength necessary to measure the test object. As an example, using measurement wavelengths of 0.50, 0.51, 0.54, and 0.70 μm, effective wavelengths of 25.5, 6.75, and 1.75 μm can be produced. Intensity data are taken with appropriate phase shifts and phase data are determined modulo 2π for each of the four measurement's wavelengths. Then modulo 2π effective wavelength phases for each of the three effective wavelengths are calculated using Eq. (15.5). Standard phase-unwrapping techniques are used to remove phase ambiguities in the 25.5-μm phase. The 25.5-μm phase is used to correct ambiguities in the 6.75-μm phase. The corrected 6.75-μm phase is then used to correct ambiguities in the 1.75-μm phase. Finally, the corrected 1.75-μm phase is used to remove ambiguities in the 0.5-μm phase. By using this technique, a larger amount of noise can be tolerated than when using the 25.5-μm phase to directly correct the 0.5-μm phase. A rule of thumb for good measurements is to keep the ratio between the longer wavelength and the wavelength being corrected between a factor of 5 and 10.

An example of a 13-μm step measured using single, two-wavelength, and multiple-wavelength techniques is shown in Fig. 15.14. These measurements were taken using an interferometric optical microscope with phase-shifting capability. Figure 15.14a shows the step measured at a wavelength of 657 nm. A two-wavelength measurement using the method of Eq. (15.5) is shown in Fig. 15.14b where the measurement wavelengths are 657 and 651 nm, producing an effective wavelength of 64 μm. The difference between two consecutive measurements using two wavelengths is shown in Fig. 15.14c where the *rms* is 7.13 nm. This means that the measurement is repeatable to within $\lambda_e/9000$ at the effective wavelength. Using the data from the 651-nm measurement to correct the phase data taken at 657 nm, the result shown in Fig. 15.14d shows uncertainties of 2π, which are caused by noise in the single-wavelength measurement. If three measurement wavelengths (657, 651, and 601 nm) are used, the corrected measurement at 657 nm shown in Fig. 15.14e is much less noisy. The repeatability (difference between two consecutive measurements) of the three-wavelength measurement of Fig. 15.14e is shown in Fig. 15.14f. The RMS of the difference is 0.67 nm, which yields a dynamic range for the measurement of almost 20,000. Thus the use of multiple wavelengths can increase the dynamic range of a measurement by a factor of 10.

15.5. HOLOGRAPHIC INTERFEROMETRY FOR NONDESTRUCTIVE TESTING

Holographic interferometry techniques have been in use for 25 years in the field of stress analysis (Hildebrand and Haines 1966; Haines and Hildebrand 1966; Vest 1979, 1980, 1982; Erf 1974, 1982; Stetson 1974, 1975a; Levitt and Stetson 1976; Sommargren 1977; Hariharan et al. 1982, 1983a, 1983b, 1986; Jones and Wykes 1983; Dändliker and Thalmann 1983, 1985; Hariharan 1984, 1985; Thalmann and Dändliker 1984, 1985a, 1985b). There are two basic techniques: static and dynamic. Static tests such as double-exposure holography and real-time holographic interferometry measure an object in two different states of applied stress and find the difference between them. In these tests, the object is assumed to be not moving during the exposure time. A single exposure of the object is made before stress is applied, and a second single measurement is made after the stress is applied. Static tests can also be performed using pulsed lasers (Vest 1979; Jones and Wykes 1983; Hariharan 1984; Hariharan and Oreb 1986; Hariharan et al. 1987), where the length of the pulse is small compared to the change in object motion. For dynamic testing, the object is excited at some vibrational frequency providing a periodic motion. A single measurement averaged over multiple periods of vibration is evaluated.

The optical layout shown in Fig. 15.15 is used for holographic nondestruc-

618   HOLOGRAPHIC AND SPECKLE TESTS

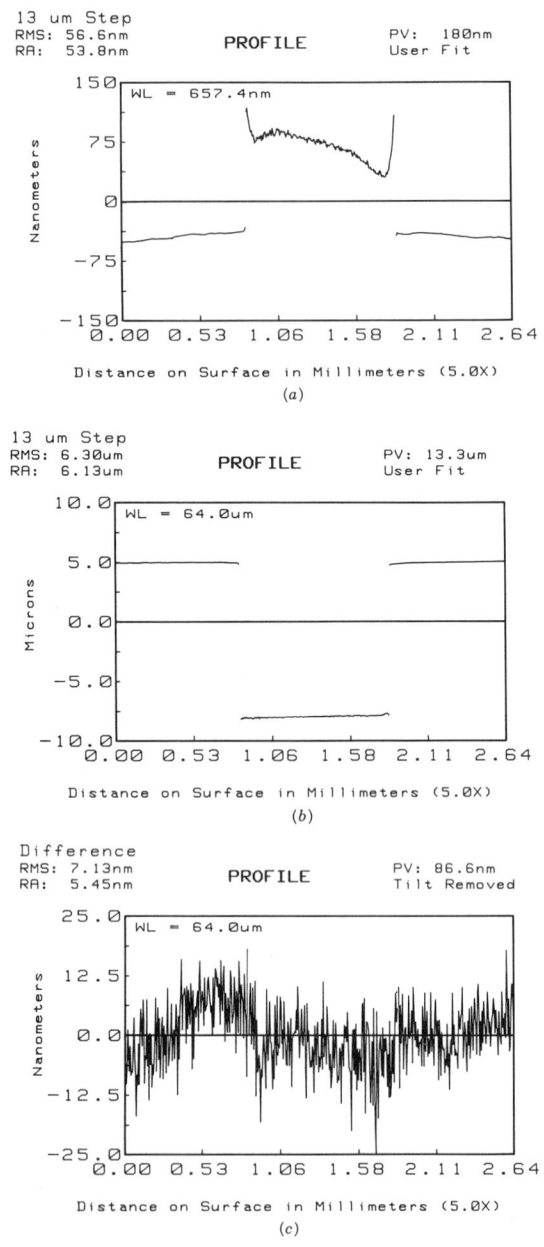

Figure 15.14. Measurement of a large step using multiple-wavelength interferometry: (a) 657 nm, (b) two-wavelength measurement using 657 and 651 nm with $\lambda_c = 64$ μm, (c) difference of consecutive two-wavelength measurements, (d) 657-nm data corrected using two-wavelength measurement, (e) three-wavelength using 657, 651, and 601 nm, and (f) difference of consecutive three-wavelength measurements.

15.5. HOLOGRAPHIC INTERFEROMETRY FOR NONDESTRUCTIVE TESTING 619

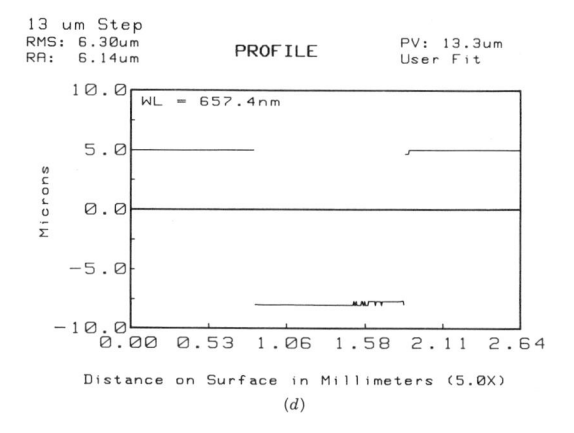

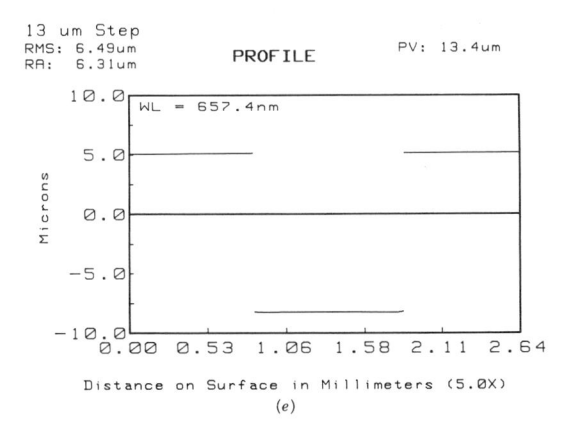

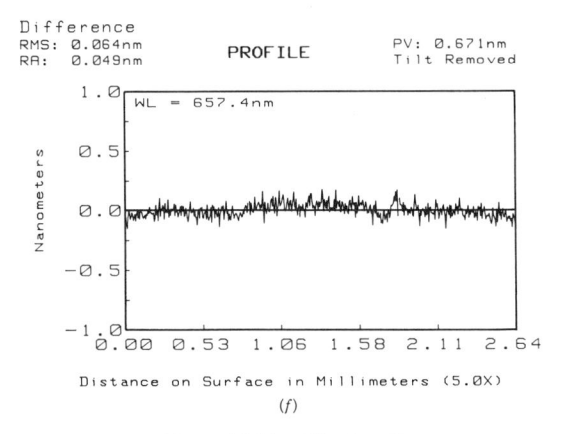

13 um Step
RMS: 6.30um
RA: 6.14um
PROFILE
PV: 13.3um
User Fit

WL = 657.4nm

(d)

13 um Step
RMS: 6.49um
RA: 6.31um
PROFILE
PV: 13.4um

WL = 657.4nm

(e)

Difference
RMS: 0.064nm
RA: 0.049nm
PROFILE
PV: 0.671nm
Tilt Removed

WL = 657.4nm

(f)

Figure 15.14. (*Continued*)

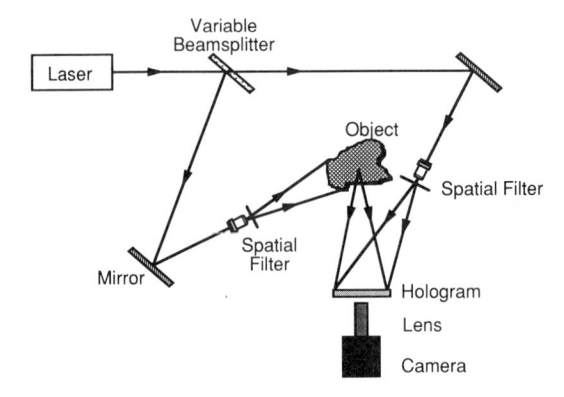

Figure 15.15. Holographic interferometer for nondestructive testing.

tive testing to measure out-of-plane displacement. It can be modified to measure in-plane displacement as well as out-of-plane displacement and to do shearing holographic interferometry measurements. A laser beam is split into two paths using a variable-density beam splitter to control the ratio of the amount of light in one beam with respect to the other beam. The object beam illuminates the object with a diverging beam of light. The object, usually diffuse, scatters the light, and some of this light is incident upon the hologram plane. The other beam is a reference beam directly incident upon the hologram plane. The angle between the object beam and the reference beam at the hologram plane will determine the spacing of the interference fringes in the hologram. Because the hologram is simply the interference of two beams of light, the difference between the path lengths of the object and reference beams must be within the coherence length of the laser being used. If an argon ion laser with an etalon or if a single-mode helium–neon laser is used, a coherence length of many meters is attainable, whereas if a multimode helium–neon laser is used, the path lengths must be within a few centimeters of one another. To make an efficient hologram, the reference beam should be six to eight times brighter than the object beam at the hologram plane, the polarizations of the two beams should be in the plane of incidence, and the angle between the object and reference beams needs to be small enough to produce resolvable interference fringes in the recording material.

Holographic interferometry techniques can be used to measure thermal changes in optical elements, changes in optical surface shape due to mounting, changes in deformable mirror shapes, and to study vibrational modes of optical elements, mounts, or entire optical systems. These techniques can be used with diffuse (ground) surfaces, specularly reflecting surfaces, and transmissive optical components. One particular application of holographic nondestructive testing is the determination of the mechanical and thermal properties of large un-

15.5. HOLOGRAPHIC INTERFEROMETRY FOR NONDESTRUCTIVE TESTING

worked mirror blanks before any time and money are spent to put optical surfaces on the blanks (Van Deelen and Nisenson 1969). Both qualitative fringe data and quantitative displacement data may be obtained.

15.5.1. Static Holographic Interferometry

For a static holographic measurement of an object, a hologram of the object is made while the object is in one stress state. In order for static measurements to work, the object must not move while the hologram is being made. After the object stress state is changed, either a second hologram is exposed as in double-exposure holographic interferometry or the object is observed through the hologram as in real-time holographic interferometry. In both cases, there is a secondary interference between the wavefront generated before the change in applied stress and the wavefront generated after the change in applied stress. For double-exposure holography both wavefronts are stored in the hologram. When the hologram is illuminated with the reference wavefront, both wavefronts are reconstructed. When these wavefronts are viewed through the hologram, cosinusoidal secondary interference fringes are visible. The secondary interference fringes correspond directly to the amount of object displacement between the two exposures. If there is no change in the object between exposures, a single interference fringe will be produced. Each fringe in the secondary interference pattern indicates one wave of displacement along a direction bisecting the illumination and viewing directions of the object.

For real-time holographic interferometry, one of the two wavefronts is stored in the hologram and the other wavefront is produced live by the test object. Once the hologram is made and developed, care must be made to ensure that the hologram is replaced in the same location so that the wavefront stored in the hologram can be interfered with the live wavefront. Real-time holographic interferometry is very similar to the use of the holographic test plate. Fringe location, fringe shape, and test sensitivity are the same for real-time techniques as they are with double-exposure techniques. If the object is in motion, static techniques can be used to provide cosinusoidal fringes as long as a pulsed laser or high-speed shutter is used to freeze the motion. The exposure time must be short enough that the object does not move during exposure.

Figure 15.16 shows the recording geometry indicating the object displacement vector **L** and the sensitivity vector **K**. The sensitivity vector is defined as the direction along which the object displacement is measured. The direction of the sensitivity vector can vary across the object surface if the field of view is large and if the source illuminating the object and the viewing position are not located at infinity. Mathematically, the secondary interference fringes for a static holographic measurement at a single point in the viewing plane can be written as

$$I = I_0(1 + \gamma \cos \Delta\phi), \tag{15.8}$$

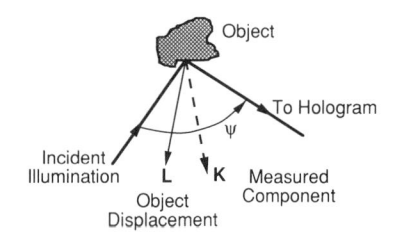

Figure 15.16. Recording geometry showing sensitivity vector and displacement vector.

where I_0 is the dc intensity, γ is the fringe visibility, and $\Delta\phi$ is the phase of the secondary interference fringes corresponding to the difference (displacement) between the two object states. The phase difference can be written as

$$\Delta\phi = \mathbf{K} \cdot \mathbf{L}. \tag{15.9}$$

The displacement of the object at a point x, y in the direction of the sensitivity vector is given by

$$D(x, y) = \frac{\Delta\phi(x, y)\lambda}{4\pi \cos(\psi/2)}, \tag{15.10}$$

where λ is the illumination wavelength, and ψ is the angle between the illumination and viewing directions. The displacement measurement is usually a combination of in-plane (along the surface of object) and out-of-plane (perpendicular to the object surface) displacement. Specific setups can be made to measure only one of these components. If all three components of displacement (x, y, and z) are desired, three measurements must be made (Pryputniewicz and Stetson 1976; Stetson 1979, 1990; Nakadate et al. 1981; Kakunai et al. 1985; Hariharan et al. 1987). As long as the change in the object shape is small, the secondary interference fringes will be localized at the object. If the applied stress causes the object to move (rigid-body translation or rotation) as well as deform, the secondary interference fringes may not be localized at the object (Vest 1979; Hariharan 1984).

An example showing the use of static holographic interferometry to test a coupling flange from a helicopter is shown in Fig. 15.17. This part was stressed between exposures by applying a static force between the rim and the center of the part. In addition to displacement measurement and stress analysis, real-time or double-exposure holographic interferometry can be used to contour objects with two wavelengths, two indices of refraction, or two different angles of object illumination (Hariharan 1984).

15.5.2. Dynamic Holographic Interferometry

If the object is excited in a periodic motion, a measurement can be made using a single exposure that averages over many vibrational periods creating a *time-*

15.5. HOLOGRAPHIC INTERFEROMETRY FOR NONDESTRUCTIVE TESTING 623

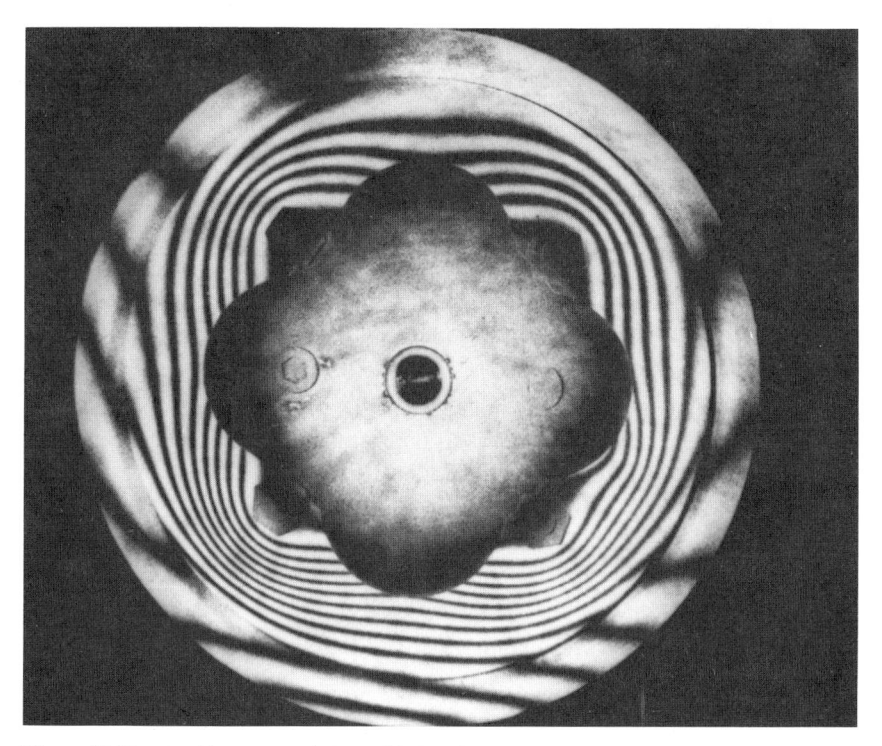

Figure 15.17. Double-exposure holographic interferometry fringes of helicopter coupling flange. (Courtesy of K. A. Stetson, United Technologies Research Center.)

average hologram (Powell and Stetson 1965; Stetson and Powell 1965; Vest 1979; Hariharan 1984). The secondary interference fringes at a single point in the viewing plane for this case can be written as

$$I = I_0 \gamma |M(\Omega)|^2, \quad (15.11)$$

where $M(\Omega)$ is the secondary fringe function due to the object motion. When the object is moving sinusoidally, the secondary fringe function becomes

$$M(\Omega) = J_0(\Omega), \quad (15.12)$$

where J_0 is a zero-order Bessel function, and $\Omega = \mathbf{K} \cdot \mathbf{L} \propto A/\lambda$ for a vibration amplitude A and illumination wavelength λ. Note that this result is independent of the vibration excitation frequency. The time-average secondary interference fringes for a sinusoidal vibration become

$$I = I_0 \gamma J_0^2(\Omega). \quad (15.13)$$

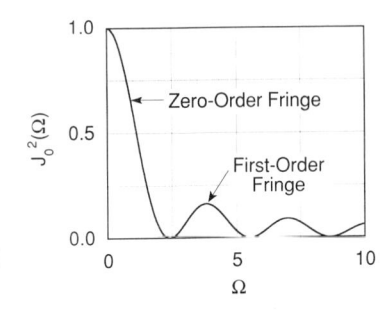

Figure 15.18. Time-average holography (Bessel function) fringes.

These fringes are shown in Fig. 15.18. For positions on the object that are stationary, the intensity is a maximum for the zero-order fringe. As the amplitude of the object movement increases, the intensity of the secondary interference fringes is reduced substantially. Figure 15.19 shows an example of a helicopter gear excited at 5812 Hz and tested using time-average holography. It is obvious from four fringe orders visible that the intensity of higher-order fringes diminishes rapidly.

15.5.3. Holographic Interferometry with Phase Measurement

Until recently, holographic techniques have provided only qualitative data in the form of fringe patterns requiring a skilled operator to interpret. With the advent of TV camera frame grabbers for personal computers and fast CPUs, quantitative data using phase-shifting interferometry techniques can be generated in less than a minute. To obtain phase data from object displacement, a phase shifter is placed in one beam of the interferometer shown in Fig. 15.15. Standard phase-shifting techniques (as described in Chapter 14) can be used with holographic interferometry techniques that produce cosinusoidal fringes to generate a phase map of $\Delta\phi(x, y)$ corresponding to the object displacement $D(x, y)$ (Hariharan et al. 1982, 1983a, 1983b, 1986; Dändliker and Thalmann 1983, 1985; Hariharan 1984, 1985; Thalmann and Dändliker 1984, 1985a, 1985b). An example of a phase measurement using real-time holographic interferometry to measure changes due to thermal variations in two translation stages made of different materials is shown in Fig. 15.20. A hologram was made of the two stages, and the phase measurement was performed after letting the stages sit at room temperature for a couple of hours. Because of the different materials used to make the stages, one stage is much more stable than the other.

Quantitative data can also be extracted from time-average vibration fringes using phase-shifting techniques (Neumann et al. 1970; Stetson 1982; Oshida et al. 1983; Nakadate and Saito 1985; Stetson and Brohinsky 1988). Three separate phase measurements need to be taken (Stetson and Brohinsky 1988). The first one is made with the object vibrating and by shifting the relative phase between the object and reference beams as in standard phase-shifting interfer-

15.5. HOLOGRAPHIC INTERFEROMETRY FOR NONDESTRUCTIVE TESTING

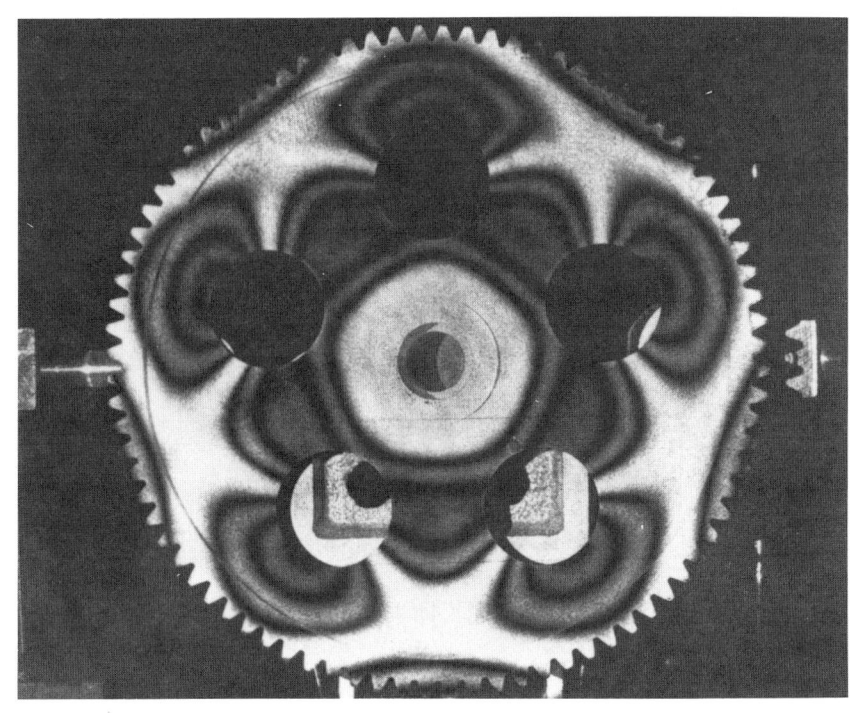

Figure 15.19. Time-average holographic fringes of helicopter gear excited at 5812 Hz. (Courtesy of K. A. Stetson, United Technologies Research Center.)

ometry. The second measurement is made by applying a vibration of the same frequency as the object vibration to a piezo-electric translator (PZT) in the reference beam of the interferometer. A bias phase between the object and reference vibration is added such that the relative phase difference between the vibrations is $+\pi/3$. Relative phase shifts between the object and reference beams are then applied and standard phase-shifting methods are used to calculate the phase. A third measurement is taken such that the relative phase difference between the object and reference vibrations is $-\pi/3$. Assuming a sinusoidal object vibration and 90° relative phase shifts for the phase calculations, one of the total of 12 frames of data recorded can be written as

$$I_{ji} = I_0[1 + \gamma \cos(\phi + \delta_i)J_0(\Omega + \beta_j)], \qquad (15.14)$$

where $\delta_i = 0$, $\pi/2$, π, and $3\pi/4$, and $\beta_j = -\pi/3$, 0, and $\pi/3$. The amplitude of the vibration can then be calculated using

$$\Omega = \tan^{-1}\left[\frac{1}{\sqrt{3}}\left(\frac{H_1 - H_3}{2H_2 - H_1 - H_3}\right)\right], \qquad (15.15)$$

```
STAGES
 rms: 0.909      Interval: 1.000    Units: um
 p-v: 5.559                         pupil: 100%
```

(a)

```
STAGES
 rms: 0.909      DISPLACEMENT       Units: um
 p-v: 5.559                         pupil: 100%
```

(b)

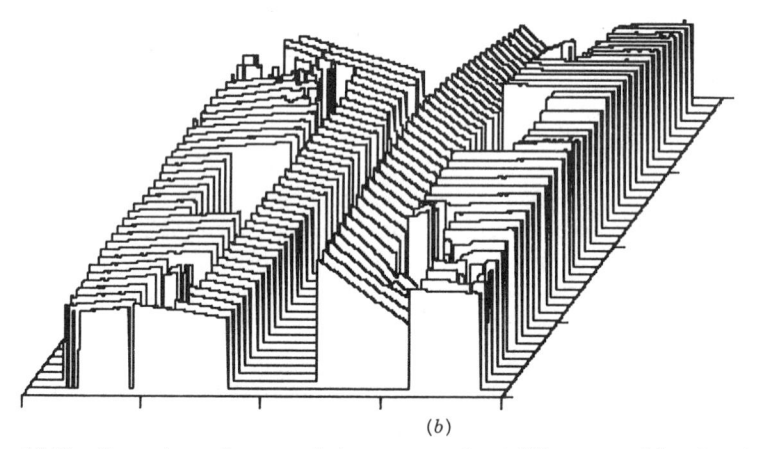

Figure 15.20. Comparison of two translation stages made of different materials using phase-shifting interferometry techniques with real-time holographic interferometry. (a) Two-dimensional contour plot where the contour interval is 1 μm and (b) isometric contour plot.

where

$$H_1 = (I_{11} - I_{13})^2 + (I_{12} - I_{14})^2 = 4I_0^2\gamma^2 J_0^2(\Omega - \pi/3), \quad (15.16)$$

$$H_2 = (I_{21} - I_{23})^2 + (I_{22} - I_{24})^2 = 4I_0^2\gamma^2 J_0^2(\Omega), \text{ and} \quad (15.17)$$

$$H_3 = (I_{31} - I_{33})^2 + (I_{32} - I_{34})^2 = 4I_0^2\gamma^2 J_0^2(\Omega + \pi/3). \quad (15.18)$$

Equation (15.15) assumes the form of $\cos^2$ for the fringes. Because of this, a look-up table is necessary to find the difference between the $J_0^2(\Omega)$ and $\cos^2(\Omega)$ functions (Stetson and Brohinsky 1988). The error due to the difference between the $J_0^2(\Omega)$ and $\cos^2(\Omega)$ functions is dependent on the fringe order which can be determined from the H_2 measurement.

When the object under observation is not stable enough to take consecutive exposures with different phase shifts, phase information can be obtained from a single interferogram using the Fourier transform technique (Kreis 1986, 1987). For this technique, the Fourier transform of the fringe pattern is found, one diffraction order is filtered out and then shifted to zero frequency, and then this single order is inverse Fourier transformed to obtain the phase distribution. This technique requires sinusoidal fringes and enough straight "tilt" fringes to separate orders. Multiple frames of data with different phase shifts for phase measurements can also be taken simultaneously using multiple cameras.

15.6. SPECKLE INTERFEROMETRY AND TV HOLOGRAPHY

When a laser beam is scattered off of a diffuse surface, the scattered light has a grainy appearance. The grains are an interference phenomenon known as *speckle*. The statistics of the speckle distribution depend on the statistics of the object surface. If imaged by a lens, the speckles are said to be subjective, and the smallest speckle in the image will have a size equal to the Airy disk 2.44 $\lambda F^\#$ generated by the optical system, where $F^\#$ is the working F number of the system (Goodman 1975a). The intensity distribution and the statistics of the speckle pattern are an indication of the roughness of the surface used to generate the speckle pattern (Goodman 1975a, 1975b; Erf 1978; Fujii and Lit 1978). A speckle pattern generated by an object surface can be thought of as the object's fingerprint. When the object is perturbed in some way, the speckle pattern will change in a predictable way. Two different types of techniques are speckle photography and speckle interferometry (Ennos 1975; Stetson 1975b; Pryputniewicz 1985; Huntley 1989). Both techniques involve the comparison of two or more speckle patterns. Speckle interferometry either includes a reference beam or compares the object with itself to enable measurement of the phase change in the speckles. In speckle interferometry, it is assumed that the speckles from one speckle pattern to another are correlated so that they do not shift by more than the diameter of a speckle between exposures. In contrast, speckle photography looks at the correlation between two speckle patterns where the fringes arise from a translation between exposures and can be used to measure larger displacements. The speckles from a small area of the two speckle patterns (translated with respect to one another) generate Young's fringes in the Fourier plane. For the rest of this discussion we are going to concentrate upon speckle interferometry.

To aid in understanding how speckle interferometry works, consider a diffuse object placed in the object beam of a Twyman–Green interferometer and a photographic plate placed at the image of the object with an $F^\#$ large enough to produce visible speckle. After the speckle pattern is recorded and the photographic plate replaced in the original position, no light will get through the plate because the speckles line up with those stored in the photographic plate. This is similar to a single fringe using the holographic test plate. As the object is tilted, fringes will be seen that correspond to the tilt of the object. These fringes are modulated by a speckle pattern and appear noisy. The secondary interference fringes are also known as *speckle correlation fringes* because they correspond to the correlation between the speckles in the two fringe patterns (Jones and Wykes 1983; Løkberg and Slettemoen 1987).

Speckle interferometry techniques can be used for the same applications as holographic interferometry techniques (Archbold et al. 1970; Butters and Leendertz 1971; Macovski et al. 1971; Biedermann and Ek 1975; Ennos 1975; Stetson 1975b; Chiang and Asundi 1981; Jaisingh and Chiang 1981; Jones and Wykes 1983; Creath 1985a, 1985b, 1986a, 1986b; Nakadate 1986; Stetson and Brohinsky 1985; Pryputniewicz 1985; Robinson and Williams 1986; Brdicko et al. 1987; Løkberg and Slettemoen 1987; Robinson 1987). Using a single illumination wavelength, static and dynamic measurements of object displacement provide the same results as holographic nondestructive testing. The speckle interferometry nondestructive testing techniques can measure only a single component of object displacement. This component, in the direction of the sensitivity vector, is described in the section on holographic interferometry for nondestructive testing. Two-wavelength techniques can also be used to perform a reduced-sensitivity test equivalent to two-wavelength holography or interferometry.

15.6.1. Relationship between Holographic and Speckle Interferometry

Holographic and speckle interferometry are very similar. In both cases, interference between an object and test beam are recorded. For speckle interferometry, the speckles are usually resolved by the recording system; however, this is not a necessary condition (Creath 1985a). The major difference between the techniques is that for speckle techniques an intermediate holographic recording is not required. Electronic storage and processing can take the place of the intermediate hologram in real-time techniques. Another difference between speckle and holographic techniques is that the object and reference beams are usually in line with one another in a speckle system; whereas, for holographic interferometry, they can have any angle between them (Leith and Upatnieks 1962). This difference is necessary because speckle techniques have evolved to use lower-resolution recording media (such as TV cameras) than holographic interferometry. An in-line speckle interferometry setup for the measurement of

15.6. SPECKLE INTERFEROMETRY AND TV HOLOGRAPHY

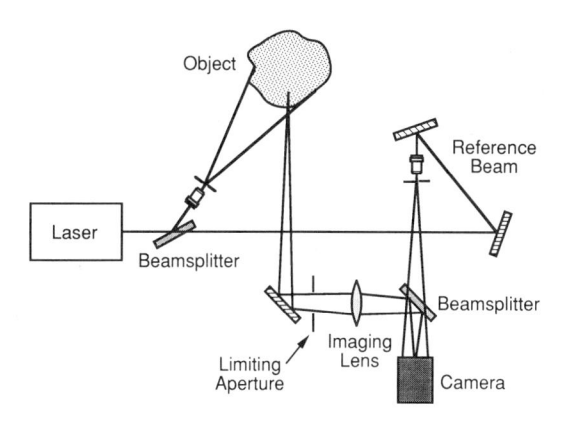

Figure 15.21. Speckle interferometry setup similar to that for making an in-line hologram.

out-of-plane displacement is shown in Fig. 15.21. The optical system is similar to those described for holographic test techniques where the path lengths must be adjusted to be within the coherence length of the source, the polarizations of the two beams must be such that they are in the plane of incidence at the viewing plane, and the speckle should be recorded in a plane conjugate to the test surface so that the interference corresponds directly to changes in the object surface. Speckle interferometry techniques also utilize a small aperture that determines the size of the speckle at the recording plane.

Essentially, speckle interferometry techniques can perform holographic interferometry without the need to make a hologram. A major speckle interferometry technique is electronic speckle-pattern interferometry (ESPI) (Archbold et al. 1970; Butters and Leendertz 1971; Macovski et al. 1971; Biedermann and Ek 1975; Ennos 1975; Jones and Wykes 1983; Løkberg and Slettemoen 1987). This technique uses a TV camera as the recording device and processes interferograms in either electronics or a computer or array processor. It is sometimes also known as *TV holography*. A slightly different variation on ESPI that provides improved fringe contrast has been coined electro-optic holography (EOH) by Stetson and Brohinsky (Stetson and Brohinsky 1985, 1986, 1987; Bushman 1989; Feit 1989; Stetson 1989; Stetson et al. 1989). EOH is very promising as a TV holography system. Phase measurements can be applied to ESPI and EOH to provide quantitative displacement maps of the same data obtained in holographic interferometry.

15.6.2. Electronic Speckle-Pattern Interferometry

ESPI uses an optical setup similar to that of Fig. 15.21, where a TV camera, CCD camera, or detector array is placed at the image of the test surface (Archbold et al. 1970; Butters and Leendertz 1971; Macovski et al. 1971; Biedermann and Ek 1975; Ennos 1975; Jones and Wykes 1983; Løkberg and Slettemoen 1987). All of the processing is performed in electronics, and results are displayed in real-time on a monitor. Because standard TV signals are generated, the results of a test can be stored on videotape for later viewing and processing. The aperture size is adjusted so that the speckle will be resolved by the camera. (When the speckles are reduced in size, the effect is to reduce the contrast of the secondary interference fringes.) For static ESPI measurements, a speckle interferogram of the object is recorded by the camera and stored electronically. A single interferogram can be written as

$$I = I_0 (1 + \gamma \cos \phi), \tag{15.19}$$

where I_0 is the dc intensity, γ is the visibility, and ϕ is the phase of the interference between the reference beam and the speckle pattern scattered by the object. After the stress applied to the object is changed, Eq. (15.19) becomes

$$I = I_0(1 + \gamma \cos \phi'), \tag{15.20}$$

where $\phi' = \phi + \Delta\phi$, and $\Delta\phi$ is the phase change. The stored interferogram [Eq. (15.19)] is then subtracted from exposures recorded after a change in applied stress [Eq. (15.20)] to the object, and the difference is squared to yield

$$I^2 = 4I_0^2\gamma^2(\sin^2 \phi)[\sin^2 (\Delta\phi/2)]. \tag{15.21}$$

This equation shows that there are fringes due to object displacement $\Delta\phi$ as well as fringes due to the phase of speckles ϕ resulting from the interference between the reference and speckled object beams. The $\sin^2 \phi$ term causes modulation of the speckle and makes the fringes noticeably noisy.

Dynamic time-average measurements can be made with ESPI using a single frame of data. For a vibrating object, the signal at the camera for a single detector point averaged over many cycles of vibration is given by

$$I = I_0[1 + \gamma \cos \phi M(\Omega)], \tag{15.22}$$

where ϕ is the phase difference between the object and reference beams, and Ω is proportional to the amplitude of the object displacement. In order to process this signal, the dc component is filtered out, and then the signal is rectified and squared to yield

$$I = I_0\gamma (\cos^2 \phi)|M(\Omega)|^2. \tag{15.23}$$

Note that this expression has a factor of $\cos \phi$, which is not included in the time-average holographic vibration fringes given by Eq. (15.11). The $\cos^2 \phi$

15.6. SPECKLE INTERFEROMETRY AND TV HOLOGRAPHY

term is due to the phase of the speckles and causes the secondary fringes to be noisy. For a sinusoidal object motion, Eq. (15.23) can be rewritten as

$$I = I_0 \gamma \, (\cos^2 \phi) J_0^2(\Omega). \tag{15.24}$$

Because the processing for static and dynamic measurements is performed in electronics, fringe data can be obtained with ESPI at TV frame rates (25 or 30 frames per second). This speed enables measurements to be made even when the object is not very stable.

Examples of fringes obtained using ESPI are shown in Figs. 15.22 and 15.23. Fringes for a single frame of time-average data for a vibrating flat plate are shown in Fig. 15.22a. By averaging a number of statistically independent fringe patterns obtained by changing the angle of illumination on the object, the speckle noise in the fringe pattern can be reduced as shown in Fig. 15.22b (Creath 1985c). Figure 15.23 shows a time-average interferogram of a car body vibrating at 110 Hz (Malmo and Vikhagen 1988). This data was taken with the car sitting in a parking lot. The body was covered with retroreflective tape and excited using a loud speaker on the seat. A commercially available ESPI instrument with a 7-mW helium–neon laser was used for the measurements.

15.6.3. Electro-Optic Holography

A major disadvantage of ESPI is that, after processing the interference fringe data, there are terms proportional to the phase of the speckle produced by the interference between the reference beam and the speckle in the object beam. These terms cause noisy-looking fringes. By processing the data differently,

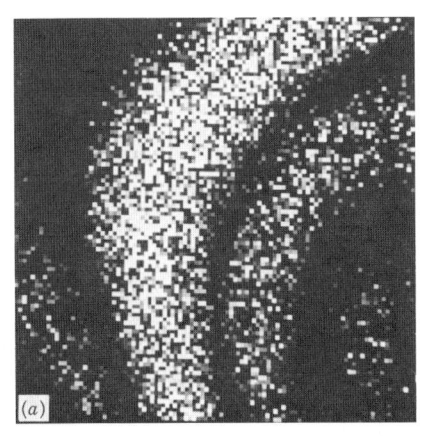

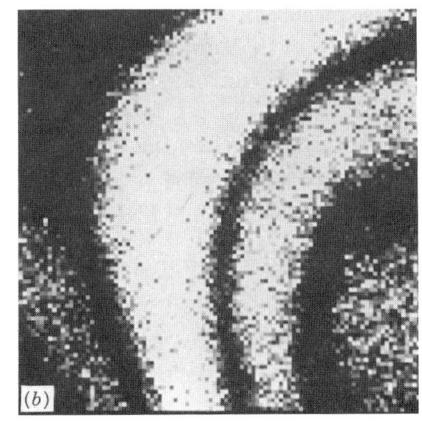

Figure 15.22. ESPI interferogram of vibration mode of a flat plate using (a) single interferogram and (b) with speckle averaging. (From Creath 1985c.)

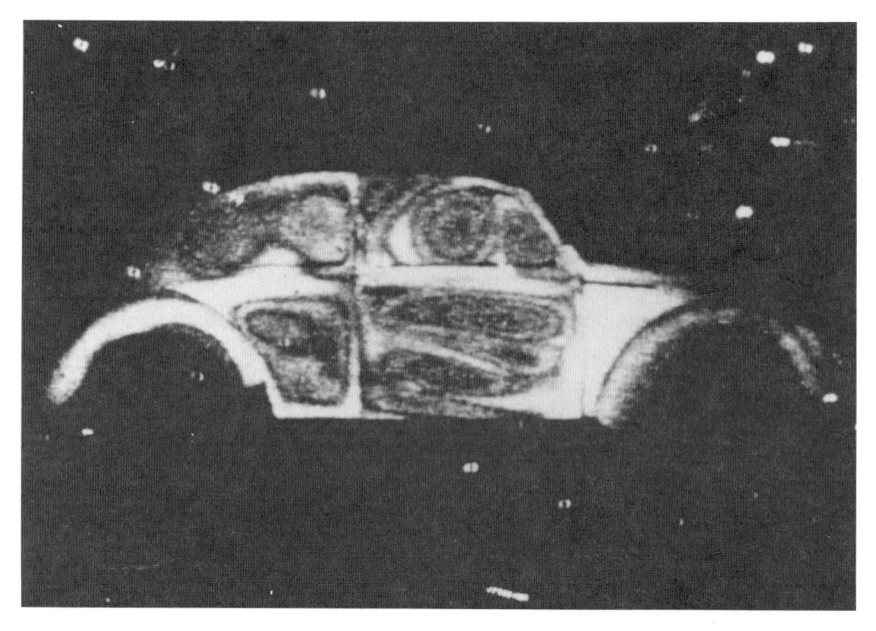

Figure 15.23. ESPI interferogram of car body vibrating at 110 Hz taken in parking lot. (From Malmo and Vikhagen 1988.)

Stetson and Brohinsky have shown that the speckle term can be removed (Stetson and Brohinsky 1985, 1986, 1987; Bushman 1989; Feit 1989; Stetson 1989; Stetson et al. 1989). The electro-optic holography (EOH) system uses the same optical setup as ESPI; however, one mirror in the reference beam is mounted on a PZT to provide relative phase shifts between the object and reference beams. Data can be taken at TV frame rates and stored on videotape. The major difference of EOH from ESPI is the use of a pipeline image processor to handle a large number of images at once instead of the simple electronics of ESPI, which process a single frame at a time.

For static measurements, this technique utilizes four frames of data with 90° relative phase shifts between the object and reference, which are stored in an array processor. One of these frames of data is written as

$$I_i = I_0[1 + \gamma \cos(\phi + \delta_i)], \qquad (15.25)$$

where $\delta_i = 0$, $\pi/2$, π, and $3\pi/4$ are the applied phase shifts for I_1, I_2, I_3, and I_4. After the applied stress on the object is changed, four more frames of data are recorded with 90° relative phase shifts:

$$I'_i = I_0[1 + \gamma \cos(\phi' + \delta_i)], \qquad (15.26)$$

15.6. SPECKLE INTERFEROMETRY AND TV HOLOGRAPHY

where $\phi' = \phi + \Delta\phi$. The eight frames of data are combined in a pipeline image processing system to yield

$$I = \sqrt{[(I_1 - I_3) + (I'_1 - I'_3)]^2 + [(I_2 - I_4) + (I'_2 - I'_4)]^2}$$
$$= 4I_0\gamma \cos(\Delta\phi/2). \tag{15.27}$$

This calculation produces fringes proportional to the phase change due to the object displacement. There is no extra term due to the speckles. The calculation involves only addition, subtraction, and the application of a look-up table to find the square root of the sum of the squares of two numbers. This is easily accomplished using an array processor. As the applied stress to the object changes, four new frames need to be obtained. As long as the applied stress is slowly varying, the secondary interference fringes can be calculated after each video frame is recorded and 90° relative phase shifts are applied between consecutive video frames. Thus, these fringes can be calculated and displayed at TV frame rates using an image-processing system.

For a dynamic measurement using EOH, four time-average frames of data are taken while the object is vibrating with 90° relative phase shifts. One of these frames is written as

$$I_i = I_0[1 + \gamma \cos(\phi + \delta_i)M(\Omega)], \tag{15.28}$$

where the phase shifts δ_i are 0, $\pi/2$, π, and $3\pi/4$. These four frames can then be combined to obtain secondary fringes using

$$I = \sqrt{(I_1 - I_3)^2 + (I_2 - I_4)^2} = 2I_0\gamma|M(\Omega)|, \tag{15.29}$$

which can be written as

$$I = \sqrt{(I_1 - I_3)^2 + (I_2 - I_4)^2} = 2I_0\gamma|J_0(\Omega)| \tag{15.30}$$

for a sinusoidal object vibration. Note that the speckle modulation term is not present with this technique.

An example of fringes produced with EOH is shown in Fig. 15.24. Figure 15.24a shows a rectangular plate undergoing static deflection, and Fig. 15.24b shows a time-average vibration mode of the same plate. EOH can also be used for shearography (shearing speckle interferometry) by changing the optical system (Stetson 1989).

15.6.4. Speckle Interferometry/TV Holography with Phase Measurement

A technique similar to the two-wavelength-interferometry method can be used for quantitative measurement of object displacement or deformation in EPSI

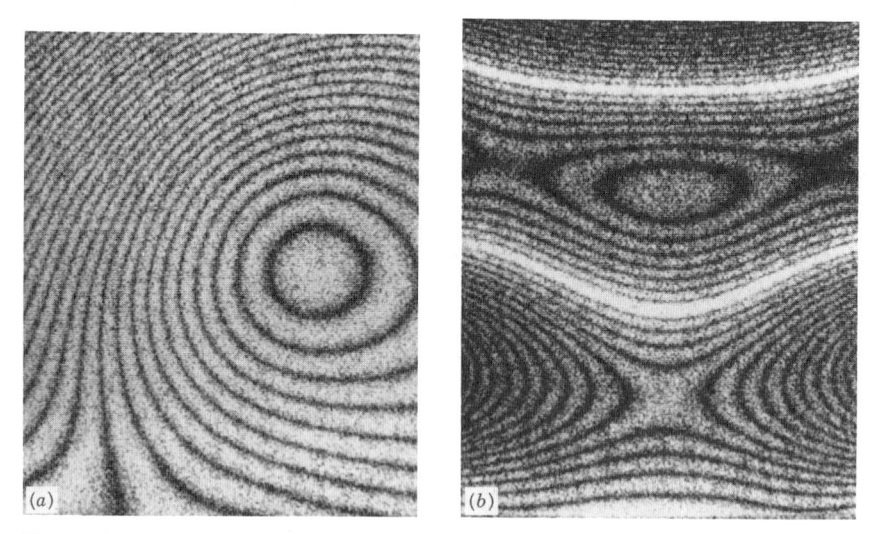

Figure 15.24. EOH interferograms of (*a*) rectangular plate undergoing static deflection and (*b*) time-average vibration mode of same plate. (Courtesy of K. A. Stetson, United Technologies Research Center.)

(Nakadate and Saito 1985; Creath 1985b, 1985a; Robinson and Williams 1986). Using a single measurement wavelength, intensity data with multiple phase shifts are taken for an object. After the stress applied to the object is changed, a second set of intensity data are taken. Modulo 2π phases of the interference pattern are then calculated using standard phase-shifting techniques for each object state. The phase due to the change in the object state is found by taking the difference of these two phases

$$\Delta\phi = \phi' - \phi, \qquad (15.31)$$

where ϕ and ϕ' are the phases before and after the applied force, and $\Delta\phi$ is the phase due to change in the object position and shape. $\Delta\phi$ can also be thought of as the phase due to the secondary interference fringes produced in double-exposure holographic interferometry. The phase can also be calculated using an equation similar to Eq. (15.6). An alternative phase calculation that requires data at only two different phase shifts for displacement/deformation measurement (a total of four frames of data, two before and two after) instead of three or more has been developed by Kerr et al. (1990).

In EOH the eight frames of data taken for a static measurement [Eqs. (15.25) and (15.26)] can be used to calculate the phase change $\Delta\phi$ due to the displacement of the object,

$$\Delta\phi = \tan^{-1}\left(\frac{C_1 - C_2}{C_3 - C_4}\right), \qquad (15.32)$$

where the quantities C_i are given by

$$C_1 = [(I_1 - I_3) + (I'_1 - I'_3)]^2 + [(I_2 - I_4) + (I'_2 - I'_4)]^2, \quad (15.33)$$

$$C_2 = [(I_1 - I_3) - (I'_1 - I'_3)]^2 + [(I_2 - I_4) - (I'_2 - I'_4)]^2, \quad (15.34)$$

$$C_3 = [(I_1 - I_3) + (I'_2 - I'_4)]^2 + [(I_2 - I_4) + (I'_1 - I'_3)]^2, \quad (15.35)$$

$$C_4 = [(I_1 - I_3) - (I'_2 - I'_4)]^2 + [(I_2 - I_4) - (I'_1 - I'_3)]^2. \quad (15.36)$$

These calculations only involve simple calculations and the use of look-up tables, which can be done in an array processor. Potentially, the modulo 2π phase corresponding to the object displacement can be calculated in real time at video frame rates. Quantitative measurements of time-average interferograms can be made using the techniques described for holographic nondestructive testing in Section 15.5.3 and Eqs. (15.14) through (15.18) (Pryputniewicz and Stetson 1989).

Acknowledgements. The authors acknowledge the support of WYKO Corporation during the preparation of this manuscript.

REFERENCES

Archbold, E., J. M. Burch, and A. E. Ennos, "Recording of In-Plane Surface Displacement by Double-Exposure Speckle Photography," *Opt. Acta*, **17**(12), 883-898 (1970).

Arnold, S. M., "E-Beam Written Computer Generated Holograms," Final Report for *AFSOR Contract No. F4962-80-C-0029*, Honeywell Corporate Technology Center, Bloomington, MN, August 1983.

Arnold, S. M., "How to Test an Asphere with a Computer Generated Hologram," *Proc. SPIE*, **1052,** 191-197 (1989).

Bartelt, H. O. and K. D. Forster, "Computer Generated Holograms with Reduced Phase Errors," *Opt. Commun.*, **26**(1), 12-16 (1978).

Biedermann, K., "Information Storage Materials for Holography and Optical Data Processing," *Opt. Acta*, **22**(2), 103-124 (1975).

Biedermann, K. and L. Ek, "A Recording and Display System for Hologram Interferometry with Low Resolution Imaging Devices," *J. Phys. E, Sic. Instrum.*, **8**(7), 571–576 (1975).

Birch, K. G. and F. J. Green, "The Application of Computer Generated Holograms to Testing Optical Elements," *J. Phys., D, Appl. Phys.*, **5**(11), 1982–1992 (1972).

Brdicko, J., M. D. Olson, and C. R. Hazell, "Theory for Surface Displacement and Strain Measurements by Laser Speckle Interferometry," *Opt. Acta*, **25**(10), 963–989 (1987).

Broder-Bursztyn, F. and D. Malacara-Hernández, "Holographic Interferometer to Test Optical Surfaces," *Appl. Opt.*, **14**(9), 2280–2282 (1975).

Bushman, T., "Development of a Holographic Computing System," *Proc. SPIE*, **1162**, 66–77 (1989).

Butters, J. N. and J. A. Leendertz, "Speckle Pattern and Holographic Techniques in Engineering Metrology," *Opt. Laser Tech.*, **3**(1), 26–30 (1971).

Caulfield, H. J., P. Mueller, D. Dvore, and A. Epstein, "Computer Holograms for Optical Testing," *Proc. SPIE*, **306**, 154–157 (1981).

Cheng, Y.-Y. and J. C. Wyant, "Two-Wavelength Phase Shifting Interferometry," *Appl. Opt.*, **23**(24), 4539–4543 (1984).

Cheng, Y.-Y. and J. C. Wyant, "Multiple-Wavelength Phase-Shifting Interferometry," *Appl. Opt.*, **24**(6), 804–807 (1985).

Chiang, F. P. and A. K. Asundi, "White Light Speckle Method with Tandem Phase Plates for 3-D Displacement and Deformation Measurement on Curved Surfaces," *Appl. Opt.*, **20**(13), 2167–2169 (1981).

Creath, K., "Digital Speckle-Pattern Interferometry," Ph.D. Dissertation, Optical Sciences Center, University of Arizona, Tucson, AZ, University Microfilms, Ann Arbor, MI 1985a.

Creath, K., "Phase-Shifting Speckle Interferometry," *Appl. Opt.*, **24**(18), 3053–3058 (1985b).

Creath, K., "Averaging Double-Exposure Speckle Interferograms," *Opt. Lett.*, **10**(12), 582–584 (1985c).

Creath, K., "Measuring Step Heights Using an Optical Profiler," *Proc. SPIE*, **661**, 296–301 (1986a).

Creath, K., "Direct Measurement of Deformations Using Digital Speckle-Pattern Interferometry," *Proc. SEM Spr. Conf. on Exp. Mech.*, New Orleans, 370–377 (1986b).

Creath, K., Y.-Y. Cheng, and J. C. Wyant, "Contouring Aspheric Surfaces Using Two-Wavelength Phase-Shifting Interferometry," *Opt. Acta*, **32**(12), 1455–1464 (1985).

Creath, K. and J. C. Wyant, "Direct Phase Measurement of Aspheric Surface Contours," *Proc. SPIE*, **645**, 101–106 (1986).

Dändliker, R. and R. Thalmann, "Determination of 3-D Displacement and Strain by Holographic Interferometry for Non-Plane Objects," *Proc. SPIE*, **398**, 11–16 (1983).

Dändliker, R. and R. Thalmann, "Heterodyne and Quasi-Heterodyne Holographic Interferometry," *Opt. Eng.*, **24**(5), 824–831 (1985).

Deever, W. T., "Testing of Commercial Aspheres with Computer-Generated Holography," *J. Opt. Soc. Am.*, **65**(10), 1216–1216 (1975).

Dörband, B. and H. J. Tiziani, "Testing Aspheric Surfaces with Computer-Generated Holograms: Analysis of Adjustment Shape Errors," *Appl. Opt.*, **24**(16), 2604–2611 (1985).

Dvore, D., *Data Formats for the ARI Holowriter Facility*, Aerodyne Research Inc., Billerica, MA, March 1983.

Emmel, P. M. and K. M. Leung, "A New Instrument for Routine Optical Testing of General Aspherics," *Proc. SPIE*, **171**, 93–99 (1979).

Ennos, A., "Speckle Interferometry," in *Laser Speckle and Related Phenomena*, J. C. Dainty, Ed., Springer-Verlag, New York, 1975, pp. 203–253.

Erf, R. K., *Holographic Nondestructive Testing*, Academic Press, Orlando, 1974.

Erf, R. K., *Speckle Metrology*, Academic Press, New York 1978.

Erf, R. K., "Holographic Nondestructive Testing: Review of a Laser Inspection Tool," *Proc. SPIE*, **349**, 47–58 (1982).

Faulde, M., A. F. Fercher, R. Torge, and R. N. Wilson, "Optical Testing by Means of Synthetic Holograms and Partial Lens Compensation," *Opt. Commun.*, **7**(4), 363–365 (1973).

Feit, E., "Electronic Holography for Non-Destructive Testing," *Advanced Imaging*, January, 42–45, (1989).

Fercher, A. F., "Computer Generated Holograms for Testing Optical Elements: Error Analysis and Error Compensation," *Opt. Acta*, **23**(5), 347–365 (1976).

Fercher, A. F., H. Z. Hu, and U. Vry, "Rough Surface Interferometry with a Two-Wavelength Heterodyne Speckle Interferometer," *Appl. Opt.*, **24**(14), 2181–2188 (1985).

Fercher, A. F. and M. Kriese, "Binare Synthetische Hologramme zur Prufung Aspharischer Optischer Elemente," *Optik*, **35**(2), 168–179 (1972).

Friesem, A. A., Y. Katzir, Z. Rav-Noy, and B. Sharon, "Photoconductor-Thermoplastic Devices for Holographic Nondestructive Testing," *Opt. Eng.*, **19**(5), 659–665 (1980).

Fujii, H. and J. W. Lit, "Surface Roughness Measurement Dichromatic Speckle Patterns: An Experimental Study," *Appl. Opt.*, **17**(17), 2690–2694 (1978).

Goodman, J. W., *Introduction to Fourier Optics*, McGraw-Hill, New York, 1968.

Goodman, J. W., "Statistical Properties of Laser Speckle Patterns," in *Laser Speckle and Related Phenomena*, J. C. Dainty, Ed., Springer-Verlag, Berlin 1975a, pp. 9–75.

Goodman, J. W., "Dependence of Image Speckle Contrast on Surface Roughness," *Opt. Commun.*, **14**(3), 324–327 (1975b).

Greivenkamp, J. E., "Interferometric Measurements at Eastman Kodak Company," *Proc. SPIE*, **816**, 212–227 (1987).

Haines, K. A. and B. P. Hildebrand, "Surface Deformation Measurement Using the Wavefront Reconstruction Technique," *Appl. Opt.*, **5**(4), 595–602 (1966).

Hansler, R. L., "Application of Holographic Interferometry to the Comparison of Highly Polished Reflecting Surfaces," *Appl. Opt.*, **7**(4), 711–712 (1968a).

Hansler, R. L., "A Holographic Foucault Knife-Edge Test for Optical Elements of Arbitrary Design," *Appl. Opt.*, **7**(9), 1863–1864 (1968b).

Hariharan, P., *Optical Holography*, Cambridge University Press, Cambridge, 1984.

Hariharan, P., "Quasi-Heterodyne Hologram Interferometry," *Opt. Eng.*, **24**(4), 632–638 (1985).

Hariharan, P. and B. F. Oreb, "Stroboscopic Holographic Interferometry: Applications of Digital Techniques," *Opt. Commun.*, **59**(2), 83–86 (1986).

Hariharan, P., B. F. Oreb, and N. Brown, "A Digital Phase-Measurement System for Real-Time Holographic Interferometry," *Opt. Commun.*, **41**(6), 393–396 (1982).

Hariharan, P., B. F. Oreb, and N. Brown, "Real-time Holographic Interferometry: A Microcomputer System for the Measurement of Vector Displacements," *Appl. Opt.*, **22**(6), 876–880 (1983a).

Hariharan, P., B. F. Oreb, and N. Brown, "A System for Real-Time Holographic Stress Analysis," *Proc. SPIE*, **370**, 189–194 (1983b).

Hariharan, P., B. F. Oreb, and N. Brown, "Computer-Aided Analysis of Holographic Interferograms Using the Phase-Shift Method: A Comment," *Appl. Opt.*, **25**(10), 1536 (1986).

Hariharan, P., B. F. Oreb, and C. H. Freund, "Stroboscopic Holographic Interferometry: Measurements of Vector Components of a Vibration," *Appl. Opt.*, **26**(18), 3899–3903 (1987).

Hildebrand, B. P. and K. A. Haines, "Interferometric Measurements Using the Wavefront Reconstruction Techniques," *Appl. Opt.*, **5**(1), 172–173 (1966).

Hildebrand, B. P. and K. A. Haines, "Multiple Wavelength and Multiple Source Holography Applied to Contour Generation," *J. Opt. Soc. Am.*, **57**(2), 155–162 (1967).

Hildebrand, B. P., K. A. Haines, and R. Larkin, "Holography as a Tool in the Testing of Large Apertures," *Appl. Opt.*, **6**(7), 1267–1269 (1967).

Huntley, J. M. "Speckle Photography Fringe Analysis: Assessment of Current Algorithms," *Appl. Opt.*, **28**(20), 4316–4322 (1989).

Ichioka, Y. and A. W. Lohmann, "Interferometric Testing of Large Optical Components with Circular Computer Holograms," *Appl. Opt.*, **11**(11), 2597–2602 (1972).

Jaisingh, G. K. and F.-P. Chiang, "Contouring by Laser Speckle," *Appl. Opt.*, **20**(19), 3385–3387 (1981).

Jones, R. and C. Wykes, *Holographic and Speckle Interferometry*, Cambridge University Press, Cambridge, 1983.

Kakunai, S., K. Iwata, R. Nagata, and H. Sekiguchi, "Measurement of Three Components of a Displacement Vector Using Heterodyne Holographic Interferometry," *Opt. Lasers Eng.*, **6**(4), 213–223 (1985).

Kerr, D., F. Mendoza-Santoyo, and J. R. Tyrer, "Extraction of Phase Data from Electronic Speckle Pattern Interferometric Fringes Using a Single-Phase-Step Method: A Novel Approach," *J. Opt. Soc. Am. A*, **7**(5), 820–826 (1990).

Kreis, T. M., "Digital Holographic Interference-Phase Measurement Using the Fourier-Transform Method," *J. Opt. Soc. Am. A*, **3**(6), 847–855 (1986).

Kreis, T. M., "Fourier-Transform Evaluation of Holographic Interference Patterns," *Proc. SPIE*, **814**, 365–371 (1987).

Kurtz, R. L. and R. B. Owen, "Holographic Recording Materials: A Review," *Opt. Eng.*, **14**(5), 393–401 (1975).

REFERENCES

Lam, P., J. D. Gaskill, and J. C. Wyant, "Two Wavelength Holographic Interferometer," *Appl. Opt.*, **23**(18), 3079-3081 (1984).

Lee, W. H., "Sampled Fourier Transform Hologram Generated by Computer," *Appl. Opt.*, **9**(3), 639-643 (1970).

Lee, W. H., "Binary Synthetic Holograms," *Appl. Opt.*, **13**(7), 1677-1682 (1974).

Lee, W. H., "Computer-Generated Holograms: Techniques and Applications," in *Progress in Optics*, Vol. XVI, E. Wolf, Ed., North-Holland, Amsterdam, 1978, pp. 121-232.

Lee, W. H., "Recent Developments in Computer-Generated Holograms," *Proc. SPIE*, **215**, 52-58 (1980).

Leith, E. N. and J. Upatnieks, "Reconstructed Wavefronts and Communication Theory," *J. Opt. Soc. Am.*, **52**(10), 1123-1129 (1962).

Leung, K. M., T. C. Lee, E. Bernal, and J. C. Wyant, "Two-Wavelength Contouring with the Automated Thermoplastic Holographic Camera," *Proc. SPIE*, **192**, 184-189 (1979).

Leung, K. M., J. C. Lindquist, and L. T. Shepherd, "E-Beam Computer Generated Holograms for Optical Testing," *Proc. SPIE*, **215**, 70-75 (1980).

Leung, K. M., S. M. Arnold, and J. C. Lindquist, "Using E-Beam Written Computer-Generated Holograms to Test Deep Aspheric Wavefronts," *Proc. SPIE*, **306**, 161-167 (1981).

Levitt, J. A. and K. A. Stetson, "Mechanical Vibrations: Mapping Their Phase with Hologram Interferometry," *Appl. Opt.*, **15**(1), 195-199 (1976).

Løkberg, O. J. and G. Å. Slettemoen, "Basic Electronic Speckle-Pattern Interferometry," in *Applied Optics and Optical Engineering*, Vol. X, R. R. Shannon and J. C. Wyant, Eds., Academic Press, San Diego, 1987, pp. 455-504.

Loomis, J. S., "Applications of Computer-Generated Holograms in Optical Testing," Ph.D. Dissertation, Optical Sciences Center, University of Arizona, Tucson, AZ, University Microfilms, Ann Arbor, MI, 1980a.

Loomis, J. S., "Computer-Generated Holography and Optical Testing," *Opt. Eng.*, **19**(5), 679-685 (1980b).

Lurionov, N. P., A. V. Lukin, and K. S. Mustafin, "Holographic Inspection of Shapes of Unpolished Surfaces," *Sov. J. Opt. Technol.*, **39**(3), 154-155 (1972).

MacGovern, A. J. and J. C. Wyant, "Computer Generated Holograms for Testing Optical Elements," *Appl. Opt.*, **10**(3), 619-624 (1971).

Macovski, A., S. D. Ramsey, and L. F. Schaefer, "Time-Lapse Interferometry and Contouring Using Television Systems," *Appl. Opt.*, **10**(12), 2722-2727 (1971).

Malmo, J. T. and E. Vikhagen, "Vibration Analysis of a Car Body by Means of TV Holography," *Exp. Tech.*, **12**(4), 28-30 (1988).

Nakadate, S., "Vibration Measurement Using Phase-Shifting Time-Average Holographic Interferometry," *Appl. Opt.*, **25**(22), 4155-4161 (1986).

Nakadate, S., N. Magome, T. Honda, and J. Tsujiuchi, "Hybrid Holographic Interferometer for Measuring Three-Dimensional Deformations," *Opt. Eng.*, **20**(2), 246-252 (1981).

Nakadate, S. and H. Saito, "Fringe Scanning Speckle-Pattern Interferometry," *Appl. Opt.*, **24**(14), 2172–2180 (1985).

Neumann, D. B., C. F. Jacobson, and G. M. Brown, "Holographic Technique for Determining the Phase of Vibrating Objects," *Appl. Opt.*, **9**(6), 1357–1362 (1970).

Ono, A. and J. C. Wyant, "Plotting Errors Measurement of CGH Using an Improved Interferometric Method," *Appl. Opt.*, **23**(21), 3905–3910 (1984).

Oshida, Y., K. Iwata, and R. Nagata, "Optical Heterodyne Measurement of Vibration Phase," *Opt. Lasers Eng.*, **4**(2), 67–79 (1983).

Pastor, J., "Hologram Interferometry and Optical Technology," *Appl. Opt.*, **8**(3), 525–531 (1969).

Pastor, J., G. E. Evans, and J. S. Harris, "Hologram-Interferometry: A Geometrical Approach," *Opt. Acta*, **17**(2), 81–96 (1970).

Polhemus, C., "Two-Wavelength Interferometry," *Appl. Opt.*, **12**(9), 2071–2074 (1973).

Powell, R. L. and K. A. Stetson, "Interferometric Vibration Analysis by Wavefront Reconstruction," *J. Opt. Soc. Am.*, **55**(12), 1593–1598 (1965).

Pryputniewicz, R. J., "Speckle Metrology Techniques and their Applications," *Proc. SPIE*, **556**, 90–98 (1985).

Pryputniewicz, R. J. and K. A. Stetson, "Holographic Strain Analysis: Extension of Fringe-Vector Method to Include Perspective," *Appl. Opt.*, **15**(3), 725–728 (1976).

Pryputniewicz, R. J. and K. A. Stetson, "Measurement of Vibration Patterns Using Electro-optic Holography," *Proc. SPIE*, **1162**, 456–467 (1989).

Robinson, D. W., "Holographic and Speckle Interferometry in the UK: A Review of Recent Developments," *Proc. SPIE*, **814**, 330–337 (1987).

Robinson, D. W. and D. C. Williams, "Digital Phase Stepping Speckle Interferometry," *Opt. Commun.*, **57**(1), 26–30 (1986).

Schmidt, W. and A. F. Fercher, "Holographic Generation of Depth Contour Using a Flash-Lamp-Pumped Dye Laser," *Opt. Commun.*, **3**(5), 363–365 (1971).

Schwider, J. and R. Burow, "The Testing of Aspherics by Means of Rotational-Symmetric Synthetic Holograms," *Opt. Appl.*, **6**, 83 (1976).

Schwider, J., R. Burow, and J. Grzanna, "CGH—Testing of Rotational Symmetric Aspheric in Compensated Interferometers," *Opt. Appl.*, **9**, 39 (1979).

Smith, D. C., "Testing Diamond Turned Aspheric Optics Using Computer-Generated Holographic (CGH) Interferometry," *Proc. SPIE*, **306**, 112–121 (1981).

Snow, K. and R. Vandewarker, "On Using Holograms for Test Glasses," *Appl. Opt.*, **9**(4), 822–827 (1970).

Sommargren, G. E., "Double-Exposure Holographic Interferometry Using Common-Path Reference Waves," *Appl. Opt.*, **16**(6), 1736–1741 (1977).

Stetson, K. A., "Fringe Interpretation for Hologram Interferometry of Rigid-Body Motions and Homogeneous Deformations," *J. Opt. Soc. Am.*, **64**(1), 1–10 (1974).

Stetson, K. A., "Homogeneous Deformations: Determination by Fringe Vectors in Hologram Interferometry," *Appl. Opt.*, **14**(9), 2256–2259 (1975a).

Stetson, K. A., "A Review of Speckle Photography and Interferometry," *Opt. Eng.*, **14**(5), 482–489 (1975b).

Stetson, K. A., "Use of Projection Matrices in Hologram Interferometry," *J. Opt. Soc. Am.*, **69**(12), 1705–1710 (1979).

Stetson, K. A., "Method of Vibration Measurements in Heterodyne Interferometry," *Opt. Lett.*, **7**(5), 233–234 (1982).

Stetson, K. A., "Electro-optic Holography for Real-Time Display and Quantitative Analysis of Interference Fringes," *Fringe '89, Automatic Processing of Fringe Patterns* (E. Berlin, April 25–28, 1989) and *Optical Sensing and Measurement, Proc. ICALEO, Laser Inst. Am.*, **70**, 78–85 (1989).

Stetson, K. A., "Use of Sensitivity Vector Variations to Determine Absolute Displacements in Double Exposure Hologram Interferometry," *Appl. Opt.*, **29**(4), 502–504 (1990).

Stetson, K. A. and W. R. Brohinsky, "Electro-optic Holography and Its Application to Hologram Interferometry," *Appl. Opt.*, **24**(21), 3631–3637 (1985).

Stetson, K. A. and W. R. Brohinsky, "Measurement of Phase Change in Heterodyne Interferometry: A Novel Scheme," *Appl. Opt.*, **25**(16), 2643–2644 (1986).

Stetson, K. A. and W. R. Brohinsky, "Electro-optic Holography System for Vibration Analysis and Nondestructive Testing," *Opt. Eng.*, **26**(12), 1234–1239 (1987).

Stetson, K. A. and W. R. Brohinsky, "Fringe-Shifting Technique for Numerical Analysis of Time-Average Holograms of Vibrating Objects," *J. Opt. Soc. Am. A*, **5**(9), 1472–1476 (1988).

Stetson, K. A., W. R. Brohinsky, J. Wahid, and T. Bushman, "An Electro-optic Holography System with Real-Time Arithmetic Processing," *J. Nondestructive Evaluation*, **8**(2), 69–76 (1989).

Stetson, K. A. and R. L. Powell, "Interferometric Hologram Evaluation and Real-Time Vibration Analysis of Diffuse Objects," *J. Opt. Soc. Am.*, **55**(12), 1694–1695 (1965).

Takahashi, T., K. Konno, and M. Kawai, "Some Improvements in Computer Hologram for Testing Aspheric Surface," *Proceedings of the ICO Conference on Optical Methods in Scientific and Industrial Measurements, Tokyo, 1974; Jpn. J. Appl. Phys.*, **14**, Suppl. 14-1, 247–251 (1975).

Thalmann, R. and R. Dändliker, "High Resolution Video-Processing for Holographic Interferometry Applied to Contouring and Measuring Deformations," *Proc. SPIE*, **492**, 299–306 (1984).

Thalmann, R. and R. Dändliker, "Holographic Contouring Using Electronic Phase Measurement," *Opt. Eng.*, **24**(6), 930–935 (1985a).

Thalmann, R. and R. Dändliker, "Automated Evaluation of 3-D Displacement and Strain by Quasi-Heterodyne Interferometry," *Proc. SPIE*, **599**, 141–148 (1985b).

Uhrich, C. and L. Hesselink, "Optical Surface Inspection Using Real-Time Fourier Transform Holography in Photorefractives," *Appl. Opt.*, **27**(21), 4497–4503 (1988).

Van Deelen, W. and P. Nisenson, "Mirror Blank Testing by Real-Time Holographic Interferometry," *Appl. Opt.*, **8**(5), 951–955 (1969).

Vest, C. M., *Holographic Interferometry*, Wiley, New York, 1979.

Vest, C. M., "Holographic Interferometry: Some Recent Developments," *Opt. Eng.*, **19**(5), 654–658 (1980).

Vest, C. M., "Status and Future of Holographic Nondestructive Testing," *Proc. SPIE*, **349**, 186–198 (1982).

Wyant, J. C., "Testing Aspherics Using Two-Wavelength Holography," *Appl. Opt.*, **10**(9), 2113-2118 (1971).

Wyant, J. C., "Holographic Testing of Aspheric Optical Elements," in *Proc. Ninth Congress of ICO*, National Academy of Science, Washington, D.C., 1974, pp. 643-664.

Wyant, J. C. and V. P. Bennett, "Using Computer Generated Holograms to Test Aspheric Wavefronts," *Appl. Opt.*, **11**(12), 2833-2839 (1972).

Wyant, J. C. and K. Creath, "Two-Wavelength Phase-Shifting Interferometer and Method," U.S. Patent No. 4,832,489 (1989).

Wyant, J. C. and P. K. O'Neill, "Computer Generated Hologram; Null Lens Test of Aspheric Wavefronts," *Appl. Opt.*, **13**(12), 2762-2765 (1974).

Wyant, J. C., P. K. O'Neill, and A. J. MacGovern, "Interferometric Method of Measuring Plotter Distortion," *Appl. Opt.*, **13**(7), 1549-1551 (1974).

Wyant, J. C., B. F. Oreb, and P. Hariharan, "Testing Aspherics Using Two-Wavelength Holography: Use of Digital Electronic Techniques," *Appl. Opt.*, **23**(22), 4020-4023 (1984).

Yatagai, T. and H. Saito, "Interferometric Testing with Computer-Generated Holograms: Aberration Balancing Method and Error Analysis," *Appl. Opt.*, **17**(4), 558-565 (1978).

Zelenka, J. S. and J. R. Varner, "A New Method for Generating Depth Contours Holographically," *Appl. Opt.*, **7**(10), 2107-2110 (1968).

Zelenka, J. S. and J. R. Varner, "Multiple-Index Holographic Contouring," *Appl. Opt.*, **8**(7), 1431-1434 (1969).

ADDITIONAL REFERENCES

This is not a complete listing. Emphasis has been placed on recent material.

Aleksoff, C. C., "Temporally Modulated Holography," *Appl. Opt.*, **10**(6), 1329-1341 (1971).

Aleksoff, C. C., "Temporal Modulation Techniques," in *Holographic Nondestructive Testing*, R. K. Erf, Ed., Academic Press, New York, 1974, pp. 247-263.

Archbold, E., J. M. Burch, A. E. Ennos, and P. A. Taylor, "Visual Observations of Surface Vibrational Nodal Patterns," *Nature*, **222**(5190), 163-165 (1969).

Archbold, E. and A. E. Ennos, "Displacement Measurement from Double-Exposure Laser Photographs," *Opt. Acta*, **19**(4), 253-271 (1972).

Archbold, E., A. E. Ennos, and M. S. Virdee, "Speckle Photography for Strain Measurement—A Critical Assessment," *Proc. SPIE*, **136**, 258-264 (1977).

Aver'yanova, G. I., N. P. Larinov, A. V. Lukin, K. S. Mustafin, and R. A. Rafikov, "The Testing of Large Aspherical Surfaces by Means of Synthetic Circular Holograms," *Sov. J. Opt. Technol.*, **42**(6), 347-349 (1975).

Baker, L. R., *Stress and Vibration: Recent Developments in Measurement and Analysis, Proc. SPIE*, **1084** (1989).

ADDITIONAL REFERENCES

Bartelt, H. and S. K. Case, "High-Efficiency Hybrid Computer-Generated Holograms," *Appl. Opt.,* **21**(16), 2886-2890 (1982).

Bates, R. H. T., B. R. Hunt, B. S. Robinson, W. R. Fright, and P. T. Gough, "Aspects of Speckle Interferometric Imaging," *Electronic Image Processing, IEE Proc.,* **214**, 164-168 (1982).

Biedermann, K., L. Ek, and L. Östlund, "A TV Speckle Interferometer," in *The Engineering Uses of Coherent Optics,* E. R. Robertson, Ed., Cambridge University Press, Cambridge, 1976, pp. 219-221.

Burch, J. M., "Laser Speckle Metrology," *Proc. SPIE,* **25**, 149-156 (1971).

Bryngdahl, O. and A. W. Lohmann, "Interferograms Are Image Holograms," *J. Opt. Soc. Am.,* **58**(1), 141-142 (1968).

Butters, J. N., "Some Applications of Electronic Speckle Pattern Interferometry," *IEEE Intern'l Opt. Computing Conf. Digest,* 85-89 (1975).

Butters, J. N., "Application of ESPI to NDT," *Opt. Lasers Tech.,* **9**(3), 117-123 (1977).

Butters, J. N., R. Jones, S. McKechnie, and C. Wykes, "Measurement Techniques for Quality Assurance and Control," *Conf. Proc. Electro-Opt./Laser Intern'l '80 UK,* 139-156 (1980).

Butters, J. N., R. Jones, and C. Wykes, "Electronic Speckle Pattern Interferometry," in *Speckle Metrology,* R. K. Erf, Ed., Academic Press, New York, 1978, pp. 111-158.

Butters, J. N. and J. A. Leendertz, "Holographic and Video Techniques Applied to Engineering Measurements," *Meas. Control,* **4**, 349-354 (1971).

Butters, J. N. and J. A. Leendertz, "A Double Exposure Technique for Speckle Pattern Interferometry," *J. Phys. E: Sci. Instrum.,* **4**(4), 277-279 (1971).

Butters, J. N. and J. A. Leendertz, "Optical Inspection," U.S. Patent No. 3,816,649 (1974).

Chiang, F. P., *International Conference on Photomechanics and Speckle Metrology, Proc. SPIE,* **814**, (1987).

Chen, C. W. and J. B. Breckinridge, "Holographic Twyman-Green Interferometer," *Appl. Opt.,* **21**(14), 2563-2568 (1982).

Cindrich, I. N., *Holographic Optics: Optically and Computer Generated, Proc. SPIE,* **1052** (1989).

Cindrich, I. N., *Computer and Optically Formed Holographic Optics, Proc. SPIE,* **1211** (1990).

Collier, R. J., C. B. Burckhardt, and L. H. Lin, *Optical Holography,* Academic Press, New York, 1971.

Cookson, T. J., J. N. Butters, and H. C. Pollard, "Pulsed Lasers in Electronic Speckle Pattern Interferometry," *Opt. Laser Tech,* **10**(3), 119-124 (1978).

Creath, K., "Phase-Measuring Holographic Technique Measures Object Deformations in Real Time," *Laser Focus/E-O,* 128-140 (April 1988).

Creath, K. "Quantitative Measurement Using Holographic Interferometry," *Proc. 4th International Conf. on Exp. Mechanics,* 449-453 (1988).

Creath, K., "Holographic Contour and Deformation Measurement Using a 1.4 Million Element Detector Array," *Appl. Opt.,* **28**(11), 2170-2175 (1989).

Creath, K. and G. Å. Slettemoen, "Vibration-Observation Techniques for Digital Speckle-Pattern Interferometry," *J. Opt. Soc. Am. A*, **2**(10), 1629–1636 (1985).

Dainty, J. C., *Laser Speckle and Related Phenomena*, 2nd ed., Springer-Verlag, Berlin, 1984.

Dändliker, R., "Heterodyne Holographic Interferometry," in *Progress in Optics*, Vol. XVII, E. Wolf, Ed., North-Holland, Amsterdam, 1980, pp. 1–84.

Dändliker, R., R. Thalmann, and J.-F. Willemin, "Fringe Interpolation by Two-Reference-Beam Holographic Interferometry: Reducing Sensitivity to Hologram Misalignment," *Opt. Commun.*, **42**(5), 301–306 (1982).

Dändliker, R. and J.-F. Willemin, "Measuring Micro-Vibrations by Heterodyne Speckle Interferometry," *Proc. SPIE*, **236**, 83–85 (1980).

Dändliker, R. and J.-F. Willemin, "Measuring Microvibrations by Heterodyne Speckle Interferometry," *Opt. Lett.*, **6**(4), 165–167 (1981).

den Boef, A. J., "Two-Wavelength Scanning Spot Interferometer Using Single-Frequency Diode Lasers," *Appl. Opt.*, **27**(2), 306–311 (1988).

Dukhopel, I. I. and L. G. Fedina, "Holographic Interferometer for Checking Lens Deformations," *Sov. J. Opt. Tech*, **47**(1), 17–19 (1980).

Ek, L. and K. Biedermann, "Fringe Contrast in a System for Hologram Interferometry with Low Resolution Imaging Devices," *J. Phys. E: Sci. Instrum*, **8**(8), 691–696 (1975).

Ek, L. and N.-E. Molin, "Detection of the Nodal Lines and the Amplitude of Vibration by Speckle Interferometry," *Opt. Commun.*, **2**(9), 419–424 (1971).

Ennos, A. E., "Speckle Interferometry," in *Coherent Optical Engineering*, F. T. Arecchi and V. Degiorgio, Eds., North-Holland, Amsterdam, 1977, pp. 129–149.

Fagan, W. F. and R. W. T. Preater, "Video Holographic Interferometry," *Proc. SPIE*, **746**, 58–60 (1987).

Fantone, S. D., "Holographic Interferometer for Testing Aspheric Molds and Molded Parts," *Appl. Opt.*, **22**(8), 1121–1226 (1983).

Gabel, R. A., "Reconstruction Errors in Computer Generated Binary Holograms: A Comparative Study," *Appl. Opt.*, **14**(9), 2252–2255 (1975).

Ganesan, A. R., C. Joenathan, and R. S. Sirohi, "Sharpening the Fringes in Digital Speckle Pattern Interferometry," *Appl. Opt.*, **27**(11), 2099–2100 (1988).

Gasvik, K. J., *Optical Metrology*, Wiley, Chichester, 1987.

George, N., "Speckle," *Proc. SPIE*, **243**, 124–140 (1980).

Goodman, J. W., "Some Fundamental Properties of Speckle," *J. Opt. Soc. Am.*, **66**(11), 1145–1150 (1976) and in *Coherent Optical Engineering* F. T. Arecchi and V. Degiorgio, Eds., North-Holland, Amsterdam, 1977, pp. 29–48.

Gorodestskii, A. A., N. P. Larinov, A. V. Lukin, and K. S. Mustakin, "Holographic Testing of Convex Surfaces Employing Wavefront Reversal," *Sov. J. Opt. Technol.*, **50**(12), 787–789 (1983).

Gregory, D. A., "Basic Physical Properties of Defocused Speckle Photography: A Tilt Topology Inspection Technique," *Opt. Laser Tech.*, **8**(5), 210–213 (1976).

Grover, C., *Optical Testing and Metrology, Proc. SPIE*, **661** (1986).

Grover, C., *Optical Testing and Metrology II, Proc. SPIE*, **955** (1988).

Guerri, G., G. Molesini, and G. DaCosta, "Experimental Determination of the Phase in a Speckle Pattern by Temporal Modulation of a Reference Field," *Appl. Opt.*, **23**(4), 524-526 (1984).

Gusev, V. G. and S. F. Balandin, "Measuring the Radii of Curvature of Spherical Mirrors by a Speckle Interferometry Method," *Sov. J. Opt. Technol.*, **52**(6), 358-360 (1985).

Gusev, V. G. and S. V. Lazarev, "Speckle Interference Display of Lens Decentering," *Sov. J. Opt. Technol.*, **53**(6), 315-317 (1986).

Gusev, V. G. and S. V. Lazarev, "Holographic Method of Testing Objectives and Lenses," *Sov. J. Opt. Technol.*, **53**(9), 507-509 (1986).

Hariharan, P., "Double-Exposure Speckle Interferometry with Instant Polaroid Film," *Opt. Commun.*, **26**(3), 325-326 (1978).

Hariharan, P. and B. F. Oreb, "Two-Index Holographic Contouring: Application of Digital Techniques," *Opt. Commun.*, **51**(3), 142-144 (1984).

Herbert, D. P., "Inspection of Out-of-Plane Surface Movements over Small Areas Using Electronic Speckle Pattern Interferometry," *Opt. Lasers Eng.*, **4**(4), 229-239 (1983).

Hernandez, R., M. Garcia, and J. Montilla, "Determination of the Step Height by Two-Wavelength Speckle Patterns," *Proc. ICO-11 Conf., Madrid, Spain*, 577-580 (1978).

Hinsch, K. D., "Holographic Interferometry of Surface Deformations of Transparent Fluids," *Appl. Opt.*, **17**(19), 3101-3107 (1978).

Høgmoen, K. and O. J. Løkberg, "Detection and Measurement of Small Vibrations Using Electronic Speckle Pattern Interferometry," *Appl. Opt.*, **16**(7), 1869-1875 (1977).

Høgmoen, K. and H. M. Pedersen, "Measurement of Small Vibrations Using Electronic Speckle Pattern Interferometry: Theory," *J. Opt. Soc. Am.*, **67**(11), 1578-1583 (1977).

Huff, L., *Applications of Holography, Proc. SPIE*, **523** (1985).

Huff, L., *Holography: Critical Review of Technology, Proc. SPIE*, **512** (1985).

Hung, M. Y. Y. and R. J. Pryputniewicz, *Industrial Laser Interferometry II, Proc. SPIE.*, **955** (1988).

Hurden, A. P. M., "Vibration Mode Analysis Using Electronic Speckle Pattern Interferometry," *Opt. Laser Tech.*, **14**(1), 21-25 (1982).

Hurden, A. P. M., "An Instrument for Vibration Mode Analysis Using Electronic Speckle Pattern Interferometry," *NDT Intern'l.*, **15**(3), 143-148 (June 1982).

Ishii, Y. and K. Murata, "Designing the Collimator of Off-Axis Spherical Mirror with a Computer-Generated Holographic Optical Element," *Proc. SPIE*, **813,** 121-124 (1987).

Jaerisch, W. and G. Makosch, "Interferometric Surface Mapping with Variable Sensitivity," *Appl. Opt.*, **17**(5), 740-743 (1978).

Jaroszewicz, Z., "Aberrationless Phase Difference Amplification in Holographic Interferometry," *Appl. Opt.*, **28**(18), 3882-3888 (1989).

Joenathan, C., A. R. Ganesan, and R. S. Sirohi, "Fringe Compensation in Speckle Interferometry: Application to Nondestructive Testing," *Appl. Opt.*, **25**(20), 3781-3784 (1986).

Joenathan, C., R. K. Mohanty, and R. S. Sirohi, "Lateral Shear Interferometry with Holo Shear Lens," *Opt. Commun.*, **52**(3), 153–156 (1984).

Joenathan, C., V. Parthiban, and R. S. Sirohi, "Shear Interferometry with Holographic Lenses," *Opt. Eng.*, **26**(4), 359–364 (1987).

Jones, R. and J. N. Butters, "Some Observations on the Direct Comparison of the Geometry of Two Objects Using Speckle Pattern Interferometric Contouring," *J. Phys. E: Sci. Instrum.*, **8**(3), 231–234 (1975).

Jones, R. and J. A. Leendertz, "A White Light Projection Technique for Viewing Double-Exposure Speckle Interferograms," *J. Phys. E: Sci Instrum.*, **7**(8), 616 (1974).

Jones, R. and C. Wykes, "De-correlation Effects in Speckle-Pattern Interferometry. 2. Displacement Dependent De-correlation and Applications to the Observation of Machine-Induced Strain," *Opt. Acta*, **24**(5), 533–550 (1977).

Jones, R. and C. Wykes, "The Comparison of Complex Object Geometries Using a Combination of Electronic Speckle Pattern Interferometric Difference Contouring and Holographic Illumination Elements," *Opt. Acta*, **25**(6), 449–472 (1978).

Jones, R. and C. Wykes, "General Parameters for the Design and Optimization of Electronic Speckle Pattern Interferometers," *Opt. Acta*, **28**(7), 949–972 (1981).

Jüptner, W. P. O., *Holography Techniques and Applications, Proc. SPIE*, **1026** (1988).

Koyuncu, B. and J. Cookson, "Semi-automatic Measurements of Small High-Frequency Vibrations Using Time-Averaged Electronic Speckle Pattern Interferometry," *J. Phys. E: Sci. Instrum.*, **13**(2), 206–208 (1980).

Krishna, M. R.. R. K. Mohanty, R. S. Sirohi, and M. P. Kothiyal, "Radial Speckle Shearing Interferometer and Its Engineering Applications," *Optik*, **67**(1), 85–94 (1984).

Kulkarni, V. G., "Holographic Interferometry with Wavefront Shearing," *Opt. Laser Technol.*, **11**(5), 269–273 (1979).

Kulkarni, V. G. and P. N. Puntambekar, "Holographic Interferometry for Testing Large Phase Objects," *Opt. Commun.*, **27**(1), 33–36 (1978).

Kujawinska, M. and D. W. Robinson, "Multichannel Phase-Stepped Holographic Interferometry," *Appl. Opt.*, **27**(2), 312–320 (1988).

Larinov, N. P., "Interferometer Having a Synthesized Hologram for Testing Convex and Flat Surfaces," *Sov. J. Opt. Technol.*, **52**(10), 596–600 (1985).

Larinov, N. P., A. V. Lukin, and R. A. Rafikov, "Holographic Inspection of Aspheric Surfaces," *Sov. J. Opt. Technol.*, **46**(4), 229–231 (1979).

Larinov, N. P., A. V. Lukin, and R. A. Rafikov, "Synthesized Hologram Used as a Simulator of the Primary Mirror of a Telescope," *Sov. J. Opt. Technol.*, **47**(1), 37–39 (1980).

Larinov, N. P., A. V. Lukin, and R. A. Rafikov, "Testing of Aspherical Surfaces by Means of Axial Synthesized Holograms," *Sov. J. Opt. Technol.*, **47**(11), 667–670 (1980).

Lee, S. H., *International Conference on Computer-Generated Holography, Proc. SPIE*, **437** (1983).

Lee, S. H., *Computer Generated Holography II, Proc. SPIE*, **884** (1988).

Leendertz, J. A., "Interferometric Displacement Measurement on Scattering Surfaces Utilizing Speckle Effect," *J. Phys. E: Sci. Instrum.*, **3**(3), 214–218 (1970).

Leendertz, J. A. and J. N. Butters, "An Image-Shearing Speckle-Pattern Interferometer for Measuring Bending Moments," *J. Phys. E: Sci. Instrum.*, **6**(11), 1107–1110 (1973).

Lohmann, A. W. and D. P. Paris, "Binary Fraunhofer Holograms, Generated by Computer," *Appl. Opt.*, **6**(10), 1739–1748 (1967).

Løkberg, O. J., "ESPI Applied to Measurement on Biological Objects In-vivo," *Proc. ICO-11 Conf., Madrid, Spain*, 567–570 (1978).

Løkberg, O. J., "Use of Chopped Laser Light in Electronic Speckle Pattern Interferometry," *Appl. Opt.*, **18**(14), 2377–2384 (1979).

Løkberg, O. J., "Advances and Applications of Electronic Speckle Pattern Interferometry (ESPI)," *Proc. SPIE*, **215**, 92–95 (1980).

Løkberg, O. J. and K. Høgmoen, "Use of Modulated Reference Wave in Electronic Speckle Pattern Interferometry," *J. Phys. E: Sci. Instrum.*, **9**(10), 847–851 (1976).

Løkberg, O. J. and K. Høgmoen, "Vibration Phase Mapping Using Electronic Speckle Pattern Interferometry," *Appl. Opt.*, **15**(11), 2701–2704 (1976).

Løkberg, O. J. and K. Høgmoen, "Holographic Methods Made Useful by Phase Modulated ESPI," *Proc. SPIE*, **136**, 222–225 (1977).

Løkberg, O. J., K. Høgmoen, and T. Gundersen, "Use of ESPI to Measure the Vibration of the Human Eardrum In-Vivo and Other Biological Movements," in *Holography in Medicine and Biology*, G. von Bally, Ed., Springer-Verlag, Berlin, 1979, pp. 212–217.

Løkberg, U. J., K. Høgmoen, and O. M. Holje, "Vibration Measurement on the Human Ear Drum in Vivo," *Appl. Opt.*, **18**(6), 763–765 (1979).

Løkberg, O. J., O. M. Holje, and H. M. Pedersen, "Scan Converter Memory Used in TV-Speckle Interferometry," *Opt. Laser Tech.*, **8**(1), 17–20 (1976).

Løkberg, O. J. and K. Krakhella, "Electronic Speckle Pattern Interferometry Using Optical Fibers," *Opt. Commun.*, **38**(3), 155–158 (1981).

Løkberg, O. J. and O. K. Ledang, "Vibration of Flutes Studied by Electronic Speckle Pattern Interferometry," *Appl. Opt.*, **23**(18), 3052–3056 (1984).

Løkberg, O. J. and J. T. Malmo, "Long-distance Electronic Speckle Pattern Interferometry," *Opt. Eng.*, **27**(2), 150–156 (1988).

Løkberg, O. J. and G. Å. Slettemoen, "Interferometric Comparison of Displacements by Electronic Speckle Pattern Interferometry," *Appl. Opt.*, **20**(15), 2630–2634 (1981).

Løkberg, O. J. and P. Svenke, "Design and Use of an Electronic Speckle Pattern Interferometer for Testing of Turbine Parts," *Opt. Lasers Eng.*, **2**(1), 1–12 (1981).

Lukin, A. V. and K. S. Mustafin, "Holographic Methods of Testing Aspherical Surfaces," *Sov. J. Opt. Technol.*, **46**(4), 237–244 (1979).

MacGovern, A. J., "Projected Fringes and Holography," *Appl. Opt.*, **11**(12), 2972–2974 (1972).

Macovski, A., "Efficient Holography Using Temporal Modulation," *Appl. Phys. Lett.*, **14**(5), 166–168 (1969).

Macovski, A. and S. D. Ramsey Jr., "Temporal Offset Optical Data Processing," *Opt. Commun.*, **4**(5), 319–325 (1972).

Massie, N. A., *Interferometric Metrology, Proc. SPIE*, **816** (1987).

Matsuda, K., "Lateral Shear Interferometer Using Twin Three-Beam Holograms," *Appl. Opt.*, **19**(15), 2643–2646 (1980).

Matsuda, K. and Y. Minami, "A New Type of Holographic Shearing Interferometer for Measurement of the Lateral Aberrations of a Lens," *Proc. SPIE*, **813,** 331–332 (1987).

Matsuda, K. and H. Shimazutsu, "A Holographic Interferometer Using an Oil Surface as a Reference Plane," *Proc. SPIE*, **813,** 333–334 (1987).

Matsuda, K., S. Watanabe, and T. Eiju, "Real-Time Measurement of Large Liquid Surface Deformation Using a Holographic Shearing Interferometer," *Appl. Opt.*, **24**(24), 4443–4447 (1985).

Menu, M. and M. L. Roblin, "Determination of Lens Aberrations by Speckle Interferometry," *J. Optics (Paris)*, **10**(2), 71–78 (1979).

Mercier, R., "Holographic Testing of Aspherical Surfaces," *Proc. SPIE*, **136,** 208–214 (1977).

Moran, S. E., R. L. Law, P. N. Craig, and W. M. Goldberg, "Optically Phase-Locked Electronic Speckle Pattern Interferometer," *Appl. Opt.*, **26**(3), 475–491 (1987).

Morris, G. M., *Holographic Optics II: Principles and Applications, Proc. SPIE*, **1136** (1989).

Nagy, R. B. and O. J. Løkberg, "Low Amplitude Measurement by Direct Lock-in ESPI, A Proposal," *Opt. Commun.*, **47**(1), 18–22 (1983).

Nakadate, S., "Vibration Measurement Using Phase-Shifting Speckle-Pattern Interferometry," *Appl. Opt.*, **25**(22), 4162–4167 (1986).

Nakadate, S., T. Yatagai, and H. Saito, "Electronic Speckle Pattern Interferometry Using Digital Image Processing Techniques," *Appl. Opt.*, **19**(11), 1879–1883 (1980).

Nakadate, S., T. Yatagai, and H. Saito, "Digital Speckle-Pattern Shearing Interferometry," *Appl. Opt.*, **19**(24), 4241–4246 (1980).

Nakadate, S., T. Yatagai, and H. Saito, "Computer-Aided Speckle Pattern Interferometry," *Appl. Opt.*, **22**(2), 237–243 (1983).

Neiswander, P. and G. Å. Slettemoen, "Electronic Speckle Pattern Interferometric Measurements of the Basilar Membrane in the Inner Ear," *Appl. Opt.*, **20**(24), 4271–4276 (1981).

Oreb, B. F., B. Sharon, and P. Hariharan, "Electronic Speckle Pattern Interferometry with a Microcomputer," *Appl. Opt.*, **23**(22), 3940–3941 (1984).

Ottonello, P., C. Pontiggia, and L. Rossi, "An ESPI Apparatus with Multichannel Data Acquisition," *Opt. Commun.*, **30**(1), 20–22 (1979).

Parthiban, V., C. Joenathan, and R. S. Sirohi, "Shear Interferometer Using Holographic Lenses and a Hologram," *J. Optics (India)*, **16**(1), 1–9 (1987).

Pedersen, H. M., "Intensity Correlation Metrology: A Comparative Study," *Opt. Acta*, **29**(1), 105–118 (1982).

Pedersen, H. M., O. J. Løkberg, and B. M. Forre, "Holographic Vibration Measurement Using a TV Speckle Interferometer with Silicon Target Vidicon," *Opt. Commun.*, **12**(4), 421-426 (1974).

Preater, R. W. T., "In-plane Strain Measurement of Large Rotating Structures," *Conf. Proc. Electro-Opt./Laser Intern'l '80 UK*, 133-138 (1980).

Preater, R. W. T., "Further Developments in In-plane Strain Measurement of Rotating Structures," *Proc. SPIE*, **369**, 591-594 (1982).

Preater, R. W. T., "Analysis of Rotating Component Strains Using Electronic Speckle Pattern Interferometry," *Proc. SPIE*, **473**, 40-43 (1984).

Prowe, B., "Ein Berechnungsverfahren für interferofgrammähnliche binäre Hologramme zur Prüfung optischer Bauteile," *Optik*, **63**(3), 203-212 (1983).

Pryputniewicz, R. J., *Industrial Laser Interferometry, Proc. SPIE*, **746** (1987).

Pryputniewicz, R. J., *Industrial Laser Interferometry III: Quantitative Analysis of Interferograms, Proc. SPIE*, **1162** (1989).

Quintanilla, M., S. Mar, and I. Arias, "A Contribution to the Holographic Measurement of the MTF and the Wavefront Aberration," *Atti. Fond. G. Ronchi*, **33**(2), 206-218 (1978).

Rastogi, P. K., "Comparative Holographic Interferometry: A Nondestructive Inspection System for Detection of Flaws," *Experimental Mechanics*, **25**(4), 325-337 (1985).

Reuss, D. L., "Interferometric Probing of a Cylindrical Tube Using Holography," *Appl. Opt.*, **24**(14), 2197-2205 (1985).

Rigden, J. D. and E. I. Gordon, "The Granularity of Scattered Optical Maser Light," *Proc. IRE (IEEE)*, **50**(11), 2367-2368 (1962).

Ruskanda, F. and A. Handojo, "Holographic Interferometer to Test Planoid Aspherics," *Appl. Opt.*, **27**(18), 3773-3775 (1988).

Schluter, M. and A. Nowatzyk, "In-plane Deformation Measurement by Video-Electronic Hologram Interferometry," *Opt. Acta*, **27**(6), 799-808 (1980).

Schwider, J., J. Grzanna, and R. Spolaczyk, "Testing Aspherics in Reflected Light Using Blazed Synthetic Holograms," *Opt. Acta*, **27**(5), 683-698 (1980).

Sirohi, R. S., "Computer Generated Holography," *J. Optics (India)*, **7**(4), 67-85 (1978).

Sirohi, R. S., "Speckle Shear Interferometry," *Opt. Laser Technol.*, **16**(5), 251-254 (1984).

Sirohi, R. S., H. Blume, and K.-J. Rosenbruch, "Optical Testing Using Synthetic Hoograms," *Opt. Acta*, **23**(3), 229-236 (1976).

Slettemoen, G. Å., "Optimal Signal Processing in Electronic Speckle Pattern Interferometry," *Opt. Commun.*, **23**(2), 213-216 (1977).

Slettemoen, G. Å., "General Analysis of Fringe Contrast in Electronic Speckle Pattern Interferometry," *Opt. Acta*, **26**(3), 313-327 (1979).

Slettemoen, G. Å., "Electronic Speckle Pattern Interferometric System Based on a Speckle Reference Beam," *Appl. Opt.*, **19**(4), 616-623 (1980).

Slettemoen, G. Å., "First-Order Statistics of Displayed Speckle Patterns in Electronic Speckle Pattern Interferometry," *J. Opt. Soc. Am.*, **71**(4), 474-482 (1981).

Slettemoen, G. Å. and O. J. Løkberg, "Speckle Reference ESPI: In Practice," *Appl. Opt.*, **20**(20), 3467-3469 (1981).

Slettemoen, G. A. and J. C. Wyant, "Maximal Fraction of Accepted Measurements in Phase-Shifting Speckle Interferometry: A Theoretical Study," *J. Opt. Soc. Am. A*, **3**(2), 210–214 (1986).

Smith, H. M., *Principles of Holography*, Wiley, New York, 1975.

Stetson, K. A., "A Rigorous Treatment of the Fringes of Hologram Interferometry," *Optik*, **29**(4), 386–400 (1969).

Stetson, K. A., "Problem of Defocusing in Speckle Photography, Its Connection to Hologram Interferometry, and Its Solution," *J. Opt. Soc. Am.*, **66**(11), 1267–1271 (1976).

Takahashi, T., K. Konno, M. Kawai, and M. Isshiki, "Computer Generated Holograms for Testing Aspheric Lenses," *Appl. Opt.*, **15**(2), 546–549 (1976).

Thalmann, R. and R. Dändliker, "Statistical Properties of Interference Phase Detection in Speckle Fields Applied to Holographic Interferometry," *J. Opt. Soc. Am. A*, **3**(7), 972–981 (1986).

Tichenor, D. A. and V. P. Madsen, "Computer Analysis of Holographic Interferograms for Nondestructive Testing," *Opt. Eng.*, **18**(5), 469–472 (1979).

Tiziani, H. J., "Application of Speckling for In-Plane Vibration Analysis," *Opt. Acta*, **18**(12), 891–902 (1971).

Tiziani, H. J., "A Study of the Use of Laser Speckle to Measure Small Tilts of Optically Rough Surfaces Accurately," *Opt. Commun.*, **5**(4), 271–276 (1972).

Tiziani, H. J. and J. Klenk, "Vibration Analysis by Speckle Techniques in Real Time," *Appl. Opt.*, **20**(8), 1467–1470 (1981).

Torge, R., "The Practical Application of Holographic Interferometry," *Proc. SPIE*, **381**, 217–221 (1983).

Trolinger, J. D., "Holography of Phase Objects," *Proc. SPIE*, **816**, 128–139 (1987).

Vest, C. M., *Holographic Nondestructive Testing: Critical Review of Technology, Proc. SPIE*, **604** (1986).

Vry, U. and A. F. Fercher, "Higher-Order Statistical Properties of Speckle Fields and Their Application to Rough-Surface Interferometry," *J. Opt. Soc. Am. A*, **3**(7), 988–1000 (1986).

Vukicevic, D., *Holographic Data Nondestructive Testing, Proc. SPIE*, **370** (1982).

Wang, D., J. Ke, and R. J. Pryputniewicz, *International Conference on Holography Applications, Proc. SPIE*, **673** (1986).

Willemin, J.-F. and R. Dändliker, "Measuring Amplitude and Phase of Microvibrations by Heterodyne Speckle Interferometry," *Opt. Lett.*, **8**(2), 102–104 (1983).

Winther, S. and G. Å. Slettemoen, "An ESPI Contouring Technique in Strain Analysis," *Proc. SPIE*, **473**, 44–47 (1984).

Wu, J. Z., X.-F. Dai, and Z.-X. Cai, "Analysis and Discussion of an Important Concept in Testing Aspheric Surfaces with CGH," *Proc. SPIE*, **813**, 357–358 (1987).

Wyant, J. C. and K. Creath, "Recent Advances in Interferometric Optical Testing," *Laser Focus/Electro-Optics*, 118–132 (Nov. 1985).

Wykes, C., "De-correlation Effects in Speckle-Pattern Interferometry. 1. Wavelength Change Dependent De-Correlation with Application to Contouring and Surface Roughness Measurement," *Opt. Acta*, **24**(5), 517–532 (1977).

Wykes, C., "Use of Electronic Speckle Pattern Interferometry (ESPI) in the Measurement of Static and Dynamic Surface Displacements," *Opt. Eng.*, **21**(3), 400–406 (1982).

Wykes, C., J. N. Butters, and R. Jones, "Fringe Contrast in Electronic Speckle Pattern Interferometry," *Appl. Opt.*, **20**(5), A50&721 (1981).

Wykes, C. and R. Jones, "Advances in Optical Metrology of Complex Objects," *Proc. SPIE*, **369,** 200–206 (1982).

Yamaguchi, I., "Speckle Displacement for General Object Deformation," *Proc. ICO-11 Conf., Madrid, Spain*, 573–576 (1978).

Yamaguchi, I., "Real-Time Measurement of In-Plane Translation and Tilt by Electronic Speckle Correlation," *Jap. J. Appl. Phys.*, **19**(3), L133–L136 (1980).

Yamaguchi, I., "Speckle Displacement and Decorrelation in the Diffraction and Image Fields for Small Object Deformation," *Opt. Acta*, **28**(10), 1359–1376 (1981).

Yamaguchi, I., "Fringe Formation in Speckle Photography," *J. Opt. Soc. Am. A*, **1**(1) 81–86 (1984).

Yamaguchi, I. and S.-I. Komatsu, "Theory and Applications of Dynamic Laser Speckles Due to In-plane Motion," *Opt. Acta*, **24**(7), 705–724 (1977).

Zou, Z., J. Liao, Y. Gu, and Q. Gu, "An Interferometric Method for Measuring Optical Spherical Surfaces Using Holographic Phase Conjugate Compensation," *Proc. SPIE*, **673,** 268–271 (1986).

16

Moiré and Fringe Projection Techniques

K. Creath and J. C. Wyant

16.1. INTRODUCTION

The term "moiré" is not the name of a person; in fact, it is a French word referring to "an irregular wavy finish usually produced on a fabric by pressing between engraved rollers" (*Webster's* 1981). In optics it refers to a beat pattern produced between two gratings of approximately equal spacing. It can be seen in everyday things such as the overlapping of two window screens, the rescreening of a half-tone picture, or with a striped shirt seen on television. The use of moiré for reduced sensitivity testing was introduced by Lord Rayleigh in 1874. Lord Rayleigh looked at the moiré between two identical gratings to determine their quality even though each individual grating could not be resolved under a microscope.

Fringe projection entails projecting a fringe pattern or grating on an object and viewing it from a different direction. The first use of fringe projection for determining surface topography was presented by Rowe and Welford in 1967. It is a convenient technique for contouring objects that are too coarse to be measured with standard interferometry. Fringe projection is related to optical triangulation using a single point of light and light sectioning where a single line is projected onto an object and viewed in a different direction to determine the surface contour (Case et al. 1987).

Moiré and fringe projection interferometry complement conventional holographic interferometry, especially for testing optics to be used at long wavelengths. Although two-wavelength holography (TWH) can be used to contour surfaces at any longer-than-visible wavelength, visible interferometric environmental conditions are required. Moiré and fringe projection interferometry can contour surfaces at any wavelength longer than 10–100 μm with reduced environmental requirements and no intermediate photographic recording setup.

Optical Shop Testing, Second Edition, Edited by Daniel Malacara.
ISBN 0-471-52232-5 © 1992, John Wiley & Sons, Inc.

Moiré is also a useful technique for aiding in the understanding of interferometry.

This chapter explains what moiré is and how it relates to interferometry. Contouring techniques utilizing fringe projection, projection and shadow moiré, and two-angle holography are all described and compared. Each of these techniques provides the same result and can be described by a single theory. The relationship between these techniques and holographic and conventional interferometry will be shown. Errors caused by divergent geometries are described, and applications of these techniques combined with phase measurement techniques are presented. Further information on these techniques can be found in the following books and book chapters: Varner (1974), Vest (1979), Hariharan (1984), Gasvik (1987), and Chiang (1978, 1983).

16.2. WHAT IS MOIRÉ?

Moiré patterns are extremely useful to help understand basic interferometry and interferometric test results. Figure 16.1 shows the moiré pattern (or beat pattern) produced by two identical straight-line gratings rotated by a small angle relative to each other. A dark fringe is produced where the dark lines are out of step one-half period, and a bright fringe is produced where the dark lines for one grating fall on top of the corresponding dark lines for the second grating. If the angle between the two gratings is increased, the separation between the bright and dark fringes decreases. [A simple explanation of moiré is given by Oster and Nishijima (1963).]

If the gratings are not identical straight-line gratings, the moiré pattern (bright and dark fringes) will not be straight equi-spaced fringes. The following anal-

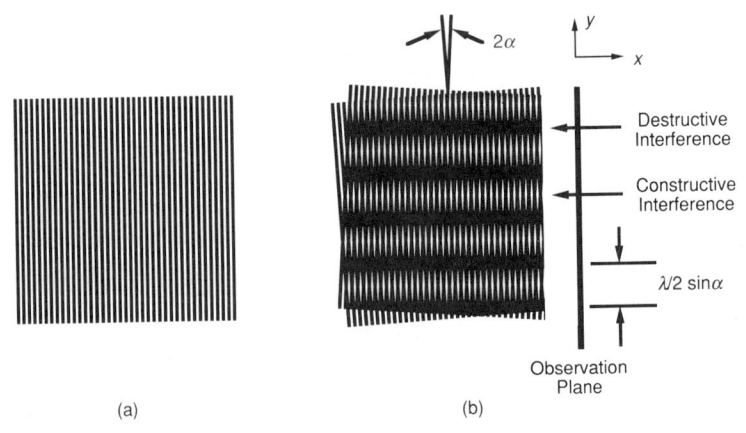

Figure 16.1. (*a*) Straight-line grating. (*b*) Moiré between two straight-line gratings of the same pitch at an angle α with respect to one another.

16.2. WHAT IS MOIRÉ?

ysis shows how to calculate the moiré pattern for arbitrary gratings. Let the intensity transmission function for two gratings $f_1(x, y)$ and $f_2(x, y)$ be given by

$$f_1(x, y) = a_1 + \sum_{n=1}^{\infty} b_{1n} \cos [n\phi_1(x, y)],$$

$$f_2(x, y) = a_2 + \sum_{m=1}^{\infty} b_{2m} \cos [m\phi_2(x, y)], \quad (16.1)$$

where $\phi(x, y)$ is the function describing the basic shape of the grating lines. For the fundamental frequency, $\phi(x, y)$ is equal to an integer times 2π at the center of each bright line and is equal to an integer plus one-half times 2π at the center of each dark line. The b coefficients determine the profile of the grating lines (i.e., square wave, triangular, sinusoidal, etc.) For a sinusoidal line profile, b_{i1} is the only nonzero term.

When these two gratings are superimposed, the resulting intensity transmission function is given by the product

$$f_1(x, y) f_2(x, y) = a_1 a_2 + a_1 \sum_{m=1}^{\infty} b_{2m} \cos [m\phi_2(x, y)]$$

$$+ a_2 \sum_{n=1}^{\infty} b_{1n} \cos [n\phi_1(x, y)]$$

$$+ \sum_{m=1}^{\infty} \sum_{n=1}^{\infty} b_{1n} b_{2m} \cos [n\phi_1(x, y)] \cos [m\phi_2(x, y)].$$

$$(16.2)$$

The first three terms of Eq. (16.2) provide information that can be determined by looking at the two patterns separately. The last term is the interesting one, and can be rewritten as

Term 4 = $\frac{1}{2} b_{11} b_{21} \cos [\phi_1(x, y) - \phi_2(x, y)]$

$$+ \frac{1}{2} \sum_{m=1}^{\infty} \sum_{n=1}^{\infty} b_{1n} b_{2m} \cos [n\phi_1(x, y) - m\phi_2(x, y)];$$

n and m both $\neq 1$

$$+ \frac{1}{2} \sum_{m=1}^{\infty} \sum_{n=1}^{\infty} b_{1n} b_{2m} \cos [n\phi_1(x, y) + m\phi_2(x, y)]x. \quad (16.3)$$

This expression shows that by superimposing the two gratings, the sum and difference between the two gratings is obtained. The first term of Eq. (16.3) represents the difference between the fundamental pattern masking up the two gratings. It can be used to predict the moiré pattern shown in Fig. 16.1. Assuming that two gratings are oriented with an angle 2α between them with the y axis of the coordinate system bisecting this angle, the two grating functions $\phi_1(x, y)$ and $\phi_2(x, y)$ can be written as

$$\phi_1(x, y) = \frac{2\pi}{\lambda_1} (x \cos \alpha + y \sin \alpha)$$

and

$$\phi_2(x, y) = \frac{2\pi}{\lambda_2} (x \cos \alpha - y \sin \alpha), \tag{16.4}$$

where λ_1 and λ_2 are the line spacings of the two gratings. Equation (16.4) can be rewritten as

$$\phi_1(x, y) - \phi_2(x, y) = \frac{2\pi}{\lambda_{\text{beat}}} x \cos \alpha + \frac{4\pi}{\bar{\lambda}} y \sin \alpha, \tag{16.5}$$

where $\bar{\lambda}$ is the average line spacing, and λ_{beat} is the beat wavelength between the two gratings given by

$$\lambda_{\text{beat}} = \frac{\lambda_1 \lambda_2}{\lambda_2 - \lambda_1}. \tag{16.6}$$

Note that this beat wavelength equation is the same as that obtained for two-wavelength interferometry as shown in Chapter 15. Using Eq. (16.3), the moiré or beat will be lines whose centers satisfy the equation

$$\phi_1(x, y) - \phi_2(x, y) = M 2\pi. \tag{16.7}$$

Three separate cases for moiré fringes can be considered. When $\lambda_1 = \lambda_2 = \lambda$, the first term of Eq. (16.5) is zero, and the fringe centers are given by

$$M\lambda = 2y \sin \alpha, \tag{16.8}$$

where M is an integer corresponding to the fringe order. As was expected, Eq. (16.8) is the equation of equi-spaced horizontal lines as seen in Fig. 16.1. The other simple case occurs when the gratings are parallel to each other with $\alpha = 0$. This makes the second term of Eq. (16.5) vanish. The moiré will then be lines that satisfy

$$M\lambda_{\text{beat}} = x. \tag{16.9}$$

16.2. WHAT IS MOIRÉ?

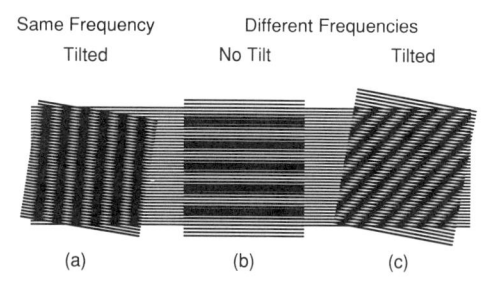

Figure 16.2. Moiré patterns caused by two straight-line gratings with (a) the same pitch tilted with respect to one another, (b) different frequencies and no tilt, and (c) different frequencies tilted with respect to one another.

These fringes are equally spaced, vertical lines parallel to the y axis. For the more general case where the two gratings have different line spacings and the angle between the gratings is nonzero, the equation for the moiré fringes will now be

$$M\bar{\lambda} = \frac{\bar{\lambda}}{\lambda_{beat}} x \cos \alpha + 2y \sin \alpha. \tag{16.10}$$

This is the equation of straight lines whose spacing and orientation is dependent on the relative difference between the two grating spacings and the angle between the gratings. Figure 16.2 shows moiré patterns for these three cases.

The orientation and spacing of the moiré fringes for the general case can be determined from the geometry shown in Fig. 16.3 (Chiang, 1983). The distance $\overline{AB}$ can be written in terms of the two grating spacings;

$$\overline{AB} = \frac{\lambda_1}{\sin(\theta - \alpha)} = \frac{\lambda_2}{\sin(\theta + \alpha)}, \tag{16.11}$$

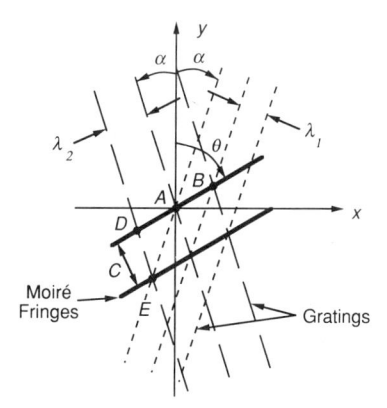

Figure 16.3. Geometry used to determine spacing and angle of moiré fringes between two gratings of different frequencies tilted with respect to one another.

where θ is the angle the moiré fringes make with the y axis. After rearranging, the fringe orientation angle θ is given by

$$\tan \theta = \tan \alpha \left(\frac{\lambda_1 + \lambda_2}{\lambda_2 - \lambda_1} \right). \quad (16.12)$$

When $\alpha = 0$ and $\lambda_1 \neq \lambda_2$, $\theta = 0°$, and when $\lambda_1 = \lambda_2$ with $\alpha \neq 0$, $\theta = 90°$ as expected. The fringe spacing perpendicular to the fringe lines can be found by equating quantities for the distance $\overline{DE}$;

$$\overline{DE} = \frac{\lambda_1}{\sin 2\alpha} = \frac{C}{\sin (\theta + \alpha)}, \quad (16.13)$$

where C is the fringe spacing or contour interval. This can be rearranged to yield

$$C = \lambda_1 \left[\frac{\sin (\theta + \alpha)}{\sin 2\alpha} \right]. \quad (16.14)$$

By substituting for the fringe orientation θ, the fringe spacing can be found in terms of the grating spacings and angle between the gratings;

$$C = \frac{\lambda_1 \lambda_2}{\sqrt{\lambda_2^2 \sin^2 2\alpha + (\lambda_2 \cos 2\alpha - \lambda_1)^2}}. \quad (16.15)$$

In the limit that $\alpha = 0$ and $\lambda_1 \neq \lambda_2$, the fringe spacing equals λ_{beat}, and in the limit that $\lambda_1 = \lambda_2 = \lambda$ and $\alpha \neq 0$, the fringe spacing equals $\lambda/(2 \sin \alpha)$. It is possible to determine λ_2 and α from the measured fringe spacing and orientation as long as λ_1 is known (Chiang 1983).

16.3. MOIRÉ AND INTERFEROGRAMS

Now that we have covered the basic mathematics of moiré patterns, let us see how moiré patterns are related to interferometry. The single grating shown in Fig. 16.1 can be thought of as a "snapshot" of a plane wave traveling to the right, where the distance between the grating lines is equal to the wavelength of light. The straight lines represent the intersection of a plane of constant phase with the plane of the figure. Superimposing the two sets of grating lines in Fig. 16.1 can be thought of as superimposing two plane waves with an angle of 2α between their directions of propagation. Where the two waves are in phase, bright fringes result (constructive interference), and where they are out of phase,

16.3. MOIRÉ AND INTERFEROGRAMS

dark fringes result (destructive interference). For a plane wave, the "grating" lines are really planes perpendicular to the plane of the figure and the dark and bright fringes are also planes perpendicular to the plane of the figure. If the plane waves are traveling to the right, these fringes would be observed by placing a screen perpendicular to the plane of the figure and to the right of the grating lines as shown in Fig. 16.1. The spacing of the interference fringes on the screen is given by Fig. (16.8), where λ is now the wavelength of light. Thus, the moiré of two straight-line gratings correctly predicts the centers of the interference fringes produced by interfering two plane waves. Since the gratings used to produce the moiré pattern are binary gratings, the moiré does not correctly predict the sinusoidal intensity profile of the interference fringes. (If both gratings had sinusoidal intensity profiles, the resulting moiré would still not have a sinusoidal intensity profile because of higher-order terms.)

More complicated gratings, such as circular gratings, can also be investigated. Figure 16.4b shows the superposition of two circular line gratings. This pattern indicates the fringe positions obtained by interfering two spherical wavefronts. The centers of the two circular line gratings can be considered the source locations for two spherical waves. Just as for two plane waves, the spacing between the grating lines is equal to the wavelength of light. When the two patterns are in phase, bright fringes are produced; and when the patterns are completely out of phase, dark fringes result. For a point on a given fringe, the difference in the distances from the two source points and the fringe point is a constant. Hence, the fringes are hyperboloids. Due to symmetry, the fringes seen on observation plane A of Fig. 16.4b must be circular. (Plane A is along the top of Fig. 16.4b and perpendicular to the line connecting the two sources as well as perpendicular to the page.) Figure 16.4c shows a binary representation of these interference fringes and represents the interference pattern obtained by interfering a nontilted plane wave and a spherical wave. (A plane wave can be thought of as a spherical wave with an infinite radius of curvature.) Figure 16.4d shows that the interference fringes in plane B are essentially straight equispaced fringes. (These fringes are still hyperbolas, but in the limit of large distances, they are essentially straight lines. Plane B is along the side of Fig. 16.4b and parallel to the line connecting the two sources as well as perpendicular to the page.)

The lines of constant phase in plane B for a single spherical wave are shown in Fig. 16.5a. (To first-order, the lines of constant phase in plane B are the same shape as the interference fringes in plane A.) The pattern shown in Fig. 16.5a is commonly called a zone plate. Figure 16.5b shows the superposition of two linearly displaced zone plates. The resulting moiré pattern of straight equi-spaced fittings illustrates the interference fringes in plane B shown in Fig. 16.4b.

Superimposing two interferograms and looking at the moiré or beat produced can be extremely useful. The moiré formed by superimposing two different

660   MOIRÉ AND FRINGE PROJECTION TECHNIQUES

Plane A

Plane B

Figure 16.4. Interference of two spherical waves. (a) Circular line grating representing a spherical wavefront. (b) Moiré pattern obtained by superimposing two circular line patterns. (c) Fringes observed in plane A. (d) Fringes observed in plane B.

16.3. MOIRÉ AND INTERFEROGRAMS 661

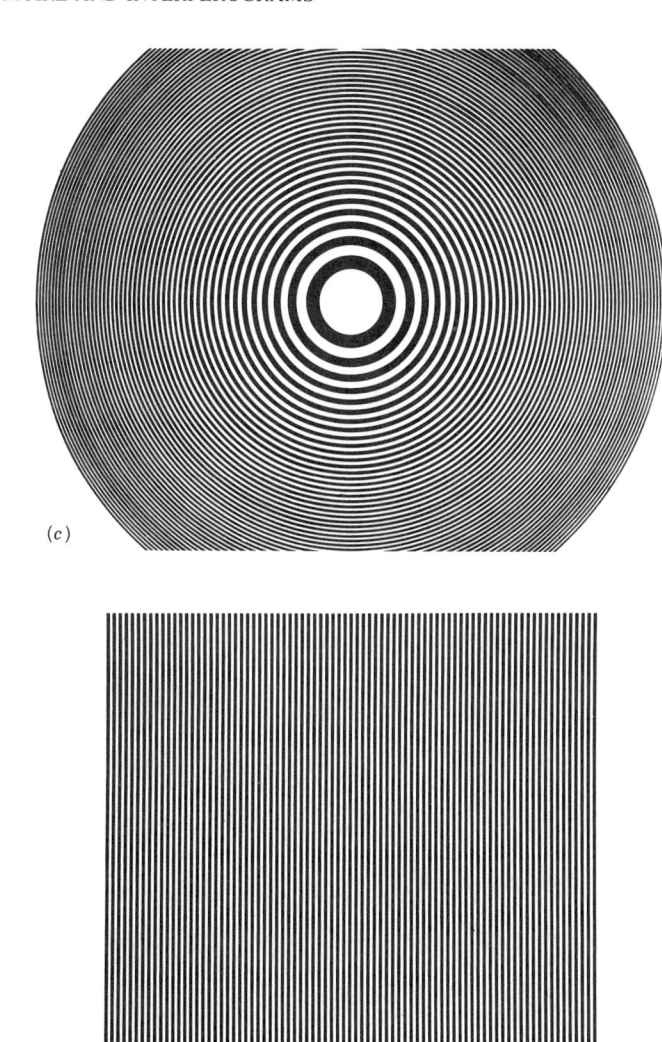

(c)

(d)

Figure 16.4. (*Continued*)

interferograms shows the difference in the aberrations of the two interferograms. For example, Fig. 16.6 shows the moiré produced by superimposing two computer-generated interferograms. One interferogram has 50 waves of tilt across the radius (Fig. 16.6a), while the second interferogram has 50 waves of tilt plus 4 waves of defocus (Fig. 16.6b). If the interferograms are aligned such that the tilt direction is the same for both interferograms, the tilt will cancel and

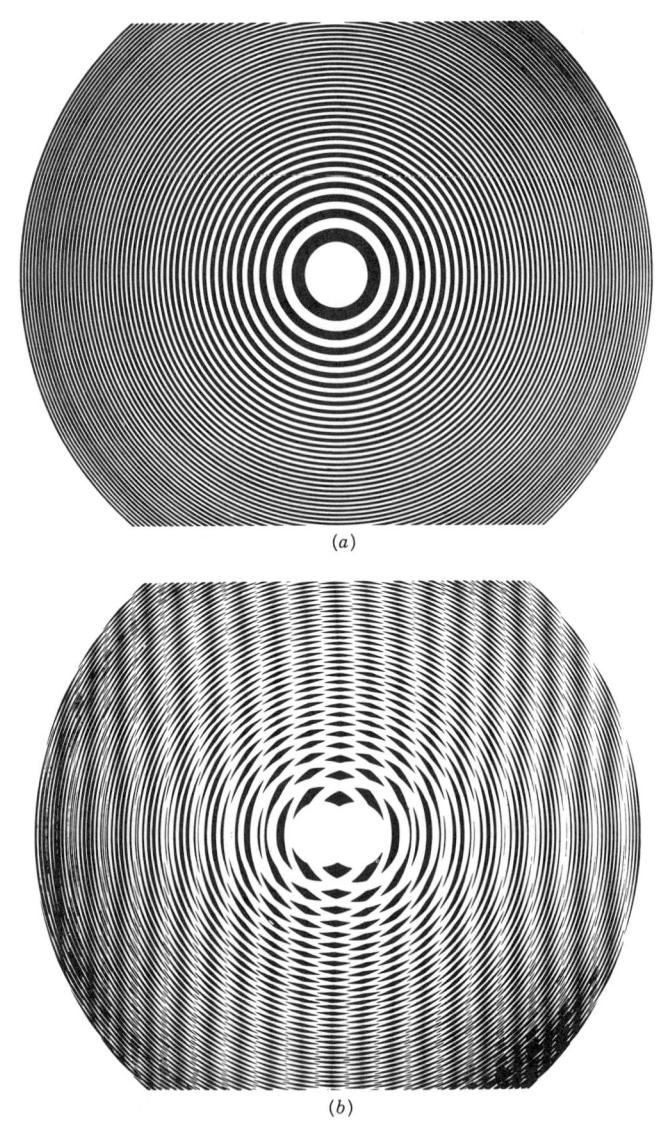

Figure 16.5. Moiré pattern produced by two zone plates. (a) Zone plate. (b) Straight-line fringes resulting from superposition of two zone plates.

only the 4 waves of defocus remain (Fig. 16.6c). In Fig. 16.6d, the two interferograms are rotated slightly with respect to each other so that the tilt will not quite cancel. These results can be described mathematically by looking at the two grating functions:

$$\phi_1(x, y) = 2\pi(50\rho \cos \phi + 4\rho^2)$$

16.3. MOIRÉ AND INTERFEROGRAMS

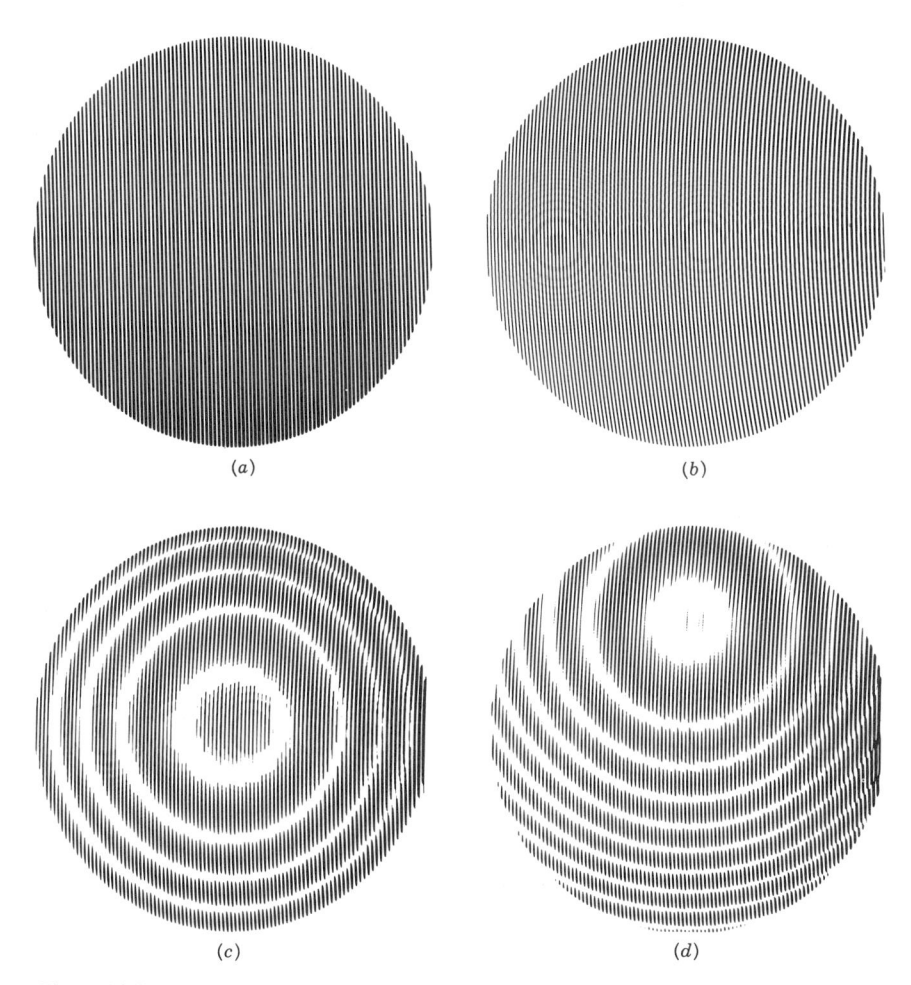

Figure 16.6. Moiré between two interferograms. (a) Interferogram having 50 waves tilt. (b) Interferogram having 50 waves tilt plus 4 waves of defocus. (c) Superposition of (a) and (b) with no tilt between patterns. (d) Slight tilt between patterns.

and

$$\phi_2(x, y) = 2\pi[50\rho \cos(\phi + \alpha)]. \tag{16.16}$$

A bright fringe is obtained when

$$50\rho[\cos\phi - \cos(\phi + \alpha)] + 4\rho^2 = M. \tag{16.17}$$

If $\alpha = 0$, the tilt cancels completely and four waves of defocus remain; otherwise, some tilt remains in the moiré pattern.

Figure 16.7 shows similar results for interferograms containing third-order aberrations. Spherical aberration with defocus and tilt is shown in Fig. 16.7d. One interferogram has 50 waves of tilt (Fig. 16.6a), and the other has 55 waves tilt, 6 waves third-order spherical aberration, and −3 waves defocus (Fig. 16.7a). Figure 16.7e shows the moiré between an interferogram having 50 waves of tilt (Fig. 16.6a) with an interferogram having 50 waves of tilt and 5 waves of coma (Fig. 16.7b) with a slight rotation between the two patterns. The moiré between an interferogram having 50 waves of tilt (Fig. 16.6a) and one having 50 waves of tilt, 7 waves third-order astigmatism, and −3.5 waves defocus (Fig. 16.7c) is shown in Fig. 16.7f. Thus, it is possible to produce simple fringe patterns using moiré. These patterns can be photocopied onto transparencies and used as a learning aid to understand interferograms obtained from third-order aberrations.

A computer-generated interferogram having 55 waves of tilt across the radius, 6 waves of spherical and −3 waves of defocus is shown in Fig. 16.7a. Figure 16.8a shows two identical interferograms superimposed with a small rotation between them. As expected, the moiré pattern consists of nearly straight equi-spaced lines. When one of the two interferograms is slipped over, the resultant moiré is shown in Fig. 16.8b. The fringe deviation from straightness in one interferogram is to the right and, in the other, to the left. Thus the sign of the defocus and spherical aberration for the two interferograms is opposite, and the moiré pattern has twice the defocus and spherical of each of the individual interferograms. When two identical interferograms given by Fig. 16.7a are superimposed with a displacement from one another, a shearing interferogram is obtained. Figure 16.9 shows vertical and horizontal displacements with and without a rotation between the two interferograms. The rotations indicate the addition of tilt to the interferograms. These types of moiré patterns are very useful for understanding lateral shearing interferograms.

Moiré patterns are produced by multiplying two intensity-distribution functions. Adding two intensity functions does not give the difference term obtained in Eq. (16.3). A moiré pattern is not obtained if two intensity functions are added. The only way to get a moiré pattern by adding two intensity functions is to use a nonlinear detector. For the detection of an intensity distribution given by $I_1 + I_2$, a nonlinear response can be written as

$$\text{Response} = a(I_1 + I_2) + b(I_1 + I_2)^2 + \cdots. \quad (16.18)$$

This produces terms proportional to the product of the two intensity distributions in the output signal. Hence, a moiré pattern is obtained if the two indivudal intensity patterns are simultaneously observed by a nonlinear detector (even if they are not multiplied before detection). If the detector produces an output linearly proportional to the incoming intensity distribution, the two intensity patterns must be multiplied to produce the moiré pattern. Since the eye

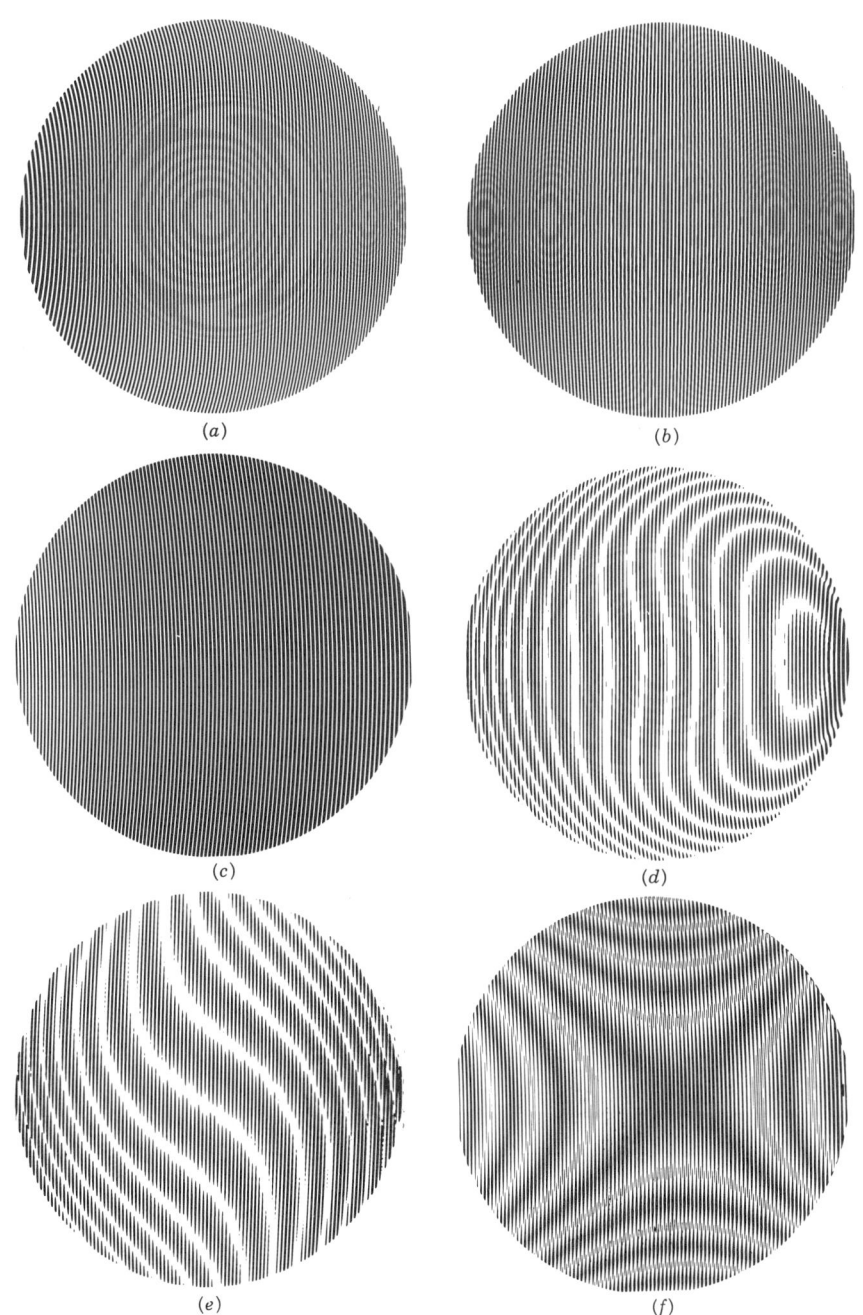

Figure 16.7. Moiré patterns showing third-order aberrations. Interferograms containing (a) 55 waves tilt, 6 waves of third-order spherical aberration, and −3 waves of defocus, (b) 50 waves tilt and 5 waves coma, and (c) 50 waves tilt, 7 waves astigmatism, and −3.5 waves of defocus. (d) Moiré pattern between Fig. 16.6a and 16.7a. (e) Moiré pattern between Fig. 16.6a and 16.7b. (f) Moiré pattern between Fig. 16.6a and 16.7c.

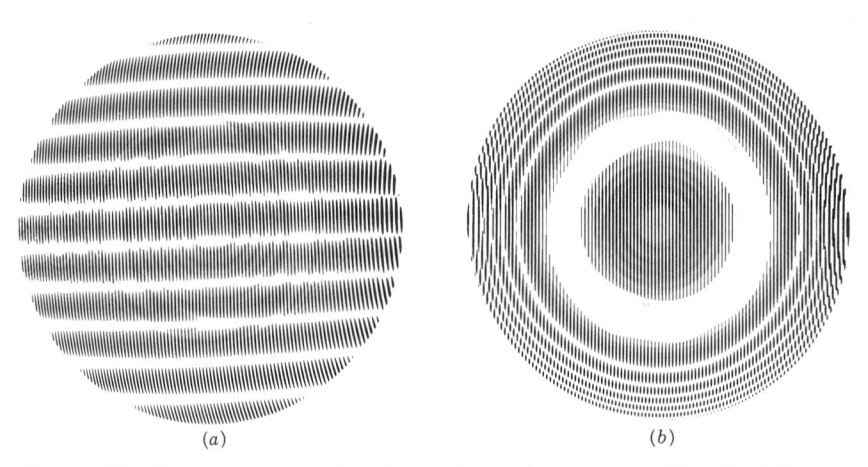

Figure 16.8. Moiré pattern by superimposing two identical interferograms (from Fig. 16.7a). (a) Both patterns having the same orientation. (b) With one pattern flipped.

is a nonlinear detector, moiré can be seen whether the patterns are added or multiplied. A good TV camera, on the other hand, will not see moiré unless the patterns are multiplied.

16.4. HISTORICAL REVIEW

Since Lord Rayleigh first noticed the phenomena of moiré fringes, moiré techniques have been used for a number of testing applications. Righi (1887) first noticed that the relative displacement of two gratings could be determined by observing the movement of the moiré fringes. The next significant advance in the use of moiré was presented by Weller and Shepherd (1948). They used moiré to measure the deformation of an object under applied stress by looking at the differences in a grating pattern before and after the applied stress. They were the first to use shadow moiré, where a grating is placed in front of a nonflat surface to determine the shape of the object behind it by using the shape of the moiré fringes. A rigorous theory of moiré fringes did not exist until the midfifties when Ligtenberg (1955) and Guild (1956, 1960) explained moiré for stress analysis by mapping slope contours and displacement measurement, respectively. Excellent historical reviews of the early work in moiré have been presented by Theocaris (1962, 1966). Books on this subject have been written by Guild (1956, 1960), Theocaris (1969), and Durelli and Parks (1970). Projection moiré techniques were introduced by Brooks and Helfinger (1969) for optical gauging and deformation measurement. Until 1970, advances in moiré techniques were primarily in stress analysis. Some of the first uses of moiré to measure surface topography were reported by Meadows *et al.* (1970), Takasaki

16.4. HISTORICAL REVIEW

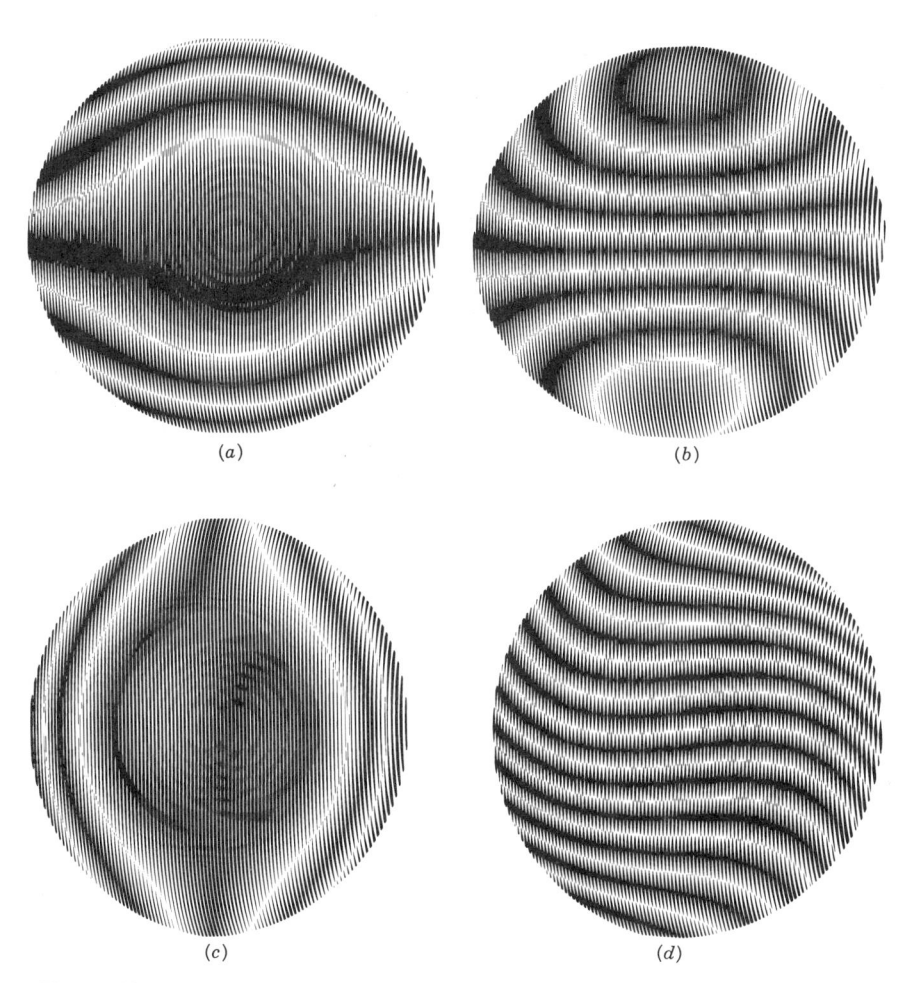

Figure 16.9. Moiré patterns formed using two identical interferograms (from Fig. 16.7a) where the two are sheared with respect to one another. (a) Vertical displacement. (b) Vertical displacement with rotation showing tilt. (c) Horizontal displacement. (d) Horizontal displacement with rotation showing tilt.

(1970), and Wasowski (1970). Moiré has also been used to compare an object to a master and for vibration analysis (Der Hovanesian and Yung 1971; Gasvik 1987). A theoretical review and experimental comparison of moiré and projection techniques for contouring is given by Benoit et al. (1975). Automatic computer fringe analysis of moiré patterns by finding fringe centers were reported by Yatagai et al. (1982). Heterodyne interferometry was first used with moiré fringes by Moore and Truax (1977), and phase measurement techniques were further developed by Perrin and Thomas (1979), Shagam (1980), and Reid

(1984b). Recent review papers on moiré techniques include Post (1982), Reid (1984a), and Halioua and Liu (1989).

The projection of interference fringes for contouring objects was first proposed by Rowe and Welford (1967). Their later work included a number of applications for projected fringes (Welford 1969) and the use of projected fringes with holography (Rowe 1971). In-depth mathematical treatments have been provided by Benoit et al. (1975) and Gasvik (1987). The relationship between projected fringe contouring and triangulation is given in a book chapter by Case et al. (1987). Heterodyne phase measurement was first introduced with projected fringes by Indebetouw (1978), and phase measurement techniques were further developed by Takeda et al. (1982), Takeda and Mutoh (1983), and Srinivasan et al. (1984, 1985).

Haines and Hildebrand first proposed contouring objects in holography using two sources (Haines and Hildebrand 1965; Hildebrand and Haines 1966, 1967). The two holographic sources were produced by changing either the angle of the illumination beam on the object or the angle of the reference beam. A small angle difference between the beams used to produce a double-exposure hologram creates a moiré in the final hologram which corresponds to topographic contours of the test object. Further insight into two-angle holography has been provided by Menzel (1974), Abramson (1976a, 1976b) and DeMattia and Fossati-Bellani (1978). The technique has also been used in speckle interferometry (Winther, 1983).

Since all of these techniques are so similar, it is sometimes hard to differentiate developments in one technique versus another. MacGovern (1972) provided a theory that linked all of these techniques together. The next part of this chapter will explain each of these techniques and then show the similarities among all of these techniques and provide a comparison to conventional interferometry.

16.5. FRINGE PROJECTION

A simple approach for contouring is to project interference fringes or a grating onto an object and then view from another direction. Figure 16.10 shows the

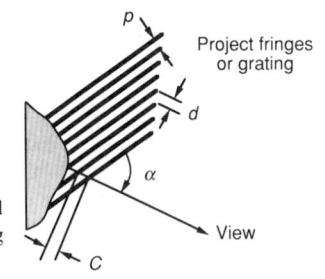

Figure 16.10. Projection of fringes or grating onto object and viewed at an angle α. p is the grating pitch or fringe spacing and C is the contour interval.

16.5. FRINGE PROJECTION

optical setup for this measurement. Assuming a collimated illumination beam and viewing the fringes with a telecentric optical system, straight equally spaced fringes are incident on the object, producing equally spaced contour intervals. The departure of a viewed fringe from a straight line shows the departure of the surface from a plane reference surface. An object with fringes projected onto it can be seen in Fig. 16.11. When the fringes are viewed at an angle α relative

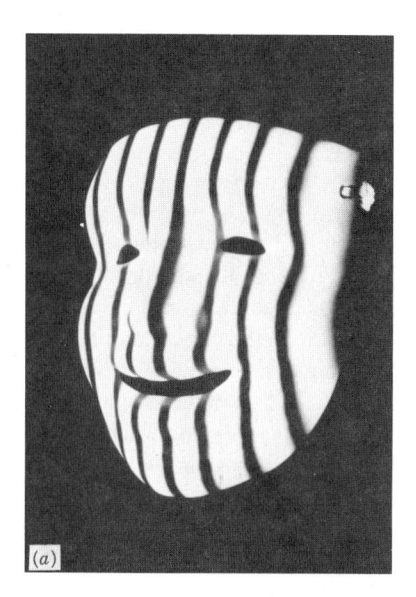

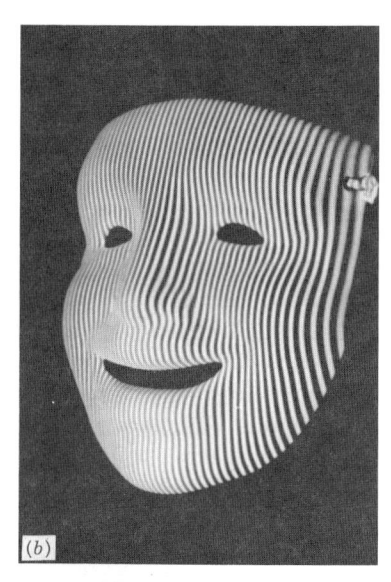

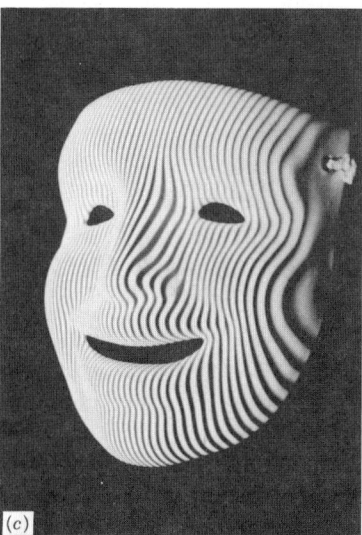

Figure 16.11. Mask with fringes projected onto it. (*a*) Coarse fringe spacing. (*b*) Fine fringe spacing. (*c*) Fine fringe spacing with an increase in the angle between illumination and viewing.

to the projection direction, the spacing of the lines perpendicular to the viewing direction will be

$$d = \frac{p}{\cos \alpha}. \tag{16.19}$$

The contour interval C (the height between adjacent contour lines in the viewing direction) is determined by the line or fringe spacing projected onto the surface and the angle between the projection and viewing directions;

$$C = \frac{p}{\sin \alpha} = \frac{d}{\tan \alpha}. \tag{16.20}$$

These contour lines are planes of equal height, and the sensitivity of the measurement is determined by α. The larger the angle α, the smaller the contour interval. If $\alpha = 90°$, then the contour interval is equal to p, and the sensitivity is a maximum. The reference plane will be parallel to the direction of the fringes and perpendicular to the viewing direction as shown in Fig. 16.12. Even though the maximum sensitivity can be obtained at 90°, this angle between the projection and viewing directions will produce a lot of unacceptable shadows on the object. These shadows will lead to areas with missing data where the object cannot be contoured. When $\alpha = 0$, the contour interval is infinite, and the measurement sensitivity is zero. To provide the best results, an angle no larger than the largest slope on the surface should be chosen.

When interference fringes are projected onto a surface rather than using a grating, the fringe spacing p is determined by the geometry shown in Fig. 16.13

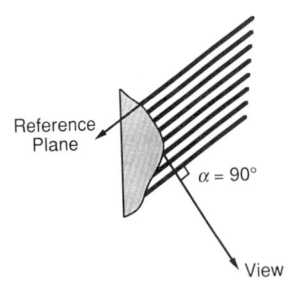

Figure 16.12. Maximum sensitivity for fringe projection with a 90° angle between projection and viewing.

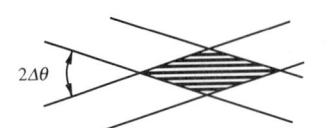

Figure 16.13. Fringes produced by two interfering beams.

16.6. SHADOW MOIRÉ

and is given by

$$p = \frac{\lambda}{2 \sin \Delta\theta}, \quad (16.21)$$

where λ is the wavelength of illumination and $2\Delta\theta$ is the angle between the two interfering beams. Substituting the expression for p into Eq. (16.20), the contour interval becomes

$$C = \frac{\lambda}{2 (\sin \Delta\theta) \sin \alpha}. \quad (16.22)$$

If a simple interferometer such as a Twyman–Green is used to generate projected interference fringes, tilting one beam with respect to the other will change the contour interval. The larger the angle between the two beams, the smaller the contour interval will be. Figures 16.11a and 16.11b show a change in the fringe spacing for interference fringes projected onto an object. The direction of illumination has been moved away from the viewing direction between Figs. 16.11b and 16.11c. This increases the angle α and the test sensitivity while reducing the contour interval. Projected fringe contouring has been covered in detail by Gasvik (1987).

If the source and the viewer are not at infinity, the fringes or grating projected onto the object will not be composed of straight, equally spaced lines. The height between contour planes will be a function of the distance from the source and viewer to the object. There will be a distortion due to the viewing of the fringes as well as due to the illumination. This means that the reference surface will not be a plane. As long as the object does not have large height changes compared to the illumination and viewing distances, a plane reference surface placed in the plane of the object can be measured first and then subtracted from subsequent measurements of the object. This enables the mapping of a plane in object space to a surface that will serve as a reference surface. If the object has large height variations, the plane reference surface may have to be measured in a number of planes to map the measured object contours to real heights. Finite illumination and viewing distances will be considered in more detail with shadow moiré in the next section.

16.6 SHADOW MOIRÉ

A simple method of moiré interferometry for contouring objects uses a single grating placed in front of the object as shown in Fig. 16.14. The grating in front of the object produces a shadow on the object that is viewed from a different direction through the grating. A low-frequency beat or moiré pattern is seen.

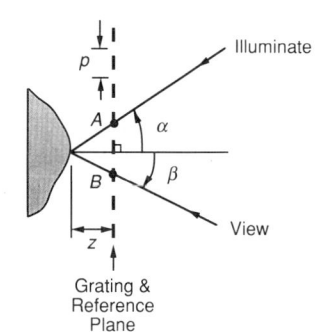

Figure 16.14. Geometry for shadow moiré with illumination and viewing at infinity, i.e., parallel illumination and viewing.

This pattern is due to the interference between the grating shadows on the object and the grating as viewed. Assuming that the illumination is collimated and that the object is viewed at infinity or through a telecentric optical system, the height z between the grating and the object point can be determined from the geometry shown in Fig. 16.14 (Meadows et al. 1970; Takasaki 1973; Chiang 1983). This height is given by

$$z = \frac{Np}{\tan \alpha + \tan \beta}, \tag{16.23}$$

where α is the illumination angle, β is the viewing angle, p is the spacing of the grating lines, and N is the number of grating lines between the points A and B (see Fig. 16.14). The contour interval in a direction perpendicular to the grating will simply be given by

$$C = \frac{p}{\tan \alpha + \tan \beta}. \tag{16.24}$$

Again, the distance between the moiré fringes in the beat pattern depends on the angle between the illumination and viewing directions. The larger the angle, the smaller the contour interval. If the high frequencies due to the original grating are filtered out, then only the moiré interference term is seen. The reference plane will be parallel to the grating. Note that this reference plane is tilted with respect to the reference plane obtained when fringes are projected onto the subject. Essentially, the shadow moiré technique provides a way of removing the "tilt" term and repositioning the reference plane. The contour interval for shadow moiré is the same as that calculated for projected fringe contouring (Eq. (16.20)) when one of the angles is zero with $d = p$. Figure 16.15 shows an object that has a grating sitting in front of it. An illumination beam is projected from one direction and viewed from another direction. Between Figs. 16.15a and 16.15b, the angles α and β have been increased. This has the effect of

16.6. SHADOW MOIRÉ

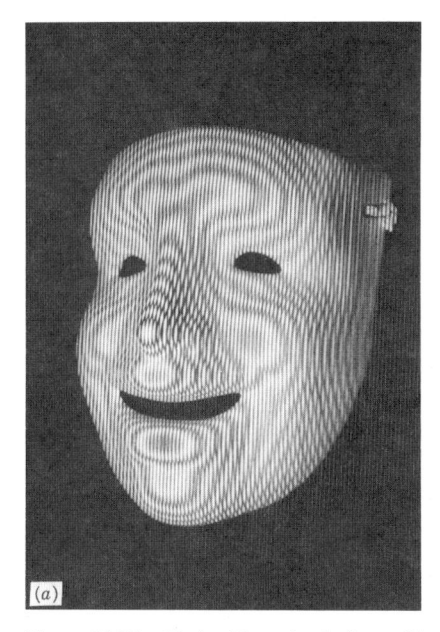

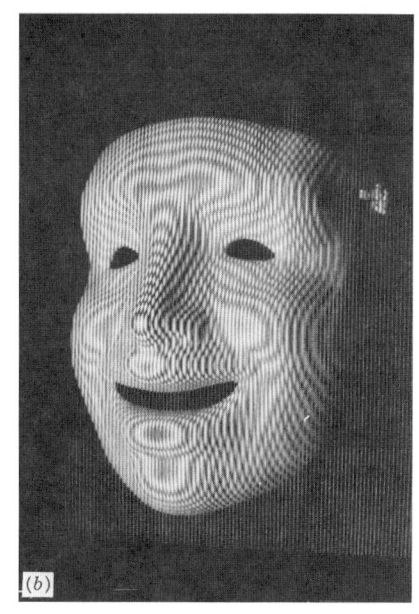

Figure 16.15. Mask with grating in front of it. (*a*) One viewing angle. (*b*) Larger viewing angle.

decreasing the contour interval, increasing the number of fringes, and rotating the reference plane slightly away from the viewer.

Most of the time, it is difficult to illuminate an entire object with a collimated beam. Therefore, it is important to consider the case of finite illumination and viewing distances. It is possible to derive this for a very general case (Meadows et al. 1970; Takasaki 1970; Bell 1985); however, for simplicity, only the case where the illumination and viewing positions are the same distance from the grating will be considered. Figure 16.16 shows a geometry where the distance between the illumination source and the viewing camera is given by *w*, and the distance between these and the grating is *l*. The grating is assumed to be close enough to the object surface so that diffraction effects are negligible. In this

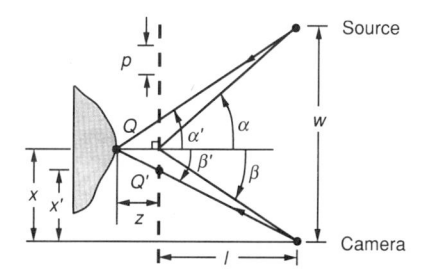

Figure 16.16. Geometry for shadow moiré with illumination and viewing at finite distances.

case the height between the object and the grating is given by

$$z = \frac{Np}{\tan \alpha' + \tan \beta'}, \qquad (16.25)$$

where α' and β' are the illumination and viewing angles at the object surface. These angles change for every point on the surface and are different from α and β in Fig. 16.16, where α and β are the illumination and viewing angles at the grating (reference) surface. The surface height can also be written as (Meadows et al. 1970; Takasaki 1973; Chiang 1983)

$$z = NC(z) = \frac{Np(l + z)}{w} = \frac{Npl}{w - Np}. \qquad (16.26)$$

This equation indicates that the height is a complex function depending on the position of each object point. Thus, the distance between contour intervals is dependent on the height of the surface and the number of fringes between the grating and the object. Individual contour lines will no longer be planes of equal height. They are now surfaces of equal height. The expression for height can be simplified by considering the case where the distance to the source and viewer is large compared to the surface height variations, $l \gg z$. Then the surface height can be expressed as

$$z = \frac{Npl}{w} = \frac{Np}{\tan \alpha + \tan \beta}. \qquad (16.27)$$

Even though the angles α and β vary from point-to-point on the surface, the sum of their tangents remains equal to w/l for all object points as long as $l \gg z$. The contour interval will be constant in this regime and will be the same as that given by Eq. (16.24).

Because of the finite distances, there is also distortion due to the viewing perspective. A point on the surface Q will appear to be at the location Q' when viewed through the grating. By similar triangles, the distances x and x' from a line perpendicular to the grating intersecting the camera location can be related using

$$\frac{x}{z + l} = \frac{x'}{l}, \qquad (16.28)$$

where x and x' are defined in Fig. 16.16. Equation (16.28) can be rearranged to yield the actual coordinate x in terms of the measured coordinate x' and the measurement geometry,

16.8. TWO-ANGLE HOLOGRAPHY

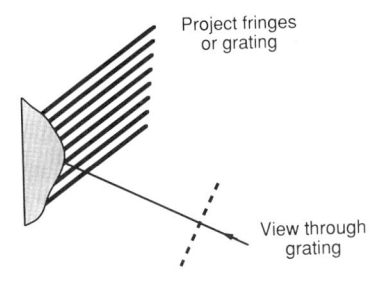

Figure 16.17. Projection moiré where fringes or a grating are projected onto a surface and viewed through a second grating.

$$x = x'\left(1 + \frac{z}{l}\right). \tag{16.29}$$

Likewise, the y coordinate can be corrected using

$$y = y'\left(1 + \frac{z}{l}\right). \tag{16.30}$$

This enables the measured surface to be mapped to the actual surface to correct for the viewing perspective. These same correction factors can be applied to fringe projection.

16.7. PROJECTION MOIRÉ

Moiré interferometry can also be implemented by projecting interference fringes or a grating onto an object and then viewing through a second grating in front of the viewer (see Fig. 16.17) (Brooks and Helfinger 1969). The difference between projection and shadow moiré is that two different gratings are used in projection moiré. The orientation of the reference plane can be arbitrarily changed by using different grating pitches to view the object. The contour interval is again given by Eq. (16.24), where d is the period of the grating in the y plane, as long as the grating pitches are matched to have the same value of d. This implementation makes projection moiré the same as shadow moiré, although projection moiré can be much more complicated than shadow moiré. A good theoretical treatment of projection moiré is given by Benoit et al. (1975).

16.8. TWO-ANGLE HOLOGRAPHY

Projected fringe contouring can also be done using holography. First a hologram of the object is made using the optical setup shown in Fig. 16.18. Then the direction of the beam illuminating the object is changed slightly. When the

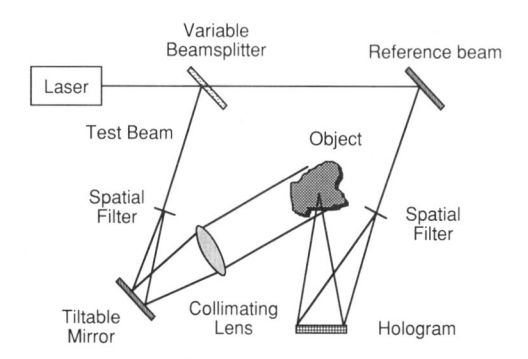

Figure 16.18. Setup for two-angle holographic interferometry.

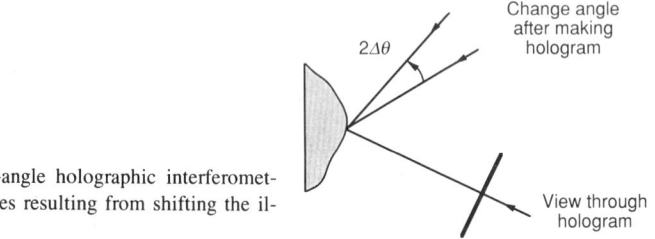

Figure 16.19. Two-angle holographic interferometric. Interference fringes resulting from shifting the illumination beam.

object is viewed through the hologram, interference fringes are seen that correspond to the interference between the wavefront stored in the hologram and the live wavefront with the tilted illumination. This process is depicted by Fig. 16.19. These fringes are exactly what would be seen if the object were illuminated with the two illumination beams simultaneously. The beams would be tilted with respect to one another by the same amount that the illumination beam was tilted after making the hologram. These fringes will look the same as those produced by projected fringe contouring and shown in Fig. 16.11. To produce straight, equally spaced fringes, the object illumination should be collimated. When collimated illumination is used, the surface contour is measured relative to a surface that is a plane. The theory of projected fringe contouring can be applied to two-angle holographic contouring yielding a contour interval given by Eq. (16.22), where $2\Delta\theta$ is the change in the angle of the object illumination. More detail on two-angle holographic contouring can be found in Haines and Hildebrand (1965), Hildebrand and Haines (1966, 1967), Vest (1979), and Hariharan (1984).

16.9. COMMON FEATURES

All of the techniques described produce fringes corresponding to contours of equal height on the object. They all have a similar contour interval determined by the fringe spacing or grating period and the angle between the illumination and viewing directions as long as the illumination and viewing are collimated. Phase shifting can be applied to any of the techniques to produce quantitative height information as long as sinusoidal fringes are present at the camera. The surface heights measured are relative to a reference surface that is a plane as long as the fringes or grating lines are straight and equally spaced at the object. The only difference between the moiré techniques and the projected fringes and two-angle holography is the change in the location of the reference plane. If the fringes are digitized or phase-measuring interferometry techniques are applied, the reference plane can be changed in the computer mathematically.

The precision of these contouring techniques depends on the number of fringes used. When the fringes are digitized using fringe-following techniques, the surface height can be determined to $\frac{1}{10}$ of a fringe. If phase measurement is used, the surface heights can be determined to $\frac{1}{100}$ of a fringe. Therefore it is advantageous to use as many fringes as possible. And because a reference plane can easily be changed in a computer, projected fringe contouring is the simplest way to contour an object interferometrically.

16.10. COMPARISON TO CONVENTIONAL INTERFEROMETRY

The measurement of surface contour can be related to making the same measurement using a Twyman–Green interferometer assuming a long effective wavelength. The loci of the lines or fringes projected onto the surface (assuming illumination and viewing at infinity) is given by

$$y = z \tan \alpha + nd, \tag{16.31}$$

where z is the height of the surface at the point y, d is the fringe spacing measured along the y axis, and n is an integer referring to fringe order number. If the same surface were tested using a Twyman–Green interferometer, a bright fringe would be obtained whenever

$$2z - y \sin \gamma = n\lambda, \tag{16.32}$$

where λ is the wavelength and γ is the tilt of the reference plane. By comparing Eqs. (16.31) and (16.32), it can be seen that they are equivalent as long as

$$d = \frac{\lambda_{\text{effective}}}{\sin \gamma} \tag{16.33}$$

and

$$\frac{2}{\sin \gamma} = \tan \alpha, \tag{16.34}$$

where $\lambda_{\text{effective}}$ is the effective wavelength. The effective wavelength can then be written as

$$\lambda_{\text{effective}} = 2C = \frac{2d}{\tan \alpha} = \frac{2p}{\cos \alpha \tan \alpha}, \qquad (16.35)$$

where C is the contour interval as defined in Eq. (16.20). Thus, contouring using these techniques is similar to measuring the object in a Twyman–Green interferometer using a source with wavelength $\lambda_{\text{effective}}$.

16.11. APPLICATIONS

These techniques can all be used for displacement measurement or stress analysis as well as for contouring objects. Displacement measurement is performed by comparing the fringe patterns obtained before and after a small movement of the object or before and after applying a load to the object. Because the sensitivity of these tests are variable, they can be used for a larger range of displacements and stresses than the holographic techniques. Differential interferometry comparing two objects or an object and a master can also be performed by comparing the two fringe patterns obtained. Finally, time-average vibration analysis can also be performed with moiré, yielding results similar to those obtained with time-average holography with a much longer effective wavelength.

Using phase-measurement techniques, the surface height relative to some reference surface can be obtained quantitatively. If the contour lines are straight and equally spaced in object space, then the reference surface will be a plane. In the computer, any plane (or surface) desired can be subtracted from the surface height to yield the surface profile relative to any plane (or surface). This is similar to viewing the contour lines through a grating (or deformed grating) to reduce their number. If the contour lines are not straight and equally spaced, the reference surface will be something other than a plane. The reference surface can be determined by placing a flat surface at the location of the object and measuring the surface height. Once this reference surface is measured, it can be subtracted from subsequent measurements to yield the surface height relative to a plane surface. Thus, with the use of phase-measuring interferometry techniques, the surface height can be made relative to any surface and transformed to surface heights relative to another surface. Taking this one step further, a master component can be compared to a number of test components to determine if their shape is within the specification. It should also be pointed out that this measurement is sensitive to a certain direction, and that there may be areas where data are missing because of shadows on the surface.

As an example, Fig. 16.20 shows the mask of Figs. 16.11 and 16.15 con-

16.11. APPLICATIONS

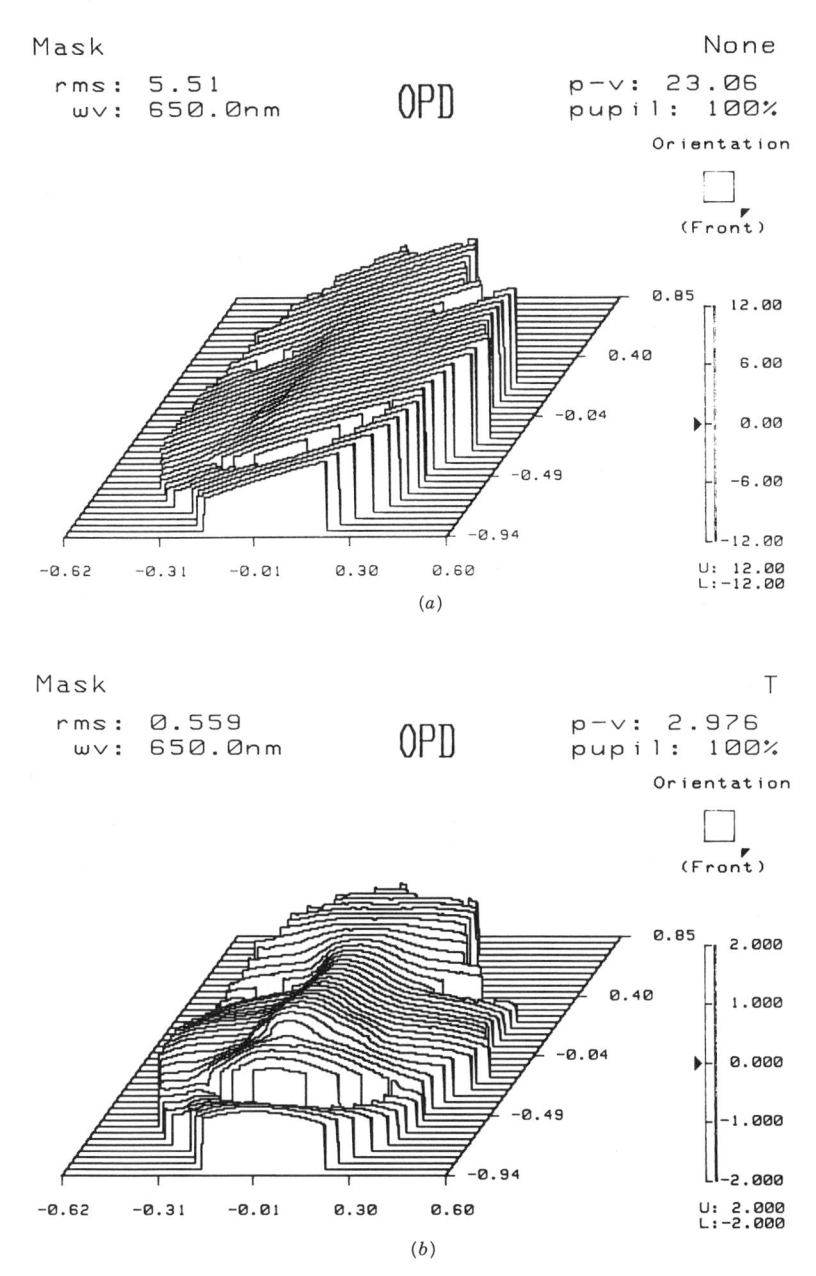

Figure 16.20. Mask measured with projected fringes and phase-measurement interferometry. (*a*) Isometric plot of measured surface height. (*b*) Isometric plot after best-fit plane removed. (*c*) Two-dimensional contour plot of measured surface height. (*d*) Two-dimensional contour plot after best-fit plane removed. Units on plots are in number of contour intervals. One contour interval is approximately 10 mm. The surface is about 150 mm in diameter.

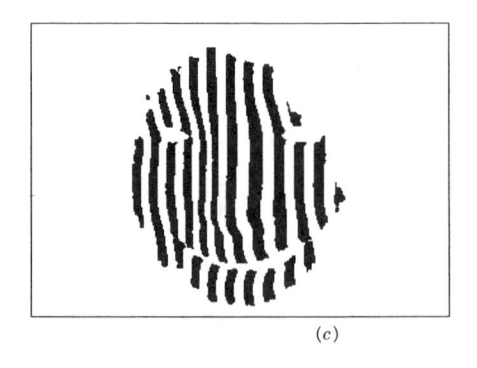

680 MOIRÉ AND FRINGE PROJECTION TECHNIQUES

```
Mask                                          None
  rms:  5.51         Interval: 2.000    p-v:  23.06
   wv:  650.0nm                        pupil:  100%
```

(c)

```
Mask                                             T
  rms:  0.559        Interval: 0.500    p-v:  2.976
   wv:  650.0nm                        pupil:  100%
```

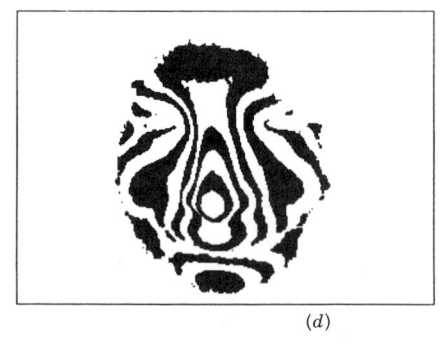

(d)

Figure 16.20. (*Continued*)

toured using fringe projection and phase-measurement interferometry. The fringes are produced using a Twyman–Green interferometer with a He–Ne laser. A high-resolution camera with 1320 × 900 pixels and a zoom lens is used to view the fringes. Surface heights are calculated using phase-measurement techniques at each detector point. A total of five interferograms were used to calculate the surface shown in Fig. 16.20a. The best-fit plane has been subtracted from the surface to yield Fig. 16.20b. In this way the reference plane has been changed. Figures 16.20c and 16.20d show two-dimensional contour maps of the object before and after the best-fit plane is removed. These contours can also be thought of as the fringes that would be viewed on the object. Figure 16.20c shows the fringes without a second grating, and Fig. 16.20d is with a second reference grating chosen to minimize the fringe spacing. The contour interval for this example is 10 mm, and the total peak-to-valley height deviation after the tilt is subtracted is about 30 mm.

16.12. SUMMARY

The techniques of projected fringe contouring, projection moiré, shadow moiré, and two-angle holographic contouring are all similar. They all involve projecting a pattern of lines or interference fringes onto an object and then viewing those contour lines from a different direction. In the case of the moiré techniques, the contour lines are viewed through a grating to reduce the total number of fringes. In all of the techniques, the surface height is measured relative to a reference surface. The reference surface will be a plane if the projected grating lines or interference fringes are straight and equally spaced at the object and viewed at infinity or with a telecentric imaging system. The use of the second grating in the moiré techniques changes the reference plane but does not affect the contour interval. The sensitivity of the techniques is a maximum when the contour lines are viewed at an angle of 90° with respect to the projection direction. Quantitative data can be obtained from any of these techniques using phase-measurement interferometry techniques. The precision of the surface-height measurement will depend on the number of fringes present. Surface-height measurements can be made with a repeatability of $\frac{1}{100}$ of a contour interval rms (root-mean-square). Thus, the number of fringes used should be as many as can easily be measured by the detection system. The contour interval can be changed to increase the number of fringes, and once the surface height is calculated, a reference surface can be subtracted in the computer to find the surface height relative to any desired surface.

Acknowledgments. The authors acknowledge the support of WYKO Corporation during the preparation of this manuscript.

REFERENCES

Abramson, N., "Sandwich Hologram Interferometry. 3: Contouring," *Appl. Opt.*, **15**(1), 200–205 (1976a).

Abramson, N., "Holographic Contouring by Translation," *Appl. Opt.*, **15**(4), 1018–1022 (1976b).

Bell, B., "Digital Heterodyne Topography," Ph.D. Dissertation, Optical Sciences Center, University of Arizona, Tuscon, AZ, 1985.

Benoit, P., E. Mathieu, J. Hormier, and A. Thomas, "Characterization and Control of Three Dimensional Objects Using Fringe Projection Techniques," *Nouv. Rev. Opt.*, **6**(2), 67-86 (1975).

Brooks, R. E. and L. O. Heflinger, "Moiré Gauging Using Optical Interference Patterns," *Appl. Opt.*, **8**(5), 935-939 (1969).

Case, S. K., J. A. Jalkio, and R. C. Kim, "3-D Vision System Analysis and Design," in *Three-Dimensional Machine Vision*, Takeo Kanade, Ed., Kluwer Academic Publishers, Norwell, MA, 1987, pp. 63-95.

Chiang, F.-P., "Moiré Methods for Contouring, Displacement, Deflection, Slope and Curvature," *Proc. SPIE*, **153**, 113-119 (1978).

Chiang, F.-P., "Moiré Methods of Strain Analysis," in *Manual on Experimental Stress Analysis*, A. S. Kobayashi, Ed., Soc. for Exp. Stress Anal., Brookfield Center, CT, 1983, pp. 51-69.

DeMattia, P. and V. Fossati-Bellani, "Holographic Contouring by Displacing the Object and the Illumination Beam," *Opt. Commun.*, **26**(1), 17-21 (1978).

Der Hovanesian, J. and Y. Y. Yung, "Moiré Contour-Sum Contour-Difference, and Vibration Analysis of Arbitrary Objects," *Appl. Opt.*, **10**(12), 2734-2738 (1971).

Dureli, A. J. and V. J. Parks, *Moiré Analysis of Strain*, Prentice-Hall, Englewood Cliffs, NJ, 1970.

Gasvik, K. J., *Optical Metrology*, Wiley, Chichester, 1987.

Guild, J., *The Interference Systems of Crossed Diffraction Gratings*, Clarendon Press, Oxford, 1956.

Guild, J., *Diffraction Gratings as Measuring Scales*, Oxford University Press, London, 1960.

Haines, K. and B. P. Hildebrand, "Contour Generation by Wavefront Reconstruction," *Phys. Lett.*, **19**(1), 10-11 (1965).

Halioua, M. and H.-C. Liu, "Optical Three-Dimensional Sensing by Phase Measuring Profilometry," *Opt. Lasers Eng.*, **11**(3), 185-215 (1989).

Hariharan, P., *Optical Holography*, Cambridge University Press, Cambridge, 1984.

Hildebrand, B. P. and K. A. Haines, "The Generation of Three-Dimensional Contour Maps by Wavefront Reconstruction," *Phys. Lett.*, **21**(4), 422-423 (1966).

Hildebrand, B. P. and K. A. Haines, "Multiple-Wavelength and Multiple-Source Holography Applied to Contour Generation," *J. Opt. Soc. Am.*, **57**(2), 155-162 (1967).

Indebetouw, G., "Profile Measurement Using Projection of Running Fringes," *Appl. Opt.*, **17**(18), 2930-2933 (1978).

Ligtenberg, F. K., "The Moiré Method," *Proc. Soc. Exp. Stress Anal. (SESA)*, **12**(2), 83-98 (1955).

MacGovern, A. J., "Projected Fringes and Holography," *Appl. Opt.*, **11**(12), 2972-2974 (1972).

Meadows, D. M., W. O. Johnson, and J. B. Allen, "Generation of Surface Contours by Moiré Patterns," *Appl. Opt.*, **9**(4), 942-947 (1970).

Menzel, E., "Comment to the Methods of Contour Holography," *Optik*, **40**(5), 557-559 (1974).

Moore, D. T. and B. E. Truax, "Phase-Locked Moiré Fringe Analysis for Automated Contouring of Diffuse Surfaces," *Appl. Opt.*, **18**(1), 91–96 (1979).

Oster, G. and Y. Nishijima, "Moiré Patterns," *Sci. Amer.*, **208**(5), 54–63 (May 1963).

Perrin, J. C. and A. Thomas, "Electronic Processing of Moiré Fringes: Application to Moiré Topography and Comparison with Photogrammetry," *Appl. Opt.*, **18**(4), 563–574 (1979).

Post, D., "Developments in Moiré Interferometry." *Opt. Eng.*, **21**(3), 458–467 (1982).

Rayleigh, Lord, "On the Manufacture and Theory of Diffraction-Gratings," *Phil. Mag. S.4*, **47**(310) 81–93 and 193–205 (1874).

Reid, G. T., "Moiré Fringes in Metrology," *Opt. Lasers Eng.*, **5**(2), 63–93 (1984a).

Reid, G. T., R. C. Rixon, and H. I. Messer, "Absolute and Comparative Measurements of Three-Dimensional Shape by Pulse Measuring Moiré Topography," *Opt. Laser Tech.*, **16**(6), 315–319 (1984b).

Righi, A., "Sui Fenomeni Che si Producono colla Sovrapposizione dei Due Reticoli e sopra Alcune Lora Applicazioni: I," *Nuovo Cim.*, **21**, 203–227 (1887).

Rowe, S. H., "Projected Interference Fringes in Holographic Interferometry," *J. Opt. Soc. Am.*, **61**(12), 1599–1603 (1971).

Rowe, S. H. and W. T. Welford, "Surface Topography of Non-Optical Surfaces by Projected Interference Fringes," *Nature*, **216**(5117), 786–787 (1967).

Shagam, R., "Heterodyne Interferometric Method for Profiling Recorded Moiré Interferograms," *Opt. Eng.*, **19**(6), 806–809 (1980).

Srinivasan, V., H. C. Liu, and M. Halioua, "Automated Phase-Measuring Profilometry of 3-D Diffuse Objects," *Appl. Opt.*, **23**(18), 3015–3108 (1984).

Srinivasan, V., H. C. Liu, and M. Halioua, "Automated Phase-Measuring Profilometry: A Phase Mapping Approach," *Appl. Opt.*, **24**(2), 185–188 (1985).

Takasaki, H., "Moiré Topography," *Appl. Opt.*, **9**(6), 1467–1472 (1970).

Takasaki, H., "Moiré Topography," *Appl. Opt.*, **12**(4), 845–850 (1973).

Takeda, M. and K. Mutoh, "Fourier Transform Profilometry for the Automatic Measurement of 3-D Object Shapes," *Appl. Opt.*, **22**(24), 3977–3982 (1983).

Takeda, M., H. Ina, and S. Kabayashi, "Fourier-Transform Method of Fringe-Pattern Analysis for Computer-Based Topography and Interferometry," *J. Opt. Soc. Am.*, **72**(1), 156–160 (1982).

Theocaris, P. S., "Moiré Fringes: A Powerful Measuring Device," *Appl. Mech. Rev.*, **15**(5), 333–339 (1962).

Theocaris, P. S., "Moiré Fringes: A Powerful Measuring Device," in *Applied Mechanics Surveys*, Spartan Books, Washington, D.C., 1966, p. 613–626.

Theocaris, P. S., *Moiré Fringes in Strain Analysis*, Pergamon Press, Oxford, 1969.

Varner, J. R., "Holographic and Moiré Surface Contouring," in *Holographic Nondestructive Testing*, R. K. Erf, Ed., Academic Press, Orlando, 1974.

Vest, C. M., *Holographic Interferometry*, Wiley, New York, 1979.

Wasowski, J., "Moiré Topographic Maps," *Opt. Commun.*, **2**(7), 321–323 (1970).

Webster's Third New International Dictionary, Merriam-Webster, Springfield, MA, 1981.

Welford, W. T., "Some Applications of Projected Interference Fringes," *Opt. Acta*, **16**(3), 371–376 (1969).

Weller, R. and B. M. Shepherd, "Displacement Measurement by Mechanical Interferometry," *Proc. Soc. Exp. Stress Anal. (SESA)*, **6**(1), 35–38 (1948).

Winther, S. and G. Å. Slettemoen, "An ESPI Contouring Technique in Strain Analysis," *Proc. SPIE*, **473**, 44–47 (1983).

Yatagai, T., M. Idesawa, Y. Yamaashi, and M. Suzuki, "Interactive Fringe Analysis System: Applications to Moiré Contourogram and Interferogram," *Opt. Eng.*, **21**(5), 901–906 (1982).

ADDITIONAL REFERENCES

Asai, K., "Contouring Method by Moiré Holography," *Jpn. J. Appl. Phys.*, **16**(10), 1805–1808 (1977).

Boehnlein, A. J. and K. G. Harding, "Adaptation of a Parallel Architecture Computer to Phase Shifted Moiré Interferometry," *Proc. SPIE*, **728**, 183–193 (1986).

Burch, J. M., "Photographic Production of Scales for Moiré Fringe Applications," in *Optics in Metrology, Brussels Colloquium, May 6–9, 1958*, P. Mollet, Ed., Pergamon Press, New York, 1960, pp. 361–368.

Cabaj, A., G. Ranninger, and G. Windischbauer, "Shadowless Moiré Topography Using a Single Source of Light," *Appl. Opt.*, **13**(4), 722–723 (1974).

Chiang, F.-P., "Techniques of Optical Signal Filtering Parallel to the Processing of Moiré-Fringe Patterns," *Exp. Mech.*, **9**(11), 523–526 (Nov. 1969).

Cline, H. E., A. S. Holik, and W. E. Lorensen, "Computer-Aided Surface Reconstruction of Interference Contours," *Appl. Opt.*, **21**(24), 4481–4488 (1982).

Cline, H. E., W. E. Lorensen, and A. S. Holik, "Automatic Moiré Contouring," *Appl. Opt.*, **23**(10), 1454–1459 (1984).

Gilbert, J. A., T. D. Dudderar, D. R. Matthys, H. S. Johnson, and R. A. Franzel, "Two-Dimensional Stress Analysis Combining High-Frequency Moiré Measurements with Finite-Element Modeling," *Exp. Tech.*, **11**(3), 24–28 (March 1987).

Halioua, M., R. S. Krishnamurthy, H.-C. Liu, and F. P. Chiang, "Automated 360° Profilometry of 3-D Diffuse Objects," *Appl. Opt.*, **24**(12), 2193–2196 (1985).

Harding, K. G., M. Michniewicz, and A. J. Boehnlein, "Small Angle Moiré Contouring," *Proc. SPIE*, **850**, 166–173 (1987).

Idesawa, M., T. Yatagai, and T. Soma, "Scanning Moiré Method and Automatic Measurement of 3-D Shapes," *Appl. Opt.*, **16**(8), 2152–2162 (1977).

Indebetouw, G., "A Simple Optical Noncontact Profilometer," *Opt. Eng.*, **18**(1), 63–66 (1979).

Jaerisch, W. and G. Makosch, "Optical Contour Mapping of Surfaces," *Appl. Opt.*, **12**(7), 1552–1557 (1973).

Kobayashi, A., Ed., *Handbook on Experimental Mechanics*, Prentice-Hall, Englewood Cliffs, NJ, 1987.

ADDITIONAL REFERENCES

Kujawinska, M., "Use of Phase-Stepping Automatic Fringe Analysis in Moiré Interferometry," *Appl. Opt.*, **26**(22), 4712–4714 (1987).

Miles, C. A. and B. S. Speight, "Recording the Shape of Animals by a Moiré Method," *J. Phys. E., Sci. Instrum.*, **8**(9), 773–776 (1975).

Pekelsky, J. R., "Automated Contour Ordering for Moiré Topograms," *Opt. Eng.*, **26**(6), 479–486 (1987).

Reid, G. T., "A Moiré Fringe Alignment Aid," *Opt. Lasers Eng.*, **4**(2), 121–126 (1983).

Schätzel, K. and G. Parry, "Real-Time Moiré Measurement of Phase Gradient," *Opt. Acta*, **29**(11), 1441–1445 (1982).

Suzuki, M. and K. Suzuki, "Moiré Topography Using Developed Recording Methods," *Opt. Lasers Eng.*, **3**(1), 59–64 (1982).

Theocaris, P. S., "Isopachic Patterns by the Moiré Method," *Exp. Mech.*, **4**(6), 153–159 (1964).

Toyooka, S. and Y. Iwaasa, "Automatic Profilometry of 3-D Diffuse Objects by Spatial Phase Detection," *Appl. Opt.*, **25**(10), 1630–1633 (1986).

Varman, P. O., "A Moiré System for Producing Numerical Data for the Profile of a Turbine Blade Using a Computer and Video Store," *Opt. Lasers Eng.*, **5**(2), 41–58 (1984).

Yatagai, T. and M. Idesawa, "Automatic Fringe Analysis for Moiré Topography," *Opt. Lasers Eng.*, **3**(1), 73–83 (1982).

17

Contact and Noncontact Profilers

K. Creath and A. Morales

17.1. INTRODUCTION

Profilers are instruments that are used to measure surface finish, surface roughness, and the geometry of small features on an object. They are different from instruments that are meant to measure large surfaces or the overall form of an object. Typically, a profiler has a maximum field of view of about 10 mm (some go up to more than 100 mm). The minimum field of view can be as small as a few nanometers. Rather than only providing qualitative data, profilers usually provide quantitative data in the form of a surface height map over the measurement area. These data can then be further analyzed to provide statistics, Fourier analysis, or critical dimension information about a surface. Stylus probes, focus sensor probes, optical and electron microscopes, and scanning probe microscopes are all examples of profiling instruments. Good reviews have been written by Bennett and Mattsson (1989) and Vorburger and Raja (1990).

Profiler types can be split into contact and noncontact devices. A contact profiler scans a probe across the surface and determines height by looking at the height variations of the probe as it is scanned. Stylus profilers are the major type of contact profiler. Noncontact devices measure surface height without coming in contact with the surface. There are many kinds of noncontact devices. Most noncontact devices are optical microscopes, although scanning probe microscopes (SPMs) are steadily gaining usefulness. Noncontact devices also include optical focus sensors that scan a point of light across the surface and relate surface height to the amount that a lens has to be moved to keep the spot in focus on the surface. A plot of the lateral and vertical measurement ranges for stylus, optical, and SPM profiling techniques is shown in Fig. 17.1 (Stedman 1989; Stedman and Lindsey 1989; Vorburger and Raja 1990). Note that these ranges are continuously changing with new advances in instruments. These ranges serve only as a guideline for a comparison of techniques.

Optical Shop Testing, Second Edition, Edited by Daniel Malacara.
ISBN 0-471-52232-5 © 1992, John Wiley & Sons, Inc.

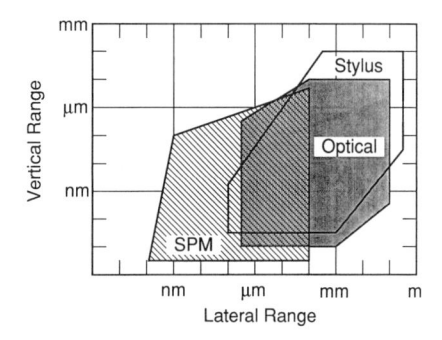

Figure 17.1. Plot of measurement ranges for stylus, optical, and SPM profilers.

Profiling systems can also be split into scanning and imaging. A scanning system measures a single data point at a time, and mechanically moves either the probe or the surface to scan a line or an area of the surface. All contact probes, as well as SPMs and focus sensors, are scanning systems. Some optical microscope systems also scan the surface. The problem with scanning systems is that there are errors due to the scanning motion. As long as these errors are systematic, they can be removed. Imaging systems almost exclusively are optical microscopes. An area on the surface is imaged onto a detector array or CCD (charge-coupled device) camera. No mechanical scanning is necessary. The main considerations for imaging systems is alignment of the sensor in the focal plane and using optics with a small amount of distortion. (Distortion is roughly equivalent to systematic mechanical scanning errors and can be corrected.)

Another major difference between profiling systems is interferometric versus noninterferometric. Most scanning systems are noninterferometric, although some scanning surfaces are interferometrically based. Surface height is determined by the height variations of the probe as it is moved across a surface. Most optical microscopes with imaging systems are interferometric.

Each of the different profiler types has different measurement properties and different advantages and disadvantages. Stylus profilers move a probe that is like a phonograph needle across the surface. Height variations in the probe are directly related to surface height. They generally only profile a single line on the surface, although there are systems available that will scan areas on the surface. Using different reference surfaces, flat as well as highly curved surfaces can easily be accommodated. These do not require any modifications to the surface; however, they may damage the surfaces of soft materials and can be very slow if a large number of points are measured in the area.

Scanning probe microscopes have only been commercially available for about 5 years. However, in that time, they have taken a considerable share of the profile measurement market. SPMs look at small areas on surfaces (up to 200 μm^2) and scan a probe very close to the surface. There are many different types of sensors. A scanning tunneling microscope (STM) senses a tunneling current between the probe and the surface. Other probes sense atomic forces, magnetic forces, thermal forces, and many other quantities. Some SPMs require coating the surface with a conducting layer, while others do not require any surface preparation. These systems typically take a couple of minutes to make a measurement.

Optical focus sensors cannot measure very smooth optical surfaces, but they can measure surfaces with intermediate roughness values like finely machined surfaces. These sensors are slow because they must focus at each measurement point by moving an optical system much in the same way that a compact disk player reads a disk. They are capable of scanning both lines and areas. Samples must reflect light back into the optical system to get a signal.

Optical microscopes can measure areas that are larger than the SPMs and smaller than the optical focus sensors and stylus probes. There are a number of different types of microscopes. Each has its advantages and disadvantages. Scanning microscopes such as the confocal microscope and the near-field scanning optical microscope can provide a higher magnification and better lateral resolution than standard optical microscopes; however, they are not well characterized. Interferometric optical microscopes are the workhorses of noncontact profiling for the measurement of surface roughness. They can measure areas of 50 μm^2 up to 5 mm^2 by simply changing the microscope objective. Just as there are different kinds of interferometers for the measurement of surface figure, there are many different kinds of interferometric microscopes. Some measure slope and others directly measure surface height. The main disadvantage of optical microscopes is that the height measured depends on the phase shift of the light upon reflection. If a single material is being measured, there is no problem. If more than one material is present in the field of view, there may be height errors that need to be accounted for. Coating the surface with a reflective material eliminates this problem. These systems are by far the fastest. They can measure the surface height over tens of thousands of data points in a few seconds.

This chapter will describe the measurement techniques and designs for the major types of instruments used for the measurement of surface roughness and small surface features. These include stylus profilers, scanning tunneling and atomic force microscopes, optical focus sensors, and interferometric optical profiles measuring slope and height. Examples of measurements will be included for some of the instruments. Other types of systems, those which are mainly used in the semiconductor industry for the measurement of critical dimensions such as electron microscopes and confocal microscopes, will not be covered in this chapter. These systems are covered in many other references (Wilson and Sheppard 1984; Reimer 1985).

17.2. STYLUS PROFILERS

The most common type of surface profiler is the stylus. These profilers work very much like a phonograph moving a small-tipped probe across the surface and sensing height variations of the tip to determine the surface height profile.

Usually the surface is moved under the stylus, but the stylus may also be moved over the surface. Styli are made of a hard material such as diamond with a tip radius of curvature between 0.1 and 25 μm. To ensure that the test surface is not damaged during measurement, the load of the stylus tip on the surface is variable. The minimum load to keep the stylus on the surface needs to be chosen so that the surface is not deformed as the stylus is moved across it. A schematic of a stylus profiler is shown in Fig. 17.2.

Because the end of the stylus is not infinitesimal in size, the output of these profilers is the convolution of the stylus tip with the surface profile. The shape of the stylus can keep it from going to the bottom of steep trenches and cause it to round off peaks on the surface. Figure 17.2 also shows the effect of the stylus on the surface profile. When the stylus tip radius is made smaller and sharper, the stylus can more easily follow the shape of the surface. However, if the tip is too sharp, the local force on the surface under the tip may be so great that the surface is locally deformed as the stylus moves over the surface. If the surface elastically deforms, it will not be damaged, but the profile may be incorrect. If the surface plastically deforms, the surface profile will be inaccurate and the surface can be permanently damaged. When a small-radius stylus is used, the scan speed must be slowed down significantly and the stylus load reduced to ensure a good measurement. The most sensitive and high-resolution stylus profilers have tip radii of tenths of a micrometer, tip loadings of milligrams, need enclosures and vibration isolation systems, and take many minutes to make a scan of a few thousand data points (Bennett and Mattsson 1989; Song and Vorburger 1991).

The lateral resolution of stylus profilers is determined by the radius of the stylus tip as well as the surface shape and the sampling interval between data points. For a stylus with a spherical tip measuring a sinusoidal surface profile, the shortest measurable wavelength (period) d of the sinusoid is given by (Bennett and Dancy 1981)

$$d = 2\pi \sqrt{ar}, \qquad (17.1)$$

where r is the stylus radius and a is the amplitude of the sinusoid. Because two samples are required to reconstruct a sinusoid, the lateral resolution will be $d/2$.

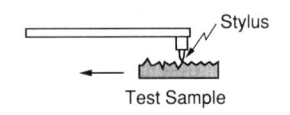

Figure 17.2. Schematic of a stylus profiler and effect of stylus shape on a measurement.

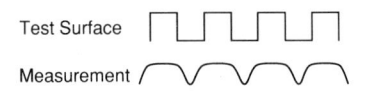

This means that for a stylus of 10-μm radius measuring 1-nm surface height variations, the lateral resolution is approximately 0.6 μm. To ensure sufficient resolution, it is best to oversample and measure at least four samples per lateral resolution element (Bennett and Mattsson 1989). Lateral resolution and transfer functions for more complex surface features can also be determined (Bennett and Dancy 1981; Al-Jumaily et al. 1987). The profile (and radius) of the stylus tip can be determined by viewing it with a scanning-electron microscope (SEM) or by scanning it over the edge of a razor blade (Vorburger and Raja 1990). Scanning-electron micrographs of different stylus tip shapes are shown in Fig. 17.3.

Most stylus profilers have reference datums of some type to ensure an accurate measurement (Vorburger and Raja 1990). The reference surface can be a skid that is moved across the surface with the stylus, or can be a separate reference surface which is scanned by a second large-radius probe in a fixed relationship to the measuring stylus. References can also be created using flexures (Vorburger and Raja 1990). Using a large-radius skid near the stylus is the easiest way to generate a reference, but it can cause errors and will remove shape and figure information. A separate reference is the most accurate but can limit the length of the scan and the measurable height variation.

Stylus profilers are normally calibrated using traceable height and roughness standards. The most common standards are step heights of chrome on glass. The step is measured periodically with the profiler to ensure calibration, and a scaling factor is calculated to apply to the profile data. Because stylus profilers are not linear over their entire height range, it is important to calibrate the instrument with a step height which is close in height to the test samples being measured. When surface roughness is being determined, it is better to use a roughness standard than a step-height standard because both lateral resolution and surface height variation need to be considered. These standards are available in a number of different types. The most common have a sinusoidal height variation with a given amplitude and a number of different spatial wavelengths. Roughness standards are also available as square-wave gratings. Because the stylus may not get down into the valleys and can round off peaks, the sinusoidal standards give more accurate indications of instrument performance at a single spatial frequency. Stylus profilers are capable of measuring surface roughness with an rms (root-mean-square) as small as 0.5 Å with lateral resolutions of 0.1 to 0.2 μm.

17.3. SCANNING PROBE MICROSCOPES

Scanning probe microscopes (SPMs) move a fine tip in close proximity to a surface. They usually scan within a few angstroms of the surface but can be in direct contact, using a very small force. The first microscope of this type was

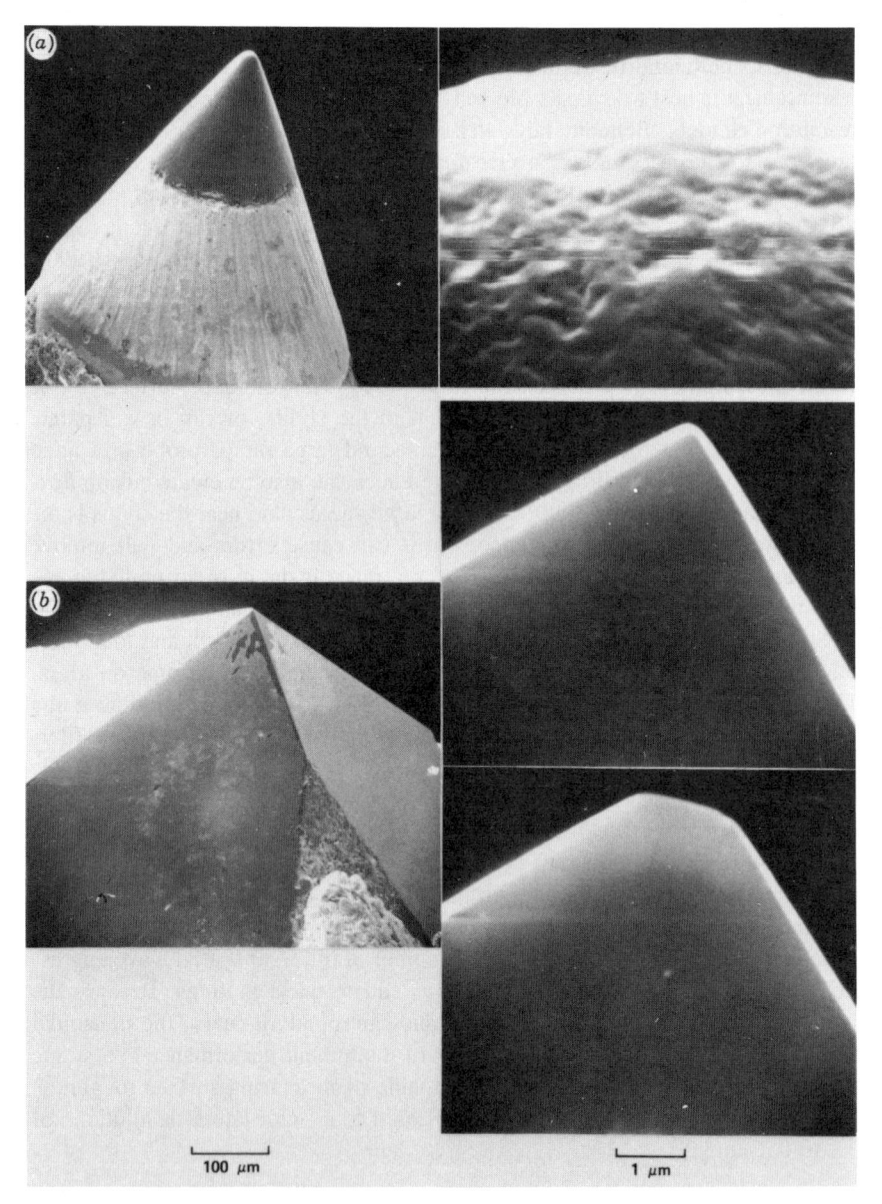

Figure 17.3. Scanning electron micrographs of (*a*) conical and (*b*) shovel-shaped pyramidal shaped styli. (Courtesy of Jean M. Bennett.)

17.3. SCANNING PROBE MICROSCOPES

a scanning tunnel microscope (STM) built by Binnig and Rohrer (1982, 1985) and won the 1986 Nobel prize in physics. Other types of SPMs are atomic force microscopes (AFM), laser force microscopes (LFM), scanning force microscopes (SFM), and magnetic force microscopes (MFM) (Wickramasinghe 1989; Pool 1990). This field is changing rapidly and new probe types are constantly being introduced. With so many different probe types, it is possible to find one that is appropriate for almost any application from integrated circuits to optical surfaces to biological objects. Reviews of SPMs have been written by Hansma and Tersoff (1987), Wickramasinghe (1989), Rugar and Hansma (1990), and Sarid (1991). The rest of this section will concentrate on the two types of SPMs: STM and AFM.

17.3.1. Scanning Tunneling Microscopes

In the STM, a tip is moved toward the test surface until a tunneling current is detected. For atomic resolution, the end of the tip has one atom interacting with the test surface. This tunneling current can only be sensed when the probe is less than 1 nm away from the surface. As the probe is moved closer to the surface, the tunneling current increases exponentially. The probe is usually scanned a few angstroms above the surface in a raster fashion using piezoelectric transducers (PZTs). To sense tunneling current, a voltage must be applied between the probe tip and the test surface. This limits STM measurements to conducting surfaces. The resolution of the image is highly dependent upon the tip geometry (van Loenen et al. 1990).

Figure 17.4 is a schematic of an STM, consisting of a fine probe tip mounted on x, y, z PZT translators. The STM can operate in either constant-current or constant-height mode. The constant-current mode uses a feedback loop to vary the height of the probe during the scan and keep the tunneling current a constant value. The other mode measures the tunneling current as a function of position to keep the tip at a constant height. Because larger height variations can be

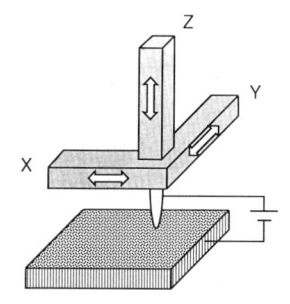

Figure 17.4. Schematic of STM.

measured, constant-current mode is most often used. The constant-height mode is faster, but it is very easy to crash the tip into the surface. Crashing not only ruins the tip, but can also harm the surface.

STMs were originally developed for atomic-resolution applications in a vacuum (Binnig and Rohrer 1982; Binnig et al. 1982). They are now being used in air and can scan areas of up to 200 μm. The scan range is determined by the range of the PZTs. Longer-range PZTs have less resolution. Just as optical microscopes have different magnifications, scan heads for different scan sizes are available. The larger scan sizes overlap the measurement range of high-magnification optical profilers and high-resolution stylus profilers. Because STMs require conducting surfaces to produce a tunneling current, surfaces such as glass cannot be measured. STMs also cannot measure metal surfaces that have formed a nonconducting oxide layer. Even with these limitations, STMs have been used by a number of people to evaluate optical surfaces (Dragoset et al. 1986; Dragoset and Vorburger 1987; Schneir et al. 1989). Figure 17.5a shows a shaded solid model plot of tracks in an optical disk coated with a conductive layer and measured using an STM.

17.3.2. Atomic Force Microscopes

Another major type of scanning probe microscope is the atomic force microscope (AFM) (Binnig et al. 1986). The AFM can profile any kind of surface. Two types of atomic forces can be used for AFMs, a repulsive force and an attractive force. In repulsive mode, the probe acts like a phonograph needle as it is moved essentially in contact with the surface. The tip is placed on the end of a cantilever, and the deflection of the cantilever is measured to determine the surface topography. A very small force is used to keep the tip against the surface. Deflection can be measured a number of ways. Alexander et al. (1989) developed a readout system that looks at the deflection of a laser beam reflected off a mirror mounted on the cantilever. Another readout technique, developed by Sarid et al. (1988), uses feedback into a diode laser from a reflection off the back of the cantilever. Figure 17.6 show schematics of both types of readout, and Fig. 17.5b shows a plot of a laser-ablated compact disk (CD) stamper measured using the diode laser readout. A comparison of stylus profiler and AFM measurements on optical surfaces can be found in Bennett et al. (1991).

The attractive force AFM is also known as the laser force microscope (LFM) (Martin et al. 1988). It will not damage a sample because it never contacts the surface. The attractive force occurs when the tip is 2–20 nm away. Because the attractive force is very small, the tip is oscillated at a high frequency, and the change in the amplitude of vibration is detected. The LFM is not capable of providing atomic resolution because the tip is scanned far from the surface; however, it still provides very good resolution.

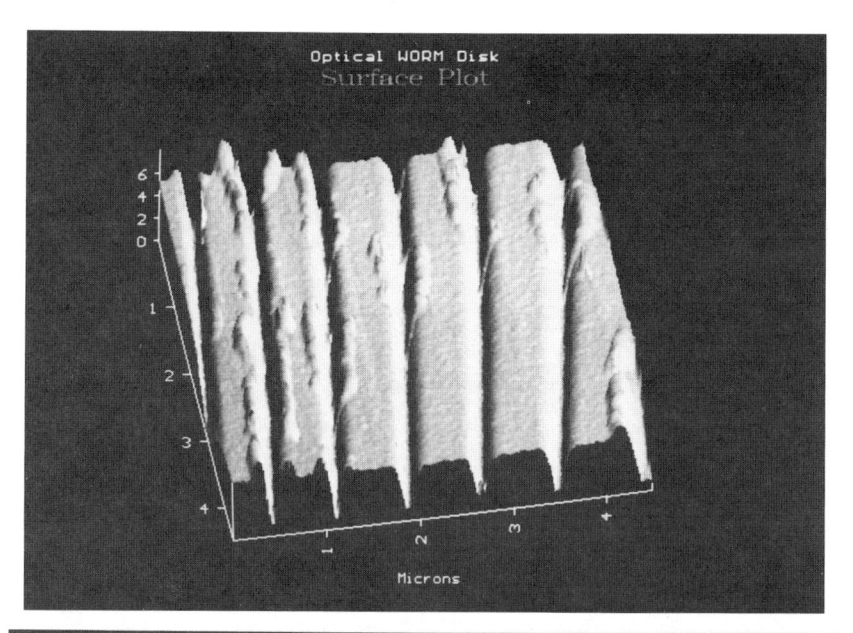

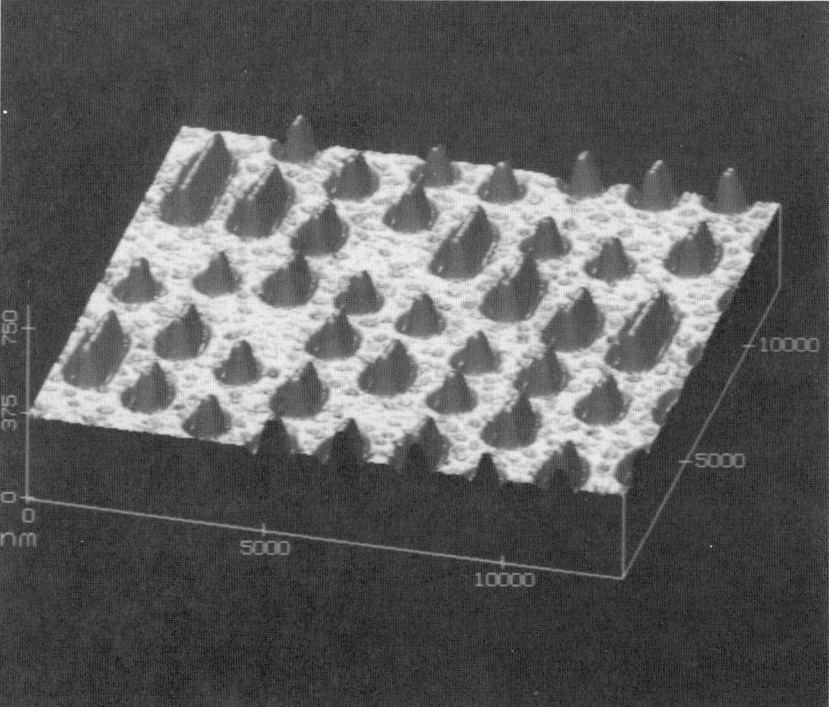

Figure 17.5. (*a*) Solid model plot of tracks in WORM (write once read many) optical disk. Sample was measured using an STM. (Data courtesy of Atomis Corporation. Photograph courtesy of WYKO Corporation.) (*b*) Solid model plot of laser oblated CD (compact disk) stamper measured using an AFM with diode-laser readout. (Courtesy of Digital Instruments.)

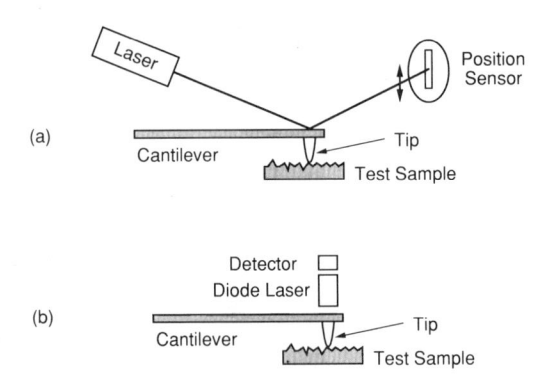

Figure 17.6. Schematic of AFM cantilever tip using (a) optical lever readout and (b) diode laser readout.

17.4. OPTICAL FOCUS SENSORS

Optical focus sensors are a noncontact means of measuring the roughness of finely machined surfaces. They measure by sensing focus at a single point on the surface and adjusting the height of the focusing lens until focus is achieved. The amount the lens is moved indicates the surface height at that data point. Either the optical head or the surface is scanned to generate a two- or three-dimensional height-profile map of the test surface. This measurement principle is very similar to the auto-focus mechanism in a compact disk player. A simple method of determining focus has been implemented in profiling instruments developed by Brodmann and Smilga (1987) and Breitmeier and Ahlers (1987). Illumination from a laser source is focused on the test surface, and the return is split at the optical axis into two parts using a prism. Each half of the beam is incident upon a split detector and the difference signal from each split detector is monitored. When the focusing lens is too high, the return beam focuses in front of the split detectors and causes a larger signal on the inner detectors; when the lens is too low, the larger signal is on the outer detectors. The sign of the difference signal will determine which side of focus the test surface is on, and is used to generate a focus error signal which moves the focusing lens to the correct position. When the focusing lens is in the correct position, both the inner and outer detectors have equal signal and the difference signal is zero. Two sets of split detectors are used to account for variations in tilt of the test surface. A sensor of this type is shown schematically in Fig. 17.7 (Brodmann and Smilga 1987). Because the focus must be adjusted to null the signal at every sampled surface point, this type of profiler can take a few minutes to generate a three-dimensional surface profiler.

The lateral resolution of optical focus sensors is limited by the size of the focus spot at the test surface, usually 1.0 to 1.5 μm in diameter. The measured

17.5. INTERFEROMETRIC OPTICAL PROFILERS

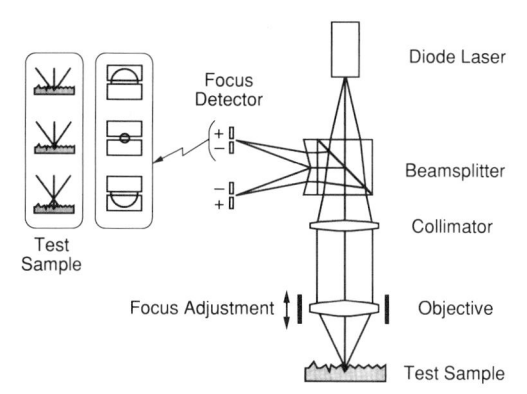

Figure 17.7. Schematics of a profilometer with an optical focus sensor.

surface height at a given sample point will be the average height of the surface over the spot size. This means that the smallest measurable features are about 2 μm. The measurement area will depend upon the sampling interval and number of data points. Another limitation of this type of profiler is that the light reflected from the test surface must get back into the sensor. If there are steep slopes on the surface, the light may get scattered out of the instrument and the signal will be lost, causing inaccurate results when only the difference signal is monitored. The height resolution of this type of profiler is related to the focusing range and the time to obtain each point. If a large height range is being measured, the movement of the focusing lens needs to be coarser to keep the time per data point the same. Otherwise, finer focusing over larger height ranges will slow the measurement time considerably.

Calibration of optical focus sensors is similar to that of stylus and SPM profilers. A traceable standard of approximately the same height or roughness as the test surface is measured and a scale factor is determined to apply to the surface profile data.

17.5. INTERFEROMETRIC OPTICAL PROFILERS

Interferometric optical profilers are noncontact instruments for the measurement of microscopic surface height profiles. They can be used to determine surface roughness or geometries of small features. All interferometric profilers are noncontact and therefore do not harm the surface under test. They are very sensitive and can measure heights with a precision of up to 0.1 Å rms (root mean square). Quantitative phase-measurement techniques are usually used to determine the phase difference between an object point and a reference. This phase is then converted into height information. All of these instruments measure heights relative to a reference surface rather than absolute distance. Because of their measurement precision and noncontact configuration, interferometric optical profilers are very good for on-line process control. All profiling instruments can be calibrated relative to an existing, traceable standard; however, most of these instruments are better than existing standards, so it is hard to determine their absolute accuracy.

The main limitation of interferometric optical profilers is due to the phase shift induced by reflection of the object beam from the test surface. Every material induces a phase shift depending on its complex index of refraction. As long as a surface is comprised of a single material and does not have a transparent coating on it, the phase shift does not cause a problem. When two dissimilar materials are side-by-side on the surface, they will have different phase shifts upon reflection and the height difference at the boundary will be incorrect. By knowing the optical constants of the different materials (Bennett 1964), it is possible to correct for this difference (Church and Lange 1986). A thin transparent layer such as an oxide layer on a metal is a harder problem. With measurements at two separate wavelengths, the thickness of a single-layer film and a profile of the substrate below it can be determined (Marcellin-Dibon 1989; Li and Talke 1990; Brophy et al. 1992). A composite material such as a ceramic needs to be coated with an opaque material (i.e., a metal) to get a good profile of the surface.

Measurement area and spatial frequency response are important considerations when determining which instrument and what magnification to use as well as whether to scan or image the surface. If the instrument you are using will not resolve the spatial wavelength or feature size you want to measure, it will not provide useful information. Also, measurement and calculation time are directly related to the number of measured data points. A larger number of data points do not necessarily add significant information. The user needs to determine the necessary measurement ranges and choose an instrument accordingly. These trade-offs have been discussed at length in the literature (Church et al. 1985; Bennett 1985; Vorburger 1987; Creath and Wyant 1988; Vorburger and Hembree, 1989; Creath 1990).

Early uses of interferometry for microprofiling were the Nomarski microscope (Nomarski 1955), multiple-beam interference (Tolansky 1960), and the FECO interferometer (Bennett 1976). However, these instruments only provided qualitative fringe data, which had to be interpolated for data between fringe centers. Beginning in 1980, the addition of phase-measurement interferometry (PMI) techniques enabled automated, quantitative measurement of phase data sampled uniformly over the interferogram. Eastman (1980) and Koliopoulos and Wyant (1980) were among the first to apply phase-measurement techniques to an interferometric microscope for measuring surface heights. Eastman based his system on a scanning Fizeau interferometer, and Koliopoulos and Wyant based theirs on a Mirau interferometer. The first instrument with sub-

17.5. INTERFEROMETRIC OPTICAL PROFILERS

angstrom precision for the measurement of optical surface finish was developed by Sommargren (1981). This instrument was based on a Wollaston prism and measured a circular path about a reference point. A system based on a differential microscope with a Wollaston prism for the measurement of step heights was developed by Makosch and his colleagues (Makosch and Solf 1981; Makosch and Drollinger 1984; Makosch 1988). This instrument directly measured surface slope, which was integrated to generate a line profile of the surface. A similar instrument using a different phase measurement technique was developed by Eastman and Zavislan (Eastman and Zavislan 1983; Zavislan and Eastman 1985). An optical profiler based on the work of Koliopoulos and Wyant (1980) utilizing Mirau, Michelson, and Linnik interferometers for the measurement of the surface roughness was developed by Wyant et al. (Wyant et al. 1984, 1986; Bhushan et al. 1985; Wyant and Prettyjohns 1986). This instrument directly measured surface heights over either a single line or an area and provided quantitative data about surface properties. A different type of surface height measurement profiler was developed independently by Huang (1983) and Downs et al. (1985). These profilers utilize a common-path interferometer where both beams are incident upon the test surface. One beam is defocused to act as a reference beam. The actual optical and phase measurement implementations differ between the two instruments. A similar instrument was later developed by Pantzer et al. (Pantzer et al. 1986; Pantzer 1987). Recent developments have included a phase-measuring Fizeau or Mirau interferometric microscope developed by Biegen and Smythe (1988), which is similar to the Wyant et al. instrument. Another recent profiler using an interferometric microscope that can measure height over a large height range using broadband illumination has been developed by Strand and co-workers (Strand 1985; Strand and Katzir 1987). It utilizes information about the fringe visibility to determine fringe order and height. An extended height range has also been produced by scanning an interferometric objective in height (Davidson et al. 1987, 1988; Dockrey and Hendricks 1989; Lee and Strand 1990). This instrument pieces together a contour map using data points that have a high degree of coherence for each height value. The height range can also be extended using two or more discrete wavelengths (Creath 1987), and the lateral range can be extended by combining multiple subaperture measurements (Cochran and Creath 1988).

Optical interferometric profiling instruments can be split into three different categories. The first category includes direct surface height measurement by imaging the surface. These instruments include those developed by Wyant et al. and Biegen and Smythe which are based on traditional Michelson, Mirau, Linnik, and Fizeau interferometers. Another category of instruments also measures surface height directly but uses concentric-beam interferometers where one of two collinear beams is defocused to provide a reference beam. These instruments all scan rather than image the object. The instruments developed by Huang, Downs et al., and Pantzer et al. fit into this category. The third category of instruments involves the use of polarization microscopes with Wollaston prisms. These scanning instruments include the Sommargren, Makosch and Solf, and Eastman and Zavislan developments. Except for the Sommargren instrument, these interferometers measure slope, which is integrated to obtain surface height. The Sommargren instrument measures height relative to a single surface point in a circular profile. The instruments in each of these categories will be discussed in more detail.

17.5.1. Traditional Interferometers

Traditional interferometric optical profilers are based on the Michelson, Mirau, Linnik, and Fizeau interferometers. Except for the work of Eastman (1980), these instruments all image the surface to provide height data over an area of the surface. They are based on standard microscopes where the objective has been replaced with an interferometric objective. Quantitative data are obtained by utilizing phase-shifting techniques that move the reference surface relative to the test surface using a piezo-electric transducer. They record a number of frames of data to calculate surface heights at each detector point. The measurement repeatability is found by subtracting two consecutive measurements. It is generally less than a few angstroms rms and can be as small as 0.1 Å rms. These instruments can either use narrow-bandwidth illumination such as a white-light source with a filter (Wyant et al. 1984, 1986) or laser illumination (Biegen and Smythe 1988). Laser illumination can be provided as either a point or an extended source. A point source is useful for the Fizeau interferometer and an extended source is needed for the Mirau interferometer. If laser illumination is used, polarization optics can be used to keep light from feeding back into the source. Schematics of commercially available implementations of two of these instruments are shown in Fig. 17.8. Figure 17.8a shows the TOPO profiler manufactured by WYKO Corporation (see reference). This instrument is based on the Wyant et al. instrument and utilizes Michelson, Mirau, and Linnik interferometric microscope objectives. Figure 17.8b shows the Maxim profiler manufactured by Zygo Corporation (see reference). This instrument is based on the Biegen and Smythe instrument and utilizes either Fizeau or Mirau interferometric objectives. Figure 17.9 shows an isometric plot of a surface with craters produced by a high-power laser. These data were taken with a 10X Mirau objective having a 4-μm lateral resolution.

The heart of these instruments is the interferometric microscope objective. Schematics of Michelson, Mirau, Linnik, and Fizeau interferometric objectives are shown in Fig. 17.10. Because these interferometers have separate reference surfaces, they are sensitive to vibration and need to be placed on an isolation table in a quiet environment. Environmental effects can significantly reduce the measurement precision. Each of the interferometric objectives has its own specific uses. The measurement conditions and magnification needed will deter-

17.5. INTERFEROMETRIC OPTICAL PROFILERS

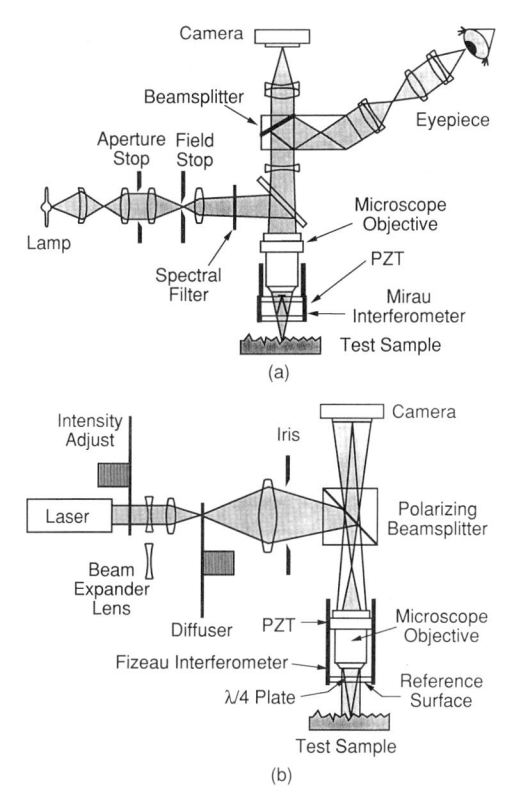

Figure 17.8. Schematics of (*a*) WYKO TOPO interferometric optical profiler and (*b*) Zygo Maxim interferometric optical profiler.

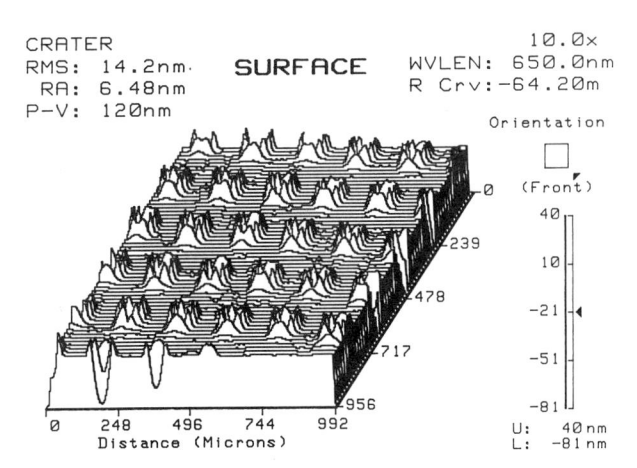

Figure 17.9. Isometric plot of surface with craters produced by a high-power laser. (Courtesy of WYKO Corporation.)

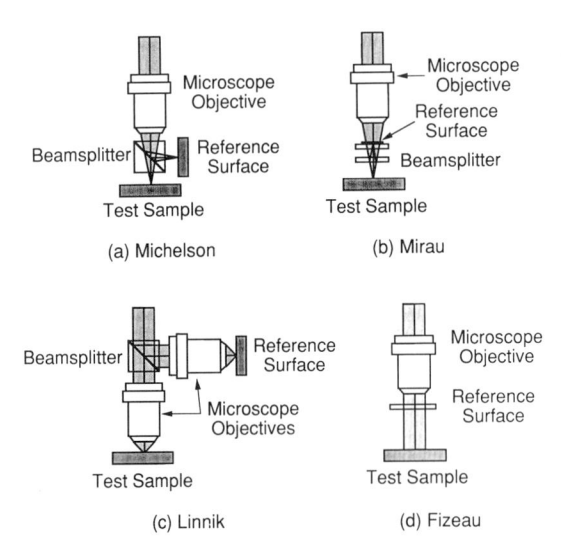

Figure 17.10. (a) Michelson, (b) Mirau, (c) Linnik, and (d) Fizeau interferometric microscope objectives.

mine which is the best objective for a particular measurement. Except for the Fizeau, all of these objectives can be adjusted to obtain white-light fringes. It is important to resolve the features on the surface you wish to measure. And no matter which objective is used, the test surface must be within the depth of field of the objective and within the coherence length of the source. This will provide high-contrast fringes and the best possible measurement. If the interferometer is adjusted so that white-light fringes are obtained when the object is at the best focal position, it makes it very easy to find the best focus and the highest contrast fringes at the same time. Because of these constraints, there are a lot of trade-offs that must be weighed in choosing an objective.

Michelson interferometers are comprised of a microscope objective, a beamsplitter, and a separate reference surface. The microscope objective must have a long working distance to fit the beamsplitter between the objective and the surface. Because of this, Michelson interferometers are only used with low-magnification objectives having low numerical apertures and long working distances. Mirau interferometers contain two small plates between the objective and the test surface. One plate contains a small reflective spot that acts as the reference surface, and the other plate is coated on one side to act as a beamsplitter. These interferometers are used for middle magnification objectives where there is not enough space for a Michelson interferometer. At lower magnifications, the reference spot obscures more of the aperture, so Miraus are not useful at magnifications of less than about 10X. For high magnifications, there is usually not enough working distance to place either a Mirau or a Michelson.

17.5. INTERFEROMETRIC OPTICAL PROFILERS

The Linnik is the most useful type of interferometer for high-magnification objectives having high numerical apertures. It is comprised of a beamsplitter, two matched microscope objectives, and a reference mirror. The entire reference arm provides path-length matching to obtain white-light fringes. The two objectives need to be matched with a beamsplitter to provide a wavefront with minimum aberration and maximum contrast fringes. The Fizeau interferometer is an unequal-path interferometer that requires a source with a long coherence length. The objective provides a collimated beam on the test surface while imaging the test surface (see Fig. 17.8d). Since interference fringes will exist over a large area of space when a long coherence length source is used, care must be taken to focus on the test surface. Extraneous fringes may get mixed up with the fringes from the surface under test and are not removable from the phase measurement because both sets of fringes will phase shift. (If a transparent surface is being measured, it is best to use another type of interferometer that requires a shorter coherence length source so that fringes are not obtained from the back surface of the object.) Fizeau objectives can be used with a large range of magnifications; however, there must be enough room for the reference surface between the objective and the test surface. With all of the interferometers except the Fizeau, there is a cone of light incident upon the test surface. Because of the obliquity of the rays at the test surface, a correction factor must be used to get accurate height information (Schulz 1954; Tolmon and Wood 1956; Gates 1956; Bruce and Thornton 1957; Ingelstam 1960; Biegen, 1989; Creath 1989; Schulz and Elssner 1991). Since these microscope systems are complicated to model, it is easiest to measure these corrections using a traceable step-height standard (Biegen 1989; Creath 1989; Schulz and Elssner 1991).

17.5.2. Concentric-Beam Interferometers

Instruments of this type have been developed by Huang (1983), Downs et al. (1985, 1989), and Pantzer et al. (1986, 1987). These interferometric optical profilers use separate areas on the test surface for the test and reference beams. The test beam is focused onto the surface. The reference beam is collinear and concentric to the test beam but defocused so that it averages over an area on the surface. They measure the height of a surface point relative to the average height in the region around the measured point. This technique will measure high-frequency roughness on the test surface while removing effects due to waviness. Low spatial frquency variations in the surface are not measured. Removing low-frequency variations may or may not be good depending on what is being measured. If a sample such as a plastic film is measured, it is advantageous to remove the low-frequency structure. If the sample is an X-ray mirror, the low-frequency data, which yields information about low angle scatter, will be lost.

The two beams in these interferometers can be generated by a number of

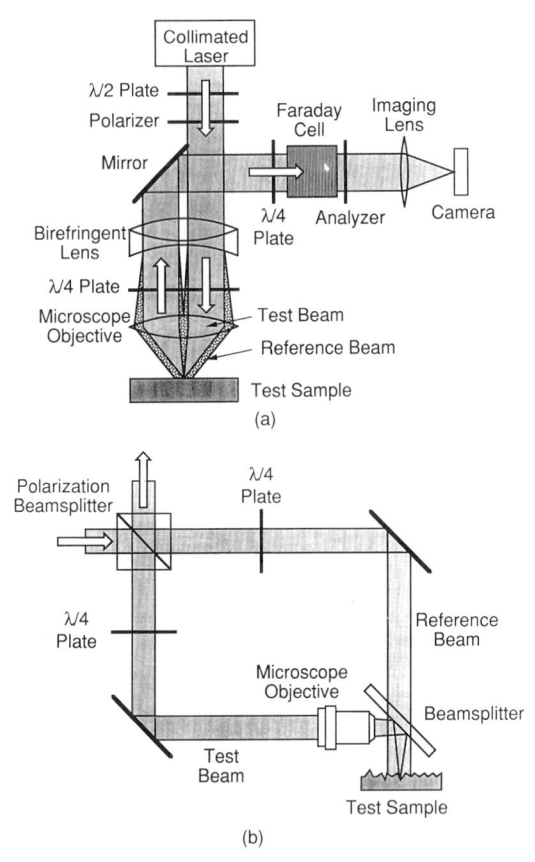

Figure 17.11. Schematics of (a) Downs et al. interferometric optical profiler and (b) Pantzer et al. interferometric optical profiler.

means. Figure 17.11 shows schematics of the systems developed by Downs et al. and Pantzer et al. Downs et al. (Fig. 17.11a) use a birefringent lens to produce the two beams. Pantzer et al. (Fig. 17.11b) use a Mach–Zehnder-like configuration to produce the two beams. Both instruments use polarization techniques, scan the test sample, and implement phase-measurement techniques to obtain a surface profile. These interferometers have the advantage that effects due to vibration and air turbulence are reduced because both beams travel the same path. They can also be used with different magnifications to provide a wide measurement range.

17.5.3. Polarization Interferometers

These common-path interferometers use a Wollaston prism and polarization techniques to split the light into two focused beams on the test surface. The

17.5. INTERFEROMETRIC OPTICAL PROFILERS

height difference between the two points on the surface is measured using phase-measurement techniques. Instruments using these techniques have been developed by Sommargren (1981), Makosch and Solf (1981), and Eastman and Zavislan (1983). Sommargren's instrument was the first of these. It provided measurements with less than 0.1 Å rms precision. The objective of Sommargren's instrument is depicted in Fig. 17.12. It consists of a Wollaston prism and a microscope objective. The instrument looks up at the sample from below. Two beams are focused onto the test surface. The test surface is rotated about one of these points, which acts as the reference. Height between this reference point and the scanning point is measured using phase-measurement techniques. As long as the rotation axis is lined up with one of the beams, this interferometer provides a highly accurate trace of the height profile. The output is a circular trace around a path with approximately a 160-μm radius to produce a 1-mm scan using a 1.8-μm spot size. This system is commercially available from Zygo Corporation (see reference). A scan of a very smooth optical surface with an rms roughness 0.86 Å is shown in Fig. 17.13. The height scale is ± 5 Å, and the peak-to-valley height variation is 5.98 Å.

The Makosch and Solf and Eastman and Zavislan systems are very similar, except that the object or objective is scanned in a linear path rather than a circular path. These instruments measure slope rather than height. Even though these instruments are similar to the Sommargren instrument (and to one another), their optical implementation and measurement techniques are different. Different magnifications can be used with different Wollaston prisms to provide a range of slopes to which the instrument is sensitive. Higher resolution is obtainable with a higher magnification, and greater slopes are measurable with a Wollaston prism which provides a smaller separation between beams. Measured slope can be directly related to power spectrum, or it can be integrated to get height information. Because of the integration, errors in mechanical motion or electronics can add up to reduce the measurement precision. The larger the number of data points, the more noise is present. However, this type of interferometer enables long scans of up to 100 mm to be made. Even though there

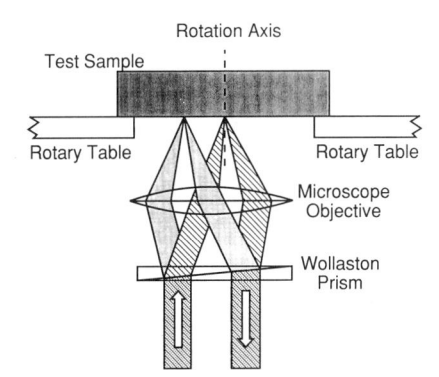

Figure 17.12. Schematic of interferometric microscope objective from Sommargren system.

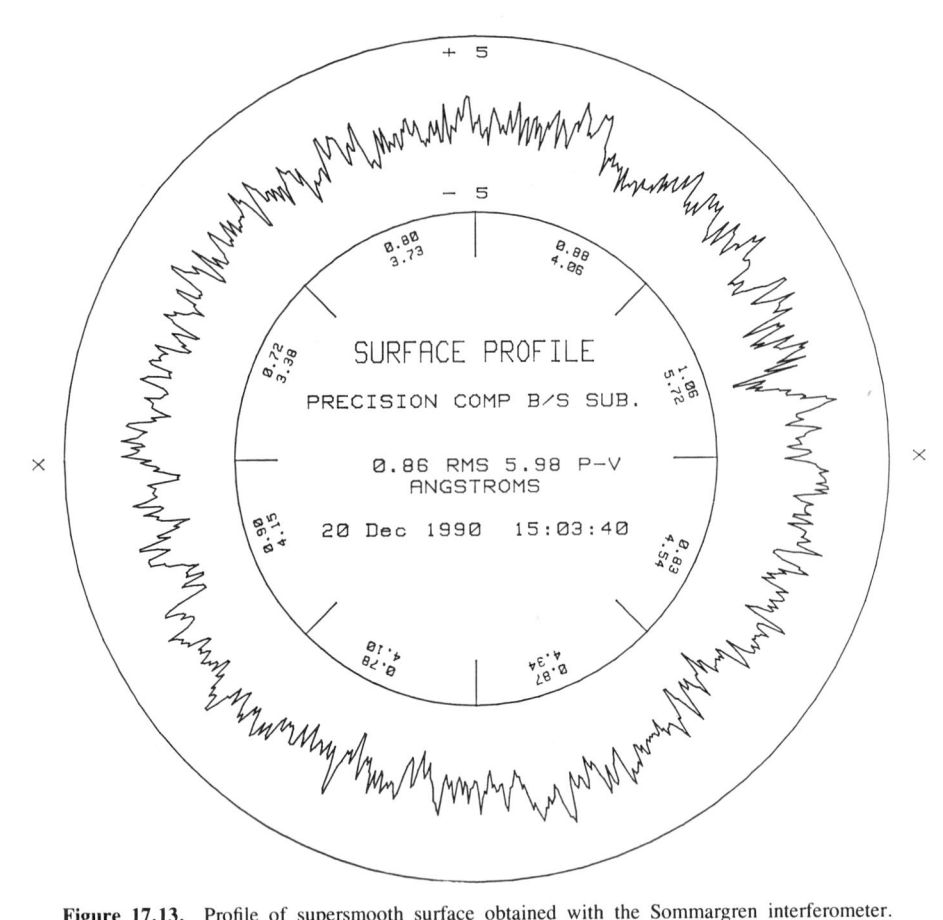

Figure 17.13. Profile of supersmooth surface obtained with the Sommargren interferometer. (Courtesy of Zygo Corporation.)

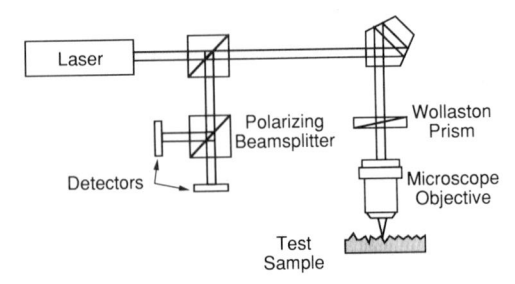

Figure 17.14. Schematic of the Eastman and Zavislan interferometer as produced by Chapman Instruments.

is more noise than in an imaging system, it has the advantage of long scan lengths. The Eastman and Zavislan instrument has been further developed by Bristow and his co-workers (Bristow and Arackellian 1987; Bristow et al. 1989) and is commercially available from Chapman Instruments (see reference). A schematic of it is shown in Fig. 17.14.

The polarization technique can be taken one step further to provide curvature data rather than slope data. This technique has recently been developed utilizing a calcite prism and two quadrant detectors to measure curvature (Glenn 1990).

17.6. COMPARISON OF PROFILERS

Each different profiler type has its own advantages, disadvantages, and preferred applications. To choose a specific instrument, the user needs to consider such things as lateral and vertical measurement range, resolution, measurement time, cost, precision, accuracy, means of calibration, and compatibility with other equipment. Stylus profilers are general-utility instruments. Rougher surfaces are measurable with the focus sensors. High-spatial-resolution applications are best accommodated by SPMs. Very smooth optical surfaces will require interferometric optical profilers or the more expensive stylus profilers. Fast measurement times are most likely obtained using imaging interferometric optical profilers. Finally, it should be note that there is a definite correlation between cost, high precision, and accuracy.

Acknowledgments. K. Creath wishes to acknowledge the support of WYKO Corporation during the preparation of this manuscript.

REFERENCES

Alexander, S., L. Hellemans, O. Marti, J. Schneir, V. Elings, P. K. Hansma, M. Longmire, and J. Gurley, "An Atomic Resolution Resolution Atomic Force Microscope Implemented Using an Optical Lever," *J. Appl. Phys.*, **65**(1), 164–167 (1989).

Al-Jumaily, G. A., S. R. Wilson, K. C. Jungling, J. R. McNeil, and J. M. Bennett, "Frequency response characteristics of a mechanical stylus profilometer," *Opt. Eng.* **26**, 953–958 (1987).

Atomis, Inc., 2946 San Pablo Av., Berkeley, CA 94702.

Bennett, J. M., "Precise Method for Measuring the Absolute Phase Change Upon Reflection," *J. Opt. Soc. Am*, **54**(5), 612–624 (1964).

Bennett, J. M., "Measurement of the rms Roughness, Autocovariance Function and Other Statistical Properties of Optical Surfaces Using a FECO Scanning Interferometer," *Appl. Opt.*, **15**(11), 2705–2721 (1976).

Bennett, J. M., "Comparison of Techniques for Measuring the Roughness of Optical Surfaces," *Opt. Eng.*, **24**(3), 380–387 (1985a).

Bennett, J. M., "Comparison of Instruments for Measuring Step Heights and Surface Profiles," *Appl. Opt.*, **24**, 3766–3772 (1985b).

Bennett, J. M. and J. H. Dancy, "Stylus Profiling Instrument for Measuring Statistical Properties of Smooth Optical Surfaces," *Appl. Opt.*, **20**(10), 1785–1802 (1981).

Bennett, J. M., V. Elings and K. Kjoller, "Precision metrology for studying optical surfaces," *Opt. Photonics News*, **2**(5), 14–18 (1991).

Bennett, J. M. and L. Mattsson, *Introduction to Surface Roughness and Scattering*, Optical Society of America, Washington, D.C., 1989.

Bhushan, B., J. C. Wyant, and C. L. Koliopoulos, "Measurement of Surface Topography of Magnetic Tapes by Mirau Interferometry," *Appl. Opt.*, **24**(10), 1489–1497 (1985).

Biegen, J. F., "Calibration Requirements for Mirau and Linnik Microscope Interferometers," *Appl. Opt*, **28**(11), 1972–1974 (1989).

Biegen, J. F. and R. A. Smythe, "High Resolution Phase Measuring Laser Interferometric Microscope for Engineering Surface Metrology," in *Proc. Vol. II, 4th International Conference on Metrology and Properties of Engineering Surfaces, April 13–15, 1988*, National Bureau of Standards, Washington, D.C., 1988 and in *Proc. SPIE*, **1009**, 35–44 (1988).

Binnig, G. and H. Rohrer, "Scanning Tunneling Microscopy," *Helv. Phys. Acta*, **55**, 726–735 (1982).

Binnig, G. and H. Rohrer, "The Scanning Tunneling Microscopy," *Sci. Amer.* (August, 1985).

Binnig, G. and H. Rohrer, Ch. Gerber, and E. Weibel, "Surface Studies by Scanning Tunneling Microscopy," *Phys. Rev. Lett.*, **49**, 57–61 (1982).

Binnig, G., C. F. Quate, and Ch. Gerber, "Atomic Force Microscope," *Phys. Rev. Lett.*, **56**, 930–933 (1986).

Breitmeier, U. and R.-J. Ahlers, "Dynamically focusing electro-optical sensor-system for microprofilometry," *Proc. SPIE*, **802**, 170–173 (1987).

Bristow, T. C. and K. Arackellian, "Surface Roughness Measurements Using a Nomarski Type Scanning Instrument," *Proc. SPIE*, **749**, 114–118 (1987).

Bristow, T. C., G. Wagner, J. R. Bietry, and R. A. Auriemma, "Surface Profile Measurements of Curved Parts," *Proc. SPIE*, **1164**, 134–141 (1989).

Brodmann R. and W. Smilga, "Evaluation of a Commercial Microtopography Sensor," *Proc. SPIE* **802**, 165–169 (1987).

Brophy, C., D. Cohen, and W. Hahn, "Simultaneous Profiles of Single-Layer Film Thickness and Surface Height Variation from Phase-Shifting Interferometry," submitted to *Applied Optics* (1992).

Bruce, C. F. and B. S. Thornton, "Obliquity Effects in Interference Microscopes," *J. Sci. Instrum*, **34**(5), 203–204 (1957).

Chapman Instruments, 50 Saginaw Dr., Rochester, NY 14623.

Church, E. L. and S. R. Lange, "Structure Effects in Optical Surface Metrology," *Proc. SPIE*, **680**, 124–140 (1986).

REFERENCES

Church, E. L., T. V. Vorburger, and J. C. Wyant, "Direct Comparison of Mechanical and Optical Measurements of the Finish of Precision Machined Optical Surfaces," *Opt. Eng.*, **24**(3), 388–395 (1985).

Cochran, E. R. and K. Creath, "Combining Multiple-Subaperture and Two-Wavelength Techniques to Extend the Measurement Limits of an Optical Surface Profiler," *Appl. Opt.*, **27**(10), 1960–1966 (1988).

Creath, K., "Step Height Measurement Using Two-Wavelength Phase-Shifting Interferometry," *Appl. Opt.*, **26**(14), 2810–2816 (1987).

Creath, K., "Calibration of Numerical Aperture Effects in Interferometric Microscope Objectives," *Appl. Opt.*, **28**(15), 3333–3338 (1989).

Creath, K., "Spatial Resolution Limits of an Optical Profiler," *Proc. 15th Congress of the ICO*, 318–319 (1990) and *Proc. SPIE*, **1319** (1990).

Creath, K. and J. C. Wyant, "Interferometric Measurement of the Roughness of Machined Parts," *Proc. SPIE*, **954**, 246–251 (1988).

Davidson, M., K. Kaufman, I. Mazor, and F. Cohen, "An Application of Interference Microscopy to Integrated Circuit Inspection and Metrology," *Proc. SPIE*, **775**, 233–247 (1987).

Davidson, M., K. Kaufman, and I. Mazor, "First Results of a Product Utilizing Coherence Probe Imaging for Wafer Inspection," *Proc. SPIE*, **921**, 100–114 (1988).

Digital Instruments, 6780 Corona Dr., Santa Barbara, CA. 93117.

Dockrey, J. W. and D. Hendricks, "The Application of Coherence Probe Microscopy for Submicron Critical Dimension Linewidth Measurement," *Proc. SPIE*, **1087**, 120–137 (1989).

Downs, M. J., W. H. McGiven, and H. J. Ferguson, "Optical System for Measuring the Profiles of Supersmooth Surfaces," *Prec. Eng.*, **7**, 211–215 (1985).

Downs, M. J., N. M. Mason, and J. C. C. Nelson, "Measurement of the Profiles of Super Smooth Surfaces Using Optical Interferometry," *Proc. SPIE*, **1009**, 14–17 (1989).

Dragoset, R. A. and T. V. Vorburger, "Scanning Tunneling Microscopy (STM) of a Diamond-Turned Surface and a Grating Replica," *Proc. SPIE*, **749**, 54–58 (1987).

Dragoset, R. A., R. D. Young, H. P. Layer, S. R. Mielczarek, E. C. Teague, and R. J. Celotta, "Scanning Tunneling Microscope Applied to Optical Surfaces," *Opt. Lett.*, **11**, 560–562 (1986).

Eastman, J. M., "The Scanning Fizeau Interferometer: An Automated Instrument for Characterizing Optical Surfaces," *Opt. Eng*, **19**(6), 810–814 (1980).

Eastman, J. M. and J. M. Zavislan, "A New Optical Surface Microprofiling Instrument," *Proc. SPIE*, **429**, 56–64 (1983).

Gates, J. W., "Fringe Spacing in Interference Microscopes," *J. Sci. Instrum.*, **33**(12), 507–507 (1956).

Glenn, P., "Robust, Sub-Angstrom Level Mid-Spatial Frequency Profilometry," *Proc. SPIE*, **1333**, 175–182 (1990).

Hansma, P. K. and J. Tersoff, "Scanning tunneling microscopy," *J. Appl. Phys.*, **61**(2), R1–R23 (1987).

Huang, C. C., "Optical Heterodyne Profilometer," *Proc. SPIE*, **429**, 65-74 (1983), and *Opt. Eng.*, **23**(4), 365-370 (1984).

Ingelstam, E., "Problems Related to the Accurate Interpretation of Microinterferograms," in *Interferometry, National Physical Laboratory Symposium No. 11*, Her Majesty's Stationery Office, London, 1960, pp. 139-163.

Koliopoulos, C. L. and J. C. Wyant, "Profilometer for Diamond-Turned Optics Using a Phase-Shifting Interferometer" (abstract only), *J. Opt. Soc. Am.*, **70**(12), 1591-1591 (1980).

Lee, B. S. and T. C. Strand, "Profilometry with a Coherence Scanning Microscope," *Appl. Opt.*, **29**(26), 3784-3788 (1990).

Li, Y. F. and F. E. Talke, "Limitations and Corrections of Optical Profilometry in Surface Characterization of Carbon Coated Magnetic Recording Disks," *Trans. ASME, J. Tribology*, **112**(4), 670-677 (1990).

Makosch, G., "LASSI—A Scanning Differential ac Interferometer for Surface Profile and Roughness Measurement," *Proc. SPIE*, **1009**, 244-253 (1988).

Makosch, G. and B. Drollinger, "Surface Profile Measurement with a Scanning Differential ac Interferometer," *Appl. Opt.*, **23**(24), 4544-4553 (1984).

Makosch, G. and B. Solf, "Surface Profiling by Electro-optical Phase Measurements," *Proc. SPIE*, **316**, 42-53 (1981).

Marcellin-Dibon, E., "The Effects of Reflection from Diverse Materials on Phase," *Project Final Report*, Optical Science Center, University of Arizona, Tucson, AZ, 85721 (1989).

Martin, Y., C. C. Williams, and H. K. Wickramasinghe, "Tip Techniques for Microcharacterization of Materials," *Scanning Microscopy*, **2**(1), 3-8 (1988).

Nomarski, G., "Microinterférométre Différentiel à Ondes Polarisées," *J. Phys. Rad.*, **16**, 9S-13S (1955).

Pantzer, D., J. Politch, and L. Ek, "Heterodyne Profiling Instrument for the Angstrom Region," *Appl. Opt.*, **25**(22), 4168-4172 (1986).

Pantzer, D., "Step Response and Spatial Resolution of an Optical Heterodyne Profiling Instrument," *Appl. Opt.*, **26**(18), 3915-3918 (1987).

Pool, R., "The Children of the STM," *Science*, **247**, 634-636 (1990).

Reimer, L., *Scanning Electron Microscopy*, Vol. 45, Springer Series in Optical Sciences, Springer-Verlag, New York, 1985.

Rugar, D. and P. Hansma, "Atomic force microscopy," *Phys. Today*, **43**(10), 23-30 (1990).

Sarid, D., *Scanning Force Microscopy*, Oxford Univ. Press, New York, 1991.

Sarid, D., D. Iams, V. Weissenberger, and L. S. Bell, "Compact scanning-force microscope using a laser diode," *Opt. Lett.*, **13**(12), 1057-1059 (1988).

Schneir, J., J. A. Dagata, H. H. Harary, C. J. Evans, A. J. Melmed, H. B. Elswijk, and J. Sauvageau, "Scanning Tunneling Microscopy of Optical Surfaces," *Proc. SPIE*, **1164**, 112-120 (1989).

Schulz, G., "Über Interferenzen Gleicher Dicke and Längenmessung mit Lichtwellen," *Ann. Physik*, **14**(6), 177-187 (1954).

REFERENCES

Schulz, G. and K.-E. Elssner, "Errors in Phase-Measurement Interferometry with High Numerical Aperture," *Appl. Opt*, **30** (1991).

Sommargren, G. E., "Optical Heterodyne Profilometry," *Appl. Opt.*, **20**(4), 610–618 (1981).

Song, J. F. and T. V. Vorburger, "Stylus profiling at high resolution and low force," *Appl. Opt.*, **30**(1), 42–50 (1991).

Stedman, M., "Limits of Surface Measurement by Optical Probes," *Proc. SPIE*, **1009**, 62–67 (1989).

Stedman, M. and K. Lindsey, "Limits of Surface Measurement by Stylus Instruments," *Proc. SPIE*, **1009**, 56–61 (1989).

Strand, T. C. and Y. Katzir, "Extended Unambiguous Range Interferometry," *Appl. Opt.*, **26**(19), 4274–4281 (1987).

Tolansky, S., *Surface Microtopography*, Wiley-Interscience, New York, 1950.

Tolmon, F. R. and J. G. Wood, "Fringe Spacing in Interference Microscopes," *J. Sci. Instrum.*, **33**(6), 236–238 (1956).

van Loenen, E. J., D. Dijkkamp, A. J. Hoeven, J. M. Lenssinck, and J. Dieleman, "Evidence for Tip Imaging in Scanning Tunneling Microscopy," *Appl. Phys. Lett.*, **56**(18), 1755–1757 (1990).

Vorburger, T. V., "Measurements of Roughness of Very Smooth Surfaces," *Annals CIRP*, **36**(2), 503–509 (1987).

Vorburger, T. V. and G. G. Hembree, "Characterization of Surface Topography," *Navy Metrology, R & D Program Conf. Report*, Dept. of the Navy Metrology Eng. Center, Corna, CA, April 1989.

Vorburger, T. V. and J. Raja, *Surface Finish and Metrology*, NISTIR 89-4088, U.S. Dept. of Commerce, National Institute of Standards and Technology, Gaithersburg, MD, June 1990.

Wickramasinghe, H. K., "Scanned-Probe Microscopes," *Sci. Amer.*, 98–105 (October, 1989).

Wilson, T. and C. J. R. Sheppard, *Theory and Practice of Scanning Optical Microscopy*, Academic Press, London, 1984.

Wyant, J. C., C. L. Koliopoulos, B. Bhushan, and O. E. George, "An Optical Profilometer for Surface Characterization of Magnetic Media," *ASLE Trans.*, **27**(2), 101–113 (1984).

Wyant, J. C., C. L. Koliopoulos, B. Bhushan, and D. Basila, "Development of a Three-Dimensional Noncontact Digital Optical Profiler," *Trans. ASME, J. Tribology*, **108**(1), 1–8 (1986).

Wyant, J. C. and K. N. Prettyjohns, "Three-Dimensional Surface Metrology Using a Computer Controlled Non-contact Instrument," *Proc. SPIE*, **661**, 292–295 (1986).

WYKO Corporation, 2650 E. Elvira Rd., Tucson, AZ 85706.

Zygo Corporation, Laurel Brook Rd., P.O. Box 448, Middlefield, CT 06455-0448.

Zavislan, J. M. and J. M. Eastman, "Microprofiling of Precision Surfaces," *Proc. SPIE*, **525**, 169–173 (1985).

ADDITIONAL REFERENCES

Almarzouk, K., "Three-Beam Interferometric Profilometer," *Appl. Opt.*, **22**(12), 1893–1897 (1983).

Barut, B. C. and P. Langenbeck, "Optical Profiling Using an Interference Microscope," *Proc. SPIE*, **954**, 101–108 (1988).

Bates, R., "Optimizing Parameters for Profiling Measurements," *Microelectron. Manuf. Testing*, **13**(14), 7–9 (1990).

Bender, J. and G. Flint, "In-process Measurement of Fast Aspherics," *Proc. SPIE*, **171**, 70–76 (1979).

Bristow, T. C. and D. Lindquist, "Surface Measurements with a Non-contact Nomarski-Profiling Instrument," *Proc. SPIE*, **816**, 106 (1987).

Brodmann, R., O. Gerstorfer, and G. Thum, "Optical Roughness Measuring Instrument for Fine-Machined Surfaces," *Opt. Eng.*, **24**(6), 408–413 (1985).

Budis, I. Y., V. Y. Orlov, and V. P. Spiridonov, "Contact Aspherometer," *Sov. J. Opt. Technol*, **47**, 106 (1980).

Cencic, B., M. Barut, and P. Langenbeck, "Optical Profiling Using an Interference Microscope," *Proc. SPIE*, **954**, 100 (1988).

Cochran, E. R. and J. C. Wyant, "Longscan Surface Profile Measurements Using a Phase-Modulated Mirau Interferometer," *Proc. SPIE*, **680**, 112–117 (1986).

Diaz-Uribe, R., A. Cornejo-Rodriguez, J. Pedraza-Contreras, O. Cardona-Nunez, and A. Cordero-Davila, "Profile Measurement of a Conic Surface, Using a He–Ne Laser and a Nodal Bench," *Appl. Opt.*, **24**(16), 2612–2615 (1985).

Diaz-Uribe, R., R. Pastrana-Sánchez, and A. Cornejo-Rodrìguez, "Profile Measurement of Aspheric Surfaces by Laser Beam Reflection," *Proc. SPIE*, **813**, 355 (1987).

Dil, J. G., P. F. Greve, and W. Mesman, "Measurement of Steep Aspheric Surfaces," *Appl. Opt.*, **17**(4), 553–557 (1978).

Egdall, M. and R. S. Breidenthal, "Large Surface Measuring Machine," *Proc. SPIE*, **416**, 62 (1983).

Ennos, A. E. and M. S. Virdee, "Precision Measurement of Surface Form by Laser Autocollimation," *Proc. SPIE*, **398**, 252 (1983).

Francini, F., G. Molesini, F. Quercioli, and B. Tiribilli, "Scanning Aspherical Surfaces with the Focus-Wavelength Encoded Profilometer," *Proc. SPIE*, **645**, 2–11 (1986).

Garrat, J. D. and S. C. Bottomley, "Technology Transfer in the Development of a Nanotopographic Instrument," *Nanotechnology IOP Publishing Ltd.*, **1**, 38–43 (1990).

Gauler, J. L., "Comparison of Two Common Methods of Surface Topography Evaluation," *Opt. Eng.*, **21**(6), 991–997 (1982).

Gorecki, Ch., G. Tribillon, and J. Mignot, "Profilometre Optique en Lumiere Blanche," *J. Opt.*, **14**(1), 19–23 (1983).

Greenleaf, A. H., "Self-Calibrating Surface Measuring Machine," *Opt. Eng.*, **22**(2), 276–280 (1983).

ADDITIONAL REFERENCES

Halioua, M. and Hsin-Chu Liu, "Optical Three-Dimensional Sensing by Phase Measuring Profilometry," *Opt. Lasers Eng.*, **11,** 185-215 (1989).

Heynacher, E., "Production and Testing of Aspheric Surfaces, State of Art in the F.R. of Germany," *Proc. SPIE*, **381,** 39 (1983).

Indebetouw, G., "A Simple Optical Noncontact Profilometer," *Opt. Eng.*, **18,** 63 (1979).

Kohli, K. S., K. N. Chopra, and R. Hradaynath, "Quantitative Evaluation of Surface Shape of Medium Size Optical Paraboloidal Reflectors," *J. Opt. (Paris)*, **10,** 89 (1979).

Leushina, T. M., "Testing of Aspherical Surfaces on a Goniometer," *Sov. J. Opt. Technol.*, **45,** 544 (1978).

Levin, B. M., T. A. Volkova, and Y. A. Myasnikov, "OP-1 Optical Flatness Meter," *Sov. J. Opt. Technol.*, **47,** 209 (1980).

Lou, D. Y., A. Martùnez, and D. Stanton, "Surface Profile Measurement with a Dual-Beam Optical System," *Appl. Opt.*, **23**(5), 746-751 (1984).

Lytle, J. D. and A. L. Palmer, "Aspheric Profile Gauging Using a Bootstrap Data Interpretation Technique," *Appl. Opt.*, **18**(7), 1064-1066 (1979).

Marx, E. and T. V. Vorburger, "Light Scattered by Random Rough Surfaces and Roughness Determination," *Proc. SPIE*, **1165,** 72-86 (1989).

Molesini, G., F. Quercioli, B. Tiribili, and M. Trivi, "Testing Reflective Optical Surfaces with a Non-contacting Probe," *Proc. SPIE*, **954,** 399 (1988).

Omar, B. A., A. J. Holloway, and D. C. Emmony, "Differential Phase Quadrature Surface Profiling Interferometer," *Appl. Opt.*, **29**(31), 4715-4719 (1990).

Payne, J. M., J. M. Hollis, and J. W. Findlay, "New Method of Measuring the Shape of Precise Antenna Reflectors," *Rev. Sci. Instrum*, **47**(1), 50-55 (1976).

Peterson, R. W., G. M. Robinson, R. A. Carlsen, C. D. Englund, P. J. Moran, and W. M. Wirth, "Interferometric Measurements of the Surface Profile of Moving Samples," *Appl. Opt.*, **23**(10), 1464-1466 (1984).

Phillips, M. J. and D. J. Whitehouse, "Some Theoretical Aspects of the Measurement of Curved Surfaces," *J. Physics E: Sci. Inst.*, **10,** 164-169 (1977).

Powell, I., "Aspheric Surface Calibrator," *Appl. Opt.*, **20**(19), 3367-3377 (1981).

Quercioli, F., B. Tiribilli, and G. Molesini, "Optical Surface Profile Transducer," *Opt. Eng.*, **27**(2), 135-142 (1988).

Rodenstock, RM-400, RM-500, and RM-600, Technical data sheets, Rodenstock, Metrology Division, P.O. Box 140440 D-8000, München 5, W. Germany.

Sasaki, O. and H. Okazaki, "Sinusoidal Phase Modulating Interferometry for Surface Profile Measurement," *App. Opt.*, **25**(18), 3137-3140 (1986).

Saul, R. S. and T. L. Williams, "An Infra-Red Inhomogeneity Scanner," *Opt. Acta*, **25,** 1149 (1978).

Singh, H. S., G. K. Sharma, and G. P. Dimri, "Checking the Surface Contour of Higher Order Deep Aspherics Using Precision Steel Balls," *J. Opt. (India)*, **14,** 69 (1985).

Stevens, D. M. and R. K. Morton, "A Unique Solution to Aspheric Measurement and Analysis as Part of a Manufacturing Process," *Proc. SPIE*, **966,** 150 (1988).

Suemoto, Y. and Y. Takeishi, "Fiber Optic Heterodyne Displacement Detection in Wide Dynamic Range," *Opt. Commun.*, **68,** 67 (1988).

Takacs, P. Z., Feng S. Ch. K., E. L. Church, S. Quian, and W. Liu, "Long Trace Profile Measurements on Cylindrical Aspheres," *Proc. SPIE*, **966,** 354 (1988).

Talystep and Talysurf Technical data from Rank Taylor Hobson Inc., 411 East Jarvis Avenue, Des Plaines, Ill. 60018-5997. Ulrich Breitmeier Messtechnik GMBH, Technical data sheet UB-16, UBM instruments, Ottostrabe 2 D-7505 Ettlingen, Germany.

Tencor P-1 Long Scan Profiler Data Sheet 4/90 and 5/90, Tencor Instruments, 2400 Charleston Road, Mountain View, CA 94043.

Von Bieren, K., "Interferometry of Wave Fronts Reflected off Conical Surfaces," *Appl. Opt*, **22,** 2109 (1983).

Vorburger, T. V., D. E. Gilsinn, F. E. Scire, M. J. Mclay, C. H. W. Giauque, and E. C. Teague, "Optical Measurement of the Roughness of Sinusoidal Surfaces," *Wear*, **109,** 15–27 (1986).

Vorburger, T. V., *Surface Finish Metrology*," Short Course Notes SC-12, SPIE's OE LASA'90, January 1990.

Vustenko, V. I., G. A. Susskil, and V. G. Shkarban, "Photoelectric Aspherometer," *Sov. J. Opt. Technol.*, **52,** 153 (1985).

Williams, T. L., "A Scanning Gauge for Measuring the Form of Spherical and Aspherical Surfaces," *Opt. Acta*, **25,** 1155 (1978).

Yoshisumi, K., T. Murao, J. Masui, R. Imanaka, et. al., "Ultrahigh Accuracy 3-D Profilometer," *Appl. Opt.*, **26,** 1647 (1987).

18

Angle, Distance, Curvature, and Focal Length Measurements

J. Z. Malacara

18.1. INTRODUCTION

Geometrical measurements in the optical workshop are developed to measure lens and system parameters, not only looking for precision but also for speed and simplicity. These measurements should be made in international units and, in some cases, must be certified by a laboratory from some secondary standards. A review of this topic may be found in the article by Geiser (1965). Geometrical measurements in optical testing may be classified mainly in distance, angle, curvature, and focal length measurements.

In the production run it is preferred not to make measurements but comparisons or null measurements. This makes gauges and templates more appropriate for the shop than expensive and precise equipment.

18.2. ANGLE MEASUREMENTS

Angle measurements in the optical shop require different levels of accuracy. While for glass cutting the accuracy can be several degrees, for standard test plates less than a second of arc may be required. For every case different measurement methods are developed. For angle measurements the most common unit remains the degree, which divides the circle in 360 parts. A radian is an angle such that arc length is equal to the radius of the circle. A gradient, which is scarcely used, is an attempt to use decimal fractions of a quadrant.

18.2.1. Divided Circles and Goniometers

Protractors are used as main-angle measuring devices. Although a typical protractor has an accuracy of about 30 min, modern electronic protractors can attain

---

Optical Shop Testing, Second Edition, Edited by Daniel Malacara.
ISBN 0-471-52232-5 © 1992, John Wiley & Sons, Inc.

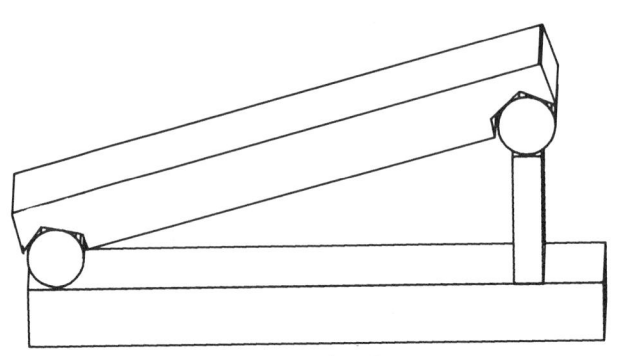

Figure 18.1. Sine plate.

up to 0.5 min of precision. During the manufacturing process, several methods are used to cut a block of glass at a given angle. An angle vise and a sine plate (Fig. 18.1) are mechanisms to both support the glass and to measure the angle. For these supporting heads the table is raised at some known distance for a known hypotenuse length. For glass machine tools a rotary table is frequently used. An accuracy of about 30 min is attainable for these instruments. Horne (1972) describes a serrated table for angle measurements up to 0.1 s.

For a semifinished glass wedge, the angle can be compared with an angle block set. Angle blocks are templates commercially available to a degree and accuracy of ±20 s, to obtain virtually any angle between 1° and 90°. By reversing the blocks, negative angles are obtained as described by Horne (1974). Angle blocks are made from hardened steel with a flat precision-polished face. This permits angle blocks in conjunction with a goniometer to be used as an angle standard. Angle blocks are sometimes certified up to ±0.1 s of arc.

Similar to angle blocks, polygons are available from 3 up to 12 faces, although Horne (1974) reports a 72-face polygon. Polygons are mostly used with an autocollimator for divided circles and goniometer calibrations.

Goniometers (Fig. 18.2) are precision spectrometers with a fixed collimator and a moving telescope. Microscopes or magnifiers are located at opposite ends in the divided circle; the mean value of the readings are calculated to compensate for centering errors. In goniometers accuracy can typically yield up to 20 s arc, although Gleiser (1965) reviews a system for checking such a circle to better than one arc second. For some precise angle measurements theodolites could be used, reaching an accuracy up to a second arc. Goniometers are used to check prisms angles; in this case the telescope with illuminated reticle is used as an autocollimator. The glass surface must be semipolished and wet; for a precision work it must be fully polished and flat.

18.2.2. Bevel Gauge

The bevel gauge consists of two straight bars hinged at their edges by a pivot, as shown in Fig. 18.3. This device may be used to generate a master prism

18.2. ANGLE MEASUREMENTS

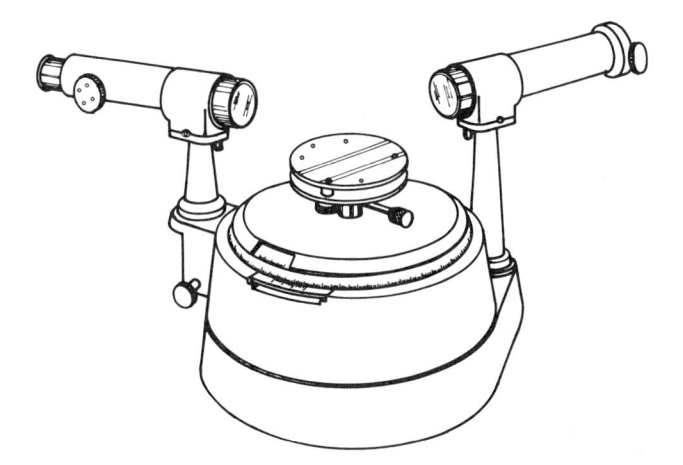

Figure 18.2. Goniometer.

whose angles are of 45° to about 20 arc second (Deve 1945). To accomplish this, first, the bevel gauge is approximately set to 45°. The second step is to find two right-angle prisms with the same size, for which an angle on each of them is close to 45°, but almost equal, as far as the bevel gauge can measure. During the measurements the gauge is held between the eye and a bright light in order to see very small spaces.

Let us now assume that the two angles that were found are equal to 46°. These two prisms are cemented with the two equal angles together, as shown in Fig. 18.3. The cemented surface is the bisector of a 92° angle. The next step is to regrind the large face until it becomes flat and the angles at the ends appear equal as measured with the bevel gauge. Under these conditions, when the two

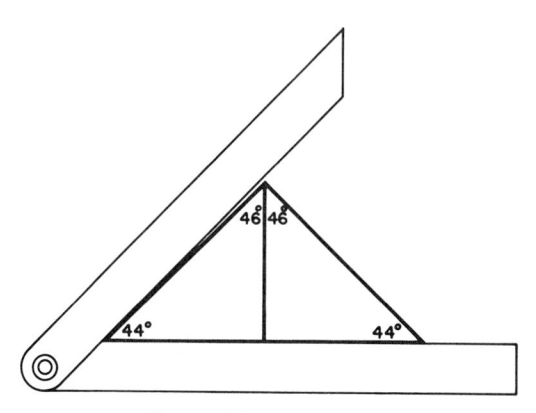

Figure 18.3. Bevel gauge.

718 ANGLE, DISTANCE, CURVATURE, AND FOCAL LENGTH MEASUREMENTS

prisms are separated, the right angle should be exactly 90° to the accuracy available with the bevel gauge. The final step is to regrind the large face of each prism until the end angles are equal. Then, these angles will be exactly 45°.

If the measured prism has a 50-mm hypotenuse, a space of 5 μm at one end represents an angle of 0.0001 rad or 20 arc seconds.

18.2.3. Autocollimator

An autocollimator is essentially a telescope focused to infinity. An illuminated reticle located at the focal plane is sent to infinity by the objective. A reflecting surface perpendicular to the beam going out images the reticle on itself. When the reflecting surface fails to be perpendicular to the beam, the reticle image is displaced at the focal plane. The amount of displacement d is

$$d = 2\alpha f \tag{18.1}$$

with α being the tilt angle for the mirror and f the focal length for the telescope (Fig. 18.4). A low-sensitivity autocollimator manufactured by Adam Hilger Ltd. is called the "Angle Dekkor."

Objective lenses are usually a corrected doublet, although sometimes a negative lens is included to form a telephoto lens to increase effective focal length, maintaining compactness. The collimating lens adjustment is critical for the final accuracy. Talbot interferometry can be used for a precise adjustment (Konthiyal and Sirohi 1987).

For observation at the focal plane, several means have been developed. Figure 18.5 (Noble 1978) illustrates some illuminated eyepieces in which the reticle is calibrated to measure the displacement. Gauss and Abbe illuminators

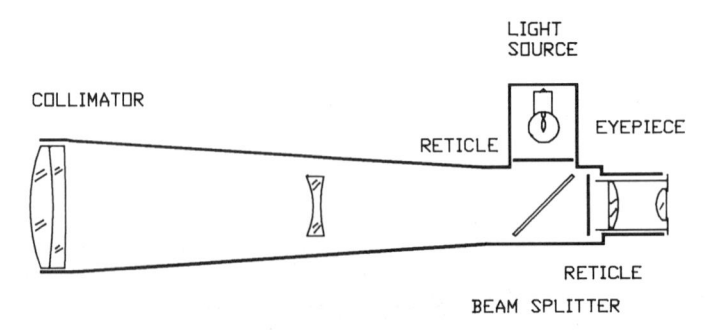

Figure 18.4. An autocollimator.

18.2. ANGLE MEASUREMENTS

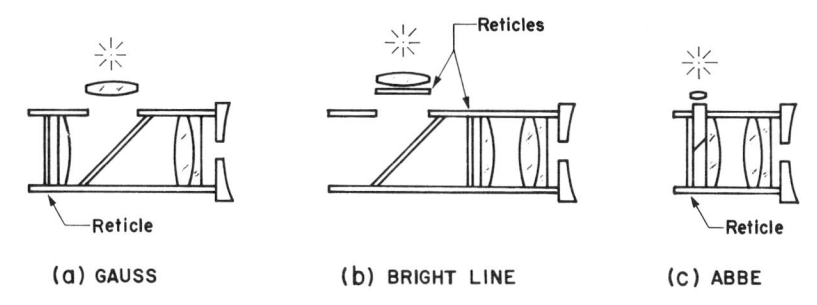

(a) GAUSS   **(b) BRIGHT LINE**   **(c) ABBE**

Figure 18.5. Illuminated eyepieces for autocollimators and traveling microscopes.

give a dark reticle in a bright field, while bright field (Carnell and Welford 1971) is more appropriate for low-reflectance surfaces. Rank (1946) modified a Gauss eyepiece to give dark field. In other systems a drum micrometer displaces a reticle to be positioned at the image plane for the first reticle. Some autocollimators have a microscope to observe the returning image. Such a system is called a microptic autocollimator.

Autocollimators are used for angle measurements in prisms and polygons as well as in other applications such as parallelism measurements in optical flats, divided circles manufacturing (Horne 1974), and, by integrating values for a scan in position, flatness for machine tool and optical beds can also be measured (Young 1967).

Some variants to the basic autocollimator include dual-axis reticle or micrometer for measurements for both axes. A moving slit can be included in the focal plane and an electronic synchronous detection system serves the purpose of indicating a null position for the return beam. Also, a position-measuring detector at the focal plane and a display indicate the angular deviation. Particularities about electronic autocollimators are described by Thurston (1986). Some new computerized versions include software for on-line data reducing.

Direct-reading autocollimators can measure over a field of about one degree, and this field is reduced with the distance to be measured. Precision in an autocollimator is limited by the method for measuring the image's centroid. In a diffraction-limited visual system, diffraction image sets the limit for the precision. Usually the effective f number is small and the image is odd-shaped, like in a prism. For a precision electronic measuring system centroid measurement is limited by the electronic detector, independent from the diffraction image itself, and can go beyond the diffraction limit. In some photoelectric systems precision is improved up to more than 10 times.

Reflecting surface in autocollimation measurements must be kept close to the objective, otherwise the reflected beam fails to enter the system, with a subsequent decrease in the range. The reflecting surface must be a high-quality one.

A curved surface has the effect of introducing another lens in the system with a change in the effective focal length (Young 1967). Irregular surfaces affect the shape for the reflecting beam too.

Several accessories are part of an autocollimator. For a single-axis angle measurement, a pentaprism is used; an optical square permits angle measurements for surfaces at right angles. Perpendicularity is measured with a pentaprism and a mirror (Fig. 18.6). A handy horizontal reference can be produced with an oil pool, and care must be taken with the surface stability. A complete description for the autocollimator is found in Hume (1965).

18.2.4. Interferometric Measurements

Interferometric methods find their main applications in measuring very small wedge angles in glass slabs (Met 1966; Leppelmeier and Mullenhoff 1970) and in parallelism measurements (Wasilik et al., 1971). Refer to Chapter 1 for further details.

Interferometric measurements of large angles may also be performed. In one method a collimated laser beam is reflected from the surfaces from a rotating glass slab. The produced fringes can be considered as coming from a Murty lateral shear interferometer (Malacara and Harris 1970). This device can be used as a secondary standard to produce angles from 0° to 360° with an accuracy within a second of arc. Further analysis of this method is done by Tentori and Celaya (1986). In another system a Michelson interferometer is used with an electronic counter to measure in a range of $\pm 5°$ with a resolution of 10^{-4} degrees (Stijns 1986; Shi and Stijns 1988). An interferometric optical sine bar for angles in the millisecond of arc was built by Chapman (1974).

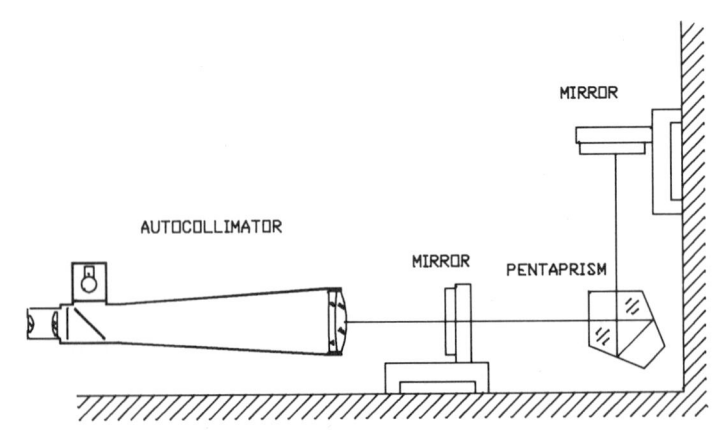

Figure 18.6. Perpendicularity measurement with an autocollimator.

18.2.5. Measurement of Angles in Prisms

A problem frequently encountered in the optical shop is precise angle measurement in the manufacturing of prisms. In most cases prism angles are 90°, 45°, and 30°. These angles are easily measured by comparison from a standard; other angles need an autocollimator or goniometer to be accurately measured.

To accurately measure angles, it is important that the prism is free of pyramidal error. Given a prism with angles A, B, and C (Fig. 18.7a), let OA be perpendicular to plane ABC. If line AP is perpendicular to segment BC, then the angle AOP is a measure for the pyramidal error. In a prism with pyramidal error the angles between the faces, as measured in planes perpendicular to the edges between these faces, adds to over 180°. To simply detect pyramidal error in a prism, Johnson (1947) and Martin (1924) suggest looking at both the refracted and the reflected image from a straight line (Fig. 18.7b). When pyramidal error is present, the line appears to be broken. A far target could be graduated to measure directly in minutes. A sensitivity of up to 3 min could be obtained.

During the milling process in a production run, a blank glass is mounted in a jig collinear to a master prism (Fig. 18.8). An autocollimator pointing to the master prism accurately sets the position for each prism face (Twyman 1957; DeVany 1968, 1971). With a carefully set diamond lap, pyramidal error is minimized. In a small quantity run angles can be checked with a bevel gauge. Visual tests for a prism in a bevel gauge can measure an error less than a minute of arc (Noble 1978).

A 90° angle in a prism can be measured with a goniometer by internal reflection (Fig. 18.9a). At the autocollimator two images are seen, and their angular separation is $2N\alpha$, α being the prism angle error, although it is not possible to know the sign for the angle error. Since the hypotenuse face has to be polished and the glass must be homogeneous, the measurement of the external angle with respect to a reference flat is preferred (Fig. 18.9b). In this case the angle error is determined by a change in the angle by tilting the prism. If the external angle is decreased and the images separate further, then the external angle is less then 90°. Conversely, if the images separate by tilting in such a way that the external angle increases, then the external angle is larger than 90°.

Several methods to determine the sign of the error have been proposed. DeVany (1978) suggests that when looking at the double image from the autocollimator, the image should be defocused inward. If the images tend to separate, then the angle in the prism is greater than 90°. Conversely, an outward defocusing will move the images closer to each other for an angle greater than 90°. Another way to eliminate the sign of the error in the angle is by introducing between the autocollimator and the prism a glass plate with a small wedge whose orientation is known. The wedge should cover only one half of the prism aperture. Ratajczyk and Bodnar (1966) suggested a different method using polarized light.

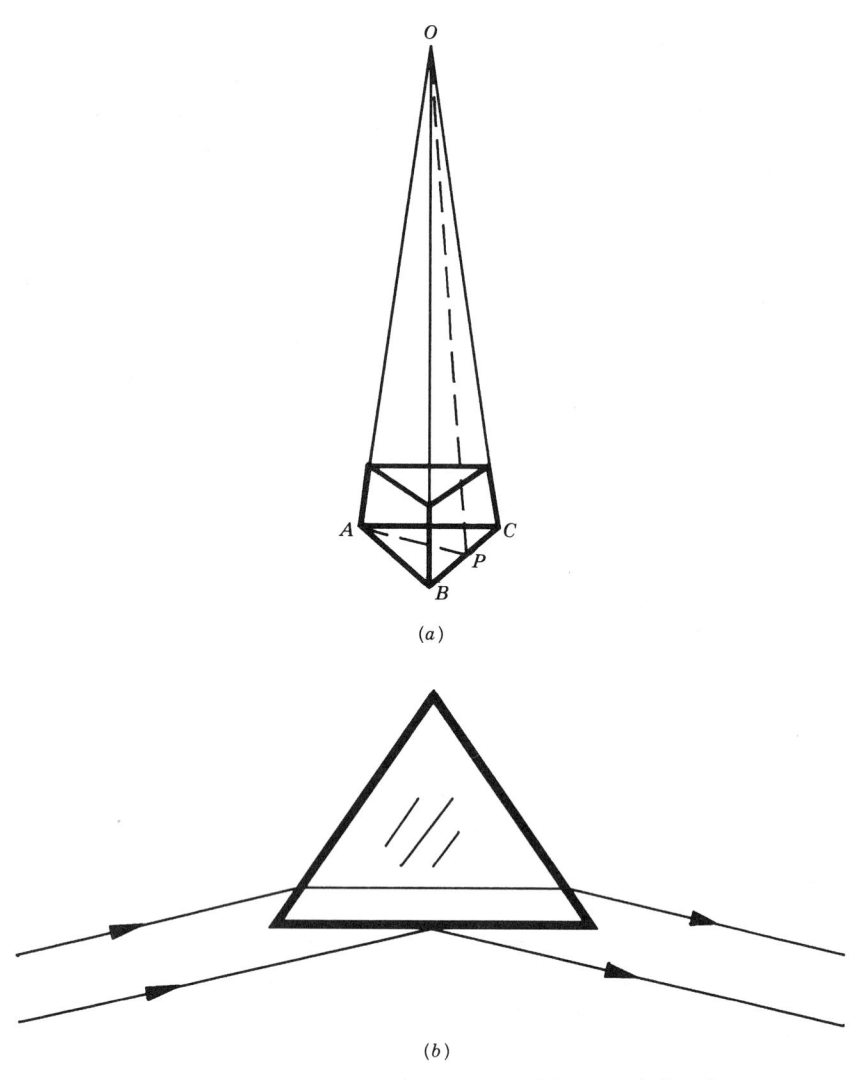

Figure 18.7. Pyramidal error in a prism. (*a*) Nature of the error. (*b*) Test for the error.

Right-angle prisms can be measured with an autocollimator with acceptable precision (Taarev 1985). With some practice perfect cubes with angles more accurate than 2 s of arc can be obtained (DeVany 1979).

An extremely simple test for the 90° angle in prisms (Johnson 1947) is performed by looking to the retroreflected image of the observer's pupil without any instrument. The shape of the image of the pupil determines the error, as shown in Fig. 18.10. The sensitivity of this test is not very high and may be

18.2. ANGLE MEASUREMENTS

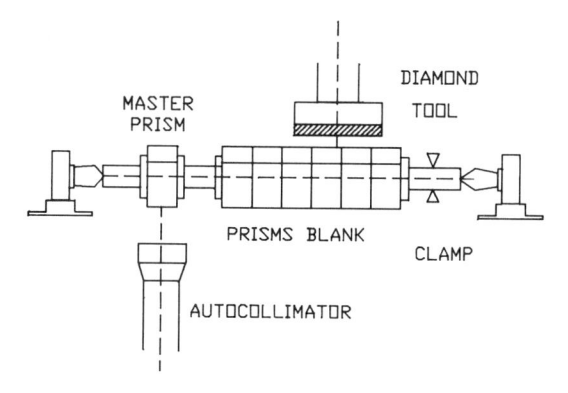

Figure 18.8. Milling prisms by replication.

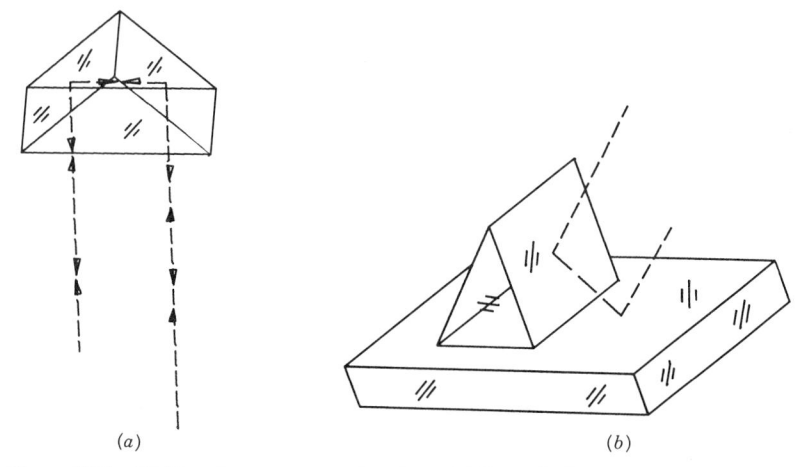

Figure 18.9. Right-angle measurement in prisms. (*a*) Internal. (*b*) External measurements.

used only as a coarse qualitative test. As shown by Malacara and Flores (1990), a small improvement in the sensitivity of this test may be obtained if a screen with a small hole is placed in front of the eye, as in Fig. 18.11. A cross centered on the small hole is painted on the front face of the screen. The observed images are as shown in Fig. 18.11. It is easy to see the similarity between this test and the Placido disc used some years ago in optometry and ophthalmology for observing irregularities in the cornea of a patient.

It is interesting to notice that, as opposed to the collimator test, there is no error sign uncertainty in the tests just described. The reason is that the observed plane is located where the two prism surfaces intersect. An improvement described by Malacara and Flores (1990), combining these simple tests with an autocollimator, is obtained with the instrument in Fig. 18.12. In this system the

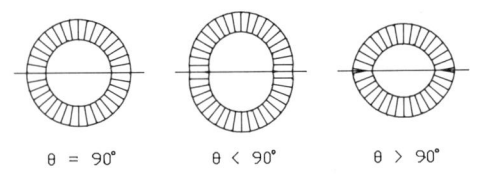

724   ANGLE, DISTANCE, CURVATURE, AND FOCAL LENGTH MEASUREMENTS

Figure 18.10. Retroreflected images of the observer's pupil on a 90° prism.

(a)

(b)

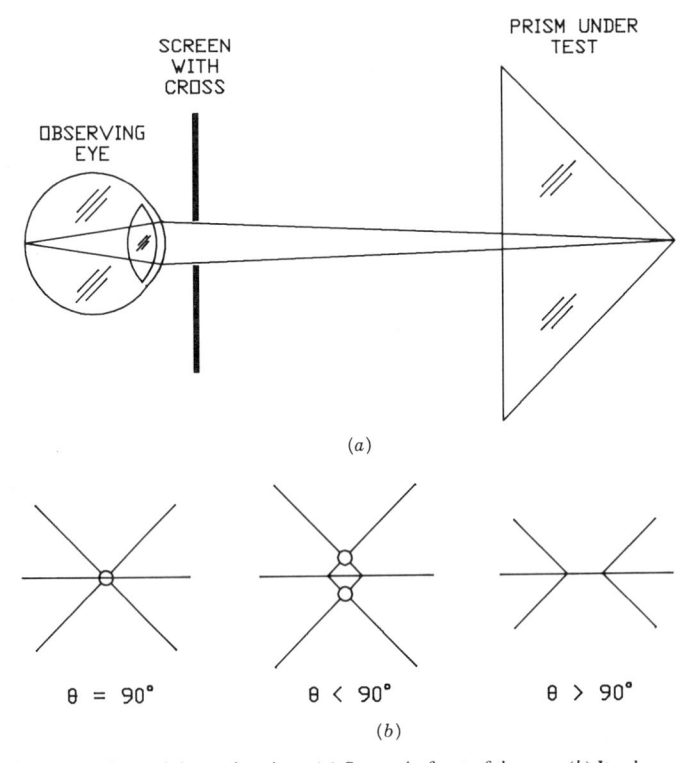

Figure 18.11. Testing a right-angle prism. (a) Screen in front of the eye. (b) Its observed images.

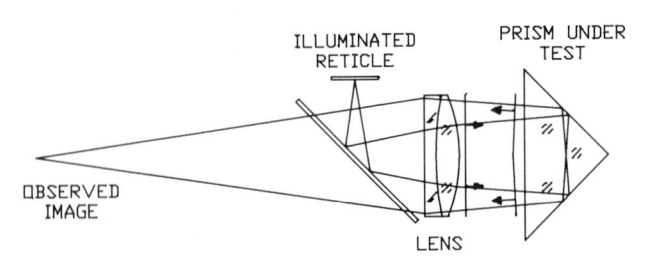

Figure 18.12. Modified autocollimator for testing the right angle in prisms without sign uncertainty.

line defining the intersection between the two surfaces is out of focus and barely visible, while the reticle is in perfect focus at the eyepiece.

18.3. DISTANCE MEASUREMENTS

All distance measurements are essentially a comparison against a master scale, and in order to maintain standards of accuracy, all masters should be referred to a single source. This source is the meter, which is defined as 1, 650, 763.73 wavelengths of the orange radiation in vacuum of the Krypton-86 isotope. The appropriate method to measure a distance depends on the required accuracy, the magnitude of the distance, and many other factors. There are so many different methods that to describe them would make a complete treatise. Here, we will make only a brief description of some commonly used methods in optical testing.

One common way of calibrating small distances is by means of gauge blocks or slip gauges. The length of these gauges is defined as the distance between the central points of opposite faces, plus about one tenth of a micron to take into account thickness irregularities. The gauges are typically calibrated by means of a Fizeau interferometer with a well-known procedure, as described by Horne (1974).

18.3.1. Interferometric Measurement of Distances

The measurement of distances by interferometric methods (Bruning 1978; Steinmetz et al. 1987; Massie and Caulfield 1987) is basically done by counting fringes in an interferometer while increasing the optical path difference. The low temporal coherence or monochromaticity of most light sources limit this procedure to short distances. Lasers, however, have a much longer coherence length due to their higher monochromaticity, which may be longer than 1 km. Their frequency stability is higher than 10^{-8}. Thus it has been possible with lasers to make interferometric distance measurements over several meters.

There are two basic approaches to interferometric distance measurements, one with a dc electrical analog and another with an ac analog. We will follow Bruning (1978) in the description of these two methods.

Figure 18.13 illustrates the dc approach. The two interfering wavefronts are flat and parallel. As the optical path difference is changed by moving one of the mirrors, an irradiance detector in the pattern will detect a sinusoidally varying signal, but the direction of the change cannot be determined. In order to determine this change, two signals in quadrature are needed. This requirement may be satisfied in many ways. In one method, illustrated in Fig. 18.13, a linearly or circularly polarized beam of light is divided at a beam splitter into the p and s components. Then each beam is reflected by a cube corner prism, each with

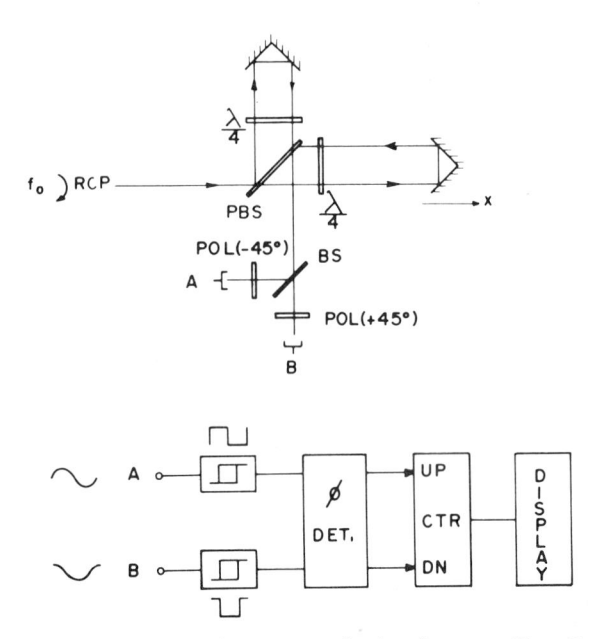

Figure 18.13. Direct current distance-measuring interferometer. (From Bruning 1978)

a $\lambda/4$ phase plate in front of it, with their axes at 45°. The returning beams will be circularly polarized and then recombined again in the beam splitter. The recombined beam of two circularly polarized beams is divided in a beam splitter and two linear polarizers are placed in front of each beam, but with their axes mutually perpendicular, at +45° and −45°. These two polarizers may be rotated to make the two irradiances equal. When the optical path difference changes by moving one of the cube corner prisms, the two beams A and B contain sinusoidally varying signals, but with a phase difference of 90°. The direction of motion of the moving prism may be sensed by determining whether the phase of A leads or lags the phase of B. This information is used to make the fringe counter go up or down. If the prism is shifted a distance x, the fringe count is

$$\Delta_{count} = \pm \left[\frac{2x}{\lambda} \right], \qquad (18.2)$$

where [ ] denotes the integer part of the argument.

This interferometer has the problem that any change in the irradiance may be easily interpreted as a fringe, although the solution could be to monitor the light source. A more serious problem is the requirement that the static interference pattern is free of fringes. Fringes may appear because of multiple reflections or turbulence.

18.3. DISTANCE MEASUREMENTS

The second method, called the ac approach, is illustrated in Fig. 18.14. The light source is a frequency stabilized He–Ne laser whose light beam is Zeeman split into two frequencies f_1 and f_2 by an application of an axial magnetic field (Burgwald and Kruger 1970). The frequency difference is several megahertz, and both beams have circular polarization, but with opposite sense. A $\lambda/4$ phase delay plate transforms the signals f_1 and f_2 into two orthogonal linearly polarized beams. A sample of this mixed signal or carrier is detected at A' by using a polarizer at 45°. The two orthogonally polarized beams with frequencies f_1 and f_2 are separated at a polarizing beam splitter. Then each beam is transformed into a circularly polarized beam by means of $\lambda/4$ phase plates; one of them becomes right handed and the other left handed. After reflection on the prisms the handiness of this polarization is changed and then the beams return to the same phase plates, where they are converted again to orthogonal linearly polarized beams. After recombination on the beam splitter, a polaroid at 45° will take the components of both means in this plane and finally they are detected at point B'.

If one of the cube corner prisms is moving, the two signals at A' and B' will be Doppler shifted in frequency by an amount Δf given by

$$\Delta f = \frac{2\dot{x}}{\lambda_2} \qquad (18.3)$$

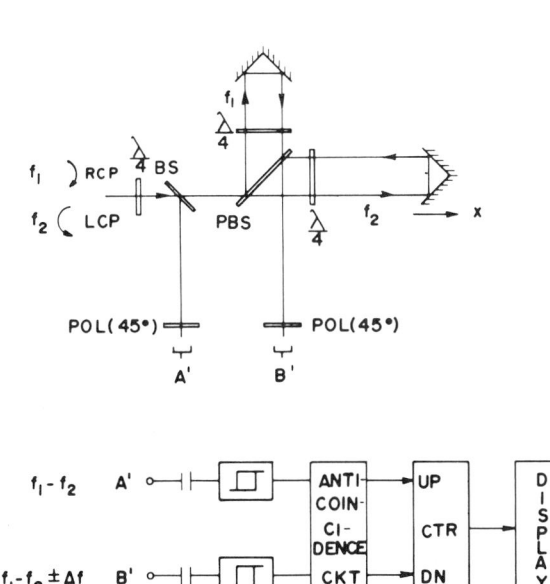

Figure 18.14. Alternating current distance-measuring interferometer. (From Bruning 1978)

where $\dot{x}$ is the cube corner prism velocity and λ_2 is the wavelength corresponding to f_2. The beating pulses due to the combination of these two signals are detected by electronic counting of both signals and detecting coincidences. If the prism moves a distance x, the number of pulses detected is given by

$$\Delta_{\text{count}} = \pm \left[\frac{2x}{\lambda_2}\right]. \tag{18.4}$$

The advantage of this method with respect to the first is that fringe counting is accomplished by processing ac electric signals, which are not subject to drift. These signals may even be processed to obtain a better signal-to-noise ratio and higher resolution (Dukes and Gordon 1970).

18.3.2. Thickness of Optical Components

The usual way of measuring lenses and glass plates thicknesses is by means of micrometers and dial gauges. The accuracy with these devices is about ± 0.01 mm. This accuracy may be increased by an order of magnitude to about ± 1.0 μm by the use of interferometry. This is done with a Michelson interferometer, comparing the thickness of the lens or glass plate with that of a calibrated reference glass plate with approximately the same thickness, made out of the same material (Tsuruta and Ichihara 1975). In this method the two mirrors of a dispersion-compensated Michelson interferometer are replaced by the lens to be tested and a reference plane parallel plate of the same material as the lens. The next step is to adjust the interferometer to produce white-light Newton rings with the front surfaces of the lens and the plate. Then the plate is translated along the arm of the interferometer, until the rear surface produces white-light rings. The displacement is the optical thickness Nt of the lens.

18.4. RADIUS OF CURVATURE MEASUREMENTS

The curvature of a spherical optical surface or the local curvature of an aspherical surface may be measured by means of mechanical or optical methods, as will be described next. Some methods measure the sagitta, some the surface slope, and some others directly the position of the center of curvature.

18.4.1. Mechanical Measurement of Radius of Curvature

Templates. The easiest way to measure the radius of curvature is by comparing it with metal templates with different radius of curvature until the best fit is obtained. The template is held against the optical surface with a bright light

18.4. RADIUS OF CURVATURE MEASUREMENTS

source in front of the observer and behind the template and the optical surface. If the surface is polished, openings close to one wavelength may be detected. If the opening is very narrow, the light becomes blue due to diffraction.

Test Plates. Another method is to use a test plate with opposite curvature as template, increasing its accuracy. The problem is that the surface has to be polished. (See Chapter 1.)

Spherometers. This is the most popular mechanical device for measuring radius of curvature. The value of the radius is calculated by measuring the sagitta (see Fig. 18.15). A classical spherometer consists of three equally spaced feet with a central moving plunger. The spherometer is first placed on top of a flat surface and then on top of the surface to be measured. The difference in the position of the central plunger is the sagitta of the spherical surface. Several practical problems may arise. One is that sharp legs may scratch the surface, thus, a steel ball is placed at the end of the legs as well as at the end of the plunger (Aldis spherometer). In this case if the measured sagitta is z, the radius of curvature R of the surface is given by (Cooke, 1964),

$$R = \frac{z}{2} + \frac{y^2}{2z} \pm r, \qquad (18.5)$$

where r is the radius of curvature of the balls. The plus sign is used for concave surfaces and the minus sign for convex surfaces.

The precision of this instrument may be obtained by differentiating Eq. (18.6), as follows:

$$\frac{dR}{dz} = \frac{1}{2} - \frac{y^2}{2z^2} \qquad (18.6)$$

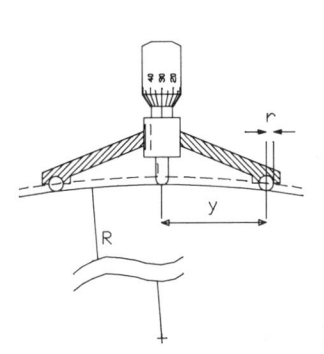

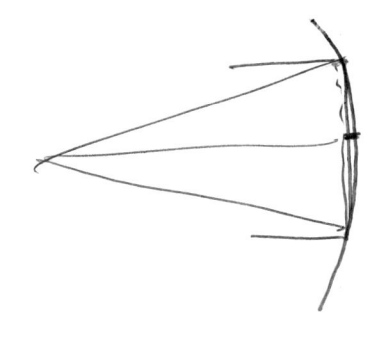

Figure 18.15. Three-leg spherometer.

Table 18.1. Spherometer Precision[a]

Radius of Sphere R (mm)	Sagitta z (mm)	Precision ΔR (mm)	Fractional Precision $\Delta R/R$
10,000	0.125	−400	−0.040
5,000	0.250	−100	−0.020
2,000	0.625	−16	−0.008
1,000	1.251	−4	−0.004
500	2.506	−1	−0.002
200	6.351	−0.15	−0.0008

[a] $y = 50$ mm; $\Delta z = 0.005$ mm.
(From Noble 1978.)

obtaining,

$$\Delta R = \frac{\Delta z}{2}\left(1 - \frac{y^2}{z^2}\right). \tag{18.7}$$

This result is valid assuming that the spherometer is perfectly built and that the dimensional parameters y and r are well known. The uncertainty comes only from the measurement of the sagitta. Noble (1978) has made an evaluation of this precision for a spherometer with $y = 50$ mm and a sagitta reading uncertainty equal to 0.005 mm, and found the results in Table 18.1. We may see that the precision is better than 2%.

An extensive analysis of the precision and accuracy of several types of spherometers is given in the book by Jurek (1977).

Another type of spherometer is the so-called ring spherometer, which has a cup instead of the three legs. The cup is flat in the upper part and has its outside and external walls with a cylindrical shape, as in Fig. 18.16. A concave surface touches the external edge of the cup, whereas the convex surface touches the

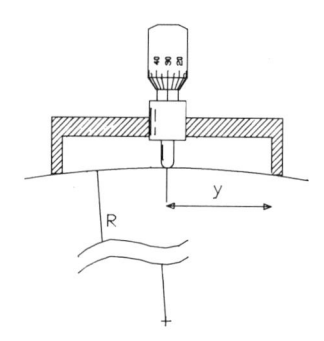

Figure 18.16. Ring spherometer.

18.4. RADIUS OF CURVATURE MEASUREMENTS

internal edge of the ring. Thus Eq. (18.5) may be used if a different value of y is used for concave and convex surfaces, and r is taken as zero. In this instrument the cups may be interchangeable, with different diameters for different surface diameters and radii of curvature. There is an averaging effect because the ring strikes only the high spots. The main advantage is that an astigmatic deformation of the surface is easily detected, but it cannot be measured. With the three-leg spherometer the astigmatic deformations cannot even be detected.

The spherometer accuracy may be improved in many ways, by different methods of taking the readings of the sagitta. One method is the Steinheil spherometer (Martin 1924) in which a mechanical device is employed to indicate the pressure between the central plunger and the surface to be measured. In the Abbe spherometer (Martin 1924) the displacement of the central plunger is measured with the aid of a scale and a reading microscope. The dial spherometer is very popular for quick measurements in industrial processes. The movement of the plunger activates the hand of a circular measuring dial.

Some modern spherometers use a differential transformer as a transducer to measure the plunger displacement. This transformer, coupled to an electronic circuit, produces a voltage linear with the plunger displacement. This voltage is then analyzed by a microprocessor. The microprocessor then calculates the radius or power in any desired units and displays it.

The bar spherometer shown in Fig. 18.17 permits the measurement of the astigmatism since it measures the curvature along any diameter. However, the accuracy of the measurement may not be as high as in the previous devices, due to tilting of the instrument. A commercial version of a small bar spher-

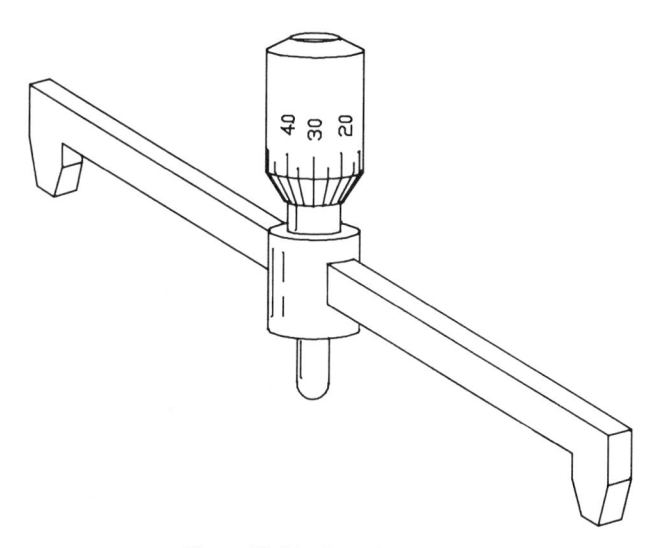

Figure 18.17. Bar spherometer.

732   ANGLE, DISTANCE, CURVATURE, AND FOCAL LENGTH MEASUREMENTS

ometer for the specific application in optometric work is the Geneva gauge. In this gauge the scale is directly calibrated in diopters since the power is linear with the sagitta, assuming that the refractive index of the glass is 1.53.

A problem indirectly related is the calculation of the sagitta from a knowledge of the radius of curvature, in order to mount a lens properly in its mechanical holder. Different graphical and algebraic procedures have been described (Zanker 1981; Foote 1981; Brixner 1982).

18.4.2. Optical Measurement of Radius of Curvature

Foucault Test. The easiest method to find the radius of curvature of a concave surface is by means of the knife-edge test (see Chapter 8) in order to locate the center of curvature. Then the distance from the knife edge to the optical surface is measured with a scale. The accuracy of this method for long radii of curvature is not very good because the scale may sag due to gravity.

Autocollimator. In the autocollimator technique (Horne 1972) the radius of curvature is determined through measurements of the slopes of the optical surface. The well-known property that a pentaprism produces a 90° deflection of a light beam, independently of small errors in its orientation, is used. The method is illustrated in Fig. 18.18 where we may see that the pentaprism travels over the optical surface to be measured, along one diameter. The first step is to center the light on the reticle of the autocollimator when the vertex of the surface is being examined. Then the pentaprism is moved outside the central part of the surface in order to measure the slope variations. From these measurements the radius of curvature and even its shape may be calculated. This method is used only for large radii of curvature, and it is equally applicable for both concave and convex surfaces.

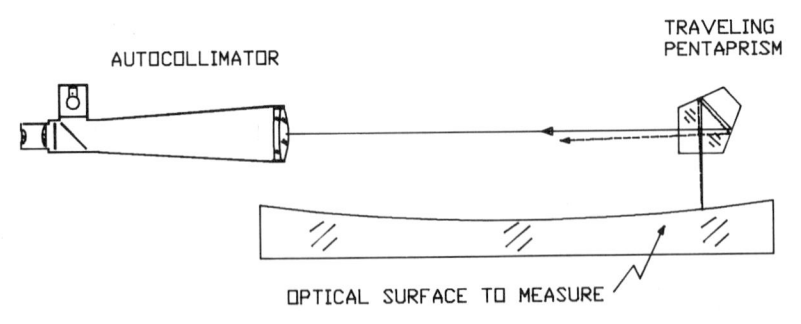

Figure 18.18. Autocollimator and pentaprism used to determine radius of curvature by measuring surface slopes.

18.4. RADIUS OF CURVATURE MEASUREMENTS

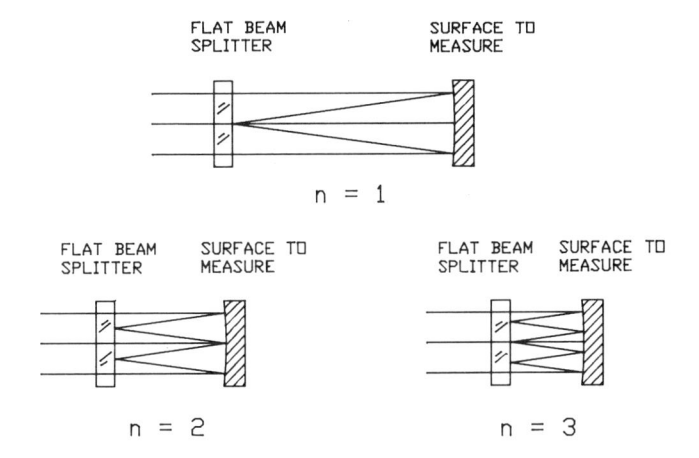

Figure 18.19. Confocal cavity arrangements used to measure radius of curvature.

Confocal Cavity Technique. The optical cavity technique (Gerchman and Hunter 1979, 1980) permits the interferometric measurement of very long radii of curvature with an accuracy of 0.01%. The method consists in forming the cavity of a Fizeau interferometer (see Chapter 1) as illustrated in Fig. 18.19. This is a confocal cavity of nth order, where n is the number of times the path is folded. The radius of curvature is approximately $2n$ times the cavity length Z_n, but for a higher accuracy values listed in Table 18.2 should be used.

Traveling Microscope. This is one of the most popular methods for measuring the radius of curvature of small concave optical surfaces, with short radius of curvature. As shown in Fig. 18.20, a point light source is produced at the front focus of a microscope objective. This light source illuminates the concave op-

Table 18.2. Constants Relating Cavity Length to Radius of Curvature $Z_n = C_n R$

n	C_n
1	0.5
2	0.25
3	0.1464466
4	0.0954915
5	0.0669873
6	0.0495156
7	0.0380603
8	0.0301537

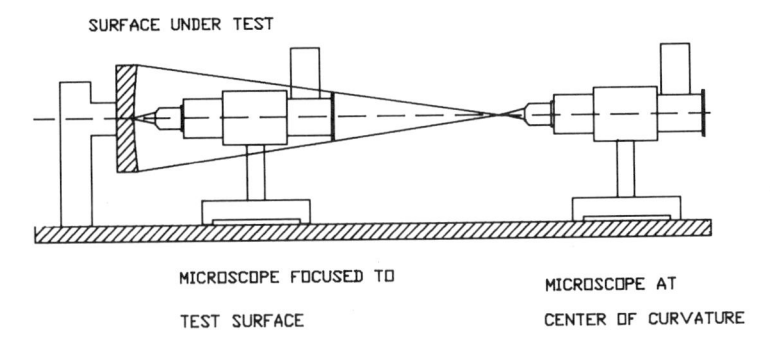

Figure 18.20. Traveling microscope to measure radii of curvature.

tical surface to be measured, near its center of curvature. Then this concave surface forms an image, also close to its center of curvature. This image is observed with the same microscope used to illuminate the surface. During this procedure the microscope is focused both at the center of curvature and at the surface to be measured. A sharp image of the light source is observed at both places. The radius of curvature is the distance between these two positions for the microscope.

This distance traveled by the microscope may be measured on a vernier scale, obtaining a precision of about ± 0.1 mm. If a bar micrometer is used, the precision may be increased by an order of magnitude. In this case two small convex buttons are required, one fixed to the microscope carriage and the other to the stationary part of the bench. They must face each other when the microscope carriage is close to the optical bench fixed component.

Carnell and Welford (1971) describe a method that requires only one measurement. The microscope is focused only at the center of curvature. Then the radius of curvature is measured by inserting a bar micrometer with one end touching the vertex of the optical surface and the other end is adjusted until it is observed in focus on the microscope. Accuracies of a few microns are obtained with this method.

In order to focus the microscope properly, the image of an illuminated reticle must fall after reflection on the same reticle itself, as in the Gauss eyepiece in Fig. 18.5. The reticle and its image appear as dark lines in a bright field. The focusing accuracy may be increased with a dark field. Carnell and Welford obtained a dark field with two reticles, as in Fig. 18.5, one illuminated with bright lines, and the other with dark lines.

A convex surface may also be measured with this method, if a well-corrected lens with a conjugate longer than the radius of curvature of the surface under test is used. Another alternative for measuring convex surfaces is by inserting an optical device with prisms in front of the microscope, as described by Jurek (1977).

Some practical aspects of the traveling microscope are examined by Rank (1946), who obtained a dark field at focus with an Abbe eyepiece, which introduces the illumination with a small prism. This method has been implemented using a laser light source by O'Shea and Tilstra (1988).

Additional optical methods to measure the radius of curvature of a spherical surface have been described. Evans (1971, 1972a, 1972b) determines the radius by measuring the lateral displacements on a screen of a laser beam reflected on the optical surface, when this optical surface is laterally displaced. Cornejo-Rodriguez and Cordero-Dávila (1980), Klingsporn (1979), and Diaz-Uribe et al. (1986) rotate the surface about its center of curvature in a nodal bench.

18.5. FOCAL LENGTH MEASUREMENTS

There are two definitions for the focal length of an optical system. One is the back focal length, which is the distance from the last surface of the system to the focus. The other is the distance from the principal plane to the focus. The back focal length is easily measured following the same procedure used for measuring the radius of curvature, using a microscope and the lens bench.

18.5.1. Nodal Slide Bench

In an optical system in air the principal points (intersection of the principal plane and the optical axis) coincide with the nodal points. Thus to locate this point we may use the well-known property that small rotations of the lens about an axis, perpendicular to the optical axis and passing through the nodal point, do not produce any lateral shift of the image. The instrument used to perform this procedure, shown in Fig. 18.21, is called an optical nodal slide bench (Kingslake 1932). This bench has provision for slowly moving the lens under test longitudinally, in order to find the nodal point.

The bench is illuminated with a collimated light source and the image produced by the lens under test is examined with a microscope. The lens is then given small movements about a vertical axis, as it is being displaced longitudinally. This procedure is stopped until a point is found in which the image does not move laterally while rotating the lens. This axis of rotation is the nodal point. Then the distance from the nodal point to the image is the effective focal length.

18.5.2. Focimeters

A focimeter is an instrument designed to measure the focal length of lenses in a simple manner. A light source illuminates a reticle and a convergent lens with focal length f is placed at a distance x from the reticle. The lens to be measured

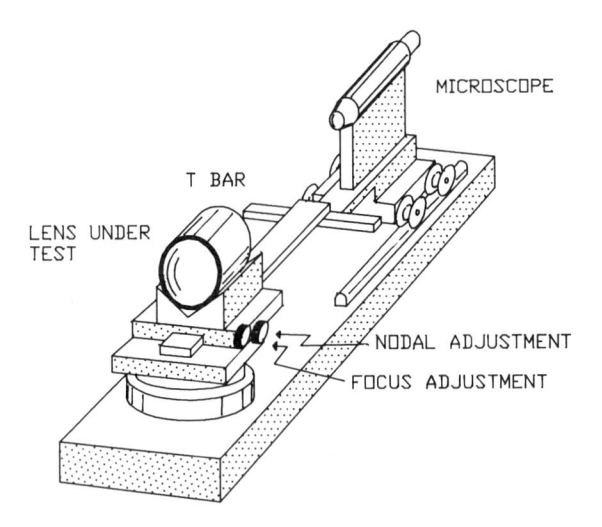

Figure 18.21. Nodal slide bench.

is placed at a distance d from the convergent lens. The magnitude of x is variable and adjusted until the light beam going out from the lens under test becomes collimated. This collimation is verified by means of a small telescope in front of this lens, focused to infinity. The values of d and the focal length f are chosen to be equal. Then the back focal length f_b of the lens under test is given by

$$\frac{1}{f_b} = \frac{1}{d} - \frac{x}{d^2}. \tag{18.8}$$

As we see, the power of the lens being measured is linear with distance x. There are many variations of this instrument. Some modern focimeters measure the lateral deviation of a light ray from the optical axis (transverse aberration), as in Fig. 18.22, when a defocus is introduced (Evans 1971, 1972a, 1972b;

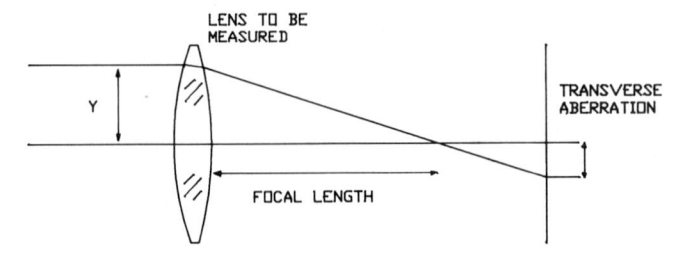

Figure 18.22. Focal length determination by transverse aberration measurements.

Bouchard and Cogno 1982). This method is mainly used in some modern automatic focimeters for optometric applications. To measure the transverse aberration a position sensing detector is frequently used.

18.5.3. Other Focal Length Measurements

A clever method used to automatically find the position of the focus has been described by Howland and Proll (1970). They used optical fibers to illuminate the lens in an autocollimating configuration and the location of the image was also determined using optical fibers.

REFERENCES

Bouchaud, P. and J. A. Cogno, "Automatic Method for Measuring Simple Lens Power," *Appl. Opt.*, **21**, 3068 (1982).

Brixner, B., "Easier Way to Find the Sagittal Depth: Comments," *Appl. Opt.*, **21**, 976 (1982).

Bruning, J., "Fringe Scanning," in *Optical Shop Testing*, 1st ed., D. Malacara, Ed., Wiley, New York, 1978.

Burgwald, G. M. and W. P. Kruger, "An Instant-On Laser for Length Measurements," *Hewlett-Packard J.*, **21**, 2 (1970).

Carnell, K. H. and W. T. Welford, "A Method for Precision Spherometry of Concave Surfaces," *J. Phys.*, **54**, 1060-1062 (1971).

Chapman, G. D., "Interferometric Angular Measurement," *Appl. Opt.*, **13**, 1646-1651 (1974).

Cooke, F., "The Bar Spherometer," *Appl. Opt.*, **3**, 87-88 (1964).

Cornejo-Rodriguez, A. and A. Cordero-Dávila, "Measurement of Radii of Curvature of Convex and Concave Surfaces Using a Nodal Bench and a He-Ne Laser," *Appl. Opt.*, **19**, 1743-1745 (1980).

DeVany, A. S., "Making and Testing Right Angle and Dove Prisms," *Appl. Opt.*, **7**, 1085-1087 (1968).

DeVany, A. S., "Reduplication of a Penta-Prism Angle Using Master Angle Prisms and Plano Interferometer," *Appl. Opt.*, **10**, 1371-1375 (1971).

DeVany, A. S., "Testing Glass Reflecting-Angles of Prisms," *Appl. Opt.*, **17**, 1661-1662 (1978).

DeVany, A. S., "Near Perfect Optical Square," *Appl. Opt.*, **18**, 1284-1286 (1979).

Deve, C., *Optical Workshop Principles* (Trans. by T. L. Tippell), Hilger and Watts, London, 1945.

Diaz-Uribe, R., J. Pedraza-Contreras, O. Cardona-Nuñez, A Cordero-Dávila, and A. Cornejo-Rodriguez, "Cylindrical Lenses: Testing and Radius of Curvature Measurement," *Appl. Opt.*, **25**, 1707-1709 (1986).

Dukes, J. N. and G. B. Gordon, "A Two-Hundred-Foot Yardstick with Graduations Every Microinch," *Hewlett-Packard J.*, **21**, 2 (1970).

738 ANGLE, DISTANCE, CURVATURE, AND FOCAL LENGTH MEASUREMENTS

Evans, J. D., "Method for Approximating the Radius of Curvature of Small Concave Spherical Mirrors Using a He-Ne Laser," *Appl. Opt.*, **10**, 995-996 (1971).

Evans, J. D., "Equations for Determining the Focal Length of On-Axis Parabolic Mirrors by He-Ne Laser Reflection," *Appl. Opt.*, **11**, 712-714 (1972a).

Evans, J. D., "Error Analysis to: Method for Approximating the Radius of Curvature of Small Concave Spherical Mirrors Using a He-Ne Laser," *Appl. Opt.*, **11**, 945-946 (1972b).

Foote, V. S., "Easier Way to Find the Sagitta Depth," *Appl. Opt.*, **20**, 2605-2605 (1981).

Geiser, R. D., "Precision and Accuracy," in *Applied Optics and Optical Engineering*, Vol. I, R. Kingslake, Ed., Academic Press, New York, 1965, Chap. 11.

Gerchman, M. C. and G. C. Hunter, "Differential Technique for Accurately Measuring the Radius of Curvature of Long Radius Concave Optical Surfaces," *Proc. SPIE*, **192**, 75-84 (1979).

Gerchman, M. C. and G. C. Hunter, "Differential Technique for Accurately Measuring the Radius of Curvature of Long Radius Concave Optical Surfaces," *Opt. Eng.*, **19**, 843-848 (1980).

Horne, D. F., *Optical Production Technology*, Adam Hilger, London, and Crane Russak, New York, 1972, Chap. XI.

Horne, D. F., *Dividing, Ruling and Mask Making*, Adam Hilger, London, 1974, Chap. VII.

Howland, B. and A. F. Proll, "Apparatus for the Accurate Determination of Flange Focal Distance," *Appl. Opt.*, **11**, 1247-1251 (1970).

Hume, K. J., *Metrology with Autocollimators*, Hilger and Watts, London, 1965.

Johnson, B. K., *Optics and Optical Instruments*, Dover, New York, 1947, Chaps. II and VIII.

Jurek, B., *Optical Surfaces*, Elsevier Scientific, New York, 1977.

Kingslake, R., "A New Bench for Testing Photographic Lenses," *J. Opt. Soc. Am.*, **22**, 207-222 (1932).

Klingsporn, P. E., "Use of a Laser Interferometric Displacement-Measuring System for Noncontact Positioning of a Sphere on a Rotation Axis through Its Center and for Measuring the Spherical Contour," *Appl. Opt.*, **18**, 2881-2890 (1979).

Kothiyal, M. P., and R. S. Sirohi, "Improved Collimation Testing Using Talbot Interferometry," *Appl. Opt.*, **26**, 4056-4057, (1987).

Leppelmier, G. W. and D. J. Mullenhoff, "A Technique to Measure the Wedge Angle of Optical Flats," *Appl. Opt.*, **9**, 509-510 (1970).

Malacara, D. and O. Harris, "Interferometric Measurement of Angles," *Appl. Opt.*, **9**, 1630-1633 (1970).

Malacara, D. and R. Flores, "A Simple Test for the 90 Degrees Angle in Prisms," *Proc. SPIE, 1332*, 36 (1990).

Martin, L. C., *Optical Measuring Instruments*, Blackie and Sons, London, 1924.

Massie, N. A. and J. Caulfield, "Absolute Distance Interferometry," *Proc. SPIE*, **816**, 149-157 (1987).

Met, V., "Determination of Small Wedge Angles Using a Gas Laser," *Appl. Opt.*, **5,** 1242–1244 (1986).

Noble, R. E., "Some Parameter Measurements," in *Optical Shop Testing*, 1st ed., D. Malacara, Ed., Wiley, New York, 1978.

O'Shea, D. C. and S. A. Tilstra, "Non Contact Measurements of Refractive Index and Surface Curvature," *Proc. SPIE*, **966,** 172–176 (1988).

Rank, D. H., "Measurement of the Radius of Curvature of Concave Spheres," *J. Opt. Soc. Am.*, **36,** 108–110 (1946).

Ratajczyk, F. and Z. Bodner, "An Autocollimation Measurement of the Right Angle Error with the Help of Polarized Light," *Appl. Opt.*, **5,** 755–758 (1966).

Steinmetz, C., R. Burgoon, and J. Harris, "Accuracy Analysis and Improvements for the Hewlett-Packard Laser Interferometer System," *Proc. SPIE*, **816,** 79–94 (1987).

Stijns, E., "Measuring Small Rotation Rates with a Modified Michelson Interferometer," *Proc. SPIE*, **661,** 264–266 (1986).

Tareev, A. M., "Testing the Angles of High-Precision Prisms by Means of an Autocollimator and a Mirror Unit," *Sov. J. Opt. Technol.*, **52,** 50–52 (1985).

Tentori, D. and M. Celaya, "Continuous Angle Measurement with a Jamin Inteferometer," *Appl. Opt.*, **25,** 215–220 (1986).

Thurston, T. H., "Specifying Electronic Autocollimators," *Proc. SPIE*, **661,** 399–401, (1986).

Tsuruta, T. and Y. Ichihara, "Accurate Measurement of Lens Thickness by Using White-Light Fringes," *Jpn. J. Appl. Phys.*, **14,** Suppl. 14-1, 369–372 (1975).

Twyman, F., *Prisms and Lens Making*, 2nd ed., Hilger and Watts, London, 1957.

Wasilik, J. H., T. V. Blomquist, and C. S. Willett, "Measurement of Parallelism of the Surfaces of a Transparent Sample Using Two-Beam Nonlocalized Fringes Produced by a Laser," *Appl. Opt.*, **10,** 2107–2112 (1971).

Williams, T. L., "A Scanning Gauge for Measuring the Form of Spherical and Aspherical Surfaces," *Opt. Acta*, **25,** 1155–1166 (1978).

Young, A. W., "Optical Workshop Instruments," in *Applied Optics and Optical Engineering*, Vol. 4, R. Kingslake, Ed., Academic Press, New York, 1967, Chap. 7.

Zanker, A., "Easy Way to Find the Saggita Depth," *Appl. Opt.*, **20,** 725–726 (1981).

ADDITIONAL REFERENCES

Barton, N. P., "Measurement of Roof and Other Prisms," *Proc. SPIE*, **163,** 121, (1979).

Bayle, A. and J. Espiard, "Sur la Construction des Grandes Telescopes d'Astronomie," *Nouv. Rev. Opti. Appl.*, **3,** 67–73 (1972).

Bergman, T. G. and J. L. Thompson, "An Interference Method for Determining the Degree of Parallelism of (Laser) Surfaces," *Appl. Opt.*, **7,** 923–925 (1968).

Bernardo, L. M. and O. D. D. Soares, "Evaluation of the Focal Distance of a Lens by Talbot Interferometry," *Appl. Opt.*, **27,** 296–301 (1988).

Bogdanov, A. P. et al., "Automated Testing of Multifaced Prisms," *Sov. J. Opt. Technol.*, **45**, 405, (1978).

Bondarenko, I. D., "Methods of Testing the Fabrication Quality of Corner Reflectors," *Sov. J. Opt. Technol.*, **52**, 430–435 (1985).

Chang, Chon-Wen and D. C. Su, "An Improved Technique of Measuring the Focal Length of a Lens," *Opt. Commun.*, **73**, 257–262 (1989).

Chapman, G. D.,"Optical Metrology in Length and Mechanical Standards," *Proc. SPIE*, **661**, 242–248 (1986).

Diaz-Uribe, R. and A. Cornejo-Rodriguez, "Conic Constant and Paraxial Radius of Curvature Measurements for Conic Surfaces," *Appl. Opt.*, **25**, 3731–3734 (1986).

Dil, J. G., P. F. Greve, and W. Mesman, "Measurement of Steep Aspheric Surfaces," *Appl. Opt.*, **17**, 553–557 (1978).

Dorofeeva, M. V., and N. A. Kashkarova, "Methods for Aligning a Block of Dove Prisms in the Near IR," *Sov. J. Opt. Technol.*, **47**, 231–232 (1980).

Farcinade, M., "Measurement of the Radii and Local Distortions of Apparatus for Inspecting Spectacle Lenses," *Mes. Regul. Autom.*, **43**, 45 (1978).

Glatt, I. and O. Kafri, "Determination of the Focal Length of Nonparaxial Lenses by Moire Deflectometry," *Appl. Opt.*, **26**, 2507–2508 (1987).

Gurari, M. L., A. P. Golikov, and S. I. Prytkov, "Measurement of the Curvature Radii and Local Distortions of Mirror Surfaces," *Sov. J. Opt. Technol.*, **45**, 585 (1978).

Gusev, V. G., and S. F. Balandin, "Measuring the Radii of Curvature of Spherical Mirrors by a Speckle Interferometry Method," *Sov. J. Opt. Technol.*, **52**, 358–360 (1985).

Kahn, H., J. Groot, and A. Beltz, "Electronic Measuring Technique to Simplify Testing of Mirrors," *Opt. Laser Technol.*, **14**, 303–307 (1982).

Kasana, R. S. and K. J. Rosenbruch, "The Use of a Plane Parallel Glass Plate for Determining the Lens Parameters," *Opt. Commun.*, **46**, 69–73 (1983).

Kessler, D. and R. V. Shack, "Dynamical Optical Tests of a High-Speed Polygon," *Appl. Opt.*, **20**, 1015–1019 (1981).

Kolomiltsov, and I. V. Novikova, "Interference Angle Gauge for Certification of Autocollimators," *Sov. J. Opt. Technol.*, **50**, 97–100 (1987).

Langlois, P., R. A. Lessard, and A. Boibon, "Real-Time Curvature Radii Measurements Using Diffraction Edge Waves," *Appl. Opt.*, **24**, 1107–1112 (1985).

Marlatov, I. V., and I. V. Lavrent'eva, "Error of the IZS-7 Spherometer," *Sov. J. Opt. Technol.*, **53**, 616–617 (1986).

Mumzhiu, A. M., "A Complex of Facilities for Testing Spectacle Optics," *Biomed. Eng.*, **12**, 48 (1978).

Nikolov, A. D., A. S. Dimitrov, and P. A. Kralchevsky, "Accuracy of the Differential-Interferometric Measurements of Curvature," *Opt. Acta*, **33**, 1359–1368 (1986).

Nakano, Y. and K. Murata, "Talbot Interferometry for Measuring the Focal Length of a Lens," *Appl. Opt.*, **24**, 3162–3166 (1985).

O'Shea, D. and S. A. Tilstra, "Non-Contact Measurements of Refractive Index and Surface Curvature," *Proc. SPIE*, **966**, 172–176 (1988).

Patson, G. E., "A Method for Checking Focal Length While Grinding," *Sky Telesc.*, **26**, 358-360 (1963).

Pernick, B. J. and B. Hyman, "Least Squares Technique for Determining Principal Plane Location and Focal Length," *Appl. Opt.*, **26**, 2938-2939 (1987).

Saunders, J. B., "Suggested Arrangement of Mirrors to Form Multiple Reference Angles," *J. Opt. Soc. Am.*, **51**, 859-862 (1961).

Sen, D. and P. N. Puntambekar, "Shearing Interferometers for Testing Corner Curves and Right Angle Prisms," *Appl. Opt.*, **5**, 1009-1014 (1966).

Shannon, R. R., "The Testing of Complete Objectives," in *Applied Optics and Optical Engineering*, Vol. III, R. Kingslake, Ed., Academic Press, New York, 1965, Chap. 5.

Shi, P. and E. Stijns, "New Optical Method for Measuring Small Angle Rotations," *Appl. Opt.*, **27**, 4342-4344 (1988).

Soares, O. D. D. and J. F. Fernandez, "Laser Aided Spherometer," *Proc. SPIE*, **954**, 234-240 (1988).

Steinmetz, C. R., "Sub-micron Position Measurement and Control on Precision Machine Tools with Laser Interferometry," *Precision Eng.*, **12**(1), 12-24 (1990).

Strakun, G. I., A. M. Mumzhiu, and I. A. Mitsevich, "Automatic Dioptometer with a Projection Readout System," *Sov. J. Opt. Technol.*, **45**, 356 (1978).

Talim, S. P., "Measurement of the Refractive Index of a Prism by a Critical Angle Method," *Opt. Acta*, **25**, 157-165 (1978).

Tareev, A. M., "Improving the Testing Efficiency of the Angles of Rectangular Prisms During their Manufacturing Process," *Sov. J. Opt. Technol.*, **46**, 603 (1979).

Tareev, A. M., "Testing of Optical Prisms for Angular Errors," *Sov. J. Opt. Technol.*, **50**, 316-318 (1983).

Tew, E. J., Jr., "Measurement Techniques Used in the Optical Workshop," *Appl. Opt.*, **5**, 695-700 (1966).

Tewari, R. D., A. M. Ghodgaonkar, and K. Ramani, "Measurement of the Apex Angle of a 60° Prism," *Opt. Eng.*, **22**, 371-372 (1983).

U.S. Department of Defense, *Military Handbook 141* (Mil. HDBK-141), 1963.

Appendix 1

An Optical Surface and Its Characteristics

A1.1. DEFINITION OF AN OPTICAL SURFACE

Aspherical surfaces are extremely important in optical systems and have been studied and described by many authors, for example, by Schulz (1988). Of special interest are surfaces with rotational symmetry, which may be defined by means of the following relation, taking the z axis as the axis of revolution:

$$z = \frac{cS^2}{1 + [1 - (K+1)c^2S^2]^{1/2}} + A_1 S^4 + A_2 S^6 + A_3 S^8 + A_4 S^{10}, \quad (A1.1)$$

where $S^2 = x^2 + y^2$ and $c = 1/r = 1/$radius of curvature. Also, A_1, A_2, A_3, and A_4 are the aspheric deformation constants, and K is a function of the eccentricity of a conic surface ($K = -e^2$), called the conic constant. If the A_i are all zero, the surface is a conic surface of revolution, according to the following:

Hyperboloid	$K < -1$
Paraboloid	$K = -1$
Ellipse rotated about its major axis (prolate spheroid or ellipsoid)	$-1 < K < 0$
Sphere	$K = 0$
Ellipse rotated about its minor axis (oblate spheroid)	$K > 0$

As it is easy to see, the conic constant is not defined for a flat surface ($c = 0$).

For conic surfaces of revolution there is an expression for z somewhat simpler than the general expression (A1.1):

$$z = \frac{1}{K+1} \left[r - \sqrt{r^2 - (K+1)S^2} \right]. \quad (A1.2)$$

This serves for all conics except the paraboloid, where

$$z = \frac{S^2}{2r}. \quad (A1.3)$$

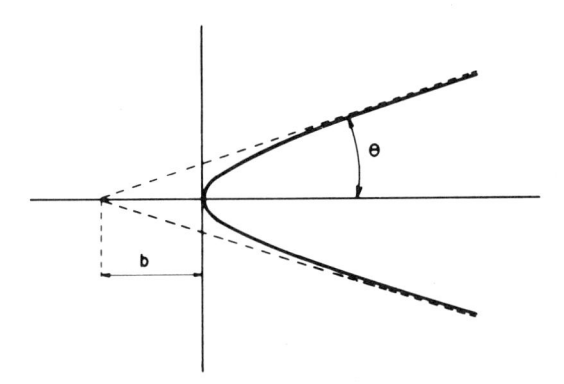

Figure A1.1. Axicon surface.

A surface of the axicon type that has the shape of a cone may be represented by means of a hyperboloid with an extremely large curvature, as shown in Fig. A1.1, thus obtaining

$$K = -(1 + \tan^2 \theta) < -1,$$

and

$$c = \frac{1}{(K + 1)b}.$$

(A1.4)

A1.2. PARAMETERS FOR CONIC SURFACES

The positions of the foci for the conic surfaces are functions of r and K and are given by the following relations, as illustrated in Fig. A1.2:

$$d_1 = \frac{r}{K + 1}, \tag{A1.5}$$

$$d_2 = \frac{r}{K + 1}(2\sqrt{K}), \tag{A1.6}$$

$$d_3, d_4 = \frac{r}{K + 1}(1 \pm \sqrt{-K}), \tag{A1.7}$$

$$d_5 = \frac{r}{2}, \tag{A1.8}$$

$$d_6, d_7 = \frac{r}{K + 1}(\sqrt{-K} \pm 1). \tag{A1.9}$$

A1.3. SOME USEFUL EXPANSIONS OF z

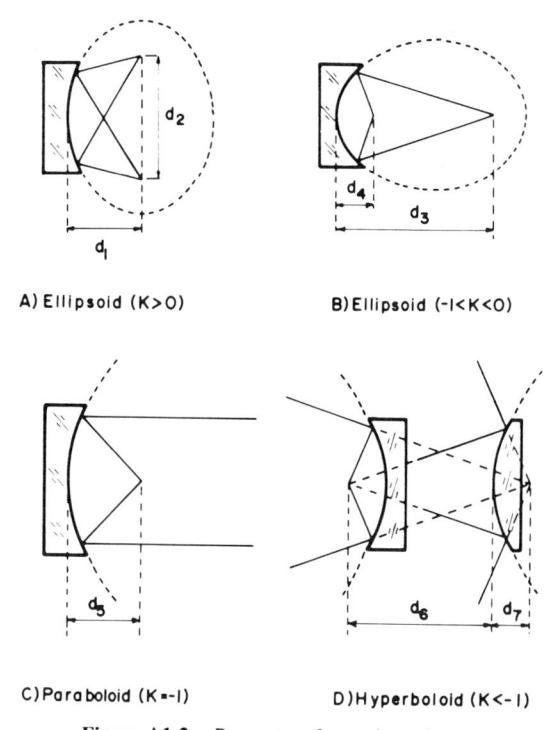

A) Ellipsoid (K>0)

B) Ellipsoid (-1<K<0)

C) Paraboloid (K=-1)

D) Hyperboloid (K<-1)

Figure A1.2. Parameters for conic surfaces.

A1.3. SOME USEFUL EXPANSIONS OF z

Sometimes it is convenient to consider an aspheric or conic optical surface as the sum of its osculating sphere plus some deformation terms, as follows:

$$z = \frac{cS^2}{1 + (1 - c^2 S^2)^{1/2}} + B_1 S^4 + B_2 S^6 + B_3 S^8 + B_4 S^{10}, \quad (A1.10)$$

where

$$B_1 = A_1 + \frac{[(K+1) - 1]c^3}{8}, \quad (A1.11)$$

$$B_2 = A_2 + \frac{[(K+1)^2 - 1]c^5}{16}, \quad (A1.12)$$

$$B_3 = A_3 + \frac{5[(K+1)^3 - 1]c^7}{128}, \quad (A1.13)$$

$$B_4 = A_4 + \frac{7vt[(K+1)^4 - 1]c^9}{256}; \quad (A1.14)$$

or at other times this expression is preferred:

$$z = D_2 S^2 + D_4 S^4 + D_6 S^6 + D_8 S^8 + D_{10} S^{10}, \quad (A1.15)$$

where

$$D_2 = \frac{c}{2}, \quad (A1.16)$$

$$D_4 = \frac{c}{2}\left(\frac{c}{2}\right)^2 + B_1 = \frac{c^3}{8} + B_1, \quad (A1.17)$$

$$D_6 = c\left(\frac{c}{2}\right)^4 + B_2 = \frac{c^5}{16} + B_2, \quad (A1.18)$$

$$D_8 = \frac{5c}{2}\left(\frac{c}{2}\right)^6 + B_3 = \frac{5c^7}{128} + B_3, \quad (A1.19)$$

$$D_{10} = \frac{14c}{2}\left(\frac{c}{2}\right)^2 + B_4 = \frac{7c^9}{256} + B_4, \quad (A1.20)$$

A1.4. ABERRATION OF THE NORMALS TO THE SURFACE

A normal to the optical surface intersects the optical axis at a distance Z_n from the vertex of the surface. To compute this distance we need to know the value of the derivative of z with respect to S, which is given by

$$\frac{dz}{dS} = \frac{cS}{[1 - (K + 1)C^2 S^2]^{1/2}} + 4A_1 S^3 + 6A_2 S^5 + 8A_3 S^7 + 10A_4 S^9.$$

$$(A1.21)$$

The distance Z_n is then given by

$$Z_n = \frac{S}{dz/dS} + z; \quad (A1.22)$$

and, as shown by Buchroeder et al. (1972), for conic surfaces this expression becomes

$$Z_n = \frac{1}{c} - Kz. \quad (A1.23)$$

A1.5. SPHERICAL ABERRATION OF AN ASPHERICAL SURFACE

Equation (A1.22) may be approximated by

$$Z_n = \frac{1}{c} - \frac{(Kc^3 + 8A_1)S^2}{2c^2}. \tag{A1.24}$$

A1.5. SPHERICAL ABERRATION OF AN ASPHERICAL SURFACE

A system of k-centered reflective or refractive aspherical surfaces has a transverse third-order spherical aberration given by

$$\text{Sph}^T = \sum_{j=1}^{k} \text{Sph}_j \tag{A1.25}$$

with

$$\text{Sph}_j = \frac{y(N_{-1} - N)}{-2N_k u_k} \left\{ (8A_1 + Kc^3)y^3 + \frac{N_{-1}}{N^2} \right.$$

$$\left. \cdot [(N + N_{-1})u_{-1} + N_{-1}yc](yc + u_1)^2 \right\}, \tag{A1.26}$$

where u_{-1} and u are the slopes of the incident and refracted or reflected rays, N_{-1} and N are the refractive indices, and y is the height of the ray on the surface. The sign convention is in Fig. A1.3. For a reflective surface we may write N_{-1}

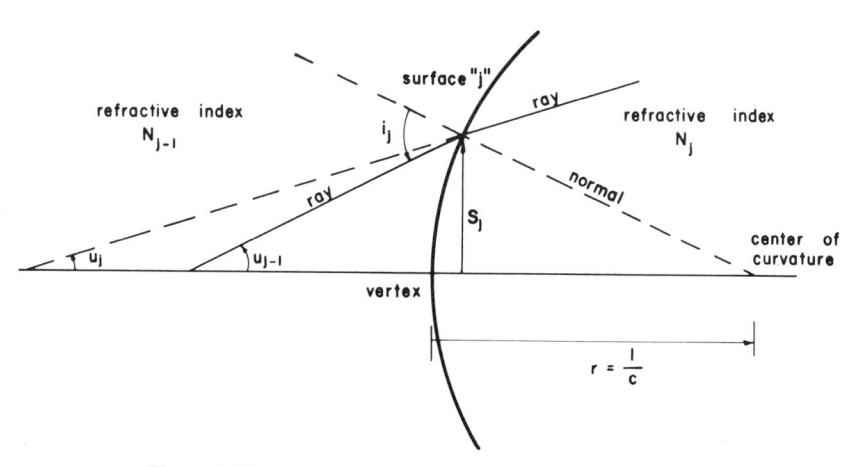

Figure A1.3. Sign convention (all quantities shown are positive).

$= -N = 1$, hence the transverse spherical aberration becomes

$$\text{Sph}_j = \frac{y}{N_k u_k} \left\{ (8A_1 + Kc^3)y^3 + (yc + u_{-1})^2 yc \right\}. \tag{A1.27}$$

Considering now the particular case of a mirror tested at its center of curvature ($N_k = -1$, $u_{-1} = 0$, and $u = -yc$), the transverse spherical aberration is given by

$$\text{Sph} = \frac{-(8A_1 + Kc^3)y^3}{c}, \tag{A1.28}$$

while with a point source at infinity and the image at the focus ($u = -2yc$) we have

$$\text{Sph}_f = \frac{-(8A_1 + Kc^3 + c^3)y^3}{2c}. \tag{A1.29}$$

A1.6. COMA OF A CONCAVE MIRROR

The expression for the third-order sagittal coma with the pupil at the optical surface is independent of the asphericity of the mirror and is given by

$$\text{Coma}_s = (yc + u)cyh \tag{A1.30}$$

where h is the image height (or deviation from the optical axis). It is quite interesting that, if a surface is tested near the center of curvature ($u = -yc$), no third-order coma is introduced if the light source is placed slightly off axis.

A1.7. ASTIGMATISM OF A CONCAVE MIRROR

The Petzval surface of a concave reflecting optical surface depends only on the curvature of the surface and has a curvature $1/\rho_p$, given by

$$\frac{1}{\rho_p} = 2c. \tag{A1.31}$$

We restrict ourselves to the case of an optical surface and its pupil at the same place, since this is the most interesting configuration in optical testing. Then we can show that the sagittal surface is always flat and that the tangential surface has a curvature $1/\rho_T$, given by

$$\frac{1}{\rho_T} = -4c. \tag{A1.32}$$

The expression for the third-order transverse sagittal astigmatism (pupil at the optical surface) as measured on the Petzval surface is

$$\text{Ast}_s = cuh^2. \tag{A1.33}$$

The tangential astigmatism on the Petzval surface has three times the magnitude of the sagittal astigmatism. The difference between these two aberrations is a residual transverse aberration, given by

$$TA_{\text{ast}} = \text{Ast}_t - \text{Ast}_s = 2\text{Ast}_s. \tag{A1.34}$$

Thus, when testing with the light source slightly off axis, the apparent astigmatic difference between the tangential and the sagittal wavefront profiles is

$$W = -\frac{2}{l}\int_0^y \text{Ast}_s \, dy \tag{A1.35}$$

where l is the distance from the surface to the image ($y = ul$). It may be shown that

$$W = -cu^2h^2 = \frac{-ch^2y^2}{l^2}. \tag{A1.36}$$

Therefore an apparent astigmatism equal to $W/2$ will be found on the surface under test. The tangential curvature appears stronger than the sagittal curvature. Notice that, if the surface is tested at the center of curvature, l is the radius of curvature and h is half the separation between the point source and the image.

A1.8. CAUSTIC PRODUCED BY AN ASPHERIC SURFACE

When testing an aspheric surface at its center of curvature, it is useful at times to know the pertinent dimensions for the caustic, which may be derived by the method to be explained. The wavefront $W(S)$ reflected from the aspheric surface can be written as (see Fig. A1.4)

$$W(S) = \frac{Kc^3}{4}S^4 + \frac{(\Delta L)c^2}{2}S^2 \tag{A1.37}$$

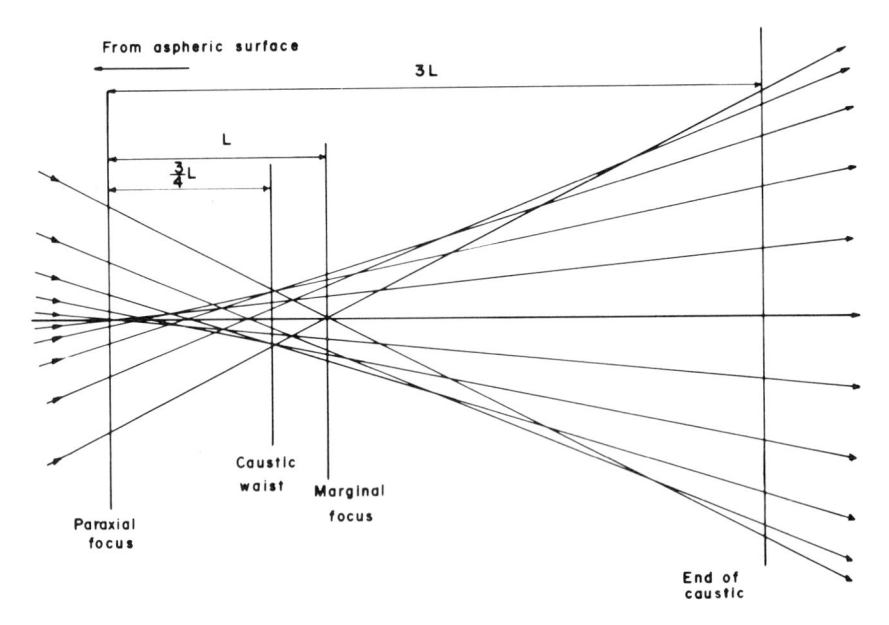

Figure A1.4. Caustic produced by an aspheric surface.

where ΔL is the distance of the plane under consideration from the paraxial focus. The distance ΔL from the paraxial to the marginal focus can be found by means of the condition

$$\left(\frac{dW}{dS}\right)_{S=S_{\max}} = 0 \tag{A1.38}$$

where $S_{\max}$ is the semidiameter of the surface under test. Thus we obtain

$$(\Delta L)_{\text{marginal focus}} = L = -KcS_{\max}^2. \tag{A1.39}$$

The distance ΔL to the end of the caustic from the paraxial focus is found with the condition

$$\left(\frac{d^2W}{dS^2}\right)_{S=S_{\max}} = 0, \tag{A1.40}$$

thus obtaining

$$(\Delta L)_{\text{end of caustic}} = 3L. \tag{A1.41}$$

A1.9. OFF-AXIS PARABOLOIDS

The distance ΔL from the waist of the caustic to the paraxial focus is obtained with the condition

$$\left(\frac{dW}{dS}\right)_{S=S_{\max}} = -\left(\frac{dW}{dS}\right)_{S=\sigma}, \qquad (A1.42)$$

where σ is the value of S that gives a maximum or minimum for dW/dS or, equivalently, $(d^2W/dS^2)_{S=\sigma} = 0$. Then it is possible to obtain

$$(\Delta L)_{\text{waist of caustic}} = \tfrac{3}{4}L. \qquad (A1.43)$$

The diameter w of the waist, or circle of least confusion, is given by

$$w = \frac{2}{c}\left(\frac{dW}{dS}\right)_{S=S_{\max}}, \qquad (A1.44)$$

using the value of ΔL for the waist of the caustic. The result is

$$w = -\tfrac{1}{2} Kc^2 S_{\max}^3. \qquad (A1.45)$$

The diameter at the beginning of the caustic (paraxial focus) is equal to $4w$, and at the end of the caustic it is equal to $8w$.

A1.9. OFF-AXIS PARABOLOIDS

An off-axis paraboloid like that in Fig. A1.5, whose axis is tilted by an angle θ, may be considered in a first approximation as a toroidal surface if the diameter is small compared with its radius of curvature. The tangential curvature c_t is given by (Malacara 1990)

$$c_t = \frac{\cos^3 \theta}{2F} \qquad (A1.46)$$

where F is the paraboloid focal length. The sagittal curvature c_s is given by

$$c_s = \frac{\cos \theta}{2F} \qquad (A1.47)$$

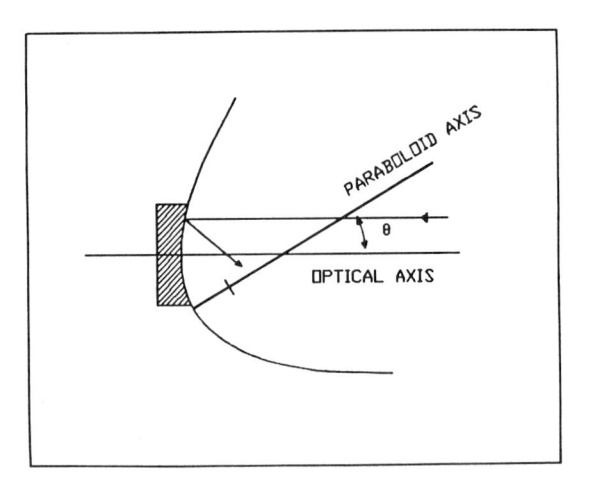

Figure A1.5. Off-axis paraboloid.

since the on-axis vertex curvature c of the paraboloid is given by

$$c = \frac{1}{2F}, \qquad (A1.48)$$

we may see that

$$c_t c^2 = c_s^2, \qquad (A1.49)$$

which is true not only for paraboloid but for any conic, as shown by Menchaca and Malacara (1984).

If the diameter of the off-axis paraboloid is not very small relative to its focal length, an additional comalike term has to be considered. With larger apertures even a triangular astigmatism term appears (Malacara, 1990).

A1.10. SPHERO-CYLINDRICAL AND TOROIDAL SURFACES

A sphero-cylindrical surface may be represented by the expression (Menchara and Malacara, 1986)

$$z = \frac{c_s x^2 + c_t y^2}{1 + \left[1 - \frac{(c_s x^2 + c_t y^2)^2}{(x^2 + y^2)^{1/2}}\right]}. \qquad (A1.50)$$

If the sagittal and tangential curvatures c_s and c_t are equal to c, this expression becomes identical to (A1.1). This expression is completely symmetrical in its form, in x and y.

A toroidal surface has the shape of a donut, and it may be represented by

$$z = ([(r^2 - y^2)^{1/2} + R - r]^2 - x^2)^{1/2} + R. \qquad (A1.51)$$

This expression is not symmetrical in x and y as Eq. (A1.50) because the toroid has an axis of symmetry about an axis parallel to the y axis ($x = 0$, $z = R$), but it does not have any axis of symmetry parallel to the x axis. Here, R is the radius of curvature on the x–z plane (semidiameter of the donut), and r is the radius of curvature in the y–z plane (radius of the circular section of the donut). If the optical surface is small compared with its radius of curvature, the distinction between the sphero-cylindrical and the toroidal surfaces becomes purely academic, and they are the same for all practical purposes. They have in common that their cross sections in the x–z and y–z planes are circles.

REFERENCES

Buchroeder, R. A., L. H. Elmore, R. V. Shack, and P. N. Slater, "The Design, Construction, and Testing of the Optics for a 147-cm-Aperture Telescope," *Optical Sciences Center Technical Report*, No. 79, University of Arizona, 1972.

Feder, D. P., "Optical Calculations with Automatic Computing Machinery," *J. Opt. Soc. Am.*, **41,** 630 (1951).

Malacara, Daniel, "Some Parameters and Characteristics of an Off-Axis Paraboloid," *Opt. Eng.*, **30,** 1277 (1990).

Menchaca, C. and D. Malacara, "Directional Curvature in a Conic Mirror," *Appl. Opt.*, **23,** 3258 (1984).

Menchaca, Carmen and Daniel Malacara, "Toroidal and Sphero-Cylindrical Surfaces," *Appl. Opt.*, **25,** 3008 (1986).

Schulz, Gunter, "Aspheric Surfaces," in *Progress in Optics*, Vol. 2, E. Wolf, Ed., Elsevier Science Publisher, New York, 1988, p. 351.

Shnurr, Alvin D. and Allen Mann, "Optical Figure Characterization for Cylindrical Mirrors and Lenses," *Opt. Eng.*, **20,** 412 (1981).

ADDITIONAL REFERENCES

Briers, John D., "Best-Fit Spheres and Conics as an Aid in the Manufacture and Testing of Diamond-Turned Aspheric Optics," *Opt. Acta*, **32,** 169 (1985).

Cardona-Nuñez, O., A. Cornejo-Rodriguez, R. Diaz-Uribe, and A. Cordero-Dávila, "Conic that Best Fits an Off-Axis Conic Section," *Appl. Opt.*, **25,** 3585 (1986).

Levine, Michael A., "Complete On-Axis Focusing Formula For Any Conic Mirror," *J. Opt. Soc. Am. A.*, **3,** 2082 (1986).

Tatian, Berge, "Least-Squares Fitting of Aspheric Surfaces by a Conicoid," *Appl. Opt.*, **28,** 4687 (1989).

Appendix 2

Some Useful Null Testing Configurations

A2.1. INTRODUCTION

In this appendix we will describe some of the null testing configurations used to test some of the most common optical surfaces and optical components. Foucault, Ronchi, and many other interferometric tests can be used with these arrangements. Some of these tests are described in this book.

A2.2. FLAT AND CONCAVE SPHERICAL SURFACES

Some null test configurations appropriate for flat or concave spherical surfaces are in Fig. A2.1, with the relevant parameters and dimensions, (Ritchey, 1904).

A2.3. CONCAVE PARABOLIC OR HYPERBOLIC SURFACES

A popular configuration for testing paraboloids (see Fig. A2.2) is by autocollimation with an optical flat. The amount of spherical convexity or concavity permissible in the flat mirror used in autocollimation tests was shown by Burch (1938) to be given by

$$\delta = 64 \left(\frac{F}{D}\right)^2 \frac{\epsilon}{4Q - \frac{1}{2}}, \quad (A2.1)$$

where F is the effective focal length and D is the aperture diameter of the system under test. The symbol δ represents the depth in fringes of the spherical concavity or convexity of the "flat" mirror, and ϵ represents the tolerance, also in fringes, of the zonal effect error introduced. If the system under test is refracting, the zonal error is $-2\epsilon/(N - 1)$, where N is the refractive index. The parameter Q is defined by

$$Q = -\frac{\text{OSC}}{\sin^2 \theta} = \frac{1}{\sin^2 \theta}\left(\frac{Y}{F \sin \theta} - 1\right) \quad (A2.2)$$

756   SOME USEFUL NULL TESTING CONFIGURATIONS

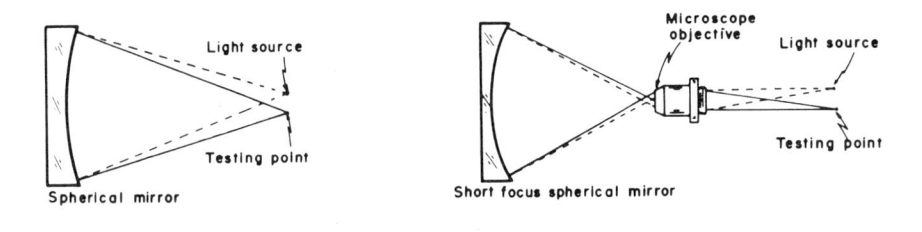

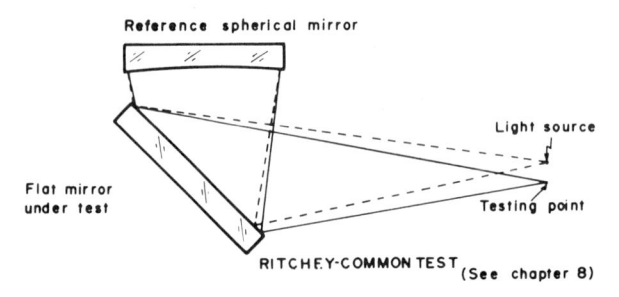

Figure A2.1. Flat and concave spherical surfaces.

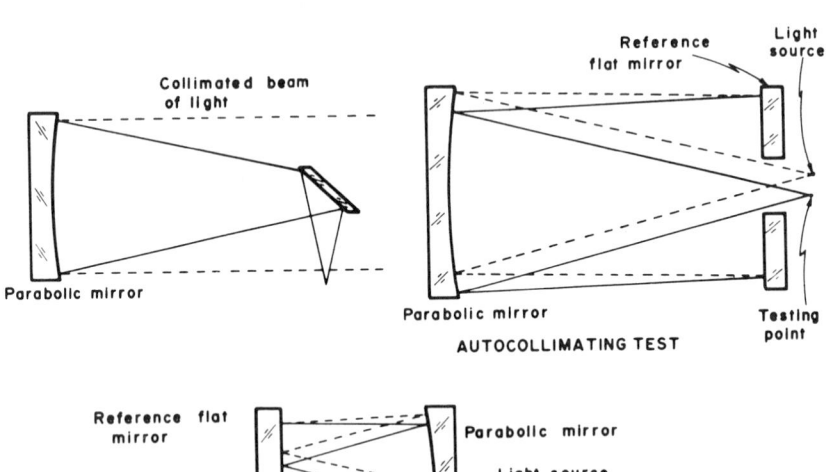

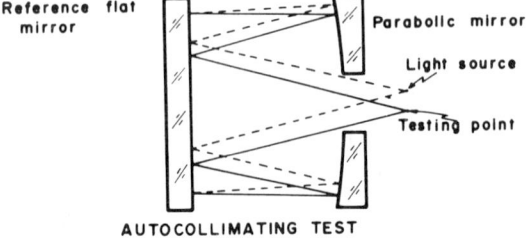

Figure A2.2. Concave parabolic surfaces.

where OSC is the "offense against the sine condition" and θ is the angle at which a marginal ray with height Y at the entrance pupil converges to the focus of the system. As Burch pointed out, a paraboloid and an aplanatic system are the two cases of practical interest: for these, $Q = \frac{1}{4}$ and $Q = 0$, respectively, giving

$$\delta = \pm 128 \left(\frac{F}{D}\right)^2 \epsilon. \tag{A2.3}$$

If the paraboloid has a large aperture, a point light source may be placed at its focus. Then, the collimated beam may be examined with another paraboloid with the same diameter but much larger focal length, as proposed by Parks (1974).

An off-axis paraboloid may be tested in many ways, as explained by Meinel and Meinel (1989). If the paraboloid is not too large, a flat mirror may be used in an autocollimating configuration, as shown in Fig. A2.3.

A2.4. CONCAVE ELLIPSOIDAL OR SPHEROIDAL SURFACES

An ellipsoidal mirror obtained by rotation of the ellipse about its major axis may be tested with conjugates at finite but different distances (Kirkham 1953), as shown in Fig. A2.3. An oblate spheroid is obtained by rotating the ellipse about its minor axis. In this case the images at the foci of the ellipse are astigmatic. A small cylindrical lens may be placed near its focus in order to correct this astigmatism if necessary (Everhart, 1966), as shown in Fig. A2.4. Several different arrangements to test oblate spheroids using refractive compensators have been proposed, as described by Rodgers and Parks (1984).

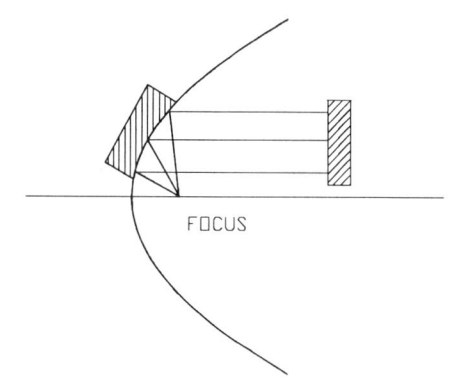

Figure A2.3. Testing an off-axis paraboloid.

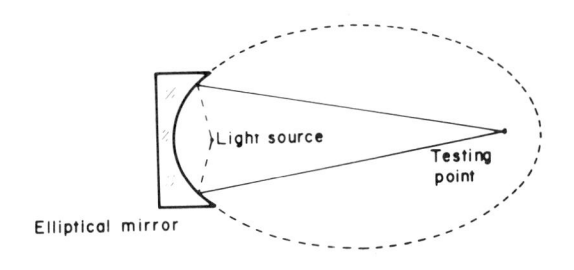

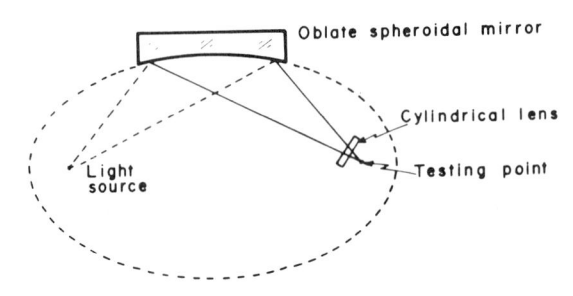

Figure A2.4. Elliptical surfaces.

A2.5. HYPERBOLOIDAL SURFACES

A convex hyperbolic surface may be tested with any of the methods shown in Fig. A2.5. The first method is the Hindle test (Hindle 1931) described in the chapter on compensators. The second method is the Silvertooth test, for concave hyperboloids (Silvertooth 1940). The third method illustrated in this figure is the Ritchey's autocollimation method for a complete telescopic system. An interesting variation of the Hindle test has been described by Robbert et al. (1979).

An inconvenience with the Hindel test is that a large spherical mirror is needed. Another solution has been proposed by Meinel and Meinel (1983a, 1983b) in order to test it from the back surface. The mirror has to be made out of fused quartz in order to have good transparency and homogeneity. There are two possible solutions, one is shown in Fig. A2.6 with the light source and the testing point at the same position. The surface has to be slightly convex, with a long radius of curvature. As pointed out by Meinel and Meinel, a better solution is obtained if the back surface is made flat and the spherical aberration is completely corrected by placing the light source and the testing point separated along the optical axis, as shown in Fig. A2.7. Many interesting variations of this test and some others may be found in a paper by Parks and Shao (1988).

A2.6. ASPHERICAL LENSES

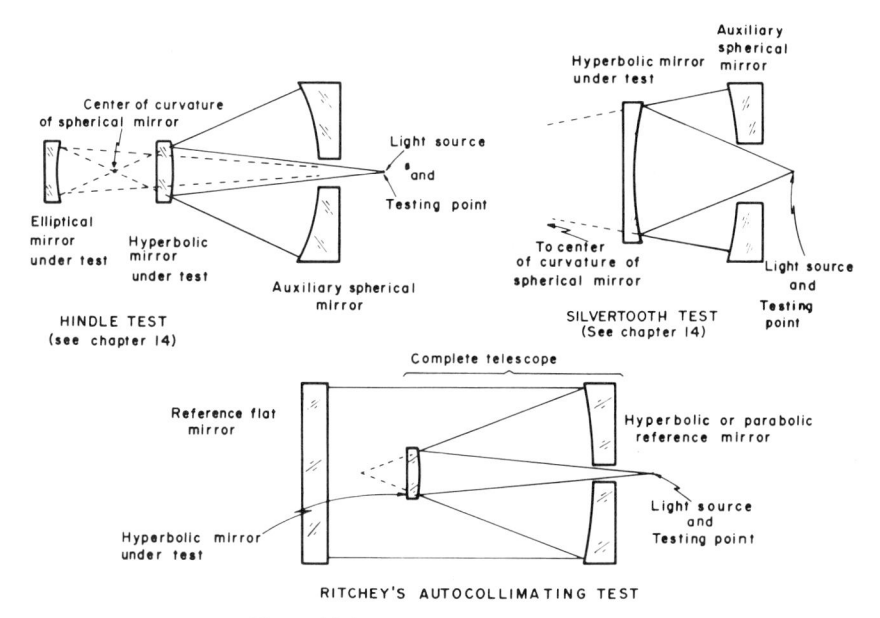

Figure A2.5. Telescope secondary mirrors.

A2.6. ASPHERICAL LENSES

A convex aspherical surface in a lens may be tested as shown in Fig. A2.8 (James and Waterworth 1965). The spherical aberration introduced by the aspherical surface is minimized by selecting an appropriate distance from the light source to the surface. This may be done by ray tracing, but an approximate

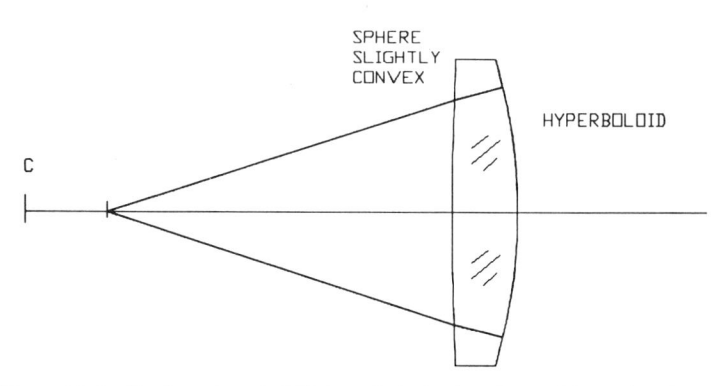

Figure A2.6. Testing a hyperboloid through the back surface, with equal conjugates.

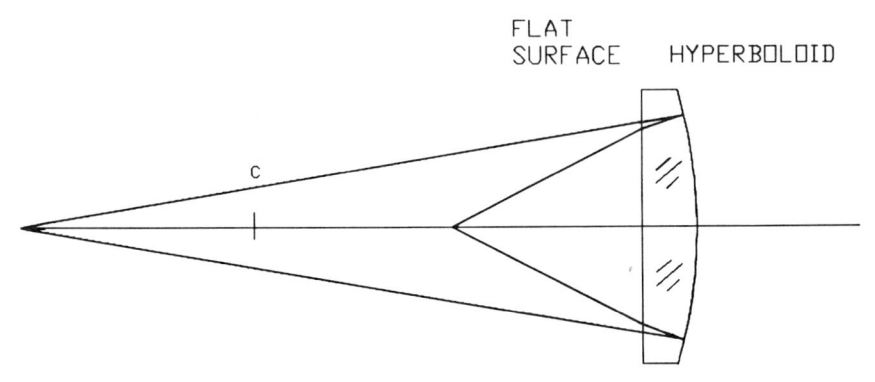

Figure A2.7. Testing a hyperboloid through the back surface, with different conjugates.

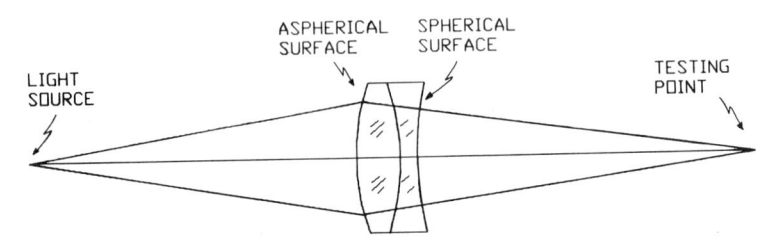

Figure A2.8. Testing of an aspherical lens.

solution may be calculated by requiring that the third-order spherical aberration in Eq. (A1.26) be zero. By making $N_{-1} = 1$, and defining a distance $l = y/u_{-1}$ this condition is

$$(8A_1 + Kc^3) + \frac{1}{N^2}[(N + 1)/l + c](c + 1/l)^2 = 0. \qquad (A2.4)$$

The spherical aberration of the second surface is eliminated by contacting this surface using oil, to another lens, whose second surface is spherical and concentric with the testing point.

A2.7. TELESCOPE REFRACTING OBJECTIVES

A telescope doublet may be easily tested by autocollimation against an optical flat, as shown in Fig. A2.9.

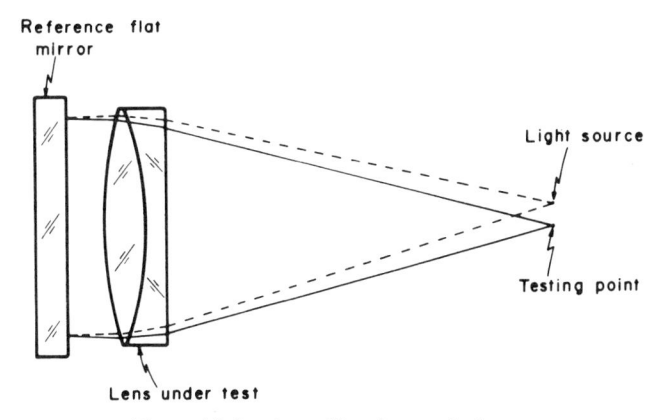

Figure A2.9. Autocollimating test for lenses.

REFERENCES

Burch, C. R., "Tolerance Permissible in Flats of Autocollimation Tests," *Mon. Not. R. Astron. Soc.*, **98,** 670 (1938).

Everhart, E., "Null Test for Wright Telescope Mirrors," *Appl. Opt.*, **5,** 717 (1966).

Hindle, J. H., "A New Test for Cassegrainian and Gregorian Secondary Mirrors," *Mont. Not. R. Astron. Soc.*, **91,** 592 (1931). Reprinted in *Amateur Telescope Making*, vol. 1, A. T. Ingalls, Ed., Scientific American, New York, 1950, p. 215.

James, W. E. and M. D. Waterworth, "A Method for Testing Aspheric Surfaces," *Opt. Acta*, **12,** 223 (1965).

Kirkham, Alan R., "The Direct Focal Test for Gregorian Secondaries," in *Amateur Telescope Making*, vol. 1, A. G. Ingalls, Ed., Scientific American, New York, 1953, p. 271.

Meinel, Aden B. and Marjorie P. Meinel, "Self-Null Corrector Test for Telescope Hyperbolic Secondaries," *Appl. Opt.*, **22,** 520 (1983a).

Meinel, Aden B. and Marjorie P. Meinel, "Self-Null Corrector Test for Telescope Hyperbolic Secondaries: Comments," *Appl. Opt.*, **22,** 2405 (1983b).

Meinel, Aden B. and Marjorie P. Meinel, "Optical Testing of Off-Axis Parabolic Segments Without Auxiliary Optical Elements," *Opt. Eng.*, **28,** 71 (1989).

Parks, Robert E., "Making and Testing an $f/0.15$ Parabola," *Appl. Opt.*, **13,** 1987 (1974).

Parks, Robert E., and L. Z. Shao, "Testing Large Hyperbolic Secondary Mirrors," *Opt. Eng.*, **27,** 1057 (1988).

Ritchey, G. W., "On the Modern Reflecting Telescope and the Making and Testing of Optical Mirrors," *Smithson. Contrib. Knowl.*, **34,** 3 (1904).

Robbert, C. R., P. R. Yoder, and L. A. Montagnino, "Typical Error Budget for Testing a High Performance Aspheric Telescope Mirror," *Proc. SPIE*, **181,** 56 (1979).

Rodgers, John M. and Robert E. Parks, "Null Tests for Oblate Spheroids," *Appl. Opt.*, **23,** 1246 (1984).

Silvertooth, W., "A Modification of the Hindle Test for Cassegrain Secondaries," *J. Opt. Soc. Am.*, **30,** 140 (1940).

ADDITIONAL REFERENCE

Mackintosh A., Ed., *Advanced Telescope Techniques, Vol. 1: Optics*, Willmann-Bell, Inc., Richmond, VA, 1987.

Additional Bibliography

Some additional bibliographic material on the general subject of optical testing are listed here. These references cover topics described in several of the chapters of this book, but many are so general that they were not mentioned in any of them.

ARTICLES AND CHAPTERS IN BOOKS

Anderson, David S., "Interferometry at the University of Arizona," *Proc. SPIE*, **816**, 158-179 (1987).

Briers, John D., "Interferometric Testing of Optical Systems and Components," *Opt. Laser Technol.*, **4**, 28-41 (1972).

Caulfield, H. John and William Friday, "Bibliography on Optical Testing," *Appl. Opt.*, **20**, 1497-1498 (1981).

Cornejo-Rodriguez, Alejandro, John H. Caulfield, and William Friday, "Testing of Optical Surfaces: A Bibliography," *Appl. Opt.*, **20**, 4148-4148 (1981).

Creath, Katherine, "WYKO Systems for Optical Metrology," *Proc. SPIE*, **816**, 111-127 (1987).

Eastman, D. R., "Interferometric Testing at Perkin Elmer," *Proc. SPIE*, **816**, 228-237 (1987).

Greivenkamp, John E., "Interferometric Measurements at Eastman Kodak Company," *Proc. SPIE*, **816**, 212-227 (1987).

Hariharan, P., "Interferometry with Lasers," in *Progress in Optics*, vol. 24, E. Wolf, Ed., Elsevier, New York, 1987, p. 105.

Hariharan, P., "Interferometric Metrology: Current Trends and Future Prospects," *Proc. SPIE*, **816**, 2-8 (1987).

Kwon, Osuk Y., "Advanced Wavefront Sensing at Lockheed," *Proc. SPIE*, **816**, 196-211 (1987).

Malacara, Daniel, Alejandro Cornejo-Rodriguez, and M. V. R. K. Murty, "Bibliography of Various Optical Testing Methods," *Appl. Opt.*, **14**, 1065-1080 (1975).

Malacara, Daniel, "Updated Optical Testing Bibliography," *Appl. Opt.*, **29**, 1384-1387 (1990).

Malacara, Daniel, "A Review of Interferogram Analysis Methods," *Proc. SPIE, 1332*, 678-689 (1990).

Ritchey, George W., "On the Modern Reflecting Telescope and the Making and Testing of Optical Mirrors," *Smithson. Contrib. Knowl.*, **34**, 3, (1904).

Rosenbruch, Klaus J., "Testing of Optical Components and Systems," *J. Optics (India)*, **13**, 23-43 (1984).

Saunders, James B., "Precision Measurements," in *Advances in Optical Techniques*, A. C. S. Van Heel, Ed., North-Holland, Amsterdam, 1967.

Schwider, J., "Advanced Evaluation Techniques in Interferometry," in *Progress in Optics*, vol. 28, E. Wolf, Ed., Pag. 273, North-Holland, Amsterdam, 1990.

Smythe-Robert A., J. A. Soobitzki, and B. E. Truax, "Recent Advances in Interferometry at Zygo," *Proc. SPIE*, **816**, 95-105 (1987).

Tiziani, Hans J., "Application of Interferometry for Testing Macro/Microgeometry of Optical Surfaces," *Proc. SPIE*, **381**, 209-216 (1983).

Tiziani, Hans J., "Current State of Optical Testing," *Proc. SPIE*, **680**, 2-5 (1986).

Tiziani, Hans J., "Optical Methods for Precision Measurements," *Opt. Quantum Elect.*, **21**, 253-282 (1989).

Wyant, James W., "Interferometric Optical Metrology: Basic Principles and New Systems," *Laser Focus*, May, 65-71 (1982).

Wyant, James C., "Interferometric Testing of Aspheric Surfaces," *Proc. SPIE*, **816**, 19-39 (1987).

Wyant, James C., "WYKO System for Optical Metrology," *Proc. SPIE*, **816**, 111-126 (1987).

Yatagai, Toyohiko, "Interferometric Testing Technology. Developments and Applications," *Proc. SPIE*, **816**, 58-78 (1987).

BOOKS

Dörband, Bernd, *Analyse Optischer Systeme Mit Hilfe von Automatischer Streifenauswertung und Strahldurchrechnung*, Universität Stuttgart, Stuttgart, 1986.

Guenther, A. H., *Optical Inteferograms, Reduction and Interpretation*, American Society for Testing and Materials, Philadelphia, PA, 1978.

Hariharan, P., *Optical Interferometry*, Academic Press, New York, 1985.

Ingalls, Albert G., *Amateur Telescope Making*, Vols. I, II, and III, Scientific American, New York, 1953, 1954, and 1956.

Mackintosh, A., Ed., *Advanced Telescope Techniques, Vol. 1: Optics*, Willmannn-Bell Inc., Richmond, VA, 1987.

Malacara, Daniel, Ed., *Selected Papers on Optical Shop Metrology*, SPIE Milestone Series Vol. MS 18, SPIE Opt. Eng. Press, Bellingham, Washington, 1990.

Massie, N. A., Ed., *Critical Reviews of Optical Science and Technology: Interferometric Metrology, Proc. SPIE Vol. 816*, SPIE Press, Bellingham, Washington, 1988.

Ronchi, Vasco, *La Prova dei Sistemi Ottici*, Nicola Zanichelli, Bologna, 1925.

Steel, William H., *Interferometry*, 2nd ed., Cambridge University Press, Cambridge, 1983.

Twyman, F., *Prism and Lens Making*, 2nd ed., Hilger and Watts, London, 1957.

Index

Abbe eyepiece, 718, 735
Abbe spherometer, 731
Aberration compensators, 427. *See also* Compensators
Aberration polynomial, 456
Aberration removal, 484
Aberrations, 78
   aberration polynomial, 456
   of aspherical surfaces, 746
   detection:
      with Foucault test, 273
      with lateral shear, 125
      with Ronchi test, 324
      with star test, 398, 400
      with Twyman–Green, 77
   in double pass interferometer, 247
   isometric plots for primary, 457
Absolute calibration of interferometers, 577
Absolute testing of flats, 43, 580
AC interferometry, 501
Air turbulence, 543
Airy pattern, 399
   defocused, 401
Aldis spherometer, 729
Aliasing, 553
Alignment fiducials, 588
Amici prism, testing with Twyman–Green interferometer, 65
Angle measurement, 715
   angle block set, 716
   Angle Dekkor, 718
   angle vise, 716
   autocollimator, 718
   bevel gauge, 716
   divided circles, 715
   goniometers, 715
   interferometric, 720
   polygons, 716
   prisms, 721
      and polygons, 719
   sine plate, 716
Aspherical lens testing, 759
Aspherical surface:
   aberration at center of curvature, 427
   astigmatism of, 748
   caustic of, 749
   coma of, 747

   definition, 743
   patterns with Ronchi test, 331
   spherical aberration of, 747
   testing, 584
      with computer hologram (CGH), 604
      by phase shifting, 553
Astigmatism, detection, 747
   with Foucault test, 272
   with lateral shear, 132
   with Ronchi test, 327
   with rotational shear, 192
   with start test, 423
   with Twyman–Green, 81
Atomic force microscope (AFM), 694
Autocollimation test, 756
   for lenses, 761
   for parabolic mirrors, 367
Autocollimator, 718, 732
   direct reading, 719
   electronic, 719
   microptic, 719
Autocorrelation method, to measure OTF, 116
Autostigmatic systems, 109
Axial symmetry, detection by star test, 423
Axicon surface, 744

Babinet compensator, lateral shear interferometer with, 164
Bar spherometer, 731
Beam splitter:
   accuracy of, in Twyman–Green, 56
   birefringent, 99
   Saunders's, 110
Bevel gauge, 716, 721
Bibliography, general, 763
Bilinear interpolating function, 487
Birefringent beam splitters, 99
Birefringent lens, 103
Burch compensator, 428, 429
Burch interferometer, 96

Calibration of interferometers:
   absolute, 577
   reference, 577
   three flat test, 580

Cassegrain system tested with star testing, 419
Caustic, 297, 748
Chromatic aberration, detection with star test, 422
Circular grid, 353
Coddington equations, 311
Coherence:
  in double pass interferometer, 247
  in lateral shear interferometer, 124
  in Newton and Fizeau interferometer, 10, 25
  in radial shear interferometer, 178
  in Ronchi test, 351
  in Twyman-Green interferometer, 58
Coma aberration, detection, 747
  with Foucault test, 275
  with lateral shear, 131
  with Ronchi test, 326
  with rotational shear, 193
  with star test, 412
  with Twyman-Green, 80
Common path interferometer, 95, 113
Compensators, 427
  Burch, 428, 429
  Couder, 428
  Dall, 430
  holographic, 599
  Offner, 434
  Ross, 429
  Shafer, 433
Computer processing in interferometry, 568
Concave conicoids, *see* Concave ellipsoidal surface; Concave hyperboloidal surface; Concave oblate spheroidal surface; Concave paraboloidal surface
Concave ellipsoidal surface, null testing configuration, 757
Concave hyperboloidal surface, testing:
  with Foucault test, 267, 289
  with Hartmann test, 370
  with Hindle test, 451
  with Offner compensator, 434
  with Platzeck-Gaviola test, 295
  with Ronchi test, 331
  with Shafer compensator, 433
  with Silvertooth test, 759
Concave mirror:
  astigmatism of, 748
  coma of, 748
Concave oblate spheroidal surface:
  null testing configuration, 757
  testing:
    with Modified Dall compensator, 444
    with Shafer compensator, 433
Concave paraboloidal surface:
  null testing configurations, 755

off-axis paraboloids, 751
tested with star testing, 418
testing:
  by autocollimation, 367
  with Dall compensator, 434
  with Foucault test, 267, 289
  with Hartmann test, 370
  with Lower test, 356
  with Offner compensator, 434
  with Platzeck-Gaviola test, 295
  with Ronchi test, 331
  with Shafer compensator, 433
Concave prolate spheroidal surface, *see* Concave ellipsoidal surface
Concave spherical surface:
  tested with star testing, 417
  testing:
    with common path interferometers, 95
    with double pass Fizeau, 252
    with Fizeau interferometer, 34
    with Foucault, 266
    with Gardner-Bennett test, 391
    with lateral shear interferometer, 152
    with Michelson test, 391
    with multipass interferometer, 257
    with multiple beam interferometer, 232
    with Newton fringes, 14
    with radial shear interoferometers, 173
    with Ronchi test, 323
    with test plates, 14
    with Twyman-Green interferometer, 67, 69
    with wire test, 290
  testing configurations, 755
Concentric beam interferometers, 703
Confocal cavity, 733
Conic constant, 743
Conic surface:
  aberration at center of curvature, 427
  aberration of the normals, 746
  definition, 743
  null testing configurations, 447, 745, 755
  parameters for, 744
Convex conicoid, *see* Convex ellipsoidal surface; Convex hyperboloidal surface; Convex paraboloidal surface
Convex ellipsoidal surface:
  null transmission configuration, 447
  testing with Hindle test, 451
Convex hyperboloidal surface:
  with compensating lens, 449
  null transmission configuration, 447, 448
  testing:
    with Hindle test, 450, 759
    with Meinel test, 448, 449, 759, 760
    with Ritchey test, 759
    with Simpson-Oland-Meckel, 451

INDEX

Convex paraboloidal surface, testing with Hindle test, 450
Convex prolate spheroidal surface, *see* Convex ellipsoidal surface
Convex spherical surface, testing:
  with Fizeau interferometer, 34
  with Newton fringes, 14
  with test plates, 14
  with Twyman–Green interferometer, 67, 69
Couder compensator, 428
Couder screen, 289
Coupled interferometer, 234
Cube corner prism, testing:
  with double pass Fizeau, 251
  with Fizeau interferometer, 30
  with Twyman–Green, 65
Curvature measurement, *see* Radius of curvature measurement
Cyclic interferometer, 141, 181

Dall compensator, 430
Detectors:
  charged coupled device (CCD), 530
  high speed, 532
  nonlinearities, 539
  in phase shifting interferometry, 546
  solid state, 530
  solid state sensors, 546
Diffraction based, lateral shear interferometer, 160
Diffraction grating (or ruling), testing, 64, 66
Diffractive beam splitters, 75
Digitizing tablets, 488
Direct interferometry, 494, 566
Distance measurement, 725
  interferometric, 574, 725
Distortion, detection by star test, 423
Divided circles, 715
Double focus interferometer, 108
Double focus systems, 102
Double pass interferometer, 107, 183
Dyson interferometer, 108
Dyson system to test microscopes, 60

E-beam recorder CGH, 604
Electronic speckle pattern interferometry, 629
Electro-optic holography, 631
Ellipsoidal surface, definition, 743. *See also* Concave ellipsoidal surface; Convex ellipsoidal surface
Equal chromatic order fringes, 221
Equivalent wavelength:
  in two wavelength holography, 612
  in two wavelength interferometry, 561

Errors, asymmetric, 580
Extended range PSI techniques, 553
Extraneous fringes, 543

Fabry–Perot interferometer, 212, 225
  Tolansky fringes in, 231
FECO interferometer, 698
Fit variance, 467
Fizeau interferometer, 1, 18, 218, 508
  collimating lens for, 36
  double passed, 249
  with laser, 27
  light source for, 25
  multipass, 258
  multiple beam, 212
  Wyko interferometer, 29
  Zygo interferometer, 28
Flat surface:
  null testing configurations, 755
  testing:
    with autocollimators, 719
    with Fizeau interferometer, 22
    with multipass, 259
    with multiple beam interferometers, 226
    with Newton fringes, 4
    with a reference sphere, 107
    with Ritchey–Common test, 309, 369
    with three flat method, 43, 580
    with Twyman–Green, 63
Focal length measurement, 735, 737
  focimeters, 735
  nodal slide bench, 735
Focimeters, 735
Foucault test, 265, 732
  computer simulated patterns, 281
  with Couder screen, 289
  geometrical theory, 270
  physical theory, 280
  practical configurations, 270
  zonal test with, 289
Fourier analysis of interferograms, 491, 566
Four step algorithm, for phase shifting interferometry, 511
Fringe contrast in Ronchi test, 346
Fringe digitization, 487
Fringe interval in multiple beam interferometer, 225
Fringe projection, 668
  contour interval, 668
  techniques, 653
Fringe scanning interferometry, 501
Fringe sharpness in Ronchi test, 346
Fringe stability, 72

Gardner–Bennett test, 391
Gas laser, coherence of, 70

Gates reversal shear interferometer, 201
Gauge blocks, 725
Gauss eyepiece, 718, 734
Gaussian beams, 410
Geneva gauge, 732
Geometrical theory:
  of Platzeck-Gaviola test, 296
  of wire test, 291
Glass plates, testing of, 64
Goniometers, 715
Gram-Schmidt orthogonalization, 480
Graphic tablets, *see* Digitizing tablets
Grazing incidence interferometer, 573
  multipass, 258
Grid of wavefront data, 486
Group refractive index, 62

Haidinger fringes, application of, 38, 41
Haidinger interferometer, 1, 36
  laser source for, 38
Hartmann test, 367
  data reduction, 384
  helical pattern screen in, 377
  implementation, 379
  null, 356
  pattern for hyperboloidal mirror, 384
  radial pattern screen in, 374
  recent developments, 384
  screens in, 374
  set up, 379
  square array screen in, 379
  theory, 370
  and transverse aberrations, 373
  wavefront determination in, 373
Heterodyne interferometry, 501, 564
Hindle test, 450, 758, 759
Hologram:
  computer generated (CGH), 604
  to correct aberrations, 73
  errors in, 601
  real, 560
  synthetic, 603
Holographic:
  multiple beam interferometer, 237
  null tests, 599, 602
  radial shear interferometer, 188
  recording media, 560
  testing of symmetrical components, 601
  wavefront storage, 601
Holographic interferometry, 617
Holographic nondestructive testing:
  double exposure, 622
  dynamic, 622
  phase measurement, 624
  real time, 622
  secondary interference fringes, 622
  static, 621

Holographic tests, 599
Holography:
  nondestructive testing, 617, 625
  recording of two wavelengths, 614
  television, 627
  two angle, 675
  two wavelength, 612
Holography, television, *see* Television holography
Houston interferometer, 71
Hyperboloidal surface, definition, 743. *See also* Concave hyperboloidal surface; Convex hyperboloidal surface

Increased accuracy in double pass interferometer, 252
Inhomogeneity testing, 29
Integrating bucket, 515
Interference pattern, imaging of, 86
Interferograms:
  analysis of, 85
  direct analysis of, 494
  evaluation, 455
  Fourier analysis of, 491
  in Twyman-Green interferometer, 77
Interferometers:
  Burch, 96
  commercially available, 28, 29, 771
  common path, 95, 113
  concentric beam, 703
  cyclic, 141, 181
  Fabry-Perot, 212, 225
  Fizeau, 1, 18, 218, 508
  Haidinger, 1, 36
  Heterodyne, 501, 564
  Jamin, 138, 203
  lateral shear, 103, 111, 123
  Linnik or Smartt, 112, 299
  Mach-Zehnder, 74, 143, 147, 180, 508, 572
  Michelson, 51, 140, 144
  multiple beam, 207
  multiple pass, 247, 255
  Murty, 151
  Newton, 1
  phase shifting, 501
  polarization, 704
  radial shear, 173
  reversal shear, 173, 198, 200
  rotational shear, 173, 191
  Saunders, 111
  Shack, 34
  Twyman-Green, 51, 508
Interferometric microscopes, 689
Interferometric optical sine bar, 720
Interferometry:
  AC, 501
  fringe scanning, 501

INDEX

heterodyne, 501
phase shifting, 501
real time, 501
Interpolating functions, 487
Irradiance function, 503
Isometric plots for primary aberrations, 457

Jamin interferometer, 138, 203

Knife-edge test, 265, 732
Koster reversal shear interferometer, 201

Laser diode, testing with Mach–Zehnder, 75, 575
Lateral shear interferometer, 103, 111, 123
 with Babinet compensator, 164
 based on diffraction, 159
 based on polarization, 162
 in collimated light, 138
 in convergent light, 144
 with double frequency grating, 162
 using lasers, 151
 with Ronchi rulings, 342
 theory of, 125
Least squares fit of polynomials, 472
Lens:
 testing:
  by autocollimation, 760
  with Foucault, 266
  with Jamin interferometer, 203
  with Twyman–Green, 67
  with wire test, 290
 testing aspherical lenses, 447, 759
Light source:
 for Fizeau interferometer, 25
 for interferometry, 586
 for star testing, 414
Linear interpolating function, 487
Linnik interferometer, 112, 299
Liquid reference flats, 21
Lower test, 356
Lyot test, 305

Mach–Zehnder interferometer, 74, 143, 147, 180, 508, 572
Matrix:
 to go from monomials to Zernike, 476
 to go from Zernike to monomials, 473
Michelson interferometer, 51, 140, 144
Michelson test, 391
Microscope objectives:
 tested with star testing, 415, 419
 testing with radial shear interferometer, 184
 testing with Twyman–Green, 60, 68
Microscopic interferometry:
 concentric beam, 703

Fizeau, 699
Linik, 699
Michelson, 699
Mirau, 699
polarization, 704
Modulation of data, 513, 524, 528, 530
Moire, 675
 applications, 678
 contour interval, 658
 and conventional interferometry, 678
 effective wavelength, 678
 fringe equation, 656
 fringe projection, 668
 historical review, 666
 and interferograms, 658
 shadow, 671
 surface height determination, 674
 techniques, 653
 theory of, 654
Monomial representation of wavefronts, 457, 470
Multiple beam interferometer, 207
 for curved surfaces, 232
 holographic, 237
 precision in, 210
Multiple pass interferometer, 247, 255
Multiple wavelength:
 phase shifting, 616
 techniques, 612
Murty interferometer, 151
 applications of, 154
Murty lateral shear interferometer, 720
Murty radial shear interferometer, 187

Newton interferometer, 1
 light sources for, 11
 measurement of aspheric surfaces, 17
 peak error, 12, 13
 source size, 10
Newton rings, 1
Nodal lens bench, 68
Nodal slide bench, 421, 735
Noise, fixed pattern, 505
Nomarski microscope, 698
Nondestructive testing, with holographic interferometry, 617
Null Ronchi–Hartmann test, 356
Null Ronchi rulings, 332
Null Ronchi test, 332
Null tests, 424, 443
 using compensators, 427
 with computer generated holograms, 604
 configurations, 755
 errors in holographic, 602
 holographic, 599, 602
Nyquist frequency, 550

Oblate spheroidal surface, definition, 743. *See also* Concave oblate spheroidal surface; Convex oblate spheroidal surface
Oblique incidence interferometer, 76
Off-axis paraboloids, 751
Offner compensator, 434
　reflective, 438
　refractive, 435
Opaque surfaces, measurement of, 18
Optical design programs, 571
Optical errors in interferometer, 543
Optical flats, materials for, 11, 13. *See also* Flat surfaces
Optical focus sensor, 689, 696
Optical frequency difference, in phase shifting interferometry, 507
Optical path, in double pass interferometer, 247
Optical sine bar, 720
Optical surface:
　characteristics, 743
　definition, 743
Optical system, for star testing, 416
Optical testing books, 763
Optical transfer function, 113
Orthogonalization, Gram-Schmidt, 480
Orthogonal polynomials, 461

Paraboloidal surface, *see also* Concave paraboloidal surface; Convex paraboloidal surface
　definition, 743
　off-axis paraboloids, 751
Parallelism measurement, 719
Parallel plate interferometer, 151
Phase conjugating interferometer, 76
Phase gratings, for Ronchi test, 348
Phase lock interferometry, 564
Phase modulation tests, 265, 299
Phase shift calibration, 533
Phase shifting:
　with Bragg cell, 509
　holographic interferometry, 624
　holographic test plate, 603
　methods of, 506
　with moving grating, 507
　with moving mirror, 506
　multiple wavelength, 616
　with polarization components, 507
　with Ronchi test, 354
　with tilted parallel plate, 507
Phase shifting algorithms:
　averaging 3 + 3, 527
　Carre, 524
　Hariharan, 527
　least squares, 522
　N step, 524
　three-step, 519
　2 + 1, 529
Phase shifting interferometry, 501
　advantages of, 504
　air turbulence in, 543
　algorithms for, 511
　errors, 524, 528
　error sources, 536
　extraneous fringes in, 543
　future trends in, 588
　heterodyne interferometry, 564
　light sources for, 586
　modulation of laser diode, 587
　phase lock interferometry, 564
　quantization errors in, 541, 542
　simultaneous, 575
　source stability, 540
　stroboscopic, 577
　sub-Nyquist, 555
　synchronous detection in, 562
　two-wavelength, 560, 614
　vibration measurement, 625
　zero crossing analysis, 562
Phase unwrapping, in phase shifting interferometry, 514, 551
Photographic objectives, testing with nodal slide bench, 421
Physical theory:
　of Foucault test, 280
　of Ronchi test, 342
　of wire test, 294
Plane Parallel plate, testing:
　with autocollimator, 719
　with double pass Fizeau, 249
　with Fizeau interferometer, 22
　with Haidinger interferometer, 36
　with lateral shear, 151, 154
　with Twyman-Green, 63
Platzeck-Gaviola test, 295
　geometrical theory, 296
Point diffraction interferometer, 112, 299
Polariscope, Savart, 99
Polarization based, lateral shear interferometer, 162
Polychromatic light, and star test, 408
Polygon, 716
　testing, 719
Polynomial fitting of wavefronts, 472
Polynomial representation of wavefronts:
　monomial representation, 457, 470
　polynomial representation, 135, 456
　with Zernike polynomials, 464
Porro prism, testing of, 65
Primary aberrations, 78

INDEX

aberration polynomial for, 456
of aspherical surfaces, 746
detection:
  with double pass, 247
  with Foucault test, 273
  with lateral shear, 129
  with Ronchi test, 324
  with star test, 398, 400
  isometric plots for, 457
Prisms:
  measuring with bevel gauge, 717
  testing:
    with autocollimator, 719, 721
    with Fizeau interferometer, 30, 42
    with Jamin interferometer, 203
    with Twyman-Green, 65
Profiling:
  contact, 687
  FECO interferometer, 698
  imaging, 687
  interferometric microscope, 689, 698, 699
  Nomarski microscope, 698
  noncontact, 687
  optical focus sensor, 689, 696
  scanning, 687
  scanning probe microscope, 688
  scanning tunneling microscope, 693, 694
  stylus, 687, 688, 689
Prolate spheroidal surface, definition, 743.
  See also Concave ellipsoidal surface;
  Convex ellipsoidal surface
PTZ, Piezo electri transducer, 506
Pyramidal error, 721

Quadratic interpolating function, 487
Quantization errors, 541

Radial pattern screen, in Hartmann test, 374
Radial shear interferometer, 173
  Brown's, 179
  double pass, 183
  fringe visibility in, 178
  Hariharan and Sen's, 179
  holographic, 188
  laser, 185
  Mach-Zehnder based, 180
  Murty's, 187
  sensitivity of, 177
  single pass, 177
  Som's, 186
  Steel's, 188
  thick lens, 188
Radius of curvature measurement, 155, 585, 728
  Abbe spherometer, 731

Aldis spherometer, 729
autocollimator, 732
bar spherometer, 731
confocal cavity technique, 733
Foucault test, 732
Geneva gauge, 732
mechanical, 728
with multipass, 260
optical methods, 732
Ring spherometer, 730
spherometers, 729
Steinheil spherometer, 731
templates, 728
test plates, 729
traveling microscope, 733
Real time interferometry, 501
Reduced sensitivity tests:
  multiple wavelengths, 612
  two wavelengths, 612
Reference flats, liquid, 21
Reflective Offner compensator, 438
Refracting compensators, comments about, 443
Refracting objectives, testing, 760
Refractive index, measurement of, 156
Refractive Offner compensator, 435
Reversal shear interferometer, 173, 198, 200
  Gate's, 201
  Koster's, 201
  Saunder's, 202
Reversing interferometer, 111
Right angle prisms, testing, 30, 65
  with double pass Fizeau, 251
Ring spherometer, 730
Ritchey-Common test, 309, 369, 756
Ronchigrams:
  for primary aberrations, 324
  for some typical deformations, 336
Ronchi-Hartmann test, null, 356
Ronchi test, 159, 321
  fringe contrast and sharpness, 346
  geometrical theory, 322, 349
  history, 321
  null, 332
  with phase gratings, 348
  phase shifting with, 354
  physical theory, 342, 349
  practical aspects, 350
  sideband, 355
  Talbot effect in, 347
Root mean square value (RMS), 467, 485
Ross compensator, 429
Rotational shear interferometer, 173, 191
  Armitage and Lohmann's, 196
  Murty and Hagerot's, 196
  with phase compensation, 197

Rotational shear interferometer
    (*Continued*)
  source size compensated, 198
  source size uncompensated, 195

Saunders interferometer, 111
  prism, 110
  reversal shear, 202
Savart polariscope, 99, 103
Scanning method to measure OTF, 113
Scanning probe microscope, 691
Scanning tunneling microscope (STM), 693
Scatter plates, 96
Screens in Hartmann test, 374
Screen tests, 367
Series interferometer, 234, 236
Shack interferometer, 34
Shadow moire, 671
Shafer compensator, 433
Sideband Ronchi test, 355
Silvertooth test, 758
Simpson–Oland–Meckel test, testing with, 451
Simultaneous PSI, 575
Sine plate, 716
Single wavelength measurements, correction of, 616
Sivertooth test, 759
Small wavefront aberrations, measurement of, 299
Smartt interferometer, 112, 299
Solid state sensors, 546
Som's radial shear interferometer, 186
Source size compensation, in rotational shear interferometer, 195, 198
Spatial coherence:
  in lateral shear interferometer, 124
  in radial shear interferometer, 178
  in Twyman–Green interferometer, 58
Spatial sampling, in phase shifting interferometry, 546, 548
Spatial synchronous interferometry, 566
Speckle interferometry, 627
Speckle tests, 599
Spherical aberration, 747
  caustic, 297
  detection:
    with Foucault test, 273
    with lateral shear, 129
    with radial shear, 176
    with Ronchi test, 326
    with star test, 421
    with Twyman–Green, 79
Spherical surface, definition, 743. *See also* Concave spherical surface; Convex spherical surface
Sphero-cylindrical surface, 752
Spheroid, definition, 743
Spheroidal concave surface, testing, 757
  of oblate, 445
Spherometer, 729
Square array screen, in Hartmann test, 379
Star test, 397
  with central osculations, 408
  detection:
    of astigmatism, 423
    of axial symmetry, 423
    of chromatic aberration, 422
    of distortion, 423
    of spherical aberration, 421
  with large aberrations, 419
  light source for, 411
  with nonnull configurations, 424
  optical system for, 416
  and polychromatic test, 408
  quantitative, 419
  with small aberrations, 411
  with small convergence beams, 411
Steel's radial shear interferometer, 188
Steinheil spherometer, 731
Strehl limit, 397
Stroboscopic PSI interferometer, 577
Subaperture testing, 584
Sub-Nyquist interferometry, 555
Surface finish measurement, by profiling, 687
Surface roughness measurement, by profiling, 687
Symmetrical optical components, testing by holography, 601
Synchronous detection, in phase shifting interferometry, 562

Talbot effect, in Ronchi test, 347
Television holography, 627
  and electronic speckle pattern, 629
  and electro-optic holography, 631
  phase measurement, 633
  and speckle interferometry, 627
Templates, 728
Temporal coherence, in Twyman–Green interferometer, 61
Test plates, 14, 729
  holographic, 602
Thick lens, radial shear interferometer, 188
Thickness measurements, 728
Thin film thickness, measurement of, 226
Tolansky fringes, 231
Toroidal surfaces, 752

INDEX

Transverse aberrations:
  for conic mirror, 332
  and wavefront deformations, 323, 371
Traveling microscope, 733
Triangular interferometer, 141, 181. *See also* Cyclic interferometer
Triangular prisms, testing of, 66
Two-wavelength:
  holography, 612
  interferometry, 612
  phase measurement, 614
  phase shifting interferometry, 560
  technique, 612
Twyman–Green interferograms, 77
Twyman–Green interferometer, 51, 508
  beam splitter for, 53
  coherence requirements in, 58
  compensation in, 55
  double passed, 247, 252
  fringe stability in, 72
  multipass, 255
  phase conjugating, 76
  spatial coherence in, 58
  temporal coherence in, 61
  testing with, 63
  unequal path, 69, 71
  uses of, 63

Unequal path interferometer, 69, 71

Variance of a wavefront, 467
Vibration errors, 542
Video fringe analysis, 488

Wavefront analysis, 484
Wavefront deformations and transverse aberrations, 323, 371

Wavefront evaluation, *see* Wavefront measurement
Wavefront fitting, 455, 472
Wavefront measurement:
  with direct interferometry, 494, 566
  with Fourier methods, 491, 566
  with Hartmann test, 367
  with lateral shear interferometer, 135
  with phase shifting interferometer, 501
  with Ronchi test, 335
  with Twyman–Green interferometer, 85
Wavefront representation:
  monomial representation, 457, 470
  polynomial representation, 135, 456
  with Zernike polynomials, 464
Wavefront variance, 467
Waves per radius, 568
Williams interferometer, 52
Wind tunnel measurements, 158
Wire test, 265, 288, 306
  geometrical theory, 291
  physical theory, 294
Wollaston prism, 101
  uses of, 106
Wolter test, 306
Wright–Schmidt system, 444
Wyko inteferometer, 29

Zernike polynomials, 461, 470, 468, 581
  properties of, 466
  wavefront representation with, 464, 482
Zernike test, 113, 299, 305
Zero crossing in phase shifting interferometry, 562
Zonal screen, 289
Zygo interferometer, 28